A First Course in the Numerical Analysis of Differential Equations

Numerical analysis presents different faces to the world. For mathematicians it is a *bona fide* mathematical theory with an applicable flavour. For scientists and engineers it is a practical, applied subject, part of the standard repertoire of modelling techniques. For computer scientists it is a theory on the interplay of computer architecture and algorithms for real-number calculations.

The tension between these standpoints is the driving force of this book, which presents a rigorous account of the fundamentals of numerical analysis both of ordinary and partial differential equations. The point of departure is mathematical, but the exposition strives to maintain a balance among theoretical, algorithmic and applied aspects of the subject.

This new edition has been extensively updated, and includes new chapters on developing subject areas: geometric numerical integration, an emerging paradigm for numerical computation that exhibits exact conservation of important geometric and structural features of the underlying differential equation; spectral methods, which have come to be seen in the last two decades as a serious competitor to finite differences and finite elements; and conjugate gradients, one of the most powerful contemporary tools in the solution of sparse linear algebraic systems.

Other topics covered include numerical solution of ordinary differential equations by multistep and Runge–Kutta methods; finite difference and finite elements techniques for the Poisson equation; a variety of algorithms to solve large, sparse algebraic systems; methods for parabolic and hyperbolic differential equations and techniques for their analysis. The book is accompanied by an appendix that presents brief back-up in a number of mathematical topics.

Professor ISERLES concentrates on fundamentals: deriving methods from first principles, analysing them with a variety of mathematical techniques and occasionally discussing questions of implementation and applications. By doing so, he is able to lead the reader to a theoretical understanding of the subject without neglecting its practical aspects. The outcome is a textbook that is mathematically honest and rigorous and provides its target audience with a wide range of skills in both ordinary and partial differential equations.

Cambridge Texts in Applied Mathematics

All titles listed below can be obtained from good booksellers or from Cambridge University Press. For a complete series listing, visit http://www.cambridge.org/uk/series/sSeries.asp?code=CTAM

Rarefied Gas Dynamics: From Basic Concepts to Actual Calculations
CARLO CERCIGNANI

Symmetry Methods for Differential Equations: A Beginner's Guide
PETER E. HYDON

High Speed Flow
C. J. CHAPMAN

Wave Motion
J. BILLINGHAM AND A. C. KING

An Introduction to Magnetohydrodynamics
P. A. DAVIDSON

Linear Elastic Waves
JOHN G. HARRIS

Vorticity and Incompressible Flow
ANDREW J. MAJDA AND ANDREA L. BERTOZZI

Infinite-Dimensional Dynamical Systems
JAMES C. ROBINSON

Introduction to Symmetry Analysis
BRIAN J. CANTWELL

Bäcklund and Darboux Transformations
C. ROGERS AND W. K. SCHIEF

Finite Volume Methods for Hyperbolic Problems
RANDALL J. LEVEQUE

Introduction to Hydrodynamic Stability
P. G. DRAZIN

Theory of Vortex Sound
M. S. HOWE

Scaling
GRIGORY ISAAKOVICH BARENBLATT

Complex Variables: Introduction and Applications (2nd Edition)
MARK J. ABLOWITZ AND ATHANASSIOS S. FOKAS

A First Course in Combinatorial Optimization
JON LEE

Practical Applied Mathematics: Modelling, Analysis, Approximation
SAM HOWISON

An Introduction to Parallel and Vector Scientific Computation
RONALD W. SHONKWILER AND LEW LEFTON

A First Course in Continuum Mechanics
OSCAR GONZALEZ AND ANDREW M. STUART

Applied Solid Mechanics
PETER HOWELL, GREGORY KOZYREFF AND JOHN OCKENDON

A First Course in the Numerical Analysis of Differential Equations

Second Edition

ARIEH ISERLES
Department of Applied Mathematics and Theoretical Physics
University of Cambridge

CAMBRIDGE
UNIVERSITY PRESS

CAMBRIDGE
UNIVERSITY PRESS

University Printing House, Cambridge CB2 8BS, United Kingdom

Cambridge University Press is part of the University of Cambridge.

It furthers the University's mission by disseminating knowledge in the pursuit of education, learning and research at the highest international levels of excellence.

www.cambridge.org
Information on this title: www.cambridge.org/9780521734905

First published 2009
4th printing 2014

A catalogue record for this publication is available from the British Library

ISBN 978-0-521-73490-5 Paperback

Contents

III Partial differential equations of evolution 347

Preface to the second edition

In an ideal world this second edition should have been written at least three years ago but, needless to say, this is not an ideal world. Annoyingly, there are just 24 hours per day, rather less annoyingly I have joyously surrendered myself to the excitements of my own research and, being rather good at finding excuses, delayed the second edition again and again.

Yet, once I braced myself, banished my writer's block and started to compose in my head the new chapters, I was taken over by the sheer pleasure of writing. Repeatedly I have found myself, as I often do, thanking my good fortune for working in this particular corner of the mathematical garden, the numerical analysis of differential equations, and striving in a small way to communicate its oft-unappreciated beauty.

The last sentence is bound to startle anybody experienced enough in the fashions and prejudices of the mathematical world. Numerical analysis is often considered neither beautiful nor, indeed, profound. Pure mathematics is beautiful if your heart goes after the joy of abstraction, applied mathematics is beautiful if you are excited by mathematics as a means to explain the mystery of the world around us. But numerical analysis? Surely, we compute only when everything else fails, when mathematical theory cannot deliver an answer in a comprehensive, pristine form and thus we are compelled to throw a problem onto a number-crunching computer and produce boring numbers by boring calculations. This, I believe, is nonsense.

A mathematical problem does not cease being mathematical just because we have discretized it. The purpose of discretization is to render mathematical problems, often approximately, in a form accessible to efficient calculation by computers. This, in particular, means rephrasing and approximating analytic statements as a finite sequence of algebraic steps. Algorithms and numerical methods are, by their very design, suitable for *computation* but it makes them neither simple nor easy as *mathematical* constructs. Replacing derivatives by finite differences or an infinite-dimensional space by a hierarchy of finite-dimensional spaces does not necessarily lead to a more fuzzy form of reasoning. We can still ask proper mathematical questions with uncompromising rigour and seek answers with the full mathematical etiquette of precise definitions, statements and proofs. The rules of the game do not change at all.

Actually, it is almost inevitable that a discretized mathematical problem is, *as a mathematical problem,* more difficult and more demanding of our mathematical ingenuity. To give just one example, it is usual to approximate a partial differential equation of evolution, an infinite-dimensional animal, in a finite-dimensional space (using, for example, finite differences, finite elements or a spectral method). This finite-dimensional approximation makes the problem tractable on a computer, a machine that can execute a finite number of algebraic operations in finite time. However, once we wish to answer the big mathematical question underlying our discourse, how well does the finite-dimensional model approximate the original equation, we are compelled to consider not one finite-dimensional system but an infinite progression of such systems, of increasing (and unbounded) dimension. In effect, we are not just approximating a single equation but an entire infinite-dimensional function space. Of course, if all you want is numbers, you can get away with hand-waving arguments or use the expertise and experience of others. But once you wish to understand honestly the term 'analysis' in 'numerical analysis', prepare yourself for real mathematical experience.

I hope to have made the case that true numerical analysis operates according to standard mathematical rules of engagement (while, needless to say, fully engaging with the algorithmic and applied parts of its inner self). My stronger claim, illustrated in a small way by the material of this book, is that numerical analysis is perhaps the most eclectic and demanding client of the entire width and breadth of mathematics. Typically in mathematics, a discipline rests upon a fairly small number of neighbouring disciplines: once you visit a mathematical library, you find yourself time and again visiting a fairly modest number of shelves. Not so in the numerical analysis of differential equations. Once you want to understand the subject in its breadth, rather than specializing in a narrow and strictly delineated subset, prepare yourself to navigate across all library shelves! This volume, being a textbook, is purposefully steering well clear of deep and difficult mathematics. However, even at the sort of elementary level of mathematical sophistication suitable for advanced undergraduates, faithful to the principle that every unusual bit of mathematics should be introduced and explained I expect the reader to identify the many and varied mathematical sources of our discourse. This opportunity to revel and rejoice in the varied mathematical origins of the subject, of pulling occasional rabbits from all kinds of mathematical hats, is what makes me so happy to work in numerical analysis. I hope to have conveyed, in a small and inevitably flawed manner, how different strands of mathematical thinking join together to form this discipline.

Three chapters have been added to the first edition to reflect the changing face of the subject. The first is on geometric numerical integration, the emerging science of the numerical computation of differential equations in a way that renders exactly their qualitative features. The second is on spectral methods, an important competitor to the more established finite difference and finite element techniques for partial differential equations. The third new chapter reviews the method of conjugate gradients for the solution of the large linear algebraic systems that occur once partial differential equations are discretized.

Needless to say, the current contents cannot reflect all the many different ideas, algorithms, methods and insights that, in their totality, make the subject of computational differential equations. Writing a textbook, the main challenge is not what to include, but what to exclude! It would have been very easy to endure the publisher's unhappiness and expand this book to several volumes, reporting on numerous exciting themes such domain decomposition, meshless methods, wavelet-based methods, particle methods, homogenization – the list goes on and on. Easy, but perhaps not very illuminating, because this is not a cookbook, a dictionary or a compendium: it is a textbook that, ideally, should form the backdrop to a lecture course. It would not have been very helpful to bury the essential didactic message under a mountain of facts, exciting and useful as they might be. The main purpose of a lecture course – and hence of a textbook – is to provide enough material, insight and motivation to prepare students for further, often independent, study. My aim on these pages has been to provide this sort of preparation.

The flowchart on p. xix displays the connectivity and logical progression of the current 17 chapters. Although it is unlikely that the entire contents of the book can be encompased in less than a year-long intensive lecture course, the flowchart is suggestive of many different ways to pick and choose material while maintaining the inner integrity and coherence of the exposition.

This is the moment to thank all those who helped me selflessly in crafting an edition better than one I could have written singlehandedly. Firstly, all those users of the first edition who have provided me with feedback, communicated errors and misprints, queried the narrative, lavished praise or extended well-deserved criticism.[1] Secondly, those of my colleagues who read parts of the draft, offered remarks (mostly encouraging but sometimes critical: I appreciated both) and frequently saved me from embarrassing blunders: Ben Adcock, Alfredo Deaño, Euan Spence, Endre Süli and Antonella Zanna. Thirdly, my friends at Cambridge University Press, in particular David Tranah, who encouraged this second edition, pushed me when a push was needed, let me get along without undue harassment otherwise and was always willing to share his immense experience. Fourthly, my copy editor Susan Parkinson, as always pedantic in the best sense of the word. Fifthly, the terrific intellectual environment in the Department of Applied Mathematics and Theoretical Physics of the University of Cambridge, in particular among my colleagues and students in the Numerical Analysis Group. We have managed throughout the years to act not only as a testing bed, and sometimes a foil, to each other's ideas but also as a milieu where it is always delightful to abandon mathematics for a break of (relatively decent) coffee and uplifting conversation on just about anything. And last, but definitely not least, my wife and best friend, Dganit, who has encouraged and helped me always, in more ways than I can count or floating-number arithmetic can bear.

And so, over to you, the reader. I hope to have managed to convey to you, even if in a small and imperfect manner, not just the raw facts that, in their totality, make up the numerical analysis of differential equations, but the beauty and the excitement of the subject.

Arieh Iserles
August 2008

[1] I wish to thank less, though, those students who emailed me for solutions to the exercises before their class assignment was due.

Preface to the first edition

Books – so we are often told – should be born out of a sense of mission, a wish to share knowledge, experience and ideas, a penchant for beauty. This book has been born out of a sense of frustration.

For the last decade or so I have been teaching the numerical analysis of differential equations to mathematicians, in Cambridge and elsewhere. Examining this extensive period of trial and (frequent) error, two main conclusions come to mind and both have guided my choice of material and presentation in this volume.

Firstly, mathematicians are different from other varieties of *homo sapiens*. It may be observed that people study numerical analysis for various reasons. **Scientists** and **engineers** require it as a means to an end, a tool to investigate the subject matter that *really* interests them. Entirely justifiably, they wish to spend neither time nor intellectual effort on the finer points of mathematical analysis, typically preferring a style that combines a cook-book presentation of numerical methods with a leavening of intuitive and hand-waving explanations. **Computer scientists** adopt a different, more algorithmic, attitude. Their heart goes after the clever algorithm and its interaction with computer architecture. Differential equations and their likes are abandoned as soon as decency allows (or sooner). They are replaced by discrete models, which in turn are analysed by combinatorial techniques. **Mathematicians,** though, follow a different mode of reasoning. Typically, mathematics students are likely to participate in an advanced numerical analysis course in their final year of undergraduate studies, or perhaps in the first postgraduate year. Their studies until that point in time would have consisted, to a large extent, of a progression of formal reasoning, the familiar sequence of axiom $\Rightarrow$ theorem $\Rightarrow$ proof $\Rightarrow$ corollary $\Rightarrow$ .... Numerical analysis does not fit easily into this straitjacket, and this goes a long way toward explaining why many students of mathematics find it so unattractive.

Trying to teach numerical analysis to mathematicians, one is thus in a dilemma: should the subject be presented purely as a mathematical theory, intellectually pleasing but arid insofar as applications are concerned or, alternatively, should the audience be administered an application-oriented culture shock that might well cause it to vote with its feet?! The resolution is not very difficult, namely to present the material in a *bona fide* mathematical manner, occasionally veering toward issues of applications and algorithmics but never abandoning honesty and rigour. It is perfectly allowable to omit an occasional proof (which might well require material outside the scope of the presentation) and even to justify a numerical method on the grounds of plausibility and a good track record in applications. But plausibility, a good track record, intuition and old-fashioned hand-waving do not constitute an honest mathematical argument and should never be presented as such.

Secondly, students should be exposed in numerical analysis to both ordinary and partial differential equations, as well as to means of dealing with large sparse algebraic systems. The pressure of many mathematical subjects and sub-disciplines is such that only a modest proportion of undergraduates are likely to take part in more than a single advanced numerical analysis course. Many more will, in all likelihood, be faced with the need to solve differential equations numerically in the future course of their professional life. Therefore, the option of restricting the exposition to ordinary differential equations, say, or to finite elements, while having the obvious merit of cohesion and sharpness of focus is counterproductive in the long term.

To recapitulate, the ideal course in the numerical analysis of differential equations, directed toward mathematics students, should be mathematically honest and rigorous and provide its target audience with a wide range of skills in both ordinary and partial differential equations. For the last decade I have been desperately trying to find a textbook that can be used to my satisfaction in such a course – in vain. There are many fine textbooks on particular aspects of the subject: numerical methods for ordinary differential equations, finite elements, computation of sparse algebraic systems. There are several books that span the whole subject but, unfortunately, at a relatively low level of mathematical sophistication and rigour. But, to the best of my knowledge, no text addresses itself to the right mathematical agenda at the right level of maturity. Hence my frustration and hence the motivation behind this volume.

This is perhaps the place to review briefly the main features of this book.

★ We cover a broad range of material: the numerical solution of ordinary differential equations by multistep and Runge–Kutta methods; finite difference and finite element techniques for the Poisson equation; a variety of algorithms for solving the large systems of sparse algebraic equations that occur in the course of computing the solution of the Poisson equation; and, finally, methods for parabolic and hyperbolic differential equations and techniques for their analysis. There is probably enough material in this book for a one-year fast-paced course and probably many lecturers will wish to cover only part of the material.

★ This is a textbook for mathematics students. By implication, it is not a textbook for computer scientists, engineers or natural scientists. As I have already argued, each group of students has different concerns and thought modes. Each assimilates knowledge differently. Hence, a textbook that attempts to be different things to different audiences is likely to disappoint them all. Nevertheless, non-mathematicians in need of numerical knowledge can benefit from this volume, but it is fair to observe that they should perhaps peruse it somewhat later in their careers, when in possession of the appropriate degree of motivation and background knowledge.

  On an even more basic level of restriction, this is a textbook, not a monograph or a collection of recipes. Emphatically, our mission is *not* to bring the exposition to the state of the art or to highlight the most advanced developments. Likewise, it is not our intention to provide techniques that cater for all possible problems and eventualities.

★ An annoying feature of many numerical analysis texts is that they display inordinately long lists of methods and algorithms to solve any one problem. Thus, not just one Runge–Kutta method but twenty! The hapless reader is left with an arsenal of weapons but, all too often, without a clue which one to use and why. In this volume we adopt an alternative approach: methods are derived from underlying principles and these principles, rather than the algorithms themselves, are at the centre of our argument. As soon as the underlying principles are sorted out, algorithmic fireworks become the least challenging part of numerical analysis – the real intellectual effort goes into the mathematical analysis.

This is not to say that issues of software are not important or that they are somehow of a lesser scholarly pedigree. They receive our attention in Chapter 6 and I hasten to emphasize that good software design is just as challenging as theorem-proving. Indeed, the proper appreciation of difficulties in software and applications is enhanced by the understanding of the analytic aspects of numerical mathematics.

★ A truly exciting aspect of numerical analysis is the extensive use it makes of different mathematical disciplines. If you believe that numerics are a mathematical cop-out, a device for abandoning mathematics in favour of something 'softer', you are in for a shock. Numerical analysis is perhaps the most extensive and varied user of a very wide range of mathematical theories, from basic linear algebra and calculus all the way to functional analysis, differential topology, graph theory, analytic function theory, nonlinear dynamical systems, number theory, convexity theory – and the list goes on and on. Hardly any theme in modern mathematics fails to inspire and help numerical analysis. Hence, numerical analysts must be open-minded and ready to borrow from a wide range of mathematical skills – this is not a good bolt-hole for narrow specialists!

In this volume we emphasize the variety of mathematical themes that inspire and inform numerical analysis. This is not as easy as it might sound, since it is impossible to take for granted that students in different universities have a similar knowledge of pure mathematics. In other words, it is often necessary to devote a few pages to a topic which, in principle, has nothing to do with numerical analysis *per se* but which, nonetheless, is required in our exposition. I ask for the indulgence of those readers who are more knowledgeable in arcane mathematical matters – all they need is simply to skip few pages . . .

★ There is a major difference between recalling and understanding a mathematical concept. Reading mathematical texts I often come across concepts that are familiar and which I have certainly encountered in the past. Ask me, however, to recite their precise definition and I will probably flunk the test. The proper and virtuous course of action in such an instance is to pause, walk to the nearest mathematical library and consult the right source. To be frank, although sometimes I pursue this course of action, more often than not I simply go on reading. I have every reason to believe that I am not alone in this dubious practice.

In this volume I have attempted a partial remedy to the aforementioned phenomenon, by adding an appendix named 'Bluffer's guide to useful mathematics'. This appendix lists in a perfunctory manner definitions and major theorems in a range of topics – linear algebra, elementary functional analysis and approximation theory – to which students should have been exposed previously but which might have been forgotten. Its purpose is neither to substitute elementary mathematical courses nor to offer remedial teaching. If you flick too often to the end of the book in search of a definition then, my friend, perhaps you had better stop for a while and get to grips with the underlying subject, using a proper textbook. Likewise, if you always pursue a virtuous course of action, consulting a proper source in each and every case of doubt, please do not allow me to tempt you off the straight and narrow.

★ Part of the etiquette of writing mathematics is to attribute material and to refer to primary sources. This is important not just to quench the vanity of one's colleagues but also to set the record straight, as well as allowing an interested reader access to more advanced material. Having said this, I entertain serious doubts with regard to the practice of sprinkling each and every paragraph in a *textbook* with copious references. The scenario is presumably that, having read the sentence '... suppose that $\boldsymbol{x} \in \mathbb{U}$, where $\mathbb{U}$ is a foliated widget [37]', the reader will look up the references, identify '[37]' with a paper of J. Bloggs in *Proc. SDW,* recognize the latter as *Proceedings of the Society of Differentiable Widgets,* walk to the library, locate the journal (which will be actually on the shelf, rather than on loan, misplaced or stolen) ... All this might not be far-fetched as far as advanced mathematics monographs are concerned but makes very little sense in an undergraduate context. Therefore I have adopted a practice whereby there are no references in the text proper. Instead, each chapter is followed by a section of 'Comments and bibliography', where we survey briefly further literature that might be beneficial to students (and lecturers).

Such sections serve a further important purpose. Some students – am I too optimistic? – might be interested and inspired by the material of the chapter. For their benefit I have given in each 'Comments and bibliography' section a brief discussion of further developments, algorithms, methods of analysis and connections with other mathematical disciplines.

★ Clarity of exposition often hinges on transparency of notation. Thus, throughout this book we use the following convention:

- lower-case lightface sloping letters $(a, b, c, \alpha, \beta, \gamma, \ldots)$ represent scalars;
- lower-case boldface sloping letters $(\boldsymbol{a}, \boldsymbol{b}, \boldsymbol{c}, \boldsymbol{\alpha}, \boldsymbol{\beta}, \boldsymbol{\gamma}, \ldots)$ represent vectors;
- upper-case lightface letters $(A, B, C, \Theta, \Phi, \ldots)$ represent matrices;
- letters in calligraphic font $(\mathcal{A}, \mathcal{B}, \mathcal{C}, \ldots)$ represent operators;
- shell capitals $(\mathbb{A}, \mathbb{B}, \mathbb{C}, \ldots)$ represent sets.

Mathematical constants like $\mathrm{i} = \sqrt{-1}$ and e, the base of natural logarithms, are denoted by roman, rather than italic letters. This follows British typesetting convention and helps to identify the different components of a mathematical formula.

As with any principle, our notational convention has its exceptions. For example, in Section 3.1 we refer to Legendre and Chebyshev polynomials by the conventional notation, P_n and T_n: any other course of action would have caused utter confusion. And, again as with any principle, grey areas and ambiguities abound. I have tried to eliminate them by applying common sense but this, needless to say, is a highly subjective criterion.

This book started out life as two sets of condensed lecture notes – one for students of Part II (the last year of undergraduate mathematics in Cambridge) and the other for students of Part III (the Cambridge advanced degree course in mathematics). The task of expanding lecture notes to a full-scale book is, unfortunately, more complicated than producing a cup of hot soup from concentrate by adding boiling water, stirring and simmering for a short while. Ultimately, it has taken the better part of a year, shared with the usual commitments of academic life. The main portion of the manuscript was written in Autumn 1994, during a sabbatical leave at the California Institute of Technology (Caltech). It is my pleasant duty to acknowledge the hospitality of my many good friends there and the perfect working environment in Pasadena.

A familiar computer proverb states that, while the first 90% of a programming job takes 90% of the time, the remaining 10% also takes 90% of the time . . . Writing a textbook follows similar rules and, back home in Cambridge, I have spent several months reading and rereading the manuscript. This is the place to thank a long list of friends and colleagues whose help has been truly crucial: Brad Baxter (Imperial College, London), Martin Buhmann (Swiss Institute of Technology, Zürich), Yu-Chung Chang (Caltech), Stephen Cowley (Cambridge), George Goodsell (Cambridge), Mike Holst (Caltech), Herb Keller (Caltech), Yorke Liu (Cambridge), Michelle Schatzman (Lyon), Andrew Stuart (Stanford), Stefan Vandewalle (Louven) and Antonella Zanna (Cambridge). Some have read the manuscript and offered their comments. Some provided software well beyond my own meagre programming skills and helped with the figures and with computational examples. Some have experimented with the manuscript upon their students and listened to their complaints. Some contributed insight and occasionally saved me from embarrassing blunders. All have been helpful, encouraging and patient to a fault with my foibles and idiosyncrasies. None is responsible for blunders, errors, mistakes, misprints and infelicities that, in spite of my sincerest efforts, are bound to persist in this volume.

This is perhaps the place to extend thanks to two 'friends' that have made the process of writing this book considerably easier: the TEX typesetting system and the `MATLAB` package. These days we take mathematical typesetting for granted but it is often forgotten that just a decade ago a mathematical manuscript would have been hand-written, then typed and retyped and, finally, typeset by publishers – each stage requiring laborious proofreading. In turn, `MATLAB` allows us a unique opportunity to turn our office into a computational-cum-graphic laboratory, to bounce ideas off the computer screen and produce informative figures and graphic displays. Not since the discovery of coffee have any inanimate objects caused so much pleasure to so many mathematicians!

The editorial staff of Cambridge University Press, in particular Alan Harvey, David Tranah and Roger Astley, went well beyond the call of duty in being helpful, friendly and cooperative. Susan Parkinson, the copy editor, has worked to the highest standards. Her professionalism, diligence and good taste have done wonders in sparing the readers numerous blunders and the more questionable examples of my hopeless wit. This is a pleasant opportunity to thank them all.

Last but never the least, my wife and best friend, Dganit. Her encouragement, advice and support cannot be quantified in conventional mathematical terms. Thank you!

I wish to dedicate this book to my parents, Gisella and Israel. They are not mathematicians, yet I have learnt from them all the really important things that have motivated me as a mathematician: love of scholarship and admiration for beauty and art.

Arieh Iserles
August 1995

Flowchart of contents

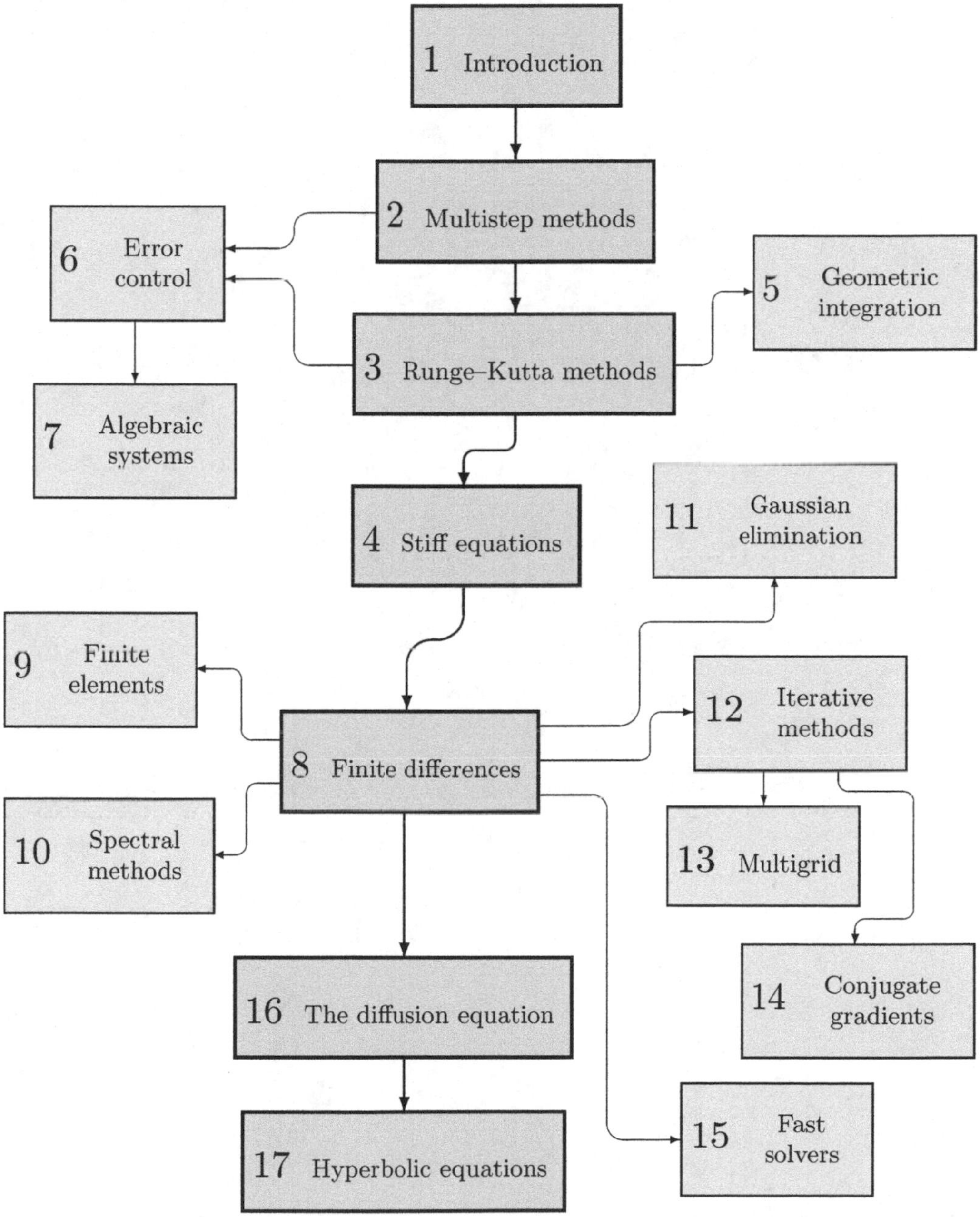

PART I

Ordinary differential equations

1

Euler's method and beyond

1.1 Ordinary differential equations and the Lipschitz condition

We commence our exposition of the computational aspects of differential equations by examining closely numerical methods for *ordinary differential equations (ODEs).* This is important because of the central role of ODEs in a multitude of applications. Not less crucial is the critical part that numerical ODEs play in the design and analysis of computational methods for *partial differential equations (PDEs).* Thus, even if your main interest is in solving PDEs, ideally you should first master computational ODEs, not just to familiarize yourself with concepts, terminology and ideas but also because (as we will see in what follows) many discretization methods for PDEs reduce the underlying problem to the computation of ODEs.

Our goal is to approximate the solution of the problem

$$\boldsymbol{y}' = \boldsymbol{f}(t, \boldsymbol{y}), \quad t \geq t_0, \qquad \boldsymbol{y}(t_0) = \boldsymbol{y}_0. \tag{1.1}$$

Here $\boldsymbol{f}$ is a sufficiently well-behaved function that maps $[t_0, \infty) \times \mathbb{R}^d$ to $\mathbb{R}^d$ and the initial condition $\boldsymbol{y}_0 \in \mathbb{R}^d$ is a given vector; $\mathbb{R}^d$ denotes here – and elsewhere in this book – the d-dimensional real Euclidean space.

The 'niceness' of $\boldsymbol{f}$ may span a whole range of desirable attributes. At the very least, we insist on $\boldsymbol{f}$ obeying, in a given vector norm $\|\cdot\|$, the *Lipschitz condition*

$$\|\boldsymbol{f}(t, \boldsymbol{x}) - \boldsymbol{f}(t, \boldsymbol{y})\| \leq \lambda \|\boldsymbol{x} - \boldsymbol{y}\| \quad \text{for all} \quad \boldsymbol{x}, \boldsymbol{y} \in \mathbb{R}^d, \ t \geq t_0. \tag{1.2}$$

Here $\lambda > 0$ is a real constant that is independent of the choice of $\boldsymbol{x}$ and $\boldsymbol{y}$ – a *Lipschitz constant.* Subject to (1.2), it is possible to prove that the ODE system (1.1) possesses a unique solution.[1] Taking a stronger requirement, we may stipulate that $\boldsymbol{f}$ is an *analytic* function – in other words, that the Taylor series of $\boldsymbol{f}$ about every $(t, \boldsymbol{y}_0) \in [0, \infty) \times \mathbb{R}^d$ has a positive radius of convergence. It is then possible to prove that the solution $\boldsymbol{y}$ itself is analytic. Analyticity comes in handy, since much of our investigation of numerical methods is based on Taylor expansions, but it is often an excessive requirement and excludes many ODEs of practical importance.

In this volume we strive to steer a middle course between the complementary vices of mathematical nitpicking and of hand-waving. We solemnly undertake to avoid any

[1] We refer the reader to the Appendix for a brief refresher course on norms, existence and uniqueness theorems for ODEs and other useful odds and ends of mathematics.

needless mention of exotic function spaces that present the theory in its most general form, whilst desisting from woolly and inexact statements. Thus, we *always* assume that $\boldsymbol{f}$ is Lipschitz and, as necessary, may explicitly stipulate that it is analytic. An intelligent reader could, if the need arose, easily weaken many of our 'analytic' statements so that they are applicable also to sufficiently-differentiable functions.

1.2 Euler's method

Let us ponder briefly the meaning of the ODE (1.1). We possess two items of information: we know the value of $\boldsymbol{y}$ at a single point $t = t_0$ and, given any function value $\boldsymbol{y} \in \mathbb{R}^d$ and time $t \geq t_0$, we can tell the slope from the differential equation. The purpose of the exercise being to guess the value of $\boldsymbol{y}$ at a new point, the most elementary approach is to use linear interpolation. In other words, we estimate $\boldsymbol{y}(t)$ by making the approximation $\boldsymbol{f}(t, \boldsymbol{y}(t)) \approx \boldsymbol{f}(t_0, \boldsymbol{y}(t_0))$ for $t \in [t_0, t_0 + h]$, where $h > 0$ is sufficiently small. Integrating (1.1),

$$\boldsymbol{y}(t) = \boldsymbol{y}(t_0) + \int_{t_0}^{t} \boldsymbol{f}(\tau, \boldsymbol{y}(\tau))\, \mathrm{d}\tau \approx \boldsymbol{y}_0 + (t - t_0)\boldsymbol{f}(t_0, \boldsymbol{y}_0). \tag{1.3}$$

Given a sequence t_0, $t_1 = t_0 + h$, $t_2 = t_0 + 2h, \ldots$, where $h > 0$ is the *time step*, we denote by $\boldsymbol{y}_n$ a numerical estimate of the exact solution $\boldsymbol{y}(t_n)$, $n = 0, 1, \ldots$ Motivated by (1.3), we choose

$$\boldsymbol{y}_1 = \boldsymbol{y}_0 + h\boldsymbol{f}(t_0, \boldsymbol{y}_0).$$

This procedure can be continued to produce approximants at t_2, t_3 and so on. In general, we obtain the recursive scheme

$$\boldsymbol{y}_{n+1} = \boldsymbol{y}_n + h\boldsymbol{f}(t_n, \boldsymbol{y}_n), \qquad n = 0, 1, \ldots, \tag{1.4}$$

the celebrated *Euler method.*

Euler's method is not only the most elementary computational scheme for ODEs and, simplicity notwithstanding, of enduring practical importance. It is also the cornerstone of the numerical analysis of differential equations of evolution. In a deep and profound sense, all the fancy multistep and Runge–Kutta schemes that we shall discuss are nothing but a generalization of the basic paradigm (1.4).

◇ **Graphic interpretation** Euler's method can be illustrated pictorially. Consider, for example, the scalar *logistic equation* $y' = y(1 - y)$, $y(0) = \frac{1}{10}$. Fig. 1.1 displays the first few steps of Euler's method, with a grotesquely large step $h = 1$. For each step we show the exact solution with initial condition $y(t_n) = y_n$ in the vicinity of $t_n = nh$ (dotted line) and the linear interpolation via Euler's method (1.4) (solid line).

The initial condition being, by definition, exact, so is the slope at t_0. However, instead of following a curved trajectory the numerical solution is piecewise-linear. Having reached t_1, say, we have moved to a wrong trajectory (i.e., corresponding to a different initial condition). The slope at t_1 is wrong – or, rather, it is the correct slope of the wrong solution! Advancing further, we might well stray even more from the original trajectory.

A realistic goal of numerical solution is not, however, to avoid errors altogether; after all, we approximate since we do not know the exact solution in the first place! An error-generating mechanism exists in every algorithm for numerical ODEs and our purpose is to understand it and to ensure that, in a given implementation, errors do not accumulate beyond a specified tolerance. Remarkably, even the excessive step $h = 1$ leads in Fig. 1.1 to a relatively modest local error. ◇

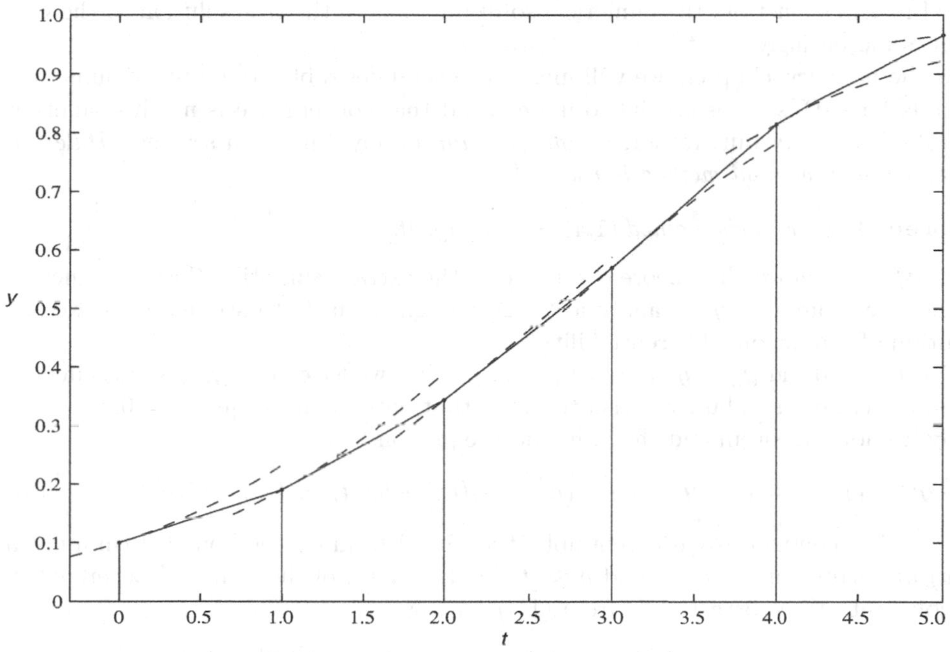

Figure 1.1 Euler's method, as applied to the equation $y' = y(1 - y)$ with initial value $y(0) = \frac{1}{10}$.

Euler's method can be easily extended to cater for variable steps. Thus, for a general monotone sequence $t_0 < t_1 < t_2 < \cdots$ we approximate as follows:

$$\boldsymbol{y}(t_{n+1}) \approx \boldsymbol{y}_{n+1} = \boldsymbol{y}_n + h_n \boldsymbol{f}(t_n, \boldsymbol{y}_n),$$

where $h_n = t_{n+1} - t_n$, $n = 0, 1, \ldots$ However, for the time being we restrict ourselves to constant steps.

How good is Euler's method in approximating (1.1)? Before we even attempt to answer this question, we need to formulate it with considerably more rigour. Thus, suppose that we wish to compute a numerical solution of (1.1) in the compact interval

$[t_0, t_0+t^*]$ with some time-stepping numerical method, not necessarily Euler's scheme. In other words, we cover the interval by an equidistant grid and employ the time-stepping procedure to produce a numerical solution. Each grid is associated with a different numerical sequence and the critical question is whether, as $h \to 0$ and the grid is being refined, the numerical solution tends to the exact solution of (1.1). More formally, we express the dependence of the numerical solution upon the step size by the notation $\boldsymbol{y}_n = \boldsymbol{y}_{n,h}$, $n = 0, 1, \ldots, \lfloor t^*/h \rfloor$. A method is said to be *convergent* if, for every ODE (1.1) with a Lipschitz function $\boldsymbol{f}$ and every $t^* > 0$ it is true that

$$\lim_{h\to 0+} \max_{n=0,1,\ldots,\lfloor t^*/h\rfloor} \|\boldsymbol{y}_{n,h} - \boldsymbol{y}(t_n)\| = 0,$$

where $\lfloor \alpha \rfloor \in \mathbb{Z}$ is the integer part of $\alpha \in \mathbb{R}$. Hence, convergence means that, for every Lipschitz function, the numerical solution tends to the true solution as the grid becomes increasingly fine.[2]

In the next few chapters we will mention several desirable attributes of numerical methods for ODEs. It is crucial to understand that convergence is not just another 'desirable' property but, rather, a *sine qua non* of any numerical scheme. *Unless it converges, a numerical method is useless!*

Theorem 1.1 *Euler's method* (1.4) *is convergent.*

Proof We prove this theorem subject to the extra assumption that the function $\boldsymbol{f}$ (and therefore also $\boldsymbol{y}$) is analytic (it is enough, in fact, to stipulate the weaker condition of continuous differentiability).

Given $h > 0$ and $\boldsymbol{y}_n = \boldsymbol{y}_{n,h}$, $n = 0, 1, \ldots, \lfloor t^*/h \rfloor$, we let $\boldsymbol{e}_{n,h} = \boldsymbol{y}_{n,h} - \boldsymbol{y}(t_n)$ denote the numerical error. Thus, we wish to prove that $\lim_{h\to 0+} \max_n \|\boldsymbol{e}_{n,h}\| = 0$.

By Taylor's theorem and the differential equation (1.1),

$$\boldsymbol{y}(t_{n+1}) = \boldsymbol{y}(t_n) + h\boldsymbol{y}'(t_n) + \mathcal{O}\big(h^2\big) = \boldsymbol{y}(t_n) + h\boldsymbol{f}(t_n, \boldsymbol{y}(t_n)) + \mathcal{O}\big(h^2\big), \tag{1.5}$$

and, $\boldsymbol{y}$ being continuously differentiable, the $\mathcal{O}\big(h^2\big)$ term can be bounded (in a given norm) uniformly for all $h > 0$ and $n \le \lfloor t^*/h \rfloor$ by a term of the form ch^2, where $c > 0$ is a constant. We subtract (1.5) from (1.4), giving

$$\boldsymbol{e}_{n+1,h} = \boldsymbol{e}_{n,h} + h[\boldsymbol{f}(t_n, \boldsymbol{y}(t_n) + \boldsymbol{e}_{n,h}) - \boldsymbol{f}(t_n, \boldsymbol{y}(t_n))] + \mathcal{O}\big(h^2\big).$$

Thus, it follows by the triangle inequality from the Lipschitz condition and the aforementioned bound on the $\mathcal{O}\big(h^2\big)$ reminder term that

$$\begin{aligned} \|\boldsymbol{e}_{n+1,h}\| &\le \|\boldsymbol{e}_{n,h}\| + h\|\boldsymbol{f}(t_n, \boldsymbol{y}(t_n) + \boldsymbol{e}_{n,h}) - \boldsymbol{f}(t_n, \boldsymbol{y}(t_n))\| + ch^2 \\ &\le (1 + h\lambda)\|\boldsymbol{e}_{n,h}\| + ch^2, \qquad n = 0, 1, \ldots, \lfloor t^*/h \rfloor - 1. \end{aligned} \tag{1.6}$$

We now claim that

$$\|\boldsymbol{e}_{n,h}\| \le \frac{c}{\lambda} h \left[(1 + h\lambda)^n - 1\right], \qquad n = 0, 1, \ldots \tag{1.7}$$

[2]We have just introduced a norm through the back door: cf. appendix subsection A.1.3.3 for an exact definition. This, however, should cause no worry, since all norms are equivalent in finite-dimensional spaces. In other words, if a method is convergent in one norm, it converges in all ...

The proof is by induction on n. When $n = 0$ we need to prove that $\|\boldsymbol{e}_{0,h}\| \le 0$ and hence that $\boldsymbol{e}_{0,h} = \boldsymbol{0}$. This is certainly true, since at t_0 the numerical solution matches the initial condition and the error is zero.

For general $n \ge 0$ we assume that (1.7) is true up to n and use (1.6) to argue that

$$\|\boldsymbol{e}_{n+1,h}\| \le (1+h\lambda)\frac{c}{\lambda}h\left[(1+h\lambda)^n - 1\right] + ch^2 = \frac{c}{\lambda}h\left[(1+h\lambda)^{n+1} - 1\right].$$

This advances the inductive argument from n to $n+1$ and proves that (1.7) is true. The constant $h\lambda$ is positive, therefore $1 + h\lambda < \mathrm{e}^{h\lambda}$ and we deduce that $(1+h\lambda)^n < \mathrm{e}^{nh\lambda}$. The index n is allowed to range in $\{0, 1, \ldots, \lfloor t^*/h \rfloor\}$, hence $(1+h\lambda)^n < \mathrm{e}^{\lfloor t^*/h \rfloor h\lambda} \le \mathrm{e}^{t^*\lambda}$. Substituting into (1.7), we obtain the inequality

$$\|\boldsymbol{e}_{n,h}\| \le \frac{c}{\lambda}(\mathrm{e}^{t^*\lambda} - 1)h, \qquad n = 0, 1, \ldots, \lfloor t^*/h \rfloor.$$

Since $c(\mathrm{e}^{t^*\lambda} - 1)/\lambda$ is independent of h, it follows that

$$\lim_{\substack{h\to 0\\ 0\le nh\le t^*}} \|\boldsymbol{e}_{n,h}\| = 0.$$

In other words, Euler's method is convergent. ■

◇ **Health warning** At first sight, it might appear that there is more to the last theorem than meets the eye – not just a proof of convergence but also an upper bound on the error. In principle this is perfectly true: the error of Euler's method is indeed always bounded by $hc\mathrm{e}^{t^*\lambda}/\lambda$. Moreover, with very little effort it is possible to demonstrate, e.g. by using the *Peano kernel theorem* (A.2.2.6), that a reasonable choice is $c = \max_{t\in[t_0,t_0+t^*]} \|\boldsymbol{y}''(t)\|$. The problem with this bound is that, unfortunately, in an overwhelming majority of practical cases it is too large by many orders of magnitude. It falls into the broad category of statements like 'the distance between London and New York is less than 47 light years' which, although manifestly true, fail to contribute significantly to the sum total of human knowledge.

The problem is not with the proof *per se* but with the insensitivity of a Lipschitz constant. A trivial example is the scalar linear equation $y' = -100y$, $y(0) = 1$. Therefore $\lambda = 100$ and, since $y(t) = \mathrm{e}^{-100t}$, $c = \lambda^2$. We thus derive the upper bound of $100h(\mathrm{e}^{100t^*} - 1)$. Letting $t^* = 1$, say, we have

$$|y_n - y(nh)| \le 2.69 \times 10^{45} h. \tag{1.8}$$

It is easy, however, to show that $y_n = (1 - 100h)^n$, hence to derive the exact expression

$$|y_n - y(nh)| = \left|(1-100h)^n - \mathrm{e}^{-100nh}\right|$$

which is smaller by many orders of magnitude than (1.8) (note that, unless nh is very small, to all intents and purposes $\mathrm{e}^{-100nh} \approx 0$).

The moral of our discussion is simple. *The bound from the proof of Theorem 1.1 must not be used in practical estimations of numerical error!* ◇

Euler's method can be rewritten in the form $\boldsymbol{y}_{n+1} - [\boldsymbol{y}_n + h\boldsymbol{f}(t_n, \boldsymbol{y}_n)] = \boldsymbol{0}$. Replacing $\boldsymbol{y}_k$ by the exact solution $\boldsymbol{y}(t_k)$, $k = n, n+1$, and expanding the first few terms of the Taylor series about $t = t_0 + nh$, we obtain

$$\begin{aligned}&\boldsymbol{y}(t_{n+1}) - [\boldsymbol{y}(t_n) + h\boldsymbol{f}(t_n, \boldsymbol{y}(t_n))]\\&= [\boldsymbol{y}(t_n) + h\boldsymbol{y}'(t_n) + \mathcal{O}(h^2)] - [\boldsymbol{y}(t_n) + h\boldsymbol{y}'(t_n)] = \mathcal{O}(h^2).\end{aligned}$$

We say that the Euler's method (1.4) is of *order* 1. In general, given an arbitrary time-stepping method

$$\boldsymbol{y}_{n+1} = \boldsymbol{\mathcal{Y}}_n(\boldsymbol{f}, h, \boldsymbol{y}_0, \boldsymbol{y}_1, \ldots, \boldsymbol{y}_n), \qquad n = 0, 1, \ldots,$$

for the ODE (1.1), we say that it is of *order* p if

$$\boldsymbol{y}(t_{n+1}) - \boldsymbol{\mathcal{Y}}_n(\boldsymbol{f}, h, \boldsymbol{y}(t_0), \boldsymbol{y}(t_1), \ldots, \boldsymbol{y}(t_n)) = \mathcal{O}(h^{p+1})$$

for every analytic $\boldsymbol{f}$ and $n = 0, 1, \ldots$ Alternatively, a method is of order p if it recovers *exactly* every polynomial solution of degree p or less.

The order of a numerical method provides us with information about its *local behaviour* – advancing from t_n to t_{n+1}, where $h > 0$ is sufficiently small, we are incurring an error of $\mathcal{O}(h^{p+1})$. Our main interest, however, is in not the local but the *global* behaviour of the method: how well is it doing in a fixed bounded interval of integration as $h \to 0$? Does it converge to the true solution? How fast? Since the local error decays as $\mathcal{O}(h^{p+1})$, the number of steps increases as $\mathcal{O}(h^{-1})$. The naive expectation is that the global error decreases as $\mathcal{O}(h^p)$, but – as we will see in Chapter 2 – it cannot be taken for granted for each and every numerical method without an additional condition. As far as Euler's method is concerned, Theorem 1.1 demonstrates that all is well and that the error indeed decays as $\mathcal{O}(h)$.

1.3 The trapezoidal rule

Euler's method approximates the derivative by a constant in $[t_n, t_{n+1}]$, namely by its value at t_n (again, we denote $t_k = t_0 + kh$, $k = 0, 1, \ldots$). Clearly, the 'cantilevering' approximation is not very good and it makes more sense to make the constant approximation of the derivative equal to the average of its values at the endpoints. Bearing in mind that derivatives are given by the differential equation, we thus obtain an expression similar to (1.3):

$$\begin{aligned}\boldsymbol{y}(t) &= \boldsymbol{y}(t_n) + \int_{t_n}^{t} \boldsymbol{f}(\tau, \boldsymbol{y}(\tau))\,\mathrm{d}\tau\\&\approx \boldsymbol{y}(t_n) + \tfrac{1}{2}(t - t_n)[\boldsymbol{f}(t_n, \boldsymbol{y}(t_n)) + \boldsymbol{f}(t, \boldsymbol{y}(t))].\end{aligned}$$

This is the motivation behind the *trapezoidal rule*

$$\boldsymbol{y}_{n+1} = \boldsymbol{y}_n + \tfrac{1}{2}h[\boldsymbol{f}(t_n, \boldsymbol{y}_n) + \boldsymbol{f}(t_{n+1}, \boldsymbol{y}_{n+1})]. \tag{1.9}$$

To obtain the order of (1.9), we substitute the exact solution,

$$\begin{aligned}
&\boldsymbol{y}(t_{n+1}) - \{\boldsymbol{y}(t_n) + \tfrac{1}{2}h[\boldsymbol{f}(t_n, \boldsymbol{y}(t_n)) + \boldsymbol{f}(t_{n+1}, \boldsymbol{y}(t_{n+1}))]\} \\
&= [\boldsymbol{y}(t_n) + h\boldsymbol{y}'(t_n) + \tfrac{1}{2}h^2\boldsymbol{y}''(t_n) + \mathcal{O}(h^3)] \\
&\quad - (\boldsymbol{y}(t_n) + \tfrac{1}{2}h\{\boldsymbol{y}'(t_n) + [\boldsymbol{y}'(t_n) + h\boldsymbol{y}''(t_n) + \mathcal{O}(h^2)]\}) = \mathcal{O}(h^3).
\end{aligned}$$

Therefore the trapezoidal rule is of order 2.

Being forewarned of the shortcomings of local analysis, we should not jump to conclusions. Before we infer that the error decays globally as $\mathcal{O}(h^2)$, we must first prove that the method is convergent. Fortunately, this can be accomplished by a straightforward generalization of the method of proof of Theorem 1.1.

Theorem 1.2 *The trapezoidal rule* (1.9) *is convergent.*

Proof Subtracting

$$\boldsymbol{y}(t_{n+1}) = \boldsymbol{y}(t_n) + \tfrac{1}{2}h\,[\boldsymbol{f}(t_n, \boldsymbol{y}(t_n)) + \boldsymbol{f}(t_{n+1}, \boldsymbol{y}(t_{n+1}))] + \mathcal{O}(h^3)$$

from (1.9), we obtain

$$\begin{aligned}
\boldsymbol{e}_{n+1,h} = \boldsymbol{e}_{n,h} + \tfrac{1}{2}h\,\{&[\boldsymbol{f}(t_n, \boldsymbol{y}_n) - \boldsymbol{f}(t_n, \boldsymbol{y}(t_n))] \\
&+ [\boldsymbol{f}(t_{n+1}, \boldsymbol{y}_{n+1}) - \boldsymbol{f}(t_{n+1}, \boldsymbol{y}(t_{n+1}))]\} + \mathcal{O}(h^3).
\end{aligned}$$

For analytic $\boldsymbol{f}$ we may bound the $\mathcal{O}(h^3)$ term by ch^3 for some $c > 0$, and this upper bound is valid uniformly throughout $[t_0, t_0 + t^*]$. Therefore, it follows from the Lipschitz condition (1.2) and the triangle inequality that

$$\|\boldsymbol{e}_{n+1,h}\| \le \|\boldsymbol{e}_{n,h}\| + \tfrac{1}{2}h\lambda\,\{\|\boldsymbol{e}_{n,h}\| + \|\boldsymbol{e}_{n+1,h}\|\} + ch^3.$$

Since we are ultimately interested in letting $h \to 0$ there is no harm in assuming that $h\lambda < 2$, and we can thus deduce that

$$\|\boldsymbol{e}_{n+1,h}\| \le \left(\frac{1 + \frac{1}{2}h\lambda}{1 - \frac{1}{2}h\lambda}\right) \|\boldsymbol{e}_{n,h}\| + \left(\frac{c}{1 - \frac{1}{2}h\lambda}\right) h^3. \tag{1.10}$$

Our next step closely parallels the derivation of inequality (1.7). We thus argue that

$$\|\boldsymbol{e}_{n,h}\| \le \frac{c}{\lambda}\left[\left(\frac{1 + \frac{1}{2}h\lambda}{1 - \frac{1}{2}h\lambda}\right)^n - 1\right] h^2. \tag{1.11}$$

This follows by induction on n from (1.10) and is left as an exercise to the reader.

Since $0 < h\lambda < 2$, it is true that

$$\frac{1 + \frac{1}{2}h\lambda}{1 - \frac{1}{2}h\lambda} = 1 + \frac{h\lambda}{1 - \frac{1}{2}h\lambda} \le \sum_{\ell=0}^{\infty} \frac{1}{\ell!}\left(\frac{h\lambda}{1 - \frac{1}{2}h\lambda}\right)^{\ell} = \exp\left(\frac{h\lambda}{1 - \frac{1}{2}h\lambda}\right).$$

Consequently, (1.11) yields

$$\|\boldsymbol{e}_{n,h}\| \le \frac{ch^2}{\lambda}\left(\frac{1 + \frac{1}{2}h\lambda}{1 - \frac{1}{2}h\lambda}\right)^n \le \frac{ch^2}{\lambda}\exp\left(\frac{nh\lambda}{1 - \frac{1}{2}h\lambda}\right).$$

This bound is true for every nonnegative integer n such that $nh \le t^*$. Therefore

$$\|\boldsymbol{e}_{n,h}\| \le \frac{ch^2}{\lambda} \exp\left(\frac{t^*\lambda}{1-\frac{1}{2}h\lambda}\right)$$

and we deduce that

$$\lim_{\substack{h\to 0 \\ 0\le nh\le t^*}} \|\boldsymbol{e}_{n,h}\| = 0.$$

In other words, the trapezoidal rule converges. ■

The number $ch^2 \exp[t^*\lambda/(1-\frac{1}{2}h\lambda)]/\lambda$ is, again, of absolutely no use in practical error bounds. However, a significant difference from Theorem 1.1 is that for the trapezoidal rule the error decays *globally* as $\mathcal{O}(h^2)$. This is to be expected from a second-order method if its convergence has been established.

Another difference between the trapezoidal rule and Euler's method is of an entirely different character. Whereas Euler's method (1.4) can be executed explicitly – knowing $\boldsymbol{y}_n$ we can produce $\boldsymbol{y}_{n+1}$ by computing a value of $\boldsymbol{f}$ and making a few arithmetic operations – this is not the case with (1.9). The vector $\boldsymbol{v} = \boldsymbol{y}_n + \frac{1}{2}h\boldsymbol{f}(t_n, \boldsymbol{y}_n)$ can be evaluated from known data, but that leaves us in each step with the task of finding $\boldsymbol{y}_{n+1}$ as the solution of the system of algebraic equations

$$\boldsymbol{y}_{n+1} - \tfrac{1}{2}h\boldsymbol{f}(t_{n+1}, \boldsymbol{y}_{n+1}) = \boldsymbol{v}.$$

The trapezoidal rule is thus said to be *implicit*, to distinguish it from the *explicit* Euler's method and its ilk.

Solving nonlinear equations is hardly a mission impossible, but we cannot take it for granted either. Only in texts on pure mathematics are we allowed to wave a magic wand, exclaim 'let $\boldsymbol{y}_{n+1}$ be a solution of ...' and assume that all our problems are over. As soon as we come to deal with actual computation, we had better specify how we plan (or our computer plans) to undertake the task of evaluating $\boldsymbol{y}_{n+1}$. This will be a theme of Chapter 7, which deals with the implementation of ODE methods. It suffices to state now that the cost of numerically solving nonlinear equations does not rule out the trapezoidal rule (and other implicit methods) as viable computational instruments. Implicitness is just one attribute of a numerical method and we must weigh it alongside other features.

◇ **A 'good' example** Figure 1.2 displays the (natural) logarithm of the error in the numerical solution of the scalar linear equation $y' = -y + 2\mathrm{e}^{-t}\cos 2t$, $y(0) = 0$ for (in descending order) $h = \frac{1}{2}$, $h = \frac{1}{10}$ and $h = \frac{1}{50}$.

How well does the plot illustrate our main distinction between Euler's method and the trapezoidal rule, namely faster decay of the error for the latter? As often in life, information is somewhat obscured by extraneous 'noise'; in the present case the error oscillates. This can be easily explained by the periodic component of the exact solution $y(t) = \mathrm{e}^{-t}\sin 2t$. Another observation is that, for both Euler's method and the trapezoidal rule, the error, twists and turns notwithstanding, does decay. This, on the face of it, can be explained by the decay of the exact solution but is an important piece of news nonetheless.

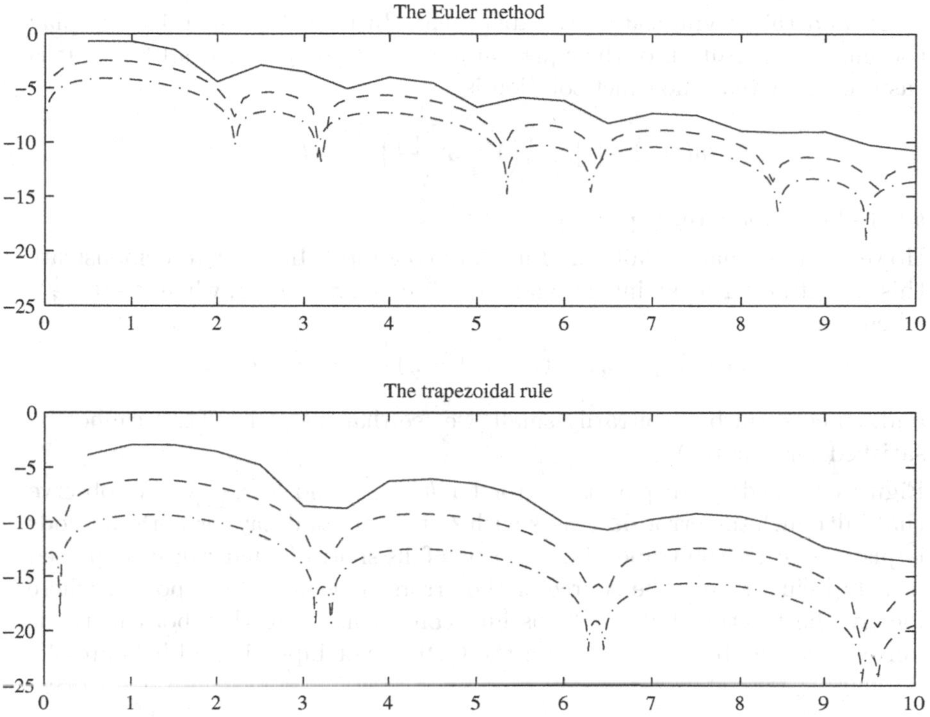

Figure 1.2 Euler's method and the trapezoidal rule, as applied to $y' = -y+2e^{-t}\cos 2t$, $y(0)=0$. The logarithm of the error, $\ln|y_n - y(t_n)|$, is displayed for $h=\frac{1}{2}$ (solid line), $h=\frac{1}{10}$ (broken line) and $h=\frac{1}{50}$ (broken-and-dotted line).

Our most pessimistic assumption is that errors might accumulate from step to step but, as can be seen from this example, this prophecy of doom is often misplaced. This is a highly nontrivial point, which will be debated at greater length throughout Chapter 4.

Factoring out oscillations and decay, we observe that errors indeed decrease with h. More careful examination verifies that they increase at roughly the rate predicted by order considerations. Specifically, for a convergent method of order p we have $\|\boldsymbol{e}\| \approx ch^p$, hence $\ln\|\boldsymbol{e}\| \approx \ln c + p\ln h$. Denoting by $\boldsymbol{e}^{(1)}$ and $\boldsymbol{e}^{(2)}$ the errors corresponding to step sizes $h^{(1)}$ and $h^{(2)}$ respectively, it follows that $\ln\|\boldsymbol{e}^{(2)}\| \approx \ln\|\boldsymbol{e}^{(1)}\| - p\ln(h^{(2)}/h^{(1)})$. The ratio of consecutive step sizes in Fig. 1.2 being five, we expect the error to decay by (at least) a constant multiple of $\ln 5 \approx 1.6094$ and $2\ln 5 \approx 3.2189$ for Euler and the trapezoidal rule respectively. The actual error decays if anything slightly faster than this. ◇

◇ **A 'bad' example** Theorems 1.1 and 1.2 and, indeed, the whole numerical ODE theory, rest upon the assumption that (1.1) satisfies the Lipschitz condition. We can expect numerical methods to underperform in the absence of (1.2), and this is vindicated by experiment. In Figs. 1.3 and 1.4 we display the numerical solution of the equation $y' = \ln 3\left(y - \lfloor y \rfloor - \frac{3}{2}\right)$, $y(0) = 0$. It is easy to verify that the exact solution is

$$y(t) = -\lfloor t \rfloor + \tfrac{1}{2}\left(1 - 3^{t-\lfloor t \rfloor}\right), \qquad t \geq 0,$$

where $\lfloor x \rfloor$ is the integer part of $x \in \mathbb{R}$.

However, the equation fails the Lipschitz condition. In order to demonstrate this, we let $m \geq 1$ be an integer and set $x = m+\varepsilon$, $z = m-\varepsilon$, where $\varepsilon \in \left(0, \frac{1}{4}\right)$. Then

$$\left|\left(x - \lfloor x \rfloor - \tfrac{3}{2}\right) - \left(z - \lfloor z \rfloor - \tfrac{3}{2}\right)\right| = \frac{1-2\varepsilon}{2\varepsilon}|x - z|$$

and, since ε can be arbitrarily small, we see that inequality (1.2) cannot be satisfied for a finite λ.

Figures 1.3 and 1.4 display the error for $h = \frac{1}{100}$ and $h = \frac{1}{1000}$. We observe that, although the error decreases with h, the rate of decay for both methods is just $\mathcal{O}(h)$: for the trapezoidal rule this falls short of what can be expected in a Lipschitz case. The source of the errors is clear: integer points, where locally the function fails the Lipschitz condition. Note that both methods perform equally badly – but when the ODE is not Lipschitz, all bets are off! ◇

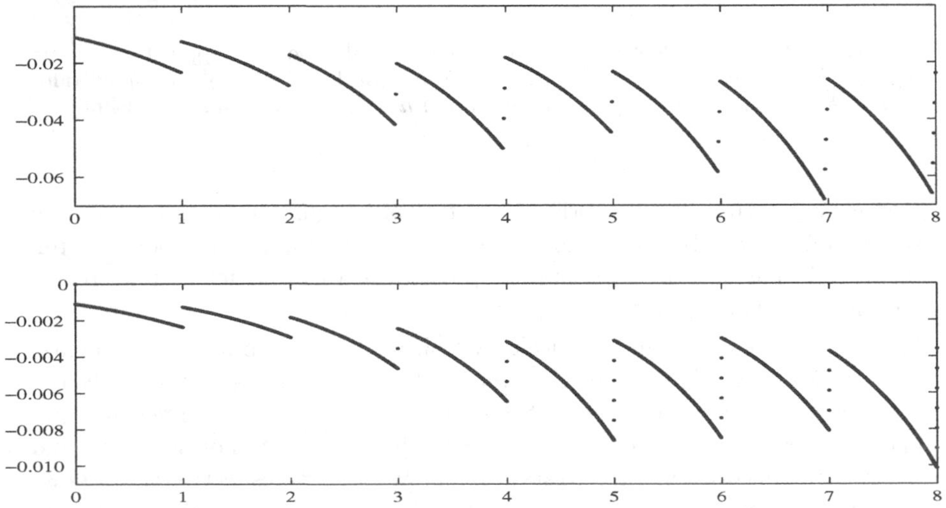

Figure 1.3 The error using Euler's method for $y' = \ln 3\left(y - \lfloor y \rfloor - \frac{3}{2}\right)$, $y(0) = 0$. The upper figure corresponds to $h = \frac{1}{100}$ and the lower to $h = \frac{1}{1000}$.

Two assumptions have led us to the trapezoidal rule. Firstly, for sufficiently small h, it is a good idea to approximate the derivative by a constant and, secondly, in choosing

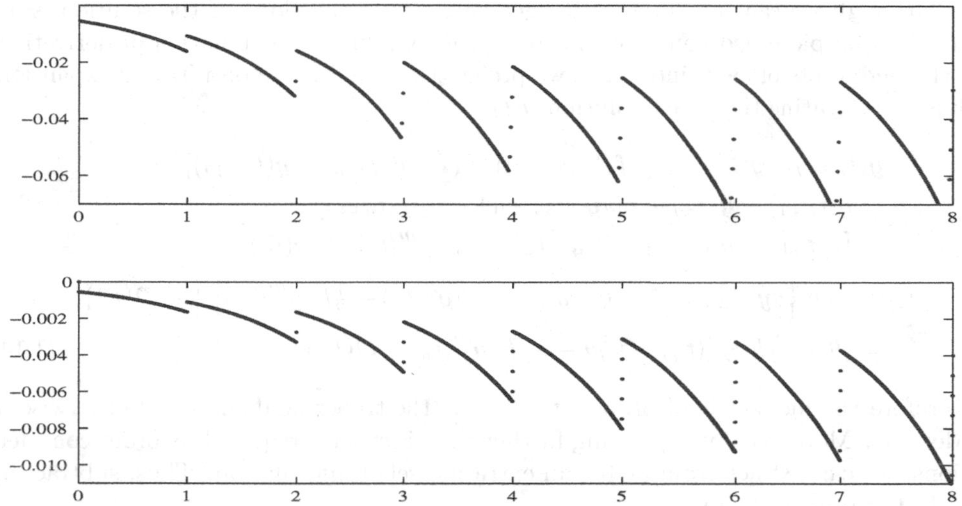

Figure 1.4 The error using the trapezoidal rule for the same equation as in Fig. 1.3. The upper figure corresponds to $h = \frac{1}{100}$ and the lower to $h = \frac{1}{1000}$.

the constant we should not 'discriminate' between the endpoints – hence the average

$$\boldsymbol{y}'(t) \approx \tfrac{1}{2}[\boldsymbol{f}(t_n, \boldsymbol{y}_n) + \boldsymbol{f}(t_{n+1}, \boldsymbol{y}_{n+1})]$$

is a sensible choice. Similar reasoning leads, however, to an alternative approximation,

$$\boldsymbol{y}'(t) \approx \boldsymbol{f}\left(t_n + \tfrac{1}{2}h, \tfrac{1}{2}(\boldsymbol{y}_n + \boldsymbol{y}_{n+1})\right), \qquad t \in [t_n, t_{n+1}],$$

and to the *implicit midpoint rule*

$$\boldsymbol{y}_{n+1} = \boldsymbol{y}_n + h\boldsymbol{f}\left(t_n + \tfrac{1}{2}h, \tfrac{1}{2}(\boldsymbol{y}_n + \boldsymbol{y}_{n+1})\right). \tag{1.12}$$

It is easy to prove that (1.12) is second order and that it converges. This is left to the reader in Exercise 1.1.

The implicit midpoint rule is a special case of the *Runge–Kutta* method. We defer the discussion of such methods to Chapter 3.

1.4 The theta method

Both Euler's method and the trapezoidal rule fit the general pattern

$$\boldsymbol{y}_{n+1} = \boldsymbol{y}_n + h[\theta\boldsymbol{f}(t_n, \boldsymbol{y}_n) + (1-\theta)\boldsymbol{f}(t_{n+1}, \boldsymbol{y}_{n+1})], \qquad n = 0, 1, \ldots, \tag{1.13}$$

with $\theta = 1$ and $\theta = \frac{1}{2}$ respectively. We may contemplate using (1.13) for any fixed value of $\theta \in [0, 1]$ and this, appropriately enough, is called a *theta method*. It is explicit for $\theta = 1$, otherwise implicit.

Although we can interpret (1.13) geometrically – the slope of the solution is assumed to be piecewise constant and provided by a linear combination of derivatives at the endpoints of each interval – we prefer the formal route of a Taylor expansion. Thus, substituting the exact solution $\boldsymbol{y}(t)$,

$$\begin{aligned}
&\boldsymbol{y}(t_{n+1}) - \boldsymbol{y}(t_n) - h[\theta \boldsymbol{f}(t_n, \boldsymbol{y}(t_n)) + (1-\theta)\boldsymbol{f}(t_{n+1}, \boldsymbol{y}(t_{n+1}))] \\
&= \boldsymbol{y}(t_{n+1}) - \boldsymbol{y}(t_n) - h[\theta \boldsymbol{y}'(t_n) + (1-\theta)\boldsymbol{y}'(t_{n+1})] \\
&= \left[\boldsymbol{y}(t_n) + h\boldsymbol{y}'(t_n) + \tfrac{1}{2}h^2\boldsymbol{y}''(t_n) + \tfrac{1}{6}h^3\boldsymbol{y}'''(t_n)\right] - \boldsymbol{y}(t_n) \\
&\quad - h\left\{\theta\boldsymbol{y}'(t_n) + (1-\theta)\left[\boldsymbol{y}'(t_n) + h\boldsymbol{y}''(t_n) + \tfrac{1}{2}h^2\boldsymbol{y}'''(t_n)\right]\right\} + \mathcal{O}(h^4) \\
&= \left(\theta - \tfrac{1}{2}\right)h^2\boldsymbol{y}''(t_n) + \left(\tfrac{1}{2}\theta - \tfrac{1}{3}\right)h^3\boldsymbol{y}'''(t_n) + \mathcal{O}(h^4). \qquad (1.14)
\end{aligned}$$

Therefore the method is of order 2 for $\theta = \frac{1}{2}$ (the trapezoidal rule) and otherwise of order one. Moreover, by expanding further than is strictly required by order considerations, we can extract from (1.14) an extra morsel of information. Thus, subtracting the last expression from

$$\boldsymbol{y}_{n+1} - \boldsymbol{y}_n - h\left[\theta\boldsymbol{f}(t_n, \boldsymbol{y}_n) + (1-\theta)\boldsymbol{f}(t_{n+1}, \boldsymbol{y}_{n+1})\right] = \boldsymbol{0},$$

we obtain for sufficiently small $h > 0$

$$\begin{aligned}
\boldsymbol{e}_{n+1} &= \boldsymbol{e}_n + \theta h[\boldsymbol{f}(t_n, \boldsymbol{y}(t_n) + \boldsymbol{e}_n) - \boldsymbol{f}(t_n, \boldsymbol{y}(t_n))] \\
&\quad + (1-\theta)h[\boldsymbol{f}(t_{n+1}, \boldsymbol{y}(t_{n+1}) + \boldsymbol{e}_{n+1}) - \boldsymbol{f}(t_{n+1}, \boldsymbol{y}(t_{n+1}))] \\
&\quad \begin{cases} -\tfrac{1}{12}h^3\boldsymbol{y}'''(t_n) + \mathcal{O}(h^4), & \theta = \tfrac{1}{2}, \\ +\left(\theta - \tfrac{1}{2}\right)h^2\boldsymbol{y}''(t_n) + \mathcal{O}(h^3), & \theta \neq \tfrac{1}{2}. \end{cases}
\end{aligned}$$

Considering $\boldsymbol{e}_{n+1}$ as an unknown, we apply the *implicit function theorem* – this is allowed since $\boldsymbol{f}$ is analytic and, for sufficiently small $h > 0$, the matrix

$$I - (1-\theta)h\frac{\partial \boldsymbol{f}(t_{n+1}, \boldsymbol{y}(t_{n+1}))}{\partial \boldsymbol{y}}$$

is nonsingular. The conclusion is that

$$\boldsymbol{e}_{n+1} = \boldsymbol{e}_n \begin{cases} -\tfrac{1}{12}h^3\boldsymbol{y}'''(t_n) + \mathcal{O}(h^4), & \theta = \tfrac{1}{2}, \\ +\left(\theta - \tfrac{1}{2}\right)h^2\boldsymbol{y}''(t_n) + \mathcal{O}(h^3), & \theta \neq \tfrac{1}{2}. \end{cases}$$

The theta method is convergent for every $\theta \in [0, 1]$, as can be verified with ease by generalizing the proofs of Theorems 1.1 and 1.2. This is is the subject of Exercise 1.1.

Why, a vigilant reader might ask, bother with the theta method except for the special values $\theta = 1$ and $\theta = \frac{1}{2}$? After all, the first is unique in conferring explicitness and the second is the only second-order theta method. The reasons are threefold. Firstly, the whole concept of order is based on the assumption that the numerical error is concentrated mainly in the leading term of its Taylor expansion. This is true as $h \to 0$, except that the step length, when implemented on a real computer, never actually tends to zero ... Thus, in very special circumstances we might wish to annihilate higher-order terms in the error expansion; for example, letting $\theta = \frac{2}{3}$ gets rid of the $\mathcal{O}(h^3)$ term while retaining the $\mathcal{O}(h^2)$ component. Secondly, the theta method is our first example of a more general approach to the design of numerical algorithms, whereby simple geometric intuition is replaced by a more formal approach based on a Taylor expansion and the implicit function theorem. Its study is a good preparation for the material of Chapters 2 and 3. Finally, the choice $\theta = 0$ is of great practical relevance. The first-order implicit method

$$\boldsymbol{y}_{n+1} = \boldsymbol{y}_n + h\boldsymbol{f}(t_{n+1}, \boldsymbol{y}_{n+1}), \qquad n = 0, 1, \ldots, \tag{1.15}$$

is called the *backward Euler's method* and is a favourite algorithm for the solution of *stiff* ODEs. We defer the discussion of stiff equations to Chapter 4, where the merits of the backward Euler's method and similar schemes will become clear.

Comments and bibliography

An implicit goal of this book is to demonstrate that the computation of differential equations is not about discretizing everything in sight by the first available finite-difference approximation and throwing it on the nearest computer. It is all about designing clever and efficient algorithms and understanding their mathematical features. The narrative of this chapter introduces us to convergence and order, the essential building blocks in this quest to understand discretization methods.

We assume very little knowledge of the analytic (as opposed to numerical) theory of ODEs throughout this volume: just the concepts of existence, uniqueness, the Lipschitz condition and (mainly in Chapter 4) explicit solution of linear initial value systems. In Chapter 5 we will be concerned with more specialized geometric features of ODEs but we take care to explain there all nontrivial issues. A brief résumé of essential knowledge is reviewed in Appendix section A.2.3, but a diligent reader will do well to refresh his or her memory with a thorough look at a reputable textbook, for example Birkhoff & Rota (1978) or Boyce & DiPrima (1986).

Euler's method, the grandaddy of all numerical schemes for differential equations, is introduced in just about every relevant textbook (e.g. Conte & de Boor, 1990; Hairer *et al.*, 1991; Isaacson & Keller, 1966; Lambert, 1991), as is the trapezoidal rule. More traditional books have devoted considerable effort toward proving, with the Euler–Maclaurin formula (Ralston, 1965), that the error of the trapezoidal rule can be expanded in odd powers of h (cf. Exercise 1.8), but it seems that nowadays hardly anybody cares much about this observation, except for its applications to Richardson's extrapolation (Isaacson & Keller, 1966).

We have mentioned in Section 1.2 the Peano kernel theorem. Its knowledge is marginal to the subject matter of this book. However, if you want to understand mathematics and learn a simple, yet beautiful, result in approximation theory, we refer to A.2.2.6 and A.2.2.7 and references therein.

Birkhoff, G. and Rota, G.-C. (1978), *Ordinary Differential Equations* (3rd edn), Wiley, New York.

Boyce, W.E. and DiPrima, R.C. (1986), *Elementary Differential Equations and Boundary Value Problems* (4th edn), Wiley, New York.

Conte, S.D. and de Boor, C. (1990), *Elementary Numerical Analysis: An Algorithmic Approach* (3rd edn), McGraw-Hill Kōgakusha, Tokyo.

Hairer, E, Nørsett, S.P. and Wanner, G. (1991), *Solving Ordinary Differential Equations I: Nonstiff Problems* (2nd edn) Springer-Verlag, Berlin.

Isaacson, E. and Keller, H.B. (1966), *Analysis of Numerical Methods,* Wiley, New York.

Lambert, J.D. (1991), *Numerical Methods for Ordinary Differential Systems,* Wiley, London.

Ralston, A. (1965), *A First Course in Numerical Analysis,* McGraw-Hill Kōgakusha, New York.

Exercises

1.1 Apply the method of proof of Theorems 1.1 and 1.2 to prove the convergence of the implicit midpoint rule (1.12) and of the theta method (1.13).

1.2 The linear system $\boldsymbol{y}' = A\boldsymbol{y}$, $\boldsymbol{y}(0) = \boldsymbol{y}_0$, where A is a symmetric matrix, is solved by Euler's method.

a Letting $\boldsymbol{e}_n = \boldsymbol{y}_n - \boldsymbol{y}(nh)$, $n = 0, 1, \ldots$, prove that

$$\|\boldsymbol{e}_n\|_2 \le \|\boldsymbol{y}_0\|_2 \max_{\lambda \in \sigma(A)} \left|(1+h\lambda)^n - \mathrm{e}^{nh\lambda}\right|,$$

where $\sigma(A)$ is the set of eigenvalues of A and $\|\cdot\|_2$ is the Euclidean matrix norm (cf. A.1.3.3).

b Demonstrate that for every $-1 \ll x \le 0$ and $n = 0, 1, \ldots$ it is true that

$$\mathrm{e}^{nx} - \tfrac{1}{2}nx^2\mathrm{e}^{(n-1)x} \le (1+x)^n \le \mathrm{e}^{nx}.$$

(*Hint: Prove first that* $1 + x \le \mathrm{e}^x$, $1 + x + \frac{1}{2}x^2 \ge \mathrm{e}^x$ *for all* $x \le 0$, *and then argue that, provided* $|a-1|$ *and* $|b|$ *are small, it is true that* $(a-b)^n \ge a^n - na^{n-1}b$.)

c Suppose that the maximal eigenvalue of A is $\lambda_{\max} < 0$. Prove that, as $h \to 0$ and $nh \to t \in [0, t^*]$,

$$\|\boldsymbol{e}_n\|_2 \le \tfrac{1}{2}t\lambda_{\max}^2\mathrm{e}^{\lambda_{\max}t}\|\boldsymbol{y}_0\|_2 h \le \tfrac{1}{2}t^*\lambda_{\max}^2\|\boldsymbol{y}_0\|_2 h.$$

d Compare the order of magnitude of this bound with the upper bound from Theorem 1.1 in the case

$$A = \begin{bmatrix} -2 & 1 \\ 1 & -2 \end{bmatrix}, \qquad t^* = 10.$$

1.3 We solve the scalar linear system $y' = ay$, $y(0) = 1$.

a Show that the 'continuous output' method

$$u(t) = \frac{1 + \frac{1}{2}a(t - nh)}{1 - \frac{1}{2}a(t - nh)} y_n, \qquad nh \le t \le (n+1)h, \quad n = 0, 1, \ldots,$$

is consistent with the values of y_n and y_{n+1} which are obtained by the trapezoidal rule.

b Demonstrate that u obeys the perturbed ODE

$$u'(t) = au(t) + \frac{\frac{1}{4}a^3(t - nh)^2}{[1 - \frac{1}{2}a(t - nh)]^2} y_n, \qquad t \in [nh, (n+1)h],$$

with initial condition $u(nh) = y_n$. Thus, prove that

$$u((n+1)h) = \mathrm{e}^{ha}\left[1 + \tfrac{1}{4}a^3 \int_0^h \frac{\mathrm{e}^{-\tau a}\tau^2 \, \mathrm{d}\tau}{(1 - \frac{1}{2}a\tau)^2}\right] y_n.$$

c Let $e_n = y_n - y(nh)$, $n = 0, 1, \ldots$. Show that

$$e_{n+1} = \mathrm{e}^{ha}\left[1 + \tfrac{1}{4}a^3 \int_0^h \frac{\mathrm{e}^{-\tau a}\tau^2 \, \mathrm{d}\tau}{(1 - \frac{1}{2}a\tau)^2}\right] e_n + \tfrac{1}{4}a^3 \mathrm{e}^{(n+1)ha} \int_0^h \frac{\mathrm{e}^{-\tau a}\tau^2 \, \mathrm{d}\tau}{(1 - \frac{1}{2}a\tau)^2}.$$

In particular, deduce that $a < 0$ implies that the error propagates subject to the inequality

$$|e_{n+1}| \le \mathrm{e}^{ha}\left(1 + \tfrac{1}{4}|a|^3 \int_0^h \mathrm{e}^{-\tau a}\tau^2 \, \mathrm{d}\tau\right)|e_n| + \tfrac{1}{4}|a|^3 \mathrm{e}^{(n+1)ha} \int_0^h \mathrm{e}^{-\tau a}\tau^2 \, \mathrm{d}\tau.$$

1.4 Given $\theta \in [0, 1]$, find the order of the method

$$\boldsymbol{y}_{n+1} = \boldsymbol{y}_n + h\boldsymbol{f}\left(t_n + (1 - \theta)h, \theta\boldsymbol{y}_n + (1 - \theta)\boldsymbol{y}_{n+1}\right).$$

1.5 Provided that $\boldsymbol{f}$ is analytic, it is possible to obtain from $\boldsymbol{y}' = \boldsymbol{f}(t, \boldsymbol{y})$ an expression for the second derivative of $\boldsymbol{y}$, namely $\boldsymbol{y}'' = \boldsymbol{g}(t, \boldsymbol{y})$, where

$$\boldsymbol{g}(t, \boldsymbol{y}) = \frac{\partial \boldsymbol{f}(t, \boldsymbol{y})}{\partial t} + \frac{\partial \boldsymbol{f}(t, \boldsymbol{y})}{\partial \boldsymbol{y}} \boldsymbol{f}(t, \boldsymbol{y}).$$

Find the orders of the methods

$$\boldsymbol{y}_{n+1} = \boldsymbol{y}_n + h\boldsymbol{f}(t_n, \boldsymbol{y}_n) + \tfrac{1}{2}h^2 \boldsymbol{g}(t_n, \boldsymbol{y}_n)$$

and

$$\boldsymbol{y}_{n+1} = \boldsymbol{y}_n + \tfrac{1}{2}h[\boldsymbol{f}(t_n, \boldsymbol{y}_n) + \boldsymbol{f}(t_{n+1}, \boldsymbol{y}_{n+1})] + \tfrac{1}{12}h^2[\boldsymbol{g}(t_n, \boldsymbol{y}_n) - \boldsymbol{g}(t_{n+1}, \boldsymbol{y}_{n+1})].$$

1.6* Assuming that $\boldsymbol{g}$ is Lipschitz, prove that both methods from Exercise 1.5 converge.

1.7 Repeated differentiation of the ODE (1.1), for analytic $\boldsymbol{f}$, yields explicit expressions for functions $\boldsymbol{g}_m$ such that

$$\frac{\mathrm{d}^m \boldsymbol{y}(t)}{\mathrm{d}t^m} = \boldsymbol{g}_m(t, \boldsymbol{y}(t)), \qquad m = 0, 1, \ldots$$

Hence $\boldsymbol{g}_0(t, \boldsymbol{y}) = \boldsymbol{y}$ and $\boldsymbol{g}_1(t, \boldsymbol{y}) = \boldsymbol{f}(t, \boldsymbol{y})$; $\boldsymbol{g}_2$ has been already defined in Exercise 1.5 as $\boldsymbol{g}$.

a Assuming for simplicity that $\boldsymbol{f} = \boldsymbol{f}(\boldsymbol{y})$ (i.e. that the ODE system (1.1) is *autonomous*), derive $\boldsymbol{g}_3$.

b Prove that the mth *Taylor method*

$$\boldsymbol{y}_{n+1} = \sum_{k=0}^{m} \frac{1}{k!} h^k \boldsymbol{g}_k(t_n, \boldsymbol{y}_n), \qquad n = 0, 1, \ldots,$$

is of order m for $m = 1, 2, \ldots$

c Let $\boldsymbol{f}(\boldsymbol{y}) = \Lambda \boldsymbol{y} + \boldsymbol{b}$, where the matrix Λ and the vector $\boldsymbol{b}$ are independent of t. Find the explicit form of $\boldsymbol{g}_m$ for $m = 0, 1, \ldots$ and thereby prove that the mth Taylor method reduces to the recurrence

$$\boldsymbol{y}_{n+1} = \left(\sum_{k=0}^{m} \frac{1}{k!} h^k \Lambda^k \right) \boldsymbol{y}_n + \left(\sum_{k=1}^{m} \frac{1}{k!} h^k \Lambda^{k-1} \right) \boldsymbol{b}, \qquad n = 0, 1, \ldots$$

1.8 Let $\boldsymbol{f}$ be analytic. Prove that, for sufficiently small $h > 0$ and an analytic function $\boldsymbol{x}$, the function

$$\boldsymbol{x}(t+h) - \boldsymbol{x}(t-h) - h\boldsymbol{f}\left(\tfrac{1}{2}(\boldsymbol{x}(t-h) + \boldsymbol{x}(t+h))\right)$$

can be expanded into power series in *odd* powers of h. Deduce that the error in the implicit midpoint rule (1.13), when applied to *autonomous* ODEs $\boldsymbol{y}' = \boldsymbol{f}(\boldsymbol{y})$ also admits an expansion in odd powers of h. (*Hint: First try to prove the statement for a scalar function f. Once you have solved this problem, a generalization should present no difficulties.*)

2

Multistep methods

2.1 The Adams method

A typical numerical method for an initial value ODE system computes the solution on a step-by-step basis. Thus, the Euler method advances the solution from t_0 to t_1 using $\boldsymbol{y}_0$ as an initial value. Next, to advance from t_1 to t_2, we *discard* $\boldsymbol{y}_0$ and employ $\boldsymbol{y}_1$ as the new initial value.

Numerical analysts, however, are thrifty by nature. Why discard a potentially valuable vector $\boldsymbol{y}_0$? Or, with greater generality, why not make the solution depend on several past values, provided that these values are available?

There is one perfectly good reason why not – the exact solution of

$$\boldsymbol{y}' = \boldsymbol{f}(t, \boldsymbol{y}), \quad t \geq t_0, \qquad \boldsymbol{y}(t_0) = \boldsymbol{y}_0 \tag{2.1}$$

is uniquely determined ($\boldsymbol{f}$ being Lipschitz) by a single initial condition. Any attempt to pin the solution down at more than one point is mathematically nonsensical or, at best, redundant. This, however, is valid only with regard to the true solution of (2.1). When it comes to computation, this redundancy becomes our friend and past values of $\boldsymbol{y}$ can be put to a very good use – provided, however, that we are very careful indeed.

Thus let us suppose again that $\boldsymbol{y}_n$ is the numerical solution at $t_n = t_0 + nh$, where $h > 0$ is the step size, and let us attempt to derive an algorithm that intelligently exploits past values. To that end, we assume that

$$\boldsymbol{y}_m = \boldsymbol{y}(t_m) + \mathcal{O}\left(h^{s+1}\right), \qquad m = 0, 1, \ldots, n + s - 1, \tag{2.2}$$

where $s \geq 1$ is a given integer. Our wish being to advance the solution from t_{n-s+1} to t_{n+s}, we commence from the trivial identity

$$\boldsymbol{y}(t_{n+s}) = \boldsymbol{y}(t_{n+s-1}) + \int_{t_{n+s-1}}^{t_{n+s}} \boldsymbol{y}'(\tau)\,\mathrm{d}\tau = \boldsymbol{y}(t_{n+s-1}) + \int_{t_{n+s-1}}^{t_{n+s}} \boldsymbol{f}(\tau, \boldsymbol{y}(\tau))\,\mathrm{d}\tau. \tag{2.3}$$

Wishing to exploit (2.3) for computational ends, we note that the integral on the right incorporates $\boldsymbol{y}$ not just at the grid points – where approximations are available – but throughout the interval $[t_{n+s-1}, t_{n+s}]$. The main idea of an *Adams method* is to use past values of the solution to approximate $\boldsymbol{y}'$ in the interval of integration. Thus, let $\boldsymbol{p}$ be an interpolation polynomial (cf. A.2.2.1–A.2.2.5) that matches $\boldsymbol{f}(t_m, \boldsymbol{y}_m)$ for $m = n, n+1, \ldots, n+s-1$. Explicitly,

$$\boldsymbol{p}(t) = \sum_{m=0}^{s-1} p_m(t) \boldsymbol{f}(t_{n+m}, \boldsymbol{y}_{n+m}),$$

where the functions

$$p_m(t) = \prod_{\substack{\ell=0\\ \ell\neq m}}^{s-1} \frac{t - t_{n+\ell}}{t_{n+m} - t_{n+\ell}} = \frac{(-1)^{s-1-m}}{m!(s-1-m)!} \prod_{\substack{\ell=0\\ \ell\neq m}}^{s-1} \left(\frac{t-t_n}{h} - \ell\right), \tag{2.4}$$

for every $m = 0, 1, \ldots, s-1$, are *Lagrange interpolation polynomials*. It is an easy exercise to verify that indeed $\boldsymbol{p}(t_m) = \boldsymbol{f}(t_m, \boldsymbol{y}_m)$ for all $m = n, n+1, \ldots, n+s-1$. Hence, (2.2) implies that $\boldsymbol{p}(t_m) = \boldsymbol{y}'(t_m) + \mathcal{O}(h^s)$ for this range of m. We now use interpolation theory from A.2.2.2 to argue that, $\boldsymbol{y}$ being sufficiently smooth,

$$\boldsymbol{p}(t) = \boldsymbol{y}'(t) + \mathcal{O}(h^s), \qquad t \in [t_{n+s-1}, t_{n+s}].$$

We next substitute $\boldsymbol{p}$ in the integrand of (2.3), replace $\boldsymbol{y}(t_{n+s-1})$ by $\boldsymbol{y}_{n+s-1}$ there and, having integrated along an interval of length h, incur an error of $\mathcal{O}(h^{s+1})$. In other words, the method

$$\boldsymbol{y}_{n+s} = \boldsymbol{y}_{n+s-1} + h\sum_{m=0}^{s-1} b_m \boldsymbol{f}(t_{n+m}, \boldsymbol{y}_{n+m}), \tag{2.5}$$

where

$$b_m = h^{-1}\int_{t_{n+s-1}}^{t_{n+s}} p_m(\tau)\,\mathrm{d}\tau = h^{-1}\int_0^h p_m(t_{n+s-1}+\tau)\,\mathrm{d}\tau, \qquad m = 0, 1, \ldots, s-1,$$

is of order $p = s$. Note from (2.4) that the coefficients $b_0, b_1, \ldots, b_{s-1}$ are independent of n and of h; thus we can subsequently use them to advance the iteration from t_{n+s} to t_{n+s+1} and so on.

The scheme (2.5) is called the s-step *Adams–Bashforth* method.

Having derived explicit expressions, it is easy to state Adams–Bashforth methods for moderate values of s. Thus, for $s = 1$ we encounter our old friend, the Euler method, whereas $s = 2$ gives

$$\boldsymbol{y}_{n+2} = \boldsymbol{y}_{n+1} + h\left[\tfrac{3}{2}\boldsymbol{f}(t_{n+1}, \boldsymbol{y}_{n+1}) - \tfrac{1}{2}\boldsymbol{f}(t_n, \boldsymbol{y}_n)\right] \tag{2.6}$$

and $s = 3$ gives

$$\boldsymbol{y}_{n+3} = \boldsymbol{y}_{n+2} + h\left[\tfrac{23}{12}\boldsymbol{f}(t_{n+2}, \boldsymbol{y}_{n+2}) - \tfrac{4}{3}\boldsymbol{f}(t_{n+1}, \boldsymbol{y}_{n+1}) + \tfrac{5}{12}\boldsymbol{f}(t_n, \boldsymbol{y}_n)\right]. \tag{2.7}$$

Figure 2.1 displays the logarithm of the error in the solution of $y' = -y^2$, $y(0) = 1$, by Euler's method and the schemes (2.6) and (2.7). The important information can be read off the y-scale: when h is halved, say, Euler's error decreases linearly, the error of (2.6) decays quadratically and (2.7) displays cubic decay. This is hardly surprising, since the order of the s-step Adams–Bashforth method is, after all, s and the global error decays as $\mathcal{O}(h^s)$.

Adams–Bashforth methods are just one instance of multistep methods. In the remainder of this chapter we will encounter several other families of such schemes. Later in this book we will learn that different multistep methods are suitable in different situations. First, however, we need to study the general theory of order and convergence.

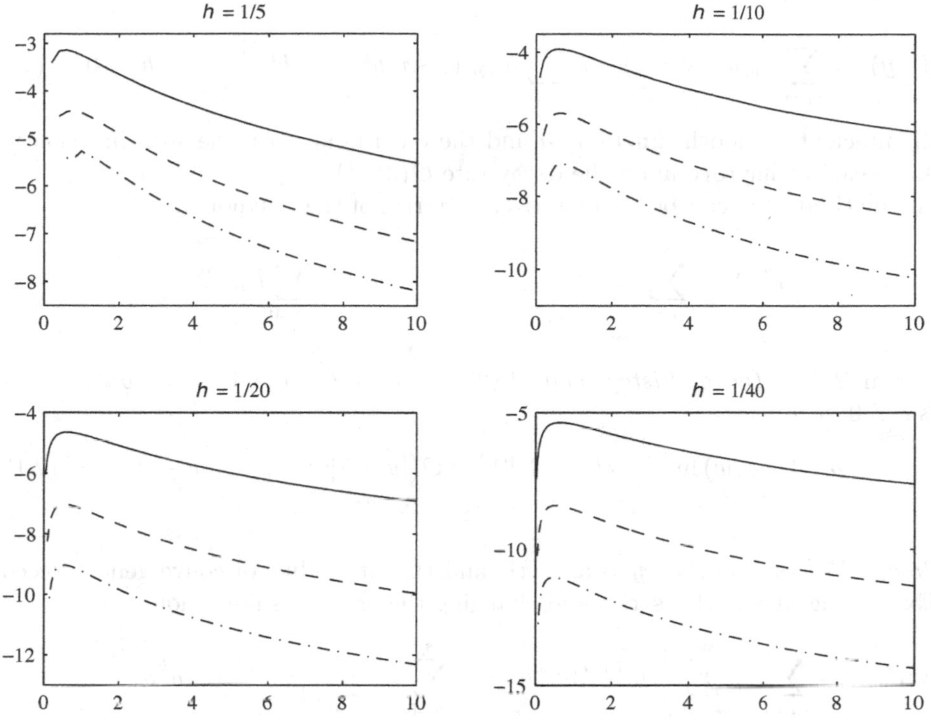

Figure 2.1 Plots of $\ln |y_n - y(t_n)|$ for the first three Adams–Bashforth methods, as applied to the equation $y' = -y^2$, $y(0) = 1$. Euler's method, (2.6) and (2.7) correspond to the solid, broken and broken-and-dotted lines respectively.

2.2 Order and convergence of multistep methods

We write a general s-step method in the form

$$\sum_{m=0}^{s} a_m \boldsymbol{y}_{n+m} = h \sum_{m=0}^{s} b_m \boldsymbol{f}(t_{n+m}, \boldsymbol{y}_{n+m}), \qquad n = 0, 1, \ldots, \tag{2.8}$$

where a_m, b_m, $m = 0, 1, \ldots, s$, are given constants, independent of h, n and the underlying differential equation. It is conventional to normalize (2.8) by letting $a_s = 1$. When $b_s = 0$ (as is the case with the Adams–Bashforth method) the method is said to be *explicit;* otherwise it is *implicit.*

Since we are about to encounter several criteria that play an important role in choosing the coefficients a_m and b_m, a central consideration is to obtain a reasonable value of the order. Recasting the definition from Chapter 1, we note that the method

(2.8) is of *order* $p \geq 1$ if and only if

$$\psi(t, \boldsymbol{y}) := \sum_{m=0}^{s} a_m \boldsymbol{y}(t+mh) - h \sum_{m=0}^{s} b_m \boldsymbol{y}'(t+mh) = \mathcal{O}\big(h^{p+1}\big), \qquad h \to 0, \tag{2.9}$$

for all sufficiently smooth functions $\boldsymbol{y}$ and there exists at least one such function for which we cannot improve upon the decay rate $\mathcal{O}\big(h^{p+1}\big)$.

The method (2.8) can be characterized in terms of the polynomials

$$\rho(w) := \sum_{m=0}^{s} a_m w^m \qquad \text{and} \qquad \sigma(w) := \sum_{m=0}^{s} b_m w^m.$$

Theorem 2.1 *The multistep method (2.8) is of order $p \geq 1$ if and only if there exists $c \neq 0$ such that*

$$\rho(w) - \sigma(w) \ln w = c(w-1)^{p+1} + \mathcal{O}\big(|w-1|^{p+2}\big), \qquad w \to 1. \tag{2.10}$$

Proof We assume that $\boldsymbol{y}$ is analytic and that its radius of convergence exceeds sh. Expanding in a Taylor series and changing the order of summation,

$$\begin{aligned}
\psi(t, \boldsymbol{y}) &= \sum_{m=0}^{s} a_m \sum_{k=0}^{\infty} \frac{1}{k!} \boldsymbol{y}^{(k)}(t) m^k h^k - h \sum_{m=0}^{s} b_m \sum_{k=0}^{\infty} \frac{1}{k!} \boldsymbol{y}^{(k+1)}(t) m^k h^k \\
&= \left(\sum_{m=0}^{s} a_m \right) \boldsymbol{y}(t) + \sum_{k=1}^{\infty} \frac{1}{k!} \left(\sum_{m=0}^{s} m^k a_m - k \sum_{m=0}^{s} m^{k-1} b_m \right) h^k \boldsymbol{y}^{(k)}(t).
\end{aligned}$$

Thus, to obtain order p it is neccesary and sufficient that

$$\begin{gathered}
\sum_{m=0}^{s} a_m = 0, \qquad \sum_{m=0}^{s} m^k a_m = k \sum_{m=0}^{s} m^{k-1} b_m, \qquad k = 1, 2, \ldots, p. \\
\sum_{m=0}^{s} m^{p+1} a_m \neq (p+1) \sum_{m=0}^{s} m^p b_m.
\end{gathered} \tag{2.11}$$

Let $w = \mathrm{e}^z$; then $w \to 1$ corresponds to $z \to 0$. Expanding again in a Taylor series,

$$\begin{aligned}
\rho(\mathrm{e}^z) - z\sigma(\mathrm{e}^z) &= \sum_{m=0}^{s} a_m \mathrm{e}^{mz} - z \sum_{m=0}^{s} b_m \mathrm{e}^{mz} \\
&= \sum_{m=0}^{s} a_m \left(\sum_{k=0}^{\infty} \frac{1}{k!} m^k z^k \right) - z \sum_{m=0}^{s} b_m \left(\sum_{k=0}^{\infty} \frac{1}{k!} m^k z^k \right) \\
&= \sum_{k=0}^{\infty} \frac{1}{k!} \left(\sum_{m=0}^{s} m^k a_m \right) z^k - \sum_{k=1}^{\infty} \frac{1}{(k-1)!} \left(\sum_{m=0}^{s} m^{k-1} b_m \right) z^k.
\end{aligned}$$

Therefore

$$\rho(\mathrm{e}^z) - z\sigma(\mathrm{e}^z) = cz^{p+1} + \mathcal{O}\big(z^{p+2}\big)$$

for some $c \neq 0$ if and only if (2.11) is true. The theorem follows by restoring $w = e^z$. ■

An alternative derivation of the order conditions (2.11) assists in our understanding of them. The map $\boldsymbol{y} \mapsto \boldsymbol{\psi}(t, \boldsymbol{y})$ is linear, consequently $\boldsymbol{\psi}(t, \boldsymbol{y}) = \mathcal{O}(h^{p+1})$, if and only if $\boldsymbol{\psi}(t, q) = 0$ for every polynomial q of degree p. Because of linearity, this is equivalent to

$$\boldsymbol{\psi}(t, q_k) = 0, \qquad k = 0, 1, \ldots, p,$$

where $\{q_0, q_1, \ldots, q_p\}$ is a basis of the $(p+1)$-dimensional space of p-degree polynomials (see A.2.1.2, A.2.1.3). Setting $q_k(t) = t^k$ for $k = 0, 1, \ldots, p$, we immediately obtain (2.11).

◇ **Adams–Bashforth revisited ...** Theorem 2.1 obviates the need for 'special tricks' such as were used in our derivation of the Adams–Bashforth methods in Section 2.1. Given any multistep scheme (2.8), we can verify its order by a fairly painless expansion into series. It is convenient to express everything in the currency $\xi := w - 1$. For example, (2.6) results in

$$\rho(w) - \sigma(w) \ln w = (\xi + \xi^2) - \left(1 + \tfrac{3}{2}\xi\right)\left(\xi - \tfrac{1}{2}\xi^2 + \tfrac{1}{3}\xi^3 + \cdots\right) = \tfrac{5}{12}\xi^3 + \mathcal{O}(\xi^4);$$

thus order 2 is validated. Likewise, we can check that (2.7) is indeed of order 3 from the expansion

$$\begin{aligned}\rho(w) - \sigma(w) \ln w &= \xi + 2\xi^2 + \xi^3 \\ &\quad - \left(1 + \tfrac{5}{2}\xi + \tfrac{23}{12}\xi^2\right)\left(\xi - \tfrac{1}{2}\xi^2 + \tfrac{1}{3}\xi^3 - \tfrac{1}{4}\xi^4 + \cdots\right) \\ &= \tfrac{3}{8}\xi^4 + \mathcal{O}(\xi^5).\end{aligned}$$

◇

Nothing, unfortunately, could be further from good numerical practice than to assess a multistep method solely – or primarily – in terms of its order. Thus, let us consider the two-step implicit scheme

$$\boldsymbol{y}_{n+2} - 3\boldsymbol{y}_{n+1} + 2\boldsymbol{y}_n = h\left[\tfrac{13}{12}\boldsymbol{f}(t_{n+2}, \boldsymbol{y}_{n+2}) - \tfrac{5}{3}\boldsymbol{f}(t_{n+1}, \boldsymbol{y}_{n+1}) - \tfrac{5}{12}\boldsymbol{f}(t_n, \boldsymbol{y}_n)\right]. \quad (2.12)$$

It is easy to ascertain that the order of (2.12) is 2. Encouraged by this – and not being very ambitious – we will attempt to use this method to solve numerically the exceedingly simple equation $y' \equiv 0$, $y(0) = 1$. A single step reads $y_{n+2} - 3y_{n+1} + 2y_n = 0$, a recurrence relation whose general solution is $y_n = c_1 + c_2 2^n$, $n = 0, 1, \ldots$, where $c_1, c_2 \in \mathbb{R}$ are arbitrary. Suppose that $c_2 \neq 0$; we need both y_0 and y_1 to launch time-stepping and it is trivial to verify that $c_2 \neq 0$ is equivalent to $y_1 \neq y_0$. It is easy to prove that the method fails to converge. Thus, choose $t > 0$ and let $h \to 0$ so that $nh \to t$. Obviously $n \to \infty$ and this implies that $|y_n| \to \infty$, which is far from the exact value $y(t) \equiv 1$.

The failure in convergence does not require, realistically, that $c_2 \neq 0$ be induced by y_1. Any calculation on a real computer introduces a roundoff error which, sooner or later, is bound to render $c_2 \neq 0$ and so bring about a geometric growth in the error of the method.

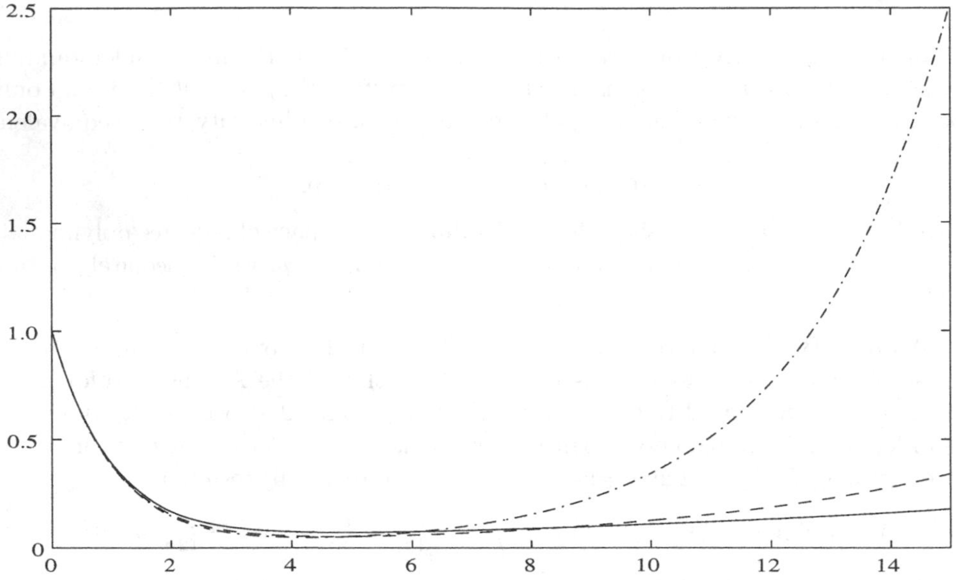

Figure 2.2 The breakdown in the numerical solution of $y' = -y$, $y(0) = 1$, by a nonconvergent numerical scheme, showing how the situation worsens with decreasing step size. The solid, broken and broken-and-dotted lines denote $h = \frac{1}{10}, \frac{1}{20}$ and $\frac{1}{40}$ respectively.

Needless to say, a method that cannot integrate the simplest possible ODE with any measure of reliability should not be used for more substantial computational ends. Nontrivial order is not sufficient to ensure convergence! The need thus arises for a criterion that allows us to discard bad methods and narrow the field down to convergent multistep schemes.

◇ **Failure to converge** Suppose that the linear equation $y' = -y$, $y(0) = 1$, is solved by a two-step, second-order method with $\rho(w) = w^2 - 2.01w + 1.01$, $\sigma(w) = 0.995w - 1.005$. As will be soon evident, this method also fails the convergence criterion, although not by a wide margin! Figure 2.2 displays three solution trajectories, for progressively decreasing step sizes $h = \frac{1}{10}, \frac{1}{20}, \frac{1}{40}$. In all instances, in its early stages the solution perfectly resembles the decaying exponential, but after a while small perturbations grow at an increasing pace and render the computation meaningless. It is a characteristic of nonconvergent methods that decreasing the step size actually makes matters worse! ◇

We say that a polynomial obeys the *root condition* if all its zeros reside in the closed complex unit disc and all its zeros of unit modulus are simple.

Theorem 2.2 (The Dahlquist equivalence theorem) *Suppose that the error in the starting values $\boldsymbol{y}_1, \boldsymbol{y}_2, \ldots, \boldsymbol{y}_{s-1}$ tends to zero as $h \to 0+$. The multistep method (2.8) is convergent if and only if it is of order $p \geq 1$ and the polynomial ρ obeys the root condition.* ■

It is important to make crystal clear that convergence is not simply another attribute of a numerical method, to be weighed alongside its other features. If a method is not convergent – and regardless of how attractive it may look – **do not use it!**

Theorem 2.2 allows us to discard method (2.12) without further ado, since $\rho(w) = (w-1)(w-2)$ violates the root condition. Of course, this method is contrived and, even were it convergent, it is doubtful whether it would have been of much interest. However, more 'respectable' methods fail the convergence test. For example, the method

$$\begin{aligned} &\boldsymbol{y}_{n+3} + \tfrac{27}{11}\boldsymbol{y}_{n+2} - \tfrac{27}{11}\boldsymbol{y}_{n+1} - \boldsymbol{y}_n \\ &= h\left[\tfrac{3}{11}\boldsymbol{f}(t_{n+3}, \boldsymbol{y}_{n+3}) + \tfrac{27}{11}\boldsymbol{f}(t_{n+2}, \boldsymbol{y}_{n+2}) + \tfrac{27}{11}\boldsymbol{f}(t_{n+1}, \boldsymbol{y}_{n+1}) + \tfrac{3}{11}\boldsymbol{f}(t_n, \boldsymbol{y}_n)\right] \end{aligned}$$

is of order 6; it is the *only* three-step method that attains this order! Unfortunately,

$$\rho(w) = (w-1)\left(w + \frac{19 + 4\sqrt{15}}{11}\right)\left(w + \frac{19 - 4\sqrt{15}}{11}\right)$$

and the root condition fails. However, note that Adams–Bashforth methods are safe for all $s \geq 1$, since $\rho(w) = w^{s-1}(w-1)$.

◇ **Analysis and algebraic conditions** Theorem 2.2 demonstrates a state of affairs that prevails throughout mathematical analysis. Thus, we desire to investigate an *analytic* condition, e.g. whether a differential equation has a solution, whether a continuous dynamical system is asymptotically stable, whether a numerical method converges. By their very nature, analytic concepts involve infinite processes and continua, hence one can expect analytic conditions to be difficult to verify, to the point of unmanageability. For all we know, the human brain (exactly like a digital computer) might be essentially an algebraic machine. It is thus an important goal in mathematical analysis to search for equivalent *algebraic* conditions. The Dahlquist equivalence theorem is a remarkable example of this: everything essentially reduces to determining whether the zeros of a polynomial reside in a unit disc, and this can be checked in a finite number of algebraic operations! In the course of this book we will encounter numerous other examples of this state of affairs. Cast your mind back to basic infinitesimal calculus and you are bound to recall further instances where analytic problems are rendered in an algebraic language. ◇

The multistep method (2.8) has $2s+1$ parameters. Had order been the sole consideration, we could have utilized all the available degrees of freedom to maximize it. The outcome, an (implicit) s-step method of order $2s$, is unfortunately not convergent for $s \geq 3$ (we have already seen the case $s = 3$). In general, it is possible to prove that the maximal order of a convergent s-step method (2.8) is at most $2\lfloor(s+2)/2\rfloor$ for implicit schemes and just s for explicit ones; this is known as the *Dahlquist first barrier*.

The usual practice is to employ orders $s+1$ and s for s-step implicit and explicit methods respectively. An easy procedure for constructing such schemes is as follows. Choose an arbitrary s-degree polynomial ρ that obeys the root condition and such that $\rho(1) = 0$ (according to (2.11), $\rho(1) = \sum a_m = 0$ is necessary for order $p \geq 1$). Dividing the order condition (2.10) by $\ln w$ we obtain

$$\sigma(w) = \frac{\rho(w)}{\ln w} + \mathcal{O}(|w-1|^p). \tag{2.13}$$

(Note that division by $\ln w$ shaves off a power of $|w-1|$ and that the singularity at $w = 1$ in the numerator and the denominator is removable.) Suppose first that $p = s+1$ and no restrictions are placed on σ. We expand the fraction in (2.13) into a Taylor series about $w = 1$ and let σ be the sth-degree polynomial that matches the series up to $\mathcal{O}(|w-1|^{s+1})$. The outcome is a convergent, s-step method of order $s+1$. Likewise, to obtain an explicit method of order s, we let σ be an $(s-1)$th-degree polynomial (to force $b_m = 0$) that matches the series up to $\mathcal{O}(|w-1|^s)$.

Let us, for example, choose $s = 2$ and $\rho(w) = w^2 - w$. Letting, as before, $\xi = w - 1$, we have

$$\begin{aligned}\frac{\rho(w)}{\ln w} &= \frac{\xi + \xi^2}{\xi - \frac{1}{2}\xi^2 + \frac{1}{3}\xi^3 + \mathcal{O}(\xi^4)} = \frac{1+\xi}{1 - \frac{1}{2}\xi + \frac{1}{3}\xi^2} + \mathcal{O}(\xi^3)\\ &= (1+\xi)\left(1 + \tfrac{1}{2}\xi - \tfrac{1}{12}\xi^2\right) + \mathcal{O}(\xi^3) = 1 + \tfrac{3}{2}\xi + \tfrac{5}{12}\xi^2 + \mathcal{O}(\xi^3).\end{aligned}$$

Thus, for quadratic σ and order 3 we truncate, obtaining

$$\sigma(w) = 1 + \tfrac{3}{2}(w-1) + \tfrac{5}{12}(w-1)^2 = -\tfrac{1}{12} + \tfrac{2}{3}w + \tfrac{5}{12}w^2,$$

whereas in the explicit case where σ is linear we have $p = 2$, and so recover, unsurprisingly, the Adams–Bashforth scheme (2.6).

The choice $\rho(w) = w^{s-1}(w-1)$ is associated with *Adams* methods. We have already seen the explicit Adams–Bashforth schemes; their implicit counterparts are Adams–Moulton methods. However, provided that we wish to maximize the order subject to convergence, without placing any extra constraints on the multistep method, Adams schemes are the most reasonable choice. After all, if – as implied in the statement of Theorem 2.2 – large zeros of ρ are bad, it makes perfect sense to drive as many zeros as we can to the origin!

2.3 Backward differentiation formulae

Classical texts in numerical analysis present several distinct families of multistep methods. For example, letting $\rho(w) = w^{s-2}(w^2-1)$ leads to s-order explicit *Nystrom* methods and and to implicit *Milne* methods of order $s+1$ (see Exercise 2.3). However, in a well-defined yet important situation, certain multistep methods are significantly better than other schemes of the type (2.8). These are the *backward differentiation formulae* (BDFs), whose importance will become apparent in Chapter 4.

An s-order s-step method is said to be a BDF if $\sigma(w) = \beta w^s$ for some $\beta \in \mathbb{R} \setminus \{0\}$.

Lemma 2.3 *For a BDF we have*

$$\beta = \left(\sum_{m=1}^{s} \frac{1}{m}\right)^{-1} \quad \textit{and} \quad \rho(w) = \beta \sum_{m=1}^{s} \frac{1}{m} w^{s-m}(w-1)^m. \tag{2.14}$$

Proof The order being $p = s$, (2.10) implies that

$$\rho(w) - \beta w^s \ln w = \mathcal{O}\left(|w-1|^{s+1}\right), \qquad w \to 1.$$

We substitute $v = w^{-1}$, hence

$$v^s \rho(v^{-1}) = -\beta \ln v + \mathcal{O}\left(|v-1|^{s+1}\right), \qquad v \to 1.$$

Since

$$\ln v = \ln[1 + (v-1)] = \sum_{m=1}^{s} \frac{(-1)^{m-1}}{m}(v-1)^m + \mathcal{O}\left(|v-1|^{s+1}\right),$$

we deduce that

$$v^s \rho(v^{-1}) = \beta \sum_{m=1}^{s} \frac{(-1)^m}{m}(v-1)^m.$$

Therefore

$$\begin{aligned}\rho(w) &= \beta v^{-s} \sum_{m=1}^{s} \frac{(-1)^m}{m}(v-1)^m = \beta \sum_{m=1}^{s} \frac{(-1)^m}{m} w^s (w^{-1}-1)^m \\ &= \beta \sum_{m=1}^{s} \frac{1}{m} w^{s-m}(w-1)^m.\end{aligned}$$

To complete the proof of (2.14), we need only to derive the explicit form of β. It follows at once by imposing the normalization condition $a_s = 1$ on the polynomial ρ. ■

The simplest BDF has been already encountered in Chapter 1: when $s = 1$ we recover the backward Euler method (1.15). The next two BDFs are

$$s = 2, \qquad \boldsymbol{y}_{n+2} - \tfrac{4}{3}\boldsymbol{y}_{n+1} + \tfrac{1}{3}\boldsymbol{y}_n = \tfrac{2}{3}h\boldsymbol{f}(t_{n+2}, \boldsymbol{y}_{n+2}), \tag{2.15}$$

$$s = 3, \qquad \boldsymbol{y}_{n+3} - \tfrac{18}{11}\boldsymbol{y}_{n+2} + \tfrac{9}{11}\boldsymbol{y}_{n+1} - \tfrac{2}{11}\boldsymbol{y}_n = \tfrac{6}{11}h\boldsymbol{f}(t_{n+3}, \boldsymbol{y}_{n+3}). \tag{2.16}$$

Their derivation is trivial; for example, (2.16) follows by letting $s = 3$ in (2.14). Therefore

$$\beta = \frac{1}{1 + \frac{1}{2} + \frac{1}{3}} = \tfrac{6}{11}$$

and

$$\rho(w) = \tfrac{6}{11}\left[w^2(w-1) + \tfrac{1}{2}w(w-1)^2 + \tfrac{1}{3}(w-1)^3\right] = w^3 - \tfrac{18}{11}w^2 + \tfrac{9}{11}w - \tfrac{2}{11}.$$

Since BDFs are derived by specifying σ, we cannot be sure that the polynomial ρ of (2.14) obeys the root condition. In fact, the root condition fails for all but a few such methods.

Theorem 2.4 *The polynomial* (2.14) *obeys the root condition and the underlying BDF method is convergent if and only if* $1 \le s \le 6$. ■

Fortunately, the 'good' range of s is sufficient for all practical considerations.

Underscoring the importance of BDFs, we present a simple example that demonstrates the limitations of Adams schemes; we hasten to emphasize that this is by way of a trailer for our discussion of *stiff* ODEs in Chapter 4.

Let us consider the linear ODE system

$$\boldsymbol{y}' = \begin{bmatrix} -20 & 10 & 0 & \cdots & 0 \\ 10 & -20 & \ddots & \ddots & \vdots \\ 0 & \ddots & \ddots & \ddots & 0 \\ \vdots & \ddots & \ddots & -20 & 10 \\ 0 & \cdots & 0 & 10 & -20 \end{bmatrix} \boldsymbol{y}, \qquad \boldsymbol{y}(0) = \begin{bmatrix} 1 \\ 1 \\ \vdots \\ 1 \\ 1 \end{bmatrix}. \tag{2.17}$$

We will encounter in this book numerous instances of similar systems; (2.17) is a handy paradigm for many linear ODEs that occur in the context of discretization of the partial differential equations of evolution.

Figure 2.3 displays the Euclidean norm of the solution of (2.17) by the second-order Adams–Bashforth method (2.6), with two (slightly) different step sizes, $h = 0.027$ (the solid line) and $h = 0.0275$ (the broken line). The solid line is indistinguishable in the figure from the norm of the true solution, which approaches zero as $t \to \infty$. Not so the norm for $h = 0.0275$: initially, it shadows the correct value pretty well but, after a while, it runs away. The whole qualitative picture is utterly false! And, by the way, things rapidly get considerably worse when h is increased: for $h = 0.028$ the norm reaches 2.5×10^4, while for $h = 0.029$ it shoots to 1.3×10^{11}.

What is the mechanism that degrades the numerical solution and renders it so sensitive to small changes in h? At the moment it suffices to state that the quality of local approximation (which we have quantified in the concept of 'order') is not to blame; taking the third-order scheme (2.7) in place of the current method would have only made matters worse. However, were we to attempt the solution of this ODE with (2.15), say, and with any $h > 0$ then the norm would tend to zero in tandem with the exact solution. In other words, methods such as BDFs are singled out by a favourable property that makes them the methods of choice for important classes of ODEs. Much more will be said about this in Chapter 4.

Comments and bibliography

There are several ways of introducing the theory of multistep methods. Traditional texts have emphasized the derivation of schemes by various interpolation formulae. The approach of Section 2.1 harks back to this approach, as does the name 'backward differentiation formula'.

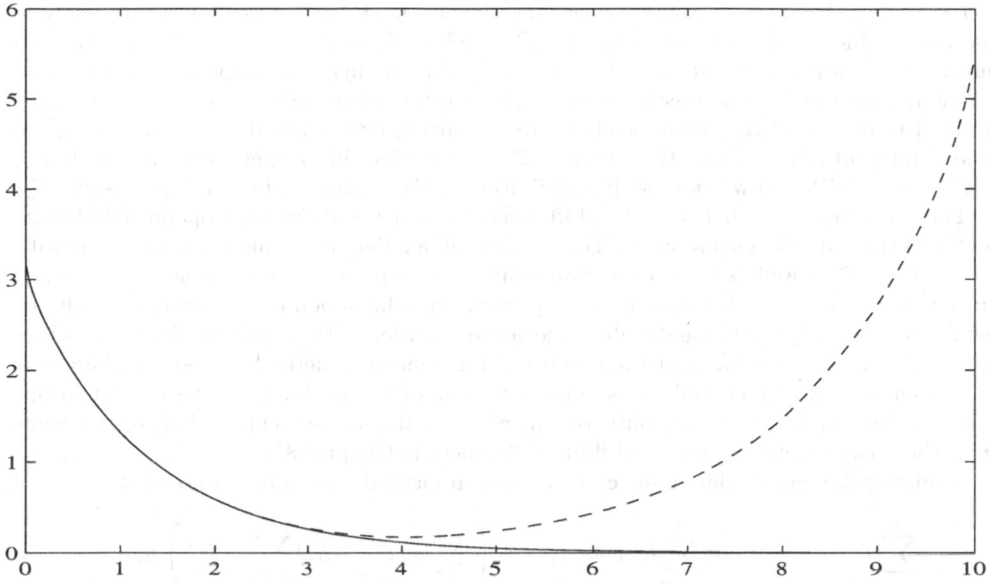

Figure 2.3 The norm of the numerical solution of (2.17) by the Adams–Bashforth method (2.6) for $h = 0.027$ (solid line) and $h = 0.0275$ (broken line).

Other books derive order conditions by sheer brute force, requiring that the multistep formula (2.8) be exact for all polynomials of degree p, since this is equivalent to requiring order p. Equation (2.8) can be expressed as a linear system of $p+1$ equations in the $2s+1$ unknowns $a_0, a_1, \ldots, a_{s-1}, b_0, b_1, \ldots, b_s$. A solution of this system yields a multistep method of the requisite order (of course, we must check it for convergence!), although this procedure does not add much to our understanding of such methods.[1] Linking order with an approximation of the logarithm, along the lines of Theorem 2.1, elucidates matters on a considerably more profound level. This can be shown by the following hand-waving argument.

Given an analytic function g, say, and a number $h > 0$, we denote $g_n^{(k)} = g^{(k)}(t_0 + hn)$, $k, n = 0, 1, \ldots$, and define two operators that map such 'grid functions' into themselves, the *shift* operator $\mathcal{E} g_n^{(k)} := g_{n+1}^{(k)}$ and the *differential* operator $\mathcal{D} g_n^{(k)} := g_n^{(k+1)}$, $k, n = 0, 1, \ldots$ (see Section 8.1). Expanding in a Taylor series about $t_0 + nh$,

$$\mathcal{E} g_n^{(k)} = \sum_{\ell=0}^{\infty} \frac{1}{\ell!} g_n^{(k+\ell)} h^\ell = \left(\sum_{\ell=0}^{\infty} \frac{1}{\ell!} (h\mathcal{D})^\ell \right) g_n^{(k)}, \qquad k, n = 0, 1, \ldots$$

Since this is true for every analytic g with a radius of convergence exceeding h, it follows that, at least formally, $\mathcal{E} = \exp(h\mathcal{D})$. The exponential of the operator, exactly like the more familiar matrix exponential, is defined by a Taylor series.

The above argument can be tightened at the price of some mathematical sophistication. The main problem with naively defining $\mathcal{E}$ as the exponential of $h\mathcal{D}$ is that, in the standard spaces beloved by mathematicians, $\mathcal{D}$ is not a *bounded* linear operator. To recover boundedness we need to resort to a more exotic space.

[1]Though low on insight and beauty, brute force techniques are occasionally useful in mathematics just as in more pedestrian walks of life.

Let $\mathbb{U} \subseteq \mathbb{C}$ be an open connected set and denote by $\mathcal{A}(\mathbb{U})$ the vector space of analytic functions defined in $\mathbb{U}$. The sequence $\{f_n\}_{n=0}^{\infty}$, where $f_n \in \mathcal{A}(\mathbb{U})$, $n = 0, 1, \ldots$, is said to converge to f *locally uniformly* in $\mathcal{A}(\mathbb{U})$ if $f_n \to f$ uniformly in every compact (i.e., closed and bounded) subset of $\mathbb{U}$. It is possible to prove that there exists a metric (a 'distance function') on $\mathcal{A}(\mathbb{U})$ that is consistent with locally uniform convergence and to demonstrate, using the Cauchy integral formula, that the operator $\mathcal{D}$ is a bounded linear operator on $\mathcal{A}(\mathbb{U})$. Hence so is $\mathcal{E} = \exp(h\mathcal{D})$, and we can justify a definition of the exponential via a Taylor series.

The correspondence between the shift operator and the differential operator is fundamental to the numerical solution of ODEs – after all, a differential equation provides us with the action of $\mathcal{D}$ as well as with a function value at a single point, and the act of numerical solution is concerned with (repeatedly) approximating the action of $\mathcal{E}$. Equipped with our new-found knowledge, we should realize that approximation of the exponential function plays (often behind the scenes) a crucial role in designing numerical methods. Later, in Chapter 4, approximations of exponentials, this time with a matrix argument, will be crucial to our understanding of important stability issues, whereas the above-mentioned correspondence forms the basis for our exposition of finite differences in Chapter 8.

Applying the operatorial approach to multistep methods, we note at once that

$$\sum_{m=0}^{s} a_m \boldsymbol{y}(t_{n+m}) - h \sum_{m=0}^{s} b_m \boldsymbol{y}'(t_{n+m}) = \left(\sum_{m=0}^{s} a_m \mathcal{E}^m - h\mathcal{D} \sum_{m=0}^{s} b_m \mathcal{E}^m \right) \boldsymbol{y}(t_n)$$

$$= [\rho(\mathcal{E}) - h\mathcal{D}\sigma(\mathcal{E})]\, \boldsymbol{y}(t_n).$$

Note that $\mathcal{E}$ and $\mathcal{D}$ commute (since $\mathcal{E}$ is given in terms of a power series in $\mathcal{D}$), and this justifies the above formula. Moreover, $\mathcal{E} = \exp(h\mathcal{D})$ means that $h\mathcal{D} = \ln \mathcal{E}$, where the logarithm, again, is defined by means of a Taylor expansion (about the identity operator $\mathcal{I}$). This, in tandem with the observation that $\lim_{h\to 0+} \mathcal{E} = \mathcal{I}$, is the basis to an alternative 'proof' of Theorem 2.1 – a proof that can be made completely rigorous with little effort by employing the implicit function theorem.

The proof of the equivalence theorem (Theorem 2.2) and the establishment of the first barrier (see Section 2.2) by Germund Dahlquist, in 1956 and 1959 respectively, were important milestones in the history of numerical analysis. Not only are these results of great intrinsic impact but they were also instrumental in establishing numerical analysis as a *bona fide* mathematical discipline and imparting a much-needed rigour to numerical thinking. It goes without saying that numerical analysis is not *just* mathematics. It is much more! Numerical analysis is first and foremost about the computation of mathematical models originating in science and engineering. It employs mathematics – and computer science – to an end. Quite often we use a computational algorithm because, although it lacks formal mathematical justification, our experience and intuition tell us that it is efficient and (hopefully) provides the correct answer. There is nothing wrong with this! However, as always in applied mathematics, we must bear in mind the important goal of casting our intuition and experience into a rigorous mathematical framework. Intuition is fallible and experience attempts to infer from incomplete data – mathematics is still the best tool of a computational scientist!

Modern texts in the numerical analysis of ODEs highlight the importance of a structured mathematical approach. The classic monograph of Henrici (1962) is still a model of clear and beautiful exposition and includes an easily digestible proof of the Dahlquist first barrier. Hairer *et al.* (1991) and Lambert (1991) are also highly recommended. In general, books on numerical ODEs fall into two categories: pre-Dahlquist and post-Dahlquist. The first category is nowadays of mainly historical and antiquarian significance.

We will encounter multistep methods again in Chapter 4. As has been already seen in Section 2.3, convergence and reasonable order are far from sufficient for the successful computation of ODEs. The solution of such *stiff* equations requires numerical methods with superior stability properties.

Much of the discussion of multistep methods centres upon their implementation. The present chapter avoids any talk of implementation issues – solution of the (mostly nonlinear) algebraic equations associated with implicit methods, error and step-size control, the choice of the starting values $\boldsymbol{y}_1, \boldsymbol{y}_2, \ldots, \boldsymbol{y}_{s-1}$. Our purpose has been an introduction to multistep schemes and their main properties (convergence, order), as well as a brief survey of the most distinguished members of the multistep methods menagerie. We defer the discussion of implementation issues to Chapters 6 and 7.

Hairer, E., Nørsett, S.P. and Wanner, G. (1991), *Solving Ordinary Differential Equations I: Nonstiff Problems* (2nd edn), Springer-Verlag, Berlin.

Henrici, P. (1962), *Discrete Variable Methods in Ordinary Differential Equations,* Wiley, New York.

Lambert, J.D. (1991), *Numerical Methods for Ordinary Differential Systems,* Wiley, London.

Exercises

2.1 Derive explicitly the three-step and four-step Adams–Moulton methods and the three-step Adams–Bashforth method.

2.2 Let $\eta(z, w) = \rho(w) - z\sigma(w)$.

a Demonstrate that the multistep method (2.8) is of order p if and only if

$$\eta(z, e^z) = cz^{p+1} + \mathcal{O}(z^{p+2}), \qquad z \to 0,$$

for some $c \in \mathbb{R} \setminus \{0\}$.

b Prove that, subject to $\partial\eta(0,1)/\partial w \neq 0$, there exists in a neighbourhood of the origin an analytic function $w_1(z)$ such that $\eta(z, w_1(z)) = 0$ and

$$w_1(z) = e^z - c\left(\frac{\partial\eta(0,1)}{\partial w}\right)^{-1} z^{p+1} + \mathcal{O}(z^{p+2}), \qquad z \to 0. \tag{2.18}$$

c Show that (2.18) is true if the underlying method is convergent. (*Hint: Express* $\partial\eta(0,1)/\partial w$ *in terms of the polynomial* ρ.)

2.3 Instead of (2.3), consider the identity

$$\boldsymbol{y}(t_{n+s}) = \boldsymbol{y}(t_{n+s-2}) + \int_{t_{n+s-2}}^{t_{n+s}} \boldsymbol{f}(\tau, \boldsymbol{y}(\tau))\,d\tau.$$

a Replace $\boldsymbol{f}(\tau, \boldsymbol{y}(\tau))$ by the interpolating polynomial $\boldsymbol{p}$ from Section 2.1 and substitute $\boldsymbol{y}_{n+s-2}$ in place of $\boldsymbol{y}(t_{n+s-2})$. Prove that the resultant explicit *Nystrom* method is of order $p = s$.

b Derive the two-step Nystrom method in a closed form by using the above approach.

c Find the coefficients of the two-step and three-step Nystrom methods by noticing that $\rho(w) = w^{s-2}(w^2 - 1)$ and evaluating σ from (2.13).

d Derive the two-step third-order implicit *Milne* method, again letting $\rho(w) = w^{s-2}(w^2 - 1)$ but allowing σ to be of degree s.

2.4 Determine the order of the three-step method

$$\boldsymbol{y}_{n+3} - \boldsymbol{y}_n = h\left[\tfrac{3}{8}\boldsymbol{f}(t_{n+3}, \boldsymbol{y}_{n+3}) + \tfrac{9}{8}\boldsymbol{f}(t_{n+2}, \boldsymbol{y}_{n+2}) + \tfrac{9}{8}\boldsymbol{f}(t_{n+1}, \boldsymbol{y}_{n+1}) + \tfrac{3}{8}\boldsymbol{f}(t_n, \boldsymbol{y}_n)\right],$$

the *three-eighths* scheme. Is it convergent?

2.5* By solving a three-term recurrence relation, calculate analytically the sequence of values $y_2, y_3, \ldots$ that is generated by the *midpoint rule*

$$\boldsymbol{y}_{n+2} = \boldsymbol{y}_n + 2h\boldsymbol{f}(t_{n+1}, \boldsymbol{y}_{n+1})$$

when it is applied to the differential equation $y' = -y$. Starting from the values $y_0 = 1$, $y_1 = 1 - h$, show that the sequence diverges as $n \to \infty$. Recall, however, from Theorem 2.1 that the root condition, in tandem with order $p \geq 1$ and suitable starting conditions, imply convergence to the true solution in a *finite* interval as $h \to 0+$. Prove that this implementation of the midpoint rule is consistent with the above theorem. (*Hint: Express the roots of the characteristic polynomial of the recurrence relation as* $\exp(\pm \sinh^{-1} h)$.)

2.6 Show that the explicit multistep method

$$\boldsymbol{y}_{n+3} + \alpha_2\boldsymbol{y}_{n+2} + \alpha_1\boldsymbol{y}_{n+1} + \alpha_0\boldsymbol{y}_n = h[\beta_2\boldsymbol{f}(t_{n+2}, \boldsymbol{y}_{n+2}) + \beta_1\boldsymbol{f}(t_{n+1}, \boldsymbol{y}_{n+1}) + \beta_0\boldsymbol{f}(t_n, \boldsymbol{y}_n)]$$

is fourth order only if $\alpha_0 + \alpha_2 = 8$ and $\alpha_1 = -9$. Hence deduce that this method cannot be both fourth order and convergent.

2.7 Prove that the BDFs (2.15) and (2.16) are convergent.

2.8 Find the explicit form of the BDF for $s = 4$.

2.9 An s-step method with $\sigma(w) = w^{s-1}(w + 1)$ and order s might be superior to a BDF in certain situations.

a Find a general formula for ρ and β, along the lines of (2.14).

b Derive explicitly such methods for $s = 2$ and $s = 3$.

c Are the last two methods convergent?

3

Runge–Kutta methods

3.1 Gaussian quadrature

The exact solution of the trivial ordinary differential equation (ODE)

$$y' = f(t), \quad t \geq t_0, \qquad y(t_0) = y_0,$$

whose right-hand side is independent of y, is $y_0 + \int_{t_0}^{t} f(\tau)\,\mathrm{d}\tau$. Since a very rich theory and powerful methods exist to compute integrals numerically, it is only natural to wish to utilize them in the numerical solution of general ODEs

$$\boldsymbol{y}' = \boldsymbol{f}(t, \boldsymbol{y}), \quad t \geq t_0, \qquad \boldsymbol{y}(t_0) = \boldsymbol{y}_0, \tag{3.1}$$

and this is the rationale behind *Runge–Kutta* methods. Before we debate such methods, it is thus fit and proper to devote some attention to the numerical calculation of integrals, a subject of significant importance on its own merit.

It is usual to replace an integral with a finite sum, a procedure known as *quadrature*. Specifically, let ω be a nonnegative function acting in the interval (a, b), such that

$$0 < \int_a^b \omega(\tau)\,\mathrm{d}\tau < \infty, \qquad \left|\int_a^b \tau^j \omega(\tau)\,\mathrm{d}\tau\right| < \infty, \quad j = 1, 2, \ldots;$$

ω is dubbed the *weight function*. We approximate as follows:

$$\int_a^b f(\tau)\omega(\tau)\,\mathrm{d}\tau \approx \sum_{j=1}^{\nu} b_j f(c_j), \tag{3.2}$$

where the numbers $b_1, b_2, \ldots, b_\nu$ and $c_1, c_2, \ldots, c_\nu$, which are independent of the function f (but, in general, depend upon ω, a and b), are called the quadrature *weights* and *nodes,* respectively. Note that we do not require a and b in (3.2) to be bounded; the choices $a = -\infty$ or $b = +\infty$ are perfectly acceptable. Of course, we stipulate $a < b$.

How good is the approximation (3.2)? Suppose that the quadrature matches the integral exactly whenever f is an arbitrary polynomial of degree $p - 1$. It is then easy to prove, e.g. by using the Peano kernel theorem (see A.2.2.6), that, for every function f with p smooth derivatives,

$$\left|\int_a^b f(\tau)\omega(\tau)\,\mathrm{d}\tau - \sum_{j=1}^{\nu} b_j f(c_j)\right| \leq c \max_{a \leq t \leq b} \left|f^{(p)}(t)\right|,$$

where the constant $c > 0$ is independent of f. Such a quadrature formula is said to be of *order* p.

We denote the set of all real polynomials of degree m by $\mathbb{P}_m$. Thus, (3.2) is of order p if it is exact for every $f \in \mathbb{P}_{p-1}$.

Lemma 3.1 *Given any distinct set of nodes $c_1, c_2, \ldots, c_\nu$, it is possible to find a unique set of weights $b_1, b_2, \ldots, b_\nu$ such that the quadrature formula (3.2) is of order $p \geq \nu$.*

Proof Since $\mathbb{P}_{\nu-1}$ is a linear space, it is necessary and sufficient for order ν that (3.2) is exact for elements of an arbitrary basis of $\mathbb{P}_{\nu-1}$. We choose the simplest such basis, namely $\{1, t, t^2, \ldots, t^{\nu-1}\}$, and the order conditions then read

$$\sum_{j=1}^{\nu} b_j c_j^m = \int_a^b \tau^m \omega(\tau)\,\mathrm{d}\tau, \qquad m = 0, 1, \ldots, \nu - 1. \tag{3.3}$$

This is a system of ν equations in the ν unknowns $b_1, b_2, \ldots, b_\nu$, whose matrix, the nodes being distinct, is a nonsingular *Vandermonde* matrix (A.1.2.5). Thus, the system possesses a unique solution and we recover a quadrature of order $p \geq \nu$. ■

The weights $b_1, b_2, \ldots, b_\nu$ can be derived explicitly with little extra effort and we make use of this in (3.14) below. Let

$$p_j(t) = \prod_{\substack{k=1\\k\neq j}}^{\nu} \frac{t - c_k}{c_j - c_k}, \qquad j = 1, 2, \ldots, \nu,$$

be Lagrange polynomials (A.2.2.3). Because

$$\sum_{j=1}^{\nu} p_j(t) g(c_j) = g(t)$$

for every polynomial g of degree $\nu - 1$, it follows that

$$\sum_{j=1}^{\nu} \int_a^b p_j(\tau)\omega(\tau)\,\mathrm{d}\tau\, c_j^m = \int_a^b \left[\sum_{j=1}^{\nu} p_j(\tau) c_j^m\right] \omega(\tau)\,\mathrm{d}\tau = \int_a^b \tau^m \omega(\tau)\,\mathrm{d}\tau$$

for every $m = 0, 1, \ldots, \nu - 1$. Therefore

$$b_j = \int_a^b p_j(\tau)\omega(\tau)\,\mathrm{d}\tau, \qquad j = 1, 2, \ldots, \nu,$$

is the solution of (3.3).

A natural inclination is to choose quadrature nodes that are equispaced in $[a, b]$, and this leads to the so-called *Newton–Cotes* methods. This procedure, however, falls far short of optimal; by making an adroit choice of $c_1, c_2, \ldots, c_\nu$, we can, in fact, double the order to 2ν.

Each weight function ω determines an *inner product* (see A.1.3.1) in the interval (a, b), namely

$$\langle f, g \rangle := \int_a^b f(\tau)g(\tau)\omega(\tau)\,\mathrm{d}\tau,$$

whose domain is the set of all functions f, g such that

$$\int_a^b [f(\tau)]^2\omega(\tau)\,\mathrm{d}\tau, \int_a^b [g(\tau)]^2\omega(\tau)\,\mathrm{d}\tau < \infty.$$

We say that $p_m \in \mathbb{P}_m$, $p_m \not\equiv 0$, is an mth *orthogonal polynomial* (with respect to the weight function ω) if

$$\langle p_m, \hat{p} \rangle = 0, \quad \text{for every} \quad \hat{p} \in \mathbb{P}_{m-1}. \tag{3.4}$$

Orthogonal polynomials are not unique, since we can always multiply p_m by a nonzero constant without violating (3.4). However, it is easy to demonstrate that *monic* orthogonal polynomials are unique. (The coefficient of the highest power of t in a monic polynomial equals unity.) Suppose that both p_m and $\tilde{p}_m$ are monic mth-degree orthogonal polynomials with respect to the same weight function. Then $p_m - \tilde{p}_m \in \mathbb{P}_{m-1}$ and, by (3.4), $\langle p_m, p_m - \tilde{p}_m \rangle = \langle \tilde{p}_m, p_m - \tilde{p}_m \rangle = 0$. We thus deduce from the linearity of the inner product that $\langle p_m - \tilde{p}_m, p_m - \tilde{p}_m \rangle = 0$, and this is possible, according to Appendix subsection A.1.3.1, only if $\tilde{p}_m = p_m$.

Orthogonal polynomials occur in many areas of mathematics; a brief list includes approximation theory, statistics, representation of groups, the theory of ordinary and partial differential equations, functional analysis, quantum groups, coding theory, combinatorics, mathematical physics and, last but not least, numerical analysis.

◇ **Classical orthogonal polynomials** Three families of weights give rise to *classical orthogonal polynomials*.

Let $a = -1$, $b = 1$ and $\omega(t) = (1 - t)^\alpha(1 + t)^\beta$, where $\alpha, \beta > -1$. The underlying orthogonal polynomials are known as *Jacobi* polynomials $P_m^{(\alpha,\beta)}$. We single out for special attention the *Legendre* polynomials P_m, which correspond to $\alpha = \beta = 0$, and the *Chebyshev* polynomials T_m, associated with the choice $\alpha = \beta = -\frac{1}{2}$. Note that for $\min\{\alpha, \beta\} < 0$ the weight function has a singularity at the endpoints ± 1. There is nothing wrong with that, provided ω is integrable in $[0, 1]$; but this is exactly the reason we require $\alpha, \beta > -1$.

The other two 'classics' are the *Laguerre* and *Hermite* polynomials. The Laguerre polynomials $L_m^{(\alpha)}$ are orthogonal with respect to the weight function $\omega(t) = t^\alpha \mathrm{e}^{-t}$, $(a, b) = (0, \infty)$, $\alpha > -1$, whereas the Hermite polynomials H_m are orthogonal in $(a, b) = \mathbb{R}$ with respect to the weight function $\omega(t) = \mathrm{e}^{-t^2}$.

Why are classical orthogonal polynomials so named? Firstly, they have been very extensively studied and occur in a very wide range of applications. Secondly, it is possible to prove that they are singled out by several properties that, in a well-defined sense, render them the 'simplest' orthogonal polynomials. For example – and do not try to prove this on your own! – $P_m^{(\alpha,\beta)}$, $L_m^{(\alpha)}$ and H_m are the only orthogonal polynomials whose derivatives are also orthogonal with some other weight function. ◇

The theory of orthogonal polynomials is replete with beautiful results which, perhaps regrettably, we do not require in this volume. However, one morsel of information, germane to the understanding of quadrature, is about the location of zeros of orthogonal polynomials.

Lemma 3.2 *All m zeros of an orthogonal polynomial p_m reside in the interval (a, b) and they are simple.*

Proof Since

$$\int_a^b p_m(\tau)\omega(\tau)\,\mathrm{d}\tau = \langle p_m, 1\rangle = 0$$

and $\omega \geq 0$, it follows that p_m changes sign at least once in (a, b). Let us thus denote by $x_1, x_2, \ldots, x_k$ all the points in (a, b) where p_m changes sign. We already know that $k \geq 1$. Let us assume that $k \leq m - 1$ and set

$$q(t) := \prod_{j=1}^{k}(t - x_j) = \sum_{i=0}^{k} q_i t^i.$$

Therefore p_m changes sign in (a, b) at exactly the same points as q and the product $p_m q$ does not change sign there at all. The weight function being nonnegative and $p_m q \not\equiv 0$, we deduce on the one hand that

$$\int_a^b p_m(\tau)q(\tau)\omega(\tau)\,\mathrm{d}\tau \neq 0.$$

On the other hand, the orthogonality condition (3.4) and the linearity of the inner product imply that

$$\int_a^b p_m(\tau)q(\tau)\omega(\tau)\,\mathrm{d}\tau = \sum_{i=0}^{k} q_i\langle p_m, t^i\rangle = 0,$$

because $k \leq m - 1$. This is a contradiction and we conclude that $k \geq m$. Since each sign-change of p_m is a zero of the polynomial and, according to the fundamental theorem of algebra, each $\hat{p} \in \mathbb{P}_m \setminus \mathbb{P}_{m-1}$ has exactly m zeros in $\mathbb{C}$, we deduce that p_m has exactly m simple zeros in (a, b). ■

Theorem 3.3 *Let $c_1, c_2, \ldots, c_\nu$ be the zeros of p_ν and let $b_1, b_2, \ldots, b_\nu$ be the solution of the Vandermonde system (3.3). Then*

(i) *The quadrature method (3.2) is of order 2ν;*

(ii) *No other quadrature can exceed this order.*

Proof Let $\hat{p} \in \mathbb{P}_{2\nu-1}$. Applying the Euclidean algorithm to the pair $\{\hat{p}, p_\nu\}$ we deduce that there exist $q, r \in \mathbb{P}_{\nu-1}$ such that $\hat{p} = p_\nu q + r$. Therefore, according to (3.4),

$$\int_a^b \hat{p}(\tau)\omega(\tau)\,\mathrm{d}\tau = \langle p_\nu, q\rangle + \int_a^b r(\tau)\omega(\tau)\,\mathrm{d}\tau = \int_a^b r(\tau)\omega(\tau)\,\mathrm{d}\tau;$$

we recall that $\deg q \le \nu - 1$. Moreover,

$$\sum_{j=1}^{\nu} b_j \hat{p}(c_j) = \sum_{j=1}^{\nu} b_j p_\nu(c_j) q(c_j) + \sum_{j=1}^{\nu} b_j r(c_j) = \sum_{j=1}^{\nu} b_j r(c_j)$$

because $p_\nu(c_j) = 0$, $j = 1, 2, \ldots, \nu$. Finally, $r \in \mathbb{P}_{\nu-1}$ and Lemma 3.1 imply

$$\int_a^b r(\tau)\omega(\tau)\,\mathrm{d}\tau = \sum_{j=1}^{\nu} b_j r(c_j).$$

We thus deduce that

$$\int_a^b \hat{p}(\tau)\omega(\tau)\,\mathrm{d}\tau = \sum_{j=1}^{\nu} b_j \hat{p}(c_j), \qquad \hat{p} \in \mathbb{P}_{2\nu-1},$$

and that the quadrature formula is of order $p \ge 2\nu$.

To prove (ii) (and, incidentally, to affirm that $p = 2\nu$, thereby completing the proof of (i)) we assume that, for some choice of weights $b_1, b_2, \ldots, b_\nu$ and nodes $c_1, c_2, \ldots, c_\nu$, the quadrature formula (3.2) is of order $p \ge 2\nu + 1$. In particular, it would then integrate exactly the polynomial

$$\hat{p}(t) := \prod_{i=1}^{\nu} (t - c_i)^2, \qquad \hat{p} \in \mathbb{P}_{2\nu}.$$

This, however, is impossible, since

$$\int_a^b \hat{p}(\tau)\omega(\tau)\,\mathrm{d}\tau = \int_a^b \left[\prod_{i=1}^{\nu} (\tau - c_i)\right]^2 \omega(\tau)\,\mathrm{d}\tau > 0,$$

while

$$\sum_{j=1}^{\nu} b_j \hat{p}(c_j) = \sum_{j=1}^{\nu} b_j \prod_{i=1}^{\nu} (c_j - c_i)^2 = 0.$$

The proof is complete. ■

The optimal methods of the last theorem are commonly known as *Gaussian quadrature* formulae.

In what follows we will require a generalization of Theorem 3.3. Its proof is left as an exercise to the reader.

Theorem 3.4 *Let $r \in \mathbb{P}_\nu$ obey the orthogonality conditions*

$$\langle r, \hat{p} \rangle = 0 \quad \textit{for every} \quad \hat{p} \in \mathbb{P}_{m-1}, \qquad \langle r, t^m \rangle \ne 0,$$

for some $m \in \{0, 1, \ldots, \nu\}$. We let $c_1, c_2, \ldots, c_\nu$ be the zeros of the polynomial r and choose $b_1, b_2, \ldots, b_\nu$ consistently with (3.3). The quadrature formula (3.2) has order $p = \nu + m$. ■

3.2 Explicit Runge–Kutta schemes

How do we extend a quadrature formula to the ODE (3.1)? The obvious approach is to integrate from t_n to $t_{n+1} = t_n + h$:

$$\boldsymbol{y}(t_{n+1}) = \boldsymbol{y}(t_n) + \int_{t_n}^{t_{n+1}} \boldsymbol{f}(\tau, \boldsymbol{y}(\tau))\,\mathrm{d}\tau = \boldsymbol{y}(t_n) + h\int_0^1 \boldsymbol{f}(t_n + h\tau, \boldsymbol{y}(t_n + h\tau))\,\mathrm{d}\tau,$$

and to replace the second integral by a quadrature. The outcome might have been the 'method'

$$\boldsymbol{y}_{n+1} = \boldsymbol{y}_n + h\sum_{j=1}^{\nu} b_j \boldsymbol{f}(t_n + c_j h,\ \boldsymbol{y}(t_n + c_j h)), \qquad n = 0, 1, \ldots,$$

except that we do not know the value of $\boldsymbol{y}$ at the nodes $t_n + c_1 h, t_n + c_2, \ldots, t_n + c_\nu h$. We must resort to an approximation!

We denote our approximation of $\boldsymbol{y}(t_n + c_j h)$ by $\boldsymbol{\xi}_j$, $j = 1, 2, \ldots, \nu$. To start with, we let $c_1 = 0$, since then the approximation is already provided by the former step of the numerical method, $\boldsymbol{\xi}_1 = \boldsymbol{y}_n$. The idea behind *explicit Runge–Kutta (ERK)* methods is to express each $\boldsymbol{\xi}_j$, $j = 2, 3, \ldots, \nu$, by updating $\boldsymbol{y}_n$ with a linear combination of $\boldsymbol{f}(t_n, \boldsymbol{\xi}_1)$, $\boldsymbol{f}(t_n + hc_2, \boldsymbol{\xi}_2)$, $\ldots$, $\boldsymbol{f}(t_n + c_{j-1}h, \boldsymbol{\xi}_{j-1})$. Specifically, we let

$$\begin{aligned}
\boldsymbol{\xi}_1 &= \boldsymbol{y}_n,\\
\boldsymbol{\xi}_2 &= \boldsymbol{y}_n + ha_{2,1}\boldsymbol{f}(t_n, \boldsymbol{\xi}_1),\\
\boldsymbol{\xi}_3 &= \boldsymbol{y}_n + ha_{3,1}\boldsymbol{f}(t_n, \boldsymbol{\xi}_1) + ha_{3,2}\boldsymbol{f}(t_n + c_2 h, \boldsymbol{\xi}_2),\\
&\ \vdots\\
\boldsymbol{\xi}_\nu &= \boldsymbol{y}_n + h\sum_{i=1}^{\nu-1} a_{\nu,i}\boldsymbol{f}(t_n + c_i h, \boldsymbol{\xi}_i),\\
\boldsymbol{y}_{n+1} &= \boldsymbol{y}_n + h\sum_{j=1}^{\nu} b_j \boldsymbol{f}(t_n + c_j h, \boldsymbol{\xi}_j).
\end{aligned} \tag{3.5}$$

The matrix $A = (a_{j,i})_{j,i=1,2,\ldots,\nu}$, where missing elements are defined to be zero, is called the *RK matrix*, while

$$\boldsymbol{b} = \begin{bmatrix} b_1\\ b_2\\ \vdots\\ b_\nu \end{bmatrix} \quad \text{and} \quad \boldsymbol{c} = \begin{bmatrix} c_1\\ c_2\\ \vdots\\ c_\nu \end{bmatrix}$$

are the *RK weights* and *RK nodes* respectively. We say that (3.5) has ν *stages*. Confusingly, sometimes the $\boldsymbol{\xi}_j$ are called 'RK stages'; elsewhere this name is reserved for $\boldsymbol{f}(t_n + c_j h, \boldsymbol{\xi}_j)$, $j = 1, 2, \ldots, s$. To avoid confusion, we henceforth desist from using the phrase 'RK stages'.

How should we choose the RK matrix? The most obvious way consists of expanding everything in sight in Taylor series about $(t_n, \boldsymbol{y}_n)$; but, in a naive rendition, this is of strictly limited utility. For example, let us consider the simplest nontrivial case, $\nu = 2$. Assuming sufficient smoothness of the vector function $\boldsymbol{f}$, we have

$$\begin{aligned}\boldsymbol{f}(t_n + c_2 h, \boldsymbol{\xi}_2) &= \boldsymbol{f}(t_n + c_2 h,\ \boldsymbol{y}_n + a_{2,1} h \boldsymbol{f}(t_n, \boldsymbol{y}_n)) \\ &= \boldsymbol{f}(t_n, \boldsymbol{y}_n) + h\left[c_2 \frac{\partial \boldsymbol{f}(t_n, \boldsymbol{y}_n)}{\partial t} + a_{2,1} \frac{\partial \boldsymbol{f}(t_n, \boldsymbol{y}_n)}{\partial \boldsymbol{y}} \boldsymbol{f}(t_n, \boldsymbol{y}_n)\right] + \mathcal{O}(h^2)\,;\end{aligned}$$

therefore the last equation in (3.5) becomes

$$\begin{aligned}\boldsymbol{y}_{n+1} &= \boldsymbol{y}_n + h(b_1 + b_2)\boldsymbol{f}(t_n, \boldsymbol{y}_n) \\ &\quad + h^2 b_2 \left[c_2 \frac{\partial \boldsymbol{f}(t_n, \boldsymbol{y}_n)}{\partial t} + a_{2,1} \frac{\partial \boldsymbol{f}(t_n, \boldsymbol{y}_n)}{\partial \boldsymbol{y}} \boldsymbol{f}(t_n, \boldsymbol{y}_n)\right] + \mathcal{O}(h^3)\,.\end{aligned} \tag{3.6}$$

We need to compare (3.6) with the Taylor expansion of the exact solution about the same point $(t_n, \boldsymbol{y}_n)$. The first derivative is provided by the ODE, whereas we can obtain $\boldsymbol{y}''$ by differentiating (3.1) with respect to t:

$$\boldsymbol{y}'' = \frac{\partial \boldsymbol{f}(t, \boldsymbol{y})}{\partial t} + \frac{\partial \boldsymbol{f}(t, \boldsymbol{y})}{\partial \boldsymbol{y}} \boldsymbol{f}(t, \boldsymbol{y}).$$

We denote the exact solution at t_{n+1}, subject to the initial condition $\boldsymbol{y}_n$ at t_n, by $\tilde{\boldsymbol{y}}$. Therefore, by the Taylor theorem,

$$\tilde{\boldsymbol{y}}(t_{n+1}) = \boldsymbol{y}_n + h\boldsymbol{f}(t_n, \boldsymbol{y}_n) + \tfrac{1}{2}h^2 \left[\frac{\partial \boldsymbol{f}(t_n, \boldsymbol{y}_n)}{\partial t} + \frac{\partial \boldsymbol{f}(t_n, \boldsymbol{y}_n)}{\partial \boldsymbol{y}} \boldsymbol{f}(t_n, \boldsymbol{y}_n)\right] + \mathcal{O}(h^3)\,.$$

Comparison with (3.6) gives us the condition for order $p \geq 2$:

$$b_1 + b_2 = 1, \qquad b_2 c_2 = \tfrac{1}{2}, \qquad a_{2,1} = c_2. \tag{3.7}$$

It is easy to verify that the order cannot exceed 2, e.g. by applying the ERK method to the scalar equation $y' = y$.

The conditions (3.7) do not define a two-stage ERK uniquely. Popular choices of parameters are displayed in the *RK tableaux*

$$\begin{array}{c|cc} 0 & & \\ \frac{1}{2} & \frac{1}{2} & \\ \hline & 0 & 1 \end{array}\,, \qquad \begin{array}{c|cc} 0 & & \\ \frac{2}{3} & \frac{2}{3} & \\ \hline & \frac{1}{4} & \frac{3}{4} \end{array} \qquad \text{and} \qquad \begin{array}{c|cc} 0 & & \\ 1 & 1 & \\ \hline & \frac{1}{2} & \frac{1}{2} \end{array}\,,$$

which are of the following form:

$$\begin{array}{c|c} \boldsymbol{c} & A \\ \hline & \boldsymbol{b}^\top \end{array}\,.$$

A naive expansion can be carried out (with substantially greater effort) for $\nu = 3$, whereby we can obtain third-order schemes. However, this is clearly not a serious contender in the technique-of-the-month competition. Fortunately, there are substantially more powerful and easier means of analysing the order of Runge–Kutta methods. We commence by observing that the condition

$$\sum_{i=1}^{j-1} a_{j,i} = c_j, \qquad j = 2, 3, \ldots, \nu,$$

is necessary for order 1 – otherwise we cannot recover the solution of $y' = y$. The simplest device, which unfortunately is valid *only* for $p \le 3$, consists of verifying the order for the scalar *autonomous* equation

$$y' = f(y), \quad t \ge t_0, \qquad y(t_0) = y_0, \tag{3.8}$$

rather than for (3.1). We do not intend here to justify the above assertion but merely to demonstrate its efficacy in the case $\nu = 3$. We henceforth adopt the 'local convention' that, unless indicated otherwise, all the quantities are evaluated at t_n, e.g. $y \sim y_n$, $f \sim f(y_n)$ etc. Subscripts denote derivatives. In the notation of (3.5), we have

$$\begin{aligned}
\xi_1 &= y \\
&\Rightarrow \quad f(\xi_1) = f; \\
\xi_2 &= y + hc_2 f \\
&\Rightarrow \quad f(\xi_2) = f(y + hc_2 f) = f + hc_2 f_y f + \tfrac{1}{2}h^2 c_2^2 f_{yy} f^2 + \mathcal{O}(h^3); \\
\xi_3 &= y + h(c_3 - a_{3,2}) f(\xi_1) + ha_{3,2} f(\xi_2) \\
&= y + (c_3 - a_{3,2}) f + ha_{3,2} f(y + hc_2 f) + \mathcal{O}(h^3) \\
&= y + hc_3 f + h^2 a_{3,2} c_2 f_y f + \mathcal{O}(h^3) \\
&\Rightarrow \quad f(\xi_3) = f(y + hc_3 f + h^2 a_{3,2} c_2 f_y f) + \mathcal{O}(h^3) \\
&\qquad\qquad = f + hc_3 f_y f + h^2 \left(\tfrac{1}{2} c_3^2 f_{yy} f^2 + a_{3,2} c_2 f_y^2 f\right) + \mathcal{O}(h^3).
\end{aligned}$$

Therefore

$$\begin{aligned}
y_{n+1} &= y + hb_1 f + hb_2 \left(f + hc_2 f_y f + \tfrac{1}{2} h^2 c_2^2 f_{yy} f^2\right) \\
&\quad + hb_3 \left[f + hc_3 f_y f + h^2 \left(\tfrac{1}{2} c_3^2 f_{yy} f^2 + a_{3,2} c_2 f_y^2 f\right)\right] + \mathcal{O}(h^4) \\
&= y_n + h(b_1 + b_2 + b_3) f + h^2 (c_2 b_2 + c_3 b_3) f_y f \\
&\quad + h^3 \left[\tfrac{1}{2}(b_2 c_2^2 + b_3 c_3^3) f_{yy} f^2 + b_3 a_{3,2} c_2 f_y^2 f\right] + \mathcal{O}(h^4).
\end{aligned}$$

Since

$$\tilde{y}' = f, \qquad \tilde{y}'' = f_y f, \qquad \tilde{y}''' = f_{yy} f^2 + f_y^2 f$$

the expansion of $\tilde{y}$ reads

$$\tilde{y}_{n+1} = y + hf + \tfrac{1}{2} h^2 f_y f + \tfrac{1}{6} h^3 \left(f_{yy} f^2 + f_y^2 f\right) + \mathcal{O}(h^4).$$

Comparison of the powers of h leads to third-order conditions, namely

$$b_1 + b_2 + b_3 = 1, \qquad b_2 c_2 + b_3 c_3 = \tfrac{1}{2}, \qquad b_2 c_2^2 + b_3 c_3^2 = \tfrac{1}{3}, \qquad b_3 a_{3,2} c_2 = \tfrac{1}{6}.$$

Some instances of third-order three-stage ERK methods are important enough to merit an individual name, for example the *classical* RK method

$$\begin{array}{c|ccc}
0 & & & \\
\frac{1}{2} & \frac{1}{2} & & \\
1 & -1 & 2 & \\
\hline
 & \frac{1}{6} & \frac{2}{3} & \frac{1}{6}
\end{array}$$

and the *Nystrom* scheme

$$\begin{array}{c|ccc} 0 & & & \\ \frac{2}{3} & \frac{2}{3} & & \\ \frac{2}{3} & 0 & \frac{2}{3} & \\ \hline & \frac{1}{4} & \frac{3}{8} & \frac{3}{8} \end{array}.$$

Fourth order is not beyond the capabilities of a Taylor expansion, although a great deal of persistence and care (or, alternatively, a good symbolic manipulator) are required. The best-known fourth-order four-stage ERK method is

$$\begin{array}{c|cccc} 0 & & & & \\ \frac{1}{2} & \frac{1}{2} & & & \\ \frac{1}{2} & 0 & \frac{1}{2} & & \\ 1 & 0 & 0 & 1 & \\ \hline & \frac{1}{6} & \frac{1}{3} & \frac{1}{3} & \frac{1}{6} \end{array}.$$

The derivation of higher-order ERK methods requires a substantially more advanced technique based upon graph theory. It is well beyond the scope of this volume (but see the comments at the end of this chapter). The analysis is further complicated by the fact that ν-stage ERKs of order ν exist only for $\nu \le 4$. To obtain order 5 we need six stages, and matters become considerably worse for higher orders.

3.3 Implicit Runge–Kutta schemes

The idea behind *implicit Runge–Kutta* (IRK) methods is to allow the vector functions $\boldsymbol{\xi}_1, \boldsymbol{\xi}_2, \ldots, \boldsymbol{\xi}_\nu$ to depend upon each other in a more general manner than that of (3.5). Thus, let us consider the scheme

$$\begin{aligned} \boldsymbol{\xi}_j &= \boldsymbol{y}_n + h \sum_{i=1}^{\nu} a_{j,i} \boldsymbol{f}(t_n + c_i h, \boldsymbol{\xi}_i), \qquad j = 1, 2, \ldots, \nu, \\ \boldsymbol{y}_{n+1} &= \boldsymbol{y}_n + h \sum_{j=1}^{\nu} b_j \boldsymbol{f}(t_n + c_j h, \boldsymbol{\xi}_j). \end{aligned} \tag{3.9}$$

Here $A = (a_{j,i})_{j,i=1,2,\ldots,\nu}$ is an arbitrary matrix, whereas in (3.5) it was strictly lower triangular. We impose the convention

$$\sum_{i=1}^{\nu} a_{j,i} = c_j, \qquad j = 1, 2, \ldots, \nu,$$

which is necessary for the method to be of nontrivial order. The ERK terminology – RK nodes, RK weights etc. – stays in place.

For general RK matrix A, the algorithm (3.9) is a system of νd coupled algebraic equations, where $\boldsymbol{y} \in \mathbb{R}^d$. Hence, its calculation faces us with a task of an altogether different magnitude than the explicit method (3.5). However, IRK schemes possess important advantages; in particular they may exhibit superior stability properties.

Moreover, as will be apparent in Section 3.4, there exists for every $\nu \geq 1$ a unique IRK method of order 2ν, a natural extension of the Gaussian quadrature formulae of Theorem 3.3.

◇ **A two-stage IRK method** Let us consider the method

$$\begin{aligned} \boldsymbol{\xi}_1 &= \boldsymbol{y}_n + \tfrac{1}{4}h\left[\boldsymbol{f}(t_n,\boldsymbol{\xi}_1) - \boldsymbol{f}(t_n + \tfrac{2}{3}h, \boldsymbol{\xi}_2)\right], \\ \boldsymbol{\xi}_2 &= \boldsymbol{y}_n + \tfrac{1}{12}h\left[3\boldsymbol{f}(t_n,\boldsymbol{\xi}_1) + 5\boldsymbol{f}(t_n + \tfrac{2}{3}h, \boldsymbol{\xi}_2)\right], \\ \boldsymbol{y}_{n+1} &= \boldsymbol{y}_n + \tfrac{1}{4}h\left[\boldsymbol{f}(t_n,\boldsymbol{\xi}_1) + 3\boldsymbol{f}(t_n + \tfrac{2}{3}h, \boldsymbol{\xi}_2)\right]. \end{aligned} \tag{3.10}$$

In tableau notation it reads

$$\begin{array}{c|cc} 0 & \frac{1}{4} & -\frac{1}{4} \\ \frac{2}{3} & \frac{1}{4} & \frac{5}{12} \\ \hline & \frac{1}{4} & \frac{3}{4} \end{array}.$$

To investigate the order of (3.10), we again assume that the underlying ODE is scalar and autonomous – a procedure that is justified since we do not intend to exceed third order. As before, the convention is that each quantity, unless explicitly stated to the contrary, is evaluated at y_n. Let $k_1 := f(\xi_1)$ and $k_2 := f(\xi_2)$. Expanding about y_n,

$$\begin{aligned} k_1 &= f + \tfrac{1}{4}hf_y(k_1 - k_2) + \tfrac{1}{32}h^2 f_{yy}(k_1 - k_2)^2 + \mathcal{O}(h^3), \\ k_2 &= f + \tfrac{1}{12}hf_y(3k_1 + 5k_2) + \tfrac{1}{288}h^2 f_{yy}(3k_1 + 5k_2)^2 + \mathcal{O}(h^3), \end{aligned}$$

therefore $k_1, k_2 = f + \mathcal{O}(h)$. Substituting this on the right-hand side of the above equations yields $k_1 = f + \mathcal{O}(h^2)$, $k_2 = f + \frac{2}{3}hf_y f + \mathcal{O}(h^2)$. Substituting again these enhanced estimates, we finally obtain

$$\begin{aligned} k_1 &= f - \tfrac{1}{6}h^2 f_y^2 f + \mathcal{O}(h^3), \\ k_2 &= f + \tfrac{2}{3}hf_y f + h^2\left(\tfrac{5}{18}f_y^2 f + \tfrac{2}{9}f_{yy}f^2\right) + \mathcal{O}(h^3). \end{aligned}$$

Consequently, on the one hand we have

$$\begin{aligned} y_{n+1} &= y_n + h(b_1 k_1 + b_2 k_2) \\ &= y + hf + \tfrac{1}{2}h^2 f_y f + \tfrac{1}{6}h^3(f_y^2 f + f_{yy}f^2) + \mathcal{O}(h^4). \end{aligned} \tag{3.11}$$

On the other hand, $y' = f$, $y'' = f_y f$, $y''' = f_y^2 f^2 + f_{yy} f^2$ and the exact expansion is

$$\tilde{y}_{n+1} = y + hf + \tfrac{1}{2}h^2 f_y f + \tfrac{1}{6}h^3(f_y^2 f + f_{yy}f^2) + \mathcal{O}(h^4),$$

and this matches (3.11). We thus deduce that the method (3.10) is of order at least 3. It is, actually, of order exactly 3, and this can be demonstrated by applying (3.10) to the linear equation $y' = y$. ◇

It is perfectly possible to derive IRK methods of higher order by employing the graph-theoretic technique mentioned at the end of Section 3.2. However, an important subset of implicit Runge–Kutta schemes can be investigated very easily and without any cumbersome expansions by an entirely different approach. This will be the theme of the next section.

3.4 Collocation and IRK methods

Let us abandon Runge–Kutta methods for a little while and consider instead an alternative approach to the numerical solution of the ODE (3.1). As before, we assume that the integration has been already carried out up to $(t_n, \boldsymbol{y}_n)$ and we seek a recipe to advance it to $(t_{n+1}, \boldsymbol{y}_{n+1})$, where $t_{n+1} = t_n + h$. To this end we choose ν distinct *collocation parameters* $c_1, c_2, \ldots, c_\nu$ (preferably in $[0, 1]$, although this is not essential to our argument) and seek a νth-degree polynomial $\boldsymbol{u}$ (with vector coefficients) such that

$$\begin{aligned} \boldsymbol{u}(t_n) &= \boldsymbol{y}_n, \\ \boldsymbol{u}'(t_n + c_j h) &= \boldsymbol{f}(t_n + c_j h, \boldsymbol{u}(t_n + c_j h)), \qquad j = 1, 2, \ldots, \nu. \end{aligned} \tag{3.12}$$

In other words, $\boldsymbol{u}$ obeys the initial condition and satisfies the differential equation (3.1) exactly at ν distinct points. A *collocation* method consists of finding such a $\boldsymbol{u}$ and setting

$$\boldsymbol{y}_{n+1} = \boldsymbol{u}(t_{n+1}).$$

The collocation method sounds eminently plausible. Yet, you will search for it in vain in most expositions of ODE methods. The reason is that we have not been entirely sincere at the beginning of this section: collocation is nothing other than a Runge–Kutta method in disguise.

Lemma 3.5 *Set*

$$q(t) := \prod_{j=1}^{\nu} (t - c_j), \qquad q_\ell(t) := \frac{q(t)}{t - c_\ell}, \qquad \ell = 1, 2, \ldots, \nu,$$

and let

$$a_{j,i} := \int_0^{c_j} \frac{q_i(\tau)}{q_i(c_i)} \, \mathrm{d}\tau, \qquad j, i = 1, 2, \ldots, \nu, \tag{3.13}$$

$$b_j := \int_0^1 \frac{q_j(\tau)}{q_j(c_j)} \, \mathrm{d}\tau, \qquad j = 1, 2, \ldots, \nu. \tag{3.14}$$

The collocation method (3.12) *is identical to the IRK method*

$$\begin{array}{c|c} \boldsymbol{c} & A \\ \hline & \boldsymbol{b}^\top \end{array}.$$

Proof According to appendix subsection A.2.2.3, the *Lagrange interpolation polynomial*

$$\boldsymbol{r}(t) := \sum_{\ell=1}^{\nu} \frac{q_\ell((t - t_n)/h)}{q_\ell(c_\ell)} \boldsymbol{w}_\ell$$

satisfies $\boldsymbol{r}(t_n + c_\ell h) = \boldsymbol{w}_\ell$, $\ell = 1, 2, \ldots, \nu$. Let us choose $\boldsymbol{w}_\ell = \boldsymbol{u}'(t_n + c_\ell h)$, $\ell = 1, 2, \ldots, \nu$. The two $(\nu - 1)$th-degree polynomials $\boldsymbol{r}$ and $\boldsymbol{u}'$ coincide at ν points and we thus conclude that $\boldsymbol{r} \equiv \boldsymbol{u}'$. Therefore, invoking (3.12),

$$\boldsymbol{u}'(t) = \sum_{\ell=1}^{\nu} \frac{q_\ell((t-t_n)/h)}{q_\ell(c_\ell)} \boldsymbol{u}'(t_n + c_\ell h) = \sum_{\ell=1}^{\nu} \frac{q_\ell((t-t_n)/h)}{q_\ell(c_\ell)} \boldsymbol{f}(t_n + c_\ell h, \boldsymbol{u}(t_n + c_\ell h)).$$

We will integrate the last expression. Since $\boldsymbol{u}(t_n) = \boldsymbol{y}_n$, the outcome is

$$\begin{aligned} \boldsymbol{u}(t) &= \boldsymbol{y}_n + \int_{t_n}^{t} \sum_{\ell=1}^{\nu} \boldsymbol{f}(t_n + c_\ell h, \boldsymbol{u}(t_n + c_\ell h)) \frac{q_\ell((\tau - t_n)/h)}{q_\ell(c_\ell)} \, \mathrm{d}\tau \\ &= \boldsymbol{y}_n + h \sum_{\ell=1}^{\nu} \boldsymbol{f}(t_n + c_\ell h, \boldsymbol{u}(t_n + c_\ell h)) \int_0^{(t-t_n)/h} \frac{q_\ell(\tau)}{q_\ell(c_\ell)} \, \mathrm{d}\tau. \end{aligned} \tag{3.15}$$

We set $\boldsymbol{\xi}_j := \boldsymbol{u}(t_n + c_j h)$, $j = 1, 2, \ldots, \nu$. Letting $t = t_n + c_j h$ in (3.15), the definition (3.13) implies that

$$\boldsymbol{\xi}_j = \boldsymbol{y}_n + h \sum_{i=1}^{\nu} a_{j,i} \boldsymbol{f}(t_n + c_i h, \boldsymbol{\xi}_i), \qquad j = 1, 2, \ldots, \nu,$$

whereas $t = t_{n+1}$ and (3.14) yield

$$\boldsymbol{y}_{n+1} = \boldsymbol{u}(t_{n+1}) = \boldsymbol{y}_n + \sum_{j=1}^{\nu} b_j \boldsymbol{f}(t_n + c_j h, \boldsymbol{\xi}_j).$$

Thus, we recover the definition (3.9) and conclude that the collocation method (3.12) is an IRK method. ■

◇ **Not every Runge–Kutta method originates in collocation** Let $\nu = 2$, $c_1 = 0$ and $c_2 = \frac{2}{3}$. Therefore

$$q(t) = t\left(t - \tfrac{2}{3}\right), \qquad q_1(t) = t - \tfrac{2}{3}, \qquad q_2(t) = t$$

and (3.13), (3.14) yield the IRK method with tableau

$$\begin{array}{c|cc} 0 & 0 & 0 \\ \frac{2}{3} & \frac{1}{3} & \frac{1}{3} \\ \hline & \frac{1}{4} & \frac{3}{4} \end{array}.$$

Given that every choice of collocation points corresponds to a unique collocation method, we deduce that the IRK method (3.10) (again, with $\nu = 2$, $c_1 = 0$ and $c_2 = \frac{2}{3}$) has no collocation counterpart. There is nothing wrong in this, except that we cannot use the remainder of this section to elucidate the order of (3.10). ◇

Not only are collocation methods a special case of IRK but, as far as actual computation is concerned, to all intents and purposes the IRK formulation (3.9) is preferable. The one advantage of (3.12) is that it lends itself very conveniently to analysis and obviates the need for cumbersome expansions. In a sense, collocation methods are the true inheritors of the quadrature formulae.

Before we can reap the benefits of the formulation (3.12), we need first to present (without proof) an important result on the estimation of error in a numerical solution. It is frequently the case that we possess a smoothly differentiable 'candidate solution' $\boldsymbol{v}$, say, to the ODE (3.1). Typically, such a solution can be produced by any of a myriad of approximation or perturbation techniques, by extending (e.g. by interpolation) a numerical solution from a grid to the whole interval of interest or by formulating 'continuous' numerical methods – the collocation (3.12) is a case in point.

Given such a function $\boldsymbol{v}$, we can calculate the *defect*

$$\boldsymbol{d}(t) := \boldsymbol{v}'(t) - \boldsymbol{f}(t, \boldsymbol{v}(t)).$$

Clearly, there is a connection between the magnitude of the defect and the error $\boldsymbol{v}(t) - \boldsymbol{y}(t)$: since $\boldsymbol{d}(t) \equiv 0$ when $\boldsymbol{v} = \boldsymbol{y}$, the exact solution, we can expect a small value of $\|\boldsymbol{d}(t)\|$ to imply that the error is small. Such a connection is important, since, unlike the error, we can evaluate the defect without knowing the exact solution $\boldsymbol{y}$.

Matters are simple for linear equations. Thus, suppose that

$$\boldsymbol{y}' = \Lambda \boldsymbol{y}, \qquad \boldsymbol{y}(t_0) = \boldsymbol{y}_0. \tag{3.16}$$

We have $\boldsymbol{d}(t) = \boldsymbol{v}'(t) - \Lambda \boldsymbol{v}(t)$ and therefore the linear inhomogeneous ODE

$$\boldsymbol{v}' = \Lambda \boldsymbol{v} + \boldsymbol{d}(t), \quad t > t_0, \qquad \boldsymbol{v}(t_0) \ \text{given}.$$

The exact solution is provided by the familiar variation-of-constants formula,

$$\boldsymbol{v}(t) = \mathrm{e}^{(t-t_0)\Lambda}\boldsymbol{v}_0 + \int_{t_0}^{t} \mathrm{e}^{(t-\tau)\Lambda}\boldsymbol{d}(\tau)\,\mathrm{d}\tau, \qquad t \ge t_0,$$

while the solution of (3.16) is, of course,

$$\boldsymbol{y}(t) = \mathrm{e}^{(t-t_0)\Lambda}\boldsymbol{y}_0, \qquad t \ge t_0.$$

We deduce that

$$\boldsymbol{v}(t) - \boldsymbol{y}(t) = \mathrm{e}^{(t-t_0)\Lambda}(\boldsymbol{v}_0 - \boldsymbol{y}_0) + \int_{t_0}^{t} \mathrm{e}^{(t-\tau)\Lambda}\boldsymbol{d}(\tau)\,\mathrm{d}\tau, \qquad t \ge t_0;$$

thus the error can be expressed completely in terms of the 'observables' $\boldsymbol{v}_0 - \boldsymbol{y}_0$ and $\boldsymbol{d}$.

It is perhaps not very surprising that we can establish a connection between the error and the defect for the linear equation (3.16) since, after all, its exact solution is known. Remarkably, the variation-of-constants formula can be rendered, albeit in a somewhat weaker form, in a nonlinear setting.

Theorem 3.6 (The Alekseev–Gröbner lemma) *Let $\boldsymbol{v}$ be a smoothly differentiable function that obeys the initial condition $\boldsymbol{v}(t_0) = \boldsymbol{y}_0$. Then*

$$\boldsymbol{v}(t) - \boldsymbol{y}(t) = \int_{t_0}^{t} \Phi(t-\tau, \boldsymbol{v}(t-\tau))\boldsymbol{d}(\tau)\,\mathrm{d}\tau, \qquad t \ge t_0, \tag{3.17}$$

where Φ is the matrix of partial derivatives of the solution of the ODE $\boldsymbol{w}' = \boldsymbol{f}(t, \boldsymbol{w})$, $\boldsymbol{w}(\tau) = \boldsymbol{v}(\tau)$, with respect to $\boldsymbol{v}(\tau)$. ■

The matrix Φ is, in general, unknown. It can be estimated quite efficiently, a practice which is useful in error control, but this ranges well beyond the scope of this book. Fortunately, we do not need to know Φ for the application that we have in mind!

Theorem 3.7 *Suppose that*

$$\int_0^1 q(\tau)\tau^j \,\mathrm{d}\tau = 0, \qquad j = 0, 1, \ldots, m-1, \tag{3.18}$$

for some $m \in \{0, 1, \ldots, \nu\}$. (The polynomial $q(t) = \prod_{\ell=1}^{\nu}(t - c_\ell)$ has been defined already in the proof of Lemma 3.5.) Then the collocation method (3.12) *is of order $\nu + m$.*[1]

Proof We express the error of the collocation method by using the Alekseev–Gröbner formula (3.17) (with t_0 replaced by t_n and, of course, the collocation solution $\boldsymbol{u}$ playing the role of $\boldsymbol{v}$; we recall that $\boldsymbol{u}(t_n) = \boldsymbol{y}_n$ and hence the conditions of Theorem 3.6 are satisfied). Thus

$$\boldsymbol{y}_{n+1} - \tilde{\boldsymbol{y}}(t_{n+1}) = \int_{t_n}^{t_{n+1}} \Phi(t_{n+1} - \tau, \boldsymbol{u}(t_{n+1} - \tau))\boldsymbol{d}(\tau) \,\mathrm{d}\tau.$$

(We recall that $\tilde{\boldsymbol{y}}$ denotes the exact solution of the ODE for the initial condition $\tilde{\boldsymbol{y}}(t_n) = \boldsymbol{y}_n$.) We next replace the integral by the quadrature formula with respect to the weight function $\omega(t) \equiv 1$, $t_n < t < t_{n+1}$, with the quadrature nodes $t_n + c_1 h$, $t_n + c_2 h$, $\ldots$, $t_n + c_\nu h$. Therefore

$$\begin{aligned} \boldsymbol{y}_{n+1} - \tilde{\boldsymbol{y}}(t_{n+1}) = \sum_{j=1}^{\nu} & b_j \Phi(t_{n+1}, t_n + c_j h, \boldsymbol{u}(t_n + c_j h))\boldsymbol{d}(t_n + c_j h) \\ & + \text{the error of the quadrature.} \end{aligned} \tag{3.19}$$

However, according to the definition (3.12) of collocation,

$$\boldsymbol{d}(t_n + c_j h) = \boldsymbol{u}'(t_n + c_j h) - \boldsymbol{f}(t_n + c_j h, \boldsymbol{u}(t_n + c_j h)) = \boldsymbol{0}, \qquad j = 1, 2, \ldots, \nu.$$

According to Theorem 3.4, the order of quadrature with the weight function $\omega(t) \equiv 1$, $0 \le t \le 1$, with nodes $c_1, c_2, \ldots, c_\nu$, is $m + \nu$. Therefore, translating linearly from $[0, 1]$ to $[t_n, t_{n+1}]$ and paying heed to the length of the latter interval, $t_{n+1} - t_n = h$, it follows that the error of the quadrature in (3.19) is $\mathcal{O}(h^{\nu+m+1})$. We thus deduce that $\boldsymbol{y}_{n+1} - \tilde{\boldsymbol{y}}(t_{n+1}) = \mathcal{O}(h^{\nu+m+1})$ and prove the theorem.[2] ■

---

[1] If $m = 0$ this means that (3.18) does not hold for any value of j and the theorem claims that the underlying collocation method is then of order ν.

[2] Strictly speaking, we have only proved that the error is at least of order $\nu + m$. However, if m is the largest integer such that (3.18) holds, then it is trivial to prove that the order cannot exceed $\nu + m$; for example, apply the collocation to the equation $y' = (\nu + m + 1)t^{\nu+m}$, $y(0) = 0$.

Corollary *Let $c_1, c_2, \ldots, c_\nu$ be the zeros of the polynomials $\tilde{P}_\nu \in \mathbb{P}_\nu$ that are orthogonal with respect to the weight function $\omega(t) \equiv 1$, $0 \le t \le 1$. Then the underlying collocation method* (3.12) *is of order 2ν.*

Proof The corollary is a straightforward consequence of the last theorem, since the definition of orthogonality (3.4) implies in the present context that (3.18) is satisfied by $m = \nu$. ■

◇ **Gauss–Legendre methods** The ν-stage order-2ν methods from the last corollary are called *Gauss–Legendre* (Runge–Kutta) methods. Note that, according to Lemma 3.2, the nodes $c_1, c_2, \ldots, c_\nu \in (0,1)$ are, as necessary for collocation, distinct. The polynomials $\tilde{P}_\nu$ can be obtained explicitly, e.g. by linearly transforming the more familiar *Legendre* polynomials P_ν, which are orthogonal with respect to the weight function $\omega(t) \equiv 1$, $-1 < t < 1$. The (monic) outcome is

$$\tilde{P}_\nu(t) = \frac{(\nu!)^2}{(2\nu)!} \sum_{k=0}^{\nu} (-1)^{\nu-k} \binom{\nu}{k} \binom{\nu+k}{k} t^k.$$

For $\nu = 1$ we obtain $\tilde{P}_1(t) = t - \frac{1}{2}$, hence $c_1 = \frac{1}{2}$. The method, which can be written in a tableau form as

$$\begin{array}{c|c} \frac{1}{2} & \frac{1}{2} \\ \hline & 1 \end{array},$$

is the familiar *implicit midpoint rule* (1.12). In the case $\nu = 2$ we have $\tilde{P}_2(t) = t^2 - t + \frac{1}{6}$, therefore $c_1 = \frac{1}{2} - \frac{\sqrt{3}}{6}$, $c_2 = \frac{1}{2} + \frac{\sqrt{3}}{6}$. The formulae (3.13), (3.14) lead to the two-stage fourth-order IRK method

$$\begin{array}{c|cc} \frac{1}{2} - \frac{\sqrt{3}}{6} & \frac{1}{4} & \frac{1}{4} - \frac{\sqrt{3}}{6} \\ \frac{1}{2} + \frac{\sqrt{3}}{6} & \frac{1}{4} + \frac{\sqrt{3}}{6} & \frac{1}{4} \\ \hline & \frac{1}{2} & \frac{1}{2} \end{array}.$$

The computation of nonlinear algebraic systems that originate in IRK methods with large ν is expensive but this is compensated by the increase in order. It is impossible to lay down firm rules, and the exact point whereby the law of diminishing returns compels us to choose a lower-order method changes from equation to equation. It is fair to remark, however, that the three-stage Gauss–Legendre is probably the largest that is consistent with reasonable implementation costs:

$$\begin{array}{c|ccc} \frac{1}{2} - \frac{\sqrt{15}}{10} & \frac{5}{36} & \frac{2}{9} - \frac{\sqrt{15}}{15} & \frac{5}{36} - \frac{\sqrt{15}}{30} \\ \frac{1}{2} & \frac{5}{36} + \frac{\sqrt{15}}{24} & \frac{2}{9} & \frac{5}{36} - \frac{\sqrt{15}}{24} \\ \frac{1}{2} + \frac{\sqrt{15}}{10} & \frac{5}{36} + \frac{\sqrt{15}}{30} & \frac{2}{9} + \frac{\sqrt{15}}{15} & \frac{5}{36} \\ \hline & \frac{5}{18} & \frac{4}{9} & \frac{5}{18} \end{array}.$$

◇

Comments and bibliography

A standard text on numerical integration is Davis & Rabinowitz (1967), while highly readable accounts of orthogonal polynomials can be found in Chihara (1978) and Rainville (1967). We emphasize that although the theory of orthogonal polynomials is of tangential importance to the subject matter of this volume, it is well worth studying for its intrinsic beauty as well as its numerous applications.

Runge–Kutta methods have been known for a long time; Runge himself produced the main idea in 1895.[3] Their theoretical understanding, however, is much more recent and associated mainly with the work of John Butcher. As is often the case with progress in computational science, an improved theory has spawned new and better algorithms, these in turn have led to further theoretical comprehension and so on. Lambert's textbook (1991) presents a readable account of Runge–Kutta methods and requires a relatively modest theoretical base. More advanced accounts can be found in Butcher (1987, 2003) and Hairer *et al.* (1991).

Let us present in a nutshell the main idea behind the graph-theoretical approach of Butcher to the derivation of the order of Runge–Kutta methods. The few examples of expansion in Sections 3.2 and 3.3 already demonstrate that the main difficulty rests in the need to differentiate composite functions repeatedly. For expositional reasons only, we henceforth restrict our attention to scalar, autonomous equations.[4] Thus,

$$\begin{aligned} & y' = f(y) \\ \Rightarrow\quad & y'' = f_y(y)f(y) \\ \Rightarrow\quad & y''' = f_{yy}(y)[f(y)]^2 + [f_y(y)]^2 f(y) \\ \Rightarrow\quad & y^{(iv)} = f_{yyy}(y)[f(y)]^3 + 4f_{yy}(y)f_y(y)[f(y)]^2 + [f_y(y)]^3 f(y) \end{aligned}$$

and so on. Although it cannot yet be seen from the above, the number of terms increases *exponentially*. This should not deter us from exploring high-order methods, since there is a great deal of redundancy in the order conditions (recall from the corollary to Theorem 2.7 that it is possible to attain order 2ν with a ν-stage method!), but we need an intelligent mechanism to express the increasingly more complicated derivatives in a compact form.

Such a mechanism is provided by *graph theory*. Briefly, a *graph* is a collection of *vertices* and *edges:* it is usual to render the vertices pictorially as solid circles, while the edges are the lines joining them.[5] For example, two simple five-vertex graphs are

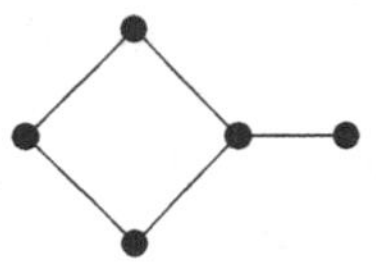

and

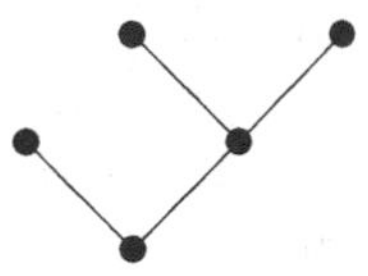

.

The *order* of a graph is the number of vertices therein: both graphs above are of order 5. We say that a graph is a *tree* if each two vertices are joined by a single path of edges. Thus the second graph is a tree, whereas the first is not. Finally, in a tree we single out one vertex and call it the *root.* This imposes a partial ordering on a *rooted tree:* the root is the lowest, its

[3] The monograph of Collatz (1966), and in particular its copious footnotes, is an excellent source on the life of many past heroes of numerical analysis.

[4] This restriction leads to loss of generality. A comprehensive order analysis should be done for systems of equations.

[5] You will have an opportunity to learn much more about graphs and their role in numerical calculations in Chapter 11.

children (i.e., all vertices that are joined to the root by a single edge) are next in line, then its children's children and so on. We adopt in our pictures the (obvious) convention that the root is always at the bottom. (Strangely, computer scientists often follow an opposite convention and place the root at the top.) Two rooted trees of the same order are said to be *equivalent* if each exhibits the same pattern of paths from its 'top' to its root – the following picture of three equivalent rooted trees should clarify this concept: the graphs

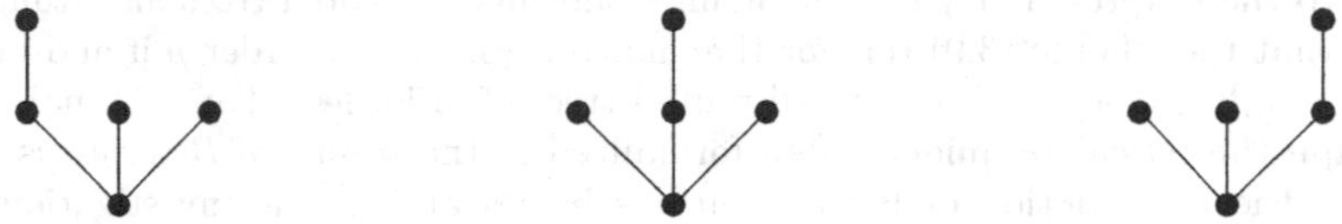

are all equivalent. We keep just one representative of each equivalence class and, hopefully without much confusion, refer to members of this reduced set as 'rooted trees'. We denote by $\gamma(\hat{t})$ the product of the order of the tree $\hat{t}$ and the orders of all possible trees that occur upon consecutive removal of the roots of $\hat{t}$. For example, for the above tree we have

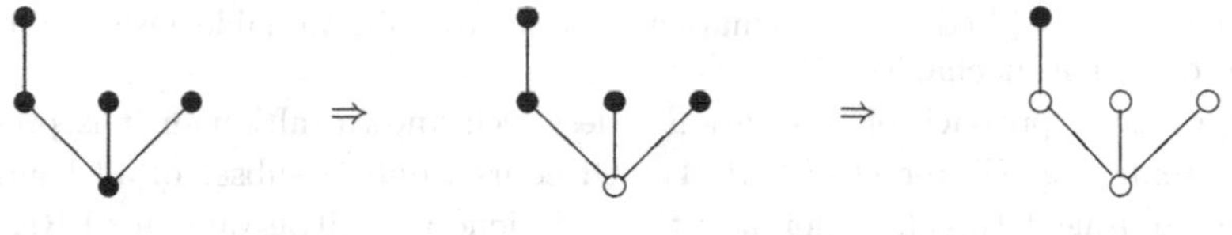

(an open circle denotes a vertex that has been removed) and $\gamma(\hat{t}) = 5 \times (2 \times 1 \times 1) \times 1 = 10$.

As we have seen above, the derivatives of y can be expressed as linear combinations of products of derivatives of f. The latter are called *elementary differentials* and they can be assigned to rooted trees according to the following rule: to each vertex of a rooted tree corresponds a derivative $f_{yy\cdots y}$, where the suffix occurs the same number of times as the number of children of the vertex, and the elementary differential corresponding to the whole tree is a product of these terms. For example,

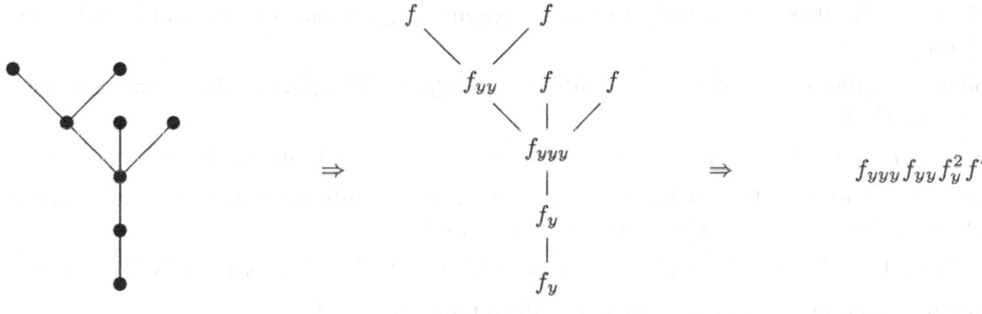

To every rooted tree there corresponds an order condition, which we can express in terms of the RK matrix A and the RK weights $\boldsymbol{b}$. This is best demonstrated by an example. We assign an index to every vertex of a tree $\hat{t}$, e.g. the tree

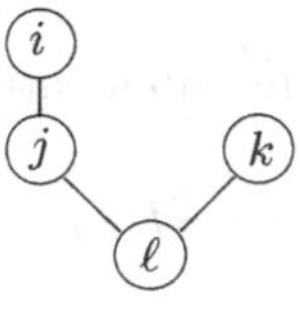

corresponds to the condition

$$\sum_{\ell,j,i,k=1}^{\nu} b_\ell a_{\ell,j} a_{\ell,k} a_{j,i} = \frac{1}{\gamma(\hat{t})} = \frac{1}{8}.$$

The general rule is clear – we multiply b_ℓ by all components $a_{q,r}$, where q and r are the indices of a parent and a child respectively, sum up for all indices ranging in $\{1, 2, \ldots, \nu\}$ and equate to the reciprocal of $\gamma(\hat{t})$. The main result linking rooted trees and Runge–Kutta methods is that the scheme (3.9) (or, for that matter, (3.5)) is of order p if and only if the above order conditions are satisfied for all rooted trees of order less than or equal to p.

The graph-theoretical technique, often formalized as the *theory of B-series,* is the standard tool in the construction of Runge–Kutta schemes and in the investigation of their properties. It is, in particular, of great importance in the investigation of the behaviour of structure-preserving Runge–Kutta methods that we will encounter in Chapter 5.

By one of these quirks of fate that make the study of mathematics so entrancing, the graph-theoretical interpretation of Runge–Kutta methods has recently acquired an unexpected application at an altogether different corner of the mathematical universe. It turns out that the abstract structure underlying this interpretation is a *Hopf algebra* of a special kind, which can be applied in mathematical physics to gain valuable insight into certain questions in quantum mechanics.

The alternative approach of collocation is less well known, although it is presented in more recent texts, e.g. Hairer *et al.* (1991). Of course, only a subset of all Runge–Kutta methods are equivalent to collocation and the technique is of little value for ERK schemes. It is, however, possible to generalize the concept of collocation to cater for *all* Runge–Kutta methods.

Butcher, J.C. (1987), *The Numerical Analysis of Ordinary Differential Equations,* John Wiley, Chichester.

Butcher, J.C. (2003), *Numerical Methods for Ordinary Differential Equations,* John Wiley, Chichester.

Chihara, T.S. (1978), *An Introduction to Orthogonal Polynomials,* Gordon and Breach, New York.

Collatz, L. (1966), *The Numerical Treatment of Differential Equations* (3rd edn), Springer-Verlag, Berlin.

Davis, P.J. and Rabinowitz, P. (1967), *Numerical Integration,* Blaisdell, London.

Hairer, E., Nørsett, S.P. and Wanner, G. (1991), *Solving Ordinary Differential Equations I: Nonstiff Problems* (2nd edn), Springer-Verlag, Berlin.

Lambert, J.D. (1991), *Numerical Methods for Ordinary Differential Systems,* Wiley, London.

Rainville, E.D. (1967), *Special Functions,* Macmillan, New York.

Exercises

3.1 Find the order of the following quadrature formulae:

a $\int_0^1 f(\tau)\,\mathrm{d}\tau = \frac{1}{6}f(0) + \frac{2}{3}f(\frac{1}{2}) + \frac{1}{6}f(1)$ (the Simpson rule);

b $\displaystyle\int_0^1 f(\tau)\,\mathrm{d}\tau = \tfrac{1}{8}f(0) + \tfrac{3}{8}f(\tfrac{1}{3}) + \tfrac{3}{8}f(\tfrac{2}{3}) + \tfrac{1}{8}f(1)$ (the three-eighths rule);

c $\displaystyle\int_0^1 f(\tau)\,\mathrm{d}\tau = \tfrac{2}{3}f(\tfrac{1}{4}) - \tfrac{1}{3}f(\tfrac{1}{2}) + \tfrac{2}{3}f(\tfrac{3}{4});$

d $\displaystyle\int_0^\infty f(\tau)\mathrm{e}^{-\tau}\,\mathrm{d}\tau = \tfrac{5}{3}f(1) - \tfrac{3}{2}f(2) + f(3) - \tfrac{1}{6}f(4).$

3.2 Let us define

$$T_n(\cos\theta) := \cos n\theta, \qquad n = 0, 1, 2, \ldots, \qquad -\pi \le \theta \le \pi.$$

a Show that each T_n is a polynomial of degree n and that the T_n satisfy the three–term recurrence relation

$$T_{n+1}(t) = 2t\,T_n(t) - T_{n-1}(t), \qquad n = 1, 2, \ldots$$

b Prove that T_n is an nth orthogonal polynomial with respect to the weight function $\omega(t) = (1-t)^{-1/2}$, $-1 < t < 1$.

c Find the explicit values of the zeros of T_n, thereby verifying the statement of Lemma 3.2, namely that all the zeros of an orthogonal polynomial reside in the open support of the weight function.

d Find b_1, b_2, c_1, c_2 such that the order of the quadrature

$$\int_{-1}^1 f(\tau)\frac{\mathrm{d}\tau}{\sqrt{1-\tau^2}} \approx b_1 f(c_1) + b_2 f(c_2)$$

is four.

(*The T_ns are known as* Chebyshev polynomials *and they have many applications in mathematical analysis. We will encounter them again in Chapter 10.*)

3.3 Construct the Gaussian quadrature formulae for the weight function $\omega(t) \equiv 1$, $0 \le t \le 1$, of orders two, four and six.

3.4 Restricting your attention to scalar autonomous equations $y' = f(y)$, prove that the ERK method with tableau

$$\begin{array}{c|cccc} 0 & & & & \\ \frac{1}{2} & \frac{1}{2} & & & \\ \frac{1}{2} & 0 & \frac{1}{2} & & \\ 1 & 0 & 0 & 1 & \\ \hline & \frac{1}{6} & \frac{1}{3} & \frac{1}{3} & \frac{1}{6} \end{array}$$

is of order 4.

3.5 Suppose that a ν-stage ERK method of order ν is applied to the linear scalar equation $y' = \lambda y$. Prove that

$$y_n = \left[\sum_{k=0}^{\nu} \frac{1}{k!}(h\lambda)^k\right]^n y_0, \qquad n = 0, 1, \ldots$$

3.6 Determine all choices of $\boldsymbol{b}$, $\boldsymbol{c}$ and A such that the two-stage IRK method

$$\begin{array}{c|c} \boldsymbol{c} & A \\ \hline & \boldsymbol{b}^\top \end{array}$$

is of order $p \geq 3$.

3.7 Write the theta method, (1.13), as a Runge–Kutta method.

3.8 Derive the three-stage Runge–Kutta method that corresponds to the collocation points $c_1 = \frac{1}{4}$, $c_2 = \frac{1}{2}$, $c_3 = \frac{3}{4}$ and determine its order.

3.9 Let $\kappa \in \mathbb{R}\setminus\{0\}$ be a given constant. We choose collocation nodes $c_1, c_2, \ldots, c_\nu$ as zeros of the polynomial $\tilde{P}_\nu + \kappa\tilde{P}_{\nu-1}$. ($\tilde{P}_m$ *is the mth-degree Legendre polynomial, shifted to the interval* $(0, 1)$. *In other words, the* $\tilde{P}_m$ *are orthogonal there with respect to the weight function* $\omega(t) \equiv 1$.)

a Prove that the collocation method (3.12) is of order $2\nu - 1$.

b Let $\kappa = -1$ and find explicitly the corresponding IRK method for $\nu = 2$.

4

Stiff equations

4.1 What are stiff ODEs?

Let us try to solve the seemingly innocent linear ODE

$$\boldsymbol{y}' = \Lambda \boldsymbol{y}, \qquad \boldsymbol{y}(0) = \boldsymbol{y}_0, \qquad \text{where} \qquad \Lambda = \begin{bmatrix} -100 & 1 \\ 0 & -\frac{1}{10} \end{bmatrix}, \tag{4.1}$$

by Euler's method (1.4). We obtain

$$\boldsymbol{y}_1 = \boldsymbol{y}_0 + h\Lambda \boldsymbol{y}_0 = (I + h\Lambda)\boldsymbol{y}_0, \qquad \boldsymbol{y}_2 = \boldsymbol{y}_1 + h\Lambda \boldsymbol{y}_1 = (I + h\Lambda)\boldsymbol{y}_1 = (I + h\Lambda)^2 \boldsymbol{y}_0$$

(where I is the identity matrix) and, in general, it is easy to prove by elementary induction that

$$\boldsymbol{y}_n = (I + h\Lambda)^n \boldsymbol{y}_0, \qquad n = 0, 1, 2, \ldots \tag{4.2}$$

Since the *spectral factorization* (A.1.5.4) of Λ is

$$\Lambda = VDV^{-1}, \qquad \text{where} \qquad V = \begin{bmatrix} 1 & 1 \\ 0 & \frac{999}{10} \end{bmatrix} \qquad \text{and} \qquad D = \begin{bmatrix} -100 & 0 \\ 0 & -\frac{1}{10} \end{bmatrix},$$

we deduce that the exact solution of (4.1) is

$$\boldsymbol{y}(t) = \mathrm{e}^{t\Lambda} = V\mathrm{e}^{tD}V^{-1}\boldsymbol{y}_0, \qquad t \geq 0, \qquad \text{where} \qquad \mathrm{e}^{tD} = \begin{bmatrix} \mathrm{e}^{-100t} & 0 \\ 0 & \mathrm{e}^{-t/10} \end{bmatrix}.$$

In other words, there exist two vectors, $\boldsymbol{x}_1$ and $\boldsymbol{x}_2$, say, dependent on $\boldsymbol{y}_0$ but not on t, such that

$$\boldsymbol{y}(t) = \mathrm{e}^{-100t}\boldsymbol{x}_1 + \mathrm{e}^{-t/10}\boldsymbol{x}_2, \qquad t \geq 0. \tag{4.3}$$

The function $g(t) = \mathrm{e}^{-100t}$ decays exceedingly fast: $g\left(\frac{1}{10}\right) \approx 4.54 \times 10^{-5}$ and $g(1) \approx 3.72 \times 10^{-44}$, while the decay of $\mathrm{e}^{-t/10}$ is a thousandfold more sedate. Thus, even for small $t > 0$ the contribution of $\boldsymbol{x}_1$ is nil to all intents and purposes and $\boldsymbol{y}(t) \approx \mathrm{e}^{-t/10}\boldsymbol{x}_2$. What about the Euler solution $\{\boldsymbol{y}_n\}_{n=0}^{\infty}$, though? It follows from (4.2) that

$$\boldsymbol{y}_n = V(I + hD)^n V^{-1}\boldsymbol{y}_0, \qquad n = 0, 1, \ldots$$

and, since

$$(I + hD)^n = \begin{bmatrix} (1 - 100h)^n & 0 \\ 0 & (1 - \frac{1}{10}h)^n \end{bmatrix},$$

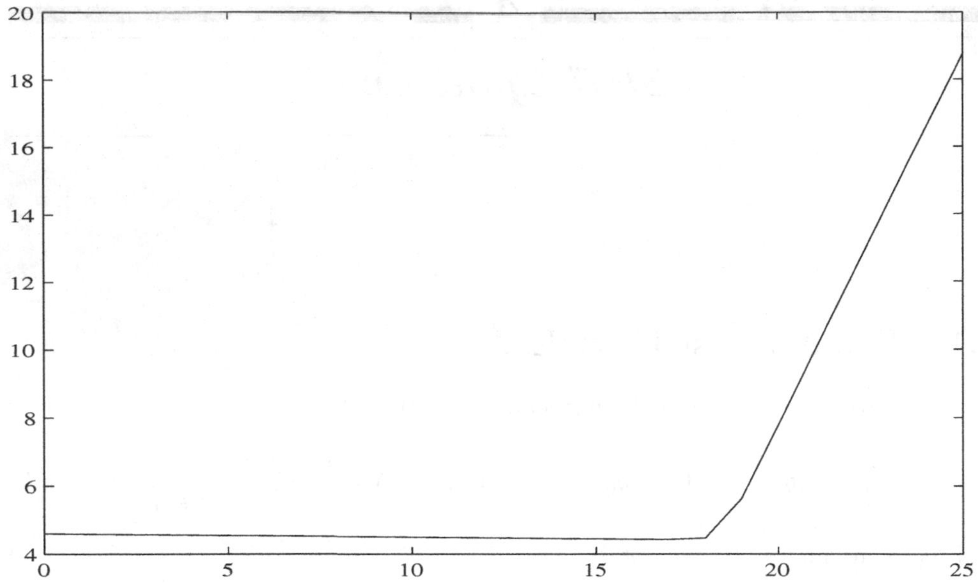

Figure 4.1 The logarithm of the Euclidean norm $\|\boldsymbol{y}_n\|$ of the Euler steps, as applied to the equation (4.1) with $h = \frac{1}{10}$ and an initial condition identical to the second (i.e., the 'stable') eigenvector. The divergence is thus entirely due to roundoff error!

it follows that

$$\boldsymbol{y}_n = (1 - 100h)^n \boldsymbol{x}_1 + (1 - \tfrac{1}{10}h)^n \boldsymbol{x}_2, \qquad n = 0, 1, \ldots \tag{4.4}$$

(it is left to the reader to prove in Exercise 4.1 that the constant vectors $\boldsymbol{x}_1$ and $\boldsymbol{x}_2$ are the same in (4.3) and (4.4)). Suppose that $h > \frac{1}{50}$. Then $|1 - 100h| > 1$ and it is a consequence of (4.4) that, for sufficiently large n, *the Euler iterates grow geometrically in magnitude,* in contrast with the asymptotic behaviour of the true solution.

Suppose that we choose an initial condition identical to an eigenvector corresponding to the eigenvalue -0.1, for example

$$\boldsymbol{y}_0 = \begin{bmatrix} 1 \\ \frac{999}{10} \end{bmatrix}.$$

Then, in exact arithmetic, $\boldsymbol{x}_1 = \boldsymbol{0}$, $\boldsymbol{x}_2 = \boldsymbol{y}_0$ and $\boldsymbol{y}_n = \left(1 - \frac{1}{10}h\right)^n \boldsymbol{y}_0$, $n = 0, 1, \ldots$; the latter converges to $\boldsymbol{0}$ as $n \to \infty$ for all reasonable values of $h > 0$ (specifically, for $h < 20$). Hence, we might hope that all will be well with the Euler method. Not so! Real computers produce roundoff errors and, unless $h < \frac{1}{50}$, sooner or later these are bound to attribute a nonzero contribution to an eigenvector corresponding to the eigenvalue -100. As soon as this occurs, the unstable component grows geometrically, as $(1 - 100h)^n$, and rapidly overwhelms the true solution.

Figure 4.1 displays $\ln \|\boldsymbol{y}_n\|$, $n = 0, 1, \ldots, 25$, with the above initial condition and the time step $h = \frac{1}{10}$. The calculation was performed on a computer equipped with the ubiquitous IEEE arithmetic,[1] which is correct (in a single algebraic operation) to about 15 decimal digits. The norm of the first 17 steps decreases at the right pace, dictated by $\left(1 - \frac{1}{10}h\right)^n = \left(\frac{99}{100}\right)^n$. However, everything then breaks down and, after just two steps, the norm increases geometrically, as $|1 - 100h|^n = 9^n$. The reader is welcome to check that the slope of the curve in Fig. 4.1 is indeed $\ln \frac{99}{100} \approx -0.0101$ initially but becomes $\ln 9 \approx 2.1972$ in the second, unstable, regime.

The choice of $\boldsymbol{y}_0$ as a 'stable' eigenvector is not contrived. Faced with an equation like (4.1) (with an arbitrary initial condition) we are likely to employ a small step size in the initial *transient* regime, in which the contribution of the 'unstable' eigenvector is still significant. However, as soon as this has disappeared and the solution is completely described by the 'stable' eigenvector, it is tempting to increase h. This must be resisted: like a malign version of the Cheshire cat, the rogue eigenvector might seem to have disappeared, but its hideous grin stays and is bound to thwart our endeavours.

It is important to understand that this behaviour has nothing to do with the local error of the numerical method; the step size is depressed not by accuracy considerations (to which we should be always willing to pay heed) but by instability.

Not every numerical method displays a similar breakdown in stability. Thus, solving (4.1) with the trapezoidal rule (1.9), we obtain

$$\boldsymbol{y}_1 = \left(\frac{I + \frac{1}{2}h\Lambda}{I - \frac{1}{2}h\Lambda}\right)\boldsymbol{y}_0, \qquad \boldsymbol{y}_2 = \left(\frac{I + \frac{1}{2}h\Lambda}{I - \frac{1}{2}h\Lambda}\right)\boldsymbol{y}_1 = \left(\frac{I + \frac{1}{2}h\Lambda}{I - \frac{1}{2}h\Lambda}\right)^2\boldsymbol{y}_0,$$

noting that since $(I - \frac{1}{2}h\Lambda)^{-1}$ and $\left(I + \frac{1}{2}h\Lambda\right)$ commute the order of multiplication does not matter, and, in general,

$$\boldsymbol{y}_n = \left(\frac{I + \frac{1}{2}h\Lambda}{I - \frac{1}{2}h\Lambda}\right)^n\boldsymbol{y}_0, \qquad n = 0, 1, \ldots \tag{4.5}$$

Substituting for Λ from (4.1) and factorizing, we deduce, in the same way as for (4.4), that

$$\boldsymbol{y}_n = \left(\frac{1 - 50h}{1 + 50h}\right)^n\boldsymbol{x}_1 + \left(\frac{1 - \frac{1}{20}h}{1 + \frac{1}{20}h}\right)^n\boldsymbol{x}_2, \qquad n = 0, 1, \ldots$$

Thus, since

$$\left|\frac{1 - 50h}{1 + 50h}\right|, \left|\frac{1 - \frac{1}{20}h}{1 + \frac{1}{20}h}\right| < 1$$

for every $h > 0$, we deduce that $\lim_{n\to\infty} \boldsymbol{y}_n = \boldsymbol{0}$. This recovers the correct asymptotic behaviour of the ODE (4.1) (cf. (4.3)) regardless of the size of h.

In other words, the trapezoidal rule does not require any restriction in the step size to avoid instability. We hasten to say that this does not mean, of course, that *any* h is suitable. It is necessary to choose $h > 0$ small enough to ensure that the local error is within reasonable bounds and the exact solution is adequately approximated. However, there is no need to decrease h to a minuscule size to prevent rogue components of the solution growing out of control.

[1] The current standard of computer arithmetic on workstations and personal computers.

The equation (4.1) is an example of a *stiff ODE*. Several attempts at a rigorous definition of stiffness appear in the literature, but it is perhaps more informative to adopt an operative (and slightly vague) designation. Thus, we say that an ODE system

$$\boldsymbol{y}' = \boldsymbol{f}(t, \boldsymbol{y}), \quad t \geq t_0, \qquad \boldsymbol{y}(t_0) = \boldsymbol{y}_0, \tag{4.6}$$

is *stiff* if its numerical solution by some methods requires (perhaps in a portion of the solution interval) a significant depression of the step size to avoid instability. Needless to say this is not a proper mathematical definition, but then we are not aiming to prove theorems of the sort 'if a system is stiff then ...'. The main importance of the above concept is in helping us to choose and implement numerical methods – a procedure that, anyway, is far from an exact science!

We have already seen the most important mechanism generating stiffness, namely, that modes with vastly different scales and 'lifetimes' are present in the solution. It is sometimes the practice to designate the quotient of the largest and the smallest (in modulus) eigenvalues of a linear system (and, for a general system (4.6), the eigenvalues of the Jacobian matrix) as the *stiffness ratio*. The stiffness ratio of (4.1) is 10^3. This concept is helpful in elucidating the behaviour of many ODE systems and, in general, it is a safe bet that if (4.6) has a large stiffness ratio then it is stiff. Having said this, it is also valuable to stress the shortcomings of linear analysis and emphasize that the stiffness ratio might fail to elucidate the behaviour of a nonlinear ODE system.

A large proportion of the ODEs that occur in practice are stiff. Whenever equations model several processes with vastly different rates of evolution, stiffness is not far away. For example, the differential equations of chemical kinetics describe reactions that often proceed on very different time scales (think of the difference in time scales of corrosion and explosion); a stiffness ratio of 10^{17} is quite typical. Other popular sources of stiffness are control theory, reactor kinetics, weather prediction, mathematical biology and electronics: they all abound with phenomena that display variation at significantly different time scales. The world record, to the author's knowledge, is held, unsurprisingly perhaps, by the equations that describe the cosmological Big Bang: the stiffness ratio is 10^{31}.

One of the main sources of stiff equations is numerical analysis itself. As we will see in Chapter 16, parabolic partial differential equations are often approximated by large systems of stiff ODEs.

4.2 The linear stability domain and A-stability

Let us suppose that a given numerical method is applied with a constant step size $h > 0$ to the scalar linear equation

$$y' = \lambda y, \quad t \geq 0, \qquad y(0) = 1, \tag{4.7}$$

where $\lambda \in \mathbb{C}$. The exact solution of (4.7) is, of course, $y(t) = \mathrm{e}^{\lambda t}$, hence $\lim_{t\to\infty} y(t) = 0$ if and only if $\operatorname{Re} \lambda < 0$. We say that the *linear stability domain* $\mathcal{D}$ of the underlying numerical method is the set of all numbers $h\lambda \in \mathbb{C}$ such that $\lim_{n\to\infty} y_n = 0$. In other words, $\mathcal{D}$ is the set of all $h\lambda$ for which the correct asymptotic behaviour of (4.7) is recovered, provided that the latter equation is stable.[2]

Let us commence with Euler's method (1.4). We obtain the solution sequence identically to the derivation of (4.2),

$$y_n = (1 + h\lambda)^n, \qquad n = 0, 1, \ldots \tag{4.8}$$

Therefore $\{y_n\}_{n=0,1,\ldots}$ is a geometric sequence and $\lim_{n\to\infty} y_n = 0$ if and only if $|1 + h\lambda| < 1$. We thus conclude that

$$\mathcal{D}_{\text{Euler}} = \{z \in \mathbb{C} \,:\, |1 + z| < 1\}$$

is the interior of a complex disc of unit radius, centred at $z = -1$ (see Fig. 4.2).

Before we proceed any further, let us ponder briefly the rationale behind this sudden interest in a humble scalar linear equation. After all, we do not need numerical analysis to solve (4.7)! However, for Euler's method and for all other methods that have been the theme of Chapters 1–3 we can extrapolate from scalar linear equations to linear ODE systems. Thus, suppose that we solve (4.1) *with an arbitrary $d \times d$ matrix* Λ. The solution sequence is given by (4.2). Suppose that Λ has a full set of eigenvectors and hence the spectral factorization $\Lambda = VDV^{-1}$, where V is a nonsingular matrix of eigenvectors and $D = \operatorname{diag}(\lambda_1, \lambda_2, \ldots, \lambda_d)$ contains the eigenvalues of Λ. Exactly as in (4.4), we can prove that there exist vectors $\boldsymbol{x}_1, \boldsymbol{x}_2, \ldots, \boldsymbol{x}_d \in \mathbb{C}^d$, dependent only on $\boldsymbol{y}_0$, not on n, such that

$$\boldsymbol{y}_n = \sum_{k=1}^{d} (1 + h\lambda_k)^n \boldsymbol{x}_k, \qquad n = 0, 1, \ldots \tag{4.9}$$

Let us suppose that the exact solution of the linear system is asymptotically stable. This happens if and only if $\operatorname{Re}\lambda_k < 0$ for all $k = 1, 2, \ldots, d$. To mimic this behaviour with Euler's method, we deduce from (4.9) that the step size $h > 0$ must be such that $|1 + h\lambda_k| < 1$, $k = 1, 2, \ldots, d$: all the products $h\lambda_1, h\lambda_2, \ldots, h\lambda_d$ must lie in $\mathcal{D}_{\text{Euler}}$. This means in practice that the step size is determined by the stiffest component of the system!

The restriction to systems with a full set of eigenvectors is made for ease of exposition only. In general, we may use a *Jordan factorization* (A.1.5.6) in place of a spectral factorization; see Exercise 4.2 for a simple example. Moreover, the analysis can be extended easily to inhomogeneous systems $\boldsymbol{y}' = \Lambda\boldsymbol{y} + \boldsymbol{a}$, and this is illustrated by Exercise 4.3.

The importance of $\mathcal{D}$ ranges well beyond linear systems. Given a nonlinear ODE system

$$\boldsymbol{y}' = \boldsymbol{f}(t, \boldsymbol{y}), \quad t \geq t_0, \qquad \boldsymbol{y}(t_0) = \boldsymbol{y}_0,$$

where $\boldsymbol{f}$ is differentiable with respect to $\boldsymbol{y}$, it is usual to require that in the nth step

$$h\lambda_{n,1},\ h\lambda_{n,2},\ \ldots,\ h\lambda_{n,d} \in \mathcal{D},$$

[2]Our interest in (4.7) with $\operatorname{Re}\lambda > 0$ is limited, since the exact solution rapidly becomes very large. However, for nonlinear equations there is an intense interest, which we will not pursue in this volume, in those equations for which a counterpart of λ, namely the Liapunov exponent, is positive.

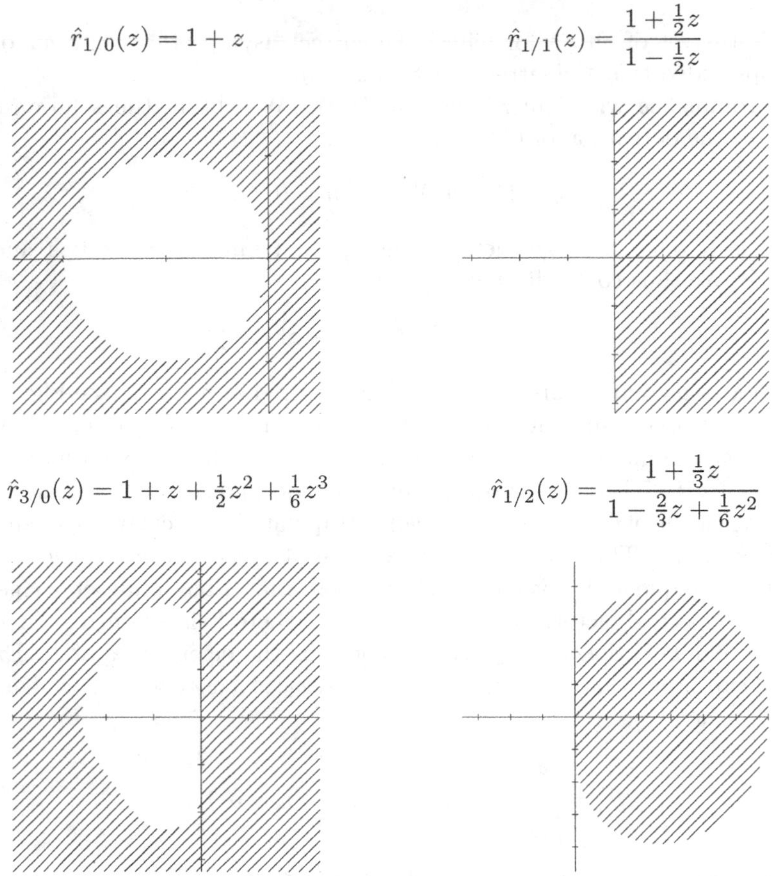

Figure 4.2 Stability domains (the unshaded areas) for various rational approximations. Note that $\hat{r}_{1/0}$ corresponds to the Euler method, while $\hat{r}_{1/1}$ corresponds both to the trapezoidal rule and the implicit midpoint rule. The $\hat{r}_{\alpha/\beta}$ notation is introduced in Section 4.3.

where the complex numbers $\lambda_{n,1}, \lambda_{n,2}, \ldots, \lambda_{n,d}$ are the eigenvalues of the *Jacobian matrix* $J_n := \partial \boldsymbol{f}(t_n, \boldsymbol{y}_n)/\partial \boldsymbol{y}$. This is based on the assumption that the local behaviour of the ODE is modelled well by the variational equation $\boldsymbol{y}' = \boldsymbol{y}_n + J_n(\boldsymbol{y} - \boldsymbol{y}_n)$. We hasten to emphasize that this practice is far from exact. Naive translation of any linear theory to a nonlinear setting can be dangerous and the correct approach is to embrace a nonlinear framework from the outset. Although in its full generality this ranges well beyond the material of this book, we provide a few pointers to modern nonlinear stability theory in Chapter 5.

Let us continue our investigation of linear stability domains. Replacing Λ by λ from (4.7) in (4.5) and bearing in mind that $y_0 = 1$, we obtain

$$y_n = \left(\frac{1 + \frac{1}{2}h\lambda}{1 - \frac{1}{2}h\lambda}\right)^n, \qquad n = 0, 1, \ldots \tag{4.10}$$

Again, $\{y_n\}_{n=0,1,\ldots}$ is a geometric sequence. Therefore, we obtain for the linear stability domain in the case of the trapezoidal rule,

$$\mathcal{D}_{\mathrm{TR}} = \left\{ z \in \mathbb{C} : \left| \frac{1+\frac{1}{2}z}{1-\frac{1}{2}z} \right| < 1 \right\}.$$

It is trivial to verify that the inequality within the braces is *identical* to $\operatorname{Re} z < 0$. In other words, *the trapezoidal rule mimics the asymptotic stability of linear ODE systems without any need to decrease the step size,* a property that we have already noticed in a special example in Section 4.1.

The latter feature is of sufficient importance to deserve a name of its own. We say that a method is *A-stable* if

$$\mathbb{C}^- := \{z \in \mathbb{C} : \operatorname{Re} z < 0\} \subseteq \mathcal{D}.$$

In other words, whenever a method is A-stable, we can choose the step size h (at least, for linear systems) on accuracy considerations only, without paying heed to stability constraints.

The trapezoidal rule is A-stable, whilst Euler's method is not. As is evident from Fig. 4.2, the graph labelled $\hat{r}_{1/2}(z)$ – but not the one labelled $\hat{r}_{3/0}(z)$ – corresponds to an A-stable method. It is left to the reader to ascertain in Exercise 4.4 that the theta method (1.13) is A-stable if and only if $0 \le \theta \le \frac{1}{2}$.

4.3 A-stability of Runge–Kutta methods

Applying the Runge–Kutta method (3.9) to the linear equation (4.7), we obtain

$$\xi_j = y_n + h\lambda \sum_{i=1}^{\nu} a_{j,i}\xi_i, \qquad j = 1, 2, \ldots, \nu.$$

Denote

$$\boldsymbol{\xi} := \begin{bmatrix} \xi_1 \\ \xi_2 \\ \vdots \\ \xi_\nu \end{bmatrix}, \qquad \mathbf{1} := \begin{bmatrix} 1 \\ 1 \\ \vdots \\ 1 \end{bmatrix} \in \mathbb{R}^\nu;$$

then $\boldsymbol{\xi} = \mathbf{1}y_n + h\lambda A\boldsymbol{\xi}$ and the exact solution of this linear algebraic system is

$$\boldsymbol{\xi} = (I - h\lambda A)^{-1}\mathbf{1}y_n.$$

Therefore, assuming that $I - h\lambda A$ is nonsingular,

$$y_{n+1} = y_n + h\lambda \sum_{j=1}^{\nu} b_j\xi_j = \left[1 + h\lambda \boldsymbol{b}^\top (I - h\lambda A)^{-1}\mathbf{1}\right] y_n, \qquad n = 0, 1, \ldots \tag{4.11}$$

We denote by $\mathbb{P}_{\alpha/\beta}$ the set of all rational functions $\hat{p}/\hat{q}$, where $\hat{p} \in \mathbb{P}_\alpha$ and $\hat{q} \in \mathbb{P}_\beta$.

Lemma 4.1 *For every Runge–Kutta method* (3.9) *there exists* $r \in \mathbb{P}_{\nu/\nu}$ *such that*

$$y_n = [r(h\lambda)]^n, \qquad n = 0, 1, \ldots \tag{4.12}$$

Moreover, if the Runge–Kutta method is explicit then $r \in \mathbb{P}_\nu$.

Proof It follows at once from (4.11) that (4.12) is valid with

$$r(z) := 1 + z\boldsymbol{b}^\top (I - zA)^{-1}\mathbf{1}, \qquad z \in \mathbb{C}, \tag{4.13}$$

and it remains to verify that r is indeed a rational function (a polynomial for an explicit scheme) of the stipulated type.

We represent the inverse of $I - zA$ using a familiar formula from linear algebra,

$$(I - zA)^{-1} = \frac{\mathrm{adj}(I - zA)}{\det(I - zA)},$$

where $\mathrm{adj}\, C$ is the *adjugate* of the $\nu \times \nu$ matrix C: the (i, j)th entry of the adjugate (also known as the 'adjunct' and abbreviated in the same way) is the determinant of the (j, i)th principal minor, multiplied by $(-1)^{i+j}$. Since each entry of $I - zA$ is linear in z, we deduce that each element of $\mathrm{adj}(I - zA)$, being (up to a sign) a determinant of a $(\nu - 1) \times (\nu - 1)$ matrix, is in $\mathbb{P}_{\nu-1}$. We thus conclude that

$$\boldsymbol{b}^\top \mathrm{adj}(I - zA)\mathbf{1} \in \mathbb{P}_{\nu-1},$$

therefore $\det(I - zA) \in \mathbb{P}_\nu$ implies $r \in \mathbb{P}_{\nu/\nu}$.

Finally, if the method is explicit then A is strictly lower triangular and $I - zA$ is, regardless of $z \in \mathbb{C}$, a lower triangular matrix with ones along the diagonal. Therefore $\det(I - zA) \equiv 1$ and r is a polynomial. ■

Lemma 4.2 *Suppose that an application of a numerical method to the linear equation* (4.7) *produces a geometric solution sequence,* $y_n = [r(h\lambda)]^n$, $n = 0, 1, \ldots$, *where* r *is an arbitrary function. Then*

$$\mathcal{D} = \{z \in \mathbb{C} : |r(z)| < 1\}. \tag{4.14}$$

Proof This follows at once from the definition of the set $\mathcal{D}$. ■

Corollary *No explicit Runge–Kutta (ERK) method* (3.5) *can be A-stable.*

Proof Given an ERK method, Lemma 4.1 states that the function r is a polynomial and (4.13) implies that $r(0) = 1$. No polynomial, except for the constant function $r(z) \equiv c \in (-1, 1)$, may be uniformly bounded by the value unity in $\mathbb{C}^-$, and this excludes A-stability. ■

For both Euler's method and the trapezoidal rule we have observed already that the solution sequence obeys the conditions of Lemma 4.2. This is hardly surprising, since both methods can be written in a Runge–Kutta formalism.

◇ **The function r for specific IRK schemes** Let us consider the methods

$$\begin{array}{c|cc} 0 & \frac{1}{4} & -\frac{1}{4} \\ \frac{2}{3} & \frac{1}{4} & \frac{5}{12} \\ \hline & \frac{1}{4} & \frac{3}{4} \end{array} \qquad \text{and} \qquad \begin{array}{c|cc} \frac{1}{3} & \frac{5}{12} & -\frac{1}{12} \\ 1 & \frac{3}{4} & \frac{1}{4} \\ \hline & \frac{3}{4} & \frac{1}{4} \end{array}.$$

We have already encountered both in Chapter 3: the first is (3.10), whereas the second corresponds to collocation at $c_1 = \frac{1}{3}$, $c_2 = 1$.

Substitution into (4.13) confirms that the function r is identical for the two methods:

$$r(z) = \frac{1 + \frac{1}{3}z}{1 - \frac{2}{3}z + \frac{1}{6}z^2}. \tag{4.15}$$

To check A-stability we employ (4.14). Representing $z \in \mathbb{C}$ in polar coordinates, $z = \rho e^{i\theta}$, where $\rho > 0$ and $|\theta + \pi| < \frac{1}{2}\pi$, we query whether $|r(\rho e^{i\theta})| < 1$. This would be equivalent to

$$\left|1 + \tfrac{1}{3}\rho e^{i\theta}\right|^2 < \left|1 - \tfrac{2}{3}\rho e^{i\theta} + \tfrac{1}{6}\rho^2 e^{2i\theta}\right|^2$$

and hence to

$$1 + \tfrac{2}{3}\rho\cos\theta + \tfrac{1}{9}\rho^2 < 1 - \tfrac{4}{3}\rho\cos\theta + \rho^2\left(\tfrac{1}{3}\cos 2\theta + \tfrac{4}{9}\right) - \tfrac{2}{9}\rho^3\cos\theta + \tfrac{1}{36}\rho^4.$$

Rearranging terms, the condition for $\rho e^{i\theta} \in \mathcal{D}$ becomes

$$2\rho\left(1 + \tfrac{1}{9}\rho^2\right)\cos\theta < \tfrac{1}{3}\rho^2(1 + \cos 2\theta) + \tfrac{1}{36}\rho^4 = \tfrac{2}{3}\rho^2\cos^2\theta + \tfrac{1}{36}\rho^4,$$

and this is obeyed for all $z \in \mathbb{C}^-$ since $\cos\theta < 0$ for all such z. Both methods are therefore A-stable.

A similar analysis can be applied to the Gauss–Legendre methods of Section 3.4, but the calculations become increasingly labour intensive for large values of ν. Fortunately, we are just about to identify a few shortcuts that render this job significantly easier. ◇

Our first observation is that there is no need to check every $z \in \mathbb{C}^-$ to verify that a given rational function r originates in an A-stable method (such an r is called *A-acceptable*).

Lemma 4.3 *Let r be an arbitrary rational function that is not a constant. Then $|r(z)| < 1$ for all $z \in \mathbb{C}^-$ if and only if all the poles of r have positive real parts and $|r(it)| \le 1$ for all $t \in \mathbb{R}$.*

Proof If $|r(z)| < 1$ for all $z \in \mathbb{C}^-$ then, by continuity, $|r(z)| \le 1$ for all $z \in \operatorname{cl}\mathbb{C}^-$. In particular, r is not allowed to have poles in the closed left half-plane and $|r(it)| \le 1$, $t \in \mathbb{R}$.

To prove the converse we note that, provided its poles reside to the right of $i\mathbb{R}$, the rational function r is analytic in the closed set $\operatorname{cl}\mathbb{C}^-$. Therefore, and since r is not constant, it attains its maximum along the boundary. In other words $|r(\mathrm{i}t)| \le 1$, $t \in \mathbb{R}$, implies $|r(z)| < 1$, $z \in \mathbb{C}^-$, and the proof is complete. ■

The benefits of the lemma are apparent in the case of the function (4.15): the poles reside at $2 \pm \mathrm{i}\sqrt{2}$, hence at the open right half-plane. Moreover $|r(\mathrm{i}t)| \le 1$, $t \in \mathbb{R}$, is equivalent to

$$\left|1 + \tfrac{1}{3}\mathrm{i}t\right|^2 \le \left|1 - \tfrac{2}{3}\mathrm{i}t - \tfrac{1}{6}t^2\right|^2, \qquad t \in \mathbb{R},$$

and hence to

$$1 + \tfrac{1}{9}t^2 \le 1 + \tfrac{1}{9}t^2 + \tfrac{1}{36}t^4, \qquad t \in \mathbb{R}.$$

The gain is even more spectacular for the two-stage Gauss–Legendre method, since in this case

$$r(z) = \frac{1 + \frac{1}{2}z + \frac{1}{12}z^2}{1 - \frac{1}{2}z + \frac{1}{12}z^2}$$

(although it is possible to evaluate this from the RK tableau in Section 3.4, a considerably easier derivation follows from the proof of the corollary to Theorem 4.6). Since the poles $3 \pm \mathrm{i}\sqrt{3}$ are in the open right half-plane and $|r(\mathrm{i}t)| \equiv 1$, $t \in \mathbb{R}$, the method is A-stable.

Our next result focuses on the kind of rational functions r likely to feature in (4.12).

Lemma 4.4 *Suppose that the solution sequence $\{y_n\}_{n=0}^{\infty}$, which is produced by applying a method of order p to the linear equation* (4.7) *with a constant step size, obeys* (4.12). *Then necessarily*

$$r(z) = \mathrm{e}^z + \mathcal{O}\left(z^{p+1}\right), \qquad z \to 0. \tag{4.16}$$

Proof Since $y_{n+1} = r(h\lambda)y_n$ and the exact solution, subject to the initial condition $y(t_n) = y_n$, is $\mathrm{e}^{h\lambda}y_n$, the relation (4.16) follows from the definition of order. ■

We say that a function r that obeys (4.16) is of *order p*. This should not be confused with the order of a numerical method: it is easy to construct pth-order methods with a function r whose order exceeds p, in other words, methods that exhibit superior order when applied to linear equations.

The lemma narrows down considerably the field of rational functions r that might occur in A-stability analysis. The most important functions exploit all available degrees of freedom to increase the order.

Theorem 4.5 *Given any integers α, $\beta \ge 0$, there exists a unique function $\hat{r}_{\alpha/\beta} \in \mathbb{P}_{\alpha/\beta}$ such that*

$$\hat{r}_{\alpha/\beta} = \frac{\hat{p}_{\alpha/\beta}}{\hat{q}_{\alpha/\beta}}, \qquad \hat{q}_{\alpha/\beta}(0) = 1$$

and $\hat{r}_{\alpha/\beta}$ is of order $\alpha+\beta$. The explicit forms of the numerator and the denominator are respectively

$$\begin{aligned}\hat{p}_{\alpha/\beta}(z) &= \sum_{k=0}^{\alpha}\binom{\alpha}{k}\frac{(\alpha+\beta-k)!}{(\alpha+\beta)!}z^k,\\ \hat{q}_{\alpha/\beta}(z) &= \sum_{k=0}^{\beta}\binom{\beta}{k}\frac{(\alpha+\beta-k)!}{(\alpha+\beta)!}(-z)^k = \hat{p}_{\beta/\alpha}(-z).\end{aligned} \tag{4.17}$$

Moreover $\hat{r}_{\alpha/\beta}$ is (up to a rescaling of the numerator and the denominator by a nonzero multiplicative constant) the only member of $\mathbb{P}_{\alpha/\beta}$ of order $\alpha+\beta$, and no function in $\mathbb{P}_{\alpha/\beta}$ may exceed this order. ■

The functions $\hat{r}_{\alpha/\beta}$ are called *Padé approximations* to the exponential. Most of the functions r that have been encountered so far are of this kind; thus (compare with (4.8), (4.10) and (4.15))

$$\hat{r}_{1/0}(z) = 1+z, \qquad \hat{r}_{1/1} = \frac{1+\frac{1}{2}z}{1-\frac{1}{2}z}, \qquad \hat{r}_{1/2}(z) = \frac{1+\frac{1}{3}z}{1-\frac{2}{3}z+\frac{1}{6}z^2}.$$

Padé approximations can be classified according to whether they are A-acceptable. Obviously, we need $\alpha \le \beta$ otherwise $\hat{r}_{\alpha/\beta}$ cannot be bounded in $\mathbb{C}^-$. Surprisingly, the latter condition is not sufficient. It is not difficult to prove, for example, that $\hat{r}_{0/3}$ is not A-acceptable!

Theorem 4.6 (The Wanner–Hairer–Nørsett theorem) *The Padé approximation $\hat{r}_{\alpha/\beta}$ is A-acceptable if and only if $\alpha \le \beta \le \alpha+2$.* ■

Corollary *The Gauss–Legendre IRK methods are A-stable for every $\nu \ge 1$.*

Proof We know from Section 3.4 that a ν-stage Gauss–Legendre method is of order 2ν. By Lemma 4.1 the underlying function r belongs to $\mathbb{P}_{\nu/\nu}$ and, by Lemma 4.4, it approximates the exponential function to order 2ν. Therefore, according to Theorem 4.5, $r = \hat{r}_{\nu/\nu}$, a function that is A-acceptable by Theorem 4.6. It follows that the Gauss–Legendre method is A-stable. ■

4.4 A-stability of multistep methods

Attempting to extend the definition of A-stability to the multistep method (2.8), we are faced with a problem: the implementation of an s-step method requires the provision of s values and only one of these is supplied by the initial condition. We will see in Chapter 7 how such values are derived in realistic computation. Here we adopt the attitude that a stable solution of the linear equation (4.7) is required *for all possible values of $y_1, y_2, \ldots, y_{s-1}$*. The justification of this pessimistic approach is that otherwise, even were we somehow to choose 'good' starting values, a small perturbation (e.g., a roundoff error) might well divert the solution trajectory toward instability. The reasons are similar to those already discussed in Section 4.1 in the context of the Euler method.

Let us suppose that the method (2.8) is applied to the solution of (4.7). The outcome is

$$\sum_{m=0}^{s} a_m y_{n+m} = h\lambda \sum_{m=0}^{s} b_m y_{n+m}, \qquad n = 0, 1, \ldots,$$

which we write in the form

$$\sum_{m=0}^{s} (a_m - h\lambda b_m) y_{n+m} = 0, \qquad n = 0, 1, \ldots \tag{4.18}$$

The equation (4.18) is an example of a *linear difference equation,*

$$\sum_{m=0}^{s} g_m x_{n+m} = 0, \qquad n = 0, 1, \ldots, \tag{4.19}$$

and it can be solved similarly to the more familiar linear differential equation

$$\sum_{m=0}^{s} g_m x^{(m)} = 0, \qquad t \geq t_0,$$

where the superscript indicates differentiation m times. Specifically, we form the characteristic polynomial

$$\eta(w) := \sum_{m=0}^{s} g_m w^m.$$

Let the zeros of η be $w_1, w_2, \ldots, w_q$, say, with multiplicities $k_1, k_2, \ldots, k_q$ respectively, where $\sum_{i=1}^{q} k_i = s$. The general solution of (4.19) is

$$x_n = \sum_{i=1}^{q} \left(\sum_{j=0}^{k_i - 1} c_{i,j} n^j \right) w_i^n, \qquad n = 0, 1, \ldots \tag{4.20}$$

The s constants $c_{i,j}$ are uniquely determined by the s starting values $x_0, x_1, \ldots, x_{s-1}$.

Lemma 4.7 *Let us suppose that the zeros (as a function of w) of*

$$\eta(z, w) := \sum_{m=0}^{s} (a_m - b_m z) w^m, \qquad z \in \mathbb{C},$$

are $w_1(z), w_2(z), \ldots, w_{q(z)}(z)$, while their multiplicities are $k_1(z), k_2(z), \ldots, k_{q(z)}(z)$ respectively. The multistep method (2.8) is A-stable if and only if

$$|w_i(z)| < 1, \quad i = 1, 2, \ldots, q(z) \qquad \textit{for every} \qquad z \in \mathbb{C}^-. \tag{4.21}$$

Proof As for (4.20), the behaviour of y_n is determined by the magnitude of the numbers $w_i(h\lambda)$, $i = 1, 2, \ldots, q(h\lambda)$. If all reside inside the complex unit disc then their powers decay faster than any polynomial in n, therefore $y_n \to 0$. Hence, (4.21) is sufficient for A-stability.

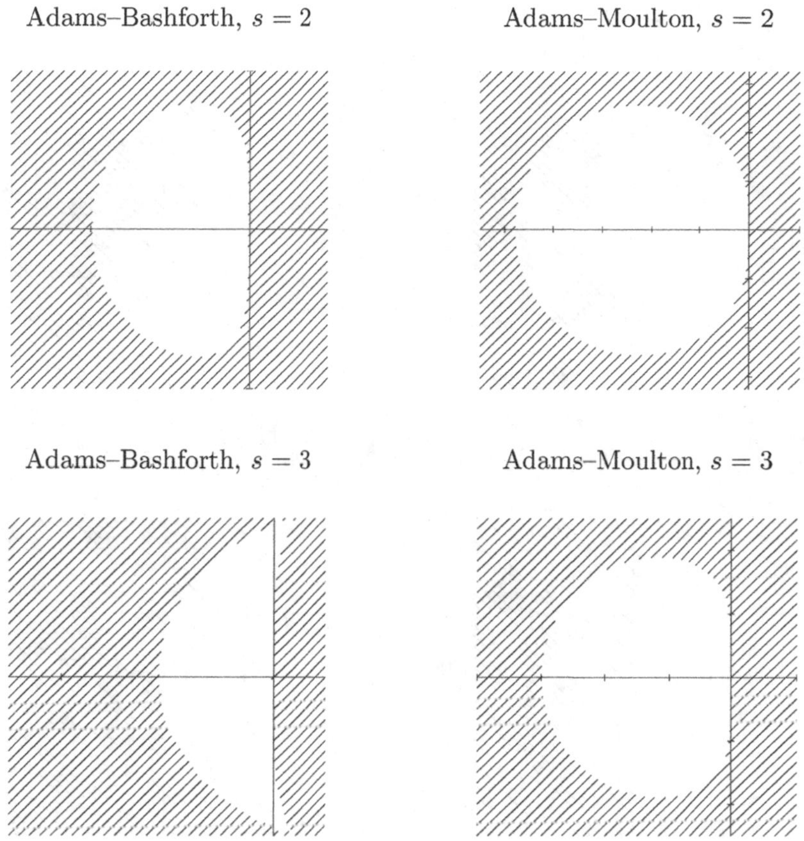

Figure 4.3 Linear stability domains $\mathcal{D}$ of Adams methods, explicit on the left and implicit on the right.

However, if $|w_1(h\lambda)| \geq 1$, say, then there exist starting values such that $c_{1,0} \neq 0$; therefore it is impossible for y_n to tend to zero as $n \to \infty$. We deduce that (4.21) is necessary for A-stability and so conclude the proof. ■

Instead of a single geometric component in (4.11), we have now a linear combination of several (in general, s) components to reckon with. This is the *quid pro quo* for using $s - 1$ starting values in addition to the initial condition, a practice whose perils have been highlighted already in the introduction to Chapter 2. According to Exercise 2.2, if a method is convergent then one of these components approximates the exponential function to the same order as the order of the method: this is similar to Lemma 4.4. However, the remaining zeros are purely parasitic: we can attribute no meaning to them so far as approximation is concerned.

Fig. 4.3 displays the linear stability domains of Adams methods, all at the same scale. Notice first how small they are and that they are reduced in size for the larger

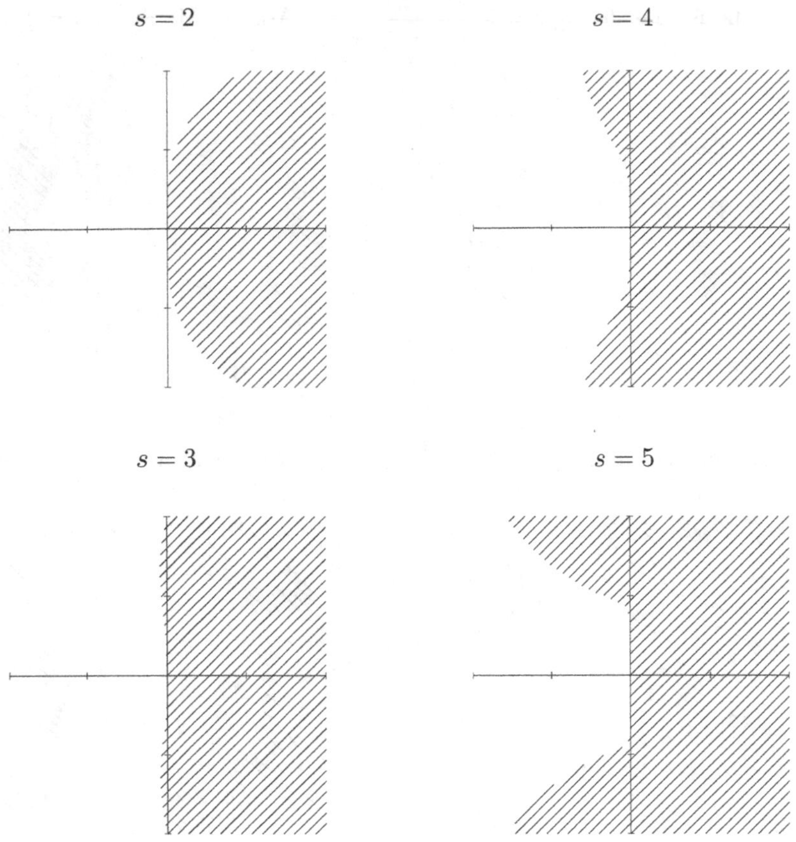

Figure 4.4 Linear stability domains $\mathcal{D}$ of BDF methods of orders $s = 2, 3, 4, 5$, shown at the same scale. Note that only $s = 2$ is A-stable.

s value. Next, pay attention to the difference between the explicit Adams–Bashforth and the implicit Adams–Moulton. In the latter case the stability domain, although not very impressive compared with those for other methods of Section 4.3, is substantially larger than for the explicit counterpart. This goes some way toward explaining the interest in implicit Adams methods, but more important reasons will be presented in Chapter 6.

However, as already mentioned in Chapter 2, Adams methods were never intended to cope with stiff equations. After all, this was the motivation for the introduction of backward differentiation formulae in Section 2.3. We turn therefore to Fig. 4.4, which displays linear stability domains for BDF methods – and are disappointed ... True, the set $\mathcal{D}$ is larger than was the case for, say, the Adams–Moulton method. However, only the two-step method displays any prospects of A-stability.

Let us commence with the good news: the BDF is indeed A-stable in the case $s = 2$. To demonstrate this we require two technical lemmas, which will be presented with a comment in lieu of a complete proof.

Lemma 4.8 *The multistep method* (2.8) *is A-stable if and only if* $b_s > 0$ *and*

$$|w_1(\mathrm{i}t)|,\ |w_2(\mathrm{i}t)|,\ \ldots,\ |w_{q(\mathrm{i}t)}(\mathrm{i}t)| \le 1, \qquad t \in \mathbb{R},$$

where $w_1, w_2, \ldots, w_{q(z)}$ *are the zeros of* $\eta(z, \cdot\,)$ *from Lemma 4.7.*

Proof On the face of it, this is an exact counterpart of Lemma 4.3: $b_s > 0$ implies analyticity in $\mathrm{cl}\,\mathbb{C}^-$ and the condition on the moduli of zeros extends the inequality on $|r(z)|$. This is deceptive, since the zeros of $\eta(z, \cdot\,)$ do not reside in the complex plane but in an s-sheeted *Riemann surface* over $\mathbb{C}$. This does not preclude the application of the maximum principle, except that somewhat more sophisticated mathematical machinery is required. ■

Lemma 4.9 (The Cohn–Schur criterion) *Both zeros of the quadratic* $\alpha w^2 + \beta w + \gamma$, *where* $\alpha, \beta, \gamma \in \mathbb{C}$, $\alpha \ne 0$, *reside in the closed complex unit disc if and only if*

$$|\alpha| \ge |\gamma|, \quad \big||\alpha|^2 - |\gamma|^2\big| \ge |\alpha\bar{\beta} - \beta\bar{\gamma}| \quad \textit{and} \quad \alpha = \gamma \ne 0 \;\Rightarrow\; |\beta| \le 2|\alpha|. \qquad (4.22)$$

Proof This is a special case of a more general result, the *Cohn–Lehmer–Schur* criterion. The latter provides a finite algorithm to check whether a given complex polynomial (of any degree) has all its zeros in any closed disc in $\mathbb{C}$. ■

Theorem 4.10 *The two-step BDF* (2.15) *is A-stable.*

Proof We have

$$\eta(z, w) = (1 - \tfrac{2}{3}z)w^2 - \tfrac{4}{3}w + \tfrac{1}{3}.$$

Therefore $b_2 = \frac{2}{3}$ and the first A-stability condition of Lemma 4.8 is satisfied. To verify the second condition we choose $t \in \mathbb{R}$ and use Lemma 4.9 to ascertain that neither of the moduli of the zeros of $\eta(\mathrm{i}t, \cdot\,)$ exceeds unity. Consequently $\alpha = 1 - \frac{2}{3}\mathrm{i}t$, $\beta = -\frac{4}{3}$, $\gamma = \frac{1}{3}$ and we obtain

$$|\alpha|^2 - |\gamma|^2 = \tfrac{4}{9}(2 + t^2) > 0$$

and

$$(|\alpha|^2 - |\gamma|^2)^2 - |\alpha\bar{\beta} - \beta\bar{\gamma}|^2 = \tfrac{16}{81}t^4 \ge 0.$$

Consequently, (4.22) is satisfied and we deduce A-stability. ■

Unfortunately, not only the 'positive' deduction from Fig. 4.4 is true. The absence of A-stability in the BDF for $s \ge 2$ (of course, $s \le 6$, otherwise the method would not be convergent and we would never use it!) is a consequence of a more general and fundamental result.

Theorem 4.11 (The Dahlquist second barrier) *The highest order of an A-stable multistep method* (2.8) *is 2.* ■

Comparing the Dahlquist second barrier with the corollary to Theorem 4.6, it is difficult to escape the impression that multistep methods are inferior to Runge–Kutta methods when it comes to A-stability. This, however, does not mean that they should not be used with stiff equations! Let us look again at Fig. 4.4. Although the cases $s = 3, 4, 5$ fail A-stability, it is apparent that for each stability domain $\mathcal{D}$ there exists $\alpha \in (0, \pi]$ such that the infinite wedge

$$\mathcal{V}_\alpha := \left\{\rho e^{i\theta} \,:\, \rho > 0,\ |\theta + \pi| < \alpha\right\} \subseteq \mathbb{C}^-$$

belongs to $\mathcal{D}$. In other words, provided that all the eigenvalues of a linear ODE system reside in $\mathcal{V}_\alpha$, no matter how far away they are from the origin, there is no need to depress the step size in response to stability restrictions. Methods with $\mathcal{V}_\alpha \subseteq \mathcal{D}$ are called A(α)-stable.[3] All BDF methods for $s \le 6$ are A(α)-stable: in particular $s = 3$ corresponds to $\alpha = 86°2'$; as Fig. 4.4 implies, almost all the region $\mathbb{C}^-$ resides in the linear stability domain.

Comments and bibliography

Different aspects of stiff equations and A-stability form the theme of several monographs of varying degrees of sophistication and detail. Gear (1971) and Lambert (1991) are the most elementary, whereas Hairer & Wanner (1991) is a compendium of just about everything known in the subject area *circa* 1991. (No text, however, for obvious reasons, abbreviates the phrase 'linear stability domain'...)

Before we comment on a few themes connected with stability analysis, let us mention briefly two topics which, while tangential to the subject matter of this chapter, deserve proper reference. Firstly, the functions $\hat{r}_{\alpha/\beta}$, which have played a substantial role in Section 4.3, are a special case of general Padé approximation. Let f be an arbitrary function that is analytic in the neighbourhood of the origin. The function $\hat{r} \in \mathbb{P}_{\alpha/\beta}$ is said to be an $[\alpha/\beta]$ *Padé approximant* of f if

$$\hat{r}(z) = f(z) + \mathcal{O}\left(z^{\alpha+\beta+1}\right), \qquad z \to 0.$$

Padé approximations possess a beautiful theory and have numerous applications, not just in the more obvious fields – the approximation of functions, numerical analysis etc. – but also in analytic number theory: they are a powerful tool in many transcendentality proofs. Baker & Graves-Morris (1981) presented a useful account of the Padé theory. Secondly, the Cohn–Schur criterion (Lemma 4.9) is a special case of a substantially more general body of knowledge that allows us to locate the zeros of polynomials in specific portions of the complex plane by a finite number of operations on the coefficients (Marden, 1966). A familiar example is the *Routh–Hurwitz* criterion, which tests whether all the zeros reside in $\mathbb{C}^-$ and is an important tool in control theory.

The characterization of all A-acceptable Padé approximations to the exponential function was the subject of a long-standing conjecture. Its resolution in 1978 by Gerhard Wanner, Ernst Hairer and Syvert Nørsett introduced the novel technique of *order stars* and was one of the great heroic tales of modern numerical mathematics. This technique can be also used to prove a far-reaching generalization of Theorem 4.11, as well as many other interesting results in the numerical analysis of differential equations. A comprehensive account of order stars features in Iserles & Nørsett (1991).

As far as A-stability for multistep equations is concerned, Theorem 4.11 implies that not much can be done. One obvious alternative, which has been mentioned in Section 4.4,

[3]Numerical analysts, being (mostly) human, tend to express α in degrees rather than radians.

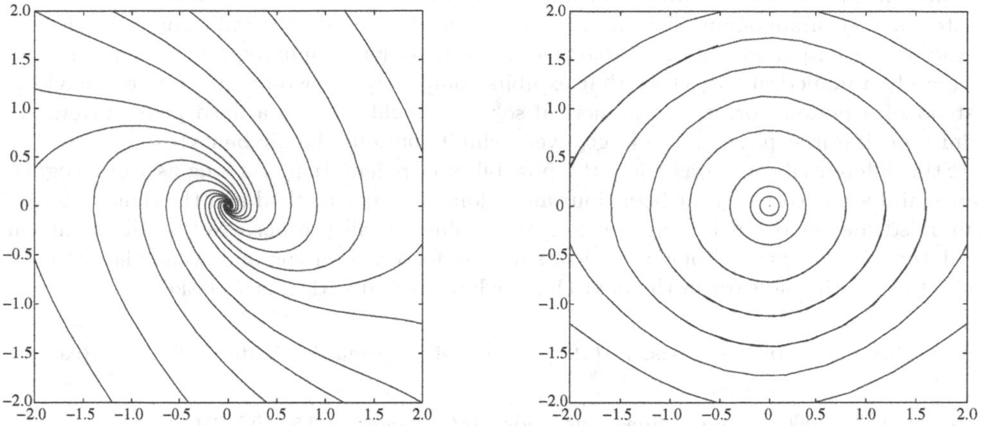

Figure 4.5 Phase planes for the damped oscillator $y'' + y' + \sin y = 0$ (on the left) and the undamped oscillator $y'' + \sin y = 0$ (on the right).

is to relax the stability requirement, in which case the order barrier disappears altogether. Another possibility is to combine the multistep rationale with the Runge–Kutta approach and possibly to incorporate higher derivatives as well. The outcome, a *general linear method* (Butcher, 2006), circumvents the barrier of Theorem 4.11.

We have mentioned in Section 4.2 that the justification of the linear model, which has led us into the concept of A-stability, is open to question when it comes to nonlinear equations. It is, however, a convenient starting point. The stability analysis of discretized nonlinear ODEs is these days a thriving industry! One model of nonlinear stability analysis is addressed in the next chapter but we make no pretence that it represents anything but a taster for a considerably more extensive theory.

And this is a convenient moment for a confession. Stiff ODEs might seem 'difficult' and indeed have been considered as such for a long time. Yet, once you get the hang of them, use the right methods and take care of stability issues, you are highly unlikely ever to go wrong. To understand why is this so and to get yourself in the right frame of mind for the next chapter, examine the phase plane of the damped nonlinear oscillator $y'' + y' + \sin y = 0$ on the left of Fig. 4.5.[4] (Of course, we convert this second-order ODE into a system of two coupled first-order ODEs $y_1' = y_2$, $y_2' = -\sin y_1 - y_2$.) No matter where we start within the displayed range, the destination is the same, the origin. Now, applying a numerical method means that our next step is typically misdirected to a neighbouring trajectory in the phase plane, but it is obvious from the figure that the flow itself is 'self correcting'. Unless we are committing errors which are both large and biased, a hallmark of an unstable method, ultimately our global picture will be at the very least of the right qualitative character: the numerical trajectory will tend to the origin. Small errors will correct themselves, provided that the method is stable enough.

Compare this with the undamped nonlinear oscillator $y'' + \sin y = 0$ on the right of Fig. 4.5. Except when it starts at the origin, in which case not much happens, the flow

[4]This system is not stiff but even this gentle damping is sufficient to convey our point, while a *real* stiff system, e.g. $y'' + 1000y' + \sin y = 0$, would have led to a plot that was considerably less intelligible.

(again, within the range of displayed initial values) progresses in periodic orbits. Now, no matter how accurate our method and no matter how stable it is, small errors can 'kick' us to the wrong trajectory; and repeated 'kicks', no matter how minute, are likely to produce ultimately a numerical trajectory that exhibits completely the wrong qualitative behaviour. Instead of a periodic orbit, the numerical solution might tend to a fixed point, diverge to infinity or, if our step size is too large, even exhibit spurious chaotic behaviour.

Stiff differential equations allow the possibility of redemption. As long as you recognise your sinful ways, correct your behaviour and adopt the right method and the right step size, your misdemeanours wil be forgiven and your solution will prosper. Not so the nonlinear oscillator $y'' + \sin y = 0$. Your numerical sins stay forever with you and accumulate forever. Or at least until you learn in the next chapter how to deal with this situation.

Baker, G.A. and Graves-Morris, P. (1981), *Padé Approximants,* Addison–Wesley, Reading, MA.

Butcher, J.C. (2006), General linear methods, *Acta Numerica* **15**, 157–256.

Gear, C.W. (1971), *Numerical Initial Value Problems in Ordinary Differential Equations,* Prentice–Hall, Englewood Cliffs, NJ.

Hairer, E. and Wanner, G. (1991), *Solving Ordinary Differential Equations II: Stiff Problems and Differential-Algebraic Equations,* Springer-Verlag, Berlin.

Iserles, A. and Nørsett, S.P. (1991), *Order Stars,* Chapman & Hall, London.

Lambert, J.D. (1991), *Numerical Methods for Ordinary Differential Systems,* Wiley, London.

Marden, M. (1966), *Geometry of Polynomials,* American Mathematical Society, Providence, RI.

Exercises

4.1 Let $\boldsymbol{y}' = \Lambda \boldsymbol{y}$, $\boldsymbol{y}(t_0) = \boldsymbol{y}_0$, be solved (with a constant step size $h > 0$) by a one-step method with a function r that obeys the relation (4.12). Suppose that a nonsingular matrix V and a diagonal matrix D exist such that $\Lambda = VDV^{-1}$. Prove that there exist vectors $\boldsymbol{x}_1, \boldsymbol{x}_2, \ldots, \boldsymbol{x}_d \in \mathbb{R}^d$ such that

$$\boldsymbol{y}(t_n) = \sum_{j=1}^{d} \mathrm{e}^{t_n \lambda_j} \boldsymbol{x}_j, \qquad n = 0, 1, \ldots,$$

and

$$\boldsymbol{y}_n = \sum_{j=1}^{d} [r(h\lambda)]^n \boldsymbol{x}_j, \qquad n = 0, 1, \ldots,$$

where $\lambda_1, \lambda_2, \ldots, \lambda_d$ are the eigenvalues of Λ. Deduce that the values of $\boldsymbol{x}_1$ and of $\boldsymbol{x}_2$, given in (4.3) and (4.4) are identical.

4.2* Consider the solution of $\boldsymbol{y}' = \Lambda \boldsymbol{y}$ where

$$\Lambda = \begin{bmatrix} \lambda & 1 \\ 0 & \lambda \end{bmatrix}, \qquad \lambda \in \mathbb{C}^-.$$

a Prove that

$$\Lambda^n = \begin{bmatrix} \lambda^n & n\lambda^{n-1} \\ 0 & \lambda^n \end{bmatrix}, \qquad n = 0, 1, \ldots$$

b Let g be an arbitrary function that is analytic about the origin. The 2×2 matrix $g(\Lambda)$ can be defined by substituting powers of Λ into the Taylor expansion of g. Prove that

$$g(t\Lambda) = \begin{bmatrix} g(t\lambda) & tg'(t\lambda) \\ 0 & g(t\lambda) \end{bmatrix}.$$

c By letting $g(z) = \mathrm{e}^z$ prove that $\lim_{t\to\infty} \boldsymbol{y}(t) = \boldsymbol{0}$.

d Suppose that $\boldsymbol{y}' = \Lambda \boldsymbol{y}$ is solved with a Runge–Kutta method, using a constant step size $h > 0$. Let r be the function from Lemma 4.1. Letting $g = r$, obtain the explicit form of $[r(h\Lambda)]^n$, $n = 0, 1, \ldots$

e Prove that if $h\lambda \in \mathcal{D}$, where $\mathcal{D}$ is the linear stability domain of the Runge–Kutta method, then $\lim_{n\to\infty} \boldsymbol{y}_n = \boldsymbol{0}$.

4.3* This question is concerned with the relevance of the linear stability domain to the numerical solution of *inhomogeneous* linear systems.

a Let Λ be a nonsingular matrix. Prove that the solution of $\boldsymbol{y}' = \Lambda\boldsymbol{y} + \boldsymbol{a}$, $\boldsymbol{y}(t_0) = \boldsymbol{y}_0$, is

$$\boldsymbol{y}(t) = \mathrm{e}^{(t-t_0)\Lambda}\boldsymbol{y}_0 + \Lambda^{-1}[\mathrm{e}^{(t-t_0)\Lambda} - I]\boldsymbol{a}, \qquad t \ge t_0.$$

Thus, deduce that if Λ has a full set of eigenvectors and all its eigenvalues reside in $\mathbb{C}^-$ then $\lim_{t\to\infty} \boldsymbol{y}(t) = -\Lambda^{-1}\boldsymbol{a}$.

b Assuming for simplicity's sake that the underlying equation is scalar, i.e. $y' = \lambda y + a$, $y(t_0) = y_0$, prove that a single step of the Runge–Kutta method (3.9) results in

$$y_{n+1} = r(h\lambda)y_n + q(h\lambda), \qquad n = 0, 1, \ldots,$$

where r is given by (4.13) and

$$q(z) := ha\boldsymbol{b}^\top(I - zA)^{-1}\boldsymbol{1} \in \mathbb{P}_{(\nu-1)/\nu}, \qquad z \in \mathbb{C}.$$

c Deduce, by induction or otherwise, that

$$y_n = [r(h\lambda)]^n y_0 + \left\{\frac{[r(h\lambda)]^n - 1}{r(h\lambda) - 1}\right\} q(h\lambda), \qquad n = 0, 1, \ldots$$

d Assuming that $h\lambda \in \mathcal{D}$, prove that $\lim_{n\to\infty} y_n$ exists and is bounded.

4.4 Determine all values of θ such that the theta method (1.13) is A-stable.

4.5 Prove that for every ν-stage explicit Runge–Kutta method (3.5) of order ν it is true that

$$r(z) = \sum_{k=0}^{\nu} \frac{1}{k!} z^k, \qquad z \in \mathbb{C}.$$

4.6 Evaluate explicitly the function r for the following Runge–Kutta methods:

$$\textbf{a}\quad \begin{array}{c|cc} 0 & 0 & 0 \\ \frac{2}{3} & \frac{1}{3} & \frac{1}{3} \\ \hline & \frac{1}{4} & \frac{3}{4} \end{array}, \qquad \textbf{b}\quad \begin{array}{c|cc} \frac{1}{6} & \frac{1}{6} & 0 \\ \frac{5}{6} & \frac{2}{3} & \frac{1}{6} \\ \hline & \frac{1}{2} & \frac{1}{2} \end{array}, \qquad \textbf{c}\quad \begin{array}{c|ccc} 0 & 0 & 0 & 0 \\ \frac{1}{2} & \frac{1}{4} & \frac{1}{4} & 0 \\ 1 & 0 & 1 & 0 \\ \hline & \frac{1}{6} & \frac{2}{3} & \frac{1}{6} \end{array}.$$

Are these methods A-stable?

4.7 Prove that the Padé approximation $\hat{r}_{0/3}$ is not A-acceptable.

4.8 Determine the order of the two-step method

$$\boldsymbol{y}_{n+2} - \boldsymbol{y}_n = \tfrac{2}{3}h\left[\boldsymbol{f}(t_{n+2}, \boldsymbol{y}_{n+2}) + \boldsymbol{f}(t_{n+1}, \boldsymbol{y}_{n+1}) + \boldsymbol{f}(t_n, \boldsymbol{y}_n)\right], \quad n = 0, 1, \ldots$$

Is it A-stable?

4.9 The two-step method

$$\boldsymbol{y}_{n+2} - \boldsymbol{y}_n = 2h\boldsymbol{f}(t_{n+1}, \boldsymbol{y}_{n+1}), \qquad n = 0, 1, \ldots \tag{4.23}$$

is called the *explicit midpoint rule*.

a Denoting by $w_1(z)$ and $w_2(z)$ the zeros of the underlying function $\eta(z, \cdot)$, prove that $w_1(z)w_2(z) \equiv -1$ for all $z \in \mathbb{C}$.

b Show that $\mathcal{D} = \emptyset$.

c We say that $\tilde{\mathcal{D}}$ is a *weak linear stability domain* of a numerical method if, when applied to the scalar linear test equation, it produces a uniformly bounded solution sequence. (It is easy to see that $\tilde{\mathcal{D}} = \mathrm{cl}\,\mathcal{D}$ for most methods of interest.) Determine explicitly $\tilde{\mathcal{D}}$ for the method (4.23).

The method (4.23) will feature again in Chapters 16 and 17, in the guise of the *leapfrog* scheme.

4.10 Prove that if the multistep method (2.8) is convergent then $0 \in \partial\tilde{\mathcal{D}}$.

5

Geometric numerical integration

5.1 Between quality and quantity

If mathematics is the language of science and engineering, differential equations form much of its grammar. A myriad of facts originating in the laboratory, in an astronomical observatory or on a field trip, flashes of enlightenment and sudden comprehension, the poetry of nature and the miracle of the human mind can all be phrased in the language of mathematical models coupling the behaviour of a physical phenomenon with its rate of change: differential equations. No wonder, therefore, that research into differential equations is so central to contemporary mathematics. Mathematical disciplines from functional analysis to algebraic geometry, from operator theory and harmonic analysis to differential geometry, algebraic topology, analytic function theory, spectral theory, nonlinear dynamical systems and beyond are, once you delve into their origins and ramifications, mostly concerned with adding insight into the great mystery of differential equations.

Modern mathematics is extraordinarily useful in deriving a wealth of qualitative information about differential equations, information that often has profound physical significance. Yet, except for particularly simple situations, it falls short of actually providing the solution in an explicit form. The task of fleshing out numbers on the mathematical bones falls to numerical analysis. And here looms danger ... The standard rules of engagement of numerical analysis are simple: deploy computing power and algorithmic ingenuity to minimize error. Yet it is possible that, in our quest for the best *quantity,* we might sacrifice *quality.* Features of the exact solution that have been derived with a great deal of mathematical ingenuity (and which might have important significance in applications) might well be lost in our quest to derive the most accurate solution with the least computing effort.

Painting with a broad brush, as one is bound to do in a textbook, we can distinguish two kinds of qualitative feature of a time-evolving differential equation, the dynamic and the geometric. The dynamic attributes of a differential equation have to do with the ultimate destination of its solution. As time increases to infinity will the solution tend to a fixed point? Will it be periodic? Or will it exhibit more 'exotic' behaviour, e.g. chaos? The geometric characteristics of a differential equation, however, typically refer to features which are invariant in time. Typical *invariants* include first integrals – thus, some differential equations conserve energy, angular momentum or (as we will see below) orthogonality. Other invariants are more elaborate and cannot be easily phrased just in terms of the solution trajectory, yet they often have deep mathematical

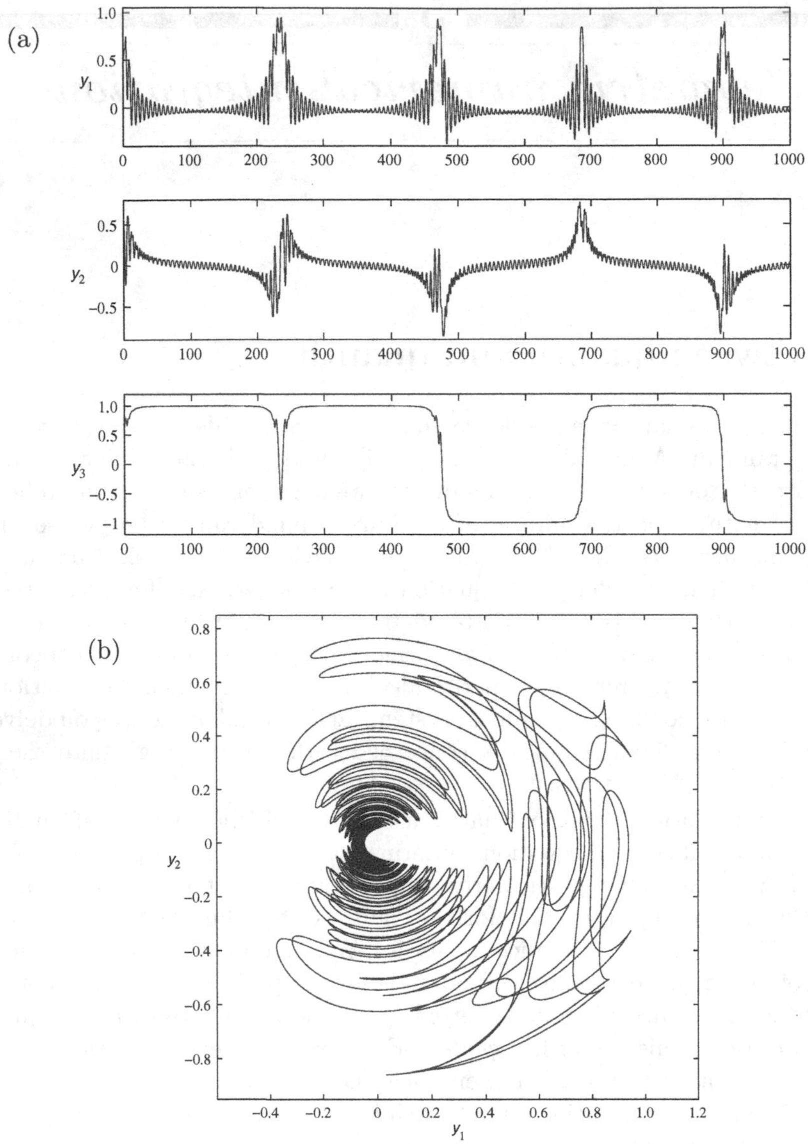

Figure 5.1 (a) The solution of the ODE system (5.1) for $t \leq 1000$ with initial value $\boldsymbol{y}(0) = (\frac{\sqrt{3}}{3}, \frac{\sqrt{3}}{3}, \frac{\sqrt{3}}{3})$ and (b) the phase plane (y_1, y_2).

and physical significance. A case in point, upon which we will elaborate at greater length later, is the conservation of symplectic form by *Hamiltonian systems.*

An innocent-looking ODE system exhibiting a wealth of dynamical and geometric features is

$$\begin{aligned} y_1' &= y_2 y_3 \sin t - y_1 y_2 y_3, \\ y_2' &= -y_1 y_3 \sin t + \tfrac{1}{20} y_1 y_3, \end{aligned} \qquad (5.1)$$

$$y_3' = y_1^2 y_2 - \tfrac{1}{20} y_1 y_2,$$

whose solution is displayed in Fig. 5.1. The solution is bounded, highly oscillatory and clearly switches between two modes that are suggestively periodic. Are these modes periodic? Are the switches chaotic? Good questions, but the system (5.1) has been designed solely for the purpose of our exposition and not much is known about it, except for one feature that can be proved with ease. Since $y_1 y_1' + y_2 y_2' + y_3 y_3' = 0$ it follows at once that as t increases the Euclidean norm $\boldsymbol{y}(t) = [y_1^2(t) + y_2^2(y) + y_3^2(t)]^{1/2}$ remains constant. In other words, the solution of (5.1) evolves on the two-dimensional unit sphere embedded in $\mathbb{R}^3$. It makes sense, at least intuitively, to compute it while respecting this feature, but simple numerical experiments using methods from Chapters 2 and 3 mostly exhibit a drift away from the sphere.

There is *a priori* no reason whatsoever why numerical methods should respect invariants or have the correct asymptotic behaviour. Does it matter? It depends and indeed is heavily sensitive to the nature of the application that our numerical solution is attempting to elucidate. Often the correct rendition of qualitative features is of lesser importance or an optional extra, but sometimes it is absolutely essential that we model the geometry or dynamics correctly. An obvious example is when the entire purpose of the computation is to shed light on the asymptotic behaviour of the solution as $t \to \infty$. In that case we are concerned very little with errors committed in finite time, but we cannot allow *any* infelicities insofar as the dynamics is concerned. Another example occurs when the conservation of a geometric feature is central to the entire purpose of the computation.

◇ **Isospectral flows** On the face of it, there is little about the ODE system

$$y_1' = 2y_4^2, \quad y_2' = 2y_5^2 - 2y_4^2, \quad y_3' = -2y_5^2, \quad y_4' = (y_2 - y_1)y_4, \quad y_5' = (y_3 - y_2)y_5$$

that meets the eye. However, once we arrange the five unknowns in a symmetric tridiagonal matrix

$$Y = \begin{bmatrix} y_1 & y_4 & 0 \\ y_4 & y_2 & y_5 \\ 0 & y_5 & y_3 \end{bmatrix},$$

we can rewrite the system in the form

$$Y' = B(Y)Y - YB(Y), \qquad \text{where} \qquad B(Y) = \begin{bmatrix} 0 & y_4 & 0 \\ -y_4 & 0 & y_5 \\ 0 & -y_5 & 0 \end{bmatrix}.$$

The solution of this matrix ODE stays symmetric and tridiagonal for every $t \ge 0$ but, more remarkably, it has a striking feature: the eigenvalues stay put as the solution evolves!

Had this been true just for one innocent-looking ODE system, this might have merited little interest. However, our system can be generalized, whence it becomes of considerably greater interest. Thus, let Y_0 be an arbitrary real symmetric $d \times d$ matrix and suppose that the Lipschitz function B maps such a matrix into real, skew-symmetric $d \times d$ matrices. The matrix ODE system

$$Y' = B(Y)Y - YB(Y), \quad t \geq 0, \qquad Y(0) = Y_0, \tag{5.2}$$

is said to be *isospectral:* the eigenvalues of $Y(t)$ coincide with these of Y_0 for all $t \geq 0$. The proof is important because, as we will see later, it can be readily translated into a numerical method. Thus, we seek a solution of the form $Y(t) = Q(t)Y_0Q^{-1}(t)$, where $Q(t)$ is a $d \times d$ matrix function. Since $\mathrm{d}Q^{-1}/\mathrm{d}t = -Q^{-1}Q'Q^{-1}$, substitution into (5.2) readily confirms that this is indeed the case, provided that Q itself satisfies the differential equation

$$Q' = B(QY_0Q^{-1})Q, \quad t \geq 0, \qquad Q(0) = I. \tag{5.3}$$

Therefore the matrices $Y(t)$ and Y_0 share the same eigenvalues.

Actually, this is not the end of the story! Let $Z(t) = Q(t)Q^\top(t)$. Direct differentiation and the skew-symmetry of B imply that Z obeys the matrix ODE

$$Z' = Q'Q^\top + Q(Q^\top)' = Q'Q^\top + Q(Q')^\top = BQQ^\top + QQ^\top B^\top = BZ - ZB$$

with the initial condition $Z(0) = I$. But the only possible solution of this equation is $Z(t) \equiv I$, and we thus deduce that $QQ^\top = I$. *In other words, the solution of* (5.3) *is an orthogonal matrix!* We file this important fact for future use, noting for the present the implication that $Y = QY_0Q^{-1} = QY_0Q^\top$ is indeed symmetric.

It is possible to show that for some choices of the matrix function B the solution of (5.2) invariably tends to a fixed point $\hat{Y}$ as $t \to \infty$ and also that $\hat{Y}$ is a diagonal matrix. Because of our discussion, it is clear that the diagonal of $\hat{Y}$ consists of the eigenvalues of Y_0 and, conceivably, we could solve (5.2) as a means to their computation. However, for this approach to make sense, it is crucial that our numerical method renders the eigenvalues correctly. The bad news is that *all* the methods that we have mentioned so far in this book are unequal to this task! ◇

Think again about the example of isospectral flows. A numerical method is bound to commit an error: this is part and parcel of a numerical solution. Our requirement, though, is that (within roundoff) this error is nil insofar as eigenvalues are concerned!

Isospectral flows are but one example of cases where the conservation of 'geometry' is an issue. Many other invariants are important for physical or mathematical reasons. Moreover, the distinction between dynamics and geometry in long-time integration is fairly moot. In important cases it is possible to prove that the maintenance of a geometric feature guarantees the computation of the correct dynamics.

The part of the numerical analysis of differential equations concerned with computation in a way that respects dynamic and geometric features is called 'geometric numerical integration' (GNI). This is a fairly new theory, which has already led to an important change of focus, more in the numerical analysis of ODEs than in the computation of PDEs (where the theory is much more incomplete). In this chapter we restrict our narrative to three examples of GNI in action. A more comprehensive treatment of the subject must be relegated to specialized monographs.

We have mentioned two reasons why it might be good to conserve dynamic or geometric features: their intrinsic mathematical importance (recall eigenvalues and isospectral flows) and their significance in applications. Intriguingly, there is a third reason, and it has to do with numerical analysis itself. It is possible to prove for important categories of equations that, once certain geometric invariants are respected under discretization, numerical error accumulates much more slowly. This becomes very important in long-term computations.

5.2 Monotone equations and algebraic stability

We have already seen in Chapter 4 a simple linear model concerned with the conservation of the dynamics. To employ the terminology of the current chapter, we observed that A-stable methods render correctly the dynamics of linear ODE systems $\boldsymbol{y}' = A\boldsymbol{y}$ when all the eigenvalues of the matrix A reside in the left half-plane. In this section we present a simple model for the analysis of computational dynamics in a nonlinear setting.

Let $\langle \cdot\,, \cdot \rangle$ be an inner product in $\mathbb{C}^d$ and $\|\cdot\|$ the corresponding norm. We say that the ODE system

$$\boldsymbol{y}' = \boldsymbol{f}(t, \boldsymbol{y}), \qquad t \geq 0, \tag{5.4}$$

is *monotone* (with respect to the given inner product) if the function $\boldsymbol{f}$ satisfies the inequality

$$\operatorname{Re}\langle \boldsymbol{u} - \boldsymbol{v}, \boldsymbol{f}(t, \boldsymbol{u}) - \boldsymbol{f}(t, \boldsymbol{v})\rangle \leq 0, \qquad t \geq 0, \quad \boldsymbol{u}, \boldsymbol{v} \in \mathbb{C}^d. \tag{5.5}$$

The importance of monotonicity follows from the next result.

Lemma 5.1 *Subject to the monotonicity condition* (5.5)*, the ODE* (5.4) *is dissipative: given two solutions, $\boldsymbol{u}$ and $\boldsymbol{v}$, say, with initial conditions $\boldsymbol{u}(0) = \boldsymbol{u}_0$ and $\boldsymbol{v}(0) = \boldsymbol{v}_0$, the function $\|\boldsymbol{u}(t) - \boldsymbol{v}(t)\|$ decreases monotonically for $t \geq 0$.*

Proof Let $\phi(t) = \frac{1}{2}\|\boldsymbol{u}(t) - \boldsymbol{v}(t)\|^2$. It then follows from (5.4) and (5.5) that

$$\begin{aligned}
\phi'(t) &= \frac{1}{2}\frac{\mathrm{d}}{\mathrm{d}t}\langle \boldsymbol{u}(t) - \boldsymbol{v}(t), \boldsymbol{u}(t) - \boldsymbol{v}(t)\rangle \\
&= \frac{1}{2}\langle \boldsymbol{u}'(t) - \boldsymbol{v}'(t), \boldsymbol{u}(t) - \boldsymbol{v}(t)\rangle + \frac{1}{2}\langle \boldsymbol{u}(t) - \boldsymbol{v}(t), \boldsymbol{u}'(t) - \boldsymbol{v}'(t)\rangle \\
&= \langle \boldsymbol{u}(t) - \boldsymbol{v}(t), \boldsymbol{f}(t, \boldsymbol{u}(t)) - \boldsymbol{f}(t, \boldsymbol{v}(t))\rangle \leq 0.
\end{aligned}$$

This proves the lemma. ■

The intuitive interpretation of Lemma 5.1 is that different solution trajectories of (5.4) never depart from each other. From the dynamical point of view, this means that small perturbations remain forever small.

◇ **Even scalar equations can be interesting!** Consider the scalar equation $y' = \frac{1}{8} - y^3$ and its fairly mild perturbation $y' = \frac{1}{8} + \frac{1}{6}y - y^3$. It is easy to

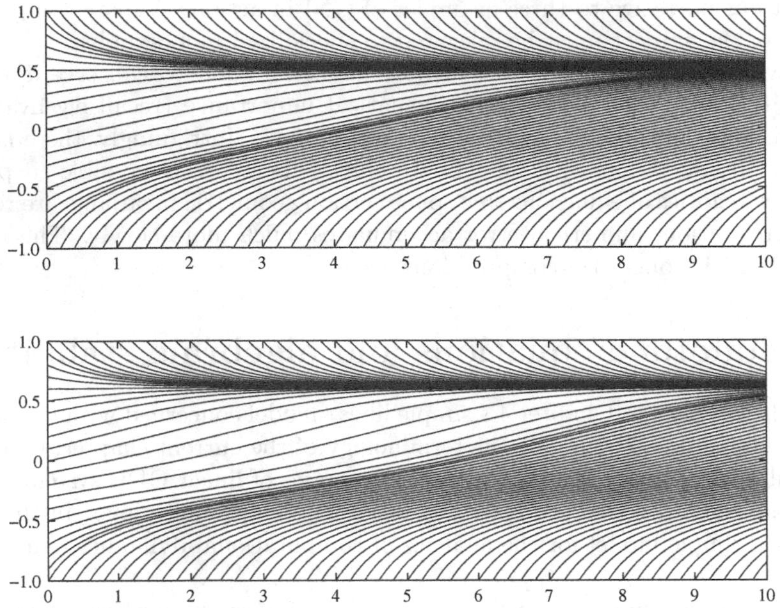

Figure 5.2 Solution trajectories for a monotone (top plot) scalar cubic equation and its nonmonotone perturbation.

see that, while the first obeys (5.4) and is monotone, this is not true for the second equation.

Solution trajectories for a range of initial values are plotted for both equations in Fig. 5.2. On the face of it, they are fairly similar and it is easy to verify that in both cases all solutions approach a unique fixed point for $t \gg 1$. Yet, while for the monotone equation the trajectories *always* bunch up, it is easy to discern in the bottom plot examples of trajectories that depart from each other even if only for a while.

This behaviour has intriguing implications once these equations are discretized. A numerical solution bears an error which, no matter how small (unless we are extraordinarily lucky and there is no error at all!), means that our next step resides on a nearby trajectory. Now, if that trajectory does not take us from the correct solution *and* if the numerical method is 'stable' (in a sense which, for the time being, we leave vague), this does not matter much. If, however, the new trajectory takes us further from the correct one, it is possible that the next step will land us on yet another trajectory, even more remote from the correct one, and so on: the error cascades and in short order the solution loses its accuracy. $\diamond$

Clearly, it is important to examine whether, once a differential equation satisfies (5.4) and is monotone, the numerical methods of Chapters 2 and 3 conform with Lemma 5.1

and therefore possess the 'stability' mentioned in the last example. In what follows we address this issue insofar as Runge–Kutta methods are concerned. The treatment of multistep methods within this context is much more complicated and outside the scope of this book.

We say that the Runge–Kutta method (3.9) is *algebraically stable* if, subject to the inequality (5.5), it produces dissipative solutions. In other words, if $\boldsymbol{u}_n$ and $\boldsymbol{v}_n$ are separate solution sequences (with the same step sizes), corresponding to the initial values $\boldsymbol{u}_0$ and $\boldsymbol{v}_0$ respectively, then, necessarily,

$$\|\boldsymbol{u}_{n+1} - \boldsymbol{v}_{n+1}\| \le \|\boldsymbol{u}_n - \boldsymbol{v}_n\|, \qquad n = 0, 1, \ldots. \tag{5.6}$$

The main new object in our analysis (and in the rest of this chapter) is a matrix which, thanks to its surprising ubiquity in many different corners of GNI, is usually referred informally as 'the famous matrix M': its elements are given by

$$m_{k,\ell} = b_k a_{k,\ell} + b_\ell a_{\ell,k} - b_k b_\ell, \qquad k, \ell = 1, 2, \ldots, \nu,$$

where $a_{k,\ell}$ and b_k are the RK matrix elements and the RK weights of the method (3.9). Note that the $\nu \times \nu$ matrix M is symmetric.

Theorem 5.2 *If the matrix M is positive semidefinite and the weights $b_1, b_2, \ldots, b_\nu$ are nonnegative then the Runge–Kutta method* (3.9) *is algebraically stable.*

Proof We need to look at detail at a single step of the method (3.9), applied at t_n to the initial vectors $\boldsymbol{u}_n$ and $\boldsymbol{v}_n$. We denote the internal stages by $\boldsymbol{r}_1, \boldsymbol{r}_2, \ldots, \boldsymbol{r}_\nu$ and $\boldsymbol{s}_1, \boldsymbol{s}_2, \ldots, \boldsymbol{s}_\nu$ respectively, and let

$$\boldsymbol{\rho}_j = \boldsymbol{u}_n + h\sum_{i=1}^{\nu} a_{j,i}\boldsymbol{r}_i, \quad \boldsymbol{\sigma}_j = \boldsymbol{v}_n + h\sum_{i=1}^{\nu} a_{j,i}\boldsymbol{s}_i, \qquad j = 1, 2, \ldots, \nu. \tag{5.7}$$

Thus,

$$\boldsymbol{r}_j = \boldsymbol{f}(t_n + c_j h, \boldsymbol{\rho}_j), \quad \boldsymbol{s}_j = \boldsymbol{f}(t_n + c_j h, \boldsymbol{\sigma}_j), \qquad j = 1, 2, \ldots, \nu \tag{5.8}$$

and

$$\boldsymbol{u}_{n+1} = \boldsymbol{u}_n + h\sum_{j=1}^{\nu} b_j \boldsymbol{r}_j, \qquad \boldsymbol{v}_{n+1} = \boldsymbol{v}_n + h\sum_{j=1}^{\nu} b_j \boldsymbol{s}_j. \tag{5.9}$$

We need to prove that the conditions of the theorem imply the inequality (5.6), namely that

$$\|\boldsymbol{u}_{n+1} - \boldsymbol{v}_{n+1}\|^2 - \|\boldsymbol{u}_n - \boldsymbol{v}_n\|^2 \le 0. \tag{5.10}$$

But, by (5.9),

$$\begin{aligned}
\|\boldsymbol{u}_{n+1} - \boldsymbol{v}_{n+1}\|^2 &= \left\langle \boldsymbol{u}_n - \boldsymbol{v}_n + h\sum_{j=1}^{\nu} b_j(\boldsymbol{r}_j - \boldsymbol{s}_j),\ \boldsymbol{u}_n - \boldsymbol{v}_n + h\sum_{j=1}^{\nu} b_j(\boldsymbol{r}_j - \boldsymbol{s}_j) \right\rangle \\
&= \left\langle \boldsymbol{u}_n - \boldsymbol{v}_n + h\sum_{j=1}^{\nu} b_j\boldsymbol{d}_j,\ \boldsymbol{u}_n - \boldsymbol{v}_n + h\sum_{j=1}^{\nu} b_j\boldsymbol{d}_j \right\rangle \\
&= \|\boldsymbol{u}_n - \boldsymbol{v}_n\|^2 + 2h\,\mathrm{Re}\left\langle \boldsymbol{u}_n - \boldsymbol{v}_n, \sum_{j=1}^{\nu} b_j\boldsymbol{d}_j \right\rangle + h^2\left\|\sum_{j=1}^{\nu} b_j\boldsymbol{d}_j\right\|^2,
\end{aligned}$$

where $\boldsymbol{d}_j = \boldsymbol{r}_j - \boldsymbol{s}_j$, $j = 1, 2, \ldots, \nu$. Thus, (5.10) is equivalent to

$$2\,\mathrm{Re}\left\langle \boldsymbol{u}_n - \boldsymbol{v}_n, \sum_{j=1}^{\nu} b_j \boldsymbol{d}_j \right\rangle + h \left\| \sum_{j=1}^{\nu} b_j \boldsymbol{d}_j \right\|^2 \leq 0. \tag{5.11}$$

Using (5.7) to replace $\boldsymbol{u}_n$ and $\boldsymbol{v}_n$ by

$$\boldsymbol{\rho}_j - h \sum_{i=1}^{\nu} a_{j,i} \boldsymbol{r}_i \qquad \text{and} \qquad \boldsymbol{\sigma}_j - h \sum_{i=1}^{\nu} a_{j,i} \boldsymbol{s}_i$$

respectively, we obtain

$$\begin{aligned}\mathrm{Re}\left\langle \boldsymbol{u}_n - \boldsymbol{v}_n, \sum_{j=1}^{\nu} b_j \boldsymbol{d}_j \right\rangle &= \sum_{j=1}^{\nu} b_j \left\langle \boldsymbol{\rho}_j - \boldsymbol{\sigma}_j - h \sum_{i=1}^{\nu} a_{j,i} \boldsymbol{d}_i, \boldsymbol{d}_j \right\rangle \\ &= \sum_{j=1}^{\nu} b_j \,\mathrm{Re}\,\langle \boldsymbol{\rho}_j - \boldsymbol{\sigma}_j, \boldsymbol{d}_j \rangle - h \sum_{j=1}^{\nu} \sum_{i=1}^{\nu} b_j a_{j,i} \,\mathrm{Re}\,\langle \boldsymbol{d}_i, \boldsymbol{d}_j \rangle.\end{aligned}$$

By our assumption, though, the system (5.4) is monotone and, using (5.8) and the nonnegativity of the weights, it follows that

$$\sum_{j=1}^{\nu} b_j \,\mathrm{Re}\,\langle \boldsymbol{\rho}_j - \boldsymbol{\sigma}_j, \boldsymbol{d}_j \rangle = \sum_{j=1}^{\nu} b_j \,\mathrm{Re}\,\langle \boldsymbol{\rho}_j - \boldsymbol{\sigma}_j, \boldsymbol{f}(t_n + c_j h, \boldsymbol{\rho}_j) - \boldsymbol{f}(t_n + c_j h, \boldsymbol{\sigma}_j) \rangle \leq 0,$$

consequently

$$\mathrm{Re}\left\langle \boldsymbol{u}_n - \boldsymbol{v}_n, \sum_{j=1}^{\nu} b_j \boldsymbol{d}_j \right\rangle \leq -h \sum_{i=1}^{\nu} \sum_{j=1}^{\nu} b_j a_{j,i} \,\mathrm{Re}\,\langle \boldsymbol{d}_j, \boldsymbol{d}_i \rangle$$

and, swapping indices,

$$\mathrm{Re}\left\langle \boldsymbol{u}_n - \boldsymbol{v}_n, \sum_{j=1}^{\nu} b_j \boldsymbol{d}_j \right\rangle \leq -h \sum_{i=1}^{\nu} \sum_{j=1}^{\nu} b_i a_{i,j} \,\mathrm{Re}\,\langle \boldsymbol{d}_i, \boldsymbol{d}_j \rangle.$$

Therefore,

$$\begin{aligned}&2\,\mathrm{Re}\left\langle \boldsymbol{u}_n - \boldsymbol{v}_n, \sum_{j=1}^{\nu} b_j \boldsymbol{d}_j \right\rangle + h \left\| \sum_{j=1}^{\nu} b_j \boldsymbol{d}_j \right\|^2 \\ &\leq h \sum_{i=1}^{\nu} \sum_{j=1}^{\nu} (b_i b_j - b_j a_{j,i} - b_i a_{i,j}) \,\mathrm{Re}\,\langle \boldsymbol{d}_i, \boldsymbol{d}_j \rangle = -h \sum_{i=1}^{\nu} \sum_{j=1}^{\nu} m_{i,j} \,\mathrm{Re}\,\langle \boldsymbol{d}_i, \boldsymbol{d}_j \rangle.\end{aligned}$$

We deduce that (5.11), and hence (5.6), are true if

$$\sum_{i=1}^{\nu} \sum_{j=1}^{\nu} m_{i,j} \,\mathrm{Re}\,\langle \boldsymbol{d}_i, \boldsymbol{d}_j \rangle \geq 0, \qquad \boldsymbol{d}_1, \boldsymbol{d}_2, \ldots, \boldsymbol{d}_\nu \in \mathbb{C}^d.$$

Recall our assumption that the matrix M is positive semidefinite. Therefore it can be written in the form $M = W\Lambda W^\top$, where W is orthogonal, and where Λ is diagonal and $\lambda_k = \Lambda_{k,k} \geq 0$, $k = 1, 2, \ldots, \nu$. Since $m_{i,j} = \sum_{k=1}^{\nu} \lambda_k w_{i,k} w_{j,k}$, $i, j = 1, 2, \ldots, \nu$, we deduce that

$$\sum_{i=1}^{\nu}\sum_{j=1}^{\nu} m_{i,j} \operatorname{Re}\langle \boldsymbol{d}_i, \boldsymbol{d}_j\rangle = \sum_{i=1}^{\nu}\sum_{j=1}^{\nu}\sum_{k=1}^{\nu} \lambda_k w_{i,k} w_{j,k} \operatorname{Re}\langle \boldsymbol{d}_i, \boldsymbol{d}_j\rangle$$

$$= \sum_{k=1}^{\nu} \lambda_k \operatorname{Re}\left\langle \sum_{i=1}^{\nu} w_{i,j}\boldsymbol{d}_i, \sum_{j=1}^{\nu} w_{j,k}\boldsymbol{d}_j \right\rangle = \sum_{k=1}^{\nu} \lambda_k \left\| \sum_{j=1}^{\nu} w_{j,k}\boldsymbol{d}_j \right\|^2 \geq 0.$$

This completes the proof: since the above argument applies to all monotone equations, the Runge–Kutta method in question is indeed algebraically stable. ■

Which RK methods can satisfy the conditions of Theorem 5.2? Definitely not explicit methods, since then $m_{k,k} = -b_k^2$, $k = 1, 2, \ldots, \nu$ and $\sum_{k=1}^{\nu} b_k = 1$ (necessary for order $p \geq 1$), which in tandem are inconsistent with the positive semidefiniteness of M.

But we do not need Theorem 5.2 in order to rule out explicit methods! A special case of a monotone equation is the scalar test equation (4.7) with $\operatorname{Re}\lambda < 0$, the cornerstone of the linear stability analysis of Chapter 4. Therefore, for an algebraically stable method it is necessary that the complex left half-plane resides within the linear stability domain: precisely the definition of A-stability! We thus deduce that only A-stable methods are candidates for algebraic stability.

Yet, algebraic stability is a stronger concept than A-stability. For example, the three-stage method

$$\begin{array}{c|ccc} 0 & 0 & 0 & 0 \\ \frac{1}{2} & \frac{5}{24} & \frac{1}{3} & -\frac{1}{24} \\ 1 & \frac{1}{6} & \frac{2}{3} & \frac{1}{6} \\ \hline & \frac{1}{6} & \frac{2}{3} & \frac{1}{6} \end{array}$$

is of order 4 (prove!) and A-stable (prove!). However, it is a matter of trivial calculation to demonstrate that the matrix

$$M = \frac{1}{36}\begin{bmatrix} -1 & 1 & 0 \\ 1 & 0 & -1 \\ 0 & -1 & 1 \end{bmatrix}$$

is not positive semidefinite.

Given an RK method

$$\begin{array}{c|c} \boldsymbol{c} & A \\ \hline & \boldsymbol{b}^\top \end{array},$$

we say that it is $B(r)$ if

$$\sum_{i=1}^{\nu} b_i c_i^{k-1} = \frac{1}{k}, \qquad k = 1, 2, \ldots, r$$

and $C(r)$ if

$$\sum_{j=1}^{\nu} a_{i,j}c_j^{k-1} = \frac{c_i^k}{k}, \qquad i = 1,2,\ldots,\nu, \quad k = 1,2,\ldots,r.$$

Lemma 5.3 *If $c_1,\ldots,c_\nu$ are distinct and a Runge–Kutta method is both $B(2\nu)$ and $C(\nu)$ then $M = O$, the zero matrix.*

Proof The Vandermonde matrix V, where $v_{k,\ell} = c_\ell^{k-1}$, $k,\ell = 1,2,\ldots,\nu$, is nonsingular (A.1.2.3). Therefore $M = O$ if and only if $\tilde{M} = O$, where $\tilde{M} = V^\top MV$. But, using the conditions $B(2\nu)$ and $C(\nu)$ where necessary,

$$\begin{aligned}
\tilde{m}_{k,\ell} &= \sum_{i=1}^{\nu}\sum_{j=1}^{\nu} c_i^{k-1}m_{i,j}c_j^{\ell-1} = \sum_{i=1}^{\nu}\sum_{j=1}^{\nu} c_i^{k-1}(b_ia_{i,j}+b_ja_{j,i}-b_ib_j)c_j^{\ell-1} \\
&= \sum_{i=1}^{\nu} b_ic_i^{k-1}\sum_{j=1}^{\nu} a_{i,j}c_j^{\ell-1} + \sum_{j=1}^{\nu} b_jc_j^{\ell-1}\sum_{i=1}^{\nu} a_{j,i}c_i^{k-1} - \sum_{i=1}^{\nu} b_ic_i^{k-1}\sum_{j=1}^{\nu} b_jc_j^{\ell-1} \\
&= \frac{1}{\ell}\sum_{i=1}^{\nu} b_ic_i^{k+\ell-1} + \frac{1}{k}\sum_{j=1}^{\nu} b_jc_j^{k+\ell-1} - \frac{1}{k\ell} = \left(\frac{1}{\ell}+\frac{1}{k}\right)\frac{1}{k+\ell} - \frac{1}{k\ell} = 0
\end{aligned}$$

for all $k,\ell = 1,2,\ldots,\nu$. Hence $\tilde{M} = O$, and so $M = O$. ■

Corollary *The Gauss–Legendre methods from Chapter 3 are algebraically stable for all $\nu \geq 1$.*

Proof We recall that each ν-stage Gauss–Legendre RK is a collocation method of order 2ν. In particular, the underlying quadrature formula is itself of order 2ν (it is the Gaussian quadrature of Theorem 3.3). This implies that

$$\sum_{i=1}^{\nu} b_ic_i^{k-1} = \int_0^1 x^{k-1}\,\mathrm{d}x = \frac{1}{k}, \qquad k = 1,2,\ldots,2\nu,$$

and hence that the Runge–Kutta method is $B(2\nu)$. Moreover, according to (3.13),

$$a_{k,\ell} = \int_0^{c_k} \frac{q_\ell(\tau)}{q_\ell(c_\ell)}\,\mathrm{d}\tau, \qquad k,\ell = 1,2,\ldots,\nu,$$

where $q(t) = \prod_{j=1}^{\nu}(t-c_j)$ and $q_\ell(t) = q(t)/(t-c_\ell)$. Therefore

$$\sum_{j=1}^{\nu} a_{i,j}c_j^{k-1} = \int_0^{c_i}\left(\sum_{j=1}^{\nu}\frac{q_j(\tau)}{q_j(c_j)}c_j^{k-1}\right)\mathrm{d}\tau.$$

Using an argument similar to that in the proof of Lemma 3.5, the integrand is the Lagrange interpolation polynomial of τ^{k-1}. Therefore, since $k \leq \nu$, it equals τ^{k-1} and so

$$\sum_{j=1}^{\nu} a_{i,j}c_j^{k-1} = \int_0^{c_i}\tau^{k-1}\,\mathrm{d}\tau = \frac{c_i^k}{k}, \qquad i,k = 1,\ldots,\nu.$$

Therefore the condition $C(\nu)$ is met and now we can use Lemma 5.3 to argue that $M = O$.

It remains to prove that the weights $b_1, b_2, \ldots, b_\nu$ are nonnegative. Let $k = 1, 2, \ldots, \nu$ and $f(x) = [q_k(x)/q_k(c_k)]^2$. Since f is a polynomial of degree $2\nu - 2$, it is integrated exactly by Gaussian quadrature. Moreover, $f(c_k) = 0$ and $f(c_\ell) = 0$ for $\ell \neq k$. Therefore

$$b_k = \sum_{\ell=1}^{\nu} b_\ell f(c_\ell) = \int_0^1 f(\tau)\,\mathrm{d}\tau > 0.$$

We deduce that the Gauss–Legendre RK method is algebraically stable. ■

5.3 From quadratic invariants to orthogonal flows

Our concern in this section is with differential equations endowed with a quadratic invariant. Specifically, we consider systems (5.4) such that, for every initial value $\boldsymbol{y}_0$,

$$\boldsymbol{y}^\top(t) S \boldsymbol{y}(t) \equiv \boldsymbol{y}_0^\top S \boldsymbol{y}_0, \qquad t \geq 0, \tag{5.12}$$

where S is a nonzero symmetric $d \times d$ matrix. (We restrict our attention to real equations, while mentioning in passing that generalization to a complex setting is straightforward.)

The invariant (5.12) means that the solution of the differential equation is restricted to a lower-dimensional manifold in $\mathbb{R}^d$. If all the eigenvalues of S are of the same sign then this manifold is a generalized ellipsoid but we will not explore this issue further.

We commence our numerical analysis of quadratic invariants by asking which Runge–Kutta methods produce a solution consistent with (5.12), in other words with

$$\boldsymbol{y}_{n+1}^\top S \boldsymbol{y}_{n+1} = \boldsymbol{y}_n^\top S \boldsymbol{y}_n, \qquad n = 0, 1, \ldots \tag{5.13}$$

The framework is surprisingly similar to that of the last section and, indeed, we will use similar ideas and notation: if the truth be told, we have already done all the heavy lifting in the proof of Theorem 5.2.

Theorem 5.4 *Suppose that a Runge–Kutta method is applied to an ODE* (5.4) *with quadratic invariant* (5.12). *If $M = O$ then the method satisfies* (5.13) *and is consistent with the invariant.*

Proof We denote the internal stages of the method by $\boldsymbol{r}_1, \boldsymbol{r}_2, \ldots, \boldsymbol{r}_\nu$ and let $\boldsymbol{\rho}_k = \boldsymbol{f}(t_n + c_k h, \boldsymbol{y}_n + h\sum_{j=1}^{\nu} a_{k,j}\boldsymbol{r}_j)$, $k = 1, 2, \ldots, \nu$. Similarly to the proof of Theorem 5.2, we calculate

$$\begin{aligned}\boldsymbol{y}_{n+1}^\top S \boldsymbol{y}_{n+1} &= \left(\boldsymbol{y}_n + h\sum_{k=1}^{\nu} b_k \boldsymbol{r}_k\right)^\top S \left(\boldsymbol{y}_n + h\sum_{\ell=1}^{\nu} b_\ell \boldsymbol{r}_\ell\right) \\ &= \boldsymbol{y}_n^\top S \boldsymbol{y}_n + h\left(\sum_{k=1}^{\nu} b_k \boldsymbol{r}_k^\top S \boldsymbol{y}_n + \boldsymbol{y}_n^\top S \sum_{\ell=1}^{\nu} b_\ell \boldsymbol{r}_\ell\right) + h^2 \sum_{k=1}^{\nu}\sum_{\ell=1}^{\nu} b_k b_\ell \boldsymbol{r}_k^\top S \boldsymbol{r}_\ell.\end{aligned}$$

Letting $\boldsymbol{y}_n = \boldsymbol{\rho}_k - h\sum_{\ell=1}^{\nu} a_{k,\ell}\boldsymbol{r}_\ell$, we have

$$\boldsymbol{r}_k^\top S\boldsymbol{y}_n = \boldsymbol{r}_k^\top S\boldsymbol{\rho}_k - h\sum_{\ell=1}^{\nu} a_{k,\ell}\boldsymbol{r}_k^\top S\boldsymbol{r}_\ell, \qquad k = 1, 2, \ldots, \nu.$$

However, differentiating (5.12) and using the symmetry of S yields

$$\boldsymbol{0} = \boldsymbol{y}^\top(t)S\boldsymbol{y}'(t) = \boldsymbol{y}^\top(t)S\boldsymbol{f}(t, \boldsymbol{y}(t)), \qquad t \geq 0.$$

The above identity still holds when we replace $\boldsymbol{y}(t)$ by an arbitrary vector $\boldsymbol{x} \in \mathbb{R}^d$, because $\boldsymbol{y}_0 \in \mathbb{R}^d$ is itself arbitrary. In particular, it is true for $\boldsymbol{x} = \boldsymbol{\rho}_k$ and, since $\boldsymbol{r}_k = \boldsymbol{f}(t_n + c_k h, \boldsymbol{\rho}_k)$, we deduce that $\boldsymbol{r}_k^\top S\boldsymbol{\rho}_k = \boldsymbol{\rho}_k^\top S\boldsymbol{r}_k = 0$. Therefore

$$\boldsymbol{r}_k^\top S\boldsymbol{y}_n = -h\sum_{\ell=1}^{\nu} a_{k,\ell}\boldsymbol{r}_k^\top S\boldsymbol{r}_\ell, \qquad k = 1, 2, \ldots, \nu,$$

and, by the same token,

$$\boldsymbol{y}_n^\top S\boldsymbol{r}_\ell = -h\sum_{k=1}^{\nu} a_{\ell,k}\boldsymbol{r}_k^\top S\boldsymbol{r}_\ell, \qquad \ell = 1, 2, \ldots, \nu.$$

Assembling all this gives

$$\begin{aligned}\boldsymbol{y}_{n+1}^\top S\boldsymbol{y}_{n+1} &= \boldsymbol{y}_n^\top S\boldsymbol{y}_n - h^2\sum_{k=1}^{\nu}\sum_{\ell=1}^{\nu}(b_k a_{k,\ell} + b_\ell a_{\ell,k} - b_k b_\ell)\boldsymbol{r}_k^\top S\boldsymbol{r}_\ell \\ &= \boldsymbol{y}_n^\top S\boldsymbol{y}_n - h^2\sum_{k=1}^{\nu}\sum_{\ell=1}^{\nu} m_{k,\ell}\boldsymbol{r}_k^\top S\boldsymbol{r}_\ell = \boldsymbol{y}_n^\top S\boldsymbol{y}_n\end{aligned}$$

and, as required, we have recovered (5.13). ■

We deduce from the corollary to Lemma 5.3 that Gauss–Legendre methods conserve quadratic invariants.

Linear invariants are trivially satisfied by all multistep and Runge–Kutta methods, yet they are not terribly interesting. Quadratic invariants are probably the simplest conservation laws that have deeper significance in applications and they include the important case of differential equations evolving on a sphere (e.g. the system (5.1)).

A profound generalization of (5.12) is represented by matrix ODEs of the form

$$Y' = A(t, Y)Y, \quad t \geq 0, \qquad Y(0) = Y_0 \in \mathrm{O}(d), \tag{5.14}$$

where A is a Lipschitz function, taking $[0, \infty) \times \mathrm{O}(d)$ to $\mathfrak{so}(d)$; here $\mathrm{O}(d)$ is the set of $d \times d$ real orthogonal matrices while $\mathfrak{so}(d)$ denotes the set of $d \times d$ skew-symmetric matrices. (The system (5.3) is an example.) It follows at once from the skew-symmetry of $A(t, Y)$ that

$$\frac{\mathrm{d}}{\mathrm{d}t}Y^\top Y = Y'^\top Y + Y^\top Y' = Y^\top A^\top(t, Y)Y + Y^\top A(t, Y)Y = O.$$

Therefore $Y^\top(t)Y(t) \equiv I$ and we deduce that the solution of (5.14) is an orthogonal matrix. This justifies the name of *orthogonal flow,* which we bestow on (5.14). Note that the invariant $Y^\top Y = I$ generalizes (5.12) since it represents a set of $\frac{1}{2}d(d+1)$ quadratic invariants.

Orthogonal flows feature in numerous applications, underlying the centrality of orthogonal matrices in mathematical physics (every physical law *must* be invariant with respect to rotation of the frame of reference, and this corresponds to multiplication by an orthogonal matrix) and in numerical algebra (because working with orthogonal matrices is the safest and best-conditioned strategy in computation-intensive settings). Furthermore, being able to solve (5.14) while respecting orthogonality affords us with a powerful tool that can be applied to many other problems. Recall, for example, the *isospectral flow* (5.2). As we have already seen, its solution can be represented in the form $Y(t) = Q(t)Y_0Q^\top(t)$, where the matrix Q is a solution of an orthogonal flow. Thus, solve (5.3) orthogonally and, subject to simple manipulation, you have an isospectral solution of (5.2). Another example when an equation is (to use a technical term) 'acted' upon by an orthogonal flow is presented by the three-dimensional system

$$\boldsymbol{y}' = \boldsymbol{a}(t,\boldsymbol{y}) \times \boldsymbol{y}, \tag{5.15}$$

where $\boldsymbol{b} \times \boldsymbol{c}$ is the *vector product* of $\boldsymbol{b} \in \mathbb{R}^3$ and $\boldsymbol{c} \in \mathbb{R}^3$: those unaware (or, more likely, forgetful) of vector analysis might just use the formula

$$\boldsymbol{b} \times \boldsymbol{c} = (b_2c_3 - b_3c_2)\boldsymbol{e}_1 - (b_1c_3 - b_3c_1)\boldsymbol{e}_2 + (b_1c_2 - b_2c_1)\boldsymbol{e}_3,$$

where $\boldsymbol{e}_1, \boldsymbol{e}_2, \boldsymbol{e}_3 \in \mathbb{R}^3$ are unit vectors. Verify that an alternative way of writing the system (5.15) is

$$\boldsymbol{y}' = \begin{bmatrix} 0 & -a_3(t,\boldsymbol{y}) & a_2(t,\boldsymbol{y}) \\ a_3(t,\boldsymbol{y}) & 0 & -a_1(t,\boldsymbol{y}) \\ -a_2(t,\boldsymbol{y}) & a_1(t,\boldsymbol{y}) & 0 \end{bmatrix} \boldsymbol{y} = A(t,\boldsymbol{y})\boldsymbol{y} \tag{5.16}$$

and note that the 3×3 matrix function A is skew-symmetric! We have already seen an example, namely the system (5.1), for which

$$\boldsymbol{a}^\top(t,\boldsymbol{y}) = [\, -\tfrac{1}{20}y_1 \quad -y_1y_2 \quad -y_3 \sin t \,].$$

Solutions of (5.15) evolve on a sphere: if $\|\boldsymbol{y}(0)\| = 1$ then a unit norm is maintained by the solution $\boldsymbol{y}(t)$ for all $t \geq 0$. This follows at once from the skew-symmetry of A, replicating the argument that we used to analyse the ODE system (5.1). Thus,

$$\frac{1}{2}\frac{\mathrm{d}}{\mathrm{d}t}\|\boldsymbol{y}\|^2 = \boldsymbol{y}^\top\boldsymbol{y}' = \boldsymbol{y}^\top A(\boldsymbol{y})\boldsymbol{y} = 0.$$

Now, given any two points $\boldsymbol{\alpha}$ and $\boldsymbol{\beta}$ on the unit sphere there exists a matrix $R \in \mathrm{O}(3)$ such that $\boldsymbol{\beta} = R\boldsymbol{\alpha}$. This justifies the following construction: we seek a sufficiently smooth function Q such that $Q(t) \in \mathrm{O}(d)$ and $\boldsymbol{y}(t) = Q(t)\boldsymbol{y}_0$. Substitution into (5.16) demonstrates easily that $Q' = A(t, Q\boldsymbol{y}_0)Q$, $Q(0) = I$, which fits the pattern (5.14) of orthogonal flows.

Suppose that we have a numerical method that is guaranteed to respect the orthogonal structure of an orthogonal flow. Then immediately, and at no extra cost (well, almost none) we have a method that respects the geometric structure of all equations that are 'acted' upon by orthogonality, e.g. isospectral flows and equations on spheres. This motivates a strong interest in discretization methods with this feature.

Which Runge–Kutta methods can solve (5.14) while keeping the solution orthogonal? No prizes for guessing: the condition is again $M = O$ and the proof is identical to that of Theorem 5.4: just replace bold-faced by capital letters and S by I, and everything follows in short order.

◇ **Lie-group equations** Think for a moment about the set of all $d \times d$ real orthogonal matrices $\mathrm{O}(d)$. Such a set has two important features, analytic and algebraic. From the analytic standpoint it is a *manifold,* a portion of $\mathbb{R}^{d^2}$ (since $d \times d$ matrices can be embedded in $\mathbb{R}^{d^2}$) which can be locally linearized and such that the resulting 'linearizing mappings' can be smoothly stitched together. Algebraically, it is a group: if $U, V \in \mathrm{O}(d)$ then $U^{-1}, UV \in \mathrm{O}(d)$ and it is easy to verify the standard group axioms. A manifold endowed with a group structure is called a *Lie group.*

Lie groups are an important mathematical concept and their applications range from number theory all the way to mathematical physics. Perhaps their most important use is as a powerful searchlight to illuminate and analyse symmetries of differential equations. Many Lie groups, like the *orthogonal group* $\mathrm{O}(d)$ or the *special linear group* $\mathrm{SL}(d)$ of all $d \times d$ real matrices with unit determinant, are composed of matrices.

Numerous differential equations of interest evolve on Lie groups: an orthogonal flow is just one example. Such equations can be characterized in a manner similar to (5.14). Dispensing altogether with proofs, we consider all differentiable curves $X(t)$ evolving on a matrix Lie group $\mathcal{G}$ and passing through the identity $I \in \mathcal{G}$. Each such curve can be written in the form $X(t) = I + tA + \mathcal{O}(t^2)$. We denote the set of all such As by $\mathfrak{g}$. It is possible to prove that $\mathfrak{g}$ is a linear space, closed under a skew-symmetric operation which, in the case of matrix Lie groups, is the familiar commutation operation: if $A, B \in \mathfrak{g}$ then $[A, B] = AB - BA \in \mathfrak{g}$. Such a set is known as a *Lie algebra.* In particular, the Lie algebra corresponding to $\mathrm{O}(d)$ is $\mathfrak{so}(d)$ (can you prove it?), while the Lie algebra of $\mathrm{SL}(d)$ comprises the set of $d \times d$ real matrices with zero trace.

It is possible to prove that an equation of the form (5.14) evolves in $\mathcal{G}$, provided that $Y_0 \in \mathcal{G}$ and $A : [0, \infty) \times \mathcal{G} \to \mathfrak{g}$. It is seductive to believe, thus, that a Runge–Kutta method is bound to respect *any* Lie-group structure, provided that $M = O$. Unfortunately, this is not true and so we may not infer in this manner from orthogonal flows to general Lie-group equations. Methods that, by design, respect an arbitrary Lie-group structure are outside the scope of this book, although we comment upon them further later in this chapter. ◇

5.4 Hamiltonian systems

A huge number of ODE systems in applications ranging from mechanics to molecular dynamics, fluid mechanics, quantum mechanics, image processing, celestial mechanics, nuclear engineering and beyond can be formulated as *Hamiltonian equations*

$$\begin{aligned} \boldsymbol{p}' &= -\frac{\partial H(\boldsymbol{p},\boldsymbol{q})}{\partial \boldsymbol{q}}, \\ \boldsymbol{q}' &= \frac{\partial H(\boldsymbol{p},\boldsymbol{q})}{\partial \boldsymbol{p}}. \end{aligned} \tag{5.17}$$

Here the scalar function H is the *Hamiltonian energy*. Both $\boldsymbol{p}$ and $\boldsymbol{q}$ are vector functions of d variables. In typical applications, d is the number of degrees of freedom of a mechanical system while $\boldsymbol{q}$ and $\boldsymbol{p}$ correspond to generalized positions and momenta respectively.

Lemma 5.5 *The Hamiltonian energy $H(\boldsymbol{p}(t),\boldsymbol{q}(t))$ remains constant along the solution trajectory.*

Proof By straightforward differentiation of $H(\boldsymbol{p}(t),\boldsymbol{q}(t))$ and substitution of the ODEs (5.17) we obtain

$$\frac{\mathrm{d}}{\mathrm{d}t}H(\boldsymbol{p},\boldsymbol{q}) = \left(\frac{\partial H}{\partial \boldsymbol{p}}\right)^{\top}\boldsymbol{p}' + \left(\frac{\partial H}{\partial \boldsymbol{q}}\right)^{\top}\boldsymbol{q}' = -\left(\frac{\partial H}{\partial \boldsymbol{p}}\right)^{\top}\left(\frac{\partial H}{\partial \boldsymbol{q}}\right) + \left(\frac{\partial H}{\partial \boldsymbol{q}}\right)^{\top}\left(\frac{\partial H}{\partial \boldsymbol{p}}\right) = 0.$$

■

As a consequence of the lemma, Hamiltonian systems evolve along surfaces of constant Hamiltonian energy H, and this is demonstrated vividly in Fig. 5.3, where we display phase planes of the equations $y'' + \sin y = 0$ and $y'' + y\sin y = 0$. Both are examples of *harmonic oscillators*

$$y'' + a(y) = 0$$

and each can be easily converted into a Hamiltonian system with a single degree of freedom by letting $p = y'$, $q = y$ and $H(p,q) = \frac{1}{2}p^2 - \int_0^q a(\xi)\,\mathrm{d}\xi$. The figures indicate a great deal of additional structure, in particular that (except when the initial value is a fixed point of the equation) the motion is periodic. This is true for many, although by no means all, Hamiltonian systems.

By this stage we might be tempted to utilize the lesson we have learnt from the previous section. We have an invariant (and one important enough to deserve the grand name of 'Hamiltonian energy'): let us seek numerical methods to preserve it! However, rushing headlong into this course of action will be a mistake, because Hamiltonian systems (5.17) have another geometric feature, which is even more important: they are *symplectic*.

Before we define symplecticity it is a good idea to provide some geometric intuition, hence see Fig. 5.4. Given an autonomous differential equation $\boldsymbol{y}' = \boldsymbol{f}(\boldsymbol{y})$, we say that the *flow map* $\varphi_t(\boldsymbol{y}_0)$ is the function taking the initial value $\boldsymbol{y}_0$ to the vector $\boldsymbol{y}(t)$.

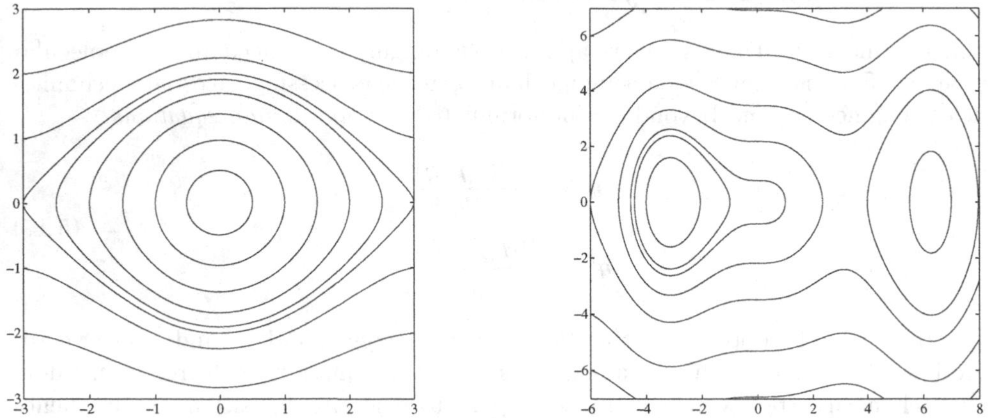

Figure 5.3 Phase planes of two nonlinear harmonic oscillators, $y'' + \sin y = 0$ (on the left) and $y'' + y \sin y = 0$ (on the right).

The definition of a flow map can be extended from vectors in $\mathbb{R}^d$ to measurable sets $\Omega \subset \mathbb{R}^d$ (roughly speaking, a subset of $\mathbb{R}^d$ is measurable if its volume is well defined). Thus,

$$\boldsymbol{\varphi}_t(\Omega) = \{\boldsymbol{y}(t) \;:\; \boldsymbol{y}(0) \in \Omega\}.$$

Let $\Omega = \{\boldsymbol{y} \in \mathbb{R}^2 \;:\; (y_1 - \frac{8}{5})^2 + y_2^2 \le \frac{2}{5}\}$. In Fig. 5.4 we display $\boldsymbol{\varphi}_t(\Omega)$ for $t = 0, 1, \ldots, 6$, for the Hamiltonian equation $y'' + \sin y = 0$. The blobs march clockwise and become increasingly distorted. However, *their area stays constant!* This is a one-degree-of-freedom manifestation of symplecticity: the flow map for Hamiltonian systems with $d = 1$ is area-preserving.

With greater generality (and a moderate amount of hand-waving) we say that a function $\boldsymbol{\varphi} : \Omega \to \mathbb{R}^{2d}$, where $\Omega \subseteq \mathbb{R}^{2d}$, is *symplectic* if $\Phi^\top(\boldsymbol{y}) J \Phi(\boldsymbol{y}) = J$ for every $\boldsymbol{y} \in \Omega$, where

$$\Phi(\boldsymbol{y}) = \frac{\partial \boldsymbol{\varphi}(\boldsymbol{y})}{\partial \boldsymbol{y}} \qquad \text{and} \qquad J = \begin{bmatrix} O & I \\ -I & O \end{bmatrix}.$$

The interpretation of symplecticity (and here hand waving comes in!) is as follows. If $d = 1$ then it corresponds to the preservation of the (oriented) area of a measurable set Ω. If $d \ge 2$, the situation is somewhat more complicated: defying intuition, area does not translate into volume! Instead, define the d two-dimensional sets

$$\Omega_k = \left\{ \begin{bmatrix} \omega_k \\ \omega_{k+d} \end{bmatrix} \boldsymbol{\omega} \in \Omega \right\}, \qquad k = 1, \ldots, d.$$

Then symplecticity corresponds to the conservation of

$$\operatorname{area} \Omega_1 + \operatorname{area} \Omega_2 + \cdots + \operatorname{area} \Omega_d.$$

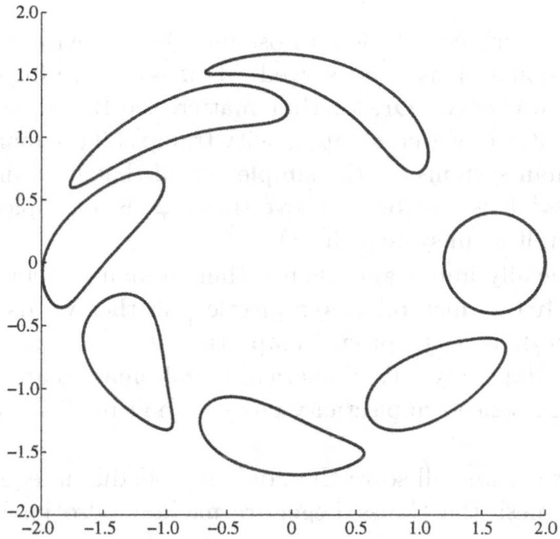

Figure 5.4 The blob story: a disc at the phase plane of the nonlinear pendulum equation $y'' + \sin y = 0$, centred at $(\frac{8}{5}, 0)$ with radius $\frac{2}{5}$, is mapped by unit time intervals with the flow map. The outcome is a progression of blobs, all having the same area.

Theorem 5.6 (The Poincaré theorem) *If H is twice continuously differentiable then the flow map φ_t of the Hamiltonian system* (5.17) *is symplectic.*

Proof It is convenient to rewrite the Hamiltonian system (5.17) in the form

$$\boldsymbol{y}' = J^{-1}\nabla H(\boldsymbol{y}), \qquad \text{where} \qquad \boldsymbol{y} = \begin{bmatrix} \boldsymbol{p} \\ \boldsymbol{q} \end{bmatrix}. \tag{5.18}$$

We denote the Jacobian of $\varphi(\boldsymbol{y})$ by $\Phi_t(\boldsymbol{y})$ and observe that

$$\frac{\mathrm{d}\varphi_t(\boldsymbol{y})}{\mathrm{d}t} = J^{-1}\nabla H(\varphi_t(\boldsymbol{y})) \quad \text{implies that} \quad \frac{\mathrm{d}\Phi_t(\boldsymbol{y})}{\mathrm{d}t} = J^{-1}\nabla^2 H(\varphi_t(\boldsymbol{y}))\Phi_t(\boldsymbol{y}).$$

Therefore

$$\begin{aligned} \frac{\mathrm{d}}{\mathrm{d}t}(\Phi_t^\top J\Phi_t) &= \left(\frac{\mathrm{d}\Phi_t}{\mathrm{d}t}\right)^\top J\Phi_t + \Phi_t^\top J\left(\frac{\mathrm{d}\Phi_t}{\mathrm{d}t}\right) \\ &= \Phi_t^\top \nabla^2 H(\varphi_t)J^{-\top}J\Phi_t + \Phi_t^\top JJ^{-1}\nabla^2 H(\varphi_t)\Phi_t \\ &= -\Phi_t^\top \nabla^2 H(\varphi_t)\Phi_t + \Phi_t^\top \nabla^2 H(\varphi_t)\Phi_t = O, \end{aligned}$$

since $J^{-\top}J = -I$. Therefore

$$\Phi_t^\top J\Phi_t \equiv \Phi_0^\top J\Phi_0 = J,$$

because $\phi_0 = I$. The theorem follows. ∎

On the face of it, symplecticity is an obscure concept: why should we care that a sum of two-dimensional areas is conserved? Is it not more important that (5.17) conserves Hamiltonian energy? Or, for that matter, the $(2d)$-dimensional volume of Ω? (Yes, it conserves it.) However, symplecticity trumps all the many other geometric features of Hamiltonian systems for the simple reason that it is, in a deep sense, the same as Hamiltonicity! It is possible to prove that if φ_t is a symplectic map then it is the flow of some Hamiltonian system (5.17).

This becomes crucially important when a Hamiltonian system is discretized by a numerical method. If this method is symplectic (in other words, the function $\boldsymbol{\psi}_h$, where $\boldsymbol{y}_{n+1} = \boldsymbol{\psi}_h(\boldsymbol{y}_n)$, is a symplectic map) then it is an exact solution of *some* Hamiltonian system. Hopefully, this 'numerical Hamiltonian' shares enough properties of the original system, hence symplecticity ensures good rendition of other geometric features.

To demonstrate this, we will solve the nonlinear pendulum equation $y'' + \sin y = 0$ with two RK schemes: the Gauss–Legendre method (also known as the implicit midpoint rule)

$$\begin{array}{c|c} \frac{1}{2} & \frac{1}{2} \\ \hline & 1 \end{array}, \tag{5.19}$$

which is symplectic and of order 2; and the Nystrom method

$$\begin{array}{c|ccc} 0 & & & \\ \frac{2}{3} & \frac{2}{3} & & \\ \frac{2}{3} & 0 & \frac{2}{3} & \\ \hline & \frac{1}{4} & \frac{3}{8} & \frac{3}{8} \end{array},$$

of order 3 but, alas, not symplectic. The solutions produced by both methods, employing a constant step size $h = \frac{1}{10}$, are displayed in Fig. 5.5 and, on the face of it, they look virtually identical. However, the entire picture changes once we examine the conservation of Hamiltonian energy in long-term integration.

In Fig. 5.6 we plot the numerical Hamiltonian energy produced by both methods in 10 000 steps of size $h = \frac{1}{10}$. As rendered by the Nystrom method, the energy slopes rapidly and soon leaves the plot altogether: this is obviously wrong. The implicit midpoint rule, however, produces an energy which, although not constant, is almost so. It oscillates within a very tight band centred on the exact constant energy $-\cos 1$ corresponding to an intial value $\boldsymbol{y}(0) = [1, 0]^\top$. Thus, although the numerical energy is not constant, it is almost so, and this behaviour persists for a very long time.

The lesson of the humble nonlinear pendulum is not over, however. In Fig. 5.7 we plot the absolute errors accumulated by the Nystrom and implicit midpoint methods, again applied with a constant step size $h = \frac{1}{10}$. We note that the error of Nystrom is larger, although the method is of higher order. More careful examination of the figure illuminates the reason underlying this surprising difference. While the error in the Nystrom method accumulates quadratically, the implicit midpoint rule yields linear error growth. Of course, a figure proves nothing yet it does manifest

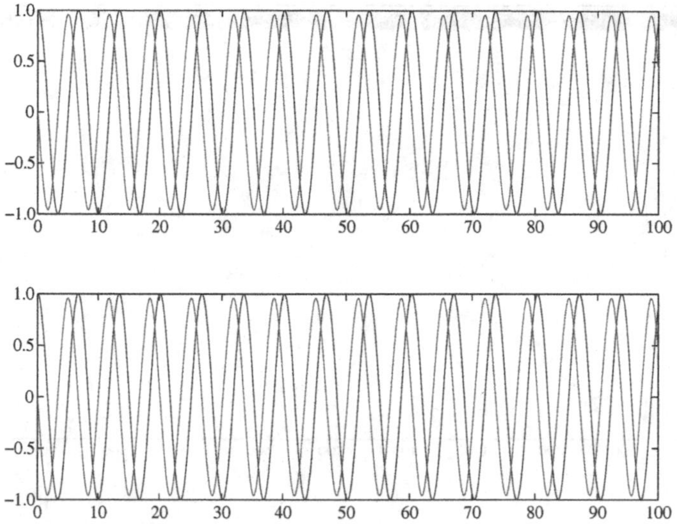

Figure 5.5 Numerical solution, with constant step $h = \frac{1}{10}$, of $y'' + \sin y = 0$ with the implicit midpoint rule (upper plot) and with the Nystrom method (lower plot).

behaviour that can be analysed and proved. Symplectic methods in general accumulate error more slowly. Unfortunately, the mathematical techniques underlying this phenomenon, i.e. the Kolmogorov–Arnold–Moser theory, backward error analysis and the theory of modulated Fourier expansions, are beyond the scope of this textbook.

Symplectic methods thus possess a number of important advantages ranging beyond the formal conservation of the symplectic invariant. Yet, there is a catch. To reap the advantages of numerical symplecticity, we must solve the equation with a constant step size, in defiance of all the words of wisdom and error-control strategies that you will read in Chapters 6 and 7. (There do exist, as a matter of fact, strategies for variable-step implementations which maintain the benefits of symplecticity, but they are fairly complicated and of limited utility.)

So far, except for an *ex cathedra* claim that the implicit midpoint method (5.19) is symplectic, we have said absolutely nothing with regard to identifying symplecticity. As in the case of algebraic stability and of the conservation of quadratic invariants, we need a clear, easily verifiable, criterion to tell us whether a Runge–Kutta method is symplectic. Careful readers of this chapter might suspect by now that the famous matrix M is just about to make its appearance, and such readers will not be disappointed.

Theorem 5.7 *If $M = O$ then the Runge–Kutta method is symplectic.*

Proof There are several ways of proving this assertion, e.g. by using exterior products or a generating function. Here we limit ourselves to familiar tools and method-

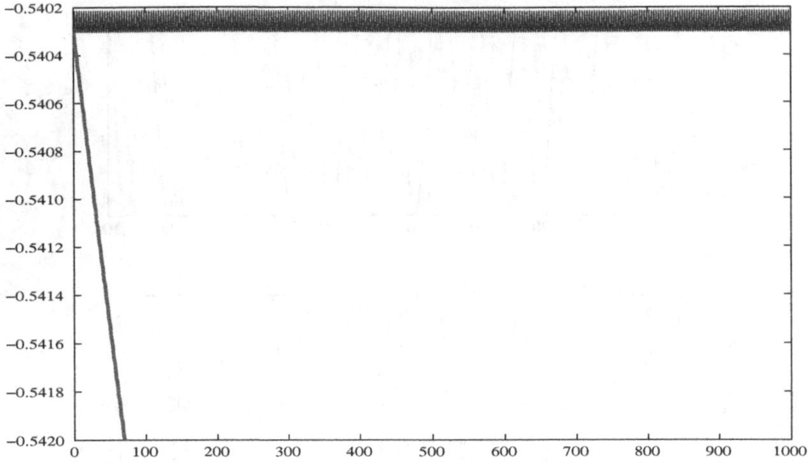

Figure 5.6 The Hamiltonian energy $\frac{1}{2}y_{n,2}^2 - \cos y_{n,1}$, as rendered by the implicit midpoint rule (the narrow band extending across the top of the figure) and the Nystrom method (the steeply sloping line).

ology: indeed, under several layers of makeup, replicate the proof of Theorem 5.4. Thus, we apply the RK method

$$\boldsymbol{\xi}_k = \boldsymbol{f}\left(t_n + c_k h,\ \boldsymbol{y}_n + h\sum_{\ell=1}^{\nu} a_{k,l}\boldsymbol{\xi}_\ell\right), \qquad k = 1, 2, \ldots, \nu,$$

$$\boldsymbol{y}_{n+1} = \boldsymbol{y}_n + h\sum_{k=1}^{\nu} b_k \boldsymbol{\xi}_k$$

to a Hamiltonian system written in the form (5.18). Letting $\Psi_n = \partial\boldsymbol{y}_n/\partial\boldsymbol{y}_0$, symplecticity means that

$$\Psi_{n+1}^\top J \Psi_{n+1} = \Psi_n^\top J \Psi_n, \qquad n = 0, 1, \ldots \tag{5.20}$$

Let

$$\Xi_k = \frac{\partial\boldsymbol{\xi}_k}{\partial\boldsymbol{y}_0}, \quad G_k = \boldsymbol{\nabla}^2 H\left(t_n + c_k h,\ \boldsymbol{y}_n + h\sum_{\ell=1}^{\nu} a_{k,\ell}\boldsymbol{\xi}_\ell\right), \qquad k = 1, 2, \ldots, \nu$$

and assume, to make matters simpler, that the symmetric matrices $G_1, \ldots, G_\nu$ are nonsingular. Now

$$\Psi_{n+1} = \Psi_n + h\sum_{k=1}^{\nu} b_k \Xi_k$$

and therefore

$$\Psi_{n+1}^\top J \Psi_{n+1} = \left(\Psi_n + h\sum_{k=1}^{\nu} b_k \Xi_k\right)^\top J\left(\Psi_n + h\sum_{k=1}^{\nu} b_k \Xi_k\right)$$

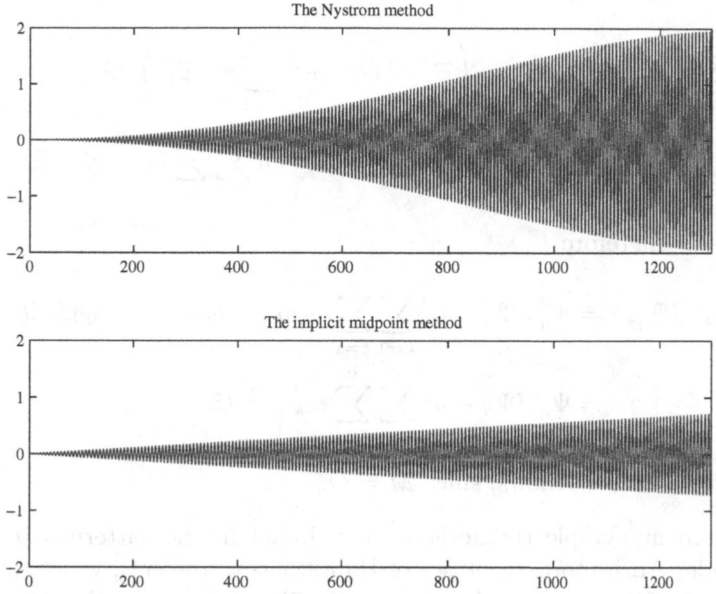

Figure 5.7 The absolute error for the nonlinear pendulum, as produced by the Nystrom method and the implicit midpoint method.

$$- \Psi_n^\top J \Psi_n + h \sum_{k=1}^{\nu} b_k \Xi_k^\top J \Psi_n + h \sum_{\ell=1}^{\nu} b_\ell \Psi_n^\top J \Xi_\ell + h^2 \sum_{k=1}^{\nu} \sum_{\ell=1}^{\nu} b_k b_\ell \Xi_k^\top J \Xi_\ell .$$

By direct differentiation in the RK method and using the special form of (5.18),

$$\Xi_k = J^{-1} G_k \left(\Psi_n + h \sum_{\ell=1}^{\nu} a_{k,\ell} \Xi_\ell \right),$$

thus

$$\Psi_n = G_k^{-1} J \Xi_k - h \sum_{\ell=1}^{\nu} a_{k,\ell} \Xi_l .$$

Therefore,

$$\begin{aligned} \sum_{k=1}^{\nu} b_k \Xi_k^\top J \Psi_n &= \sum_{k=1}^{\nu} b_k \Xi_k^\top J \left(G_k^{-1} J \Xi_k - h \sum_{\ell=1}^{\nu} a_{k,\ell} \Xi_\ell \right) \\ &= \sum_{k=1}^{\nu} b_k \Xi_k^\top J G_k^{-1} J \Xi_k - h \sum_{k=1}^{\nu} \sum_{\ell=1}^{\nu} b_k a_{k,\ell} \Xi_k^\top J \Xi_\ell . \end{aligned}$$

Likewise,

$$\sum_{\ell=1}^{\nu} b_\ell \Psi_n^\top J \Xi_\ell = \sum_{\ell=1}^{\nu} b_\ell \left(\Xi_k^\top J^\top G_\ell^{-1} - h \sum_{k=1}^{\nu} a_{\ell,k} \Xi_k^\top \right) J \Xi_\ell$$
$$= -\sum_{\ell=1}^{\nu} b_\ell \Xi_\ell^\top J G_\ell^{-1} J \Xi_\ell - h \sum_{k=1}^{\nu} \sum_{\ell=1}^{\nu} b_\ell a_{\ell,k} \Xi_k^\top J \Xi_\ell,$$

since $J^\top = -J$. Therefore

$$\Psi_{n+1}^\top J \Psi_{n+1} = \Psi_n^\top J \Psi_n - h^2 \sum_{k=1}^{\nu} \sum_{\ell=1}^{\nu} (b_k a_{k,\ell} + b_\ell a_{\ell,k} - b_k b_\ell) \Xi_k^\top J \Xi_\ell$$
$$= \Psi_n^\top J \Psi_n - h^2 \sum_{k=1}^{\nu} \sum_{\ell=1}^{\nu} m_{k,\ell} \Xi_k^\top J \Xi_\ell$$

and symplecticity (5.20) follows, since $M = O$. ■

There are many symplectic methods that do not fit the pattern (3.9) of Runge–Kutta methods. An important subset of Hamiltonian problems, which deserves specialized methods of its own, includes systems (5.17) with *separable* Hamiltonian energy,

$$H(\boldsymbol{p}, \boldsymbol{q}) = T(\boldsymbol{p}) + V(\boldsymbol{q}),$$

in other words,

$$\boldsymbol{p}' = -\frac{\partial V(\boldsymbol{q})}{\partial \boldsymbol{q}}, \qquad \boldsymbol{q}' = \frac{\partial T(\boldsymbol{p})}{\partial \boldsymbol{p}}. \tag{5.21}$$

Such systems are ubiquitous in mechanics, where T and V correspond to the kinetic and potential energy respectively of a mechanical system.

It is possible to discretize (5.21) using *two* distinct Runge–Kutta methods: one applied to the $\boldsymbol{p}'$ equation and the other to the $\boldsymbol{q}'$ equation. The great benefit of this approach is that there exist corresponding *partitioned RK methods,* which are both symplectic and explicit, making symplectic computation considerably more affordable.

Another useful technique, providing a means of deriving higher-order symplectic methods from lower-order ones, is *composition.* We illustrate it with a simple example, without any proof. Recall that the implicit midpoint rule, that is, the one-stage Gauss–Legendre Runge–Kutta method

$$\boldsymbol{y}_{n+1} = \boldsymbol{y}_n + h\boldsymbol{f}(\tfrac{1}{2}(\boldsymbol{y}_n + \boldsymbol{y}_{n+1})), \tag{5.22}$$

is symplectic. Unfortunately, it is of order 2 and in practice we might wish to employ a higher-order method. The *Yoshida method* involves three steps of (5.22), according to the pattern

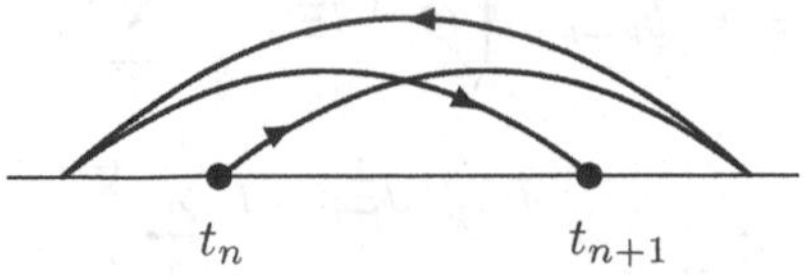

Thus, we advance from t_n with a step αh, say, where $\alpha > 1$, then turn and time-step backwards with a step $(2\alpha - 1)h$ and, finally, advance again with a step αh. By the end of this 'Yoshida shuffle', we are at t_{n+1}. Moreover, provided that we were clever enough to choose $\alpha = 1/(2 - \sqrt[3]{2})$, by the end of the journey we have a fourth-order symplectic method!

There is a notable absentee at our feast: multistep methods. Indeed, these methods are of little use when the conservation of geometric structure is at issue. This, needless to say, does not detract from their many other uses in the numerical solution of ODEs.

Comments and bibliography

In an ideal world, we would have kicked off with a chapter on computational dynamics, followed by one on differential algebraic equations. Then we would have laid down meticulously, in the language of differential geometry, the mathematical foundations of geometric numerical integration (GNI) and followed this with chapters on Lie-group methods and on Hamiltonian systems. But then, in an ideal mathematical world, on Planet Pedant, all books have at least 100 chapters and undergraduate studies extend for 15 years. Down on Planet Earth we are forced to compromise, condense and occasionally even hand-wave. Thus, if your impression by the end of this chapter is that you know enough of GNI to comprehend what it is roughly all about but not enough to claim real expertise, we have struck the right note.

The understanding that computational dynamics is important was implicit in numerical ODE research from the early days. True, as we saw in Chapter 4, the standard classical stability model was linear but there was the realization that this was just an initial step. The monotone model (5.5) was the first framework for the rigorous numerical analysis of nonlinear ODEs. As with most other fundamental ideas in numerical ODEs, it was formulated by Germund Dahlquist, who proceeded to analyse multistep methods in this setting; essentially, he proved that A-stability is sufficient for the stable solution of monotone equations by multistep methods. Insofar as Runge–Kutta methods are concerned, numerical lore has it that in 1975 John Butcher heard for the first time of the monotone model, at a conference in Scotland. He then proved Theorem 5.2 during the flight from London back to his native New Zealand: a triumph of the mathematical mind over the discomforts of long-haul travel.

The monotone model was a convenient focus for nonlinear stability analysis for a decade, until numerical analysts (together with everybody else) became aware of nonlinear dynamical systems. This has led to many more powerful models for nonlinear behaviour, e.g.

$$\langle \boldsymbol{u}, \boldsymbol{f}(\boldsymbol{u})\rangle \le \alpha - \beta\|\boldsymbol{u}\|, \qquad \boldsymbol{u} \in \mathbb{R}^d, \tag{5.23}$$

where $\alpha, \beta > 0$. If (5.4) satisfies this inequality and $g(t) = \frac{1}{2}\|\boldsymbol{y}(t)\|^2$ then

$$g'(t) = \langle \boldsymbol{y}(t), \boldsymbol{f}(\boldsymbol{y}(t))\rangle \le \alpha - \beta\|\boldsymbol{y}(t)\|^2 = \alpha - 2\beta g(t).$$

Hence $g \ge 0$ satisfies the differential inequality $g' \le -2\beta g + \alpha$; thus

$$g(t) \le \frac{\alpha}{2\beta} + \left[g(0) - \frac{\alpha}{2\beta}\right] \mathrm{e}^{-2\beta t}, \qquad t \ge 0$$

and $g(t) \le \max\{\alpha/(2\beta), g(0)\}$. Therefore the solution $\boldsymbol{y}(t)$ evolves within a bounded ball in $\mathbb{R}^d$. However, within this ball there is a great deal of freedom for the solution to do things strange and wonderful: like a monotone equation it might tend to a fixed point, but there is nothing to stop it from being periodic, quasi-periodic or chaotic. So, what is the magic condition which makes Runge–Kutta methods respect (5.23)? If your guess is $M = O$, you are right on the money ... The monograph of Stuart and Humphries (1996) is a detailed compendium of the computational dynamics of numerical methods and includes a long list of different nonlinear stability models.

Numerical analysts found it natural to adopt the ideas of computational dynamics, since they chimed with standard numerical theory. This was emphatically not the case with GNI. With the single honourable exception of Feng Kang and his group at the Chinese Academy of Sciences in Beijing, numerical analysts were too besotted with accuracy as the main organizing principle of computation to realize that qualitative and geometric attributes are important too. And important they were, since researchers in quantum chemistry, celestial mechanics and reactor physics had been using rudimentary GNI methods for decades, compelled by the nature of their differential equations. Organized, concerted research into GNI commenced only in the late 1980s although, in fairness, it soon became the mainstay of contemporary ODE research (GNI methods for PDEs are at a more tentative stage).

An alternative to the manifold-hugging methods of Section 5.4 is the formalism of *differential-algebraic equations* (DAEs). Without striving at generality, a typical DAE might be of the form

$$\boldsymbol{y}' = \boldsymbol{f}(\boldsymbol{y},\boldsymbol{x}), \quad \mathbf{0} = \boldsymbol{g}(\boldsymbol{y},\boldsymbol{x}), \quad t \ge 0, \qquad \boldsymbol{y}(0) = \boldsymbol{y}_0 \in \mathbb{R}^{d_1}, \quad \boldsymbol{x}(0) = \boldsymbol{x}_0 \in \mathbb{R}^{d_2}, \tag{5.24}$$

where the Jacobian $\partial \boldsymbol{g}/\partial \boldsymbol{x}$ is nonsingular. One interpetation of (5.24) is that the solution evolves on the manifold determined by the level set $\boldsymbol{g} = 0$. Alternatively, establishing a connection with the material of Section 5.4, DAEs can be interpreted as the limiting case of the ODEs

$$\boldsymbol{y}' = \boldsymbol{f}(\boldsymbol{y},\boldsymbol{x}), \quad \varepsilon\boldsymbol{x}' = \boldsymbol{g}(\boldsymbol{y},\boldsymbol{x}), \quad t \ge 0, \qquad \boldsymbol{y}(0) = \boldsymbol{y}_0 \in \mathbb{R}^{d_1}, \quad \boldsymbol{x}(0) = \boldsymbol{x}_0 \in \mathbb{R}^{d_2}$$

for $\varepsilon \to 0$; in other words, DAEs are stiff equations with infinite stiffness. Following this logic leads us to BDF methods (see Lemma 2.3), which are in a sense ideally adjusted to 'infinite stiffness' and which can be extended to the DAEs (5.24). A good reference on DAEs is Hairer *et al.* (1991).

The narrative of Section 5.4 is centred around orthogonal flows, but we have commented already on the considerably more general framework of Lie groups. Thus, assume that a matrix function Y satisfies the ODE (5.14), except that $Y_0 \in \mathcal{G}$ and the matrix function A maps $\mathcal{G}$ to its Lie algebra $\mathfrak{g}$. An important fact about matrix Lie groups and Lie algebras is that if $X \in \mathfrak{g}$ then $\mathrm{e}^X \in \mathcal{G}$; the exponential of a matrix was defined in Chapter 2. Now suppose that, in place of (5.14), we formulate an equation evolving in a Lie algebra. Unlike $\mathcal{G}$, the Lie algebra $\mathfrak{g}$ is a *linear space!* As long as we restrict ourselves to linear combinations and to the computation of matrix commutators (recall that $\mathfrak{g}$ is closed under commutation), we cannot go wrong and, no matter what we do, we will stay in a Lie algebra – whence exponentiation takes us back to the Lie group and our numerical solution stays in $\mathcal{G}$. The challenge is thus to reformulate (5.14) in a Lie-algebraic setting, and it is answered by the 'dexpinv' equation

$$\begin{aligned} \Omega' &= A(t,\mathrm{e}^{\Omega}Y_0) - \tfrac{1}{2}[\Omega, A(t,\mathrm{e}^{\Omega}Y_0)] + \tfrac{1}{12}[\Omega,[\Omega,A(t,\mathrm{e}^{\Omega}Y_0)]] \\ &\quad - \tfrac{1}{720}[\Omega,[\Omega,[\Omega,[\Omega,A(t,\mathrm{e}^{\Omega}Y_0)]]]] + \cdots, \qquad t \ge 0, \quad \Omega(0) = O. \end{aligned} \tag{5.25}$$

Note that here we are indeed using only the permitted operations of commutation and linear combination. Once we have computed Ω, we have $Y(t) = \mathrm{e}^{\Omega(t)}Y_0$. The idea, known as the Runge–Kutta–Munthe-Kaas method, abbreviated to the somewhat more melodic acronym RKMK, is to apply a Runge–Kutta method to an appropriately truncated equation (5.25). As an example, consider the Nystrom method of Section 3.2. Applied directly to the Lie-group equation (5.14), it reads

$$\begin{aligned}
\Xi_1 &= hA(t_n, Y_n)Y_n, \\
\Xi_2 &= hA(t_n + \tfrac{2}{3}h, Y_n + \tfrac{2}{3}\Xi_1)(Y_n + \tfrac{2}{3}h\Xi_1), \\
\Xi_3 &= hA(t_n + \tfrac{2}{3}h, Y_n + \tfrac{2}{3}\Xi_2)(Y_n + \tfrac{2}{3}h\Xi_2), \\
Y_{n+1} &= Y_n + \tfrac{1}{4}\Xi_1 + \tfrac{3}{8}\Xi_2 + \tfrac{3}{8}\Xi_3,
\end{aligned}$$

and there is absolutely no reason why should it evolve in the Lie group. However, at the Lie-algebra level, when applied to (5.25) the same method becomes

$$\begin{aligned}
& & \Xi_1 &= hA(t_n, Y_n), & F_1 &= \Xi_1, \\
\Theta_2 &= -\tfrac{2}{3}\Xi_1, & \Xi_2 &= hA(t_n + \tfrac{2}{3}h, \mathrm{e}^{\Theta_2}Y_n), & F_2 &= \Xi_2 - \tfrac{1}{2}[\Theta_2, \Xi_2], \\
\Theta_3 &= -\tfrac{2}{3}\Xi_2, & \Xi_3 &= hA(t_n + \tfrac{2}{3}h, \mathrm{e}^{\Theta_3}Y_n), & F_3 &= \Xi_3 - \tfrac{1}{2}[\Theta_3, \Xi_3],
\end{aligned}$$

$$Y_{n+1} = \mathrm{e}^{F_1/4 + 3F_2/8 + 3F_3/8}Y_n.$$

The method is explicit, of order 3, requires just three function evaluations of A and is guaranteed to stay in any Lie group. This is but one example of the many Lie-group methods reviewed in Iserles *et al.* (2000).

Hamiltonian equations are central to research into mechanics and most relevant ODEs in this area are Hamiltonian, although often phrased in the equivalent Lagrangian formulation. Marsden & Ratiu (1999) is a good introduction to this fascinating area but beware: you will need to master some differential-geometric formalism to understand what is going on! In this volume we have tried to avoid any mention of differential geometry beyond that which a reasonable undergraduate at a reasonable university would have encountered, but any serious treatment of this subject is bound to employ more sophisticated terminology. This is the moment to remind long-suffering students that 'heavy-duty' mathematical formalism might be tough to master but, once understood, makes life much easier! Proving Theorem 5.7 with *exterior products* would have been easy, virtually a repeat of the proof of Theorem 5.4.

An early, yet very readable exposition of the numerical aspects of Hamiltonian equations is Sanz-Serna & Calvo (1994). The most comprehensive and authoritative treatment of the subject is Hairer *et al.* (2006), while Leimkuhler & Reich (2004) focuses on the vital connection between numerical theory and the practical applications of Hamiltonian systems. The satisfactory implementation of a numerical method for a difficult problem consists of much more than just pulling an algorithm off the shelf. It is imperative to understand the application just as much as we understand the computation, in order to ask the right questions, ascertain the correct requirements and ultimately produce a computational solution that really addresses the problem at hand.

Runge–Kutta methods are but one (exceedingly effective) means for computing Hamiltonian equations symplectically. The list below gives a selection of other techniques that have attracted much attention in the last few years.

- The generating-function method is natural within the differential-geometric Hamiltonian formalism (which is precisely why we do not propose to explain it here). Its disadvantage is its lesser generality: essentially, each Hamiltonian requires a separate expansion. This has an important application to the production of Ph.D. dissertations and scientific papers, less so to practical computation.

- We have already mentioned partitioned Runge–Kutta methods. An elementary example of such methods, applicable to Hamiltonians of the form $H(\boldsymbol{p}, \boldsymbol{q}) = \frac{1}{2}\boldsymbol{p}^\top \boldsymbol{p} + V(\boldsymbol{q})$, is the *Störmer–Verlet method*, to which we return in Chapter 16 in a different context (and with an abbreviated name, the Störmer method; Verlet proved that it is symplectic):

$$\boldsymbol{q}_{n+1} - 2\boldsymbol{q}_n + \boldsymbol{q}_{n-1} + h^2 \frac{\partial V(\boldsymbol{q}_n)}{\partial \boldsymbol{q}} = 0. \qquad (5.26)$$

Now, before you exclaim 'but this is a multistep method!' or query where the $\boldsymbol{p}$ variables have gone, note first that our Hamiltonian equations $\boldsymbol{p}' = \partial V(\boldsymbol{q})/\partial \boldsymbol{q}$, $\boldsymbol{q}' = \boldsymbol{p}$ easily yield the second-order system $\boldsymbol{q}'' + \partial V(\boldsymbol{q})/\partial \boldsymbol{q} = 0$, which is solved by (5.26). Moreover, the multistep scheme (5.26) can be written in a one-step formulation. Thus, defining the numerical momenta as

$$\boldsymbol{p}_n = \frac{\boldsymbol{q}_{n+1} - \boldsymbol{q}_{n-1}}{2h} \qquad \text{and} \qquad \boldsymbol{p}_{n+1/2} = \frac{\boldsymbol{q}_{n+1} - \boldsymbol{q}_n}{h},$$

we can convert (5.26) into $\boldsymbol{p}_{n+1/2} = \boldsymbol{p}_{n-1/2} - h\partial V(\boldsymbol{q}_n)/\partial \boldsymbol{q}$. But $\boldsymbol{p}_{n-1/2} + \boldsymbol{p}_{n+1/2} = 2\boldsymbol{p}_n$ and eliminating $\boldsymbol{p}_{n-1/2}$ from these two expressions leads to the one-step explicit scheme

$$\begin{aligned} \boldsymbol{p}_{n+\frac{1}{2}} &= \boldsymbol{p}_n - \tfrac{1}{2}h \frac{\partial V(\boldsymbol{q}_n)}{\partial \boldsymbol{q}}, \\ \boldsymbol{q}_{n+1} &= \boldsymbol{q}_n + h\boldsymbol{p}_{n+1/2}, \\ \boldsymbol{p}_{n+1} &= \boldsymbol{p}_{n+\frac{1}{2}} - \tfrac{1}{2}h \frac{\partial V(\boldsymbol{q}_{n+1})}{\partial \boldsymbol{q}}, \end{aligned} \qquad (5.27)$$

a partitioned second-order symplectic Runge–Kutta method. The Störmer–Verlet method is probably the most popular symplectic integrator in quantum chemistry and celestial mechanics, as well as an ideal testing bed for all the different phenomena, tools and tricks of the trade of computational Hamiltonian dynamics (Hairer *et al.*, 2003).

- There is much more to composition methods than our brief mention of the Yoshida trick at the end of Section 5.4. It is possible to employ similar ideas to boost further the order of symplectic methods and reduce their error. The benefits of this approach are not restricted to the Hamiltonian setting and similar ideas have been applied to the conservation of volume and to the conservation of arbitrary first integrals of differential systems (McLachlan & Quispel, 2002).
- An altogether different approach to the solution of Hamiltonian problems is provided by *variational integrators* (Marsden & West, 2001). We have mentioned already the Lagrangian formulation of Hamiltonian problems: essentially, it is possible to convert a Hamiltonian problem into a variational one, of the kind that will be considered in Chapter 9. Now, variational integrators are finite element methods that act within the Lagrangian formulation in a manner which, back in the Hamiltonian realm, is symplectic. This is a very flexible approach with many advantages.

An important spin off of GNI is a numerical theory of highly oscillatory differential equations. Once an ODE oscillates very rapidly, standard methods (no matter how stable) force us to use step sizes which are smaller than the shortest period, and this can impose huge costs on the calculation. (If you do not believe me, try solving the humble *Airy equation* $y'' + ty = 0$ for large t with any `Matlab` ODE solver.) Geometric numerical integration methods have completely revolutionised the computation of such problems, but this is work in progress.

Dekker, K. and Verwer, J.G. (1984), *Stability of Runge–Kutta Methods for Stiff Nonlinear Differential Equations*, North-Holland, Amsterdam.

Hairer, E. and Wanner, G. (1991), *Solving Ordinary Differential Equations II: Stiff Problems and Differential-Algebraic Equations,* Springer-Verlag, Berlin.

Hairer, E., Lubich, C. and Wanner, G. (2003), Geometric numerical integration illustrated by the Störmer–Verlet method, *Acta Numerica* **12**, 399–450.

Hairer, E., Lubich, C. and Wanner, G. (2006), *Geometric Numerical Integration* (2nd edn), Springer Verlag, Berlin.

Iserles, A., Munthe-Kaas, H.Z., Nørsett, S.P. and Zanna, A. (2000), Lie-group methods, *Acta Numerica* **9**, 215–365.

Leimkuhler, B. and Reich, S. (2004), *Simulating Hamiltonian Dynamics,* Cambridge University Press, Cambridge.

Marsden, J.E. and Ratiu, T.S. (1999), *Introduction to Mechanics and Symmetry: A Basic Exposition of Classical Mechanical Systems* (2nd edn), Springer Verlag, New York.

Marsden, J.E. and West, M. (2001), Discrete mechanics and variational integrators, *Acta Numerica* **10**, 357–514.

McLachlan, R.I. and Quispel, G.R.W. (2002), Splitting methods, *Acta Numerica* **11**, 341–434.

Sanz-Serna, J.M. and Calvo, M.P. (1994), *Numerical Hamiltonian Problems,* Chapman & Hall, London.

Stuart, A.M. and Humphries, A.R. (1996), *Dynamical Systems and Numerical Analysis,* Cambridge University Press, Cambridge.

Exercises

5.1 Consider the linear ODE with variable coefficients

$$\boldsymbol{y}' = A(t)\boldsymbol{y}, \qquad t \geq 0,$$

where A is a real $d \times d$ matrix function.

a Prove that the above ODE is monotone if and only if all the eigenvalues $\mu_1(t), \ldots, \mu_d(t)$ of the symmetric matrix $B(t) = \frac{1}{2}[A(t) + A^\top(t)]$ are nonpositive.

b Assuming for simplicity that $A(t) \equiv A$, a constant matrix, demonstrate by a counterexample that it is not enough for all the eigenvalues of A to reside in the closed left complex half-plane for the equation to be monotone.

c Again let A be a constant matrix and assume further that its eigenvalues are all real. (This is not necessary for our statement but renders the proof much easier.) Prove that, provided that the ODE is monotone, all the eigenvalues of A are in the closed left complex half-plane $\mathrm{cl}\,\mathbb{C}^-$.
(*You might commence by expanding an eigenvector of A in the basis of eigenvectors of $B = \frac{1}{2}(A + A^\top)$.*)

5.2 Let $\boldsymbol{y}' = A\boldsymbol{y}$, where A is a constant $d \times d$ matrix. As in the previous exercise, let $B = \frac{1}{2}(A + A^\top)$. The *spectral abscissa* of B is $\mu[B] = \max\{\lambda : \lambda \in \sigma(B)\}$. (Thus the previous exercise amounted to proving that $\mu[B] \le 0$ is necessary and sufficient for monotonicity.)

a Prove that the function $\phi(t) = \|\boldsymbol{y}(t)\|^2$, where $\|\cdot\|$ is the standard Euclidean norm, obeys the differential inequality $\phi' \le 2\mu[B]\phi$, and thereby deduce that $\|\boldsymbol{y}(t)\| \le \mathrm{e}^{t\mu[B]}\|\boldsymbol{y}(0)\|$.

b Prove that $\alpha = \mu[B]$ is the least possible constant such that $\|\boldsymbol{y}(t)\| \le \mathrm{e}^{t\alpha}\|\boldsymbol{y}(0)\|$ for all possible initial values $\boldsymbol{y}(0)$.

5.3 For which of the following three-stage Runge–Kutta methods is it true that the matrix M is positive semidefinite?

a The fifth-order Radau IA method

$$\begin{array}{c|ccc} 0 & \frac{1}{9} & -\frac{1}{18} - \frac{\sqrt{6}}{18} & -\frac{1}{18} + \frac{\sqrt{6}}{18} \\ \frac{3}{5} - \frac{\sqrt{6}}{10} & \frac{1}{9} & \frac{11}{45} + \frac{7\sqrt{6}}{360} & \frac{11}{45} - \frac{43\sqrt{6}}{360} \\ \frac{3}{5} + \frac{\sqrt{6}}{10} & \frac{1}{9} & \frac{11}{45} + \frac{43\sqrt{6}}{360} & \frac{11}{45} - \frac{7\sqrt{6}}{360} \\ \hline & \frac{1}{9} & \frac{4}{9} + \frac{\sqrt{6}}{36} & \frac{4}{9} - \frac{\sqrt{6}}{36} \end{array},$$

b The fourth-order Lobatto IIIB method

$$\begin{array}{c|ccc} 0 & \frac{1}{6} & -\frac{1}{6} & 0 \\ \frac{1}{2} & \frac{1}{6} & \frac{1}{3} & 0 \\ 1 & \frac{1}{6} & \frac{5}{6} & 0 \\ \hline & \frac{1}{6} & \frac{2}{3} & \frac{1}{6} \end{array},$$

c The fourth-order Lobatto IIIC method

$$\begin{array}{c|ccc} 0 & \frac{1}{6} & -\frac{1}{3} & \frac{1}{6} \\ \frac{1}{2} & \frac{1}{6} & \frac{5}{12} & -\frac{1}{12} \\ 1 & \frac{1}{6} & \frac{2}{3} & \frac{1}{6} \\ \hline & \frac{1}{6} & \frac{2}{3} & \frac{1}{6} \end{array}.$$

5.4 Consider the ODE $\boldsymbol{y}' = S(\boldsymbol{y})\nabla g(\boldsymbol{y})$, where S is a $d \times d$ skew-symmetric matrix function and g is a continuously differentiable scalar function. Prove that g is a first integral of this equation, i.e. that $g(\boldsymbol{y}(t))$ stays constant for all $t \ge 0$.

(*This 'skew-gradient equation' is at the root of certain discretization methods that can be made to respect an arbitrary first integral g.*)

5.5 Our point of departure is the matrix differential equation

$$Y' = BY + YB^\top, \quad t \ge 0, \qquad Y(0) = Y_0, \tag{5.28}$$

where the matrix B has zero trace, $\sum_{k=1}^{d} b_{k,k} = 0$.

a Prove that the solution of (5.28) can be expressed in the form $Y(t) = V(t)Y_0V^\top(t)$, $t \geq 0$, where the matrix V is the solution of the ODE

$$V' = BV, \quad t \geq 0, \qquad V(0) = I. \tag{5.29}$$

b Using the fact that the trace of B is zero, prove that the determinant is an invariant of (5.29), namely that $\det V(t) \equiv 1$.

c Deduce that $\det Y(t) \equiv \det Y_0$ for all $t \geq 0$.

5.6 Consider again equation (5.29), recalling that the trace of B is zero and that the exact solution has unit determinant for all $t \geq 0$. Assume that we are solving it with a Runge–Kutta method (3.9).

a Suppose that the rational function r is given by the formula (4.13). Prove that $Y_{n+1} = r(hB)Y_n$, $n = 0, 1, \ldots$

b Deduce that the condition for $\det Y_{n+1} = \det Y_n$ is that

$$\prod_{k=1}^{d} r(h\lambda_k) = 1,$$

where $\lambda_1, \ldots, \lambda_d$ are the eigenvalues of B.
(You may assume that B has a full set of eigenvectors.)

c Supposing that the RK method is of order $p \geq 1$, prove that

$$\prod_{k=1}^{d} r(h\lambda_k) = 1 + ch^{p+1}\sum_{k=1}^{d}\lambda_k^{p+1} + \mathcal{O}(h^{p+2}), \qquad c \neq 0.$$

d Provided that $d \geq 3$, demonstrate that there exists a matrix B, consistent with our assumptions, for which $\det Y_{n+1} \neq \det Y_n$ for sufficiently small step size $h > 0$.

5.7 The solution of the linear matrix ODE

$$Y' = A(t)Y, \quad t \geq 0, \qquad Y(0) = Y_0 \in \mathcal{G},$$

evolves in the Lie group $\mathcal{G}$, subject to the assumption that $A(t) \in \mathfrak{g}$, $t \geq 0$, where $\mathfrak{g}$ is the corresponding Lie algebra.

a Consider the method

$$Y_{n+1} = \exp\left(\int_{t_n}^{t_{n+1}} A(\tau)\,\mathrm{d}\tau\right) Y_n, \qquad n = 0, 1, \ldots,$$

where $\exp(\cdots)$ is the standard matrix exponential. Prove that the method is of order 2 and that $Y_n \in \mathcal{G}$, $n = 0, 1, \ldots$

b Suppose that the integral above is discretized by Gaussian quadrature with a single node,

$$\int_{t_n}^{t_{n+1}} A(\tau)\,\mathrm{d}\tau \approx hA(t_n + \tfrac{1}{2}h).$$

Prove that the new method is also of order 2 and that it evolves in $\mathcal{G}$.

c* Prove that

$$Y_{n+1} = \exp\left(\int_{t_n}^{t_{n+1}} A(\tau)\,\mathrm{d}\tau - \tfrac{1}{2}\int_{t_n}^{t_{n+1}} \left(\int_{t_n}^{\tau} A(\zeta)\,\mathrm{d}\zeta, A(\tau)\right)\mathrm{d}\tau\right) Y_n$$

for $n = 0, 1, \ldots$ is a fourth-order method and, again, $Y_n \in \mathcal{G}$, $n = 0, 1, \ldots$

5.8 Show that the *Hénon–Heiles system*

$$\begin{aligned} p_1' &= -q_1 - 2q_2q_2, \\ p_2' &= -q_2 + \tfrac{2}{3}q_2, \\ q_1' &= p_1, \\ q_2' &= p_2 \end{aligned}$$

is Hamiltonian and identify explicitly its Hamiltonian energy.
(*The Hénon–Heiles system is a famous example of an ODE with chaotic solutions.*)

5.9 Let

$$\begin{array}{c|ccc} c_1 & a_{1,1} & \cdots & a_{1,\nu} \\ \vdots & \vdots & & \vdots \\ c_\nu & a_{\nu,1} & \cdots & a_{\nu,\nu} \\ \hline & b_1 & \cdots & b_\nu \end{array} \qquad \text{and} \qquad \begin{array}{c|ccc} \tilde{c}_1 & \tilde{a}_{1,1} & \cdots & \tilde{a}_{1,\tilde{\nu}} \\ \vdots & \vdots & & \vdots \\ \tilde{c}_{\tilde{\nu}} & \tilde{a}_{\tilde{\nu},1} & \cdots & \tilde{a}_{\tilde{\nu},\tilde{\nu}} \\ \hline & \tilde{b}_1 & \cdots & \tilde{b}_{\tilde{\nu}} \end{array}$$

be two Runge–Kutta methods. We apply them to the Hamiltonian system (5.17) for a separable Hamiltonian $H(\boldsymbol{p},\boldsymbol{q}) = T(\boldsymbol{p}) + V(\boldsymbol{q})$; the first method to the momenta $\boldsymbol{p}$ and the second to the positions $\boldsymbol{q}$. Thus

$$\boldsymbol{r}_k = \boldsymbol{p}_n - h\sum_{\ell=1}^{\nu} a_{k,\ell}\frac{\partial V(\boldsymbol{s}_\ell)}{\partial \boldsymbol{q}}, \qquad k = 1, 2, \ldots, \nu,$$

$$\boldsymbol{s}_k = \boldsymbol{q}_n + h\sum_{\ell=1}^{\tilde{\nu}} \tilde{a}_{k,\ell}\frac{\partial T(\boldsymbol{r}_\ell)}{\partial \boldsymbol{p}}, \qquad k = 1, 2, \ldots, \tilde{\nu},$$

$$\boldsymbol{p}_{n+1} = \boldsymbol{p}_n - h\sum_{k=1}^{\nu} b_k\frac{\partial V(\boldsymbol{s}_k)}{\partial \boldsymbol{q}},$$

$$\boldsymbol{q}_{n+1} = \boldsymbol{q}_n + h\sum_{k=1}^{\tilde{\nu}} \tilde{b}_k\frac{\partial T(\boldsymbol{r}_k)}{\partial \boldsymbol{p}}.$$

Further assuming that $T(\boldsymbol{p}) = \frac{1}{2}\boldsymbol{p}^\top \boldsymbol{p}$, prove that the Störmer–Verlet method (5.27) can be written as a partitioned RK method with

$$\begin{array}{c|cc} \frac{1}{2} & \frac{1}{2} & 0 \\ \frac{1}{2} & \frac{1}{2} & 0 \\ \hline & \frac{1}{2} & \frac{1}{2} \end{array} \qquad \text{and} \qquad \begin{array}{c|cc} 0 & 0 & 0 \\ 1 & \frac{1}{2} & \frac{1}{2} \\ \hline & \frac{1}{2} & \frac{1}{2} \end{array} .$$

5.10 The *symplectic Euler method* for the Hamiltonian system (5.17) reads

$$\boldsymbol{p}_{n+1} = \boldsymbol{p}_n - h\frac{\partial H(\boldsymbol{p}_{n+1}, \boldsymbol{q}_n)}{\partial \boldsymbol{q}}, \qquad \boldsymbol{q}_{n+1} = \boldsymbol{q}_n + h\frac{\partial H(\boldsymbol{p}_{n+1}, \boldsymbol{q}_n)}{\partial \boldsymbol{p}}.$$

a Show that this is a first-order method.

b Prove from basic principles that, as implied by its name, the method is indeed symplectic.

c Assuming that the Hamiltonian is separable, $H(\boldsymbol{p}, \boldsymbol{q}) = T(\boldsymbol{p}) + V(\boldsymbol{q})$, show that the method can be implemented explicitly.

6

Error control

6.1 Numerical software vs. numerical mathematics

There comes a point in every exposition of numerical analysis when the theme shifts from the familiar mathematical progression of definitions, theorems and proofs to the actual ways and means whereby computational algorithms are implemented. This point is sometimes accompanied by an air of anguish and perhaps disdain: we abandon the palace of the Queen of Sciences for the lowly shop floor of a software engineer. *Nothing could be further from the truth!* Devising an algorithm that fulfils its goal accurately, robustly and economically is an intellectual challenge equal to the best in mathematical research.

In Chapters 1–5 we have seen a multitude of methods for the numerical solution of the ODE system

$$\boldsymbol{y}' = \boldsymbol{f}(t, \boldsymbol{y}), \quad t \geq t_0, \qquad \boldsymbol{y}(t_0) = \boldsymbol{y}_0. \tag{6.1}$$

In the present chapter we are about to study how to incorporate a method into a *computational package*. It is important to grasp that, when it comes to software design, a time-stepping method is just one – albeit very important – component.

A good analogy is the design of a motor car. The time-stepping method is like the engine: it powers the vehicle along. A car with just an engine is useless: a multitude of other components – wheels, chassis, transmission – are essential for its operation. Now, the different parts of the system should not be optimized on their own but as a part of an integrated plan; there is little point in fitting a Formula 1 racing car engine into a family saloon. Moreover, the very goal of optimization is problem-dependent: do we want to optimize for speed? economy? reliability? marketability? In a well-designed car the right components are combined in such a way that they operate together as required, reliably and smoothly. The same is true for a computational package.

A user of a software package for ODEs, say, typically does not (and should not!) care about the particular choice of method or, for that matter, the other 'operating parts' – error and step-size control, solution of nonlinear algebraic equations, choice of starting values and of the initial step, visualization of the numerical solution etc. As far as a user is concerned, a computational package is simply a tool.

The tool designer – be it a numerical analyst or a software engineer – must adopt a more discerning view. The package is no longer a black box but an integrated system, which can be represented in the following flowchart:

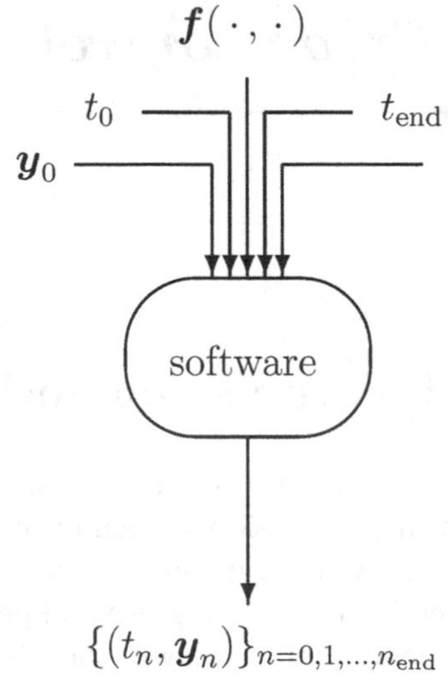

The inputs are not just the function $\boldsymbol{f}$, the starting point t_0, the initial value $\boldsymbol{y}_0$ and the endpoint t_{end} but also the error *tolerance* $\delta > 0$; we wish the numerical error in, say, the Euclidean norm to be within δ. The output is the computed solution sequence at the points $t_0 < t_1 < \cdots < t_{\text{end}}$, which, of course, are not equi-spaced.

We hasten to say that the above is actually the simplest possible model for a computational package, but it will do for expositional purposes. In general, the user might be expected to specify whether (6.1) is stiff and to express a range of preferences with regard to the form of the output. An increasingly important component in the design of a modern software package is *visualization;* it is difficult to absorb information from long lists of numbers and so its display in the form of time series, phase diagrams, Poincaré sections etc. often makes a great deal of difference. Altogether, writing, debugging, testing and documenting modern, advanced, broad-purpose software for ODEs is a highly professional and time-demanding enterprise.

In the present chapter we plan to elaborate a major component of any computational package, *the mechanism whereby numerical error is estimated* in the course of solution *and controlled by means of step size changes.* Chapter 7 is devoted to another aspect, namely solution of the nonlinear algebraic systems that occur whenever implicit methods are applied to the system (6.1).

We will describe a number of different devices for the estimation of the *local error,* i.e. the error incurred when we integrate from t_n to t_{n+1} under the assumption that $\boldsymbol{y}_n$ is 'exact'. This should not be confused with the *global error,* namely the difference between $\boldsymbol{y}_n$ and $\boldsymbol{y}(t_n)$ for all $n = 0, 1, \ldots, n_{\text{end}}$. Clever procedures for the estimation of global error are fast becoming standard in modern software packages.

The error-control devices of this chapter will be applied to three relatively simple systems of the type (6.1): the *van der Pol* equation

$$\begin{aligned} y_1' &= y_2, \\ y_2' &= (1 - y_1^2)y_2 - y_1, \end{aligned} \qquad 0 \le t \le 25, \qquad \begin{aligned} y_1(0) &= \tfrac{1}{2}, \\ y_2(0) &= \tfrac{1}{2}; \end{aligned} \tag{6.2}$$

the *Mathieu* equation

$$\begin{aligned} y_1' &= y_2, \\ y_2' &= -(2-\cos 2t)y_1, \end{aligned} \qquad 0 \le t \le 30, \qquad \begin{aligned} y_1(0) &= 1, \\ y_2(0) &= 0; \end{aligned} \tag{6.3}$$

and the *Curtiss–Hirschfelder* equation

$$y' = -50(y - \cos t), \qquad 0 \le t \le 10, \qquad y(0) = 1. \tag{6.4}$$

The first two equations are not stiff, while (6.4) is moderately so. The solution of the van der Pol equation (which is more commonly written as a second-order equation $y'' - \varepsilon(1-y^2)y' + y = 0$; here we take $\varepsilon = 1$) models electrical circuits connected with triode oscillators. It is well known that, for every initial value, the solution tends to a periodic curve (see Fig. 6.1). The Mathieu equation (which, likewise, is usually written in the second-order form $y'' + (a - b\cos 2t)y = 0$, here with $a = 2$ and $b = 1$) arises in the analysis of the vibrations of an elliptic membrane and also in celestial mechanics. Its (nonperiodic) solution remains forever bounded, without approaching a fixed point (see Fig. 6.2). Finally, the solution of the Curtiss–Hirschfelder equation (which has no known significance except as a good test case for computational algorithms) is

$$y(t) = \tfrac{2500}{2501}\cos t + \tfrac{50}{2501}\sin t + \tfrac{1}{2501}\,\mathrm{e}^{-50t}, \qquad t \ge 0,$$

and approaches a periodic curve at an exponential speed (see Fig. 6.3).

We assume throughout this chapter that $\boldsymbol{f}$ is as smooth as required.

6.2 The Milne device

Let

$$\sum_{m=0}^{s} a_m \boldsymbol{y}_{n+m} = h \sum_{m=0}^{s} b_m \boldsymbol{f}(t_{n+m}, \boldsymbol{y}_{n+m}), \qquad n = 0, 1, \ldots, \qquad a_s = 1, \tag{6.5}$$

be a given convergent multistep method of order p. The goal of assessing the local error in (6.5) is attained by employing another convergent multistep method of the same order, which we write in the form

$$\sum_{m=q}^{s} \tilde{a}_m \boldsymbol{x}_{n+m} = h \sum_{m=q}^{s} \tilde{b}_m \boldsymbol{f}(t_{n+m}, \boldsymbol{x}_{n+m}), \qquad n \ge \max\{0, -q\}, \quad \tilde{a}_s = 1. \tag{6.6}$$

Here $q \le s-1$ is an integer, which might be of either sign; the main reason for allowing negative q is that we wish to align the two methods so that they approximate at the same point t_{n+s} in the nth step. Of course, $\boldsymbol{x}_{n+m} = \boldsymbol{y}_{n+m}$ for $m = \min\{0, q\}, \min\{0, q\} + 1, \ldots, s-1$.

According to Theorem 2.1, the method (6.5), say, is of order p if and only if

$$\rho(w) - \sigma(w)\ln w = c(w-1)^{p+1} + \mathcal{O}\left(|w-1|^{p+2}\right), \qquad w \to 1,$$

where $c \neq 0$ and

$$\rho(w) = \sum_{m=0}^{s} a_m w^m, \qquad \sigma(w) = \sum_{m=0}^{s} b_m w^m.$$

By expanding ψ one term further in the proof of Theorem 2.1, it is easy to demonstrate that, provided $\boldsymbol{y}_n, \boldsymbol{y}_{n+1}, \ldots, \boldsymbol{y}_{n+s-1}$ are assumed to be error free,

$$\boldsymbol{y}(t_{n+s}) - \boldsymbol{y}_{n+s} = ch^{p+1}\boldsymbol{y}^{(p+1)}(t_{n+s}) + \mathcal{O}\left(h^{p+2}\right), \qquad h \to 0. \tag{6.7}$$

The number c is termed the *(local) error constant* of the method (6.5).[1] Let $\tilde{c}$ be the error constant of (6.6) and assume that we have selected the method in such a way that $\tilde{c} \neq c$. Therefore

$$\boldsymbol{y}(t_{n+s}) - \boldsymbol{x}_{n+s} = \tilde{c}h^{p+1}\boldsymbol{y}^{(p+1)}(t_{n+s}) + \mathcal{O}\left(h^{p+2}\right), \qquad h \to 0.$$

We subtract this expression from (6.7) and disregard the $\mathcal{O}\left(h^{p+2}\right)$ terms. The outcome is $\boldsymbol{x}_{n+s} - \boldsymbol{y}_{n+s} \approx (c - \tilde{c})h^{p+1}\boldsymbol{y}^{(p+1)}(t_{n+s})$, hence

$$h^{p+1}\boldsymbol{y}^{(p+1)}(t_{n+s}) \approx \frac{1}{c - \tilde{c}}(\boldsymbol{x}_{n+s} - \boldsymbol{y}_{n+s}).$$

Substitution into (6.7) yields an estimate of the local error, namely

$$\boldsymbol{y}(t_{n+s}) - \boldsymbol{y}_{n+s} \approx \frac{c}{c - \tilde{c}}(\boldsymbol{x}_{n+s} - \boldsymbol{y}_{n+s}). \tag{6.8}$$

This method of assessing the local error is known as the *Milne device.* Recall that our critical requirement is to maintain the local error at less than the tolerance δ. A naive approach is *error control per step,* namely to require that the *local error* κ satisfies

$$\kappa \leq \delta,$$

where

$$\kappa = \left|\frac{c}{c - \tilde{c}}\right| \|\boldsymbol{x}_{n+s} - \boldsymbol{y}_{n+s}\|$$

originates in (6.8). A better requirement, *error control per unit step,* incorporates a crude global consideration into the local estimate. It is based on the assumption that the accumulation of global error occurs roughly at a constant pace and is allied to our observation in Chapter 1 that the global error behaves like $\mathcal{O}(h^p)$. Therefore, the smaller the step size, the more stringent requirement must we place upon κ; the right inequality is

$$\kappa \leq h\delta. \tag{6.9}$$

This is the criterion that we adopt in the remainder of this exposition.

Suppose that we have executed a single time step, thereby computing a candidate solution $\boldsymbol{y}_{n+s}$. We use (6.9), where κ has been evaluated by the Milne device (or by other means), to decide whether $\boldsymbol{y}_{n+s}$ is an acceptable approximation to $\boldsymbol{y}(t_{n+s})$.

[1] The *global* error constant is defined as $c/\rho'(1)$, for reasons that are related to the theme of Exercise 2.2 but are outside the scope of our exposition.

If not, the time step is rejected: we go back to t_{n+s-1}, halve h and resume time-stepping. If, however, (6.9) holds then the new value is acceptable and we advance to t_{n+s}. Moreover, if κ is significantly smaller than $h\delta$ – for example, if $\kappa < \frac{1}{10}h\delta$ – we take this as an indication that the time step is too small (hence, wasteful) and double it. A simple scheme of this kind might be conveniently represented in flowchart form:

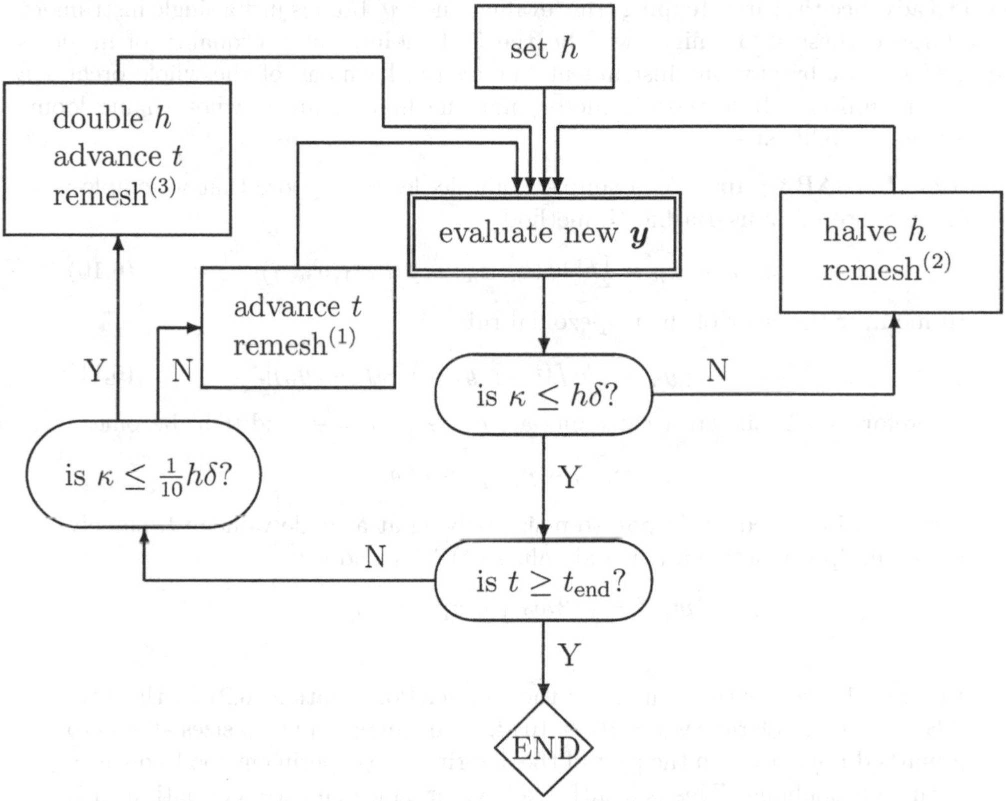

except that each box in the flowchart hides a multitude of sins! In particular, observe the need to 'remesh' the variables: multistep methods require that starting values for each step are provided on an equally spaced grid and this is no longer the case when h is amended. We need then to approximate the starting values, typically by polynomial interpolation (A.2.2.3–A.2.2.5). In each iteration we need $\hat{s} := s + \max\{0, -q\}$ vectors $\boldsymbol{y}_{n+\min\{0,q\}}, \ldots, \boldsymbol{y}_{n+s-1}$, which we rename $\boldsymbol{w}_1, \boldsymbol{w}_2, \ldots, \boldsymbol{w}_{\hat{s}}$ respectively. There are three possible cases:

(1) h **is unamended** We let $\boldsymbol{w}_j^{\text{new}} = \boldsymbol{w}_{j+1}$, $j = 1, 2, \ldots, \hat{s} - 1$, and $\boldsymbol{w}_{\hat{s}}^{\text{new}} = \boldsymbol{y}_{n+s}$.

(2) h **is halved** The values $\boldsymbol{w}_{\hat{s}-2j}^{\text{new}} = \boldsymbol{w}_{\hat{s}-j}$, $j = 0, 1, \ldots, \lfloor \hat{s}/2 \rfloor - 1$, survive, while the rest, which approximate values at the midpoints of the old grid, need to be computed by interpolation.

(3) h **is doubled** Here $\boldsymbol{w}_{\hat{s}}^{\text{new}} = \boldsymbol{y}_{n+s+1}$ and $\boldsymbol{w}_{\hat{s}-j}^{\text{new}} = \boldsymbol{w}_{\hat{s}-2j+1}$, $j = 1, 2, \ldots, \hat{s} - 1$. This requires an extra $\hat{s} - 2$ vectors, $\boldsymbol{w}_{-\hat{s}+3}, \ldots, \boldsymbol{w}_0$, which have not been defined

above. The remedy is simple, at least in principle: we need to carry forward in the previous two remeshings at least $2\hat{s}-1$ vectors to allow the scope for step-size doubling. This procedure may impose a restriction on consecutive step-size doublings.

A glance at the flowchart affirms our claim in Section 6.1 that the specific method used to advance the time-stepping (the 'evaluate new $\boldsymbol{y}$' box) is just a single instrument in a large orchestra. It might well be the first violin, but the quality of music is determined not by any one instrument but by the harmony of the whole orchestra playing in unison. It is the conductor, not the first violinist, whose name looms largest on the billboard!

◇ **The TR–AB2 pair** As a simple example, let us suppose that we employ the two-step Adams–Bashforth method

$$\boldsymbol{x}_{n+1}-\boldsymbol{x}_n=\tfrac{1}{2}h[3\boldsymbol{f}(t_n,\boldsymbol{x}_n)-\boldsymbol{f}(t_{n-1},\boldsymbol{x}_{n-1})] \tag{6.10}$$

to monitor the error of the trapezoidal rule

$$\boldsymbol{y}_{n+1}-\boldsymbol{y}_n=\tfrac{1}{2}h[\boldsymbol{f}(t_{n+1},\boldsymbol{y}_{n+1})+\boldsymbol{f}(t_n,\boldsymbol{y}_n)]. \tag{6.11}$$

Therefore $\hat{s}=2$, the error constants are $c=-\frac{1}{12}$, $\tilde{c}=\frac{5}{12}$ and (6.9) becomes

$$\|\boldsymbol{x}_{n+1}-\boldsymbol{y}_{n+1}\|\le 6h\delta.$$

Interpolation is required upon step-size halving at a single value of t, namely at the midpoint between (the old values of) t_{n-1} and t_n:

$$\boldsymbol{w}_1^{\text{new}}=\tfrac{1}{8}(3\boldsymbol{w}_2+6\boldsymbol{w}_1-\boldsymbol{y}_{n-2}).$$

Figure 6.1 displays the solution of the van der Pol equation (6.2) by the TR–AB2 pair with tolerances $\delta=10^{-3},10^{-4}$. The sequence of step sizes attests to a marked reluctance on the part of the algorithm to experiment too frequently with step-doubling. This is healthy behaviour, since an excess of optimism is bound to breach the inequality (6.9) and is wasteful. Note, by the way, how strongly the step-size sequence correlates with the size of y_2', which, for the van der Pol equation, measures the 'awkwardness' of the solution – it is easy to explain this feature from the familiar phase portrait of (6.2).

The global error, as displayed in the bottom two graphs, is of the right order of magnitude and, as we might expect, slowly accumulates with time. This is typical of non-stiff problems like (6.2).

Similar lessons can be drawn from the Mathieu equation (6.3) (see Fig. 6.2). It is perhaps more difficult to find a single characteristic of the solution that accounts for the variation in h, but it is striking how closely the step sequences for $\delta=10^{-3}$ and $\delta=10^{-4}$ correlate. The global error accumulates markedly faster but is still within what can be expected from the general theory and the accepted wisdom. The goal being to incur an error of at most δ in a unit-length interval, a final error of $(t_{\text{end}}-t_0)\delta$ is to be expected at the right-hand endpoint of the interval.

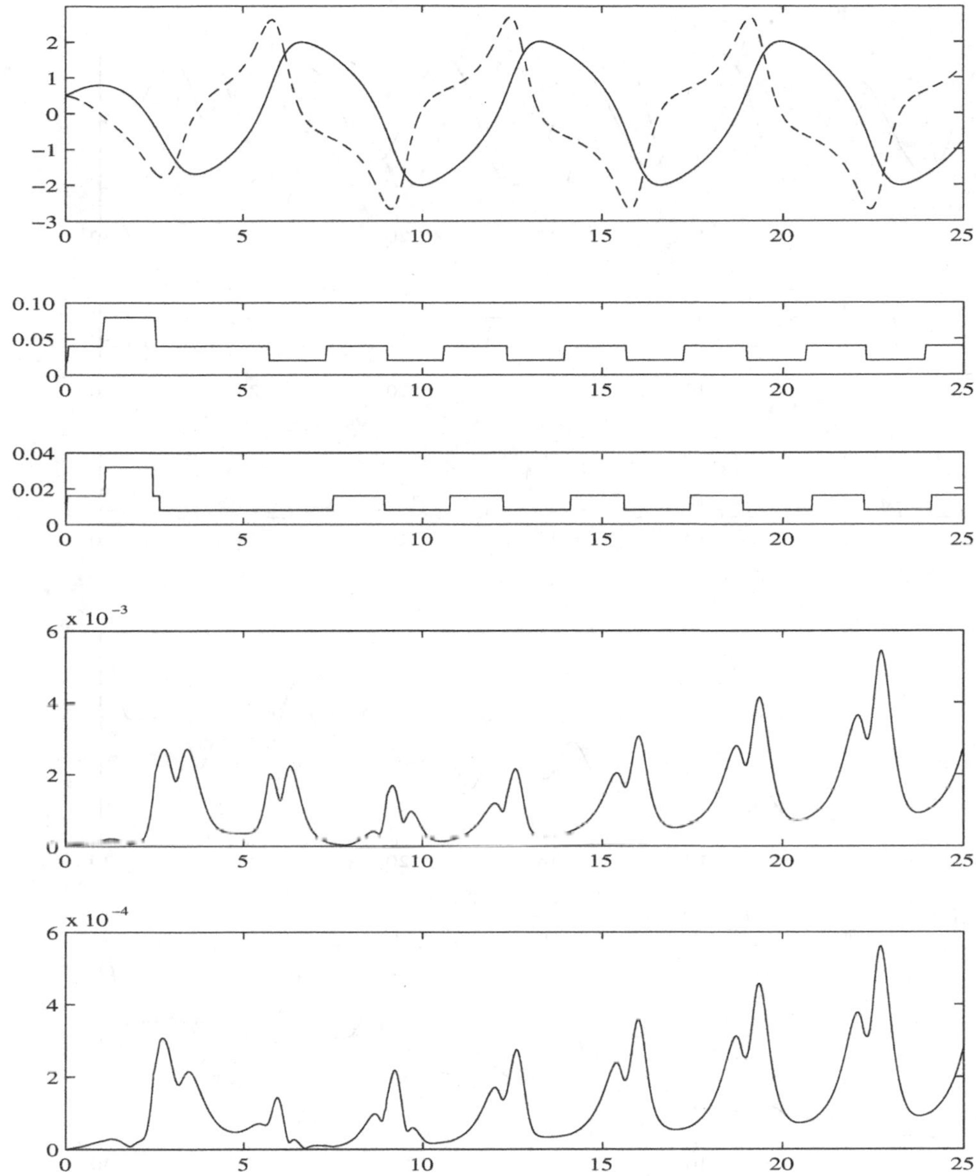

Figure 6.1 The top figure displays the two solution components of the van der Pol equation (6.2) in the interval $[0, 25]$. The other figures relate to the Milne device, applied with the pair (6.10), (6.11) to this equation. The second and third figures each feature the sequence of step sizes for tolerances δ equal to 10^{-3} and 10^{-4} respectively, while the lowest two figures show the (exact) global error for these values of δ.

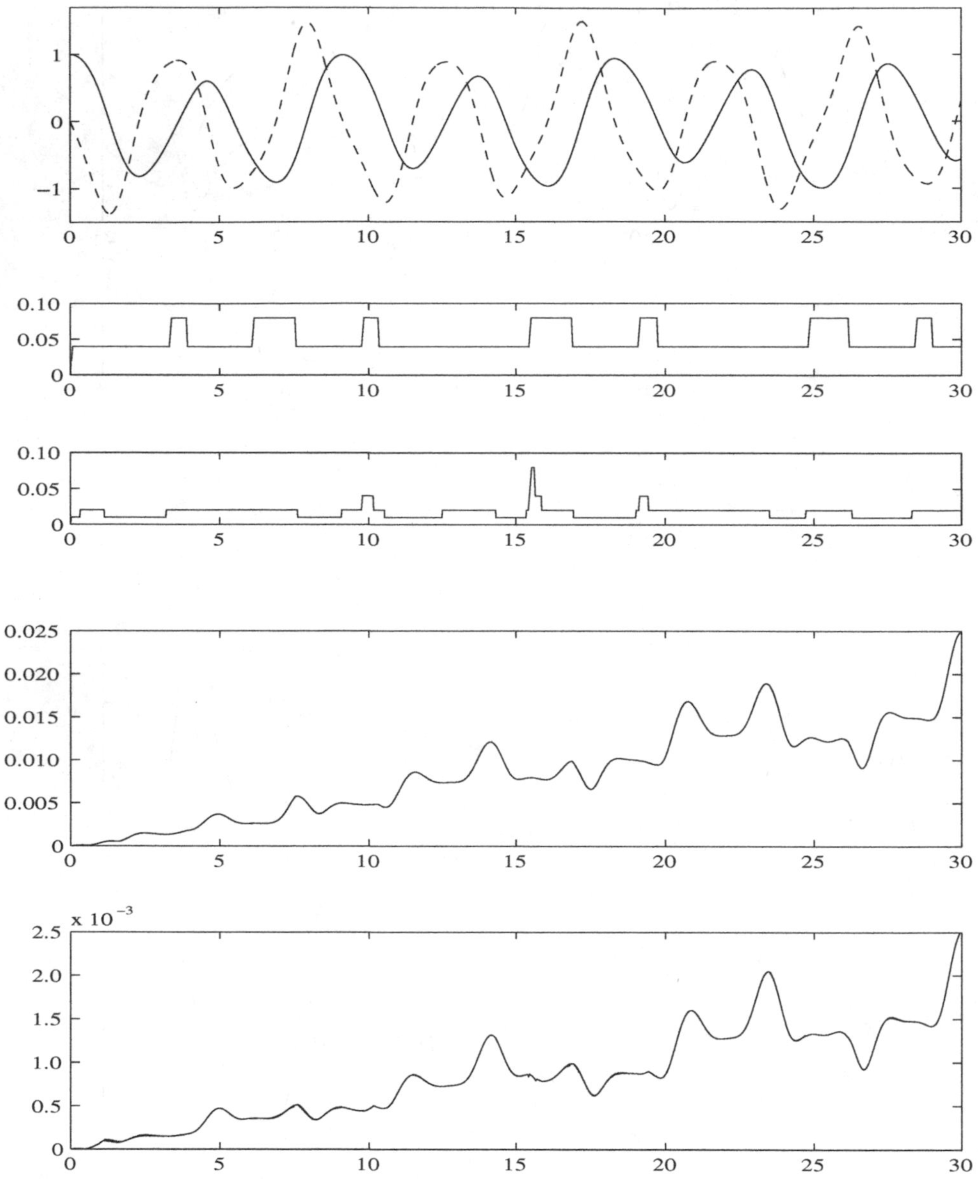

Figure 6.2 The top figure displays the two solution components of the Mathieu equation (6.3) in the interval $[0, 30]$. The other figures are concerned with the Milne device, applied with the pair (6.10), (6.11). The second and the third figures each feature the sequence of step sizes for tolerances δ equal to 10^{-3} and 10^{-4} respectively, while the lowest two figures show the (exact) global error for these values of δ.

Finally, Fig. 6.3 displays the behaviour of the TR–AB2 pair for the mildly stiff equation (6.4). The first interesting observation, looking at the step sequences, is that the step sizes are quite large, at least for $\delta = 10^{-3}$. Had we tried to solve this equation with the Euler method, we would have needed to impose $h < \frac{1}{25}$ to prevent instabilities, whereas the trapezoidal rule chugs along happily with h occasionally exceeding $\frac{1}{3}$. This is an important point to note since the stability analysis of Chapter 4 has been restricted to constant steps. It is worthwhile to record that, at least in a single computational example, A-stability allows the trapezoidal rule to select step sizes solely in pursuit of accuracy.

The accumulation of global errors displays a pattern characteristic of stiff equations. Provided that the method is adequately stable, global error does not accumulate at all and is often significantly smaller than δ! (The occasional jumps in the error in Fig. 6.3 are probably attributable to the increase in h and might well have been eliminated altogether with sufficient fine-tuning of the computational scheme.)

Note that, of course, (6.10) has exceedingly poor stability characteristics (see Fig. 4.3). This is not a handicap, since the Adams–Bashforth method is used solely for local error control. To demonstrate this point, in Fig. 6.4 we display the error for the Curtiss–Hirschfelder equation (6.4) when, in lieu of (6.10), we employ the A-stable backward differentiation formula method (2.15). Evidently, not much changes! ◇

We conclude this section by remarking again that execution of a variable-step code requires a multitude of choices and a great deal of fine-tuning. The need for brevity prevents us from discussing, for example, an appropriate procedure for the choice of the initial step size and starting values $\boldsymbol{y}_1, \boldsymbol{y}_2, \ldots, \boldsymbol{y}_{s-1}$.

6.3 Embedded Runge–Kutta methods

The comfort of a single constant whose magnitude reflects (at least for small h) the local error κ is denied us in the case of Runge–Kutta methods, (3.9). In order to estimate κ we need to resort to a different device, which again is based upon running two methods in tandem – one, of order p, to provide a candidate for solution and the other, of order $\tilde{p} \geq p+1$, to control the error.

In line with (6.5) and (6.6), we denote by $\boldsymbol{y}_{n+1}$ the candidate solution at t_{n+1} obtained from the pth-order method, whereas the solution at t_{n+1} obtained from the higher-order scheme is $\boldsymbol{x}_{n+1}$. We have

$$\left.\begin{aligned} \boldsymbol{y}_{n+1} &= \tilde{\boldsymbol{y}}(t_{n+1}) + \boldsymbol{\ell} h^{p+1} + \mathcal{O}(h^{p+2}), \qquad (6.12) \\ \boldsymbol{x}_{n+1} &= \tilde{\boldsymbol{y}}(t_{n+1}) + \mathcal{O}(h^{p+2}), \qquad (6.13) \end{aligned}\right\} \quad h \to 0,$$

where $\boldsymbol{\ell}$ is a vector that depends on the equation (6.1) (but not upon h) and $\tilde{\boldsymbol{y}}$ is the exact solution of (6.1) with initial value $\tilde{\boldsymbol{y}}(t_n) = \boldsymbol{y}_n$. Subtracting (6.13) from (6.12),

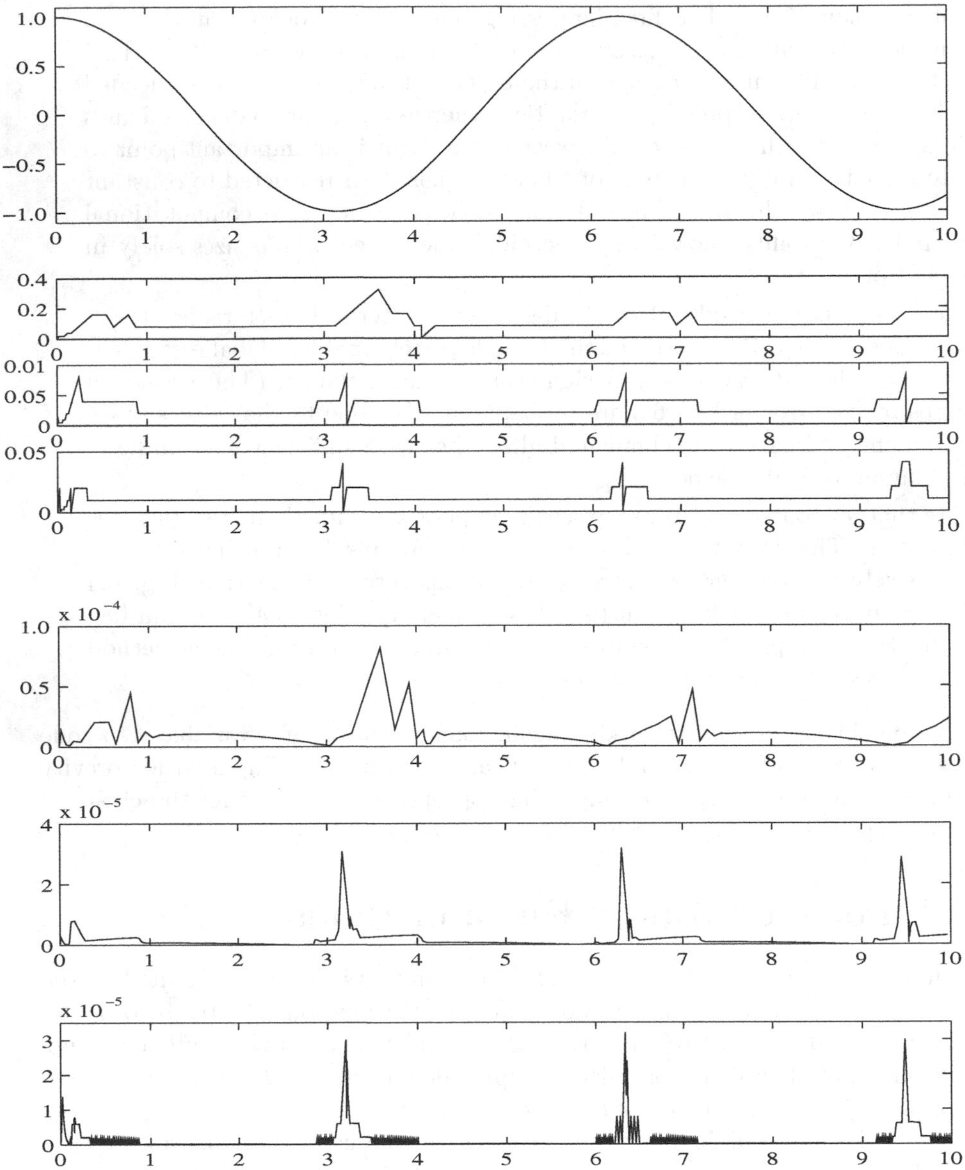

Figure 6.3 The top figure displays the solution of the Curtiss–Hirschfelder equation (6.4) in the interval $[0, 10]$. The other figures are concerned with the Milne device, applied with the pair (6.10), (6.11). The second to fourth figures feature the sequence of step sizes for tolerances δ equal to $10^{-3}, 10^{-4}$ and 10^{-5} respectively, while the bottom three figures show the (exact) global error for these values of δ.

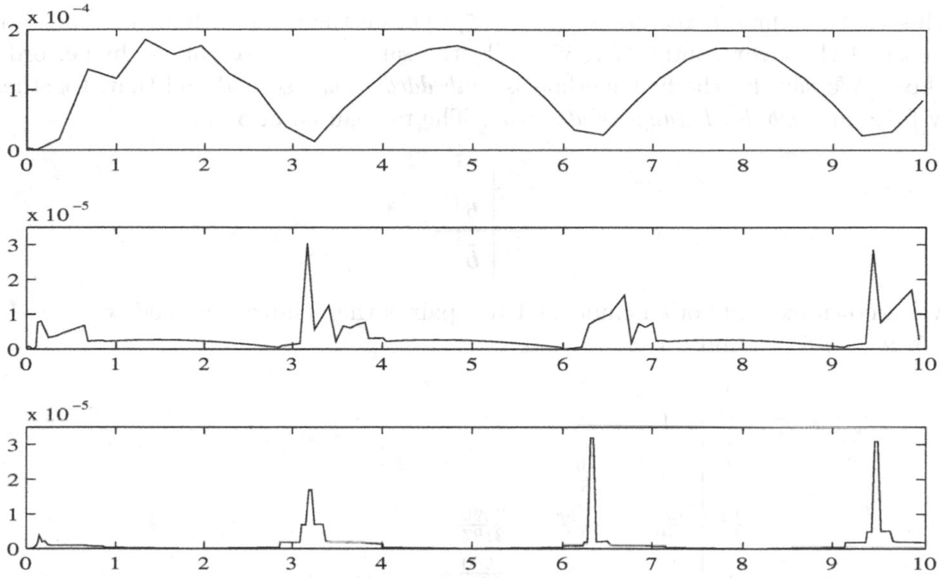

Figure 6.4 Global errors for the numerical solution of the Curtiss–Hirschfelder equation (6.4) by the TR–BDF2 pair with $\delta = 10^{-3}, 10^{-4}, 10^{-5}$.

we obtain $\boldsymbol{\ell} h^{p+1} \approx \boldsymbol{y}_{n+1} - \boldsymbol{x}_{n+1}$, the outcome being the error estimate

$$\kappa = \|\boldsymbol{y}_{n+1} - \boldsymbol{x}_{n+1}\|. \tag{6.14}$$

Once κ is available, we may proceed as in Section 6.2 except that, having opted for one-step methods, we are spared all the awkward and time-consuming minutiae of remeshing each time h is changed.

A naive application of the above approach requires the doubling, at the very least, of the expense of calculation, since we need to compute both $\boldsymbol{y}_{n+1}$ and $\boldsymbol{x}_{n+1}$. This is unacceptable since, as a rule, the cost of error control should be marginal in comparison with the cost of the main scheme.[2] However, when an ERK method (3.5) is used for the main scheme it is possible to choose the two methods in such a way that the extra expense is small. Let us thus denote by

$$\begin{array}{c|c} \boldsymbol{c} & A \\ \hline & \boldsymbol{b}^\top \end{array} \qquad \text{and} \qquad \begin{array}{c|c} \tilde{\boldsymbol{c}} & \tilde{A} \\ \hline & \tilde{\boldsymbol{b}}^\top \end{array}$$

the pth-order method and the higher-order method, of ν and $\tilde{\nu}$ stages respectively. The main idea is to choose

$$\tilde{\boldsymbol{c}} = \begin{bmatrix} \boldsymbol{c} \\ \hat{\boldsymbol{c}} \end{bmatrix}, \qquad \tilde{A} = \begin{bmatrix} A & O \\ & \hat{A} \end{bmatrix},$$

[2]Note that we have used in Section 6.2 an explicit method, the Adams–Bashforth scheme, to control the error of the implicit trapezoidal rule. This is consistent with the latter remark.

where $\hat{\boldsymbol{c}} \in \mathbb{R}^{\tilde{\nu}-\nu}$ and $\hat{A}$ is a $(\tilde{\nu}-\nu) \times \nu$ matrix, so that $\tilde{A}$ is strictly lower triangular. In this case the first ν vectors $\boldsymbol{\xi}_1, \boldsymbol{\xi}_2, \ldots, \boldsymbol{\xi}_\nu$ will be the same in both methods and the cost of the error controller is virtually the same as the cost of the higher-order method. We say the the first method is *embedded* in the second and that, together, they form an *embedded Runge–Kutta pair*. The tableau notation is

$$\begin{array}{c|c} \tilde{\boldsymbol{c}} & \tilde{A} \\ \hline & \boldsymbol{b}^\top \\ \hline & \tilde{\boldsymbol{b}}^\top \end{array}.$$

A well-known example of an embedded RK pair is the Fehlberg method, with $p = 4$, $\tilde{p} = 5$, $\nu = 5$, $\tilde{\nu} = 6$ and tableau

$$\begin{array}{c|cccccc} 0 & & & & & & \\ \frac{1}{4} & \frac{1}{4} & & & & & \\ \frac{3}{8} & \frac{3}{32} & \frac{9}{32} & & & & \\ \frac{12}{13} & \frac{1932}{2197} & -\frac{7200}{2197} & \frac{7296}{2197} & & & \\ 1 & \frac{439}{216} & -8 & \frac{3680}{513} & -\frac{845}{4104} & & \\ \frac{1}{2} & -\frac{8}{27} & 2 & -\frac{3544}{2565} & \frac{1859}{4104} & -\frac{11}{40} & \\ \hline & \frac{25}{216} & 0 & \frac{1408}{2565} & \frac{2197}{4104} & -\frac{1}{5} & \\ \hline & \frac{16}{135} & 0 & \frac{6656}{12825} & \frac{28561}{56430} & -\frac{9}{50} & \frac{2}{55} \end{array}.$$

◇ **A simple embedded RK pair** The RK pair

$$\begin{array}{c|ccc} 0 & & & \\ \frac{2}{3} & \frac{2}{3} & & \\ \frac{2}{3} & 0 & \frac{2}{3} & \\ \hline & \frac{1}{4} & \frac{3}{4} & \\ \hline & \frac{1}{4} & \frac{3}{8} & \frac{3}{8} \end{array} \qquad (6.15)$$

has orders $p = 2$, $\tilde{p} = 3$. The local error estimate becomes simply

$$\kappa = \tfrac{3}{8} \left\| \boldsymbol{f}\left(t_n + \tfrac{2}{3}h, \boldsymbol{\xi}_3\right) - \boldsymbol{f}\left(t_n + \tfrac{2}{3}h, \boldsymbol{\xi}_2\right) \right\|.$$

We applied a variable-step algorithm based on the above error controller to the problems (6.2)–(6.4), within the same framework as the computational experiments using the Milne device in Section 6.2. The results are reported in Figs 6.5–6.7.

Comparing Figs 6.1 and 6.5 demonstrates that, as far as error control is concerned, the performance of (6.15) is roughly similar to the Milne device for the TR–AB2 pair. A similar conclusion can be drawn by comparing Figs 6.2 and 6.6. On the face of it, this is also the case for our single stiff example

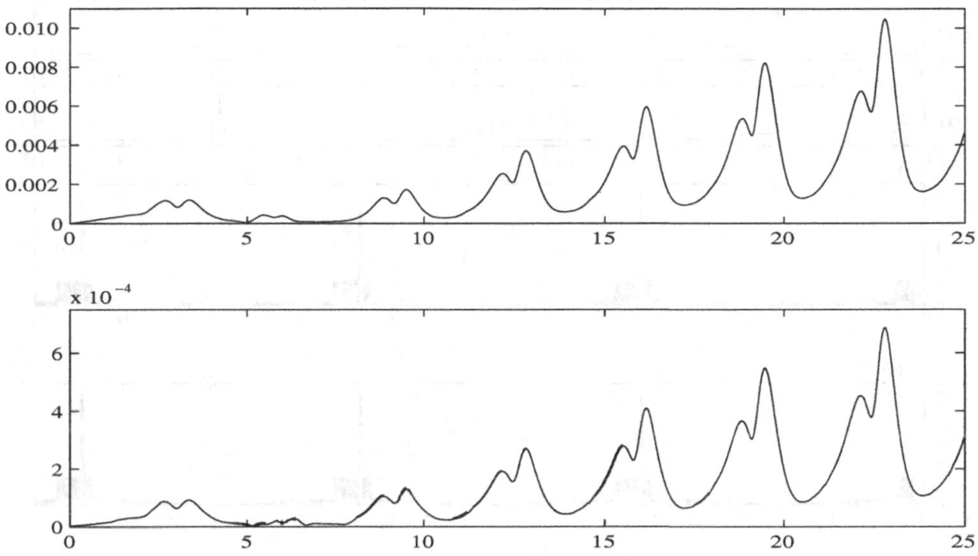

Figure 6.5 Global errors for the numerical solution of the van der Pol equation (6.2) by the embedded RK pair (6.15) with $\delta = 10^{-3}, 10^{-4}$.

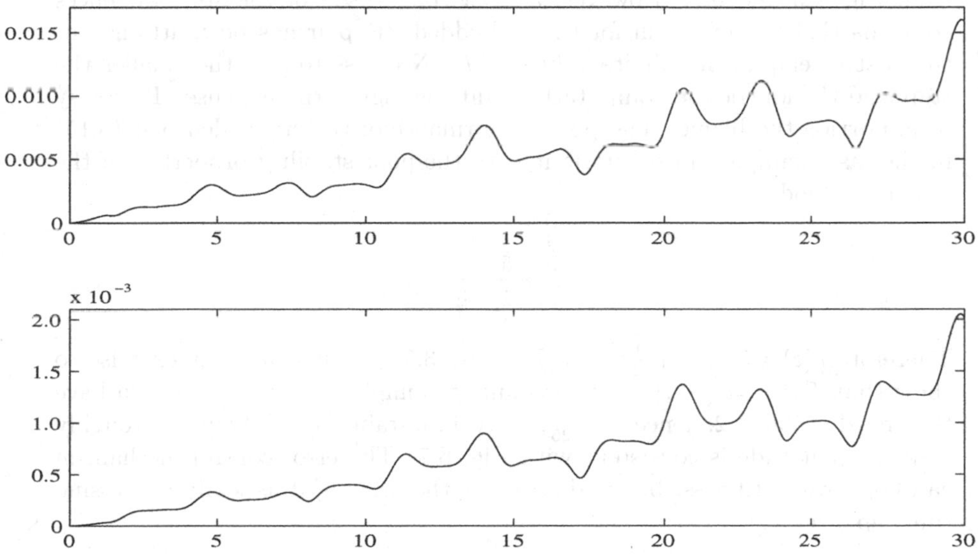

Figure 6.6 Global errors for the numerical solution of the Mathieu equation (6.3) by the embedded RK pair (6.15) with $\delta = 10^{-3}, 10^{-4}$.

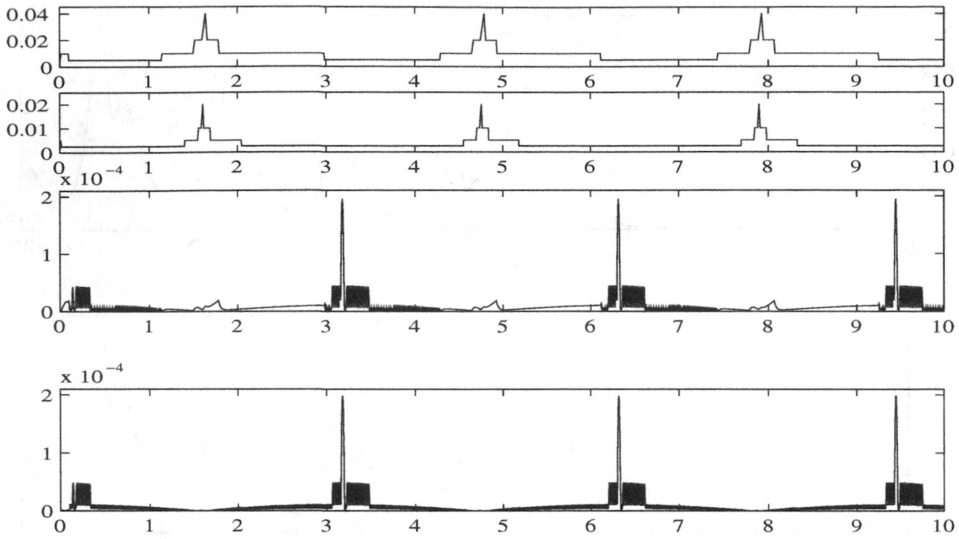

Figure 6.7 The step sequences (in the top two graphs) and global errors for the numerical solution of the Curtiss–Hirschfelder equation (6.4) by the embedded RK pair (6.15) with $\delta = 10^{-3}, 10^{-4}$.

from Fig. 6.3 and 6.7. However, a brief comparison of the step sequences confirms that the precision for the embedded RK pair has been attained at the cost of employing minute values of h. Needless to say, the smaller the step size the longer the computation and the higher the expense. It is easy to apportion the blame. The poor performance of the embedded pair (6.15) in the last example can be attributed to the poor stability properties of the 'inner' method

$$\begin{array}{c|cc} 0 & & \\ \frac{2}{3} & \frac{2}{3} & \\ \hline & \frac{1}{4} & \frac{3}{4} \end{array}.$$

Therefore $r(z) = 1 + z + \frac{1}{2}z^2$ (cf. Exercise 3.5) and it is an easy exercise to show that $\mathcal{D} \cap \mathbb{R} = (-2, 0)$. In constant-step implementation we would have thus needed $50h < 2$, hence $h < \frac{1}{25}$, to avoid instabilities. A bound of roughly similar magnitude is consistent with Fig. 6.7. The error-control mechanism can cope with stiffness, but it does so at the price of drastically depressing the step size. ◇

Exercise 6.5 demonstrates that the technique of embedded RK pairs can be generalized, at least to some extent, to cater for implicit methods, thereby rendering it more suitable for stiff equations. It is fair, though, to point out that, in practical implementations, the use of embedded RK pairs is almost always restricted to explicit Runge–Kutta schemes.

Comments and bibliography

A classical approach to error control is to integrate once with step h and to integrate again, along the same interval, with two steps of $\frac{1}{2}h$. Comparison of the two candidate solutions at the new point yields an estimate of κ. This is clearly inferior to the methods of this chapter within the narrow framework of error control. However, the 'one step, two half-steps' technique is a rudimentary example of extrapolation, which can be used *both* to monitor the error *and* to improve locally the quality of the solution (cf. Exercise 6.6). See Hairer *et al.* (1991) for an extensive description of extrapolation techniques.

We wish to mention two further means whereby local error can be monitored. The first is a general technique due to Zadunaisky, which can ride piggyback on any time-stepping method

$$\boldsymbol{y}_{n+1} = \boldsymbol{\mathcal{Y}}(\boldsymbol{f}; h; (t_0, \boldsymbol{y}_0), (t_1, \boldsymbol{y}_1), \ldots, (t_n, \boldsymbol{y}_n))$$

of order p (Zadunaisky, 1976). It proceeds as follows. We are solving numerically the ODE system (6.1) and assume the availability of $p+1$ past values, $\boldsymbol{y}_{n-p}, \boldsymbol{y}_{n-p+1}, \ldots, \boldsymbol{y}_n$. They need not correspond to equally spaced points. Let us form a pth degree interpolating polynomial $\boldsymbol{\psi}$ such that $\boldsymbol{\psi}(t_i) = \boldsymbol{y}_i$, $i = n-p, n-p+1, \ldots, n$, and consider the ODE

$$\boldsymbol{z}' = \boldsymbol{f}(t, \boldsymbol{z}) + \left[\boldsymbol{\psi}'(t) - \boldsymbol{f}(t, \boldsymbol{\psi}(t))\right], \quad t \geq t_n, \qquad \boldsymbol{z}(t_n) = \boldsymbol{y}_n. \tag{6.16}$$

Two observations are crucial. Firstly, (6.16) is merely a small perturbation of the original system (6.1): since the numerical method is of order p and $\boldsymbol{\psi}$ interpolates at $p+1$ points, it follows that $\boldsymbol{\psi}(t) = \boldsymbol{y}(t) + \mathcal{O}(h^{p+1})$, therefore, as long as $\boldsymbol{f}$ is sufficiently smooth, $\boldsymbol{\psi}'(t) - \boldsymbol{f}(t, \boldsymbol{\psi}(t)) = \mathcal{O}(h^p)$. Secondly, the *exact* solution of (6.16) is nothing other than $\boldsymbol{z} = \boldsymbol{\psi}$; this is verified at once by substitution.

We now use the underlying numerical method to approximate the solution of (6.16) at t_{n+1}, using *exactly* the same ingredients as were used in the computation of $\boldsymbol{y}_{n+1}$: the same starting values, the same approach to solving nonlinear algebraic systems, an identical stopping criterion ... The outcome is

$$\boldsymbol{z}_{n+1} = \boldsymbol{\mathcal{Y}}(\boldsymbol{g}; h; (t_0, \boldsymbol{y}_0), (t_1, \boldsymbol{y}_1), \ldots, (t_n, \boldsymbol{y}_n)),$$

where $\boldsymbol{g}(t, \boldsymbol{z}) = \boldsymbol{f}(t, \boldsymbol{z}) + [\boldsymbol{\psi}'(t) - \boldsymbol{f}(t, \boldsymbol{\psi}(t))]$. Since $\boldsymbol{g} \approx \boldsymbol{f}$, we act upon the assumption (which can be firmed up mathematically) that the error in $\boldsymbol{z}_{n+1}$ is similar to the error in $\boldsymbol{y}_{n+1}$, and this motivates the estimate

$$\kappa = \|\boldsymbol{\psi}(t_{n+1}) - \boldsymbol{z}_{n+1}\|.$$

Our second technique for error control, the *Gear automatic integration* approach, is much more than simply a device to assess the local error. It is an integrated approach to the implementation of multistep methods that not only controls the growth of the error but also helps us to choose (on a local basis) the best multistep formula out of a given range.

The actual estimate of κ in Gear's method is probably the least important detail. Recalling from (6.7) that the principal error term is of the form $ch^{p+1}\boldsymbol{y}^{(p+1)}(t_{n+s})$, we interpolate the $\boldsymbol{y}_i$ by a polynomial $\boldsymbol{\psi}$ of sufficiently high degree and replace $\boldsymbol{y}^{(p+1)}(t_{n+s})$ by $\boldsymbol{\psi}^{(p+1)}(t_{n+s})$. This, however, is only the beginning! Suppose, for example, that the underlying method is the pth-order Adams–Moulton. We subsequently form similar local-error estimates for its neighbours, Adams–Moulton methods of orders $p \pm 1$. Instead of doubling (or, if the error estimate for the pth method falls short of the tolerance, halving) the step size, we ask ourselves which of the three methods would have attained, on the basis of our estimates, the requisite tolerance δ with the largest value of h? We then switch to this method and this step size, advancing if the present step is acceptable or otherwise resuming the integration from the former point t_{n+s-1}.

This brief explanation does no justice to a complicated and sophisticated assembly of techniques, rules and tricks that makes the Gear approach the method of choice in many leading computational packages. The reader is referred to Gear (1971) and to Shampine & Gordon (1975) for details. Here we just comment on two important features. Firstly, a tremendous simplification of the tedious minutiae of interpolation and remeshing occurs if, instead of storing past values of the solution, the program deals with their finite differences, which, in effect, approximate the derivatives of $\boldsymbol{y}$ at t_{n+s-1}. This is called the *Nordsieck representation* of the multistep method (6.5). Secondly, the Gear automatic integration obviates the need for an independent (and tiresome) derivation of the requisite number of additional starting values, which is characteristic of other implementations of multistep methods (and which is often accomplished by a Runge–Kutta scheme). Instead, the integration can be commenced using a one-step method, allowing the algorithm to increase order only when enough information has accumulated for that purpose.

It might well be, gentle reader, that by this stage you are disenchanted with your prospects of programming a competitive computational package that can hold its own against the best in the field. If so, this exposition has achieved its purpose! Modern high-quality software packages require years of planning, designing, programming, debugging, testing, debugging again, documenting and testing again, by whole teams of first class experts in numerical analysis and software engineering. It is neither a job for amateurs nor an easy alternative to proving theorems.

All standard numerical software packages, for example MATLAB and symbolic packages like Maple and Mathematica that cater also for numerical calculations, have a number of well-written and well-tested ODE solvers, mostly following the ideas described in this chapter. Good and reliable software for ODEs is available from commercial companies that specialize in numerical software, e.g. IMSL and NAG, as well as from *NetLib,* a depository of free software managed by University of Tennesee at Knoxville and Oak Ridge National Laboratory (the current URL address is `http://www.netlib.org/`). General purpose software for partial differential equations is more problematic, for reasons that should be apparent later in this volume, but well-written, reliable and superbly documented packages exist for various families of such equations, often in a form suitable for specific applications such as computational fluid dynamics, electrical engineering etc.

A useful and (relatively) up-to-date guide to state-of-the-art mathematical software is available at the website `http://gams.nist.gov/`, courtesy of the (American) National Institute of Standards and Technology. An impressive source of free software is Ernst Hairer's website, `http://www.unige.ch/~hairer/software.html`. However, the number of `ftp` and websites is expanding so fast as to render a more substantive list of little lasting value. Given the volume of traffic along the information superhighway, it is likely that the ideal program for your problem exists *somewhere.* It is a moot point, however, whether it is easier to locate it or to write one of your own ...

Gear, C.W. (1971), *Numerical Initial Value Problems in Ordinary Differential Equations,* Prentice–Hall, Englewood Cliffs, NJ.

Hairer, E., Nørsett, S.P. and Wanner, G. (1991), *Solving Ordinary Differential Equations I: Nonstiff Problems* (2nd edn), Springer-Verlag, Berlin.

Shampine, L.F. and Gordon, M.K. (1975), *Computer Solution of Ordinary Differential Equations,* W.H. Freeman, San Francisco.

Zadunaisky, P.E. (1976), On the estimation of errors propagated in the numerical integration of ordinary differential equations, *Numerische Mathematik* **27**, 21–39.

Exercises

6.1 Find the error constants for the Adams–Bashforth method (2.7) and for Adams–Moulton methods with $s = 2, 3$.

6.2* Prove that the error constant of the s-step backward differentiation formula is $-\beta/(s+1)$, where β was defined in (2.14).

6.3 Instead of using (6.10) to estimate the error in the multistep method (6.11), we can use it to increase the accuracy.

a Prove that the formula (6.7) yields

$$\begin{aligned} \boldsymbol{y}(t_{n+1}) - \boldsymbol{y}_{n+1} &= -\tfrac{1}{12} h^3 \boldsymbol{y}'''(t_{n+1}) + \mathcal{O}(h^4), \\ \boldsymbol{y}(t_{n+1}) - \boldsymbol{x}_{n+1} &= -\tfrac{5}{12} h^3 \boldsymbol{y}'''(t_{n+1}) + \mathcal{O}(h^4). \end{aligned} \tag{6.17}$$

b Neglecting the $\mathcal{O}(h^4)$ terms, solve the two equations for the unknown $\boldsymbol{y}(t_{n+1})$ (in contrast with the Milne device, where we solve for $\boldsymbol{y}'''(t_{n+1})$).

c Substituting the approximate expression back into (6.7) results in a two-step implicit multistep method. Derive it explicitly and determine its order. Is it convergent? Can you identify it?[3]

6.4 Prove that the embedded RK pair

$$\begin{array}{c|ccc} 0 & & & \\ \frac{1}{2} & \frac{1}{2} & & \\ 1 & -1 & 2 & \\ \hline & 0 & 1 & \\ \hline & \frac{1}{6} & \frac{2}{3} & \frac{1}{6} \end{array}$$

combines a second-order and a third-order method.

6.5 Consider the embedded RK pair

$$\begin{array}{c|ccc} 0 & & & \\ 1 & \frac{1}{2} & \frac{1}{2} & \\ \frac{1}{2} & \frac{3}{8} & \frac{1}{8} & \\ \hline & \frac{1}{2} & \frac{1}{2} & \\ \hline & \frac{1}{6} & \frac{1}{6} & \frac{2}{3} \end{array}.$$

Note that the 'inner' two-stage method is implicit and that the third stage is explicit. This means that the added cost of error control is marginal.

[3] It is always possible to use the method (6.6) to boost the order of (6.5) except when the outcome is not convergent, as is often the case.

a Prove that the 'inner' method is of order 2, while the full three-stage method is of order 3.

b Show that the 'inner' method is A-stable. Can you identify it as a familiar method in disguise?

c Find the function r associated with the three-stage method and verify that, in line with Lemma 4.4, $r(z) = \mathrm{e}^z + \mathcal{O}(z^4)$, $z \to 0$.

6.6 Let $\boldsymbol{y}_{n+1} = \boldsymbol{\mathcal{Y}}(\boldsymbol{f}, h, \boldsymbol{y}_n)$, $n = 0, 1, \ldots$, be a one-step method of order p. We assume (consistently with Runge–Kutta methods, cf. (6.12)) that there exists a vector $\boldsymbol{\ell}_n$, independent of h, such that

$$\boldsymbol{y}_{n+1} = \tilde{\boldsymbol{y}}(t_{n+1}) + \boldsymbol{\ell}_n h^{p+1} + \mathcal{O}\left(h^{p+2}\right), \qquad h \to 0,$$

where $\tilde{\boldsymbol{y}}$ is the exact solution of (6.1) with initial condition $\tilde{\boldsymbol{y}}(t_n) = \boldsymbol{y}_n$. Let

$$\boldsymbol{x}_{n+1} := \boldsymbol{\mathcal{Y}}\left(\boldsymbol{f}, \tfrac{1}{2}h, \boldsymbol{\mathcal{Y}}\left(\boldsymbol{f}, \tfrac{1}{2}h, \boldsymbol{y}_n\right)\right).$$

Note that $\boldsymbol{x}_{n+1}$ is simply the result of traversing $[t_n, t_{n+1}]$ with the method $\boldsymbol{\mathcal{Y}}$ in two equal steps of $\frac{1}{2}h$.

a Find a real constant α such that

$$\boldsymbol{x}_{n+1} = \tilde{\boldsymbol{y}}(t_{n+1}) + \alpha \boldsymbol{\ell}_n h^{p+1} + \mathcal{O}\left(h^{p+2}\right), \qquad h \to 0.$$

b Determine a real constant β such that the linear combination $\boldsymbol{z}_{n+1} := (1 - \beta)\boldsymbol{y}_{n+1} + \beta \boldsymbol{x}_{n+1}$ approximates $\tilde{\boldsymbol{y}}(t_{n+1})$ up to $\mathcal{O}\left(h^{p+2}\right)$. (*The procedure of using the enhanced value* $\boldsymbol{z}_{n+1}$ *as the approximation at* t_{n+1} *is known as* extrapolation. *It is of widespread application.*)

c Let $\boldsymbol{\mathcal{Y}}$ correspond to the trapezoidal rule (1.9) and suppose that the above extrapolation procedure is applied to the scalar linear equation $y' = \lambda y$, $y(0) = 1$, with a constant step size h. Find a function r such that $z_{n+1} = r(h\lambda) z_n = [r(h\lambda)]^{n+1}$, $n = 0, 1, \ldots$ Is the new method A-stable?

7

Nonlinear algebraic systems

7.1 Functional iteration

From the point of view of a numerical mathematician, which we adopted in Chapters 1–5, the solution of ordinary differential equations is all about analysis – i.e. convergence, order, stability and an endless progression of theorems and proofs. The outlook of Chapter 6 parallels that of a software engineer, being concerned with the correct assembly of computational components and with choosing the step sequence dynamically. Computers, however, are engaged neither in analysis nor in algorithm design but in the real work concerned with solving ODEs, and this consists in the main of the computation of (mostly nonlinear) algebraic systems of equations.

Why not – and this is a legitimate question – use explicit methods, whether multistep or Runge–Kutta, thereby dispensing altogether with the need to calculate algebraic systems? The main reason is computational cost. This is obvious in the case of stiff equations, since, for explicit time-stepping methods, stability considerations restrict the step size to an extent that renders the scheme noncompetitive and downright ineffective. When stability questions are not at issue, it often makes very good sense to use explicit Runge–Kutta methods. The accepted wisdom is, however, that, as far as multistep methods are concerned, implicit methods should be used even for non-stiff equations since, as we will see in this chapter, the solution of the underlying algebraic systems can be approximated with relative ease.

Let us suppose that we wish to advance the (implicit) multistep method (2.8) by a single step. This entails solving the algebraic system

$$\boldsymbol{y}_{n+s} = hb_s\boldsymbol{f}(t_{n+s}, \boldsymbol{y}_{n+s}) + \boldsymbol{\gamma}, \tag{7.1}$$

where the vector

$$\boldsymbol{\gamma} = h\sum_{m=0}^{s-1} b_m\boldsymbol{f}(t_{n+m}, \boldsymbol{y}_{n+m}) - \sum_{m=0}^{s-1} a_m\boldsymbol{y}_{n+m}$$

is known. As far as implicit Runge–Kutta methods (3.9) are concerned, we need to solve at each step the system

$$\begin{aligned}
\boldsymbol{\xi}_1 &= \boldsymbol{y}_n + h[a_{1,1}\boldsymbol{f}(t_n + c_1h, \boldsymbol{\xi}_1) + a_{1,2}\boldsymbol{f}(t_n + c_2h, \boldsymbol{\xi}_2) + \cdots + a_{1,\nu}\boldsymbol{f}(t_n + c_\nu h, \boldsymbol{\xi}_\nu)],\\
\boldsymbol{\xi}_2 &= \boldsymbol{y}_n + h[a_{2,1}\boldsymbol{f}(t_n + c_1h, \boldsymbol{\xi}_1) + a_{2,2}\boldsymbol{f}(t_n + c_2h, \boldsymbol{\xi}_2) + \cdots + a_{2,\nu}\boldsymbol{f}(t_n + c_\nu h, \boldsymbol{\xi}_\nu)],\\
&\ \ \vdots\\
\boldsymbol{\xi}_\nu &= \boldsymbol{y}_n + h[a_{\nu,1}\boldsymbol{f}(t_n + c_1h, \boldsymbol{\xi}_1) + a_{\nu,2}\boldsymbol{f}(t_n + c_2h, \boldsymbol{\xi}_2) + \cdots + a_{\nu,\nu}\boldsymbol{f}(t_n + c_\nu h, \boldsymbol{\xi}_\nu)].
\end{aligned}$$

This system looks considerably more complicated than (7.1). Both, however, can be cast into a standard form, namely

$$\boldsymbol{w} = h\boldsymbol{g}(\boldsymbol{w}) + \boldsymbol{\beta}, \qquad \boldsymbol{w} \in \mathbb{R}^{\tilde{d}}, \tag{7.2}$$

where the function $\boldsymbol{g}$ and the vector $\boldsymbol{\beta}$ are known. Obviously, for the multistep method (7.1) we have $\boldsymbol{g}(\,\cdot\,) = b_m \boldsymbol{f}(t_{n+s}, \,\cdot\,)$, $\boldsymbol{\beta} = \boldsymbol{\gamma}$ and $\tilde{d} = d$. The solution of (7.2) then becomes $\boldsymbol{y}_{n+s}$. The notation is slightly more complicated for Runge–Kutta methods, although it can be simplified a great deal by using *Kronecker products*. However, we provide an example for the case $\nu = 2$ where, mercifully, no Kronecker products are required. Thus, $\tilde{d} = 2d$ and

$$\boldsymbol{g}(\boldsymbol{w}) = \begin{bmatrix} a_{1,1}\boldsymbol{f}(t_n + c_1 h, \boldsymbol{w}_1) + a_{1,2}\boldsymbol{f}(t_n + c_2 h, \boldsymbol{w}_2) \\ a_{2,1}\boldsymbol{f}(t_n + c_1 h, \boldsymbol{w}_1) + a_{2,2}\boldsymbol{f}(t_n + c_2 h, \boldsymbol{w}_2) \end{bmatrix}, \qquad \text{where} \qquad \boldsymbol{w} = \begin{bmatrix} \boldsymbol{w}_1 \\ \boldsymbol{w}_2 \end{bmatrix}.$$

Moreover,

$$\boldsymbol{\beta} = \begin{bmatrix} \boldsymbol{y}_n \\ \boldsymbol{y}_n \end{bmatrix}.$$

Provided that the solution of (7.2) is known, we then set $\boldsymbol{\xi}_j = \boldsymbol{w}_j$, $j = 1, 2$, hence

$$\boldsymbol{y}_{n+1} = \boldsymbol{y}_n + h[b_1 \boldsymbol{f}(t_n + c_1 h, \boldsymbol{w}_1) + b_2 \boldsymbol{f}(t_n + c_2 h, \boldsymbol{w}_2)].$$

Let us assume that $\boldsymbol{g}$ is nonlinear (if $\boldsymbol{g}$ were linear and the number of equations moderate, we could solve (7.2) by familiar *Gaussian elimination;* the numerical solution of large linear systems is discussed in Chapters 11–15). Our intention is to solve the algebraic system by iteration;[1] in other words, we need to make an *initial guess* $\boldsymbol{w}^{[0]}$ and provide an algorithm

$$\boldsymbol{w}^{[i+1]} = \boldsymbol{s}(\boldsymbol{w}^{[i]}), \qquad i = 0, 1, \ldots, \tag{7.3}$$

such that

(1) $\boldsymbol{w}^{[i]} \to \hat{\boldsymbol{w}}$, the solution of (7.2);

(2) the cost of each step (7.3) is small; and

(3) the progression to the limit is rapid.

The form (7.2) emphasizes two important aspects of this iterative procedure. Firstly, the vector $\boldsymbol{\beta}$ is known at the outset and there is no need to recalculate it in every iteration (7.3). Secondly, the step size h is an important parameter and its magnitude is likely to determine central characteristics of the system (7.2); since the exact solution is obvious when $h = 0$, clearly the problem is likely to be easier for small $h > 0$. Moreover, although in principle (7.2) may possess many solutions, it follows at once from the implicit function theorem that, provided $\boldsymbol{g}$ is continuously differentiable,

[1] The reader will notice that two distinct iterative procedures are taking place: time-stepping, i.e. $\boldsymbol{y}_{n+s-1} \mapsto \boldsymbol{y}_{n+s}$, and 'inner' iteration, $\boldsymbol{w}^{[i]} \mapsto \boldsymbol{w}^{[i+1]}$. To prevent confusion, we reserve the phrase 'iteration' for the latter.

nonsingularity of the Jacobian matrix $I - h\partial \boldsymbol{g}(\boldsymbol{\beta})/\partial \boldsymbol{w}$ for $h \to 0$ implies the existence of a unique solution for sufficiently small $h > 0$.

The most elementary approach to the solution of (7.2) is the *functional iteration* $\boldsymbol{s}(\boldsymbol{w}) = h\boldsymbol{g}(\boldsymbol{w}) + \boldsymbol{\beta}$, which, using (7.3) can be expressed as

$$\boldsymbol{w}^{[i+1]} = h\boldsymbol{g}(\boldsymbol{w}^{[i]}) + \boldsymbol{\beta}, \qquad i = 0, 1, \ldots \tag{7.4}$$

Much beautiful mathematics has been produced in the last few decades in connection with functional iteration, concerned mainly with the fractal nature of basins of attraction in the complex case. For practical purposes, however, we resort to the tried and trusted *Banach fixed-point theorem,* which will now be stated and proved in a formalism appropriate for the recursion (7.4).

Given a vector norm $\|\cdot\|$ and $\boldsymbol{w} \in \mathbb{R}^{\tilde{d}}$, we denote by $\mathcal{B}_\rho(\boldsymbol{w})$ the closed ball of radius $\rho > 0$ centred at $\boldsymbol{w}$:

$$\mathcal{B}_\rho(\boldsymbol{w}) = \left\{\boldsymbol{u} \in \mathbb{R}^{\tilde{d}} \,:\, \|\boldsymbol{u} - \boldsymbol{w}\| \le \rho\right\}.$$

Theorem 7.1 *Let $h > 0$, $\boldsymbol{w}^{[0]} \in \mathbb{R}^{\tilde{d}}$, and suppose that there exist numbers $\lambda \in (0,1)$ and $\rho > 0$ such that*

(i) $\|\boldsymbol{g}(\boldsymbol{v}) - \boldsymbol{g}(\boldsymbol{u})\| \le \dfrac{\lambda}{h}\|\boldsymbol{v} - \boldsymbol{u}\|$ *for every* $\boldsymbol{v}, \boldsymbol{u} \in \mathcal{B}_\rho(\boldsymbol{w}^{[0]})$;

(ii) $\boldsymbol{w}^{[1]} \in \mathcal{B}_{(1-\lambda)\rho}(\boldsymbol{w}^{[0]})$.

Then

(a) $\boldsymbol{w}^{[i]} \in \mathcal{B}_\rho(\boldsymbol{w}^{[0]})$ *for every* $i = 0, 1, \ldots$;

(b) $\hat{\boldsymbol{w}} := \lim_{i\to\infty} \boldsymbol{w}^{[i]}$ *exists, obeys equation* (7.2) *and* $\hat{\boldsymbol{w}} \in \mathcal{B}_\rho(\boldsymbol{w}^{[0]})$;

(c) *no other point in* $\mathcal{B}_\rho(\boldsymbol{w}^{[0]})$ *is a solution of* (7.2).

Proof We commence by using induction to prove that

$$\|\boldsymbol{w}^{[i+1]} - \boldsymbol{w}^{[i]}\| \le \lambda^i(1-\lambda)\rho \tag{7.5}$$

and that $\boldsymbol{w}^{[i+1]} \in \mathcal{B}_\rho(\boldsymbol{w}^{[0]})$ for all $i = 0, 1, \ldots$

Part (a) is certainly true for $i = 0$ because of condition (ii) and the definition of $\mathcal{B}_\rho(\boldsymbol{w}^{[0]})$. Now, let us assume that the statement is true for all $m = 0, 1, \ldots, i-1$. Then, by (7.2) and assumption (i),

$$\begin{aligned}\|\boldsymbol{w}^{[i+1]} - \boldsymbol{w}^{[i]}\| &= \|[h\boldsymbol{g}(\boldsymbol{w}^{[i]}) + \boldsymbol{\beta}] - [h\boldsymbol{g}(\boldsymbol{w}^{[i-1]}) + \boldsymbol{\beta}]\| \\ &= h\|\boldsymbol{g}(\boldsymbol{w}^{[i]}) - \boldsymbol{g}(\boldsymbol{w}^{[i-1]})\| \le \lambda\|\boldsymbol{w}^{[i]} - \boldsymbol{w}^{[i-1]}\|, \qquad i = 1, 2, \ldots\end{aligned}$$

This carries forward the induction for (7.5) from $i - 1$ to i.

The following sum,

$$\boldsymbol{w}^{[i+1]} - \boldsymbol{w}^{[0]} = \sum_{j=0}^{i} (\boldsymbol{w}^{[j+1]} - \boldsymbol{w}^{[j]}), \qquad i = 0, 1, \ldots,$$

telescopes; therefore, by the triangle inequality (A.1.3.3)

$$\|\boldsymbol{w}^{[i+1]} - \boldsymbol{w}^{[0]}\| = \left\| \sum_{j=0}^{i} (\boldsymbol{w}^{[j+1]} - \boldsymbol{w}^{[j]}) \right\| \le \sum_{j=0}^{i} \|\boldsymbol{w}^{[j+1]} - \boldsymbol{w}^{[j]}\|, \qquad i = 0, 1, \ldots$$

Exploiting (7.5) and summing the geometric series, we thus conclude that

$$\|\boldsymbol{w}^{[i+1]} - \boldsymbol{w}^{[0]}\| \le \sum_{j=0}^{i} \lambda^i (1-\lambda)\rho = (1-\lambda^{i+1})\rho \le \rho, \qquad i = 0, 1, \ldots$$

Therefore $\boldsymbol{w}^{[i+1]} \in \mathcal{B}_\rho(\boldsymbol{w}^{[0]})$. This completes the inductive proof and we deduce that (a) is true.

Again telescoping series, the triangle inequality and (7.5) can be used to argue that

$$\begin{aligned}\|\boldsymbol{w}^{[i+k]} - \boldsymbol{w}^{[i]}\| &= \left\| \sum_{j=0}^{k-1} (\boldsymbol{w}^{[i+j+1]} - \boldsymbol{w}^{[i+j]}) \right\| \le \sum_{j=0}^{k-1} \|\boldsymbol{w}^{[i+j+1]} - \boldsymbol{w}^{[i+j]}\| \\ &\le \sum_{j=0}^{k-1} \lambda^{i+j}(1-\lambda)\rho = \lambda^i(1-\lambda^k)\rho, \qquad i = 0, 1, \ldots, \quad k = 1, 2, \ldots\end{aligned}$$

Therefore, $\lambda \in (0,1)$ implies that for every $i = 0, 1, \ldots$ and $\varepsilon > 0$ we may choose k large enough that

$$\|\boldsymbol{w}^{[i+k]} - \boldsymbol{w}^{[i]}\| < \varepsilon.$$

In other words, $\{\boldsymbol{w}^{[i]}\}_{i=0,1,\ldots}$ is a *Cauchy sequence.*

The set $\mathcal{B}_\rho(\boldsymbol{w}^{[0]})$ being compact (i.e., closed and bounded), the Cauchy sequence $\{\boldsymbol{w}^{[i]}\}_{i=0,1,\ldots}$ converges to a limit within the set. This proves the existence of $\hat{\boldsymbol{w}} \in \mathcal{B}_\rho(\boldsymbol{w}^{[0]})$.

Finally, let us suppose that there exists $\boldsymbol{w}^\star \in \mathcal{B}_\rho(\boldsymbol{w}^{[0]})$, $\boldsymbol{w}^\star \ne \hat{\boldsymbol{w}}$, such that $\boldsymbol{w}^\star = h\boldsymbol{g}(\boldsymbol{w}^\star) + \boldsymbol{\beta}$. Then $\|\boldsymbol{w}^\star - \hat{\boldsymbol{w}}\| > 0$ implies that

$$\begin{aligned}\|\boldsymbol{w}^\star - \hat{\boldsymbol{w}}\| &= \|[h\boldsymbol{g}(\boldsymbol{w}^\star) + \boldsymbol{\beta}] - [h\boldsymbol{g}(\hat{\boldsymbol{w}}) + \boldsymbol{\beta}]\| = h\|\boldsymbol{g}(\boldsymbol{w}^\star) - \boldsymbol{g}(\hat{\boldsymbol{w}})\| \\ &\le \lambda\|\boldsymbol{w}^\star - \hat{\boldsymbol{w}}\| < \|\boldsymbol{w}^\star - \hat{\boldsymbol{w}}\|.\end{aligned}$$

This is impossible and we deduce that the fixed point $\hat{\boldsymbol{w}}$ is unique in $\mathcal{B}_\rho(\boldsymbol{w}^{[0]})$, thereby concluding the proof of the theorem. ■

If $\boldsymbol{g}$ is smoothly differentiable then, by the mean value theorem, for every $\boldsymbol{v}$ and $\boldsymbol{u}$ there exists $\tau \in (0,1)$ such that

$$\boldsymbol{g}(\boldsymbol{v}) - \boldsymbol{g}(\boldsymbol{u}) = \frac{\partial \boldsymbol{g}(\tau\boldsymbol{v} + (1-\tau)\boldsymbol{u})}{\partial \boldsymbol{w}}(\boldsymbol{v} - \boldsymbol{u}).$$

Therefore, assumption (i) of Theorem 7.1 is nothing other than a statement on the magnitude of the step size h in relation to $\|\partial \boldsymbol{g}/\partial \boldsymbol{w}\|$. In particular, if (7.2) originates in a multistep method then we need, in effect, $h|b_s| \cdot \|\partial \boldsymbol{f}(t_{n+s}, \boldsymbol{y}_{n+s})/\partial \boldsymbol{y}\| < 1$. (A similar inequality applies to Runge–Kutta methods.) The meaning of this restriction is phenomenologically similar to stiffness, as can be seen in the following example.

◇ **The trapezoidal rule and functional iteration** The iterative scheme (7.4), as applied to the trapezoidal rule (1.9), reads

$$\boldsymbol{w}^{[i+1]} = \tfrac{1}{2}h\boldsymbol{f}(t_{n+1}, \boldsymbol{w}^{[i]}) + \left[\boldsymbol{y}_n + \tfrac{1}{2}h\boldsymbol{f}(t_n, \boldsymbol{y}_n)\right], \qquad i = 0, 1, \ldots$$

Let us suppose that the underlying ODE is linear, i.e. of the form $\boldsymbol{y}' = \Lambda \boldsymbol{y}$, where Λ is symmetric. As long as we are employing the Euclidean norm, it is true that $\|\Lambda\| = \rho(\Lambda)$, the *spectral radius* of Λ (A.1.5.2). The outcome is the restriction $h\rho(\Lambda) < 2$, which imposes similar constraints on a stable implementation of the *Euler method* (1.4). Provided $\rho(\Lambda)$ is small and stiffness is not an issue, this makes little difference. However, to retain the A-stability of the trapezoidal rule for large $\rho(\Lambda)$ we must restrict $h > 0$ so drastically that all the benefits of A-stability are lost – we might just as well have used Adams–Bashforth, say, in the first place! ◇

We conclude that a useful rule of a thumb is that we may use the functional iteration (7.4) for non-stiff problems but we need an alternative when stiffness becomes an issue.

7.2 The Newton–Raphson algorithm and its modification

Let us suppose that the function $\boldsymbol{g}$ is twice continuously differentiable. We expand (7.2) about a vector $\boldsymbol{w}^{[i]}$:

$$\begin{aligned}\boldsymbol{w} &= \boldsymbol{\beta} + h\boldsymbol{g}(\boldsymbol{w}^{[i]} + (\boldsymbol{w} - \boldsymbol{w}^{[i]})) \\ &= \boldsymbol{\beta} + h\boldsymbol{g}(\boldsymbol{w}^{[i]}) + h\frac{\partial \boldsymbol{g}(\boldsymbol{w}^{[0]})}{\partial \boldsymbol{w}}(\boldsymbol{w} - \boldsymbol{w}^{[i]}) + \mathcal{O}\left(\|\boldsymbol{w} - \boldsymbol{w}^{[i]}\|^2\right).\end{aligned} \tag{7.6}$$

Disregarding the $\mathcal{O}(\|\boldsymbol{w} - \boldsymbol{w}^{[i]}\|^2)$ term, we solve (7.6) for $\boldsymbol{w} - \boldsymbol{w}^{[i]}$. The outcome,

$$\left[I - h\frac{\partial \boldsymbol{g}(\boldsymbol{w}^{[i]})}{\partial \boldsymbol{w}}\right](\boldsymbol{w} - \boldsymbol{w}^{[i]}) \approx \boldsymbol{\beta} + h\boldsymbol{g}(\boldsymbol{w}^{[i]}) - \boldsymbol{w}^{[i]},$$

suggests the iterative scheme

$$\boldsymbol{w}^{[i+1]} = \boldsymbol{w}^{[i]} - \left[I - h\frac{\partial \boldsymbol{g}(\boldsymbol{w}^{[i]})}{\partial \boldsymbol{w}}\right]^{-1}\left[\boldsymbol{w}^{[i]} - \boldsymbol{\beta} - h\boldsymbol{g}(\boldsymbol{w}^{[i]})\right], \qquad i = 0, 1, \ldots \tag{7.7}$$

This is (under a mild disguise) the celebrated *Newton–Raphson* algorithm.

The Newton–Raphson method has motivated several profound theories and attracted the attention of some of the towering mathematical minds of the twentieth century – Leonid Kantorowitz and Stephen Smale, to mention just two. We do not propose in this volume to delve into this issue, whose interest is tangential to our main theme. Instead, and without further ado, we merely comment on several features of the iterative scheme (7.7).

Firstly, as long as $h > 0$ is sufficiently small the rate of convergence of the Newton–Raphson algorithm is quadratic: it is possible to prove that there exists a constant $c > 0$ such that, for sufficiently large i,

$$\|\boldsymbol{w}^{[i+1]} - \hat{\boldsymbol{w}}\| \le c\|\boldsymbol{w}^{[i]} - \hat{\boldsymbol{w}}\|^2,$$

where $\hat{\boldsymbol{w}}$ is a solution of (7.2). This is already implicit in the fact that we have neglected an $\mathcal{O}(\|\boldsymbol{w} - \boldsymbol{w}^{[i]}\|^2)$ term in (7.6). It is important to comment that the 'sufficient smallness' of $h > 0$ is of a different order of magnitude to the minute values of $h > 0$ that are required when the functional iteration (7.4) is applied to stiff problems. It is easy to prove, for example, that (7.7) terminates in a single step when $\boldsymbol{g}$ is a linear function, regardless of any underlying stiffness (see Exercise 7.2).

Secondly, an implementation of (7.7) requires computation of the Jacobian matrix at every iteration. This is a formidable ordeal since, for a d-dimensional system, the Jacobian matrix has d^2 entries and its computation – even if all requisite formulae are available in an explicit form – is expensive.

Finally, each iteration requires the solution of a linear system of algebraic equations. It is highly unusual for such a system to be singular or ill conditioned (i.e., 'close' to singular) in a realistic computation, regardless of stiffness; the reasons, in the (simpler) case of multistep methods, are that $b_s > 0$ for all methods with reasonably large linear stability domains, the eigenvalues of $\partial \boldsymbol{f}/\partial \boldsymbol{y}$ reside in $\mathbb{C}^-$ and it is easy to prove that all the eigenvalues of the matrix in (7.7) are bounded away from zero. However, the solution of even a well-conditioned nonsingular algebraic system is a nontrivial and potentially costly task.

Both shortcomings of Newton–Raphson – the computation of the Jacobian matrix and the need to solve linear systems in each iteration – can be alleviated by using the *modified Newton–Raphson* instead. The *quid pro quo,* however, is a significant slowing-down of the convergence rate. Before we introduce the modification of (7.7), let us comment briefly on an important special case when the 'full' Newton–Raphson can (and should) be used.

A significant proportion of stiff ODEs originate in the *semi-discretization* of parabolic partial differential equations by finite difference methods (Chapter 16). In such cases the Newton–Raphson method (7.7) is very effective indeed, since the Jacobian matrix is sparse (an overwhelming majority of its elements vanish): it has just $\mathcal{O}(d)$ nonzero components and usually can be computed with relative ease. Moreover, most methods for the solution of sparse algebraic systems confer no advantage for the special form (7.9) of the modified equations, an exception being the direct factorization algorithms of Chapter 11.

◇ **The reaction–diffusion equation** A quasilinear parabolic partial differential equation with many applications in mathematical biology, chemistry and physics is the *reaction–diffusion* equation

$$\frac{\partial u}{\partial t} = \frac{\partial^2 u}{\partial x^2} + \varphi(u), \qquad 0 < x < 1, \quad t \geq 0, \tag{7.8}$$

where $u = u(x,t)$. It is given with the initial condition $u(x,0) = u_0(x)$, $0 < x < 1$, and (for simplicity) zero Dirichlet boundary conditions $u(0,t), u(1,t) \equiv 0$, $t \geq 0$.
Among the many applications of (7.8) we single out two for special mention. The choice $\varphi(u) = cu$, where $c > 0$, models the neutron density in an atom bomb (subject to the assumption that the latter is in the form of a thin uranium rod of unit length), whereas $\varphi(u) = \alpha u + \beta u^2$ (the *Fisher equation*) is used in population dynamics: the terms αu and βu^2 correspond respectively to the reproduction and interaction of a species while $\partial^2 u/\partial x^2$ models its diffusion in the underlying habitat.
A standard semi-discretization (that is, an approximation of a partial differential equation by an ODE system, see Chapter 16) of (7.8) is

$$y_k' = \frac{1}{(\Delta x)^2}(y_{k-1} - 2y_k + y_{k+1}) + \varphi(y_k), \qquad k = 1,2,\ldots,d, \qquad t \geq 0,$$

where $\Delta x = 1/(d+1)$ and $y_0, y_{d+1} \equiv 0$. Suppose that φ' is easily available, e.g. that φ is a polynomial. The Jacobian matrix

$$\left(\frac{\partial \boldsymbol{f}(t,\boldsymbol{y})}{\partial \boldsymbol{y}}\right)_{k,\ell} = \begin{cases} -\dfrac{2}{(\Delta x)^2} + \varphi'(y_k), & k = \ell, \\ \dfrac{1}{(\Delta x)^2}, & |k - \ell| = 1, \\ 0 & \text{otherwise}, \end{cases} \qquad k,\ell = 1,2,\ldots,d,$$

is fairly easy to evaluate and store. Moreover, as will be apparent in the forthcoming discussion in Chapter 11, the solution of algebraic linear systems with tridiagonal matrices is very easy and fast. We conclude that in this case there is no need to trade off the superior speed of Newton–Raphson for 'easier' alternatives. ◇

Unfortunately, most stiff systems do not share the features of the above example and for these we need to modify the Newton–Raphson iteration. This modification takes the form of a replacement of the matrix $\partial \boldsymbol{g}(\boldsymbol{w}^{[i]})/\partial \boldsymbol{w}$ by another matrix, J, say, that does not vary with i. A typical choice might be

$$J = \frac{\partial \boldsymbol{g}(\boldsymbol{w}^{[0]})}{\partial \boldsymbol{w}},$$

but it is not unusual, in fact, to retain the same matrix J for a number of time steps. In place of (7.7) we thus have

$$\boldsymbol{w}^{[i+1]} = \boldsymbol{w}^{[i]} - (I - hJ)^{-1}\left[\boldsymbol{w}^{[i]} - \boldsymbol{\beta} - h\boldsymbol{g}(\boldsymbol{w}^{[i]})\right], \qquad i = 0,1,\ldots. \tag{7.9}$$

This *modified Newton–Raphson* scheme (7.9) confers two immediate advantages. The first is obvious: we need to calculate J only once per step (or perhaps per several steps). The second is realized when the underlying linear algebraic system is solved by Gaussian elimination – the method of choice for small or moderate d. In its LU formulation (A.1.4.5), Gaussian elimination for the linear system $A\boldsymbol{x} = \boldsymbol{b}$, where $\boldsymbol{b} \in \mathbb{R}^d$, consists of two stages. Firstly, the matrix A is factorized in the form LU, where L and U are lower triangular and upper triangular matrices respectively. Secondly, we solve $L\boldsymbol{z} = \boldsymbol{b}$, followed by $U\boldsymbol{x} = \boldsymbol{z}$. While factorization entails $\mathcal{O}(d^3)$ operations (for non-sparse matrices), the solution of two triangular $d \times d$ systems requires just $\mathcal{O}(d^2)$ operations and is considerably cheaper. In the case of the iterative scheme (7.9) it is enough to factorize $A = I - hJ$ just once per time step (or once per re-evaluation of J and/or per change in h). Therefore, the cost of each single iteration goes down by an order of magnitude, as compared with the original Newton–Raphson scheme!

Of course, quadratic convergence is lost. As a matter of fact, modified Newton–Raphson is simply functional iteration except that instead of $h\boldsymbol{g}(\boldsymbol{w}) + \boldsymbol{\beta}$ we iterate the new function $h\tilde{\boldsymbol{g}}(\boldsymbol{w}) + \tilde{\boldsymbol{\beta}}$, where

$$\tilde{\boldsymbol{g}}(\boldsymbol{w}) := (I - hJ)^{-1}[\boldsymbol{g}(\boldsymbol{w}) - J\boldsymbol{w}], \qquad \tilde{\boldsymbol{\beta}} := (I - hJ)^{-1}\boldsymbol{\beta}. \tag{7.10}$$

The proof is left to the reader in Exercise 7.3.

There is nothing to stop us from using Theorem 7.1 to explore the convergence of (7.9). It follows from (7.10) that

$$\tilde{\boldsymbol{g}}(\boldsymbol{v}) - \tilde{\boldsymbol{g}}(\boldsymbol{u}) = (I - hJ)^{-1}\{[\boldsymbol{g}(\boldsymbol{v}) - \boldsymbol{g}(\boldsymbol{u})] - J(\boldsymbol{v} - \boldsymbol{u})\}. \tag{7.11}$$

Recall, however, that, subject to the sufficient smoothness of $\boldsymbol{g}$, there exists a point $\boldsymbol{z}$ on the line segment joining $\boldsymbol{v}$ and $\boldsymbol{u}$ such that

$$\boldsymbol{g}(\boldsymbol{v}) - \boldsymbol{g}(\boldsymbol{u}) = \frac{\partial \boldsymbol{g}(\boldsymbol{z})}{\partial \boldsymbol{w}}(\boldsymbol{v} - \boldsymbol{u}).$$

Given that we have chosen

$$J = \frac{\partial \boldsymbol{g}(\tilde{\boldsymbol{w}})}{\partial \boldsymbol{w}}$$

(for example, $\tilde{\boldsymbol{w}} = \boldsymbol{w}^{[0]}$), it follows from (7.11) that

$$\frac{\|\tilde{\boldsymbol{g}}(\boldsymbol{v}) - \tilde{\boldsymbol{g}}(\boldsymbol{u})\|}{\|\boldsymbol{v} - \boldsymbol{u}\|} \le \left\| \left[I - h\frac{\partial \boldsymbol{g}(\tilde{\boldsymbol{w}})}{\partial \boldsymbol{w}} \right]^{-1} \right\| \times \left\| \frac{\partial \boldsymbol{g}(\boldsymbol{z})}{\partial \boldsymbol{w}} - \frac{\partial \boldsymbol{g}(\tilde{\boldsymbol{w}})}{\partial \boldsymbol{w}} \right\|.$$

Unless the Jacobian matrix varies very considerably as a function of t, the second term on the right is likely to be small. Moreover, if all the eigenvalues of J are in $\mathbb{C}^-$ then $\|(I - hJ)^{-1}\|$ is likely also to be small; stiffness is likely to help, not hinder, this estimate! Therefore, it is possible in general to satisfy assumption (i) of Theorem 7.1 for large $\rho > 0$.

7.3 Starting and stopping the iteration

Theorem 7.1 quantifies an important point that is equally valid for every iterative method for nonlinear algebraic equations (7.2), not just for the functional iteration

(7.4) and the modified Newton–Raphson method (7.9): good performance hinges to a large extent on the quality of the starting value $\boldsymbol{w}^{[0]}$.

Provided that $\boldsymbol{g}$ is Lipschitz, condition (i) is always valid for a given $h > 0$ and sufficiently small $\rho > 0$. However, small ρ means that, to be consistent with condition (ii), we must choose an exceedingly good starting condition $\boldsymbol{w}^{[0]}$. Viewed in this light, the main purpose in replacing (7.4) by (7.9) is to allow convergence from imperfect starting values.

Even if the choice of an iterative scheme provides for a large *basin of attraction* of $\hat{\boldsymbol{w}}$ (the set of all $\boldsymbol{w}^{[0]} \in \mathbb{R}^{\tilde{d}}$ for which the iteration converges to $\hat{\boldsymbol{w}}$), it is important to commence the iteration with a good initial guess. This is true for every nonlinear algebraic system but our problem here is special – it originates in the use of a time-stepping method for ODEs. This is an important advantage.

Supposing for example that the underlying ODE method is a pth-order multistep scheme, let us recall the meaning of the solution of (7.2), namely $\boldsymbol{y}_{n+s}$. To paraphrase the last paragraph, it is an excellent policy to seek a starting condition $\boldsymbol{w}^{[0]}$ near to the vector $\boldsymbol{y}_{n+s}$. The latter is, of course, unknown, but we can obtain a good guess by using a different, *explicit,* multistep method of order p. This multistep method, called the *predictor,* provides the platform upon which the iterative scheme seeks the solution of the implicit *corrector.*

It is not enough to start an iterative procedure; we must also provide a stopping criterion, which terminates the iterative process. This might appear as a relatively straightforward task: iterate until $\|\boldsymbol{w}^{[i+1]} - \boldsymbol{w}^{[i]}\| < \varepsilon$ for a given threshold value ε (distinct from, and probably significantly smaller than, the tolerance δ that we employ in error control). However, this approach – perfectly sensible for the solution of general nonlinear algebraic systems – misses an important point: the origin of (7.2) is in a time-stepping computational method for ODEs, implemented with step size h. If convergence is slow we have two options, either to carry on iterating or to stop the procedure, abandon the current step size and commence time-stepping with a smaller value of $h > 0$. In other words, there is nothing to prevent us from using the step size both to control the error and to ensure rapid convergence of the iterative scheme.

The traditional attitude to iterative procedures, namely to proceed with perhaps thousands of iterations until convergence takes place (to a given threshold) is completely inadequate. Unless the process converges in a relatively small number of iterations – perhaps ten, perhaps fewer – the best course of action is to stop, decrease the step size and recommence time-stepping. However, this does not exhaust the range of all possible choices. Let us remember that the goal is not to solve a nonlinear algebraic system *per se* but to compute a solution of an ODE system to a given tolerance. We thus have two options.

Firstly, we can iterate for $i = 0, 1, \ldots, i_{\text{end}}$, where $i_{\text{end}} = 10$, say. After each iteration we check for convergence. Unless it is attained (within the threshold ε) we decide that h is too large and abandon the current step. This is called *iteration to convergence.*

The second option is to identify the predictor–corrector pair with the two methods (6.6) and (6.5) that were used in Chapter 6 to control the error by means of the *Milne device.* We perform just a single iteration of the corrector and substitute $\boldsymbol{w}^{[1]}$, instead of $\boldsymbol{y}_{n+s}$, into the error estimate (6.8). If $\kappa \leq h\delta$ then all is well and we let $\boldsymbol{y}_{n+s} = \boldsymbol{w}^{[1]}$. Otherwise we abandon the step. Note that we are accepting a value of $\boldsymbol{y}_{n+s}$ that solves neither the nonlinear algebraic equation nor, as a matter of fact, the implicit multistep method. This, however, is of no consequence since our $\boldsymbol{y}_{n+s}$ passes the error test – and that is all that matters! This approach is called the *PECE iteration*.[2]

The choice between PECE iteration and iteration to convergence hinges upon the relative cost of performing a single iteration and changing the step size. If the cost of changing h is negligible, we might just as well abandon the iteration unless $\boldsymbol{w}^{[1]}$, or perhaps $\boldsymbol{w}^{[2]}$ (in which case we have a PE(CE)^2 procedure), satisfies the error criterion. If, though, this cost is large we should carry on with the iteration considerably longer. Another consideration is that a PECE iteration is likely to cause severe contraction of the linear stability domain of the corrector. In particular, no such procedure can be A-stable[3] (see Exercise 7.4).

We recall the dichotomy between stiff and non-stiff ODEs. If the ODE is non-stiff then we are likely to employ the functional iteration (7.2), which costs nothing to restart. The only cost of changing h is in remeshing, which, although difficult to program, carries a very modest computational price tag. Since shrinkage of the linear stability domain is not an important issue for non-stiff ODEs, the clear conclusion is that the PECE approach is superior in this case. Moreover, if the equation is stiff then we should use the modified Newton–Raphson iteration (7.9). In order to change the step size, we need to redo the LU factorization of $I - hJ$ (since h has changed). Moreover, it is a good policy to re-evaluate J as well, unless it has already been computed at $\boldsymbol{y}_{n+s-1}$: it might well be that the failure of the iterative procedure follows from poor approximation of the Jacobian matrix. Finally, stability is definitely a crucial consideration and we should be unwilling to reconcile ourselves to a collapse in the size of the linear stability domain. All these reasons mean that the right approach is to iterate to convergence.

Comments and bibliography

The computation of nonlinear algebraic systems is as old as numerical analysis itself. This is not necessarily an advantage, since the theory has developed in many directions which, mostly, are irrelevant to the theme of this chapter.

The basic problem admits several equivalent formulations: firstly, we may regard it as finding a zero of the equation $\boldsymbol{h}_1(\boldsymbol{x}) = \boldsymbol{0}$; secondly, as computing a fixed point of the system $\boldsymbol{x} = \boldsymbol{h}_2(\boldsymbol{x})$, where $\boldsymbol{h}_2 = \boldsymbol{x} + \alpha\boldsymbol{h}_1(\boldsymbol{x})$ for some $\alpha \in \mathbb{R} \setminus \{0\}$; thirdly, as minimizing $\|\boldsymbol{h}_1(\boldsymbol{x})\|$. (With regard to the third formulation, minimization is often equivalent to the solution of a nonlinear system: provided the function $\psi : \mathbb{R}^d \to \mathbb{R}$ is continuously differentiable, the problem of finding the stationary values of ψ is equivalent to solving the system $\boldsymbol{\nabla}\psi(\boldsymbol{x}) = \boldsymbol{0}$.) Therefore, in a typical library nonlinear algebraic systems and their numerical analysis appear

[2]Predict, Evaluate, Correct, Evaluate.

[3]in a formal sense. A-stability is defined only for constant steps, whereas the whole *raison d'être* of the PECE iteration is that it is operated within a variable-step procedure. However, experience tells us that the damage to the quality of the solution in an unstable situation is genuine in a variable-step setting also.

under several headings, probably on different shelves. Good sources are Ortega & Rheinboldt (1970) and Fletcher (1987) – the latter has a pronounced optimization flavour.

Modern numerical practice has moved a long way from the old days of functional iteration and the Newton–Raphson method and its modifications. The powerful algorithms of today owe much to tremendous advances in numerical optimization, as well as to the recent realization that certain acceleration schemes for linear systems can be applied with telling effect to nonlinear problems (for example, the method of conjugate gradients, the theme of Chapter 14). However, it appears that, as far as the choice of nonlinear algebraic algorithms for practical implementation of ODE algorithms is concerned, not much has happened in the last three decades. The texts of Gear (1971) and of Shampine & Gordon (1975) represent, to a large extent, the state of the art today.

This conservatism is not necessarily a bad thing. After all, the test of the pudding is in the eating and, as far as we are aware, functional iteration (7.4) and modified Newton–Raphson (7.9), applied correctly and to the right problems, discharge their duty very well indeed. We cannot emphasize enough that the task in hand is not simply to solve an arbitrary nonlinear algebraic system but to compute a problem that arises in the calculation of ODEs. This imposes a great deal of structure, highlights the crucial importance of the parameter h and, at each iteration, faces us with the question 'Should we continue to iterate or, rather, abandon the step and decrease h?'. The transplantation of modern methods for general nonlinear algebraic systems into this framework requires a great deal of work and fine-tuning. It might well be a worthwhile project, though: there are several good dissertations here, awaiting authors!

One aspect of functional iteration familiar to many readers (and to the general public, through the agency of the mass media, exhibitions and coffee-table volumes) is the fractal sets that arise when complex functions are iterated. It is only fair to mention that, behind the façade of beautiful pictures, there lies some truly beautiful mathematics: complex dynamics, automorphic forms, Teichmüller spaces ... This, however, is largely irrelevant to the task in hand. It is a constant temptation of the wanderer in the mathematical garden to stray from the path and savour the sheer beauty and excitement of landscapes strange and wonderful. Although it may be a good idea occasionally to succumb to temptation, on this occasion we virtuously stay on the straight and narrow.

Fletcher, R. (1987), *Practical Methods of Optimization* (2nd edn), Wiley, London.

Gear, C.W. (1971), *Numerical Initial Value Problems in Ordinary Differential Equations*, Prentice–Hall, Englewood Cliffs, NJ.

Ortega, J.M. and Rheinboldt, W.C. (1970), *Iterative Solution of Nonlinear Equations in Several Variables*, Academic Press, New York.

Shampine, L.F. and Gordon, M.K. (1975), *Computer Solution of Ordinary Differential Equations*, W.H. Freeman, San Francisco.

Exercises

7.1 Let $\boldsymbol{g}(\boldsymbol{w}) = \Lambda \boldsymbol{w} + \boldsymbol{a}$, where Λ is a $d \times d$ matrix.

a Prove that the inequality (i) of Theorem 7.1 is satisfied for $\lambda = h\|\Lambda\|$ and $\rho = \infty$. Deduce a condition on h that ensures that all the assumptions of the theorem are valid.

b Let $\|\cdot\|$ be the Euclidean norm (A.1.3.3). Show that the above value of λ is the best possible, in the following sense: there exist no $\rho > 0$ and $0 < \lambda < h\|\Lambda\|$ such that

$$\|\boldsymbol{g}(\boldsymbol{v}) - \boldsymbol{g}(\boldsymbol{u})\| \le \frac{\lambda}{h}\|\boldsymbol{v} - \boldsymbol{u}\|$$

for all $\boldsymbol{v}, \boldsymbol{u} \in \mathcal{B}_\rho(\boldsymbol{w}^{[0]})$. (*Hint: Recalling that*

$$\|\Lambda\| = \max\left\{\frac{\|\Lambda\boldsymbol{x}\|}{\|\boldsymbol{x}\|} : \boldsymbol{x} \in \mathbb{R}^d,\ \boldsymbol{x} \ne \boldsymbol{0}\right\},$$

prove that for every $\varepsilon > 0$ *there exists* $\boldsymbol{x}_\varepsilon$ *such that* $\|\Lambda\| = \|\Lambda\boldsymbol{x}_\varepsilon\| / \|\boldsymbol{x}_\varepsilon\|$ *and* $\|\boldsymbol{x}_\varepsilon\| = \varepsilon$. *For any* $\rho > 0$ *choose* $\boldsymbol{v} = \boldsymbol{w}^{[0]} + \boldsymbol{x}_\rho$ *and* $\boldsymbol{u} = \boldsymbol{w}^{[0]}$.)

7.2 Let $\boldsymbol{g}(\boldsymbol{w}) = \Lambda\boldsymbol{w} + \boldsymbol{a}$, where Λ is a $d \times d$ matrix.

a Prove that the Newton–Raphson method (7.7) converges (in exact arithmetic) in a single iteration.

b Suppose that $J = \Lambda$ in the modified Newton–Raphson method (7.9). Prove that also in this case just a single iteration is required.

7.3 Prove that the modified Newton–Raphson iteration (7.9) can be written as the functional iteration scheme

$$\boldsymbol{w}^{[i+1]} = h\tilde{\boldsymbol{g}}(\boldsymbol{w}^{[i]}) + \tilde{\boldsymbol{\beta}}, \qquad i = 0, 1, \ldots,$$

where $\tilde{\boldsymbol{g}}$ and $\tilde{\boldsymbol{\beta}}$ are given by (7.10).

7.4 Let the two-step Adams–Bashforth method (2.6) and the trapezoidal rule (1.9) be respectively the predictor and the corrector of a PECE scheme.

a Applying the scheme with a constant step size to the linear scalar equation $y' = \lambda y$, $y(0) = 1$, prove that

$$y_{n+1} - \left[1 + h\lambda + \tfrac{3}{4}(h\lambda)^2\right] y_n + \tfrac{1}{4}(h\lambda)^2 y_{n-1} = 0, \qquad n = 1, 2, \ldots \quad (7.12)$$

b Prove that, unlike the trapezoidal rule itself, the PECE scheme is not A-stable. (*Hint: Let* $h|\lambda| \gg 1$. *Prove that every solution of* (7.12) *is of the form* $y_n \approx c\left[\tfrac{3}{4}(h\lambda)^2\right]^n$ *for large* n.)

7.5 Consider the PECE iteration

$$\boldsymbol{x}_{n+3} = -\tfrac{1}{2}\boldsymbol{y}_n + 3\boldsymbol{y}_{n+1} - \tfrac{3}{2}\boldsymbol{y}_{n+2} + 3h\boldsymbol{f}(t_{n+2}, \boldsymbol{y}_{n+2}),$$
$$\boldsymbol{y}_{n+3} = \tfrac{1}{11}[2\boldsymbol{y}_n - 9\boldsymbol{y}_{n+1} + 18\boldsymbol{y}_{n+2} + 6h\boldsymbol{f}(t_{n+3}, \boldsymbol{x}_{n+3})].$$

a Show that both methods are third order and that the Milne device gives an estimate $\frac{6}{17}(\boldsymbol{x}_{n+3} - \boldsymbol{y}_{n+3})$ of the error of the corrector formula.

b Let the method be applied to scalar equations, let the cubic polynomial p_{n+2} interpolate y_m at $m = n, n+1, n+2$ and let $p'_{n+2}(t_{n+2}) = f(t_{n+2}, y_{n+2})$. Verify that the predictor and corrector are equivalent to the formulae

$$
\begin{aligned}
x_{n+1} &= p_{n+2}(t_{n+3}) \\
&= p_{n+2}(t_{n+2}) + hp'_{n+2}(t_{n+2}) + \tfrac{1}{2}h^2 p''_{n+2}(t_{n+2}) + \tfrac{1}{6}h^3 p'''_{n+2}(t_{n+2}), \\
y_{n+1} &= p_{n+2}(t_{n+2}) + \tfrac{5}{11}hp'_{n+2}(t_{n+2}) - \tfrac{1}{22}h^2 p''_{n+2}(t_{n+2}) - \tfrac{7}{66}p'''_{n+2}(t_{n+2}) \\
&\quad + \tfrac{6}{11}hf(t_{n+3}, x_{n+3})
\end{aligned}
$$

respectively. These formulae make it easy to change the value of h at t_{n+2} if the Milne estimate is unacceptably large.

PART II

The Poisson equation

8
Finite difference schemes

8.1 Finite differences

The opening line of *Anna Karenina*, 'All happy families resemble one another, but each unhappy family is unhappy in its own way',[1] is a useful metaphor for the computation of ordinary differential equations (ODEs) as compared with that of *partial differential equations* (PDEs). Ordinary differential equations are a happy family; perhaps they do not resemble each other but, at the very least, we can write them in a single overarching form $\boldsymbol{y}' = \boldsymbol{f}(t, \boldsymbol{y})$ and treat them by a relatively small compendium of computational techniques. (True, upon closer examination, even ODEs are not all the same: their classification into stiff and non-stiff is the most obvious example. How many happy families will survive the deconstructing attentions of a mathematician?)

Partial differential equations, however, are a huge and motley collection of problems, each unhappy in its own way. Most students of mathematics will be aware of the classification into elliptic, parabolic and hyperbolic equations, but this is only the first step in a long journey. As soon as nonlinear – or even quasilinear – PDEs are admitted for consideration, the subject is replete with an enormous number of different problems and each problem clamours for its own brand of numerics. No textbook can (or should) cover this enormous menagerie. Fortunately, however, it is possible to distil a small number of tools that allow for a well-informed numerical treatment of several important equations and form a sound basis for the understanding of the subject as a whole.

One such tool is the classical theory of *finite differences.* The main idea in the calculus of finite differences is to replace derivatives with linear combinations of discrete function values. Finite differences have the virtue of simplicity and they account for a large proportion of the numerical methods actually used in applications. This is perhaps a good place to stress that alternative approaches abound, each with its own virtue: finite elements, spectral and pseudospectral methods, boundary elements, spectral elements, particle methods, meshless methods . . . Chapter 9 is devoted to the finite element method and Chapter 10 to spectral methods.

It is convenient to introduce finite differences in the context of real (or complex) sequences $\boldsymbol{z} = \{z_k\}_{k=-\infty}^{\infty}$ indexed by all the integers. Everything can be translated to finite sequences in a straightforward manner, except that the notation becomes more cumbersome.

[1]Leo Tolstoy, *Anna Karenina,* Translated by L. & A. Maude, Oxford University Press, London (1967).

We commence by defining the following *finite difference operators,* which map the space $\mathbb{R}^{\mathbb{Z}}$ of all such sequences into itself. Each operator is defined in terms of its action on individual elements of the sequence $\boldsymbol{z}$:

the *shift* operator,	$(\mathcal{E}\boldsymbol{z})_k$	$=$	z_{k+1};
the *forward difference* operator,	$(\Delta_+\boldsymbol{z})_k$	$=$	$z_{k+1} - z_k$;
the *backward difference* operator,	$(\Delta_-\boldsymbol{z})_k$	$=$	$z_k - z_{k-1}$;
the *central difference* operator,	$(\Delta_0\boldsymbol{z})_k$	$=$	$z_{k+\frac{1}{2}} - z_{k-\frac{1}{2}}$;
the *averaging* operator,	$(\Upsilon_0\boldsymbol{z})_k$	$=$	$\frac{1}{2}(z_{k-\frac{1}{2}} + z_{k+\frac{1}{2}})$.

The first three operators are defined for all $k = 0, \pm1, \pm2, \ldots$ Note, however, that the last two operators, Δ_0 and Υ_0, do not, as a matter of fact, map $\boldsymbol{z}$ into itself. After all, the values $z_{k+1/2}$ are meaningless for integer k. Having said this, we will soon see that, appropriately used, these operators can be perfectly well defined.

Let us assume further that the sequence $\boldsymbol{z}$ originates in the sampling of a function z, say, at equispaced points. In other words, $z_k = z(kh)$ for some $h > 0$. Stipulating (for the time being) that z is an entire function, we define

the *differential* operator,	$(\mathcal{D}\boldsymbol{z})_k$	$=$	$z'(kh)$.

Our first observation is that all these operators are *linear:* given that

$$\mathcal{T} \in \{\mathcal{E},\ \Delta_+,\ \Delta_-,\ \Delta_0,\ \Upsilon_0,\ \mathcal{D}\},$$

and that $\boldsymbol{w}, \boldsymbol{z} \in \mathbb{R}^{\mathbb{Z}}$, $a, b \in \mathbb{R}$, it is true that

$$\mathcal{T}(a\boldsymbol{w} + b\boldsymbol{z}) = a\mathcal{T}\boldsymbol{w} + b\mathcal{T}\boldsymbol{z}.$$

The superposition of finite difference operators is defined in an obvious manner, e.g.

$$\Delta_+\mathcal{E}^2 z_k = \Delta_+\left(\mathcal{E}(\mathcal{E}z_k)\right) = \Delta_+(\mathcal{E}z_{k+1}) = \Delta_+ z_{k+2} = z_{k+3} - z_{k+2}.$$

Note that we have just introduced a notational shortcut: $\mathcal{T}z_k$ stands for $(\mathcal{T}\boldsymbol{z})_k$, where $\mathcal{T}$ is an arbitrary finite difference operator.

The purpose of the calculus of finite differences is, ultimately, to approximate derivatives by linear combinations of function values along a grid. We wish to get rid of $\mathcal{D}$ by expressing it in the currency of the other operators. This, however, requires us first to define formally general functions of finite difference operators.

Because of our assumption that $z_k = z(kh)$, $k = 0, \pm1, \pm2, \ldots$, finite difference operators depend upon the parameter h. Let $g(x) = \sum_{j=0}^{\infty} a_j x^j$ be an arbitrary analytic function, given in terms of its Taylor series. Noting that

$$\mathcal{E} - \mathcal{I},\ \Upsilon_0 - \mathcal{I},\ \Delta_+,\ \Delta_-,\ \Delta_0,\ h\mathcal{D} \overset{h\to 0+}{\longrightarrow} O,$$

where $\mathcal{I}$ is the identity, we can formally expand g about $\mathcal{E} - \mathcal{I}$, $\Upsilon_0 - \mathcal{I}$, Δ_+ etc. For example,

$$g(\Delta_+)\boldsymbol{z} = \left(\sum_{j=0}^{\infty} a_j \Delta_+^j\right)\boldsymbol{z} = \sum_{j=0}^{\infty} a_j(\Delta_+^j \boldsymbol{z}).$$

It is not our intention to argue here that the above expansions converge (although they do) but merely to use them in a formal manner to define functions of operators.

◇ **The operator $\mathcal{E}^{1/2}$** What is the square root of the shift operator? One interpretation, which follows directly from the definition of $\mathcal{E}$, is that $\mathcal{E}^{1/2}$ is a 'half-shift', which takes z_k to $z_{k+1/2}$; this we can define as $z((k+\frac{1}{2})h)$. An alternative expression exploits the power series expansion

$$\sqrt{1+x} = 1 + \sum_{j=1}^{\infty} \left[\frac{(-1)^{j-1}}{2^{2j-1}}\right] \frac{(2j-2)!}{(j-1)!j!} x^j$$

to argue that

$$\mathcal{E}^{1/2} = \mathcal{I} - 2\sum_{j=1}^{\infty} \frac{(2j-2)!}{(j-1)!j!} \left[-\tfrac{1}{4}(\mathcal{E}-\mathcal{I})\right]^j.$$

Needless to say, the two definitions coincide, but the proof of this would proceed at a tangent to the theme of this chapter. Readers familiar with Newton's interpolation formula might seek a proof by interpolating $z(x+\frac{1}{2})$ on the set $\{x+jh\}_{j=0}^{\ell}$ and letting $\ell \to \infty$. ◇

Recalling the purpose of our analysis, we next express all finite difference operators in a single currency, as functions of the shift operator $\mathcal{E}$. It is trivial that $\Delta_+ = \mathcal{E} - \mathcal{I}$ and $\Delta_- = \mathcal{I} - \mathcal{E}^{-1}$, while the interpretation of $\mathcal{E}^{1/2}$ as a 'half shift' implies that $\Delta_0 = \mathcal{E}^{1/2} - \mathcal{E}^{-1/2}$ and $\Upsilon_0 = \frac{1}{2}(\mathcal{E}^{-1/2} + \mathcal{E}^{1/2})$. Finally, to express $\mathcal{D}$ in terms of the shift operator, we recall the Taylor theorem: for any analytic function z it is true that

$$\mathcal{E}z(x) = z(x+h) = \sum_{j=0}^{\infty} \frac{1}{j!}\left[\frac{\mathrm{d}^j z(x)}{\mathrm{d}x^j}\right] h^j = \left[\sum_{j=0}^{\infty} \frac{1}{j!}(h\mathcal{D})^j\right] z(x) = \mathrm{e}^{h\mathcal{D}} z(x),$$

and we deduce that $\mathcal{E} = \mathrm{e}^{h\mathcal{D}}$.[2] Formal inversion yields

$$h\mathcal{D} = \ln \mathcal{E}. \tag{8.1}$$

We conclude that, each having been expressed in terms of $\mathcal{E}$, all six finite difference operators commute. This is a useful observation since it follows that we need not bother with the order of their action whenever they are superposed.

The above operator formulae can be (formally) inverted, thereby expressing $\mathcal{E}$ in terms of Δ_+ etc. It is easy to verify that $\mathcal{E} = \mathcal{I} + \Delta_+ = (I - \Delta_-)^{-1}$. The expression for Δ_0 is a quadratic equation for $\mathcal{E}^{1/2}$,

$$(\mathcal{E}^{1/2})^2 - \Delta_0 \mathcal{E}^{1/2} - \mathcal{I} = O,$$

with two solutions, $\frac{1}{2}\Delta_0 \pm \sqrt{\frac{1}{4}\Delta_0^2 + \mathcal{I}}$. Letting $h \to 0$, we deduce that the correct formula is

$$\mathcal{E} = \left(\tfrac{1}{2}\Delta_0 + \sqrt{\mathcal{I} + \tfrac{1}{4}\Delta_0^2}\right)^2.$$

[2] We have already encountered a similar construction in Chapter 2.

We need not bother to express $\mathcal{E}$ in terms of Υ_0, since this serves no useful purpose.

Combining (8.1) with these expressions, we next write the differential operator in terms of other finite difference operators,

$$h\mathcal{D} = \ln(\mathcal{I} + \Delta_+) \tag{8.2}$$

$$h\mathcal{D} = -\ln(\mathcal{I} - \Delta_-) \tag{8.3}$$

$$h\mathcal{D} = 2\ln\left(\tfrac{1}{2}\Delta_0 + \sqrt{\mathcal{I} + \tfrac{1}{4}\Delta_0^2}\right). \tag{8.4}$$

Recall that the purpose of the exercise is to approximate the differential operator $\mathcal{D}$ and its powers (which, of course, correspond to higher derivatives). The formulae (8.2)–(8.4) are ideally suited to this purpose. For example, expanding (8.2) we obtain

$$\begin{aligned}\mathcal{D} &= \frac{1}{h}\ln(\mathcal{I} + \Delta_+) = \frac{1}{h}\left[\Delta_+ - \tfrac{1}{2}\Delta_+^2 + \tfrac{1}{3}\Delta_+^3 + \mathcal{O}(\Delta_+^4)\right]\\ &= \frac{1}{h}\left(\Delta_+ - \tfrac{1}{2}\Delta_+^2 + \tfrac{1}{3}\Delta_+^3\right) + \mathcal{O}\left(h^3\right), \qquad h \to 0,\end{aligned}$$

where we exploit the estimate $\Delta_+ = \mathcal{O}(h)$, $h \to 0$. Operating s times, we obtain an expression for the sth derivative, $s = 1, 2, \ldots,$

$$\mathcal{D}^s = \frac{1}{h^s}\left[\Delta_+^s - \tfrac{1}{2}s\Delta_+^{s+1} + \tfrac{1}{24}s(3s+5)\Delta_+^{s+2}\right] + \mathcal{O}\left(h^3\right), \qquad h \to 0. \tag{8.5}$$

The meaning of (8.5) is that the linear combination

$$\frac{1}{h^s}\left[\Delta_+^s - \tfrac{1}{2}s\Delta_+^{s+1} + \tfrac{1}{24}s(3s+5)\Delta_+^{s+2}\right] z_k \tag{8.6}$$

of the $s+3$ grid values $z_k, z_{k+1}, \ldots, z_{k+s+2}$ approximates $\mathrm{d}^s z(kh)/\,\mathrm{d}x^s$ up to $\mathcal{O}\left(h^3\right)$. Needless to say, truncating (8.5) a term earlier, for example, we obtain order $\mathcal{O}\left(h^2\right)$, whereas higher order can be obtained by expanding the logarithm further.

Similarly to (8.5), we can use (8.3) to express derivatives in terms of grid points wholly to the left,

$$\mathcal{D}^s = \frac{(-1)^s}{h^s}[\ln(\mathcal{I}-\Delta_-)]^s = \frac{1}{h^s}\left[\Delta_-^s + \tfrac{1}{2}s\Delta_-^{s+1} + \tfrac{1}{24}s(3s+5)\Delta_-^{s+2}\right] + \mathcal{O}\left(h^3\right), \quad h \to 0.$$

However, does it make much sense to approximate derivatives solely in terms of grid points that all lie to one side? Sometimes we have little choice – more about this later – but in general it is a good policy to match the numbers of points on the left and on the right. The natural candidate for this task would be the central finite difference operator Δ_0 except that now, having at last started to discuss approximation on a grid, not just operators in a formal framework, we can no longer loftily disregard the fact that $\Delta_0 \boldsymbol{z}$ is not a proper grid sequence. The crucial observation is that *even powers* of Δ_0 map the set $\mathbb{R}^{\mathbb{Z}}$ of grid sequences to itself! Thus, $\Delta_0^2 z_n = z_{n-1} - 2z_n + z_{n+1}$ and the proof for all even powers follows at once from the trivial observation that $\Delta_0^{2s} = \left(\Delta_0^2\right)^s$.

Recalling (8.4), we consider the Taylor expansion of the function $g(\xi) := \ln(\xi + \sqrt{1+\xi^2})$. By the generalized binomial theorem, we have

$$g'(\xi) = \frac{1}{\sqrt{1+\xi^2}} = \sum_{j=0}^{\infty} (-1)^j \binom{2j}{j} \left(\tfrac{1}{2}\xi\right)^{2j},$$

where $\binom{2j}{j}$ is a binomial coefficient equal to $(2j)!/(j!)^2$. Since $g(0) = 0$ and the Taylor series converges uniformly for $|\xi| < 1$, integration yields

$$g(\xi) = g(0) + \int_0^\xi g'(\tau)\,\mathrm{d}\tau = 2\sum_{j=0}^{\infty} \frac{(-1)^j}{2j+1}\binom{2j}{j}\left(\tfrac{1}{2}\xi\right)^{2j+1}.$$

Letting $\xi = \frac{1}{2}\Delta_0$, we thus deduce from (8.4) the formal expansion

$$\mathcal{D} = \frac{2}{h} g\left(\tfrac{1}{2}\Delta_0\right) = \frac{4}{h}\sum_{j=0}^{\infty} \frac{(-1)^j}{2j+1}\binom{2j}{j}\left(\tfrac{1}{4}\Delta_0\right)^{2j+1}. \tag{8.7}$$

Unfortunately, the expression (8.7) is of exactly the wrong kind – all the powers of Δ_0 therein are odd! However, since even powers of odd powers are themselves even, raising (8.7) to an even power yields

$$\begin{aligned}\mathcal{D}^{2s} = \frac{1}{h^{2s}}\bigg[&(\Delta_0^2)^s - \frac{s}{12}(\Delta_0^2)^{s+1} + \frac{s(11+5s)}{1440}(\Delta_0^2)^{s+2} \\ &- \frac{s(382+231s+35s^2)}{362880}(\Delta_0^2)^{s+3}\bigg] + \mathcal{O}(h^8), \qquad h \to 0.\end{aligned} \tag{8.8}$$

Thus, for example, the linear combination

$$\frac{1}{h^{2s}}\left[(\Delta_0^2)^s - \frac{s}{12}(\Delta_0^2)^{s+1} + \frac{s(11+5s)}{1440}(\Delta_0^2)^{s+2}\right] z_k \tag{8.9}$$

approximates $\mathrm{d}^{2s}z(kh)/\mathrm{d}x^{2s}$ to $\mathcal{O}(h^6)$.

How effective is (8.9) in comparison with (8.6)? To attain $\mathcal{O}(h^{2p})$, (8.6) requires $2s+2p$ adjacent grid points and (8.9) just $2s+2p-1$, a relatively modest saving. Central difference operators, however, have smaller error constants (see Exercises 8.3 and 8.4). More importantly, they are more convenient to use and usually lead to more tractable linear algebraic systems (see Chapters 11–15).

The expansion (8.8) is valid only for even derivatives. To reap the benefits of central differencing for odd derivatives, we require a simple, yet clever, trick. Let us thus pay attention to the averaging operator Υ_0, which has until now had only a silent part in the proceedings.

We express Υ_0 in terms of Δ_0. Since $\Upsilon_0 = \frac{1}{2}(\mathcal{E}^{1/2} + \mathcal{E}^{-1/2})$ and $\Delta_0 = \mathcal{E}^{1/2} - \mathcal{E}^{-1/2}$, it follows that

$$\begin{aligned}4\Upsilon_0^2 &= \mathcal{E} + 2\mathcal{I} + \mathcal{E}^{-1},\\ \Delta_0^2 &= \mathcal{E} - 2\mathcal{I} + \mathcal{E}^{-1}\end{aligned}$$

and, subtracting, we deduce that $4\Upsilon_0 - \Delta_0^2 = 4\mathcal{I}$. We conclude that

$$\Upsilon_0 = (\mathcal{I} + \tfrac{1}{4}\Delta_0^2)^{1/2}. \tag{8.10}$$

The main idea now is to multiply (8.7) by the identity $\mathcal{I}$, which we craftily disguise by using (8.10),

$$\mathcal{I} = \Upsilon_0 \left(\mathcal{I} + \tfrac{1}{4}\Delta_0^2\right)^{-1/2} = \Upsilon_0 \sum_{j=0}^{\infty} (-1)^j \binom{2j}{j} \left(\tfrac{1}{16}\Delta_0^2\right)^j.$$

The outcome,

$$\mathcal{D} = \frac{1}{h}(\Upsilon_0\Delta_0) \left[\sum_{j=0}^{\infty} (-1)^j \binom{2j}{j} \left(\tfrac{1}{16}\Delta_0^2\right)^j\right] \left[\sum_{i=0}^{\infty} \frac{(-1)^i}{2i+1} \binom{2i}{i} \left(\tfrac{1}{16}\Delta_0^2\right)^i\right], \tag{8.11}$$

might look messy, but has one redeeming virtue: it is constructed exclusively from *even* powers of Δ_0 and $\Upsilon_0\Delta_0$. Since

$$\Upsilon_0\Delta_0 z_k = \Upsilon_0 \left(z_{k+\frac{1}{2}} - z_{k-\frac{1}{2}}\right) = \tfrac{1}{2}\left(z_{k+1} - z_{k-1}\right),$$

we conclude that (8.11) is a linear combination of terms that reside on the grid.

The expansion (8.11) can be raised to a power but this is not a good idea, since such a procedure is wasteful in terms of grid points; an example is provided by

$$\mathcal{D}^2 = \frac{1}{h^2}(\Upsilon_0\Delta_0)^2 \left(\mathcal{I} - \tfrac{1}{3}\Delta_0^2\right) + \mathcal{O}(h^4).$$

Since

$$(\Upsilon_0\Delta_0)^2 \left(\mathcal{I} - \tfrac{1}{3}\Delta_0^2\right) z_k = \tfrac{1}{12}(-z_{k-3} + 5z_{k-2} + z_{k-1} - 10z_k + z_{k+1} + 5z_{k+2} - z_{k+3}),$$

we need seven points to attain $\mathcal{O}(h^4)$, while (8.9) requires just five points. In general, a considerably better idea is first to raise (8.7) to an odd power and then to multiply it by $\mathcal{I} = \Upsilon_0(\mathcal{I} + \frac{1}{4}\Delta_0^2)^{1/2}$. The outcome,

$$\begin{aligned} \mathcal{D}^{2s+1} = \frac{1}{h^{2s+1}}(\Upsilon_0\Delta_0) \big[&(\Delta_0^2)^s - \tfrac{1}{12}(s+2)(\Delta_0^2)^{s+1} \\ &+ \tfrac{1}{1440}(s+3)(5s+16)(\Delta_0^2)^{s+2}\big] + \mathcal{O}(h^5), \qquad h \to 0, \end{aligned} \tag{8.12}$$

lives on the grid and, other things being equal, is the recommended approximation of odd derivatives.

◇ **A simple example...** Figure 8.1 displays the (natural) logarithm of the error in the approximation of the first derivative of $z(x) = x\,\mathrm{e}^x$. The first row corresponds to the forward difference approximations

$$\frac{1}{h}\Delta_+, \qquad \frac{1}{h}\left(\Delta_+ - \tfrac{1}{2}\Delta_+^2\right) \qquad \text{and} \qquad \frac{1}{h}\left(\Delta_+ - \tfrac{1}{2}\Delta_+^2 + \tfrac{1}{3}\Delta_+^3\right),$$

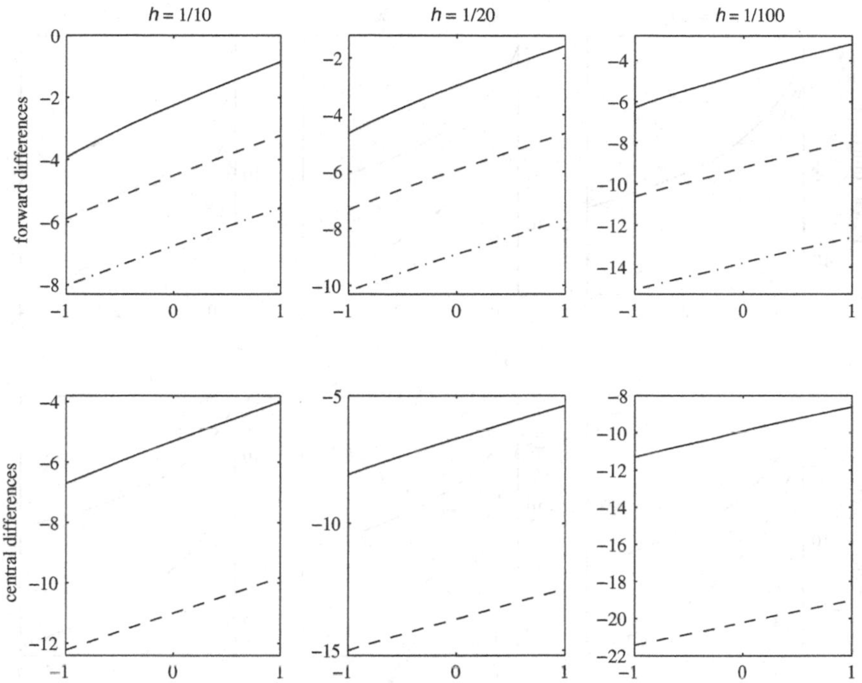

Figure 8.1 The error (on a logarithmic scale) in the approximation of z', where $z(x) = xe^x$, $-1 \le x \le 1$. Forward differences of size $\mathcal{O}(h)$ (solid line), $\mathcal{O}(h^2)$ (broken line) and $\mathcal{O}(h^3)$ (broken-and-dotted line) feature in the first row, while the second row presents central differences of size $\mathcal{O}(h^2)$ (solid line) and $\mathcal{O}(h^4)$ (broken line).

with $h = \frac{1}{10}$ and $h = \frac{1}{20}$ in the first and in the second column respectively. The second row displays the central difference approximations

$$\frac{1}{h}\Upsilon_0\Delta_0 \qquad \text{and} \qquad \frac{1}{h}\Upsilon_0\Delta_0\left(\mathcal{I} - \tfrac{1}{6}\Delta_0^2\right).$$

What can we learn from this figure? If the error behaves like ch^p, where $c \neq 0$, then its logarithm is approximately $p \ln h + \ln|c|$. Therefore, for small h, one can expect each ℓth curve in the top row to behave like the first curve, scaled by ℓ (since $p = \ell$ for the ℓth curve). This is not the case in the first two columns, since $h > 0$ is not small enough, but the pattern becomes more visible when h decreases; the reader could try $h = \frac{1}{1000}$ to confirm that this asymptotic behaviour indeed takes place. However, replacing h by $\frac{1}{2}h$ should lower each curve by an amount roughly equal to $\ln 2 \approx 0.6931$, and this can be observed by comparing the first two columns. Likewise, the curves in the third column are each lowered by about $\ln 5 \approx 1.6094$ in comparison with the second column.

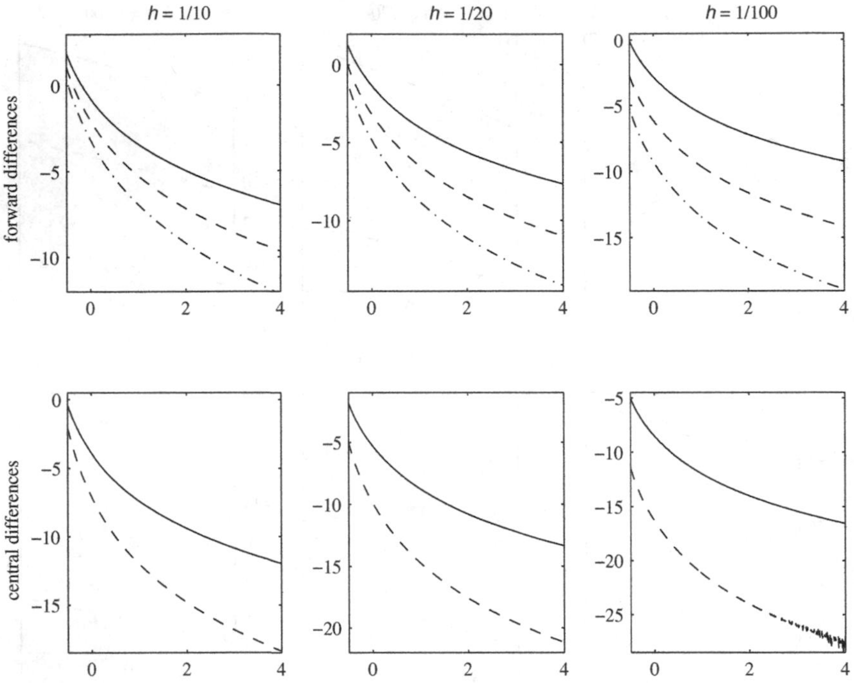

Figure 8.2 The error (on a logarithmic scale) in the approximation of z'', where $z(x) = 1/(1+x)$, $-\frac{1}{2} \le x \le 4$. For the meaning of the curves, see the caption to Fig. 8.1.

Similar information is displayed in Fig. 8.2, namely the logarithm of the error in approximating z'', where $z(x) = 1/(1+x)$, by forward differences (in the top row) and central differences (in the second row). The specific approximants

$$\frac{1}{h^2}\Delta_+^2, \qquad \frac{1}{h^2}(\Delta_+^2 - \Delta_+^3), \qquad \frac{1}{h^2}(\Delta_+^2 - \Delta_+^3 + \tfrac{11}{12}\Delta_+^4)$$

and

$$\frac{1}{h^2}\Delta_0^2, \qquad \frac{1}{h^2}(\Delta_0^2 - \tfrac{1}{12}\Delta_0^4)$$

can be easily derived from (8.6) and (8.9) respectively. The pattern is similar, except that the singularity at $x = -1$ means that the quality of approximation deteriorates at the left end of the scale; it is always important to bear in mind that estimates based on Taylor expansions break down near singularities. ◇

Needless to say, there is no bar on using several finite difference operators in a single formula (see Exercise 8.5). However, other things being equal, in such cases we usually prefer to employ central differences.

There are two important exceptions. Firstly, realistic grids do not in fact extend from $-\infty$ to ∞; this was just a convenient assumption, which has simplified the notation a great deal. Of course, we can employ finite differences on finite grids, except that the procedure might break down near the boundary. 'One-sided' finite differences possess obvious advantages in such situations. Secondly, for some PDEs the exact solution of the equation displays an innate 'preference' toward one spatial direction over the other, and in this case it is a good policy to let the approximation to the derivative reflect this fact. This behaviour is displayed by certain hyperbolic equations and we will encounter it in Chapter 17.

Finally, it is perfectly possible to approximate derivatives on non-equidistant grids. This, however, is by and large outside the scope of this book, except for a brief discussion of approximation near curved boundaries in the next section.

8.2 The five-point formula for $\nabla^2 u = f$

Perhaps the most important and ubiquitous PDE is the *Poisson* equation

$$\nabla^2 u = f, \qquad (x, y) \in \Omega, \tag{8.13}$$

where

$$\nabla^2 = \frac{\partial^2}{\partial x^2} + \frac{\partial^2}{\partial y^2},$$

$f = f(x, y)$ is a known continuous function and the domain $\Omega \subset \mathbb{R}^2$ is bounded, open and connected and has a piecewise-smooth boundary. We hasten to add that this is not the most general form of the Poisson equation – in fact we are allowed any number of space dimensions, not just two, Ω need not be bounded and its boundary, as well as the function f, can satisfy far less demanding smoothness requirements. However, the present framework is sufficient for our purpose.

Like any partial differential equation, for its solution (8.13) must be accompanied by a boundary condition. We assume the *Dirichlet* condition, namely that

$$u(x, y) = \phi(x, y), \qquad (x, y) \in \partial\Omega. \tag{8.14}$$

An implementation of finite differences always commences by inscribing a grid into the domain of interest. In our case we impose on $\mathrm{cl}\,\Omega$ a square grid $\Omega_{\Delta x}$ parallel to the axes, with an equal spacing of Δx in both spatial directions (Fig. 8.3). In other words, we choose $\Delta x > 0$, $(x_0, y_0) \in \Omega$ and let $\Omega_{\Delta x}$ be the set of all points of the form $(x_0 + k\Delta x,\ y_0 + \ell\Delta x)$ that reside in the closure of Ω. We denote

$$\begin{aligned} \boldsymbol{I}_{\Delta x} &:= \left\{(k, \ell) \in \mathbb{Z}^2 \,:\, (x_0 + k\Delta x, y_0 + \ell\Delta x) \in \mathrm{cl}\,\Omega\right\}, \\ \boldsymbol{I}^{\circ}_{\Delta x} &:= \left\{(k, \ell) \in \mathbb{Z}^2 \,:\, (x_0 + k\Delta x, y_0 + \ell\Delta x) \in \Omega\right\}, \end{aligned}$$

and, for every $(k, \ell) \in \boldsymbol{I}^{\circ}_{\Delta x}$, we let $u_{k,\ell}$ stand for the approximation to the solution $u(x_0 + k\Delta x, y_0 + \ell\Delta x)$ of the Poisson equation (8.13) at the relevant grid point. Note that, of course, there is no need to approximate grid points $(k, \ell) \in \boldsymbol{I}_{\Delta x} \setminus \boldsymbol{I}^{\circ}_{\Delta x}$, since they lie on $\partial\Omega$ and there exact values are given by (8.14).

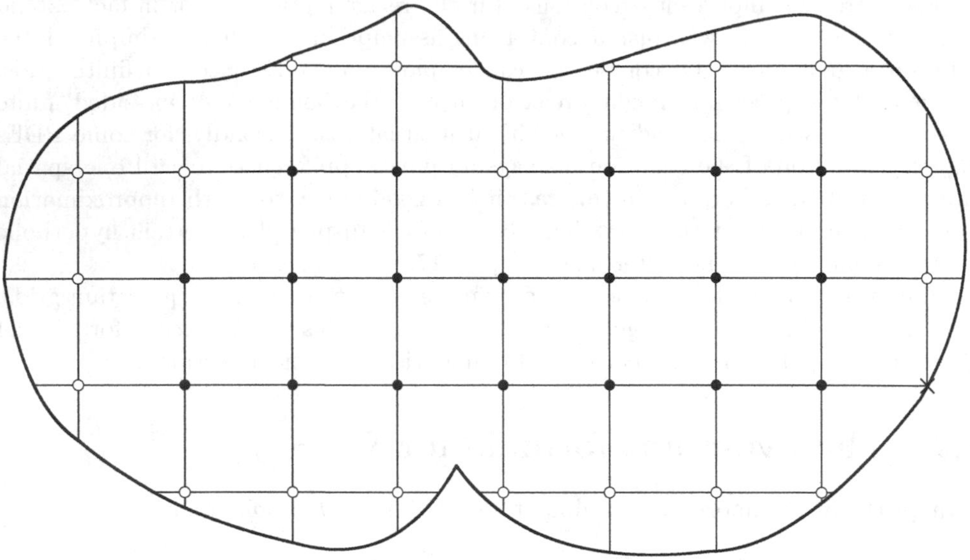

Figure 8.3 An example of a computational grid for a two-dimensional domain Ω. $\bullet$, internal points; $\circ$, near-boundary points; $\times$, a boundary point.

Wishing to approximate ∇^2 by finite differences, our first observation is that we are no longer allowed the comfort of sequences that stretch all the way to $\pm\infty$; whether in the x- or the y-direction, $\partial\Omega$ acts as an impenetrable barrier and we cannot use grid points outside $\mathrm{cl}\,\Omega$ to assist in our approximation.

Our first finite difference scheme approximates $\nabla^2 u$ at the (k,ℓ)th grid point as a linear combination of the five values $u_{k,\ell}, u_{k\pm1,\ell}, u_{k,\ell\pm1}$ and it is valid only if the immediate horizontal and vertical neighbours of (k,ℓ), namely $(k\pm1,\ell)$ and $(k,\ell\pm1)$ respectively, are in $\boldsymbol{I}_{\Delta x}$. We say, for the purposes of our present discussion, that such a point $(x_0+k\Delta x,\ y_0+\ell\Delta x)$ is an *internal point*. In general, the set $\Omega_{\Delta x}$ consists of three types of points: *boundary points,* which lie on $\partial\Omega$ and whose value is known by virtue of (8.14); internal points, which soon will be subjected to our scrutiny; and the *near-boundary points,* where we can no longer employ finite differences on an equidistant grid so that a special approach is required. Needless to say, the definition of the internal and near-boundary points changes if we employ a different configuration of points in our finite difference scheme (cf. Section 8.3).

Let us suppose that $(k,\ell)\in\boldsymbol{I}^{\circ}_{\Delta x}$ corresponds to an internal point. Following our recommendation from Section 8.1, we use central differences. Of course, our grid is now two dimensional and we can use differences in either coordinate direction. This creates no difficulty, as long as we distinguish clearly the space coordinate with respect to which our operator acts. We do this by appending a subscript, e.g. $\Delta_{0,x}$.

Let $v=v(x,y)$, $(x,y)\in\mathrm{cl}\,\Omega$, be an arbitrary sufficiently smooth function. It follows at once from (8.9) that, for every internal grid point,

$$\left.\frac{\partial^2 v}{\partial x^2}\right|_{\substack{x=x_0+k\Delta x,\\ y=y_0+\ell\Delta x}} = \frac{1}{(\Delta x)^2}\Delta_{0,x}^2 v_{k,\ell} + \mathcal{O}\big((\Delta x)^2\big),$$

$$\left.\frac{\partial^2 v}{\partial y^2}\right|_{\substack{x=x_0+k\Delta x,\\ y=y_0+\ell\Delta x}} = \frac{1}{(\Delta x)^2}\Delta_{0,y}^2 v_{k,\ell} + \mathcal{O}\big((\Delta x)^2\big),$$

where $v_{k,\ell}$ is the value of v at the (k,ℓ)th grid point. Therefore,

$$\frac{1}{(\Delta x)^2}(\Delta_{0,x}^2 + \Delta_{0,y}^2)$$

approximates ∇^2 to order $\mathcal{O}\big((\Delta x)^2\big)$. This motivates the replacement of the Poisson equation (8.13) by the *five point* finite difference scheme

$$\frac{1}{(\Delta x)^2}(\Delta_{0,x}^2 + \Delta_{0,y}^2)u_{k,\ell} = f_{k,\ell} \tag{8.15}$$

at every pair (k,ℓ) that corresponds to an internal grid point. Of course, $f_{k,\ell} = f(x_0 + k\Delta x, y_0 + \ell\Delta x)$. More explicitly, (8.15) can be written in the form

$$u_{k-1,\ell} + u_{k+1,\ell} + u_{k,\ell-1} + u_{k,\ell+1} - 4u_{k,\ell} = (\Delta x)^2 f_{k,\ell}, \tag{8.16}$$

and this motivates its name, the five-point formula. In lieu of the Poisson equation, we have a linear combination of the values of u at an (internal) grid point and at the immediate horizontal and vertical neighbours of this point.

Another way of depicting (8.16) is via a *computational stencil* (also known as a *computational molecule*). This is a pictorial representation that is self-explanatory (and becomes indispensable for more complicated finite difference schemes, which involve a larger number of points), as follows:

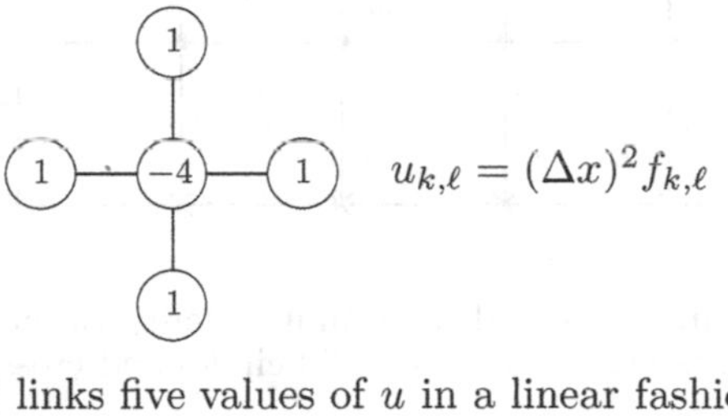

Thus, the equation (8.16) links five values of u in a linear fashion. Unless they lie on the boundary, these values are unknown. The main idea of the finite difference method is to associate with every grid point having an index in $\boldsymbol{I}_{\Delta x}^{\circ}$ (that is, every internal and near-boundary point) a single linear equation, for example (8.16). This results in a system of linear equations whose solution is our approximation $\boldsymbol{u} := (u_{k,\ell})_{(k,\ell)\in \boldsymbol{I}_{\Delta x}^{\circ}}$.

Three questions are critical to the performance of finite differences.

- Is the linear system nonsingular, so that the finite difference solution $\boldsymbol{u}$ exists and is unique?

- Suppose that a unique $\boldsymbol{u} = \boldsymbol{u}_{\Delta x}$ exists for all sufficiently small Δx, and let $\Delta x \to 0$. Is it true that the numerical solution converges to the exact solution of (8.13)? What is the asymptotic magnitude of the error?

- Are there efficient and robust means to solve the linear system, which is likely to consist of a very large number of equations?

We defer the third question to Chapters 11–15, where the theme of the numerical solution of large sparse algebraic linear systems will be debated at some length. Meantime, we address ourselves to the first two questions in the special case when Ω is a square. Without loss of generality, we let $\Omega = \{(x, y) : 0 < x, y < 1\}$. This leads to considerable simplification since, provided we choose $\Delta x = 1/(m+1)$, say, for an integer m, and let $x_0 = y_0 = 0$, all grid points are either internal or boundary (Fig. 8.4).

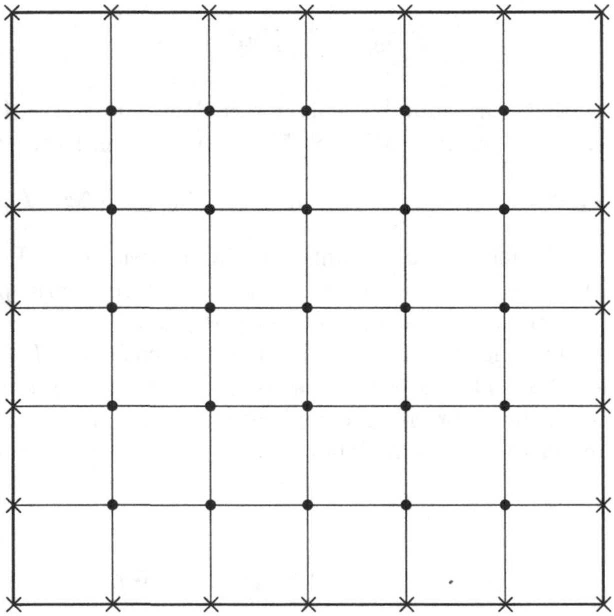

Figure 8.4 Computational grid for a unit square. As in Fig. 8.3, internal and boundary points are denoted by solid circles and crosses, respectively.

◇ **The Laplace equation** Prior to attempting to prove theorems on the behaviour of numerical methods, it is always a good practice to run a few simple programs and obtain a 'feel' for what we are, after all, trying to prove. The computer is the mathematical equivalent of a physicist's laboratory!

In this spirit we apply the five-point formula (8.15) to the *Laplace equation*

$\nabla^2 u = 0$ in the unit square $(0,1)^2$, subject to the boundary conditions

$$\begin{aligned} u(x,0) &\equiv 0, & u(x,1) &= \frac{1}{(1+x)^2+1}, & 0 \le y \le 1, \\ u(0,y) &= \frac{y}{1+y^2}, & u(1,y) &= \frac{y}{4+y^2}, & 0 \le x \le 1. \end{aligned}$$

Figure 8.5 displays the exact solution of this equation,

$$u(x,y) = \frac{y}{(1+x)^2+y^2}, \qquad 0 \le x, y \le 1,$$

as well as its numerical solution by means of the five-point formula with $m = 5$, $m = 11$ and $m = 23$; this corresponds to $\Delta x = \frac{1}{6}$, $\Delta x = \frac{1}{12}$ and $\Delta x = \frac{1}{24}$ respectively. The size of the grid halves in each consecutive numerical trial and it is evident from the figure that the error decreases by a factor of 4. This is consistent with an error decay of $\mathcal{O}((\Delta x)^2)$, which is hinted at in our construction of (8.15) and will be proved in Theorem 8.2. ◇

Recall that we wish to address ourselves to two questions. Firstly, is the linear system (8.16) nonsingular? Secondly, does its solution converge to the exact solution of the Poisson equation (8.13) as $\Delta x \to 0$? In the case of a square, both questions can be answered by employing a similar construction.

The function $u_{k,\ell}$ is defined on a two-dimensional grid and, to write the linear equations (8.16) formally in a matrix–vector notation, we need to rearrange $u_{k,\ell}$ into a one-dimensional column vector $\boldsymbol{u} \in \mathbb{R}^s$, where $s = m^2$. In other words, for any permutation $\{(k_i, \ell_i)\}_{i=1,2,\ldots,s}$ of the set $\{(k,\ell)\}_{k,\ell=1,2,\ldots,m}$ we can let

$$\boldsymbol{u} = \begin{bmatrix} u_{k_1,\ell_1} \\ u_{k_2,\ell_2} \\ \vdots \\ u_{k_s,\ell_s} \end{bmatrix}$$

and write (8.16) in the form

$$A\boldsymbol{u} = \boldsymbol{b}, \tag{8.17}$$

where A is an $s \times s$ matrix, while $\boldsymbol{b} \in \mathbb{R}^s$ includes both the inhomogeneous part $(\Delta x)^2 f_{k,\ell}$, similarly ordered, and the contribution of the boundary values. Since any permutation of the s grid points provides for a different arrangement, there are $s! = (m^2)!$ distinct ways of deriving (8.17). Fortunately, none of the features that are important to our present analysis depends on the specific ordering of the $u_{k,\ell}$.

Lemma 8.1 *The matrix A in (8.17) is symmetric and the set of its eigenvalues is*

$$\sigma(A) = \{\lambda_{\alpha,\beta} : \alpha, \beta = 1, 2, \ldots, m\},$$

where

$$\lambda_{\alpha,\beta} = -4\left\{\sin^2\left[\frac{\alpha\pi}{2(m+1)}\right] + \sin^2\left[\frac{\beta\pi}{2(m+1)}\right]\right\}, \quad \alpha, \beta = 1, 2, \ldots, m. \tag{8.18}$$

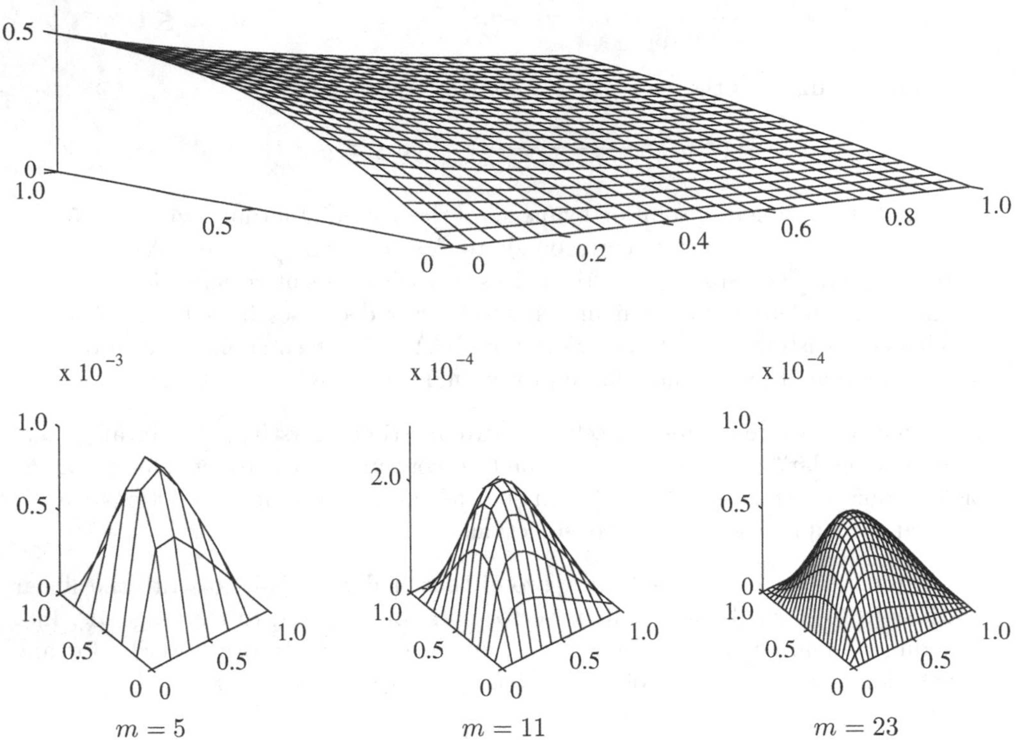

Figure 8.5 The exact solution of the Laplace equation discussed in the text and the errors of the five-point formula for $m = 5$, $m = 11$ and $m = 23$, with 25, 121 and 529 grid points respectively.

Proof To prove symmetry, we notice by inspection of (8.16) that all elements of A must be -4, 1 or 0, according to the following rule. All diagonal elements $a_{\gamma,\gamma}$ equal -4, whereas an off-diagonal element $a_{\gamma,\delta}$, $\gamma \neq \delta$, equals 1 if (i_γ, j_γ) and (i_δ, j_δ) are either horizontal or vertical neighbours and 0 otherwise. Being a neighbour is, however, a commutative relation: if (i_γ, j_γ) is a neighbour of (i_δ, j_δ) then (i_δ, j_δ) is a neighbour of (i_γ, j_γ). Therefore $a_{\gamma,\delta} = a_{\delta,\gamma}$ for all $\gamma, \delta = 1, 2, \ldots, s$.

To find the eigenvalues of A we disregard the exact way in which the matrix has been composed – after all, symmetric permutations conserve eigenvalues – and, instead, go back to the equations (8.16). Suppose that we can demonstrate the existence of a nonzero function $(v_{k,\ell})_{k,\ell=0,1,\ldots,m+1}$ such that $v_{k,0} = v_{k,m+1} = v_{0,\ell} = v_{m+1,\ell} = 0$, $k, \ell = 1, 2, \ldots, m$, and such that the homogeneous set of linear equations

$$v_{k-1,\ell} + v_{k+1,\ell} + v_{k,\ell-1} + v_{k,\ell+1} - 4v_{k,\ell} = \lambda v_{k,\ell}, \qquad k, \ell = 1, 2, \ldots, m, \qquad (8.19)$$

is satisfied for some λ. It follows that, up to rearrangement, $(v_{k,\ell})$ is an eigenvector and λ is a corresponding eigenvalue of A.

Given $\alpha, \beta \in \{1, 2, \ldots, m\}$, we let

$$v_{k,\ell} = \sin\left(\frac{k\alpha\pi}{m+1}\right)\sin\left(\frac{\ell\beta\pi}{m+1}\right), \qquad k, \ell = 0, 1, \ldots, m+1.$$

Note that, as required, $v_{k,0} = v_{k,m+1} = v_{0,\ell} = v_{m+1,\ell} = 0$, $k, \ell = 1, 2, \ldots, m$. Substituting into (8.19), we obtain

$$\begin{aligned} & v_{k-1,\ell} + v_{k+1,\ell} + v_{k,\ell-1} + v_{k,\ell+1} - 4v_{k,\ell} \\ &= \left\{\sin\left[\frac{(k-1)\alpha\pi}{m+1}\right] + \sin\left[\frac{(k+1)\alpha\pi}{m+1}\right]\right\}\sin\left(\frac{\ell\beta\pi}{m+1}\right) \qquad (8.20) \\ &\quad + \sin\left(\frac{k\alpha\pi}{m+1}\right)\left\{\sin\left[\frac{(\ell-1)\beta\pi}{m+1}\right] + \sin\left[\frac{(\ell+1)\beta\pi}{m+1}\right]\right\} - 4\sin\left(\frac{k\alpha\pi}{m+1}\right)\sin\left(\frac{\ell\beta\pi}{m+1}\right). \end{aligned}$$

We exploit the trigonometric identity

$$\sin(\theta - \psi) + \sin(\theta + \psi) = 2\sin\theta\cos\psi$$

to simplify (8.20), obtaining for the right-hand side

$$\begin{aligned} & 2\sin\left(\frac{k\alpha\pi}{m+1}\right)\cos\left(\frac{\alpha\pi}{m+1}\right)\sin\left(\frac{\ell\beta\pi}{m+1}\right) + 2\sin\left(\frac{k\alpha\pi}{m+1}\right)\sin\left(\frac{\ell\beta\pi}{m+1}\right)\cos\left(\frac{\beta\pi}{m+1}\right) \\ &\quad - 4\sin\left(\frac{k\alpha\pi}{m+1}\right)\sin\left(\frac{\ell\beta\pi}{m+1}\right) \\ &= -2\left[2 - \cos\left(\frac{\alpha\pi}{m+1}\right) - \cos\left(\frac{\beta\pi}{m+1}\right)\right]\sin\left(\frac{k\alpha\pi}{m+1}\right)\sin\left(\frac{\ell\beta\pi}{m+1}\right) \\ &= -4\left\{\sin^2\left[\frac{\alpha\pi}{2(m+1)}\right] + \sin^2\left[\frac{\beta\pi}{2(m+1)}\right]\right\}v_{k,\ell}, \qquad k, \ell = 1, 2, \ldots, m. \end{aligned}$$

Note that we have used in the last line the trigonometric identity

$$1 - \cos\theta = 2\sin^2\left(\frac{\theta}{2}\right).$$

We have thus demonstrated that (8.19) is satisfied by $\lambda = \lambda_{\alpha,\beta}$, and this completes the proof of the lemma. ■

Corollary *The matrix A is negative definite and, a fortiori, nonsingular.*

Proof We have just shown that A is symmetric, and it follows from (8.18) that all its eigenvalues are negative. Therefore (see A.1.5.1) it is negative definite and nonsingular. ■

◇ **Eigenvalues of the Laplace operator** Before we continue with the orderly flow of our exposition, it is instructive to comment on how the eigenvalues and eigenvectors of the matrix A are related to the eigenvalues and eigenfunctions of the Laplace operator ∇^2 in the unit square.

The function v, not identically zero, is said to be an *eigenfunction* of ∇^2 in a domain Ω and λ is the corresponding *eigenvalue* if v vanishes along $\partial\Omega$ and satisfies within Ω the equation $\nabla^2 v = \lambda v$. The linear system (8.19) is nothing other than a five-point discretization of this equation for $\Omega = (0,1)^2$.

The eigenfunctions and eigenvalues of ∇^2 can be evaluated easily and explicitly in the unit square. Given any two positive integers α, β, we let $v(x,y) = \sin(\alpha\pi x)\sin(\beta\pi y)$, $x, y \in [0,1]$. Note that, as required, v obeys zero Dirichlet boundary conditions. It is trivial to verify that $\nabla^2 v = -(\alpha^2+\beta^2)\pi^2 v$; hence v is indeed an eigenfunction and the corresponding eigenvalue is $-(\alpha^2+\beta^2)\pi^2$. It is possible to prove that *all* eigenfunctions of ∇^2 in $(0,1)^2$ have this form.

The vector $v_{k,\ell}$ from the proof of Lemma 8.1 can be obtained by sampling of the eigenfunction v at the grid points $\left\{\left(\frac{k}{m+1}, \frac{\ell}{k+1}\right)\right\}_{k,\ell=0,1,\ldots,m+1}$ (for $\alpha, \beta = 1, 2, \ldots, m$ only; the matrix A, unlike ∇^2, acts on a finite-dimensional space!), whereas $(\Delta x)^{-2}\lambda_{\alpha,\beta}$ is a good approximation to $-(\alpha^2+\beta^2)\pi^2$ provided α and β are small in comparison with m. Expanding $\sin^2\theta$ in a power series and bearing in mind that $(m+1)\Delta x = 1$, we readily obtain

$$\begin{aligned}\frac{\lambda_{\alpha,\beta}}{(\Delta x)^2} &= -4\left(\left\{\left[\frac{\alpha\pi}{2(m+1)}\right]^2 - \frac{1}{3}\left[\frac{\alpha\pi}{2(m+1)}\right]^4 + \cdots\right\}\right.\\ &\qquad \left. + \left\{\left[\frac{\beta\pi}{2(m+1)}\right]^2 - \frac{1}{3}\left[\frac{\beta\pi}{2(m+1)}\right]^4 + \cdots\right\}\right)\\ &= -(\alpha^2+\beta^2)\pi^2 + \tfrac{1}{12}(\alpha^4+\beta^4)\pi^4(\Delta x)^2 + \mathcal{O}\big((\Delta x)^4\big).\end{aligned}$$

Hence, $(m+1)^2\lambda_{\alpha,\beta}$ is a good approximation of an exact eigenvalue of the Laplace operator for small α, β, but the quality of approximation deteriorates rapidly as soon as $(\alpha^4+\beta^4)(\Delta x)^2$ becomes nonnegligible. ◇

Let u be the exact solution of (8.13) in a unit square. We set $\tilde{u}_{k,\ell} = u(k\Delta x, \ell\Delta x)$ and denote by $e_{k,\ell}$ the error of the five-point formula (8.15) at the (k,ℓ)th grid point, $e_{k,\ell} = u_{k,\ell} - \tilde{u}_{k,\ell}$, $k, \ell = 0, 1, \ldots, m+1$. Let the five-point equations (8.15) be represented in the matrix form (8.17) and let $\boldsymbol{e}$ denote an arrangement of $\{e_{k,\ell}\}$ into a vector in $\mathbb{R}^s$, $s = m^2$, whose ordering is identical to that of $\boldsymbol{u}$. We are measuring the magnitude of $\boldsymbol{e}$ by the Euclidean norm $\|\cdot\|$ (A.1.3.3).

Theorem 8.2 *Subject to sufficient smoothness of the function f and the boundary conditions, there exists a number $c > 0$, independent of Δx, such that*

$$\|\boldsymbol{e}\| \le c(\Delta x)^2, \qquad \Delta x \to 0. \tag{8.21}$$

Proof Since $(\Delta x)^{-2}(\Delta_{0,x}^2 + \Delta_{0,y}^2)$ approximates ∇^2 locally to order $\mathcal{O}\big((\Delta x)^2\big)$, it is true that

$$\tilde{u}_{k-1,\ell} + \tilde{u}_{k+1,\ell} + \tilde{u}_{k,\ell-1} + \tilde{u}_{k,\ell+1} - 4\tilde{u}_{k,\ell} = (\Delta x)^2 f_{k,\ell} + \mathcal{O}\big((\Delta x)^4\big) \tag{8.22}$$

for $\Delta x \to 0$. We subtract (8.22) from (8.16) and the outcome is

$$e_{k-1,\ell} + e_{k+1,\ell} + e_{k,\ell-1} + e_{k,\ell+1} - 4e_{k,\ell} = \mathcal{O}\big((\Delta x)^4\big), \qquad \Delta x \to 0,$$

or, in vector notation (and paying due heed to the fact that $u_{k,\ell}$ and $\tilde{u}_{k,\ell}$ coincide along the boundary)

$$A\boldsymbol{e} = \boldsymbol{\delta}_{\Delta x}, \tag{8.23}$$

where $\boldsymbol{\delta}_{\Delta x} \in \mathbb{R}^{m^2}$ is such that $\|\boldsymbol{\delta}_{\Delta x}\| = \mathcal{O}\big((\Delta x)^4\big)$. It follows from (8.23) that

$$\boldsymbol{e} = A^{-1}\boldsymbol{\delta}_{\Delta x}. \tag{8.24}$$

Recall from Lemma 8.1 that A is symmetric. Hence so is A^{-1} and its Euclidean norm $\|A^{-1}\|$ is the same as its spectral radius $\rho(A^{-1})$ (A.1.5.2). The latter can be computed at once from (8.18), since $\lambda \in \sigma(B)$ is the same as $\lambda^{-1} \in \sigma(B^{-1})$ for any nonsingular matrix B. Thus, bearing in mind that $(m+1)\Delta x = 1$,

$$\rho(A^{-1}) = \max_{\alpha,\beta=1,2,\ldots,m} \frac{1}{4}\left\{\sin^2\left[\frac{\alpha\pi}{2(m+1)}\right] + \sin^2\left[\frac{\beta\pi}{2(m+1)}\right]\right\}^{-1} = \frac{1}{8\sin^2\frac{1}{2}\Delta x\pi}.$$

Since

$$\lim_{\Delta x\to 0}\left[\frac{(\Delta x)^2}{8\sin^2\frac{1}{2}\Delta x\pi}\right] = \frac{1}{2\pi^2},$$

it follows that for any constant $c_1 > (2\pi^2)^{-1}$ it is true that

$$\|A^{-1}\| = \rho(A^{-1}) \le c_1(\Delta x)^{-2}, \qquad \Delta x \to 0. \tag{8.25}$$

Provided that f and the boundary conditions are sufficiently smooth,[3] u is itself sufficiently differentiable and there exists a constant $c_2 > 0$ such that $\|\boldsymbol{\delta}(\Delta x)\| \le c_2(\Delta x)^4$ (recall that $\boldsymbol{\delta}$ depends solely on the exact solution). Substituting this and (8.25) into the inequality (8.24) yields (8.21) with $c = c_1 c_2$. ■

Our analysis can be generalized to rectangles, L-shaped domains etc., provided the ratios of all sides are rational numbers (cf. Exercise 8.7). Unfortunately, in general the grid contains near-boundary points, at which the five-point formula (8.15) cannot be implemented. To see this, it is enough to look at a single coordinate direction; without loss of generality let us suppose that we are seeking to approximate ∇^2 at the point P in Fig. 8.6.

Given $z(x)$, we first approximate z'' at $P \sim x_0$ (we disregard the variable y, which plays no part in this process) as a linear combination of the values of z at P, $Q \sim x_0 - \Delta x$ and $T \sim x_0 + \tau\Delta x$. Expanding z in a Taylor series about x_0, we can easily show that

$$\frac{1}{(\Delta x)^2}\left[\frac{2}{\tau+1}z(x_0 - \Delta x) - \frac{2}{\tau}z(x_0) + \frac{2}{\tau(\tau+1)}z(x_0 + \tau\Delta x)\right]$$
$$= z''(x_0) + \tfrac{1}{3}(\tau - 1)z'''(x_0)\Delta x + \mathcal{O}\big((\Delta x)^2\big).$$

[3] We prefer not to be very specific here, since the issues raised by the smoothness and differentiability of Poisson equations are notoriously difficult. However, these requirements are satisfied in most cases of practical interest.

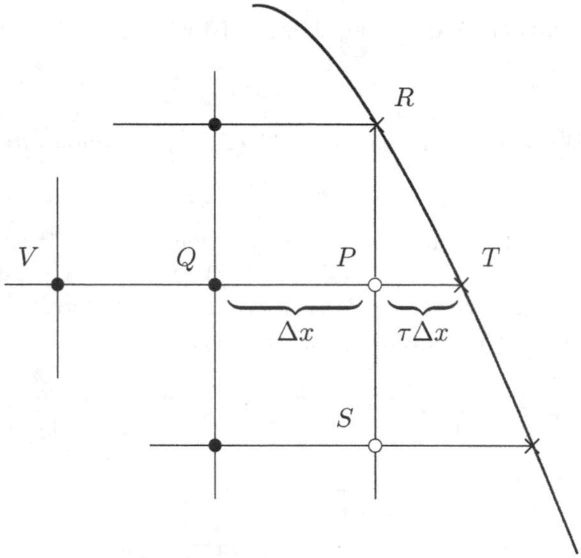

Figure 8.6 Computational grid near a curved boundary.

Unless $\tau = 1$, when everything reduces to the central difference approximation, the error is just $\mathcal{O}(\Delta x)$. To recover order $\mathcal{O}\big((\Delta x)^2\big)$, consistently with the five-point formula at internal points, we add the function value at $V \sim x_0 - 2\Delta x$ to the linear combination, whereby expansion in a Taylor series about x_0 yields

$$z''(x_0) = \frac{1}{(\Delta x)^2}\left[\frac{\tau-1}{\tau+2}z(x_0-2\Delta x) + \frac{2(2-\tau)}{\tau+1}z(x_0-\Delta x) - \frac{3-\tau}{\tau}z(x_0) + \frac{6}{\tau(\tau+1)(\tau+2)}z(x_0+\tau\Delta x)\right] + \mathcal{O}\big((\Delta x)^2\big).$$

A good approximation to $\nabla^2 u$ at P should involve, therefore, six points, P, Q, R, S, T and V. Assuming that P corresponds to the grid point (k^0, ℓ^0), say, we obtain the linear equation

$$\frac{\tau-1}{\tau+2}u_{k^0-2,\ell^0} + \frac{2(2-\tau)}{\tau+1}u_{k^0-1,\ell^0} + \frac{6}{\tau(\tau+1)(\tau+2)}u_{k^0+\tau,\ell^0} + u_{k^0,\ell^0-1} + u_{k^0,\ell^0+1} - \frac{3+\tau}{\tau}u_{k^0,\ell^0} = (\Delta x)^2 f_{k^0,\ell^0}, \tag{8.26}$$

where $(k^0+\tau, \ell^0)$ corresponds to the boundary point T, whose value is provided from the Dirichlet boundary condition. Note that if $\tau = 1$ and P becomes an internal point then this reduces to the five-point formula (8.16).

A similar treatment can be used in the y-direction. Of course, regardless of direction, we need Δx small enough that we have sufficient information to implement (8.26) or other $\mathcal{O}\big((\Delta x)^2\big)$ approximants to ∇^2 at all near-boundary points. The outcome, in tandem with (8.16) at internal points, is a linear algebraic equation for every grid point – whether internal or near-boundary – where the solution is unknown.

We will not extend Theorem 8.2 here and prove that the rate of decay of the error is $\mathcal{O}\big((\Delta x)^2\big)$ but will set ourselves a less ambitious goal: to prove that the linear algebraic system is nonsingular. First, however, we require a technical lemma that is of great applicability in many branches of matrix analysis.

Lemma 8.3 (The Geršgorin criterion) *Let $B = (b_{k,\ell})$ be an arbitrary irreducible* (A.1.2.5) *complex $d \times d$ matrix. Then*

$$\sigma(B) \subset \bigcup_{i=1}^{d} \mathbb{S}_i,$$

where

$$\mathbb{S}_i = \left\{ z \in \mathbb{C} : |z - b_{i,i}| \le \sum_{j=1,\, j\neq i}^{d} |b_{i,j}| \right\}$$

and $\sigma(B)$ is the set containing the eigenvalues of B. Moreover, $\lambda \in \sigma(B)$ may lie on $\partial\mathbb{S}_{i^0}$ for some $i^0 \in \{1, 2, \ldots, d\}$ only if $\lambda \in \partial\mathbb{S}_i$ for all $i = 1, 2, \ldots, d$. The $\mathbb{S}_i$ are known as Geršgorin discs.

Proof This is relegated to Exercise 8.8, where it is broken down into a number of easily manageable chunks. ■

There is another part of the Geršgorin criterion that plays no role whatsoever in our present discussion but we mention it as a matter of independent mathematical interest. Thus, suppose that

$$\{1, 2, \ldots, d\} = \{i_1, i_2, \ldots, i_r\} \cup \{j_1, j_2, \ldots, j_{d-r}\}$$

such that $\mathbb{S}_{i_\alpha} \cap \mathbb{S}_{i_\beta} \neq \emptyset$, $\alpha, \beta = 1, 2, \ldots, r$ and $\mathbb{S}_{i_\alpha} \cap \mathbb{S}_{j_\beta} = \emptyset$, $i = 1, 2, \ldots, r$, $j = 1, 2, \ldots, d - r$. Let $\mathcal{S} := \cup_{\alpha=1}^{r} \mathbb{S}_{i_\alpha}$. Then the set $\mathcal{S}$ includes *exactly* r eigenvalues of B.

Theorem 8.4 *Let $A\boldsymbol{u} = \boldsymbol{b}$ be the linear system obtained by employing the five-point formula* (8.16) *at internal points and the formula* (8.26) *or its reflections and extensions (catering for the case when one horizontal and one vertical neighbour are missing) at near-boundary points. Then A is nonsingular.*

Proof No matter how we arrange the unknowns into a vector, thereby determining the ordering of the rows and the columns of A, each row of A corresponds to a single equation. Therefore all the elements along the ith row vanish, except for those that feature in the linear equation that is obeyed at the grid point corresponding to this row. It follows from an inspection of (8.16) and (8.26) that $a_{i,i} < 0$, that $\sum_{j\neq i} |a_{i,j}| + a_{i,i} \le 0$ and that the inequality is sharp at a near-boundary point. (This is trivial for (8.16), while, since $\tau \in (0, 1]$, some off-diagonal components in (8.26) might be negative. Yet, simple calculation confirms that the sum of absolute values of all off-diagonal elements along a row is consistent with the above inequality. Of course, we must remember to disregard the contribution of boundary points.) It follows that the origin may not lie in the interior of the Geršgorin disc $\mathbb{S}_j$. Thus, by Lemma 8.3, $0 \in \sigma(A)$ only if $0 \in \partial\mathbb{S}_i$ for *all* rows i.

At least one equation has a neighbour on the boundary. Let this equation correspond to the i^0th row of A. Then $\sum_{j \neq i^0} |a_{i^0,j}| < |a_{i^0,i^0}|$, therefore $0 \notin \mathbb{S}_{i^0}$. We deduce that it is impossible for 0 to lie on the boundaries of all the discs $\mathbb{S}_i$; hence, by Lemma 8.3, it is not an eigenvalue. This means that A is nonsingular and the proof is complete. ■

8.3 Higher-order methods for $\nabla^2 u = f$

The Laplace operator ∇^2 has a key role in many important equations of mathematical physics, to mention just two, the parabolic *diffusion* equation

$$\frac{\partial u}{\partial t} = \nabla^2 u, \qquad u = u(x, y, t),$$

and the hyperbolic *wave* equation

$$\frac{\partial^2 u}{\partial t^2} = \nabla^2 u, \qquad u = u(x, y, t).$$

Therefore, the five-point approximation formula (8.15) is one of the workhorses of numerical analysis. This all pervasiveness, however, motivates a discussion of higher-order computational schemes.

Truncating (8.8) after two terms, we obtain in place of (8.15) the scheme

$$\frac{1}{(\Delta x)^2}\left[\Delta_{0,x}^2 + \Delta_{0,y}^2 - \tfrac{1}{12}\left(\Delta_{0,x}^4 + \Delta_{0,y}^4\right)\right] u_{k,\ell} = f_{k,\ell}. \tag{8.27}$$

More economically, this can be written as the computational stencil

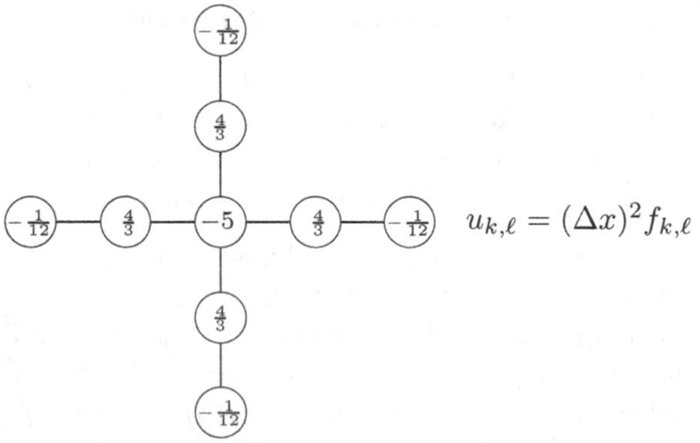

Although the error is $\mathcal{O}\big((\Delta x)^4\big)$, (8.27) is not a popular method. It renders too many points near-boundary, even in a square grid, which means that they require laborious special treatment. Worse, it gives linear systems that are considerably more expensive

to solve than, for example, those generated by the five-point scheme. In particular, the fast solvers from Sections 15.1 and 15.2 cannot be implemented in this setting.

A more popular alternative is to approximate $\nabla^2 u$ at the (k,ℓ)th grid point by means of all its eight nearest neighbours: horizontal, vertical and diagonal. This results in the *nine-point formula*

$$\frac{1}{(\Delta x)^2}\left(\Delta_{0,x}^2 + \Delta_{0,y}^2 + \tfrac{1}{6}\Delta_{0,x}^2\Delta_{0,y}^2\right) u_{k,\ell} = f_{k,\ell}, \tag{8.28}$$

more familiarly known in the computational stencil notation

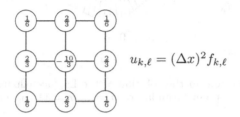

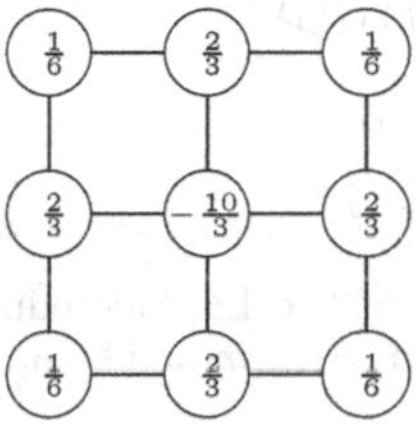

$$u_{k,\ell} = (\Delta x)^2 f_{k,\ell}$$

To analyse the error in (8.28) we recall from Section 8.1 that $\Delta_0 = \mathcal{E}^{1/2} - \mathcal{E}^{-1/2}$ and $\mathcal{E} = \mathrm{e}^{\Delta x \mathcal{D}}$. Therefore, expanding in a Taylor series in Δx,

$$\begin{aligned}\Delta_0^2 &= \mathcal{E} - 2\mathcal{I} + \mathcal{E}^{-1} = \mathrm{e}^{\Delta x\mathcal{D}} - 2\mathcal{I} + \mathrm{e}^{-\Delta x \mathcal{D}}\\ &= (\Delta x)^2\mathcal{D}^2 + \tfrac{1}{12}(\Delta x)^4\mathcal{D}^4 + \mathcal{O}\big((\Delta x)^6\big).\end{aligned}$$

Since this is valid for both spatial variables and $\mathcal{D}_x, \mathcal{D}_y$ commute, substitution into (8.28) yields

$$\begin{aligned}&\frac{1}{(\Delta x)^2}\left(\Delta_{0,x}^2 + \Delta_{0,y}^2 + \tfrac{1}{6}\Delta_{0,x}^2\Delta_{0,y}^2\right)\\ &= \frac{1}{(\Delta x)^2}\Big\{\big[(\Delta x)^2\mathcal{D}_x^2 + \tfrac{1}{12}(\Delta x)^4\mathcal{D}_x^4 + \mathcal{O}\big((\Delta x)^6\big)\big] + \big[(\Delta x)^2\mathcal{D}_y^2 + \tfrac{1}{12}(\Delta x)^4\mathcal{D}_y^4\\ &\qquad + \mathcal{O}\big((\Delta x)^6\big)\big] + \tfrac{1}{6}\big[(\Delta x)^2\mathcal{D}_x^2 + \mathcal{O}\big((\Delta x)^4\big)\big]\big[(\Delta x)^2\mathcal{D}_y^2 + \mathcal{O}\big((\Delta x)^4\big)\big]\Big\}\\ &= (\mathcal{D}_x^2 + \mathcal{D}_y^2) + \tfrac{1}{12}(\Delta x)^2(\mathcal{D}_x^2 + \mathcal{D}_y^2)^2 + \mathcal{O}\big((\Delta x)^4\big)\\ &= \nabla^2 + \tfrac{1}{12}(\Delta x)^2\nabla^4 + \mathcal{O}\big((\Delta x)^4\big).\end{aligned} \tag{8.29}$$

In other words, the error in the nine-point formula is of *exactly* the same order of magnitude as that of the five-point formula. Apparently, nothing is gained by incorporating the diagonal neighbours!

Not giving in to despair, we return to the example of Fig. 8.5 and recalculate it with the nine-point formula (8.28). The results can be seen in Fig. 8.7 and, as can be easily ascertained, *they are inconsistent with our claim that the error decays as* $\mathcal{O}\big((\Delta x)^2\big)$! Bearing in mind that Δx decreases by a factor of 2 in each consecutive graph, we would have expected the error to attenuate by a factor of 4 – instead it attenuates by a factor of 16. Too good to be true?

The reason for this spectacular behaviour is, to put it bluntly, our sloth. Had we attempted to solve a Poisson equation with a nontrivial inhomogeneous term the

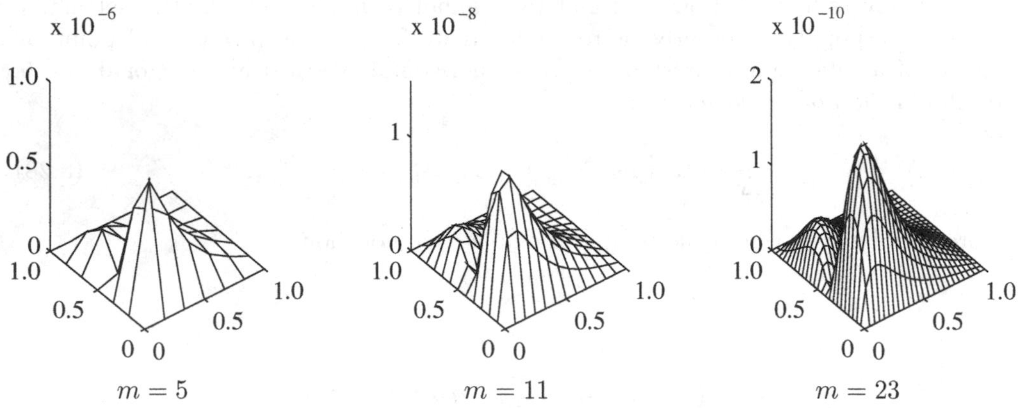

Figure 8.7 The errors in the solution of the Laplace equation from Fig. 8.5 by the nine-point formula, for $m = 5$, $m = 11$ and $m = 23$.

decrease in the error would have been consistent with our error estimate (see Fig. 8.8). Instead, we applied the nine-point scheme to the Laplace equation, which, as far as (8.28) is concerned, is a special case. We now give a hand-waving explanation of this phenomenon.

It follows from (8.29) that the nine-point scheme is an approximation of order $\mathcal{O}\big((\Delta x)^4\big)$ to the 'equation'

$$\left[\nabla^2 + \tfrac{1}{12}(\Delta x)^2\nabla^4\right] u = f. \qquad (8.30)$$

Setting aside the dependence of (8.30) on Δx and the whole matter of boundary conditions (we *are* hand-waving after all!), the operator $\mathcal{M}_{\Delta x} := \mathcal{I} + \frac{1}{12}(\Delta x)^2\nabla^2$ is invertible for sufficiently small Δx. Multiplying (8.30) by its inverse while letting $f \equiv 0$ indicates that the nine-point scheme bears an error of $\mathcal{O}\big((\Delta x)^4\big)$ when applied to the Laplace equation. This explains the rapid decay of the error in Fig. 8.7.

The Laplace equation has many applications and this superior behaviour of the nine-point formula is a matter of interest. Remarkably, the logic that has led to an explanation of this phenomenon can be extended to cater for Poisson equations as well. Thus suppose that $\Delta x > 0$ is small enough that $\mathcal{M}_{\Delta x}^{-1}$ exists, and act with this operator on both sides of (8.30). Then we have

$$\nabla^2 u = \mathcal{M}_{\Delta x}^{-1} f, \qquad (8.31)$$

a new Poisson equation for which the nine-point formula produces an error of order $\mathcal{O}\big((\Delta x)^4\big)$. The only snag is that the right-hand side differs from that in (8.13), but this is easy to put right. We replace f in (8.31) by a function $\tilde{f}$ such that

$$\tilde{f}(x, y) = f(x, y) + \tfrac{1}{12}(\Delta x)^2\nabla^2 f(x, y) + \mathcal{O}\big((\Delta x)^4\big), \qquad (x, y) \in \Omega.$$

Since $\mathcal{M}_{\Delta x}^{-1}\tilde{f} = f + \mathcal{O}\big((\Delta x)^4\big)$, the new equation (8.31) differs from (8.13) only in its

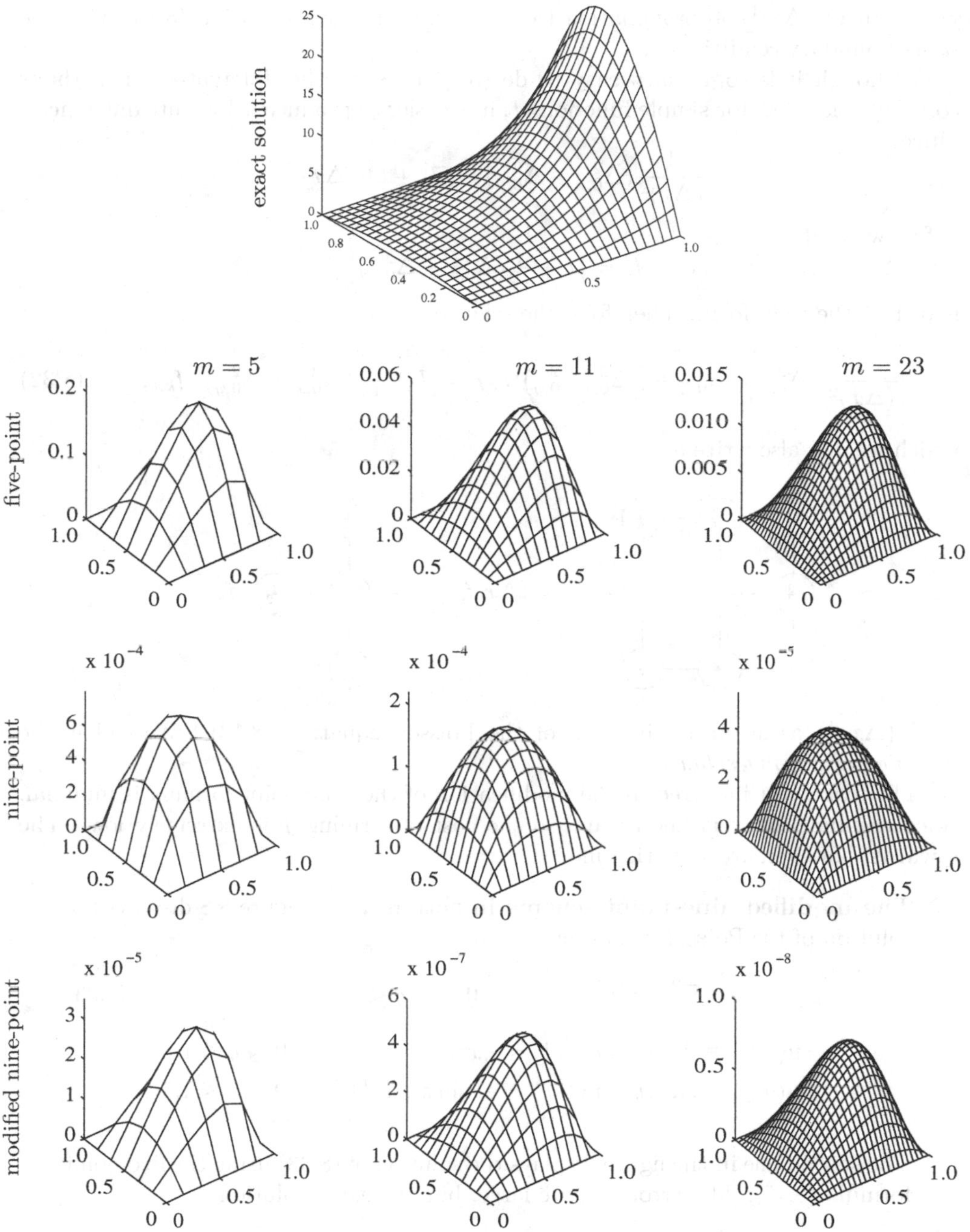

Figure 8.8 The exact solution of the Poisson equation (8.33) and the errors with the five-point, nine-point and modified nine-point schemes.

$\mathcal{O}((\Delta x)^4)$ terms. In other words, the nine-point formula, when applied to $\nabla^2 u = \tilde{f}$, yields an $\mathcal{O}((\Delta x)^4)$ approximation to the original Poisson equation (8.13) with the same boundary conditions.

Although it is sometimes easy to derive $\tilde{f}$ by symbolic differentiation, perhaps computer-assisted, for simple functions f, it is easier to produce it by finite differences. Since

$$\frac{1}{(\Delta x)^2}(\Delta_{0,x}^2 + \Delta_{0,y}^2) = \nabla^2 + \mathcal{O}((\Delta x)^2),$$

it follows that

$$\tilde{f} := \left[\mathcal{I} + \tfrac{1}{12}(\Delta_{0,x}^2 + \Delta_{0,y}^2)\right] f$$

is of just the right form. Therefore, the scheme

$$\frac{1}{(\Delta x)^2}\left(\Delta_{0,x}^2 + \Delta_{0,y}^2 + \tfrac{1}{6}\Delta_{0,x}^2\Delta_{0,y}^2\right) u_{k,\ell} = \left[\mathcal{I} + \tfrac{1}{12}\left(\Delta_{0,x}^2 + \Delta_{0,y}^2\right)\right] f_{k,\ell}, \qquad (8.32)$$

which we can also write as

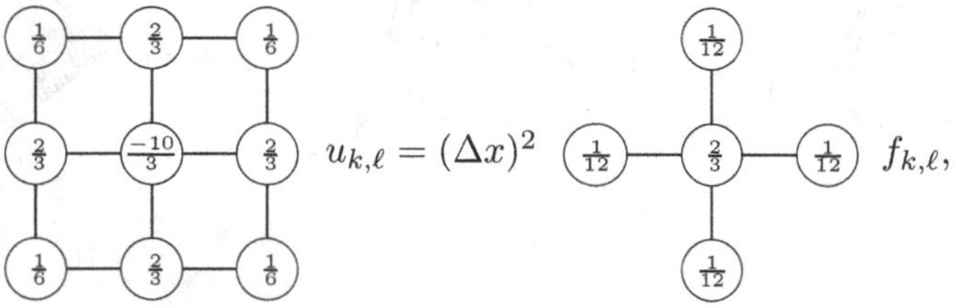

is $\mathcal{O}((\Delta x)^4)$ as an approximation of the Poisson equation (8.13). We call it the *modified nine-point scheme.*

The extra cost incurred in the modification of the nine-point formula is minimal, since the function f is known and so the cost of forming $\tilde{f}$ is extremely low. The rewards, however, are very rich indeed.

◇ **The modified nine-point scheme in action...** Figure 8.8 displays the solution of the Poisson equation

$$\nabla^2 u = x^2 + y^2, \qquad 0 < x, y < 1, \qquad (8.33)$$

$$\begin{array}{lll} u(x,0) \equiv 0, & u(x,1) = \frac{1}{2}x^2, & 0 \le y \le 1, \\ u(0,y) = \sin \pi y, & u(1,y) = \mathrm{e}^{\pi} \sin \pi y + \frac{1}{2}y^2, & 0 \le x \le 1. \end{array}$$

The second line in the figure displays the solution of (8.33) using the five-point formula (8.15). The error is quite large, but the exact solution

$$u(x,y) = \mathrm{e}^{\pi x} \sin \pi y + \tfrac{1}{2}(xy)^2, \qquad 0 \le x, y \le 1,$$

can be as much as $\mathrm{e}^{\pi} + \frac{1}{2} \approx 23.6407$, and so even the numerical solution for $m = 5$ comes within 1% of it. The most important observation, however, is that the error is attenuated by roughly a factor of 4 each time Δx is halved.

The outcome of the calculation with the nine-point formula (8.28) is displayed in the third line and, without any doubt, it is much better – about 200 times smaller than the corresponding values for the five-point scheme. There is a good reason: the error expansion of the five-point formula is

$$\frac{1}{(\Delta x)^2}(\Delta_{0,x}^2+\Delta_{0,y}^2)-\nabla^2 = \tfrac{1}{12}(\Delta x)^2(\nabla^4-2\mathcal{D}_x^2\mathcal{D}_y^2)+\mathcal{O}\big((\Delta x)^4\big), \qquad \Delta x \to 0,$$

(verification is left to the reader in Exercise 8.9) while the error expansion of the nine-point formula can be deduced at once from the above expression,

$$\frac{1}{(\Delta x)^2}\left(\Delta_{0,x}^2+\Delta_{0,y}^2+\tfrac{1}{6}\Delta_{0,x}^2\Delta_{0,y}^2\right)-\nabla^2=\tfrac{1}{12}(\Delta x)^2\nabla^4+\mathcal{O}\big((\Delta x)^4\big), \quad \Delta x \to 0.$$

As far as the Poisson equation (8.33) is concerned, we have

$$\begin{array}{rcl} (\nabla^4-\mathcal{D}_x^2\mathcal{D}_y^2)u & = & 2\pi^4 e^{\pi x}\sin \pi y, \\ \nabla^4 u & \equiv & 4, \end{array} \qquad 0 \le x, y \le 1.$$

Hence the principal error term for the five-point formula can be as much as 1127 times larger than the corresponding term for the nine-point scheme in $(0,1)^2$. This is not a general feature of the underlying numerical methods! Perhaps less striking, but nonetheless an important observation, is that the error associated with the nine-point scheme decays in Fig. 8.8 by roughly a factor of 4 with each halving of Δx, consistently with the general theory and similarly to the behaviour of the five-point formula.

The bottom line in Fig. 8.8 displays the outcome of the calculation with the modified nine-point formula (8.32). It is evident not just that the absolute magnitude of the error is vastly smaller (we do better with 25 grid points than the 'plain' nine-point formula with 529 grid points!) but also that its decay, by roughly a factor 64 whenever Δx is halved, is consistent with the expected error $\mathcal{O}\big((\Delta x)^4\big)$. ◇

Comments and bibliography

The numerical solution of the Poisson equation by means of finite differences features in many texts, e.g. Ames (1977). Diligent readers who wish first to acquaint themselves with the analytic theory of the Poisson equation (and more general elliptic equations) might consult the classical text of Agmon (1965). The best reference, however, is probably Hackbusch (1992), since it combines the analytic and numerical aspects of the subject.

The modified nine-point formula (8.32) is not as well known as it ought to be, and part of the blame lies perhaps in the original name, *Mehrstellenverfahren* (Collatz, 1966). We prefer to avoid this sobriquet in the text, to deter overenthusiastic instructors from making its spelling an issue in examinations.

Anticipating the discussion of Fourier transforms in Chapter 15, it is worth remarking briefly on the connection of the latter with finite difference operators. Denoting the Fourier transform of a function f by $\hat{f}$,[4] it is easy to prove that

$$\widehat{\mathcal{E}f}(\xi) = \mathrm{e}^{\mathrm{i}\xi h}\hat{f}(\xi), \qquad \xi \in \mathbb{C};$$

therefore

$$\left.\begin{aligned} \widehat{\Delta_+ f}(\xi) &= (\mathrm{e}^{\mathrm{i}\xi h} - 1)\hat{f}(\xi), \\ \widehat{\Delta_- f}(\xi) &= (1 - \mathrm{e}^{-\mathrm{i}\xi h})\hat{f}(\xi), \\ \widehat{\Delta_0 f}(\xi) &= (\mathrm{e}^{\mathrm{i}\xi h/2} - \mathrm{e}^{-\mathrm{i}\xi h/2})\hat{f}(\xi) = 2\mathrm{i}\left(\sin\frac{\xi h}{2}\right)\hat{f}(\xi), \\ \widehat{\Upsilon_0 f}(\xi) &= \tfrac{1}{2}(\mathrm{e}^{\mathrm{i}\xi h/2} + \mathrm{e}^{\mathrm{i}\xi h/2})\hat{f}(\xi) = \left(\cos\frac{\xi h}{2}\right)\hat{f}(\xi). \end{aligned}\right\} \quad \xi \in \mathbb{C}.$$

Likewise,

$$\widehat{\mathcal{D}f}(\xi) = \mathrm{i}\xi\hat{f}(\xi), \qquad \xi \in \mathbb{C}.$$

The calculus of finite differences can be relocated to Fourier space. For example, the Fourier transform of $h^{-2}[f(x+h) - 2f(x) + f(x-h)]$ is

$$\frac{1}{h^2}\widehat{\Delta_0^2 f}(\xi) = \frac{\mathrm{e}^{\mathrm{i}\xi h} - 2 + \mathrm{e}^{-\mathrm{i}\xi h}}{h^2}\hat{f}(\xi) = \left(-\xi^2 + \tfrac{1}{12}h^2\xi^4 - \cdots\right)\hat{f}(\xi),$$

which we recognize as the transform of $f'' - \frac{1}{12}h^2 f^{(4)} + \cdots$.

The subject matter of this chapter can be generalized in numerous directions. Much of it is straightforward intellectually, although it might require lengthy manipulation and unwieldy algebra. An example is the generalization of methods for the Poisson equation to more variables. An engineer recognizes three spatial variables, while the agreed number of spatial dimensions in theoretical physics changes by the month (or so it seems), but numerical analysis can cater for all these situations (see Exercises 8.10 and 8.11). Of course, the cost of linear algebra is bound to increase with dimension, but such considerations are easier to deal with in the framework of Chapters 11–15.

Finite differences extend to non-rectangular meshes. Although, for example, a honeycomb mesh such as the following,

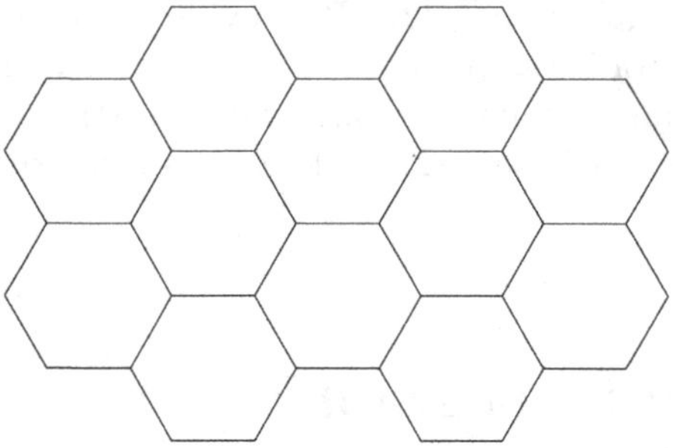

is considerably more difficult to use in practical programs, it occasionally confers advantages in dealing with curved boundaries. An example of a finite difference scheme in a honeycomb is provided by

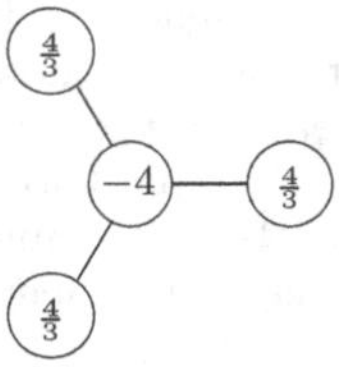

$$u_{k,\ell} = (\Delta x)^2 f_{k,\ell}.$$

[4]If you do not know the definition of the Fourier transform, skip what follows, possibly returning to it later.

A slightly more complicated grid features in Exercise 8.13. There are, however, limits to the practicability of fancy grids – Escher's tessellations and Penrose tiles are definitely out of the question!

A more wide-ranging approach to curved and complicated geometries is to use nonuniform meshes. This allows a snug fit to difficult boundaries and also opens up the possibility of 'zooming in' on portions of the set Ω where, for some reason, the solution is more problematic or error-prone, and employing a finer grid there. This is relatively easy in one dimension (see Fornberg & Sloan (1994) for a very general approach) but, as far as finite differences are concerned, virtually hopeless in several spatial dimensions. Fortunately, the finite element method, which is reviewed in Chapter 9, provides a relatively accessible means of working with multivariate nonuniform meshes.

Yet another possibility, increasingly popular because of its efficiency in parallel computing architectures, is *domain decomposition* (Chan & Mathew, 1994; Le Tallec, 1994). The main idea is to tear the set Ω into smaller subsets, where the Poisson (or another) equation can be solved on a distinct grid (and possibly even on a different computer), subsequently 'gluing' different bits together by solving smaller problems along the interfaces.

The scope of this chapter has been restricted to Dirichlet boundary conditions, but Poisson equations can be computed just as well with Neumann or mixed conditions. Moreover, finite differences and tricks like the *Mehrstellenverfahren* can be applied to other linear elliptic equations. Most notably, the computational stencil

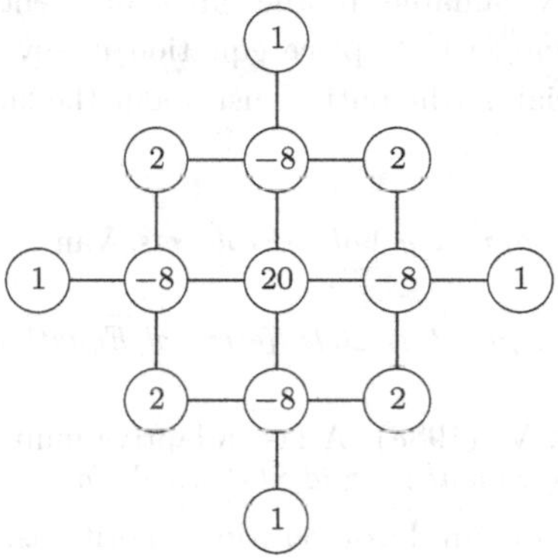

$$u_{k,\ell} = (\Delta x)^4 f_{k,\ell} \qquad (8.34)$$

represents an $\mathcal{O}\big((\Delta x)^2\big)$ method for the *biharmonic equation*

$$\nabla^4 u = f, \qquad (x, y) \in \Omega$$

(see Exercise 8.12).

However, no partial differential equation competes with the Poisson equation in regard to its importance and pervasiveness in applications. It is no wonder that there exist so many computational algorithms for (8.13), and not just finite difference, but also finite element (Chapter 9), spectral (Chapter 10), and boundary element methods. The fastest to date is the *multipole* method, originally introduced by Carrier *et al.* (1988).

This chapter would not be complete without a mention of a probabilistic method for solving the five-point system (8.15) in a special case, the Laplace equation (i.e., $f \equiv 0$). Although solution methods for linear algebraic systems belong in Chapters 11–15, we prefer to describe this method here, so that nobody mistakes it for a *viable* algorithm. Assume for example a computational grid in the shape of Manhattan Island, New York City, with streets and avenues forming rows and columns, respectively. (We need to disregard the Broadway and a few other thoroughfares that fail to conform with the grid structure, but it is the principle that matters!) Suppose that a drunk emerges from a pub at the intersection of Sixth Avenue and Twentieth Street. Being drunk, our friend turns at random in one of the four available directions and staggers ahead. Having reached the next intersection, the drunk again turns at random, and so on and so forth. Disregarding the obvious perils of muggers, New York City drivers, pollution etc., the person in question can terminate this random walk (for this is what it is in mathematical terminology) only by reaching the harbour and falling into the water – an event that is bound to take place in finite time with probability one. Depending on the exact spot where our imbibing friend takes a dip, a fine is paid to the New York Harbor Authority. In other words, we have a 'fine function' $\phi(x,y)$, defined for all (x,y) along Manhattan's waterfront, and the Harbor Authority is paid US\$ $\phi(x^0,y^0)$ if the drunk falls into the water at the point (x^0,y^0).

Suppose next that n drunks emerge from the same pub and each performs an independent meander through the city grid. Let $\tilde{u}_n$ be the average fine paid by the n drunks. It is then possible to prove that

$$\tilde{u} := \lim_{n\to\infty} \tilde{u}_n$$

is the solution of the five-point formula for the Laplace equation (with the Dirichlet boundary condition ϕ) at the grid point (Twentieth Street, Sixth Avenue).

Before trying this algorithm (hopefully, playing the part of the Harbor Authority), the reader had better be informed that the speed of convergence of $\tilde{u}_n$ to $\tilde{u}$ is $\mathcal{O}\big(1/\sqrt{n}\big)$. In other words, to obtain four significant digits we need 10^8 drunks.

This is an example of a *Monte Carlo* method (Kalos & Whitlock, 1986) and we hasten to add that such algorithms can be very valuable in other areas of scientific computing. In particular, if you are interested in solving the Laplace equation in several hundred space dimensions, a popular pastime in financial mathematics, just about the only viable approach is Monte Carlo.

Agmon, S. (1965), *Lectures on Elliptic Boundary Value Problems,* Van Nostrand, Princeton, NJ.

Ames, W.F. (1977), *Numerical Methods for Partial Differential Equations* (2nd ed.), Academic Press, New York.

Carrier, J., Greengard, L. and Rokhlin, V. (1988), A fast adaptive multipole algorithm for particle simulations, *SIAM Journal of Scientific and Statistical Computing* **9**, 669–686.

Chan, T.F. and Mathew, T.P. (1994), Domain decomposition algorithms, *Acta Numerica* **3**, 61–144.

Collatz, L. (1966), *The Numerical Treatment of Differential Equations* (3rd edn), Springer-Verlag, Berlin.

Fornberg, B. and Sloan, D.M. (1994), A review of pseudospectral methods for solving partial differential equations, *Acta Numerica* **3**, 203–268.

Hackbusch, W. (1992), *Elliptic Differential Equations: Theory and Numerical Treatment,* Springer-Verlag, Berlin.

Kalos, M.H. and Whitlock, P.A. (1986), *Monte Carlo Methods,* Wiley, New York.

Le Tallec, P. (1994), Domain decomposition methods in computational mechanics, *Computational Mechanics Advances* **1**, 121–220.

Exercises

8.1 Prove the identities

$$\Delta_- + \Delta_+ = 2\Upsilon_0\Delta_0, \qquad \Delta_-\Delta_+ = \Delta_0^2.$$

8.2 Show that formally

$$\mathcal{E} = \sum_{j=0}^{\infty} \Delta_-^j .$$

8.3 Demonstrate that for every $s \geq 1$ there exists a constant $c_s \neq 0$ such that

$$\frac{\mathrm{d}^s}{\mathrm{d}x^s} z(x) - \frac{1}{h^s}\left(\Delta_+^s - \tfrac{1}{2} s \Delta_+^{s+1}\right) z(x) = c_s \frac{\mathrm{d}^{s+2}}{\mathrm{d}x^{s+2}} z(x) h^2 + \mathcal{O}(h^3), \qquad h \to 0,$$

for every sufficiently smooth function z. Evaluate c_s explicitly for $s = 1, 2$.

8.4 For every $s \geq 1$ find a constant $d_s \neq 0$ such that

$$\frac{\mathrm{d}^{2s}}{\mathrm{d}x^{2s}} z(x) - \frac{1}{h^{2s}} \Delta_0^{2s} z(x) = d_s \frac{\mathrm{d}^{2s+2}}{\mathrm{d}x^{2s+2}} z(x) h^2 + \mathcal{O}(h^4), \qquad h \to 0,$$

for every sufficiently smooth function z. Compare the sizes of d_1 and c_2; what does this tell you about the errors in the forward difference and central difference approximations?

8.5 In this exercise we consider finite difference approximations to the derivative that use one point to the left and $s \geq 1$ points to the right of x.

a Determine constants α_j, $j = 1, 2, \ldots$, such that

$$\mathcal{D} = \frac{1}{h}\left(\beta\Delta_- + \sum_{j=1}^{\infty} \alpha_j \Delta_+^j\right),$$

where $\beta \in \mathbb{R}$ is given.

b Given an integer $s \geq 1$, show how to choose the parameter β so that

$$\mathcal{D} = \frac{1}{h}\left(\beta\Delta_- + \sum_{j=1}^{s} a_j \Delta_+^j\right) + \mathcal{O}(h^{s+1}), \qquad h \to 0.$$

8.6 Determine the order (in the form $\mathcal{O}((\Delta x)^p)$) of the finite difference approximation to $\partial^2/\partial x \partial y$ given by the computational stencil

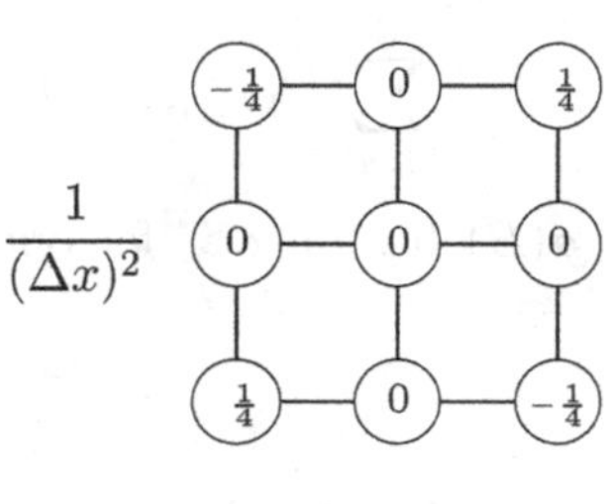

8.7 The five-point formula (8.15) is applied in an L-shaped domain of the form

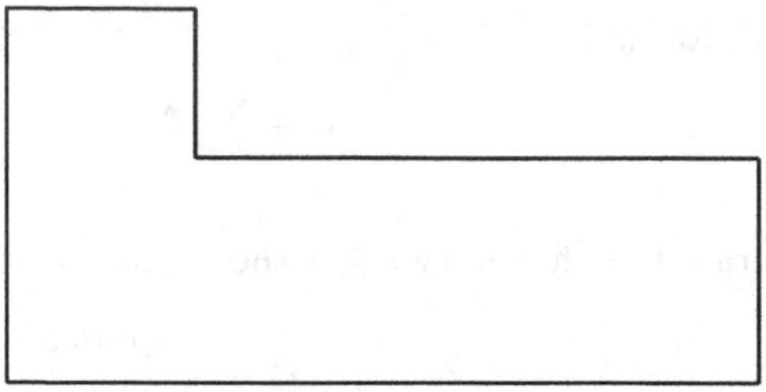

and we assume that all grid points are either internal or boundary (this is possible if the ratios of the sides are rational). Prove *without relying on Theorem 8.4 or on its method of proof* that the underlying matrix is nonsingular. (*Hint: Nothing prevents you from relying on the method of proof of Lemma 8.1.*)

8.8 In this exercise we prove, step by step, the Geršgorin criterion, which was stated in Lemma 8.3.

a Let $C = (c_{i,j})$ be an arbitrary $d \times d$ singular complex matrix. Then there exists $\boldsymbol{x} \in \mathbb{C}^d \setminus \{\mathbf{0}\}$ such that $C\boldsymbol{x} = \mathbf{0}$. Choose $\ell \in \{1, 2, \ldots, d\}$ such that

$$|x_\ell| = \max_{j=1,2,\ldots,d} |x_j| > 0.$$

By considering the ℓth row of $C\boldsymbol{x}$, prove that

$$|c_{\ell,\ell}| \le \sum_{j=1,\, j\neq\ell}^{d} |c_{\ell,j}|. \tag{8.35}$$

b Let B be a $d \times d$ matrix and choose $\lambda \in \sigma(B)$, where $\sigma(B)$ is the set containing the eigenvalues of B. Substituting $C = B - \lambda I$ in (8.35) prove that $\lambda \in \mathbb{S}_\ell$ (the Geršgorin discs $\mathbb{S}_i$ were defined in Lemma 8.3). Hence deduce that

$$\sigma(B) \subset \bigcup_{i=1}^{d} \mathbb{S}_i.$$

c Suppose that the matrix B is *irreducible* (A.1.2.5) and that the inequality (8.35) holds as an *equality* for some $\ell \in \{1, \ldots, d\}$; show that this equality implies that

$$|c_{k,k}| = \sum_{j=1,\, j\neq k}^{d} |c_{k,j}|, \qquad k = 1, 2, \ldots, d.$$

Deduce that if $\lambda \in \sigma(B)$ lies on $\partial\mathbb{S}_\ell$ for one ℓ then $\lambda \in \partial\mathbb{S}_k$ for all $k = 1, 2, \ldots, d$.

8.9 Prove that

$$\frac{1}{(\Delta x)^2}(\Delta_{0,x}^2 + \Delta_{0,y}^2) - \nabla^2 = \tfrac{1}{12}(\Delta x)^2(\nabla^4 - 2\mathcal{D}_x^2\mathcal{D}_y^2) + \mathcal{O}((\Delta x)^4), \qquad \Delta x \to 0.$$

8.10 Consider the d-dimensional Laplace operator

$$\nabla^2 = \sum_{j=1}^{d} \frac{\partial^2}{\partial x_j^2}.$$

Prove a d-dimensional generalization of (8.15),

$$\frac{1}{(\Delta x)^2} \sum_{j=1}^{d} \Delta_{0,x_j}^2 = \nabla^2 + \mathcal{O}\big((\Delta x)^2\big).$$

8.11* Let ∇^2 again denote the d-dimensional Laplace operator and set

$$\mathcal{L}_{\Delta x} = -\tfrac{2}{3}\mathcal{I} + \tfrac{2}{3}\sum_{j=1}^{d} \Delta_{0,x_j}^2 + \tfrac{2}{3}\prod_{j=1}^{d}\left(\mathcal{I} + \tfrac{1}{2}\Delta_{0,x_j}^2\right),$$

$$\mathcal{M}_{\Delta x} = \mathcal{I} + \tfrac{1}{12}\sum_{j=1}^{d} \Delta_{0,x_j}^2,$$

where $\mathcal{I}$ is the identity operator.

a Prove that

$$\frac{1}{(\Delta x)^2}\mathcal{L}_{\Delta x} = \nabla^2 + \tfrac{1}{12}(\Delta x)^2\nabla^4 + \mathcal{O}\big((\Delta x)^4\big), \qquad \Delta x \to 0$$

and

$$\mathcal{M}_{\Delta x} = \mathcal{I} + \tfrac{1}{12}(\Delta x)^2\nabla^2 + \mathcal{O}\big((\Delta x)^4\big), \qquad \Delta x \to 0.$$

b Deduce that the method

$$\mathcal{L}_{\Delta x} u_{k_1,k_2,\ldots,k_d} = (\Delta x)^2 \mathcal{M}_{\Delta x} f_{k_1,k_2,\ldots,k_d}$$

solves the d-dimensional Poisson equation $\nabla^2 u = f$ with an order-$\mathcal{O}\big((\Delta x)^4\big)$ error. (*This is the multivariate generalization of the modified nine-point formula* (8.32).)

8.12 Prove that the computational stencil (8.34) for the solution of the biharmonic equation $\nabla^4 u = f$ is equivalent to the finite difference representation

$$(\Delta_{0,x}^2 + \Delta_{0,y}^2)^2 u_{k,\ell} = (\Delta x)^4 f_{k,\ell},$$

and thereby deduce the order of the error.

8.13* Find the equivalent of the five-point formula in a computational grid of equilateral triangles,

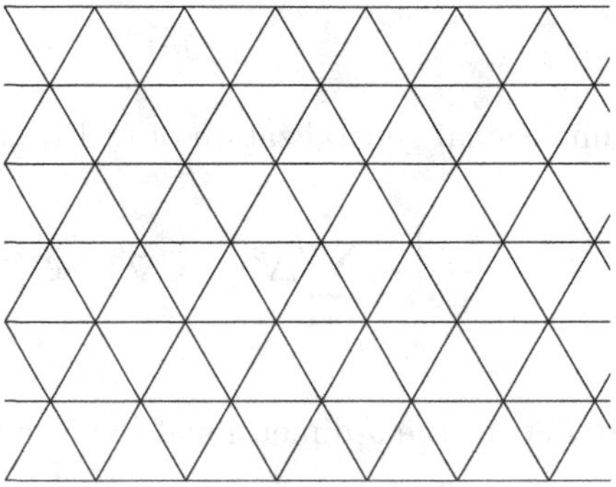

Your scheme should couple each internal grid point with its six nearest neighbours. What is the order of the error?

9

The finite element method

9.1 Two-point boundary value problems

The *finite element method* (FEM) presents to all those who were weaned on finite differences an entirely new outlook on the computation of a numerical solution for differential equations. Although it is often encapsulated in a few buzzwords – 'weak solution', 'Galerkin', 'finite element functions' – an understanding of the FEM calls not just for a different frame of mind but also for the comprehension of several principles. Each principle is important but it is their combination that makes the FEM into such an effective computational tool.

Instead of commencing our exposition from the deep end, let us first examine in detail a simple example, the Poisson equation in just *one* space variable. In principle, such an equation is $u'' = f$, but this is clearly too trivial for our purposes since it can be readily solved by integration. Instead, we adopt a more ambitious goal and examine linear *two-point boundary value problems*

$$-\frac{\mathrm{d}}{\mathrm{d}x}\left[a(x)\frac{\mathrm{d}u}{\mathrm{d}x}\right] + b(x)u = f, \qquad 0 \le x \le 1, \tag{9.1}$$

where a, b and f are given functions, a is differentiable and $a(x) > 0$, $b(x) \ge 0$, $0 < x < 1$. Any equation of the form (9.1) must be specified in tandem with proper initial or boundary data. For the time being, we assume Dirichlet boundary conditions

$$u(0) = \alpha, \qquad u(1) = \beta. \tag{9.2}$$

Two-point boundary problems (9.1) abound in applications, e.g. in mechanics, and their numerical solution is of independent interest. However, in the context of this section, our main motivation is to use them as a paradigm for a general linear boundary value problem and as a vehicle for the description of the FEM.

Throughout this chapter we extensively employ the terminology of linear spaces, inner products and norms. The reader might wish to consult appendix section A.2.1 for the relevant concepts and definitions.

Instead of estimating the solution of (9.1) on a grid, a practice that has underlain the discourse of previous chapters, we wish to approximate u by a linear combination of functions in a finite-dimensional space. We choose a function φ_0 that obeys the boundary conditions (9.2) and a set of m linearly independent functions $\varphi_1, \varphi_2, \ldots, \varphi_m$ that satisfy the zero boundary conditions $\varphi_\ell(0) = \varphi_\ell(1) = 0$, $\ell = 1, 2, \ldots, m$. In addition, these functions need to satisfy certain smoothness conditions, but it is best to leave this question in abeyance for a while. Our goal is to represent u approximately in the form

$$u_m(x) = \varphi_0(x) + \sum_{\ell=1}^{m} \gamma_\ell \varphi_\ell(x), \qquad 0 \le x \le 1, \tag{9.3}$$

where $\gamma_1, \gamma_2, \ldots, \gamma_m$ are real constants. In other words, let

$$\mathring{\mathbb{H}}_m := \operatorname{Sp}\{\varphi_1, \varphi_2, \ldots, \varphi_m\},$$

the *span* of $\varphi_1, \varphi_2, \ldots, \varphi_m$, be the set of all linear combinations of these functions. Since the φ_ℓ, $\ell = 1, 2, \ldots, m$, are linearly independent, $\mathring{\mathbb{H}}_m$ is an m-dimensional linear space. An alternative phrasing of (9.3) is that we seek

$$u_m - \varphi_0 \in \mathring{\mathbb{H}}_m$$

such that u_m approximates, in some sense, the solution of (9.1). Note that every member of $\mathring{\mathbb{H}}_m$ obeys zero boundary conditions. Hence, by design, u_m is bound to satisfy the boundary conditions (9.2). For future reference, we record this first principle of the FEM.

- *Approximate the solution in a finite-dimensional space.*

What might we mean, however, by the phrase 'approximates the solution of (9.1)'? One possibility, that we have already seen in Chapter 3, is *collocation:* we choose $\gamma_1, \gamma_2, \ldots, \gamma_m$ so as to satisfy the differential equation (9.1) at m distinct points in $[0, 1]$. Here, though, we shall apply a different and more general line of reasoning. For any choice of $\gamma_1, \gamma_2, \ldots, \gamma_m$ consider the *defect*

$$d_m(x) := -\frac{\mathrm{d}}{\mathrm{d}x}\left[a(x)\frac{\mathrm{d}u_m(x)}{\mathrm{d}x}\right] + b(x)u_m(x) - f(x), \qquad 0 < x < 1.$$

Were u_m the solution of (9.1), the defect would be identically zero. Hence, the nearer d_m is to the zero function, the better we can expect our candidate solution to be.

In the fortunate case when d_m itself lies in the space $\mathring{\mathbb{H}}_m$ for all $\gamma_1, \gamma_2, \ldots, \gamma_m$, the problem is simple. The zero function in an inner product space is identified by being orthogonal to all members of that space. Hence, we equip $\mathring{\mathbb{H}}_m$ with an inner product $\langle\,\cdot\,,\,\cdot\,\rangle$ and seek parameters $\gamma_0, \gamma_1, \ldots, \gamma_m$ such that

$$\langle d_m, \varphi_k \rangle = 0, \qquad k = 1, 2, \ldots, m. \tag{9.4}$$

Since $\{\varphi_1, \varphi_2, \ldots, \varphi_m\}$ is, by design, a basis of $\mathring{\mathbb{H}}_m$, it follows from (9.4) that d_m is orthogonal to all members of $\mathring{\mathbb{H}}_m$, hence that it is the zero function. The orthogonality conditions (9.4) are called the *Galerkin equations.*

In general, however, $d_m \notin \mathring{\mathbb{H}}_m$ and we cannot expect to solve (9.1) exactly from within the finite-dimensional space. Nonetheless, the principle of the last paragraph still holds good: we wish d_m to obey the orthogonality conditions (9.4). In other words, the goal that underlies our discussion is to render the defect orthogonal to the finite-dimensional space $\mathring{\mathbb{H}}_m$.[1] This, of course, represents valid reasoning only if $\mathring{\mathbb{H}}_m$ approximates well the infinite-dimensional linear space of all the candidate solutions of (9.1); more about this later. The second principle of the FEM is thus as follows.

- *Choose the approximation so that the defect is orthogonal to the space* $\mathring{\mathbb{H}}_m$.

Using the representation (9.3) and the linearity of the differential operator, the defect becomes

$$d_m = -\left[(a\varphi_0')' + \sum_{\ell=1}^{m} \gamma_\ell (a\varphi_\ell')'\right] + b\left[\varphi_0 + \sum_{\ell=1}^{m} \gamma_\ell \varphi_\ell\right] - f,$$

and substitution in (9.4) results, after an elementary rearrangement of terms, in

$$\sum_{\ell=1}^{m} \gamma_\ell [\langle -(a\varphi_\ell')', \varphi_k\rangle + \langle b\varphi_\ell, \varphi_k\rangle] = \langle f, \varphi_k\rangle - [\langle -(a\varphi_0')', \varphi_k\rangle + \langle b\varphi_0, \varphi_k\rangle], \qquad k = 1, 2, \ldots, m. \tag{9.5}$$

On the face of it, (9.4) is a linear system of m equations for the m unknowns $\gamma_1, \gamma_2, \ldots, \gamma_m$. However, before we rush to solve it, we must first perform a crucial operation, integration by parts.

Let us suppose that $\langle \cdot\,, \cdot \rangle$ is the standard Euclidean inner product over functions (the L_2 inner product; see A.2.1.4),

$$\langle v, w\rangle = \int_0^1 v(\tau)w(\tau)\,\mathrm{d}\tau.$$

It is defined over all functions v and w such that

$$\int_0^1 |v(\tau)|^2\,\mathrm{d}\tau, \quad \int_0^1 |w(\tau)|^2\,\mathrm{d}\tau \; < \; \infty. \tag{9.6}$$

Hence, (9.5) assumes the form

$$\sum_{\ell=1}^{m} \gamma_\ell \left(-\int_0^1 \{-[a(\tau)\varphi_\ell'(\tau)]'\varphi_k(\tau)\}\,\mathrm{d}\tau + \int_0^1 b(\tau)\varphi_\ell(\tau)\varphi_k(\tau)\,\mathrm{d}\tau \right)$$
$$= \int_0^1 f(\tau)\varphi_k(\tau)\,\mathrm{d}\tau - \left(-\int_0^1 \{-[a(\tau)\varphi_0'(\tau)]'\varphi_k(\tau)\}\,\mathrm{d}\tau + \int_0^1 b(\tau)\varphi_0(\tau)\varphi_k(\tau)\,\mathrm{d}\tau \right), \qquad k = 1, 2, \ldots, m.$$

Since for $k = 1, 2, \ldots, m$ the function φ_k vanishes at the endpoints, integration by parts may be carried out with great ease:

$$\int_0^1 \{-[a(\tau)\varphi_\ell'(\tau)]'\}\,\varphi_k(\tau)\,\mathrm{d}\tau = -a(\tau)\varphi_\ell(\tau)\varphi_k(\tau)\Big|_0^1 + \int_0^1 a(\tau)\varphi_\ell'(\tau)\varphi_k'(\tau)\,\mathrm{d}\tau$$
$$= \int_0^1 a(\tau)\varphi_\ell'(\tau)\varphi_k'(\tau)\,\mathrm{d}\tau, \qquad \ell = 0, 1, \ldots, m.$$

[1] Collocation fits easily into this formulation, provided the inner product is properly defined (see Exercise 9.1).

The outcome,

$$\sum_{\ell=1}^{m} a_{k,\ell}\gamma_\ell = \int_0^1 f(\tau)\varphi_k(\tau)\,\mathrm{d}\tau - a_{k,0}, \qquad k = 1, 2, \ldots, m, \tag{9.7}$$

where

$$a_{k,\ell} := \int_0^1 [a(\tau)\varphi'_\ell(\tau)\varphi'_k(\tau) + b(\tau)\varphi_\ell(\tau)\varphi_k(\tau)]\,\mathrm{d}\tau, \quad k = 1, 2, \ldots, m, \quad \ell = 0, 1, \ldots, m,$$

is a form of the Galerkin equations suitable for numerical work.

The reason why (9.7) is preferable to (9.5) lies in the choice of the functions φ_ℓ, which will be discussed soon. The whole point is that good choices of the basis functions (within the FEM framework) possess quite poor smoothness properties – in fact, the more we can lower the differentiability requirements, the wider the class of desirable basis functions that we might consider. Integration by parts takes away one derivative, hence we need no longer insist that the φ_ℓ are twice-differentiable in order to satisfy (9.6). Even once-differentiability is, in fact, too strong. Since the value of an integral is independent of the values that the integrand assumes on a *finite* set of points, it is sufficient to choose basis functions that are *piecewise differentiable* in $[0, 1]$. We will soon see that it is this lowering of the smoothness requirements through the agency of an integration by parts that confers important advantages on the finite element method. This is therefore our third principle.

- *Integrate by parts to depress to the maximal extent possible the differentiability requirements of the space* $\mathring{\mathbb{H}}_m$.

The importance of lowered smoothness and integration by parts ranges well beyond the FEM and numerical analysis.

◇ **Weak solutions** Let us pause and ponder for a while the meaning of the term ‘exact solution of a differential equation’, which we are using so freely and with such apparent abandon on these pages. Thus, suppose that we have an equation of the form $\mathcal{L}\boldsymbol{u} = \boldsymbol{f}$, where $\mathcal{L}$ is a differential operator – ordinary or partial, in one or several dimensions, linear or nonlinear –, provided with the right boundary and/or initial conditions. An obvious candidate for the term ‘exact solution’ is a function $\boldsymbol{u}$ that obeys the equation and the ‘side’ conditions – the *classical* solution. However, there is an alternative. Suppose that we are given an infinite-dimensional linear space $\mathring{\mathbb{H}}$ that is rich enough to include in its closure all functions of interest. Given a ‘candidate function’ $\boldsymbol{v} \in \mathring{\mathbb{H}}$, we define the defect as $\boldsymbol{d}(\boldsymbol{v}) := \mathcal{L}\boldsymbol{v} - \boldsymbol{f}$ and say that $\boldsymbol{v}$ is the *weak solution* of the differential equation if $\langle \boldsymbol{d}(\boldsymbol{v}), \boldsymbol{w}\rangle = 0$ for every $\boldsymbol{w} \in \mathring{\mathbb{H}}$.

On the face of it, we have not changed anything much – the naive point of view is that if the defect is orthogonal to the whole space then it is the zero function, hence $\boldsymbol{v}$ obeys the differential equation in the classical sense. This is a fallacy, inherited from a finite-dimensional intuition! The whole point is that, astutely integrating by parts, we are usually able to lower the differentiability requirements of $\mathring{\mathbb{H}}$ to roughly half those in the original equation. In other words, it is entirely possible for an equation to possess a weak solution that is neither a classical solution nor, indeed, can even be subjected to the action of the differential operator $\mathcal{L}$ in a naive way.

The distinction between classical and weak solutions makes little sense in the context of initial value problems for ODEs, since there the Lipschitz condition ensures the existence of a unique classical solution (which, of course, is also a weak solution). This is not the case with boundary value problems, and much of modern PDE analysis hinges upon the concept of a weak solution. $\diamond$

Suppose that the coefficients $a_{k,\ell}$ in (9.7) have been evaluated, whether explicitly or by means of quadrature (see Section 3.1). The system (9.7) comprises m linear equations in m unknowns and our next step is to employ a computer to solve it, thereby recovering the coefficients $\gamma_1, \gamma_2, \ldots, \gamma_m$ that render the best (in the sense of the underlying inner product) linear combination (9.3).

To introduce the FEM, we need another crucial ingredient, namely a specific choice of the set $\mathring{\mathbb{H}}_m$. There are, in principle, two objectives that we might seek in this choice. On the one hand, we might wish to choose the $\varphi_1, \varphi_2, \ldots, \varphi_m$ that, in some sense, are the most 'dense' in the infinite-dimensional space inhabited by the exact (weak) solution of (9.1). In other words, we might wish

$$\|u_m - u\| = \langle u_m - u, u_m - u \rangle^{1/2}$$

to be the smallest possible in $(0,1)$. This is a perfectly sensible choice, which results in the *spectral methods* that will be considered in the next chapter. Here we adopt a different goal. The dimension of the finite-dimensional space from which we are seeking the solution will be often very large, perhaps not in the particular case of (9.7), when $m \approx 100$ is at the upper end of what is reasonable, but definitely so in a multivariate case. To implement (9.7) (or its multivariate brethren) we need to calculate approximately m^2 integrals and solve a linear system of m equations. This might be a very expensive process and thus a second reasonable goal is to choose $\varphi_1, \varphi_2, \ldots, \varphi_m$ so as to make this task much easier.

The penultimate and crucial principle of the FEM is thus designed to save computational cost.

- *Choose each function φ_k so that it vanishes along most of* $(0,1)$, *thereby ensuring that $\varphi_k \varphi_\ell \equiv 0$ for most choices of $k, \ell = 1, 2, \ldots, m$.*

In other words, each function φ_k is supported on a relatively small set $\mathbb{E}_k \subset (0,1)$, say, and $\mathbb{E}_k \cap \mathbb{E}_\ell = \emptyset$ for as many $k, \ell = 1, 2, \ldots, m$ as possible.

Recall that, by virtue of integration by parts, we have narrowed the differentiability requirements so much that piecewise linear functions are perfectly acceptable in $\mathring{\mathbb{H}}_m$.

Setting $h = 1/(m+1)$, we choose

$$\varphi_k(x) = \begin{cases} 1 - k + \dfrac{x}{h}, & (k-1)h \le x \le kh, \\ 1 + k - \dfrac{x}{h}, & kh \le x \le (k+1)h, \\ 0, & |x - kh| \ge h, \end{cases} \qquad k = 1, 2, \ldots, m.$$

In other words, $\varphi_k = \psi(x/h - k)$, $k = 1, 2, \ldots, m$, where ψ is the *chapeau function* (also known as the *hat function*), represented as follows:

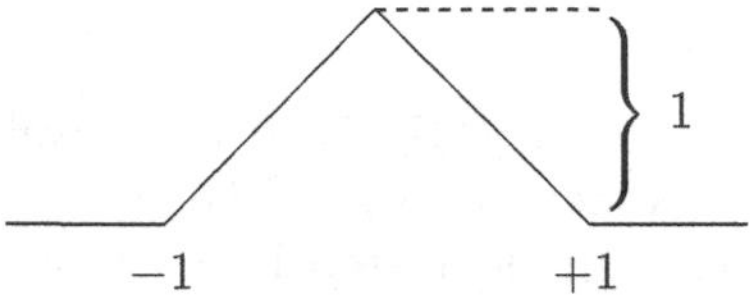

It can also be written in the form

$$\psi(x) = (x+1)_+ - 2(x)_+ + (x-1)_+ = (1 - |x|)_+, \qquad x \in \mathbb{R}, \tag{9.8}$$

where

$$(t)_+ := \begin{cases} t, & t \ge 0, \\ 0, & t < 0, \end{cases} \qquad t \in \mathbb{R}.$$

The advantages of this cryptic notation will become clear later. At present we draw attention to the fact that $\mathbb{E}_k = ((k-1)h, (k+1)h)$, and therefore

$$\mathbb{E}_k \cap \mathbb{E}_\ell = \begin{cases} ((k-1)h, (k+1)h), & k = \ell, \\ ((k-1)h, kh), & k = \ell + 1, \\ (kh, (k+1)h), & k = \ell - 1, \\ \emptyset, & |k - \ell| \ge 2, \end{cases} \qquad k, \ell = 1, 2, \ldots, m.$$

Therefore the matrix in (9.7) becomes tridiagonal. This means firstly that we need to evaluate just $\mathcal{O}(m)$, rather than $\mathcal{O}(m^2)$, integrals and secondly that the solution of triangular linear systems is very easy indeed (see Section 11.1).

◇ **Spectral methods vs. the FEM** We attempt to solve the equation

$$-\frac{\mathrm{d}}{\mathrm{d}x}\left[(1+x)\frac{\mathrm{d}u}{\mathrm{d}x}\right] + \frac{1}{1+x}u = \frac{2}{1+x}, \tag{9.9}$$

accompanied by the Dirichlet boundary conditions

$$u(0) = 0, \qquad u(1) = 1,$$

using two distinct choices of the functions $\varphi_1, \varphi_2, \ldots, \varphi_m$. Firstly, we let

$$\varphi_k(x) = \sin k\pi x, \qquad 0 \le x \le 1, \quad k = 1, 2, \ldots, m \tag{9.10}$$

and force the boundary conditions by setting

$$\varphi_0(x) = \sin \frac{\pi x}{2}, \qquad 0 \le x \le 1.$$

This is an example of a spectral method, although we hasten to confess that it is blatantly biased, since the boundary conditions are not of the right sort for spectral methods: more even-handed treatment of this construct must await Chapter 10.

The sprinter in FEM colours is the piecewise linear approximation that we have just described, i.e.,

$$\varphi_k(x) = \psi((m+1)x - k), \qquad 0 \le x \le 1 \qquad k = 1, 2, \ldots, m. \tag{9.11}$$

Boundary conditions are recovered by choosing

$$\varphi_0(x) = \psi((m+1)(x-1)), \qquad 0 \le x \le 1;$$

note that $\varphi_0(0) = 0$, $\varphi_0(1) = 1$, as required. Of course, the support of φ_0 extends beyond $(0, 1)$ but this is of no consequence, since its integration is restricted to the interval $(m/(m+1), 1)$. The errors for (9.10) and (9.11) are displayed in Figs. 9.1 and 9.2, respectively. At a first glance, both methods are performing well but the FEM has a slight edge, because of the large wiggles in Fig. 9.1 near the endpoints. This *Gibbs effect* is the penalty that we endure for attempting to approximate the non-periodic solution

$$u(x) = \frac{2x}{1+x}, \qquad 0 \le x \le 1,$$

with trigonometric functions. If we disregard the vicinity of the endpoints, the spectral method performs marginally better.

The error decays roughly quadratically in both figures, the latter consistently with the estimate $\|u_m - u\| = \mathcal{O}(m^{-2})$ (which, as we will see in Section 9.2, happens to be the correct order of magnitude). Had we played to the strengths of spectral methods by taking periodic boundary conditions, the error would have decayed at an *exponential* speed and the FEM would have been left out of the running altogether. Leaping to the defence of the FEM, we remark that it took more than a hundredfold in terms of computer time to produce Fig. 9.1 in comparison with Fig. 9.2. ◇

All the essential ingredients that together make the FEM are in now place, except for one. To clarify it, we describe an alternative methodology leading to the Galerkin equations (9.7).

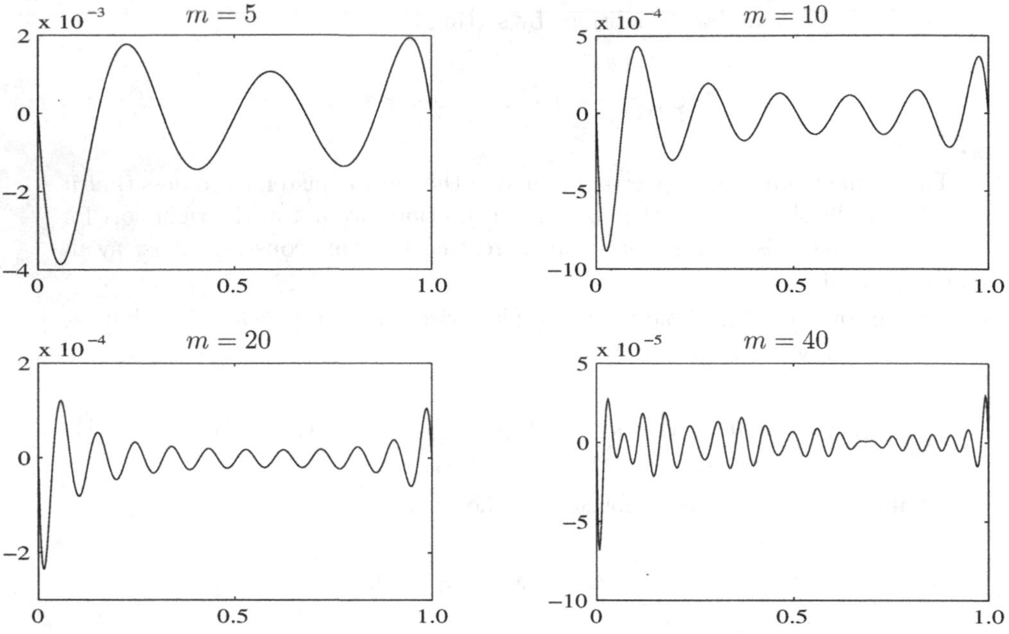

Figure 9.1 The error in (9.3) when equation (9.9) is solved using the spectral method (9.10).

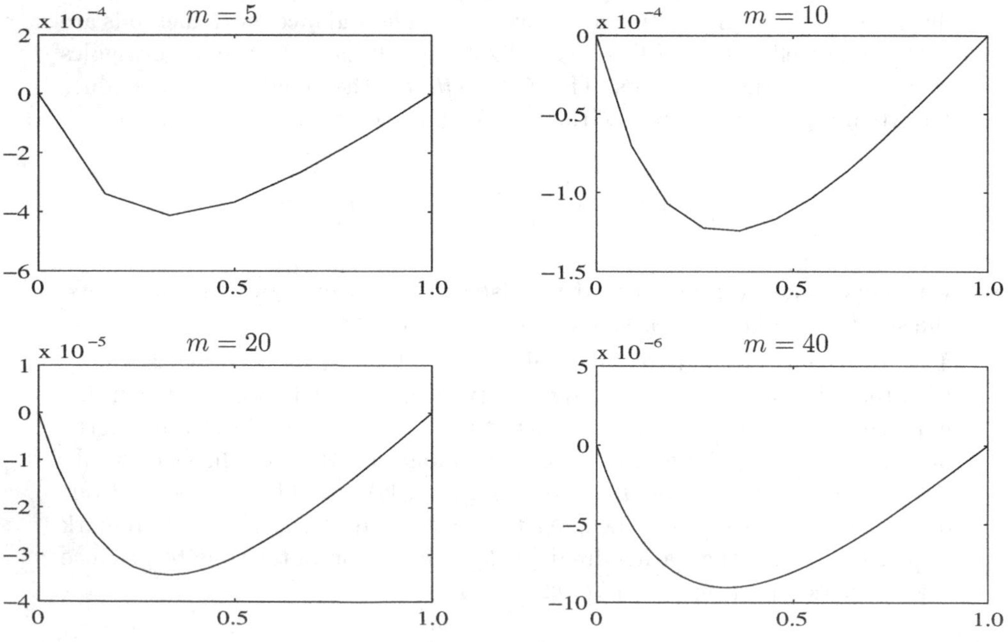

Figure 9.2 The error in (9.3) when the equation (9.9) is solved using the FEM (9.11).

Many differential equations of practical interest start their life as *variational problems* and only subsequently are converted to a more familiar form by use of the *Euler–Lagrange* equations. This gives little surprise to physicists, since the primary truth about physical models is not that derivatives (velocity, momentum, acceleration etc.) are somehow linked to the state of a system but that they arrange themselves according to the familiar principles of least action and least expenditure of energy. It is only mathematical ingenuity that renders this in the terminology of differential equations!

In a general variational problem we are given a *functional* $\mathcal{J} : \mathbb{H} \to \mathbb{R}$, where $\mathbb{H}$ is some function space, and we wish to find a function $u \in \mathbb{H}$ such that

$$\mathcal{J}(u) = \min_{v\in\mathbb{H}} \mathcal{J}(v).$$

Let us consider the following variational problem. Three functions, a, b and f are given in the interval $(0,1)$ in which we stipulate $a(x) > 0$ and $b(x) \geq 0$. The space $\mathbb{H}$ consists of all functions v that obey the boundary conditions (9.2) and

$$\int_0^1 v^2(\tau)\,\mathrm{d}\tau,\ \int_0^1 [v'(\tau)]^2\,\mathrm{d}\tau < \infty,$$

and we let

$$\mathcal{J}(v) := \int_0^1 \left\{ a(\tau)[v'(\tau)]^2 + b(\tau)[v(\tau)]^2 - 2f(\tau)v(\tau) \right\} \mathrm{d}\tau, \qquad v \in \mathbb{H}. \tag{9.12}$$

It is possible to prove that $\inf_{v\in\mathbb{H}} \mathcal{J}(v) > -\infty$ (see Exercise 9.7). Moreover, since the space $\mathbb{H}$ is complete,[2] every infimum is attainable within it and the operations inf and min become equal. Hence our variational problem always possesses a solution.

The space $\mathbb{H}$ is not a *linear* function space since (unless $\alpha = \beta = 0$) it is not closed under addition or multiplication by a scalar. However, choose an arbitrary $u \in \mathbb{H}$ and let $\mathring{\mathbb{H}} = \{v - u \,:\, v \in \mathbb{H}\}$. Therefore, all functions in $\mathring{\mathbb{H}}$ obey *zero boundary conditions.* Unlike $\mathbb{H}$, the set $\mathring{\mathbb{H}}$ is a linear space and it is trivial to prove that each function $v \in \mathbb{H}$ can be written in a unique way as $v = u + w$, where $w \in \mathring{\mathbb{H}}$. We denote this by $\mathbb{H} = u + \mathring{\mathbb{H}}$ and say that $\mathbb{H}$ is an *affine* space.

Let us suppose that $u \in \mathbb{H}$ minimizes $\mathcal{J}$. In other words, and bearing in mind that $\mathbb{H} = u + \mathring{\mathbb{H}}$,

$$\mathcal{J}(u) \leq \mathcal{J}(u+v), \qquad v \in \mathring{\mathbb{H}}. \tag{9.13}$$

We choose $v \in \mathring{\mathbb{H}} \setminus \{0\}$ and a real number $\varepsilon \neq 0$. Then

$$\begin{aligned}\mathcal{J}(u+\varepsilon v) &= \int_0^1 \left[a(u'+\varepsilon v')^2 + b(u+\varepsilon v) - 2f(u+\varepsilon v)\right]\,\mathrm{d}\tau \\ &= \int_0^1 \left\{a\left[(u')^2 + 2\varepsilon u'v' + \varepsilon^2(v')^2\right] + b\left(u^2 + 2\varepsilon uv + \varepsilon^2 v^2\right) - 2f(u+\varepsilon v)\right\}\mathrm{d}\tau\end{aligned}$$

[2] Unless you know functional analysis, do not try to prove this – accept it as an act of faith... This is perhaps the place to mention that $\mathbb{H}$ is rich enough to contain all piecewise differentiable functions in $(0,1)$ but, in order to be complete, it must contain many other functions as well.

$$= \int_0^1 \left[a(u')^2 + bu^2 - 2fu\right] \mathrm{d}\tau + 2\varepsilon \int_0^1 (au'v' + buv - fv) \,\mathrm{d}\tau$$
$$+ \varepsilon^2 \int_0^1 \left[a(v')^2 + bv^2\right] \mathrm{d}\tau$$
$$= \mathcal{J}(u) + 2\varepsilon \int_0^1 (au'v' + buv - fv) \,\mathrm{d}\tau + \varepsilon^2 \int_0^1 \left[a(v')^2 + bv^2\right] \mathrm{d}\tau. \qquad (9.14)$$

To be consistent with (9.13), we require, replacing v by εv,

$$2\varepsilon \int_0^1 (au'v' + buv) \,\mathrm{d}\tau + \varepsilon^2 \int_0^1 \left[a(v')^2 + bv^2\right] \mathrm{d}\tau \ge 0.$$

As $|\varepsilon| > 0$ can be made arbitrarily small, we can make the second term negligible, thereby deducing that

$$\varepsilon \int_0^1 (au'v' + buv - fv) \,\mathrm{d}\tau \ge 0.$$

Recall that no assumptions have been made with regard to ε, except that it is nonzero and that its magnitude is adequately small. In particular, the inequality is valid when we replace ε by $-\varepsilon$, and we therefore deduce that

$$\int_0^1 \left[a(\tau)u'(\tau)v'(\tau) + b(\tau)u(\tau)v(\tau) - f(\tau)v(\tau)\right] \mathrm{d}\tau = 0. \qquad (9.15)$$

We have just proved that (9.15) is necessary for u to be the solution of the variational problem, and it is easy to demonstrate that it is also sufficient. Thus, assuming that the identity is true, (9.14) (with $\varepsilon = 1$) gives

$$\mathcal{J}(u + v) = \mathcal{J}(u) + \int_0^1 \left[a(v')^2 + bv^2\right] \mathrm{d}\tau \ge \mathcal{J}(u), \qquad v \in \mathring{\mathbb{H}}.$$

Since $\mathbb{H} = u + \mathring{\mathbb{H}}$, it follows that u indeed minimizes $\mathcal{J}$ in $\mathbb{H}$.

Identity (9.15) possesses a further remarkable property: it is the weak form (in the Euclidean norm) of the two-point boundary value problem (9.1). This is easy to ascertain using integration by parts in the first term. Since $v \in \mathring{\mathbb{H}}$, it vanishes at the endpoints and (9.15) becomes

$$\int_0^1 \left\{- \left[a(\tau)u'(\tau)\right]' + b(\tau)u(\tau) - f(\tau)\right\} v(\tau) \,\mathrm{d}\tau = 0, \qquad v \in \mathring{\mathbb{H}}.$$

In other words, the function u is a solution of the variational problem (9.12) *if and only if it is the weak solution of the differential equation* (9.1).[3] We thus say that the two-point boundary value problem (9.1) is the *Euler–Lagrange* equation of (9.12).

Traditionally, variational problems have been converted into their Euler–Lagrange counterparts but, so far as obtaining a numerical solution is concerned, we may attempt to approximate (9.12) rather than (9.1). The outcome is the *Ritz method.*

[3] This proves, incidentally, that the solution of (9.12) is unique, but you may try to prove uniqueness directly from (9.14).

Let $\varphi_1, \varphi_2, \ldots, \varphi_m$ be linearly independent functions in $\mathring{\mathbb{H}}$ and choose an arbitrary $\varphi_0 \in \mathbb{H}$. As before, $\mathring{\mathbb{H}}_m$ is the m-dimensional linear space spanned by φ_k, $k = 1, 2, \ldots, m$. We seek a minimum of $\mathcal{J}$ in the m-dimensional affine space $\varphi_0 + \mathring{\mathbb{H}}_m$. In other words, we seek a vector $\boldsymbol{\gamma} = [\ \gamma_1 \quad \gamma_2 \quad \cdots \quad \gamma_m\ ]^\top \in \mathbb{R}^m$ that minimizes

$$\mathcal{J}_m(\boldsymbol{\delta}) := \mathcal{J}\left(\varphi_0 + \sum_{\ell=1}^{m} \delta_\ell \varphi_\ell\right), \qquad \boldsymbol{\delta} \in \mathbb{R}^m.$$

The functional $\mathcal{J}_m$ acts on just m variables and its minimization can be accomplished, by well-known rules of calculus, by letting the gradient equal zero. Since $\mathcal{J}_m$ is quadratic in its variables,

$$\mathcal{J}_m(\boldsymbol{\delta}) = \int_0^1 \left[a\left(\varphi_0' + \sum_{\ell=1}^{m} \delta_\ell \varphi_\ell'\right)^2 + b\left(\varphi_0 + \sum_{\ell=1}^{m} \delta_\ell \varphi_\ell\right)^2 - 2f\left(\varphi_0 + \sum_{\ell=1}^{m} \delta_\ell \varphi_\ell\right)\right] \mathrm{d}\tau,$$

the gradient is easy to calculate. Thus,

$$\frac{1}{2}\frac{\partial \mathcal{J}_m(\boldsymbol{\delta})}{\partial \delta_k} = \sum_{\ell=1}^{m} \int_0^1 (a\varphi_\ell'\varphi_k' + b\varphi_\ell\varphi_\ell)\,\mathrm{d}\tau + \int_0^1 (a\varphi_0'\varphi_k' + b\varphi_0\varphi_k)\,\mathrm{d}\tau - \int_0^1 f\varphi_k\,\mathrm{d}\tau.$$

Letting $\partial \mathcal{J}_m/\partial \delta_k = 0$ for $k = 1, 2, \ldots, m$ recovers exactly the form (9.7) of the Galerkin equations.

A careful reader will observe that setting the gradient to zero is merely a necessary condition for a minimum. For sufficiency we require in addition that the Hessian matrix $\left(\partial^2 \mathcal{J}_m/\partial \delta_k \partial \delta_j\right)_{k,j=1}^m$ is nonnegative definite. This is easy to prove (see Exercise 9.6).

What have we gained from the Ritz method? On the face of it, not much except for some additional insight, since it results in the same linear equations as the Galerkin method. This, however, ceases to be true for many other equations; in these cases Ritz and Galerkin result in genuinely different computational schemes. Moreover the variational formulation provides us with an important clue about how to deal with more complicated boundary conditions.

There is an important mismatch between the boundary conditions for variational problems and those for differential equations. Each differential equation requires the right amount of boundary data. For example, (9.1) requires two conditions, of the form

$$\alpha_{0,i}u(0) + \alpha_{1,i}u'(0) + \beta_{0,i}u(1) + \beta_{1,i}u'(1) = \gamma_i, \qquad i = 1, 2,$$

such that

$$\operatorname{rank}\begin{bmatrix} \alpha_{0,1} & \alpha_{1,1} & \beta_{0,1} & \beta_{1,1} \\ \alpha_{0,2} & \alpha_{1,2} & \beta_{0,2} & \beta_{1,2} \end{bmatrix} = 2.$$

Observe that (9.2) is a simple special case. However, a variational problem happily survives with less than a full complement of boundary data. For example, (9.12) can be defined with just a single boundary value, $u(0) = \alpha$, say. The rule is to replace in the Euler–Lagrange equations each 'missing' boundary condition by a *natural* boundary condition. For example, we complement $u(0) = \alpha$ with $u'(1) = 0$. (The proof is virtually identical to the reasoning that led us from (9.12) to the corresponding Euler–Lagrange equation (9.1), except that we need to use the natural boundary condition when integrating by parts.)

In the Ritz method we traverse the avenue connecting variational problems and differential equations in the opposite direction, from (9.1) to (9.15), say. This means that, whenever the two-point boundary value problem is provided with a natural boundary condition, we disregard it in the formation of the space $\mathbb{H}$. In other words, the function φ_0 need obey only the *essential* boundary conditions that survive in the variational problem. The *quid pro quo* for the disappearance of, say, $u'(1) = 0$ is that we need to add φ_{m+1} (defined consistently with (9.11)) to our space and an extra equation, for $k = m + 1$, to the linear system (9.7); otherwise, by default, we are imposing $u(1) = 0$, which is wrong.

◇ **A natural boundary condition** Consider the equation

$$-u'' + u = 2e^{-x}, \qquad 0 \le x \le 1, \tag{9.16}$$

given in tandem with the boundary conditions

$$u(0) = 0, \qquad u'(1) = 0.$$

The exact solution is easy to find: $u(x) = xe^{-x}$, $0 \le x \le 1$.

Fig. 9.3 displays the error in the numerical solution of (9.16) using the piecewise linear chapeau functions (9.8). Note that there is no need to provide the 'boundary function' φ_0 at all. It is evident from the figure that the algorithm, as expected, is clever enough to recover the correct natural boundary condition at $x = 1$. Another observation, which the figure shares with Fig. 9.2, is that the decay of the error is consistent with $\mathcal{O}(m^{-2})$.

Why not, one may ask, impose the natural boundary condition at $x = 1$? The obvious reason is that we cannot employ a chapeau function for that purpose, since its derivative will be discontinuous at the endpoint. Of course, we might instead use a more complicated function but, unsurprisingly, such functions complicate matters needlessly. ◇

A natural boundary condition is just one of several kinds of boundary data that undergo change when differential equations are solved with the FEM. We do not wish to delve further into this issue, which is more than adequately covered in specialized texts. However, and to remind the reader of the need for proper respect towards boundary data, we hereby formulate our last principle of the FEM.

- *Retain only essential boundary conditions.*

Throughout this section we have identified several principles that combine to give the finite element method. Let us repeat them with some reformulation and

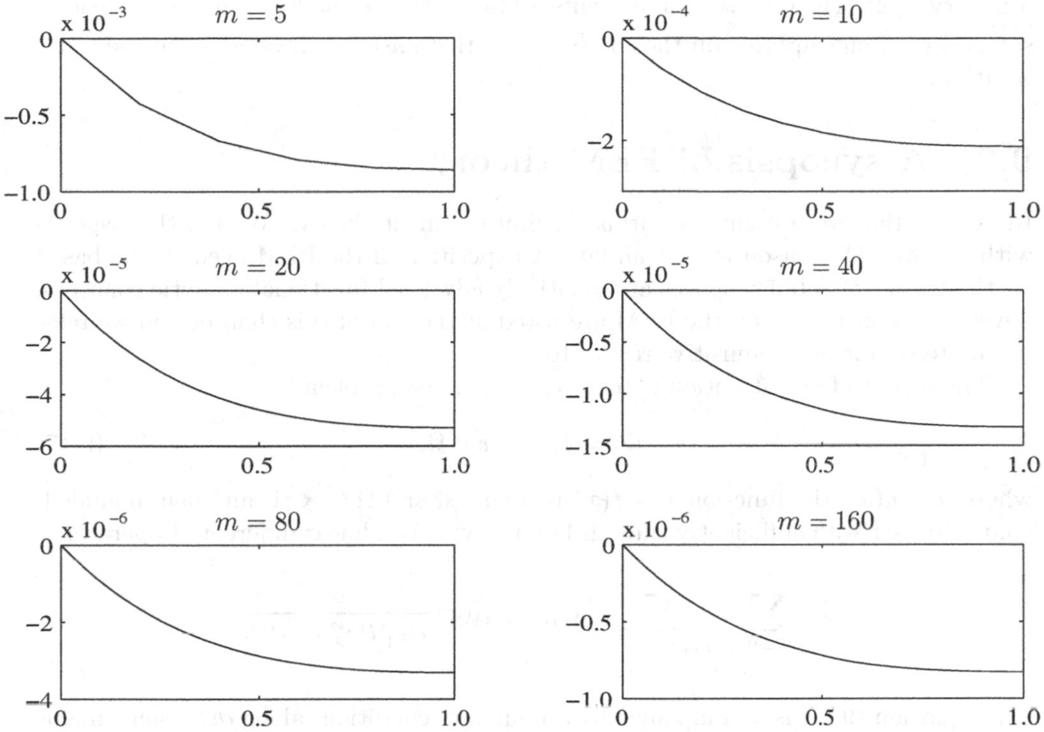

Figure 9.3 The error in the solution of the equation (9.16) with boundary data $u(0) = 0$, $u'(1) = 0$, by the Ritz–Galerkin method with chapeau functions.

also some reordering.

- *Approximate the solution in a finite-dimensional space* $\varphi_0 + \mathring{\mathbb{H}}_m \subset \mathbb{H}$.

- *Retain only essential boundary conditions.*

- *Choose the approximant so that the defect is orthogonal to* $\mathring{\mathbb{H}}_m$ *or, alternatively, so that a variational problem is minimized in* $\mathring{\mathbb{H}}_m$.

- *Integrate by parts to depress to the maximal extent possible the differentiability requirements of the space* $\mathring{\mathbb{H}}_m$.

- *Choose each function in a basis of* $\mathring{\mathbb{H}}_m$ *in such a way that it vanishes along much of the spatial domain of interest, thereby ensuring that the intersection between the supports of most of the basis functions is empty.*

Needless to say, there is much more to the FEM than these five principles. In particular, we wish to specify $\mathring{\mathbb{H}}_m$ so that for sufficiently large m the numerical solution converges to the exact (weak) solution of the underlying equation – and, preferably, converges fairly fast. This is a subject that has attracted enough research to fill many a library shelf. The next section presents a brief review of the FEM in a more general setting, with an emphasis on the choice of $\mathring{\mathbb{H}}_m$ that ensures convergence to the exact solution.

9.2 A synopsis of FEM theory

In this section we present an outline of finite element theory. We mostly dispense with proofs. The reason is that an honest exposition of the FEM needs to be based on the theory of Sobolev spaces and relatively advanced functional-analytic concepts. Several excellent texts on the FEM are listed at the end of this chapter and we refer the more daring and inquisitive reader to these.

The object of our attention is the boundary value problem

$$\mathcal{L}u = f, \qquad \boldsymbol{x} \in \Omega, \tag{9.17}$$

where $u = u(\boldsymbol{x})$, the function $f = f(\boldsymbol{x})$ is bounded and $\Omega \subset \mathbb{R}^d$ is an open, bounded, connected set with sufficiently smooth boundary; $\mathcal{L}$ is a linear differential operator,

$$\mathcal{L} = \sum_{k=0}^{2\nu} \sum_{\substack{i_1+i_2+\cdots+i_d=k \\ i_1,i_2,\ldots,i_d\geq 0}} c_{i_1,i_2,\ldots,i_d}(\boldsymbol{x}) \frac{\partial^k}{\partial x_1^{i_1} \partial x_2^{i_2} \cdots \partial x_d^{i_d}}.$$

The equation (9.17) is accompanied by ν boundary conditions along $\partial\Omega$ – some might be essential, others natural, but we will not delve further into this issue.

Let $\mathbb{H}$ be the affine space of all functions which act in Ω, whose νth derivative is square-integrable[4] and which obey all *essential* boundary conditions along $\partial\Omega$. We let $\mathring{\mathbb{H}} = \mathbb{H} - u$, where $u \in \mathbb{H}$ is arbitrary, and note that $\mathring{\mathbb{H}}$ is a linear space of functions that satisfy zero boundary conditions.

We equip ourselves with the Euclidean inner product

$$\langle v, w \rangle = \int_\Omega v(\boldsymbol{\tau}) w(\boldsymbol{\tau}) \, \mathrm{d}\boldsymbol{\tau}, \qquad v, w \in \mathbb{H},$$

and the inherited Euclidean norm

$$\|v\| = \{\langle v, v \rangle\}^{1/2}, \qquad v \in \mathbb{H}.$$

Note that we have designed $\mathring{\mathbb{H}}$ so that terms of the form $\langle \mathcal{L}v, w \rangle$ make sense for every $v, w \in \mathbb{H}$, but this is true only subject to integration by parts ν times, to depress the degree of derivatives inside the integral from 2ν down to ν. If $d \geq 2$ we need to use various multivariate counterparts of integration by parts, of which perhaps the most useful are the *divergence theorem*

$$\int_\Omega \nabla \cdot [a(\boldsymbol{x}) \nabla v(\boldsymbol{x})] w(\boldsymbol{x}) \, \mathrm{d}\boldsymbol{x} = \int_{\partial\Omega} a(\boldsymbol{s}) w(\boldsymbol{s}) \frac{\partial v(\boldsymbol{s})}{\partial n} \, \mathrm{d}\boldsymbol{s} - \int_\Omega a(\boldsymbol{x}) [\nabla v(\boldsymbol{x})] \cdot [\nabla w(\boldsymbol{x})] \, \mathrm{d}\boldsymbol{x},$$

[4] As we have already seen in Section 9.1, this does not mean that the νth derivative exists everywhere in Ω.

and *Green's formula*

$$\int_\Omega [\nabla^2 v(\boldsymbol{x})]w(\boldsymbol{x})\,\mathrm{d}\boldsymbol{x} + \int_\Omega [\nabla v(\boldsymbol{x})]\cdot[\nabla w(\boldsymbol{x})]\,\mathrm{d}\boldsymbol{x} = \int_{\partial\Omega} \frac{\partial v(\boldsymbol{s})}{\partial n} w(\boldsymbol{s})\,\mathrm{d}\boldsymbol{s}.$$

Here $\nabla = \begin{bmatrix} \partial/\partial x_1 & \partial/\partial x_2 & \cdots & \partial/\partial x_d \end{bmatrix}^\top$, while $\partial/\partial n$ is the derivative in the direction of the outward normal to the boundary $\partial\Omega$.[5] Both the divergence theorem and the Green formula are special cases of *Stokes's theorem,* which is outside the scope of our exposition.

Given a linear differential operator $\mathcal{L}$ from (9.17), we define a bilinear form $\tilde{a}(\,\cdot\,,\,\cdot\,)$ such that $\tilde{a}(v,w) = \langle \mathcal{L}v, w\rangle$ for sufficiently smooth functions v and w (i.e. $v \in C^{2\nu}(\mathrm{cl}\,\Omega)$, $w \in \mathbb{H}$) and note that $\tilde{a}(v,w)$, unlike $\langle \mathcal{L}v, w\rangle$, remains meaningful when $v, w \in \mathbb{H}$. The operator $\mathcal{L}$ is said to be

self-adjoint	if $\tilde{a}(v,w) = \tilde{a}(w,v)$ for all $v, w \in \mathring{\mathbb{H}}$;
elliptic	if $\tilde{a}(v,v) > 0$ for all $v \in \mathring{\mathbb{H}}$; and
positive definite	if it is both self-adjoint and elliptic.

An important example of a positive definite operator is

$$\mathcal{L} = -\sum_{i=1}^{d} \frac{\partial}{\partial x_i} \sum_{j=1}^{d} b_{i,j}(\boldsymbol{x}) \frac{\partial}{\partial x_j}, \tag{9.18}$$

where the matrix $B(\boldsymbol{x}) = (b_{i,j}(\boldsymbol{x}))$, $i,j = 1,2,\ldots,d$, is symmetric and positive definite for every $\boldsymbol{x} \in \Omega$. To prove this we use a variant of the divergence theorem. Since $w \in \mathring{\mathbb{H}}$, it vanishes along the boundary $\partial\Omega$ and it is easy to verify that

$$\begin{aligned} \langle \mathcal{L}v, w\rangle &= -\int_\Omega \left\{ \sum_{i=1}^{d} \frac{\partial}{\partial x_i} \left[\sum_{j=1}^{d} b_{i,j}(\boldsymbol{x}) \frac{\partial v(\boldsymbol{x})}{\partial x_j} \right] \right\} w(\boldsymbol{x})\,\mathrm{d}\boldsymbol{x} \\ &= \int_\Omega \sum_{i=1}^{d} \sum_{j=1}^{d} \left[\frac{\partial v(\boldsymbol{x})}{\partial x_i} \right] b_{i,j}(\boldsymbol{x}) \left[\frac{\partial w(\boldsymbol{x})}{\partial x_j} \right] \mathrm{d}\boldsymbol{x}. \end{aligned} \tag{9.19}$$

Note that, while the formal term $\langle \mathcal{L}v, w\rangle$ above requires v to be twice differentiable, integration by parts converts the integral into a form in which $v, w \in \mathring{\mathbb{H}}$ is allowed: this is precisely our bilinear form $\tilde{a}$. Since $b_{i,j} \equiv b_{j,i}$, $i,j = 1,2,\ldots,d$, we deduce that the last expression is symmetric in v and w. Therefore $\tilde{a}(v,w) = \tilde{a}(w,v)$ and $\mathcal{L}$ is self-adjoint. To prove ellipticity we let $w = v \not\equiv 0$ in (9.19); then

$$\tilde{a}(v,v) = \int_\Omega \sum_{i=1}^{d} \sum_{j=1}^{d} \left[\frac{\partial v(\boldsymbol{x})}{\partial x_i} \right] b_{i,j}(\boldsymbol{x}) \left[\frac{\partial v(\boldsymbol{x})}{\partial x_j} \right] \mathrm{d}\boldsymbol{x} > 0$$

[5] By rights, this means that the Laplace operator should be denoted by $\nabla^\top\nabla$, $\nabla\cdot\nabla$ or div grad, rather than ∇^2 (a distinction which becomes crucial in algebraic topology), and that, faithful to our convention, we should really use boldface to remind ourselves that ∇ is a vector. Regretfully, and with a heavy sigh, pedantry yields to convention.

by definition of the positive definiteness of matrices (A.1.3.5).

Note that both the negative of the Laplace operator, $-\nabla^2$, and the one-dimensional operator

$$-\frac{\mathrm{d}}{\mathrm{d}x}\left[a(x)\frac{\mathrm{d}}{\mathrm{d}x}\right]+b(x), \tag{9.20}$$

where $a(x) > 0$ and $b(x) \geq 0$ in the interval of interest, are special cases of (9.18); therefore they are positive definite.

Whenever a differential operator $\mathcal{L}$ is positive definite, we can identify the differential equation (9.17) with a variational problem, thereby setting the stage for the Ritz method.

Theorem 9.1 *Provided that the operator $\mathcal{L}$ is positive definite,* (9.17) *is the Euler–Lagrange equation of the variational problem*

$$\mathcal{J}(v) := \tilde{a}(v,v) - 2\langle f, v\rangle, \qquad v \in \mathbb{H}. \tag{9.21}$$

The weak solution of $\mathcal{L}u = f$ is therefore the unique minimum of $\mathcal{J}$ in $\mathbb{H}$.[6]

Proof We generalize an argument that has already been set out in Section 9.1 for the special case of the two-point boundary value problem (9.20).

Because of ellipticity, the variational functional $\mathcal{J}$ possesses a minimum (see Exercise 9.7). Let us denote a local minimum by $u \in \mathbb{H}$. Therefore, for any given $v \in \mathring{\mathbb{H}}$ and sufficiently small $|\varepsilon|$ we have

$$\mathcal{J}(u) \leq \mathcal{J}(u+\varepsilon v) = \tilde{a}(u+\varepsilon v, u+\varepsilon v) - 2\langle f, u+\varepsilon v\rangle.$$

The form $\tilde{a}$ being linear, this results in

$$\mathcal{J}(u) \leq [\tilde{a}(u,u) - 2\langle f,u\rangle] + \varepsilon[\tilde{a}(v,u) + \tilde{a}(u,v) - 2\langle f,v\rangle] + \varepsilon^2\tilde{a}(v,v)$$

and self-adjointness together with linearity yield

$$\mathcal{J}(u) \leq \mathcal{J}(u) + 2\varepsilon[\tilde{a}(u,v) - \langle f,v\rangle] + \varepsilon^2\tilde{a}(v,v).$$

In other words,

$$2\varepsilon[\tilde{a}(v,v) - \langle f,v\rangle] + \varepsilon^2\tilde{a}(v,v) \geq 0 \tag{9.22}$$

for all $v \in \mathring{\mathbb{H}}$ and sufficiently small $|\varepsilon|$.

Suppose that u is not a weak solution of (9.17). Then there exists $v \in \mathring{\mathbb{H}}$, $v \not\equiv 0$, such that $\tilde{a}(u,v) - \langle f,v\rangle \neq 0$. We may assume without loss of generality that this inner product is negative, otherwise we replace v by $-v$. It follows that, choosing sufficiently small $\varepsilon > 0$, we may render the expression on the left of (9.22) negative. Since this is forbidden by the inequality, we deduce that no such $v \in \mathring{\mathbb{H}}$ exists and u is indeed a weak solution of (9.17).

[6] An unexpected (and very valuable) consequence of this theorem is the existence and uniqueness of the solution of (9.17) in $\mathbb{H}$. Therefore Theorem 9.1 – like much of the material in this section – is relevant to both the analytic and numerical aspects of elliptic PDEs.

Assume, though, that $\mathcal{J}$ has several local minima in $\mathring{\mathbb{H}}$ and denote two such distinct functions by u_1 and u_2. Repeating our analysis with $\varepsilon = 1$ whilst replacing v by $u_2 - u_1 \in \mathring{\mathbb{H}}$ results in

$$\begin{aligned}\mathcal{J}(u_2) &= \mathcal{J}(u_1 + (u_2 - u_1)) \\ &= \mathcal{J}(u_1) + 2[\tilde{a}(u_1, u_2 - u_1) - \langle f, u_2 - u_1\rangle] + \tilde{a}(u_2 - u_1, u_2 - u_1). \qquad (9.23)\end{aligned}$$

We have just proved that $\tilde{a}(u_1, u_2 - u_1) - \langle f, u_2 - u_1\rangle = 0$, since u_1 locally minimizes $\mathcal{J}$. Moreover $\mathcal{L}$ is elliptic and $u_2 \neq u_1$, therefore $\tilde{a}(u_2 - u_1, u_2 - u_1) > 0$. Substitution into (9.23) yields the contradictory inequality $\mathcal{J}(u_1) < \mathcal{J}(u_1)$, thereby leading us to the conclusion that $\mathcal{J}$ possesses a single minimum in $\mathring{\mathbb{H}}$. ■

◇ **When is a zero really a zero?** An important yet subtle point in the theory of function spaces is the identity of the zero function. In other words, when are u_1 and u_2 really different? Suppose for example that u_2 is the same as u_1, except that it has a different value at just one point. This, clearly, will pass unnoticed by our inner product, which consists of integrals. In other words, if u_1 is a minimum of $\mathcal{J}$ (and a weak solution of (9.17)), then so is u_2; in this sense there is no uniqueness. In order to be distinct *in the sense of the function space* $\mathring{\mathbb{H}}$, u_1 and u_2 need to satisfy $\|u_2 - u_1\| > 0$. In the language of measure theory, they must differ on a set of positive Lebesgue measure.

The truth, seldom spelt out in elementary texts, is that a normed function space (i.e., a linear function space equipped with a norm) sometimes consists not of functions but of *equivalence classes* of functions: u_1 and u_2 are in the same equivalence class if $\|u_2 - u_1\| = 0$ (that is, if $u_2 - u_1$ is of measure zero). This is an artefact of function spaces defined on continua that has no counterpart in the more familiar vector spaces such as $\mathbb{R}^d$. Fortunately, as soon as this point is comprehended, we can, like everybody else, go back to our habit of referring to the members of $\mathring{\mathbb{H}}$ as 'functions'. ◇

The *Ritz method* for (9.17) (where $\mathcal{L}$ is presumed positive definite) is a straightforward generalization of the corresponding algorithm from the last section. Again, we choose $\varphi_0 \in \mathbb{H}$, let m linearly independent vectors $\varphi_1, \varphi_2, \ldots, \varphi_m \in \mathring{\mathbb{H}}$ span a finite-dimensional linear space $\mathring{\mathbb{H}}$ and seek a vector $\boldsymbol{\gamma} = \begin{bmatrix}\gamma_1 & \gamma_2 & \cdots & \gamma_m\end{bmatrix}^\top \in \mathbb{R}^m$ that will minimize

$$\mathcal{J}_m(\boldsymbol{\delta}) := \mathcal{J}\left(\varphi_0 + \sum_{\ell=1}^{m} \delta_\ell \varphi_\ell\right), \qquad \boldsymbol{\delta} \in \mathbb{R}^m.$$

We set the gradient of $\mathcal{J}_m$ to $\mathbf{0}$, and this results in the m linear equations (9.7), where

$$a_{k,\ell} = \tilde{a}(\varphi_k, \varphi_\ell), \qquad k = 1, 2, \ldots, m, \quad \ell = 0, 1, \ldots, m. \qquad (9.24)$$

Incidentally, the self-adjointness of $\mathcal{L}$ means that $a_{k,\ell} = a_{\ell,k}$, $k, \ell = 1, 2, \ldots, m$. This saves roughly half the work of evaluating integrals. Moreover, the symmetry of a matrix often simplifies the task of its numerical solution.

The general *Galerkin method* is also an easy generalization of the algorithm presented in Section 9.1 for the ODE (9.1). Again, we seek γ such that

$$\tilde{a}\left(\varphi_0 + \sum_{\ell=1}^{m} \gamma_\ell \varphi_\ell, \varphi_k\right) - \langle f, \varphi_k \rangle = 0, \qquad k = 1, 2, \ldots, m. \tag{9.25}$$

In other words, we endeavour to approximate a weak solution from a finite-dimensional space.

We have stipulated that $\mathcal{L}$ is linear, and this means that (9.25) is, again, nothing other than the linear system (9.7) with coefficients defined by (9.24). However, (9.25) makes sense even for nonlinear operators.

The existence and uniqueness of the solution of the Ritz–Galerkin equations (9.7) has already been addressed in Theorem 9.1. Another important statement is the *Lax–Milgram theorem,* which requires more than ellipticity but considerably less than self-adjointness. Moreover, it also provides a most valuable error estimate.

Given any $v \in \mathbb{H}$, we let

$$\|v\|_H := \left(\|v\|^2 + [\tilde{a}(v, v)]^{1/2}\right).$$

It is possible to prove that $\|\cdot\|_H$ is a norm – in fact, this is a special case of the famed *Sobolev norm* and it is the correct way of measuring distances in $\mathbb{H}$. We say that the bilinear form $\tilde{a}$ is

bounded if there exists $\delta > 0$ such that $|\tilde{a}(v, w)| \le \delta \|v\|_H \times \|w\|_H$ for every $v, w \in \mathbb{H}$; and

coercive if there exists $\kappa > 0$ such that $\tilde{a}(v, v) \ge \kappa \|v\|_H^2$ for every $v \in \mathbb{H}$.

Theorem 9.2 (The Lax–Milgram theorem) *Let $\mathcal{L}$ be linear, bounded and coercive and let $\mathbb{V}$ be a closed linear subspace of $\mathring{\mathbb{H}}$. There exists a unique $\tilde{u} \in \varphi_0 + \mathbb{V}$ such that*

$$\tilde{a}(\tilde{u}, v) - \langle f, v \rangle = 0, \qquad v \in \mathbb{V}$$

and

$$\|\tilde{u} - u\|_H \le \frac{\delta}{\kappa} \inf \{\|v - u\|_H : v \in \varphi_0 + \mathbb{V}\}, \tag{9.26}$$

where $\varphi_0 \in \mathbb{H}$ is arbitrary and u is a weak solution of (9.17) in $\mathbb{H}$. ■

The inequality (9.26) is sometimes called the *Céa lemma.*

The space $\mathbb{V}$ need not be finite dimensional. In fact, it could be the space $\mathring{\mathbb{H}}$ itself, in which case we would deduce from the first part of the theorem that the weak solution of (9.17) exists and is unique. Thus, exactly like Theorem 9.1, the Lax–Milgram theorem can be used for analytic, as well as numerical, ends.

A proof of the coercivity and boundedness of $\mathcal{L}$ is typically much more difficult than a proof of its positive definiteness. It suffices to say here that, for most domains of interest, it is possible to prove that the operator $-\nabla^2$ satisfies the conditions of the Lax–Milgram theorem. An essential step in this proof is the *Poincaré inequality:* there exists a constant c, dependent only on Ω, such that

$$\|v\| \le c \left\| \sum_{i=1}^{d} \frac{\partial v}{\partial x_i} \right\|, \qquad v \in \mathring{\mathbb{H}}\,.$$

As far as the FEM is concerned, however, the error estimate (9.26) is the most valuable consequence of the theorem. On the right-hand side we have a constant, δ/κ, which is independent of the choice of $\mathring{\mathbb{H}}_m = \mathbb{V}$ and of the norm of the *distance* of the exact solution u from the affine space $\varphi_0 + \mathring{\mathbb{H}}_m$. Of course, $\inf_{v \in \varphi_0 + \mathring{\mathbb{H}}_m} \|v - u\|_H$ is unknown, since we do not know u. The one piece of information, however, that is definitely true about u is that it lies in $\mathbb{H} = \varphi_0 + \mathring{\mathbb{H}}$. Therefore the distance from u to $\varphi_0 + \mathring{\mathbb{H}}_m$ can be bounded in terms of the distance of an *arbitrary* member $w \in \varphi_0 + \mathring{\mathbb{H}}$ from $\varphi_0 + \mathring{\mathbb{H}}_m$. The final observation is that φ_0 makes no difference to our estimates and we hence deduce that, subject to linearity, boundedness and coercivity, the estimation of the error in the Galerkin method can be replaced by an approximation-theoretical problem: given a function $w \in \mathring{\mathbb{H}}$ find the distance $\inf_{v \in \mathring{\mathbb{H}}_m} \|w - v\|_H$.

In particular, the question of the *convergence* of the FEM reduces, subject to the conditions of Theorem 9.2, to the following question in approximation theory. Suppose that we have an infinite sequence of linear spaces $\mathring{\mathbb{H}}_{m_1}, \mathring{\mathbb{H}}_{m_2}, \ldots \subset \mathring{\mathbb{H}}$, where $\dim \mathring{\mathbb{H}}_{m_i} = m_i$ and the sequence $\{m_i\}_{i=1}^{\infty}$ ascends monotonically to infinity. Is it true that

$$\lim_{i \to \infty} \|u_{m_i} - u\|_H = 0,$$

where u_{m_i} is the Galerkin solution in the space $\varphi_0 + \mathring{\mathbb{H}}_{m_i}$? In the light of the inequality (9.26) and of our discussion, a sufficient condition for convergence is that for every $v \in \mathring{\mathbb{H}}$ it is true that

$$\lim_{i \to \infty} \inf_{w \in \mathring{\mathbb{H}}_{m_i}} \|v - w\|_H = 0. \tag{9.27}$$

It now pays to recall, when talking of the FEM, that the spaces $\mathring{\mathbb{H}}_{m_i}$ are spanned by functions with small support. In other words, each $\mathring{\mathbb{H}}_{m_i}$ possesses a basis

$$\varphi_1^{[i]}, \varphi_2^{[i]}, \ldots, \varphi_{m_i}^{[i]} \in \mathring{\mathbb{H}}_{m_i}$$

such that each $\varphi_j^{[i]}$ is supported on the open set $\mathbb{E}_j^{[i]} \subset \mathring{\mathbb{H}}_{m_i}$ and $\mathbb{E}_k^{[i]} \cap \mathbb{E}_\ell^{[i]} = \emptyset$ for most choices of $k, \ell = 1, 2, \ldots, m_i$. In practical terms, this means that the d-dimensional set Ω needs to be partitioned as follows:

$$\mathrm{cl}\,\Omega = \bigcup_{\alpha=1}^{n_i} \mathrm{cl}\,\Omega_\alpha^{[i]}, \qquad \text{where} \qquad \Omega_\alpha^{[i]} \cap \Omega_\beta^{[i]} = \emptyset, \quad \alpha \neq \beta.$$

Each $\Omega_\alpha^{[i]}$ is called an *element,* hence the name 'finite element method'. We allow each support $\mathbb{E}_j^{[i]}$ to extend across a small number of elements. Hence, $\mathbb{E}_k^{[i]} \cap \mathbb{E}_\ell^{[i]}$

consists exactly of the sets $\Omega_\alpha^{[i]}$ (and possibly their boundaries) that are shared by both supports. This implies that an overwhelming majority of intersections is empty.

Recall the solution of (9.7) using chapeau functions. In that case $m_i = i$, $n_i = i+1$,

$$\varphi_k^{[i]} = \psi\left(x - \frac{k}{i+1}\right), \qquad k = 1, 2, \ldots, i$$

(ψ having been defined in (9.8)),

$$\Omega_\alpha^{[i]} = \left(\frac{\alpha - 1}{i+1}, \frac{\alpha}{i+1}\right), \qquad \alpha = 1, 2, \ldots, i+1,$$

and

$$\mathbb{E}_j^{[i]} = \Omega_{j-1}^{[i]} \cup \Omega_j^{[i]}, \qquad j = 1, 2, \ldots, i.$$

Further examples, in two spatial dimensions, feature in Section 9.3.

What are reasonable requirements for a 'finite element space' $\mathring{\mathbb{H}}_{m_i}$? Firstly, of course, $\mathring{\mathbb{H}}_{m_i} \subset \mathring{\mathbb{H}}$, and this means that all members of the set must be sufficiently smooth to be subjected to the weak form (i.e., after integration by parts) of action by $\mathcal{L}$. Secondly, each set $\Omega_\alpha^{[i]}$ must contain functions $\varphi_j^{[i]}$ of sufficient number and variety to be able to approximate well arbitrary functions; recall (9.27). Thirdly, as i increases and the partition is being refined, we wish to ensure that the diameters of all elements ultimately tend to zero. It is usual to express this as the requirement that $\lim_{i\to\infty} h_i = 0$, where

$$h_i = \max_{\alpha=1,2,\ldots,n_i} \operatorname{diam} \Omega_\alpha^{[i]}$$

is the *diameter* of the ith partition.[7] This does not mean that we need to refine all elements at an equal speed – an important feature of the FEM is that it lends itself to local refinement, and this confers important practical advantages. Our fourth and final requirement is that, as $i \to \infty$, the geometry of the elements does not become too 'difficult': in practical terms, the elements are likely to be polytopes (for example, polygons in $\mathbb{R}^2$) and we wish all their angles to be bounded away from zero as $i \to \infty$.

The latter two conditions are relatively simple to formulate and enforce, but the first two require further attention and elaboration. As far as the smoothness of $\varphi_j^{[i]}$, $j = 1, 2, \ldots, m_i$, is concerned, the obvious difficulty is likely to be smoothness *across element boundaries,* since it is in general easy to specify arbitrarily smooth functions within each $\Omega_\alpha^{[i]}$. However, 'approximability' of the finite element space $\mathring{\mathbb{H}}_{m_i}$ is all about what happens *inside* each element.

Our policy in the remainder of this chapter is to use elements $\Omega_\alpha^{[i]}$ that are all linear translates of the same 'master element', in the same way as the chapeau function (9.8) is defined in the interval $[-1, 1]$ and then translated to arbitrary intervals. Specifically, for $d = 2$ our interest centres on the translates of triangular elements (not necessarily all with identical angles) and quadrilateral elements. We choose functions $\varphi_j^{[i]}$ that are

[7]The quantity $\operatorname{diam} \mathbb{U}$, where $\mathbb{U}$ is a bounded set, is defined as the least radius of a ball into which this set can be inscribed. It is called the *diameter* of $\mathbb{U}$.

polynomial within each element – obviously the question of smoothness is relevant only across element boundaries. Needless to say, our refinement condition $\lim_{i\to\infty} h_i = 0$ and our ban on arbitrarily acute angles are strictly enforced.

We say that the space $\mathring{\mathbb{H}}_{m_i}$ is of *smoothness q* if each function $\varphi_j^{[i]}$, $j = 1, 2, \ldots, m_i$, is $q - 1$ times smoothly differentiable in Ω and q times differentiable inside each element $\mathbb{E}_\alpha^{[i]}$, $\alpha = 1, 2, \ldots, n_i$. (The latter requirement is automatically satisfied within our framework, since we have already required all functions $\varphi_j^{[i]}$ to be polynomials. It is stated for the sake of conformity with more general finite element spaces.) Furthermore, the space $\mathring{\mathbb{H}}_{m_i}$ is of *accuracy p* if, within each element $\Omega_\alpha^{[i]}$, the functions $\varphi_j^{[i]}$ span the set $\mathbb{P}_p^d[\boldsymbol{x}]$ of all d-dimensional polynomials of *total degree p*. The latter encompasses all functions of the form

$$\sum_{\substack{\ell_1+\cdots+\ell_d\le p\\ \ell_1,\ldots,\ell_d\ge 0}} c_{\ell_1,\ell_2,\ldots,\ell_d} x_1^{\ell_1} x_2^{\ell_2} \cdots x_d^{\ell_d},$$

where $c_{\ell_1,\ell_2,\ldots,\ell_d} \in \mathbb{R}$ for all $\ell_1, \ell_2, \ldots, \ell_d$.

Let us illustrate the above concepts for the case of (9.1) with chapeau functions. Firstly, each translate of (9.8) is continuous throughout $\Omega = (0, 1)$ but not differentiable throughout the whole interval, hence $q - 1 = 0$ and we deduce a smoothness $q = 1$. Secondly, each element $\Omega_\alpha^{[i]}$ is the support of both $\varphi_{\alpha-1}^{[i]}$ and $\varphi_\alpha^{[i]}$ (with obvious modification for $\alpha = 1$ and $\alpha = i + 1$). Both are linear functions, the first increasing, with slope $+i$, and the second decreasing with slope $-i$. Hence linear independence allows the conclusion that every linear function can be expressed inside $\Omega_\alpha^{[i]}$ as a linear combination of $\varphi_{\alpha-1}^{[i]}$ and $\varphi_\alpha^{[i]}$. Since $\mathbb{P}_1^1[\boldsymbol{x}] = \mathbb{P}_1[x]$, the set of all univariate linear functions, it follows that $\mathring{\mathbb{H}}_{m_1}$ is of accuracy $p = 1$.

Much effort has been spent in the last few pages in arguing that there is an intimate connection between smoothness, accuracy and the error estimate (9.26). Unfortunately, this is as far as we can go without venturing into much deeper waters of functional analysis – except for stating, without any proof, a theorem that quantifies this connection in explicit terms.

Theorem 9.3 *Let $\mathcal{L}$ obey the conditions of Theorem 9.2 and suppose that we solve equation* (9.17) *by the FEM, subject to the aforementioned restrictions (the shape of the elements, $\lim_{m\to\infty} h_i = 0$ etc.), with smoothness and accuracy $q = p \ge \nu$ (ν is half the number of derivatives in $\mathcal{L}$, cf.* (9.18)*). Then there exists a constant $c > 0$, independent of i, such that*

$$\|u_m - u\|_H \le c h_i^{p+1-\nu} \|u^{(p+1)}\|, \qquad i = 1, 2, \ldots \tag{9.28}$$

■

Returning to the chapeau functions and their solution of the two-point boundary value equation (9.1), we use the inequality (9.28) to confirm our impression from Figs 9.2 and 9.3, namely that the error is $\mathcal{O}(h_i)$.

Theorem 9.3 is just a sample of the very rich theory of the FEM. Error bounds are available in a variety of norms (often with more profound significance to the underlying problem than the Euclidean norm) and subject to different conditions. However, inequality (9.28) is sufficient for the applications to the Poisson equation in the next section.

9.3 The Poisson equation

As we saw in the last section, the operator $\mathcal{L} = -\nabla^2$ is positive definite, being a special case of (9.18). Moreover, we have claimed that, for most realistic domains Ω, it is coercive and bounded. The coefficients (9.24) of the Ritz–Galerkin equations are simply given by

$$a_{k,\ell} = \int_\Omega (\nabla\varphi_k)\cdot(\nabla\varphi_\ell)\,\mathrm{d}\boldsymbol{x}, \qquad k,\ell = 1,2,\ldots,m. \tag{9.29}$$

Letting $d = 2$, we assume that the boundary $\partial\Omega$ is composed of a finite number of straight segments and partition Ω into triangles. The only restriction is that no vertex of one triangle may lie on an edge of another; vertices must be shared. In other words, a configuration like

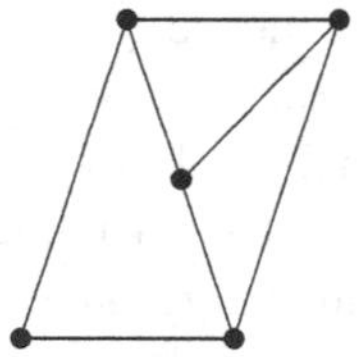

(where the position of a vertex is emphasized by '•') is not allowed. Figs 9.5 and 9.6 display a variety of triangulations that conform with this rule.

In light of (9.28), we require for convergence that $p, q \geq 1$, where p and q are the accuracy and smoothness respectively. This is similar to the situation that we have already encountered in Section 9.1 and we propose to address it with a similar remedy, namely by choosing $\varphi_1, \varphi_2, \ldots, \varphi_m$ as *piecewise linear* functions. Each function in $\mathbb{P}_1^2$ can be represented in the form

$$g(x,y) = \alpha + \beta x + \gamma y \tag{9.30}$$

for some $\alpha, \beta, \gamma \in \mathbb{R}$. Each function φ_k supported by an element Ω_j, $j = 1,2,\ldots,n$, is consequently of the form (9.30). Thus, to be accurate to order $p = 1$, each element must support at least three linearly independent functions. Recall that smoothness $q = 1$ means that every linear combination of the functions $\varphi_1, \varphi_2, \ldots, \varphi_m$ is continuous in Ω and this, obviously, need be checked only across element boundaries.

We have already seen in Section 9.1 one construction that provides both for accuracy $p = 1$ and for continuity with piecewise linear functions. The idea is to choose a basis of piecewise linear functions that vanish at all the vertices, except that at each vertex one function equals $+1$. Chapeau functions are an example of such *cardinal functions* and they have counterparts in $\mathbb{R}^2$. Fig. 9.4 displays three examples of *pyramid functions,* the planar cardinal functions, within their support (the set of all values of the argument for which they are nonzero). Unfortunately, it also demonstrates that using cardinal functions in $\mathbb{R}^2$ is, in general, a poor idea. The number of elements in each support may change from vertex to vertex and the description of each cardinal function, although easy in principle, is quite messy and inconvenient for practical work.

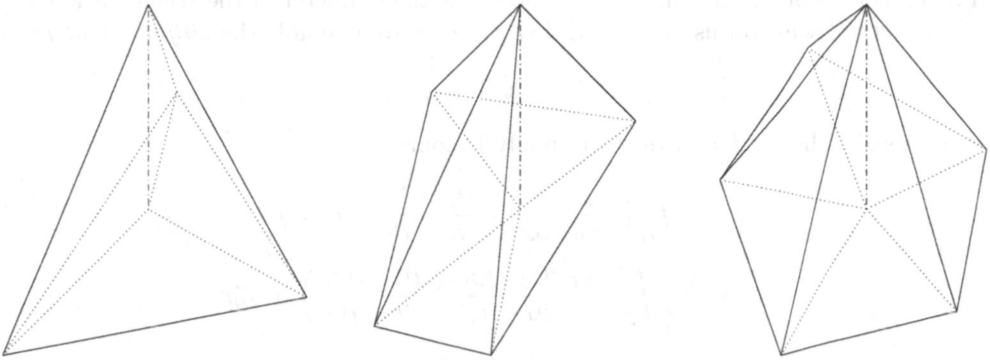

Figure 9.4 Pyramid functions for different configurations of vertices.

The correct procedure is to represent the approximation inside each Ω_j by data at its vertices. As long as we adopt this approach, how many different triangles meet at each vertex is of no importance and we can apply the same algorithm to all elements. Let the triangle in question be

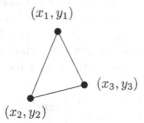

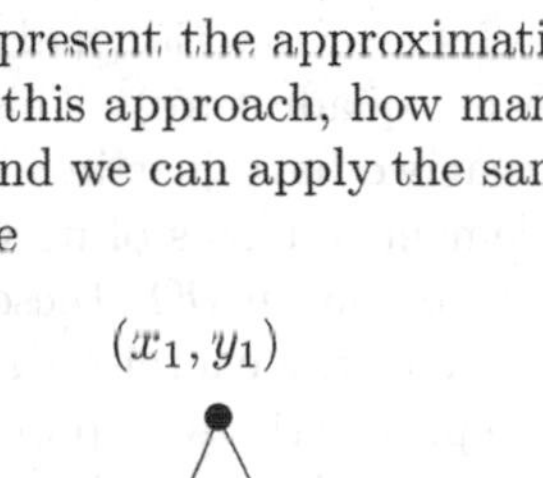

We determine the piecewise linear approximation s by interpolating at the three vertices. According to (9.30), this results in the linear system

$$\alpha + x_\ell\beta + y_\ell\gamma = g_\ell, \qquad \ell = 1, 2, 3,$$

where g_ℓ is the interpolated value at (x_ℓ, y_ℓ). Since the three vertices are not collinear, the system is nonsingular and can be solved with ease. This procedure (which, formally, is completely equivalent to the use of cardinal functions) ensures accuracy of order $p = 1$.

We need to prove that the above approach produces a function that is continuous throughout Ω, since this is equivalent to $q = 1$, the required degree of smoothness. This, however, follows from our construction. Recall that we need to prove continuity only across element boundaries. Suppose, without loss of generality, that the line segment joining (x_1, y_1) and (x_2, y_2) is not part of $\partial\Omega$ (otherwise there would be nothing to prove). The function s reduces along a straight line to a *linear function in one variable,* hence it is determined uniquely by interpolation of the two values g_1 and g_2 at the endpoints. Since these endpoints are shared by the triangle that adjoins along this edge, it follows that s is continuous there. A similar argument extends to all internal edges of the triangulation. We conclude that $p = q = 1$, hence the error (in a correct norm) decays like $\mathcal{O}(h)$, where h is the diameter of the triangulation.

A practical solution using the FEM requires us to assemble the *stiffness matrix*

$$A = (a_{k,\ell})_{k,\ell=1}^{m}.$$

The dimension being $d = 2$, (9.29) formally becomes

$$\begin{aligned}
a_{k,\ell} &= \iint_{\Omega} \left(\frac{\partial \varphi_k}{\partial x} \frac{\partial \varphi_\ell}{\partial x} + \frac{\partial \varphi_k}{\partial y} \frac{\partial \varphi_\ell}{\partial y} \right) \mathrm{d}x\, \mathrm{d}y \\
&= \sum_{j=1}^{n} \iint_{\Omega_j} \left(\frac{\partial \varphi_k}{\partial x} \frac{\partial \varphi_\ell}{\partial x} + \frac{\partial \varphi_k}{\partial y} \frac{\partial \varphi_\ell}{\partial y} \right) \mathrm{d}x\, \mathrm{d}y \\
&= \sum_{j=1}^{n} a_{k,\ell,j}, \qquad k, \ell = 1, 2, \ldots, m.
\end{aligned}$$

Inside the jth element the quantity $a_{k,\ell,j}$ vanishes, unless both φ_k and φ_ℓ are supported there. In the latter case, each is a linear function and, at least in principle, all integrals can be calculated (probably using quadrature). This, however, fails to take account of the subtle change of basis that we have just introduced in our characterization of the approximant inside each element in terms of its values on the vertices. Of course, except for vertices that happen to lie on $\partial\Omega$, these values are unknown and their computation is the whole purpose of the exercise. *The values at the vertices are our new unknowns* and we thereby rephrase the Ritz problem as follows: out of all possible piecewise linear functions that are consistent with our partition (i.e. linear inside each element), find the one that minimizes the functional

$$\begin{aligned}
\mathcal{J}(v) &= \iint_{\Omega} \left[\left(\frac{\partial v}{\partial x} \right)^2 + \left(\frac{\partial v}{\partial y} \right)^2 \right] \mathrm{d}x\, \mathrm{d}y - 2 \iint_{\Omega} f v \, \mathrm{d}x\, \mathrm{d}y \\
&= \sum_{j=1}^{n} \iint_{\Omega_j} \left[\left(\frac{\partial v}{\partial x} \right)^2 + \left(\frac{\partial v}{\partial y} \right)^2 \right] \mathrm{d}x\, \mathrm{d}y - 2 \sum_{j=1}^{n} \iint_{\Omega_j} f v \, \mathrm{d}x\, \mathrm{d}y.
\end{aligned}$$

Inside each Ω_j the function v is linear, $v(x, y) = \alpha_j + \beta_j x + \gamma_j y$, and explicitly

$$\iint_{\Omega_j} \left[\left(\frac{\partial v}{\partial x} \right)^2 + \left(\frac{\partial v}{\partial y} \right)^2 \right] \mathrm{d}x\, \mathrm{d}y = (\beta_j^2 + \gamma_j^2) \operatorname{area} \Omega_j.$$

As far as the second integral is concerned, we usually discretize it by quadrature and this, again, results in a function of α_j, β_j and γ_j.

With a little help from elementary analytic geometry, this can be expressed in terms of the values of v at the vertices. Let these be v_1, v_2, v_3, say, and assume that the corresponding (inner) angles of the triangle are $\theta_1, \theta_2, \theta_3$ respectively, where $\theta_1 + \theta_2 + \theta_3 = \pi$. Letting $\sigma_k = 1/(2\tan\theta_k)$, $k = 1, 2, 3$, we obtain

$$\iint_{\Omega_j} \left[\left(\frac{\partial v}{\partial x}\right)^2 + \left(\frac{\partial v}{\partial y}\right)^2\right] \mathrm{d}x\,\mathrm{d}y = \begin{bmatrix} v_1 & v_2 & v_3 \end{bmatrix} \begin{bmatrix} \sigma_2+\sigma_3 & -\sigma_3 & -\sigma_2 \\ -\sigma_3 & \sigma_1+\sigma_3 & -\sigma_1 \\ -\sigma_2 & -\sigma_1 & \sigma_1+\sigma_2 \end{bmatrix} \begin{bmatrix} v_1 \\ v_2 \\ v_3 \end{bmatrix}.$$

Meting out a similar treatment to the second integral and repeating this procedure for all elements in the triangulation, we finally represent the variational functional, acting on piecewise linear functions, in the form

$$\mathcal{J}(v) = \tfrac{1}{2}\boldsymbol{v}^\top \tilde{A}\boldsymbol{v} - \boldsymbol{f}^\top \boldsymbol{v}, \tag{9.31}$$

where $\boldsymbol{v}$ is the vector of the values of the function v at the m internal vertices (the number of such vertices is the same as the dimension of the space – why?). The $m \times m$ *stiffness matrix* $\tilde{A} = (\tilde{a}_{k,\ell})_{k,\ell=1}^{m}$ is assembled from the contributions of individual vertices. Obviously, $\tilde{a}_{k,\ell} = 0$ unless k and ℓ are indices of neighbouring vertices. The vector $\boldsymbol{f} \in \mathbb{R}^m$ is constructed similarly, except that it also contains the contributions of boundary vertices.

Setting the gradient of (9.31) to zero results in the linear algebraic system

$$\tilde{A}\boldsymbol{v} = \boldsymbol{f},$$

which we need to solve, e.g. by the methods of Chapters 11–15.

Our extensive elaboration of the construction of (9.31) illustrates the point that it is substantially more difficult to work with the FEM than with finite differences. The extra effort, however, is the price that we pay for extra flexibility.

Figure 9.5 displays the solution of the Poisson equation (8.33) on three meshes. These meshes are *hierarchical* – each is constructed by refining the previous one – and of increasing fineness. The graphs on the left display the meshes, while the shapes on the right are the numerical solutions as constructed from linear pieces (compare with the exact solution at the top of Fig. 8.8).

The advantages of the FEM are apparent if we are faced with difficult geometries and, even more profoundly, when it is known *a priori* that the solution is likely to be more problematic in part of the domain of interest and we wish to 'zoom in' on the triangulation there. For example, suppose that a Poisson equation is given in a domain with a *re-entrant corner* (for example, an L-shaped domain). We can expect the solution to be more difficult near such a corner and it is a good policy to refine the triangulation there.

As an example, let us consider the equation

$$\nabla^2 u + 2\pi^2 \sin \pi x \sin \pi y = 0, \qquad (x, y) \in \Omega = (-1, 1)^2 \setminus [0, 1]^2, \tag{9.32}$$

with zero Dirichlet boundary conditions along $\partial\Omega$. The exact solution is simply $u(x, y) = \sin \pi x \sin \pi y$.

Figure 9.6 displays the triangulations and underlying numerical solutions for three meshes that are increasingly refined. The triangulation is substantially finer near the re-entrant corner, as it should be, but perhaps the most important observation is that this does not require more effort than, say, the uniform tessellation of Fig. 9.5. In fact, both figures were produced by an identical program, but with different input! Although writing such a program is more of a challenge than coding finite differences, the rewards are very rich indeed . . .

The error in Figs. 9.5 and 9.6 is consistent with the bound $\|u_m - u\|_H \le ch\|u''\|$ (where h is the diameter of the triangulation), and this, in turn, is consistent with (9.28). In particular, its rate of decay (as a function of h) in Fig. 9.5 is similar to those of the five-point formula and the (unmodified) nine-point formula in Fig. 8.8. At first glance, this might perhaps seem contradictory; did we not state in Chapter 8 that the error of the five-point formula (8.15) is $\mathcal{O}(h^2)$? True enough, except that here we have been using different criteria to measure the error. Suppose, thus, that $u_{k,\ell} \approx u(k\Delta x, \ell\Delta x) + c_{k,\ell}h^2$ at all the grid points. Provided that $h = \mathcal{O}(m^{-1/2})$ (note that the number m means here the *total* number of variables in the whole grid), that there are $\mathcal{O}(m)$ grid points and that the error coefficients $c_{k,\ell}$ are of roughly similar order of magnitude, it is easy to verify that

$$\left\{ \frac{1}{m} \sum_{(k,\ell)\text{ in the grid}} [u_{k,\ell} - u(k\Delta x, \ell\Delta x)]^2 \right\}^{1/2} = \mathcal{O}(h).$$

This corresponds to the Euclidean norm in the finite element space. Although the latter is distinct from the Sobolev norm $\|\cdot\|_H$ of inequality (9.28), our argument indicates why the two error estimates are similar.

As was the case with finite difference schemes, the aforementioned accuracy sometimes falls short of that desired. This motivates a discussion of function bases having superior smoothness and accuracy properties. In one dimension this is straightforward, at least on the conceptual level: we need to replace piecewise linear functions with *splines,* functions that are kth-degree polynomials, say, in each element and possess $k-1$ smooth derivatives in the whole interval of interest. A convenient basis for kth-degree splines is provided by *B-splines,* which are distinguished by having the least possible support, extending across $k+1$ consecutive elements. In a general partition $\xi_0 < \xi_1 < \cdots < \xi_n$, say, a kth degree B-spline is defined explicitly by the formula

$$B_j^{[k]}(x) = \sum_{\ell=j}^{k+j+1} \left(\sum_{i=j,\, i\neq\ell}^{k+j+1} \frac{1}{\xi_i - \xi_\ell} \right) (x - \xi_\ell)_+^k .$$

Comparison with (9.8) ascertains that chapeau functions are nothing than linear B-splines.

The task in hand is more complicated in the case of two-dimensional triangulation, because of our dictum that everything needs to be formulated in terms of function values in an individual element and across its boundary. As a matter of fact, we have used only the values at the boundary – specifically, at the vertices – but this is about to change.

A general quadratic in $\mathbb{R}^2$ is of the form

$$s(x,y) = \alpha + \beta x + \gamma y + \delta x^2 + \eta xy + \zeta y^2;$$

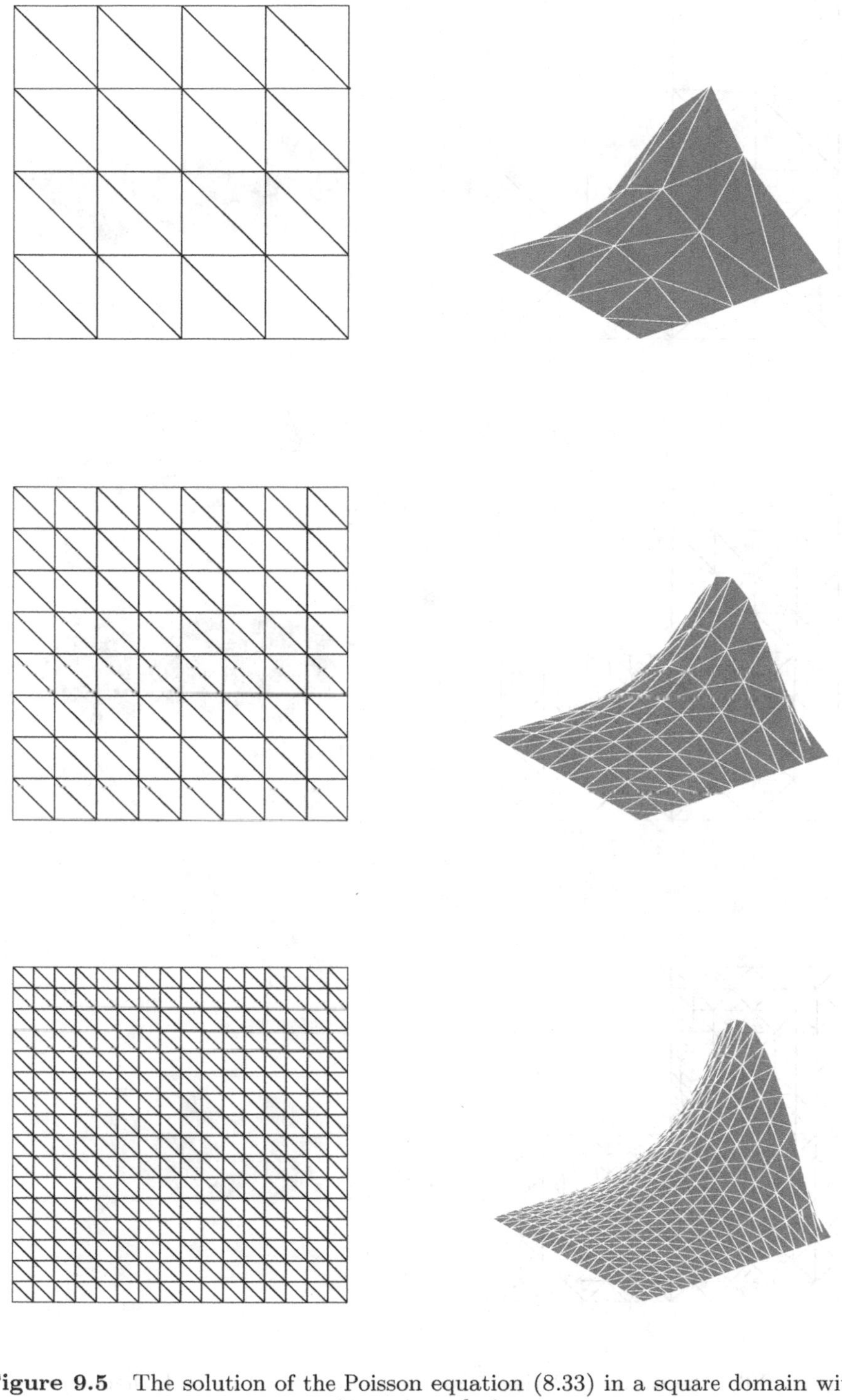

Figure 9.5 The solution of the Poisson equation (8.33) in a square domain with various triangulations.

Figure 9.6 The solution of the Poisson equation (9.32) in an L-shaped domain with various triangulations.

we note that it has six parameters. Likewise, a general cubic has ten parameters (verify!). We need to specify the correct number of interpolation points in the (closed) triangle. Two choices that give orders of accuracy $p = 2$ and $p = 3$ are

and

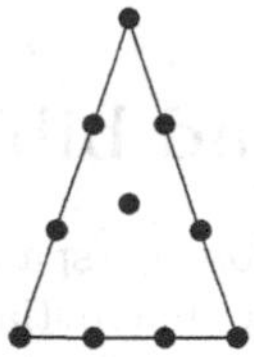

respectively. Unfortunately, their smoothness q is just 1 since, although a unique univariate quadratic or cubic, respectively, can be fitted along each edge (hence ensuring continuity), a tangental derivative might well be discontinuous. A superior interpolation pattern is

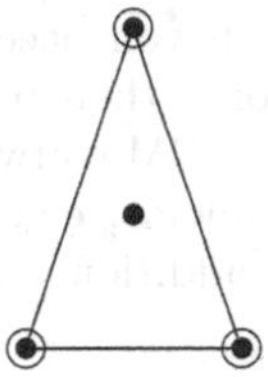

(9.33)

where '◉' means that we interpolate both function values and both spatial derivatives. We require altogether ten data items, and this is exactly the number of degrees of freedom in a bivariate cubic. Moreover, it is possible to show that the *Hermite* interpolation of both function values and (directional) derivatives along each edge results in both function and derivative smoothness there, hence $q = 2$.

Interpolation patterns like (9.33) are indispensable when, instead of the Laplace operator we consider the *biharmonic operator* ∇^4, since then $\nu = 2$ and we need $q \geq 2$ (see Exercise 9.5).

We conclude this chapter with a few words on piecewise linear interpolation with *quadrilateral* elements. The main problem in this context is that the bivariate linear function has three parameters – exactly right for a triangle but problematic in a quadrilateral. Recall that we must place interpolation points so as to attain continuity in the whole domain, and this means that at least two such points must reside along each edge. The standard solution of this conundrum is to restrict one's attention to rectangles (aligned with the axes) and interpolate with functions of the form

$$s(x, y) = s_1(x)s_2(y), \qquad \text{where} \qquad s_1(x) := \alpha + \beta x, \quad s_2(y) := \gamma + \delta y.$$

Obviously, piecewise linear functions are a proper subset of the functions s, but now we have four parameters, just right for interpolating at the four corners:

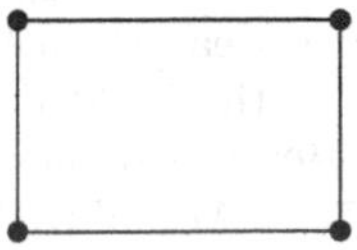

Along both horizontal edges s_2 is constant and s_1 is uniquely specified by the values at the corners. Therefore, the function s along each horizontal edge is independent of the interpolated values elsewhere in the rectangle. Since an identical statement is true for the vertical edges, we deduce that the interpolant is continuous and that $q = 1$.

Comments and bibliography

Weak solutions and Sobolev spaces are two inseparable themes that permeate the modern theory of linear elliptic differential equations (Agmon, 1965; Evans, 1998; John, 1982). The capacity of the FEM to fit snugly into this framework is not just a matter of æsthetics. Also, as we have had a chance to observe in this chapter, it provides for truly powerful error estimates and for a computational tool that can cater for a wide range of difficult situations – curved geometries, problems with internal interfaces, solutions with singularities ... Yet, the FEM is considerably less popular in applications than the finite difference method. The two reasons are the considerably more demanding theoretical framework and the more substantial effort required to program the FEM. If all you need is to solve the Poisson equation in a square, say, with nice boundary conditions, then probably there is absolutely no need to bother with the FEM (unless off-the-shelf FEM software is available), since finite differences will do perfectly well. More difficult problems, e.g. the equations of elasticity theory, the Navier–Stokes equations etc. justify the additional effort involved in mastering and using finite elements.

It is legitimate, however, to query how genuine weak solutions are. Anybody familiar with the capacity of mathematicians to generalize from the mundane yet useful to the beautiful yet useless has every right to feel sceptical. The simple answer is that they occur in many application areas, in linear as well as nonlinear PDEs and in variational problems. Moreover, seemingly 'nice' problems often have weak solutions. For a simple example, borrowed from Gelfand & Fomin (1963), we turn to the calculus of variations. Let

$$\mathcal{J}(v) := \int_{-1}^{1} v^2(\tau)[2\tau - v'(\tau)]^2 \, d\tau, \qquad v(-1) = 0, \quad v(1) = 1;$$

this is a nice cosy problem which, needless to say, should have a nice cosy solution. And it does! The exact solution can be written down explicitly,

$$u(x) = \begin{cases} x^2, & 0 \le x \le 1, \\ 0, & -1 \le x \le 0. \end{cases}$$

However, the underlying Euler–Lagrange equation is

$$y\left(4x^2 + 2y - y'^2 - yy''\right) = 0 \tag{9.34}$$

and includes a second derivative, while the function u fails to be twice differentiable at the origin. The solution of (9.34) exists only in a weak sense!

Lest the last example sounds a mite artificial (and it is – artificiality is the price of simplicity!), let us add that many equations of profound interest in applications can be investigated only in the context of weak solutions and Sobolev spaces. A thoroughly modern applied mathematician must know a great deal of mathematical analysis.

An unexpected luxury for students of the FEM is the abundance of excellent books in the subject, e.g. Axelsson & Barker (1984); Brenner & Scott (2002); Hackbusch (1992); Johnson (1987); Mitchell & Wait (1977). Arguably, the most readable introductory text is Strang & Fix (1973) – and it is rare for a book in a fast-moving subject to stay at the top of the hit parade for more than 30 years! The most comprehensive exposition of the subject, short of research papers and specialized monographs, is Ciarlet (1976). The reader is referred to this FEM feast for a thorough and extensive exposition of themes upon which we have touched briefly – error bounds, the design of finite elements in multivariate spaces – and many themes that have not been mentioned in this chapter. In particular, we encourage interested readers to consult more advanced monographs on the generalization of finite element functions to $d \geq 3$, on the attainment of higher smoothness conditions and on elements with curved boundaries. Things are often not what they seem to be in Sobolev spaces and it is always worthwhile, when charging the computational ramparts, to ensure adequate pure-mathematical covering fire.

These remarks will not be complete without mentioning recent work that blends concepts from the finite element, finite difference and spectral methods. A whole new menagerie of concepts has emerged in the last two decades: boundary element methods, the h-p formulation of the FEM, hierarchical bases... Only the future will tell how much will survive and find its way into textbooks, but these are exciting times at the frontiers of the FEM.

Agmon, S. (1965), *Lectures on Elliptic Boundary Value Problems,* Van Nostrand, Princeton, NJ.

Axelsson, O. and Barker, V.A. (1984), *Finite Element Solution of Boundary Value Problems: Theory and Computation,* Academic Press, Orlando, FL.

Brenner, S.C. and Scott, L.R. (2002), *The Mathematical Theory of Finite Element Methods* (2nd edn), Springer-Verlag, New York.

Ciarlet, P.G. (1976), *Numerical Analysis of the Finite Element Method,* North-Holland, Amsterdam.

Evans, L.C. (1998), *Partial Differential Equations,* American Mathematical Society, Providence, RI.

Gelfand, I.M. and Fomin, S.V. (1963), *Calculus of Variations,* Prentice–Hall, Englewood Cliffs, NJ.

Hackbusch, W. (1992), *Elliptic Differential Equations: Theory and Numerical Treatment,* Springer-Verlag, Berlin.

John, F. (1982), *Partial Differential Equations* (4th edn), Springer-Verlag, New York.

Johnson, C. (1987), *Numerical Solution of Partial Differential Equations by the Finite Element Method,* Cambridge University Press, Cambridge.

Mitchell, A.R. and Wait, R. (1977), *The Finite Element Method in Partial Differential Equations,* Wiley, London.

Strang, G. and Fix, G.J. (1973), *An Analysis of the Finite Element Method,* Prentice–Hall, Englewood Cliffs, NJ.

Exercises

9.1 Demonstrate that in the interval $[t_n, t_{n+1}]$ the collocation method (3.12) finds an approximation to the weak solution of the ordinary differential system

$\boldsymbol{y}' = \boldsymbol{f}(t, \boldsymbol{y})$, $\boldsymbol{y}(t_n) = \boldsymbol{y}_n$, from the space $\mathbb{P}_\nu$ of νth-degree polynomials, provided that we employ the inner product

$$\langle \boldsymbol{v}, \boldsymbol{w} \rangle = \sum_{j=1}^{\nu} \boldsymbol{v}(t_n + c_j h)^\top \boldsymbol{w}(t_n + c_j h),$$

where $h = t_{n+1} - t_n$. (*Strictly speaking,* $\langle \cdot\,, \cdot \rangle$ *is a* semi-inner product, *since it is not true that* $\langle \boldsymbol{v}, \boldsymbol{v} \rangle = 0$ *implies* $\boldsymbol{v} \equiv \boldsymbol{0}$.)

9.2 Find explicitly the coefficients $a_{k,\ell}$, $k, \ell = 1, 2, \ldots, m$, for the equation $-y'' + y = f$, assuming that the space $\mathring{\mathbb{H}}_m$ is spanned by chapeau functions on an equidistant grid.

9.3 Suppose that the equation (9.1) is solved by the Galerkin method with chapeau functions on a non-equidistant grid. In other words, we are given $0 = t_0 < t_1 < t_2 < \cdots < t_m < t_{m+1} = 1$ such that each φ_j is supported in (t_{j-1}, t_{j+1}), $j = 1, 2, \ldots, m$. Prove that the linear system (9.7) is nonsingular. (*Hint: Use the Geršgorin criterion (Lemma 8.3).*)

9.4 Let a be a given positive univariate function and

$$\mathcal{L} := \frac{\partial^2}{\partial x^2} \left[a(x) \frac{\partial^2}{\partial x^2} \right].$$

Assuming zero Dirichlet boundary conditions, prove that $\mathcal{L}$ is positive definite in the Euclidean norm.

9.5 Prove that the biharmonic operator ∇^4, acting in a parallelepiped in $\mathbb{R}^d$, is positive definite in the Euclidean norm.

9.6 Let $\mathcal{J}$ be given by (9.21), suppose that the operator $\mathcal{L}$ is positive definite and let

$$\mathcal{J}_m(\boldsymbol{\delta}) := \mathcal{J}\left(\varphi_0 + \sum_{\ell=1}^{m} \delta_\ell \varphi_\ell \right), \qquad \boldsymbol{\delta} \in \mathbb{R}^m.$$

Prove that the matrix

$$\left(\frac{\partial^2 \mathcal{J}_m(\boldsymbol{\delta})}{\partial \delta_k \partial \delta_\ell} \right)_{k,\ell=1,2,\ldots,m}$$

is positive definite, thereby deducing that the solution of the Ritz equations is indeed the global minimum of $\mathcal{J}_m$.

9.7 Let $\mathcal{L}$ be an elliptic differential operator and f a given bounded function.

a Prove that the numbers

$$c_1 := \min_{\substack{v \in \mathring{\mathbb{H}} \\ \|v\|=1}} \tilde{a}(v, v) \qquad \text{and} \qquad c_2 := \max_{\substack{v \in \mathring{\mathbb{H}} \\ \|v\|=1}} \langle f, v \rangle$$

are bounded and that $c_1 > 0$.

b Given $w \in \mathbb{H}$, prove that

$$\tilde{a}(w,w) - 2\langle f, w\rangle \geq c_1\|w\|^2 - 2c_2\|w\|.$$

(*Hint: Write* $w = \kappa v$, *where* $\|v\| = 1$ *and* $|\kappa| = \|w\|$.)

c Deduce that

$$\tilde{a}(w,w) - 2\langle f, w\rangle \geq -\frac{c_2^2}{c_1}, \qquad w \in \mathbb{H},$$

thereby proving that the functional $\mathcal{J}$ from (9.21) has a bounded minimum.

9.8 Find explicitly a cardinal piecewise linear function (a pyramid function) in a domain partitioned into equilateral triangles (cf. the graph in Exercise 8.13).

9.9 Nine interpolation points are specified in a rectangle:

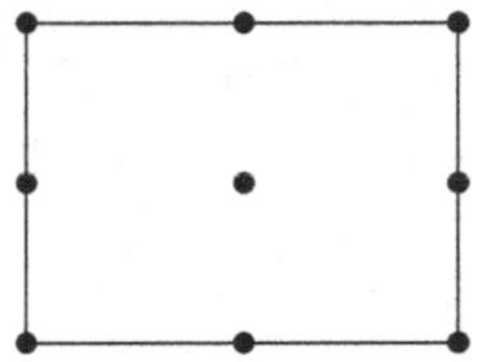

Prove that they can be interpolated by a function of the form $s(x, y) = s_1(x)s_2(y)$, where both s_1 and s_2 are quadratics. Find the orders of the accuracy and of the smoothness of this procedure.

9.10 The Poisson equation is solved in a square partition by the FEM in the manner described in Section 9.3. In each square element the approximant is the function $s(x, y) = s_1(x)s_2(y)$, where s_1 and s_2 are linear, and it is being interpolated at the vertices. Derive explicitly the entries $\tilde{a}_{k,\ell}$ of the stiffness matrix $\tilde{A}$ from (9.31).

9.11 Prove that the four interpolatory conditions specified at the vertices of the three-dimensional *tetrahedral* element

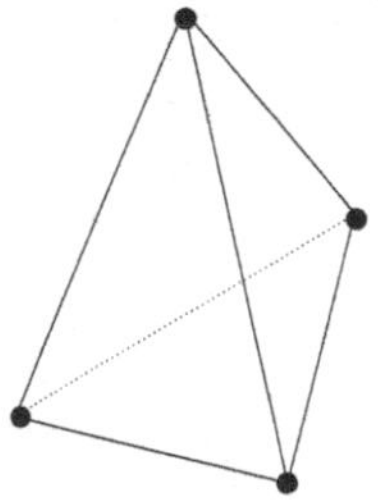

can be satisfied by a piecewise linear function.

10

Spectral methods

10.1 Sparse matrices vs. small matrices

In the previous two chapters we have introduced methods based on completely different principles: finite differences rest upon the replacement of derivatives by linear combinations of function values but the idea behind finite elements is to approximate an infinite-dimensional expansion of the solution in a finite-dimensional space. Yet the implementation of either approach ultimately leads to the solution of a system of algebraic equations. The bad news about such a system is that it tends to be very large indeed; the good news is that it is highly structured, usually very sparse, hence lending itself to effective algorithms for the solution of sparse linear algebraic systems, the theme of Chapters 11–15.

In other words, both finite differences and finite elements converge fairly slowly (hence the matrices are large) but the weak coupling between the variables results in sparsity and in practice algebraic systems can be computed notwithstanding their size. Once we formulate the organizing principle of both kinds of method in this manner, it immediately suggests an enticing alternative: methods that produce *small* matrices in the first place. Although we are giving up sparsity, the much smaller size of the matrices renders their solution affordable.

How do we construct such 'small matrix' methods? The large size of the matrices in Chapters 8 and 9 was caused by slow convergence of the underlying approximations, which resulted in a large number of parameters (grid points or finite element functions). Thus, the key is to devise approximation methods that exhibit considerably faster convergence, hence requiring much smaller number of parameters.

Before we thus approximate solutions of differential equations, we need to look at the approximation of functions. In Fig. 10.1 we display the error incurred when the function e^{-x} is approximated in $[-1,1]$ by piecewise linear functions (the chapeau functions of Chapter 9) with equally spaced nodes k/N, $k=-N/2,\ldots,N/2$; here and elsewhere in this chapter $N\geq 2$ is an even integer. The local error of piecewise linear approximation is $\mathcal{O}(N^{-2})$ and so we expect this to be roughly divided by four each time N is doubled. This is indeed confirmed by the figure. Likewise, in Fig. 10.2 we display the error when the function $\mathrm{e}^{-\cos\pi x}$ is approximated by cheapau functions. Again, the error decays fairly predictably: the reason why we are so keen on this figure will become clear later.

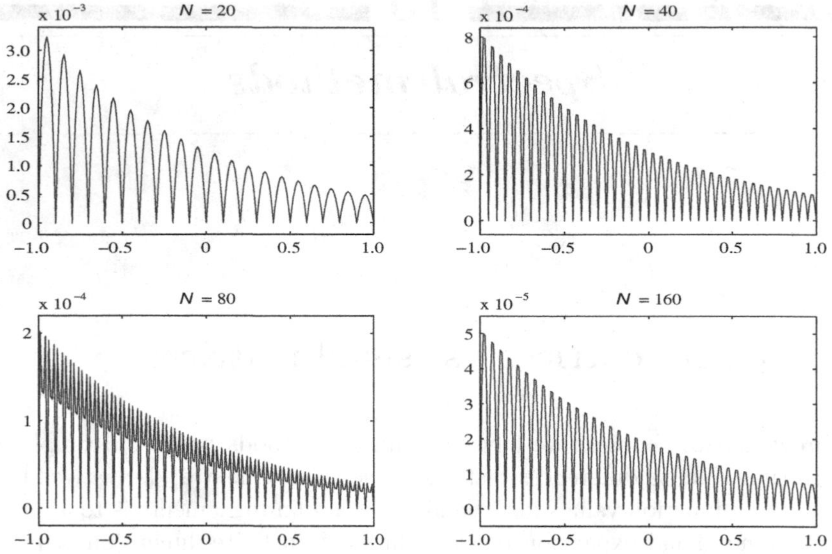

Figure 10.1 The error in approximating the function e^{-x} in $[-1,1]$ by piecewise linear functions with N degrees of freedom.

We will compare the chapeau-function approximation

$$f(x) \approx \sum_{n=-N/2+1}^{N/2} \psi(Nn - x) f\left(\frac{2n}{N}\right)$$

(see (9.8) for the definition of the chapeau function ψ) with the truncated Fourier approximation:

$$f(x) \approx \varphi_N(x) = \sum_{n=-N/2+1}^{N/2} \hat{f}_n e^{i\pi n x},$$
$$\text{where} \qquad \hat{f}_n = \frac{1}{2}\int_{-1}^{1} f(\tau) e^{-i\pi n \tau}\, d\tau, \qquad n \in \mathbb{Z}. \tag{10.1}$$

Before looking at numerical results, it is useful to recall a basic fact on Fourier series and their convergence.

Theorem 10.1 (The de la Vallée Poussin theorem) *If the function f is Riemann integrable and $\hat{f}_n = \mathcal{O}(n^{-1})$ for $|n| \gg 1$ then $\varphi_N(x) = f(x) + \mathcal{O}(N^{-1})$ as $N \to \infty$ for every point $x \in (-1,1)$ where f is Lipschitz.*

Note that if f is smoothly differentiable then, integrating by parts,

$$\hat{f}_n = -\frac{(-1)^n}{2i\pi n}[f(1) - f(-1)] - \frac{1}{2i\pi n}\widehat{f'}_n = \mathcal{O}(n^{-1}), \qquad |n| \gg 1.$$

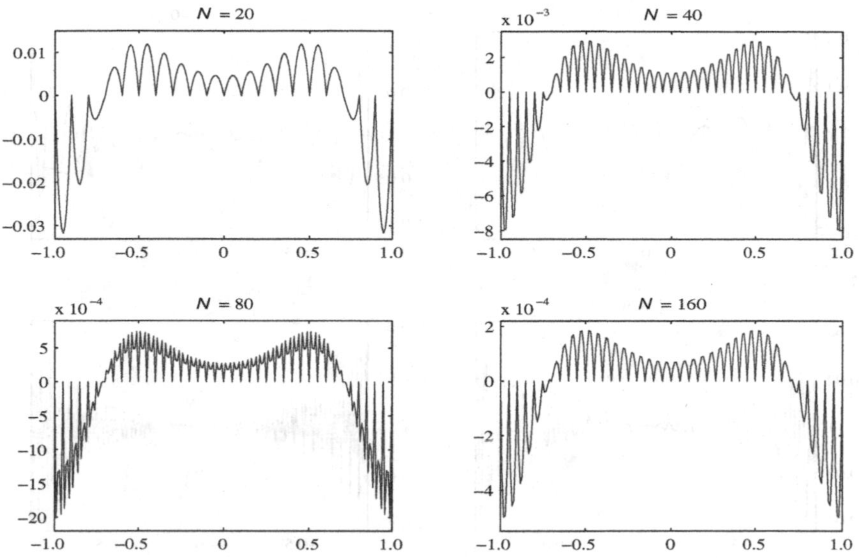

Figure 10.2 The same as Fig. 10.1, except that now we approximate the periodic function $e^{-\cos \pi x}$.

Since such a function f is Lipschitz in $(-1, 1)$, we deduce that φ_N converges to f there. (It follows from standard theorems of calculus that this convergence is uniform in every closed subinterval.) However, we can guarantee the convergence of $\mathcal{O}(N^{-1})$, though this is very slow. Even more importantly, there is nothing to make φ_N converge to f at the endpoints. As a matter of fact, it is possible to show that $\varphi_N(\pm 1) \to \frac{1}{2}[f(-1) + f(1)]$: unless f is periodic, we fail to converge to the correct function values at the endpoints. (Not a great surprise, since φ_N itself is periodic.) This implies, in addition, that the error is likely to be unacceptably large near ± 1, where the approximation oscillates wildly, a phenomenon known as *the Gibbs effect.*

All this is vividly illustrated by Fig. 10.3. Note that we have plotted the error only in the subinterval $[-\frac{9}{10}, \frac{9}{10}]$, since $\varphi_N(\pm 1) \to \cosh 1$, an altogether wrong value. But even in the open interval $(-1, 1)$ the news is not good: the convergence of $\mathcal{O}(N^{-1})$ is excruciatingly slow. Doubling the number of points increases the accuracy barely by a factor of 2.

We thus approach our second function, $e^{-\cos \pi x}$, with very modest expectations. Yet, even brief examination of Fig. 10.4 (where again we plot in $[-1, 1]$) reveals something truly amazing. Taking $N = 10$ gives an accuracy comparable to $N = 160$ with chapeau functions (compare Fig. 10.2). Doubling N results in ten significant digits, while for $N = 30$ we have exhausted the accuracy of MATLAB computer arithmetic and the plot displays randomness, a tell-tale sign of roundoff error. All this is true not just inside the interval but also on the boundary. This is precisely the rapid convergence that we have sought!

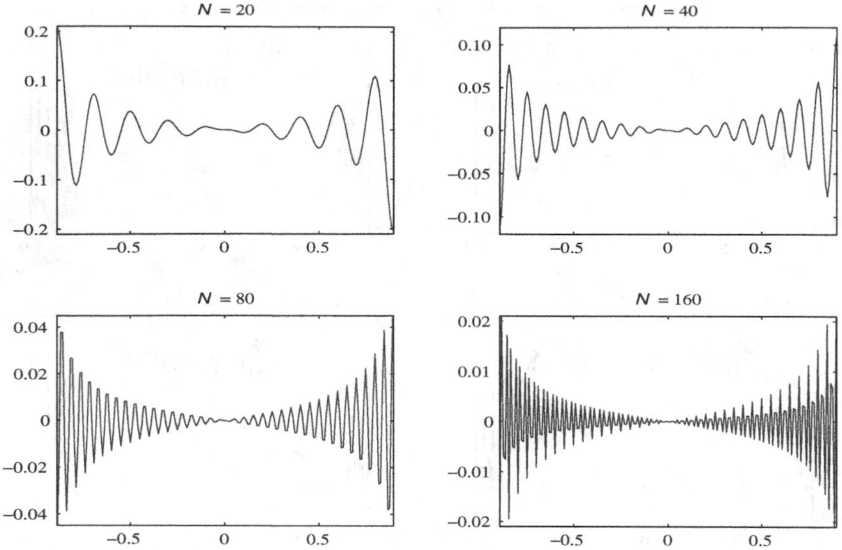

Figure 10.3 The error in approximating the function e^{-x} in $[-\frac{9}{10}, \frac{9}{10}]$ by Fourier expansion with N terms.

So what makes the Fourier series approximation to $e^{-\cos \pi x}$ so effective in comparison to e^{-x}? The brief answer is *periodicity*. In general, suppose that f is an analytic function in $[-1, 1]$ that can be extended analytically to a closed complex domain Ω such that $[-1, 1] \subset \Omega$ and to its boundary. In addition, we stipulate that f is periodic with period 2. Therefore $f^{(m)}(-1) = f^{(m)}(1)$ for all $m = 0, 1, \ldots$ We again integrate by parts, but this time we do not stop with f':

$$\hat{f}_n = -\frac{1}{2\pi i n}\widehat{f'}_n = \left(-\frac{1}{2\pi i n}\right)^2 \widehat{f''}_n = \left(-\frac{1}{2\pi i n}\right)^2 \widehat{f'''}_n = \cdots.$$

We thus have

$$\hat{f}_n = \left(-\frac{1}{2\pi i n}\right)^m \widehat{f^{(m)}}_n, \qquad m = 0, 1, \ldots \tag{10.2}$$

How large is $|\widehat{f^{(m)}}_n|$? Letting γ be the positively oriented boundary of Ω and denoting by $\alpha^{-1} > 0$ the minimal distance betweeen γ and $[-1, 1]$, the Cauchy theorem of complex analysis states that

$$f^{(m)}(x) = \frac{m!}{2\pi i}\int_\gamma \frac{f(z)\,dz}{(z-m)^{m+1}}, \qquad x \in [-1, 1]:$$

therefore, letting $\kappa = \max\{|f(z)| : z \in \gamma\} < \infty$,

$$|f^{(m)}(x)| \le \frac{m!}{2\pi}\int_\gamma \frac{|f(z)|\,|dz|}{|z-x|^{m+1}} \le \frac{\kappa \,\text{length}\,\gamma}{2\pi} m!\,\alpha^{m+1}.$$

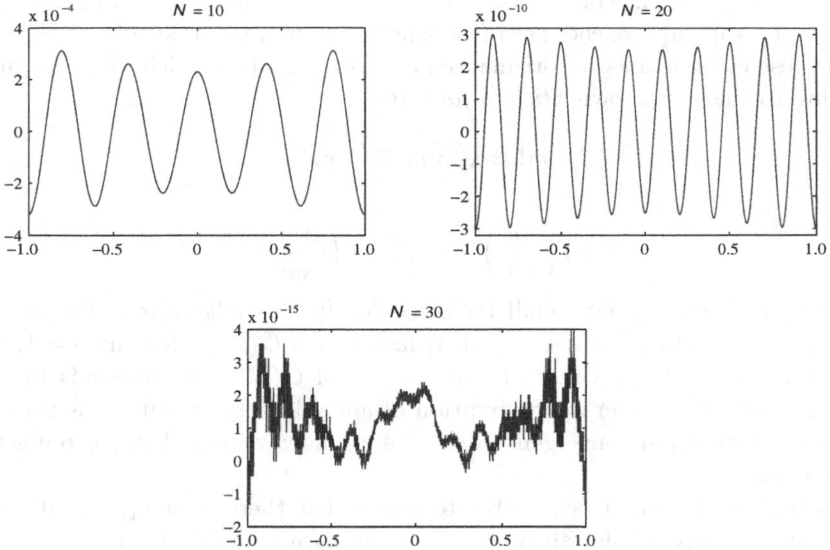

Figure 10.4 The same as Fig. 10.3, except that now we approximate the periodic function $e^{-\cos \pi x}$ and employ smaller values of N.

It follows that we can bound $|\widehat{f^{(m)}}_n| \le cm!\alpha^m$, $m = 0, 1, \ldots$, for some $c > 0$. Consequently, using (10.2) and the above upper bound of $|\widehat{f^{(m)}}_n|$,

$$\begin{aligned}|\varphi_N(x) - f(x)| &= \left|\varphi_N(x) - \sum_{n=-\infty}^{\infty} \hat{f}_n e^{i\pi n x}\right| \le \sum_{n=-\infty}^{-N/2} |\hat{f}_n| + \sum_{n=N/2+1}^{\infty} |\hat{f}_n| \\ &= \sum_{n=-\infty}^{-N/2} \frac{|\widehat{f^{(m)}}_n|}{(-2\pi n)^m} + \sum_{n=N/2+1}^{\infty} \frac{|\widehat{f^{(m)}}_n|}{(2\pi n)^m} \\ &\le \frac{cm!\alpha^m}{(2\pi)^m}\left[\frac{1}{(N/2)^m} + 2\sum_{n=N/2+1}^{\infty} \frac{1}{n^m}\right].\end{aligned}$$

However, for any $r = 1, 2, \ldots$,

$$\sum_{n=r}^{\infty} \frac{1}{n^m} \le \int_r^{\infty} \frac{d\tau}{\tau^m} = \frac{1}{m-1} r^{-m+1},$$

and we deduce that

$$|\varphi_N(x) - f(x)| \le cm!\left(\frac{\alpha}{2\pi}\right)^m \left[\frac{1}{(N/2)^m} + \frac{2}{m-1}\frac{1}{(N/2+1)^m}\right] \le 3cm!\left(\frac{\alpha}{\pi N}\right)^m. \tag{10.3}$$

We have a competition: while $\alpha/(\pi N)$ can be made as small as we want for large N, so that $[\alpha/(\pi N)]^m$ approaches zero fast when m grows, factorial $m!$ rapidly becomes large in these circumstances. Fortunately, this is one contest which the good guys win. According to the well-known *Stirling formula,*

$$m! \approx \sqrt{2\pi} m^{m+1/2} \mathrm{e}^{-m},$$

we have

$$m! \left(\frac{\alpha}{\pi N}\right)^m \approx \sqrt{2\pi m} \left(\frac{\alpha m}{\pi \mathrm{e} N}\right)^m$$

and the latter becomes very small for large N. It thus follows from (10.3) that the error $|\varphi_N - f|$ decays pointwise in $[-1,1]$ faster than $\mathcal{O}(N^{-p})$ for any $p = 1, 2, \ldots$

Since, in our setting, a rate of convergence of $\mathcal{O}(N^{-p})$ corresponds to order p, we deduce that the Fourier approximation of analytic periodic functions is of infinite order. Such very rapid convergence deserves a name: we say that φ_N tends to f at *spectral speed.*

As a matter of fact, it is possible to prove that there exist $c_1, \omega > 0$ such that $|\varphi_N(x) - f(x)| \le c_1 \mathrm{e}^{-\omega N}$ for all $N = 0, 1, \ldots$, uniformly in $[-1,1]$. Thus, convergence is at least at an exponential rate, and this explains the extraordinary accuracy evident in Fig. 10.4.

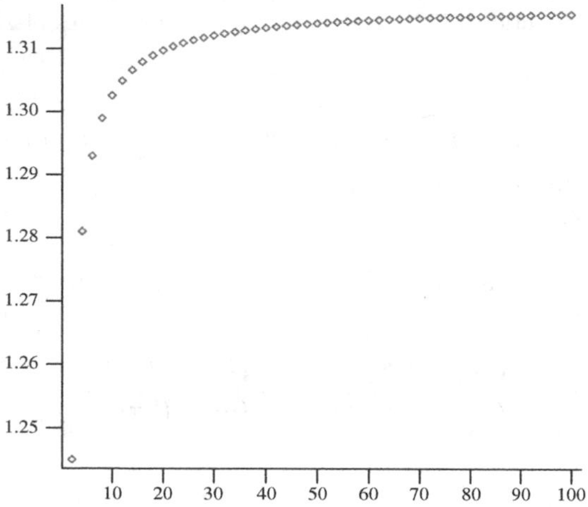

Figure 10.5 Scaled logarithm of Fourier coefficients $-(\log|\hat{f}_n|)/n$ for $n = 1, 2, \ldots, 100$ and the periodic function $f(x) = (1 + \frac{1}{2}\cos \pi x)^{-1}$.

As we have seen, spectral convergence is all about the fast decay of Fourier coefficients. In Fig. 10.5 we illustrate this with the function $f(x) = (1 + \frac{1}{2}\cos \pi x)^{-1}$. It is easy to check that it is indeed periodic and that it can be extended analytically away from $[-1,1]$, its nearest singularities residing at $\pm \log(2 + \sqrt{3})/(\pi \mathrm{i})$. The coefficients (we need to compute $\hat{f}_n$ only for $n \ge 0$, since f is even and $\hat{f}_{-n} = \hat{f}_n$) decay *very fast:* $\hat{f}_{100} \approx 7.37 \times 10^{-58}$ (and we needed more than 120 significant digits to compute this!) and the plot indicates that $\hat{f}_n \approx \mathrm{e}^{-1.32n}$.

We have accumulated enough evidence to make the case that Fourier expansions converge exceedingly fast for periodic functions. The challenge now is to utilize this behaviour in the design of discretization methods that lead to relatively small linear algebraic systems.

10.2 The algebra of Fourier expansions

We denote by $\mathcal{A}$ the set of all complex-valued functions f that are analytic in $[-1,1]$, are periodic there with period 2, and can be extended analytically into the complex plane; such functions, as we saw in the last section, have rapidly convergent Fourier expansions.

What sort of animal is $\mathcal{A}$? It is a linear space: if $f, g \in \mathcal{A}$ and $a \in \mathbb{C}$ then $f+g, af \in \mathcal{A}$. Moreover, identifying functions in $\mathcal{A}$ with their (convergent) Fourier expansion, given by

$$f(x) = \sum_{n=-\infty}^{\infty} \hat{f}_n \mathrm{e}^{\mathrm{i}\pi n x}, \qquad g(x) = \sum_{n=-\infty}^{\infty} \hat{g}_n \mathrm{e}^{\mathrm{i}\pi n x},$$

implies that

$$f(x) + g(x) = \sum_{n=-\infty}^{\infty} (\hat{f}_n + \hat{g}_n) \mathrm{e}^{\mathrm{i}\pi n x}, \qquad af(x) = \sum_{n=-\infty}^{\infty} a\hat{f}_n \mathrm{e}^{\mathrm{i}\pi n x}. \tag{10.4}$$

Thus, the algebra of $\mathcal{A}$ can be easily expressed in terms of Fourier coefficients. Moreover, simple calculation confirms that $\mathcal{A}$ is also closed with regard to multiplication:

$$f(x)g(x) = \sum_{n=-\infty}^{\infty} \left(\sum_{m=-\infty}^{\infty} \hat{f}_{n-m} \hat{g}_m \right) \mathrm{e}^{\mathrm{i}\pi n x}. \tag{10.5}$$

The inner convergent infinite sum above is called the *convolution* of the complex sequences $\hat{\boldsymbol{f}} = \{\hat{f}_n\}$ and $\hat{\boldsymbol{g}} = \{\hat{g}_n\}$ and is written as

$$\hat{\boldsymbol{f}} * \hat{\boldsymbol{g}} = \hat{\boldsymbol{h}}, \qquad \text{where} \qquad \hat{h}_n = \sum_{m=-\infty}^{\infty} \hat{f}_{n-m} \hat{g}_m, \quad n \in \mathbb{Z}. \tag{10.6}$$

Therefore $f(x)g(x) = (f * g)(x)$, where the relationship between the Fourier series is expressed at the level of functions by $h = f * g$.

Our calculus with Fourier series requires, in the context of the numerical analysis of differential equations, a means of differentiating functions. This is fairly straightforward: $f \in \mathcal{A}$ implies that $f' \in \mathcal{A}$ and

$$f'(x) = \mathrm{i}\pi \sum_{n=-\infty}^{\infty} n \hat{f}_n \mathrm{e}^{\mathrm{i}\pi n}.$$

All this extends to higher derivatives. This is the moment to recall that, in our setting, the sequence $\{\hat{f}_n\}$ decays faster than $\mathcal{O}(n^{-p})$ for $p \in \mathbb{Z}_+$, and this provides an alternative demonstration that all derivatives of f have rapidly convergent Fourier expansions.

◇ **A simple spectral method** How does all this help in our quest to compute differential equations? Consider the two-point boundary value problem

$$y'' + a(x)y' + b(x)y = f(x), \quad -1 \le x \le 1, \quad y(-1) = y(1), \tag{10.7}$$

where $a, b, f \in \mathcal{A}$. Substituting Fourier expansions and using (10.4) and (10.5), we obtain an infinite-dimensional system of linear algebraic equations

$$-\pi^2 n^2 \hat{y}_n + \mathrm{i}\pi \sum_{m=-\infty}^{\infty} m\hat{a}_{n-m}\hat{y}_m + \sum_{m=-\infty}^{\infty} \hat{b}_{n-m}\hat{y}_m = \hat{f}_n, \qquad n \in \mathbb{Z}, \tag{10.8}$$

for the unknowns $\hat{\boldsymbol{y}}$. Knowing that the sequences $\hat{\boldsymbol{a}}, \hat{\boldsymbol{b}}, \hat{\boldsymbol{f}}$ decay at spectral speed, we can truncate (10.8) into the N-dimensional system

$$\begin{gathered} -\pi^2 n^2 \hat{y}_n + \mathrm{i}\pi \sum_{m=-N/2+1}^{N/2} m\hat{a}_{n-m}\hat{y}_m + \sum_{m=-N/2+1}^{N/2} \hat{b}_{n-m}\hat{y}_m = \hat{f}_n \\ \text{for} \quad n = -N/2+1, \ldots, N/2. \end{gathered} \tag{10.9}$$

In the language of signal processing, we are approximating the solution by a *band-limited function,* one that can be described as a linear combination of a finite number of Fourier coefficients.

Note that, to avoid a needless clutter of notation, we denote both the exact Fourier coefficients in (10.8) and their N-dimensional approximation in (10.9) by $\hat{y}_n$. This should not cause any confusion.

The matrix of system (10.9) is in general dense, but our theory predicts that fairly small values of N, hence very small matrices, are sufficient for high accuracy.

By a way of example, we will choose $a(x) = f(x) = \cos \pi x$ and $b(x) = \sin 2\pi x$. This, incidentally, leads to a sparse matrix, because a and b contain just two nonzero Fourier harmonics each. Yet, the purpose of this example is not to investigate matrices but to illustrate the rate of convergence. Fig. 10.6 shows that $N = 16$ yields an accuracy of more than ten digits, while for $N = 22$ we have already hit the buffers of computer arithmetic and roundoff error. Needless to say, the direct solution of a 22×22 linear algebraic system with Gaussian elimination is so fast that it is pointless to seek an alternative. ◇

In the last example we were prescient enough to choose functions a, b and f with known Fourier coefficients. This is not the case for most realistic scenarios and, if we really expect spectral methods for differential equations to be a serious competitor to finite differences and finite elements, we must have an effective means of computing $\hat{f}_n$ for $n = -N/2+1, \ldots, N/2$.

Fortunately, Fourier coefficients can be computed very accurately indeed with remarkably small computational cost. In general, suppose that $h \in \mathcal{A}$ and we wish to

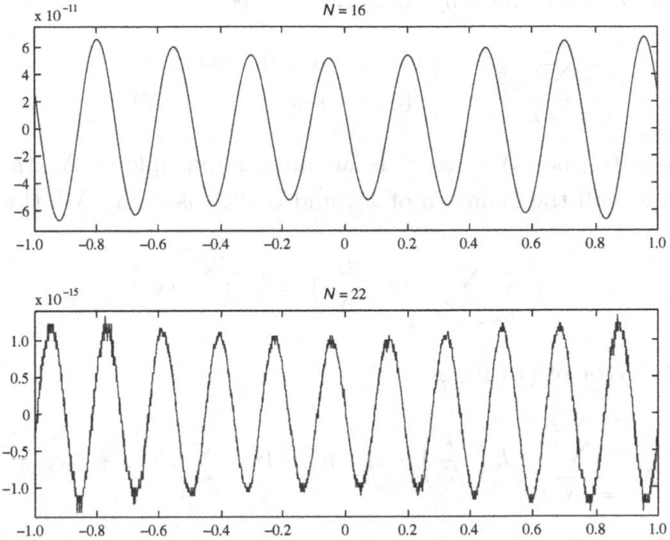

Figure 10.6 The error in solving (10.7) with $a(x) = f(x) = \cos\pi x$ and $b(x) = \sin 2\pi x$ using the spectral method (10.9) with $N = 16$ and $N = 22$ coefficients respectively.

compute its integral in $[-1,1]$. We do so by means of the deceptively simple Riemann sum

$$\int_{-1}^{1} h(\tau)\,\mathrm{d}\tau \approx \frac{2}{N} \sum_{k=-N/2+1}^{N/2} h\left(\frac{2k}{N}\right). \tag{10.10}$$

Let

$$\omega_N = \mathrm{e}^{2\pi\mathrm{i}/N}$$

be the Nth primitive *root of unity.* Substituting its Fourier expansion into (10.10) in place of h, we obtain

$$\begin{aligned}\frac{2}{N} \sum_{k=-N/2+1}^{N/2} h\left(\frac{2k}{N}\right) &= \frac{2}{N} \sum_{k=-N/2+1}^{N/2} \sum_{n=-\infty}^{\infty} \hat{h}_n \mathrm{e}^{2\pi\mathrm{i}nk/N} \\ &= \frac{2}{N} \sum_{n=-\infty}^{\infty} \hat{h}_n \sum_{k=-N/2+1}^{N/2} \omega_N^{nk} = \frac{2}{N} \sum_{n=-\infty}^{\infty} \hat{h}_n \sum_{k=0}^{N-1} \omega_N^{n(k+1-N/2)} \\ &= \frac{2}{N} \sum_{n=-\infty}^{\infty} \hat{h}_n \omega_N^{-n(N/2-1)} \sum_{k=0}^{N-1} \omega_N^{nk}.\end{aligned}$$

Since $\omega_N^N = 1$, it follows at once by summing the geometric series that

$$\sum_{k=0}^{N-1} \omega_N^{kn} = \begin{cases} N, & n \equiv 0 \pmod{N}, \\ 0, & n \not\equiv 0 \pmod{N}. \end{cases}$$

Moreover, if $n \equiv 0 \pmod{N}$, i.e. n is an integer multiple of N, then necessarily $\omega_N^{-n(N/2-1)} = 1$ (recall the definiton of ω_N and that N is even). We thus deduce that

$$\frac{2}{N} \sum_{k=-N/2+1}^{N/2} h\left(\frac{2k}{N}\right) = 2 \sum_{r=-\infty}^{\infty} \hat{h}_{Nr};$$

consequently the error of (10.10) is

$$\frac{2}{N} \sum_{k=-N/2+1}^{N/2} h\left(\frac{2k}{N}\right) - \int_{-1}^{1} h(\tau)\,\mathrm{d}\tau = \sum_{r=1}^{\infty} (\hat{h}_{Nr} + \hat{h}_{Nr}).$$

We now recall that $h \in \mathcal{A}$, hence its Fourier coefficients decay at a spectral rate, and deduce that the error of (10.10) also decays spectrally as a function of N.

In particular, letting $h(x) = f(x)\mathrm{e}^{-\mathrm{i}\pi m x}$, we obtain a spectral method for the calculation of Fourier coefficients. Specifically, given that we wish to evaluate the N coefficients $\hat{f}_n$, $n = -N/2+1, \ldots, N/2$, say, we need to compute

$$\hat{f}_n \approx \frac{2}{N} \sum_{k=-N/2+1}^{N/2} f\left(\frac{2k}{N}\right) \omega_N^{-nk}, \qquad n = -N/2+1, \ldots, N/2. \tag{10.11}$$

To calculate (10.11) we first evaluate h at N equidistant points on a grid in $[-1, 1]$ and then multiply the outcome with a matrix whose elements are $(2/N)\omega_N^{-nk}$. On the face of it, such a multiplication requires $\mathcal{O}(N^2)$ operations. However, the special structure of this matrix lends itself to perhaps the most remarkable computational algorithm ever, the *fast Fourier transform* (FFT), the subject of our next section. The outcome is a method that computes the leading N Fourier coefficients (up to spectrally small error) in just $\mathcal{O}(N \log_2 N)$ operations.

10.3 The fast Fourier transform

Let N be a positive integer and denote by Π_N the set of all complex sequences $\boldsymbol{x} = \{x_j\}_{j=-\infty}^{\infty}$ which are periodic with period N, i.e., which are such that $x_{j+N} = x_j$, $j \in \mathbb{Z}$. It is an easy matter to demonstrate that Π_N is a linear space of dimension N over the complex numbers $\mathbb{C}$ (see Exercise 10.3).

Recall that $\omega_N = \exp(2\pi\mathrm{i}/N)$ stands for the *primitive root of unity* of degree N. A *discrete Fourier transform (DFT)* is a linear mapping $\mathcal{F}_N$ defined for every $\boldsymbol{x} \in \Pi_N$ by

$$\boldsymbol{y} = \mathcal{F}_N \boldsymbol{x} \qquad \text{where} \qquad y_j = \frac{1}{N} \sum_{\ell=0}^{N-1} \omega_N^{-j\ell} x_\ell, \quad j \in \mathbb{Z}. \tag{10.12}$$

Lemma 10.2 *The DFT $\mathcal{F}_N$, as defined in (10.12), maps Π_N into itself. The mapping is invertible and*

$$\boldsymbol{x} = \mathcal{F}_N^{-1}\boldsymbol{y} \qquad \text{where} \qquad x_\ell = \sum_{j=0}^{N-1} \omega_N^{j\ell} y_j, \quad \ell \in \mathbb{Z}. \tag{10.13}$$

Moreover, $\mathcal{F}_N$ is an isomorphism of Π_N onto itself (A.1.4.2, A.2.1.9).

Proof Since ω_N is a root of unity, it is true that $\omega_N^N = \omega_N^{-N} = 1$. Therefore it follows from (10.12) that

$$y_{j+N} = \frac{1}{N}\sum_{\ell=0}^{N-1} \omega_N^{-(j+N)\ell} x_\ell = \frac{1}{N}\sum_{\ell=0}^{N-1} \omega_N^{-j\ell} x_\ell = y_j, \qquad j \in \mathbb{Z},$$

and we deduce that $\boldsymbol{y} \in \Pi_N$. Therefore $\mathcal{F}_N$ indeed maps elements of Π_N into elements of Π_N.

To prove the stipulated form of the inverse, we denote $\boldsymbol{w} := \mathcal{F}_N\boldsymbol{x}$, where $\boldsymbol{x}$ was defined in (10.13). Our first observation is that if $\boldsymbol{y} \in \Pi_N$ then it is also true that $\boldsymbol{x} \in \Pi_N$ (just change the minus to a plus in the above proof). Moreover, also changing the order of summation,

$$\begin{aligned} w_m &= \frac{1}{N}\sum_{\ell=0}^{N-1} \omega_N^{-m\ell} x_\ell = \frac{1}{N}\sum_{\ell=0}^{N-1} \omega_N^{-m\ell} \left(\sum_{j=0}^{N-1} \omega_N^{j\ell} y_j\right) \\ &= \frac{1}{N}\sum_{j=0}^{N-1} \left(\sum_{\ell=0}^{N-1} \omega_N^{(j-m)\ell}\right) y_j, \qquad m \in \mathbb{Z}. \end{aligned}$$

Within the parentheses is a geometric series that can be summed explicitly. If $j \neq m$ then we have

$$\sum_{\ell=0}^{N-1} \omega_N^{(j-m)\ell} = \frac{1-\omega_N^{(j-m)N}}{1-\omega_N^{j-m}} = 0,$$

because $\omega_N^{sN} = 1$ for every $s \in \mathbb{Z}\setminus\{0\}$, whereas in the case $j = m$ we obtain

$$\sum_{\ell=0}^{N-1} \omega_N^{(j-m)N} = \sum_{\ell=0}^{N-1} 1 = N.$$

We thus conclude that $w_m = y_m$, $m \in \mathbb{Z}$, hence that $\boldsymbol{w} = \boldsymbol{y}$. Therefore, (10.13) indeed describes the inverse of $\mathcal{F}_N$.

The existence of an inverse for every $\boldsymbol{x} \in \Pi_N$ shows that $\mathcal{F}_N$ is an isomorphism. To conclude the proof and demonstrate that this DFT maps Π_N onto itself, we suppose that there exists a $\tilde{\boldsymbol{y}} \in \Pi_N$ that cannot be the destination of $\mathcal{F}_N\boldsymbol{x}$ for any $\boldsymbol{x} \in \Pi_N$. Let us define $\tilde{\boldsymbol{x}}$ in terms of (10.13). Then, as we have already observed, $\tilde{\boldsymbol{x}} \in \Pi_N$ and it follows from our proof that $\tilde{\boldsymbol{y}} = \mathcal{F}_N\tilde{\boldsymbol{x}}$. Therefore $\tilde{\boldsymbol{y}}$ is in the range of $\mathcal{F}_N$, in contradiction to our assumption, and we conclude that the DFT $\mathcal{F}_N$ is, indeed, an isomorphism of Π_N *onto* itself. ■

It is of interest to mention an alternative proof that $\mathcal{F}_N$ is onto. According to a classical theorem from linear algebra, a linear mapping T from a finite-dimensional linear space V to itself is onto if and only if its kernel $\ker T$ consists just of the zero element of the space ($\ker T$ is the set of all $\boldsymbol{w} \in V$ such that $T\boldsymbol{w} = \boldsymbol{0}$; see A.1.4.2). Letting $\boldsymbol{x} \in \ker \mathcal{F}_N$, (10.12) yields

$$\sum_{\ell=0}^{N-1} \omega_N^{-j\ell} x_\ell = 0, \qquad j = 0, 1, \ldots, N-1. \tag{10.14}$$

This is a homogeneous linear system of N equations in the N unknowns $x_0, \ldots, x_{N-1}$. Its matrix is a *Vandermonde* matrix (A.1.2.5) and it is easy to prove that its determinant satisfies

$$\det \begin{bmatrix} 1 & \omega_N^{-1} & \cdots & \omega_N^{-(N-1)} \\ 1 & \omega_N^{-2} & \cdots & \omega_N^{-2(N-1)} \\ \vdots & \vdots & & \vdots \\ 1 & \omega_N^{-N} & \cdots & \omega_N^{-N(N-1)} \end{bmatrix} = \prod_{\ell=1}^{N} \prod_{j=0}^{\ell-1} (\omega_N^{-\ell} - \omega_N^{-j}) \neq 0.$$

Therefore the only possible solution of (10.14) is $x_0 = x_1 = \cdots = x_{N-1} = 0$ and we deduce that $\ker \mathcal{F}_N = \boldsymbol{0}$ and the mapping is onto Π_N.

◇ **Applications of the DFT** It is difficult to overstate the importance of the discrete Fourier transform in a multitude of applications ranging from numerical analysis to control theory, from computer science to coding theory signal processing, time series analysis ... Later, both in this chapter and in Chapter 15, we will employ it to provide a fast solution to discretized differential equations.

We commence with the issue that motivated us at the first place, the computation of Fourier coefficients. Note that $\omega_N^{nN/2} = (-1)^n$ (verify!) implies that, in (10.11),

$$\hat{f}_n = 2(-1)^n \omega_N^{-n} \times \frac{1}{N} \sum_{\ell=0}^{N} f\left(\frac{2\ell + 2 - N}{N}\right) \omega^{-n\ell}.$$

Since f is periodic, this immediately establishes a connection between the approximation of Fourier coefficients in (10.11) and the DFT.

This can be extended without difficulty to functions f defined in the interval $[a, b]$, where $a < b$, that are periodic and of period $b - a$. The *Fourier transform* of f is the sequence $\{\hat{f}_n\}_{n=-\infty}^{\infty}$, where

$$\hat{f}_n = \frac{1}{b-a} \int_a^b f(\tau) \exp\left(-\frac{2\mathrm{i}\pi n\tau}{b-a}\right) \mathrm{d}\tau, \qquad n \in \mathbb{Z}. \tag{10.15}$$

Fourier transforms feature in numerous branches of mathematical analysis and its applications: the library shelf labelled 'harmonic analysis' is, to a very large extent, devoted to Fourier transforms and their ramifications. More to the point, as far as the subject matter of this book is concerned they are crucial in the stability analysis of numerical methods for PDEs of evolution (see Chapters 15 and 16).

The computation of Fourier transforms is not the only application of the DFT, although arguably it is the most important. Other applications include the computation of conformal mappings, interpolation by trigonometric polynomials, the incomplete factoring of polynomials, the fast multiplication of large integers ... In Chapter 15 we utilize it in the solution of specially structured linear algebraic systems that occur in methods for the calculation of the Poisson equation with Dirichlet boundary conditions in a square. ◇

On the face of it, the evaluation of the DFT (10.12) (or of its inverse) requires $\mathcal{O}(N^2)$ operations since, owing to periodicity, it is obtained by multiplying a vector in $\mathbb{C}^N$ by a $N \times N$ complex matrix. It is one of the great wonders of computational mathematics, however, that this operation count can be reduced a very great deal. Let us assume for simplicity that $N = 2^n$, where n is a nonnegative integer. It is convenient to replace $\mathcal{F}_N$ by the mapping $\mathcal{F}_N^* := N\mathcal{F}_N$; clearly, if we can compute $\mathcal{F}_N^*\boldsymbol{x}$ cheaply then just $\mathcal{O}(N)$ operations will convert the result to $\mathcal{F}_N\boldsymbol{x}$.

Let us define, for every $\boldsymbol{x} \in \Pi_N$, 'even' and 'odd' sequences

$$\boldsymbol{x}^{[\mathrm{e}]} := \{x_{2j}\}_{j=-\infty}^{\infty} \qquad \text{and} \qquad \boldsymbol{x}^{[\mathrm{o}]} := \{x_{2j+1}\}_{j=-\infty}^{\infty}.$$

Since $\boldsymbol{x}^{[\mathrm{e}]}, \boldsymbol{x}^{[\mathrm{o}]} \in \Pi_{N/2}$, we can make the mappings

$$\boldsymbol{y}^{[\mathrm{e}]} = \mathcal{F}_{N/2}^*\boldsymbol{x}^{[\mathrm{e}]} \qquad \text{and} \qquad \boldsymbol{y}^{[\mathrm{o}]} = \mathcal{F}_{N/2}^*\boldsymbol{x}^{[\mathrm{o}]}.$$

Let $\boldsymbol{y} = \mathcal{F}_N^*\boldsymbol{x}$. Then, by (10.12),

$$\begin{aligned} y_j &= \sum_{\ell=0}^{N-1} \omega_N^{-j\ell} x_\ell = \sum_{\ell=0}^{2^n-1} \omega_{2^n}^{-j\ell} x_\ell \\ &= \sum_{\ell=0}^{2^{n-1}-1} \omega_{2^n}^{-2j\ell} x_{2\ell} + \sum_{\ell=0}^{2^{n-1}-1} \omega_{2^n}^{-j(2\ell+1)} x_{2\ell+1}, \qquad j = 0, 1, \ldots, 2^n - 1. \end{aligned}$$

However,

$$\omega_{2^n}^{2s} = \omega_{2^{n-1}}^{s}, \qquad s \in \mathbb{Z};$$

therefore

$$\begin{aligned} y_j &= \sum_{j=0}^{2^{n-1}-1} \omega_{2^{n-1}}^{-j\ell} x_{2\ell} + \omega_{2^n}^{-j} \sum_{j=0}^{2^{n-1}-1} \omega_{2^{n-1}}^{-j\ell} x_{2\ell+1} \\ &= y_j^{[\mathrm{e}]} + \omega_{2^n}^{-j} y_j^{[\mathrm{o}]}, \qquad j = 0, 1, \ldots, 2^n - 1. \end{aligned} \qquad (10.16)$$

In other words, provided that $\boldsymbol{y}^{[\mathrm{e}]}$ and $\boldsymbol{y}^{[\mathrm{o}]}$ are already known, we can synthesize them into $\boldsymbol{y}$ in $\mathcal{O}(N)$ operations.

Incidentally – and before proceeding any further – we observe that the number of operations can be reduced significantly by exploiting the identity $\omega_{2^s}^{-s} = -1$, $s \geq 1$. Hence, (10.16) yields

$$\left.\begin{aligned} y_j &= y_j^{[\mathrm{e}]} + \omega_{2^n}^{-j} y_j^{[\mathrm{o}]}, \\ y_{j+2^{n-1}} &= y_{j+2^{n-1}}^{[\mathrm{e}]} + \omega_{2^n}^{-j-2^{n-1}} y_{j+2^{n-1}}^{[\mathrm{o}]} = y_j^{[\mathrm{e}]} - \omega_{2^n}^{-j} y_j^{[\mathrm{o}]}, \end{aligned}\right\} \quad j = 0, 1, \ldots, 2^{n-1} - 1$$

(recall that $\boldsymbol{y}^{[\mathrm{e}]}, \boldsymbol{y}^{[\mathrm{o}]} \in \Pi_{2^{n-1}}$, therefore $y_{j+2^{n-1}}^{[\mathrm{e}]} = y_j^{[\mathrm{e}]}$ etc.). In other words, to combine $\boldsymbol{y}^{[\mathrm{e}]}$ and $\boldsymbol{y}^{[\mathrm{o}]}$ we need to form just 2^{n-1} products $\omega_{2^{n-1}}^{-j} y_j^{[\mathrm{o}]}$, subsequently adding or subtracting them, as required, from $y_j^{[\mathrm{e}]}$ for $j = 0, 1, \ldots, 2^{n-1}$.

All this, needless to say, is based on the premise that $\boldsymbol{y}^{[\mathrm{e}]}$ and $\boldsymbol{y}^{[\mathrm{o}]}$ are known, which, as things stand, is false. Having said this, we can form, in a similar fashion to that above, $\boldsymbol{y}^{[\mathrm{e}]}$ from $\mathcal{F}_{N/4}^{*}\boldsymbol{x}^{[\mathrm{ee}]}, \mathcal{F}_{N/4}^{*}\boldsymbol{x}^{[\mathrm{eo}]} \in \Pi_{N/4}$, where

$$\boldsymbol{x}^{[\mathrm{ee}]} = \{x_{4j}\}_{j=-\infty}^{\infty}, \qquad \boldsymbol{x}^{[\mathrm{eo}]} = \{x_{4j+2}\}_{j=-\infty}^{\infty}.$$

Likewise, $\boldsymbol{y}^{[\mathrm{o}]}$ can be obtained from two transforms of length $N/4$. This procedure can be iterated until we reach transforms of unit length, which, of course, are the variables themselves.

Practical implementation of this procedure, the famous *fast Fourier transform* (FFT), proceeds from the other end: we commence from 2^n transforms of length 1 and synthesize them into 2^{n-1} transforms of length 2. These are, in turn, combined into 2^{n-2} transforms of length 2^2, then into 2^{n-3} transforms of length 2^3 and so on, until we reach a single transform of length 2^n, the object of this whole exercise.

Assembling 2^{n-s+1} transforms of length 2^{s-1} into 2^{n-s} transforms of double the length costs $\mathcal{O}(2^{n-s} \times 2^s) = \mathcal{O}(N)$ operations. Since there are n such 'layers', the total expense of the FFT is a multiple of $2^n n = N \log_2 N$ operations. For large values of N this results in a very significant saving in comparison with naive matrix multiplication.

The order of assembly of one set of transforms into new transforms of twice the length is important – we do not just combine any two arbitrary 'strands'! The correct arrangement is displayed in Fig. 10.7 in the case $n = 4$ and the general rule is obvious. It can be best understood by expressing the index ℓ in a binary representation, but we choose not to dwell further on this.

It is elementary to generalize the DFT (10.12) to two (or more) dimensions. Thus, let $\{x_{k,j}\}_{k,j=-\infty}^{\infty}$ be N-periodic in each index, $x_{k+N,j} = x_{k,j} = x_{k,j+N}$ for $k, j \in \mathbb{Z}$. We set

$$\boldsymbol{y} = \mathcal{F}_N \boldsymbol{x} \qquad \text{where} \qquad y_{k,j} = \frac{1}{N^2} \sum_{\ell=0}^{N-1} \sum_{m=0}^{N-1} \omega_N^{-(k\ell+jm)} x_{\ell,m}, \quad k, j \in \mathbb{Z}.$$

The FFT can be extended to this case by acting separately on each index. This leads to an algorithm bearing the price tag of $\mathcal{O}\big(N^2 \log_2 N\big)$ operations.

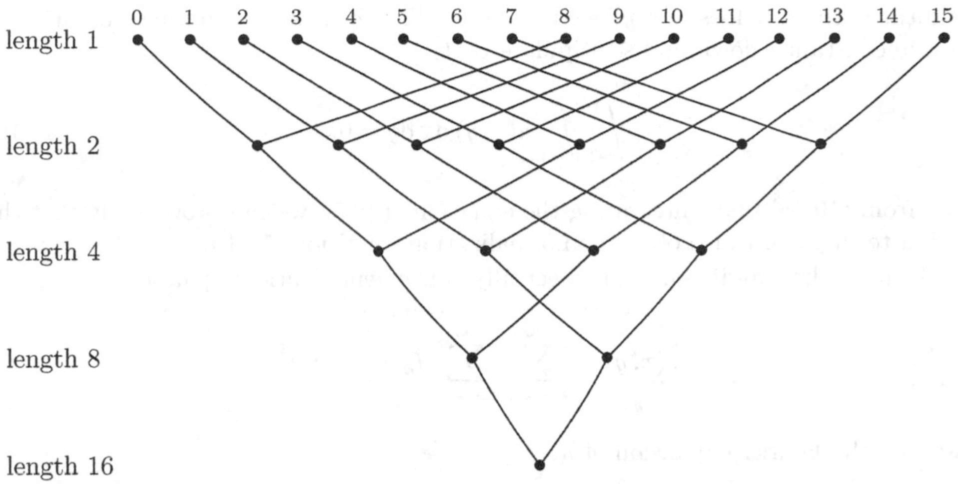

Figure 10.7 The assembly pattern in FFT for $N = 16 = 2^4$.

10.4 Second-order elliptic PDEs

Spectral methods owe their success to the confluence of three bits of mathematical magic:

- the spectral convergence of Fourier expansions of analytic periodic functions;
- the spectral convergence of a DFT approximation to Fourier coefficients of analytic periodic functions; and
- the low-cost calculation of a DFT by the fast Fourier transform.

In the previous two chapters we were concerned with solving the Poisson equation $\nabla^2 u = f$ in a square, and it is tempting to test a spectral method in this case. However, as we will see, this would not be very representative, since the special structure of a Poisson equation with periodic boundary conditions confers an unfair advantage on spectral methods.

Specifically, consider the Poisson equation

$$\nabla^2 u = f, \qquad -1 \le x, y \le 1, \tag{10.17}$$

where the analytic function f obeys the periodic boundary conditons $f(-1, y) = f(1, y)$, $-1 \le y \le 1$ and $f(x, -1) = f(x, 1)$, $-1 \le x \le 1$. We equip (10.17) with the periodic boundary conditions

$$\begin{aligned} u(-1, y) &= u(1, y), \quad u_x(-1, y) = u_x(1, y), \qquad -1 \le y \le 1, \\ u(x, -1) &= u(x, 1), \quad u_y(x, -1) = u_y(x, 1), \qquad -1 \le x \le 1, \end{aligned} \tag{10.18}$$

but a moment's reflection demonstrates that they define the solution of (10.17) only up to an additive constant: if $u(x, y)$ solves this equation and obeys periodic boundary conditions, then so does $u(x,y)+c$ for any $c \in \mathbb{R}$. We thus need another condition to pin the constant c down and so stipulate that

$$\int_{-1}^{1}\int_{-1}^{1} u(x,y)\,\mathrm{d}x\,\mathrm{d}y = 0. \tag{10.19}$$

Note from (10.18) that, integrating the equation (10.17), we can prove easily that the forcing term f must also obey the normalization condition (10.19).

We have the two-dimensional spectrally convergent Fourier expansion

$$f(x,y) = \sum_{k=-\infty}^{\infty}\sum_{\ell=-\infty}^{\infty} \hat{f}_{k,\ell}\mathrm{e}^{\mathrm{i}\pi(kx+\ell y)}$$

and seek the Fourier expansion of u,

$$u(x,y) = \sum_{k=-\infty}^{\infty}\sum_{\ell=-\infty}^{\infty} \hat{u}_{k,\ell}\mathrm{e}^{\mathrm{i}\pi(kx+\ell y)},$$

whose existence is justified by the periodic boundary conditions (10.18). Note that the normalization condition (10.19) amounts to $\hat{u}_{0,0}=0$. Therefore

$$\nabla^2 u(x,y) = -\pi^2\sum_{k=-\infty}^{\infty}\sum_{\ell=-\infty}^{\infty} (k^2+\ell^2)\hat{u}_{k,\ell}\mathrm{e}^{\mathrm{i}\pi(kx+\ell y)},$$

together with (10.17), implies at once that

$$\hat{u}_{k,\ell} = -\frac{1}{(k^2+\ell^2)\pi^2}\hat{f}_{k,\ell}, \qquad k,\ell\in\mathbb{Z}, \quad (k,\ell)\neq(0,0).$$

We have obtained the Fourier coefficients of the solution in an explicit form, without any need to solve linear algebraic equations. The explanation is simple: what we have done is to recreate in a numerical setting the familiar technique of the separation of variables. The trigonometric functions $\varphi_{k,\ell}(x,y)=\mathrm{e}^{\mathrm{i}\pi(kx+\ell y)}$ are eigenfunctions of the Laplace operator, $\nabla^2\varphi_{k,\ell}=-\pi^2(k^2+\ell^2)\varphi_{k,\ell}$, and they obey periodic boundary conditions.

A fairer impression is gained by considering a case where spectral methods do not enjoy an unfair advantage. Therefore, let us examine the second-order linear elliptic PDE,

$$\nabla\cdot(a\nabla u) = f, \qquad -1\le x,y\le 1, \tag{10.20}$$

where the positive analytic function a is periodic, as is the forcing term f. We again impose the periodic boundary conditions (10.18) and the normalization condition (10.19).

Writing

$$\nabla\cdot(a\nabla u) = a\nabla^2 u + a_x u_x + a_y u_y,$$

we thus have

$$-\pi^2\left(\sum_{k=-\infty}^{\infty}\sum_{\ell=-\infty}^{\infty}\hat{a}_{k,\ell}\mathrm{e}^{\mathrm{i}\pi(kx+\ell y)}\right)\left(\sum_{k=-\infty}^{\infty}\sum_{\ell=-\infty}^{\infty}(k^2+\ell^2)\hat{u}_{k,\ell}\mathrm{e}^{\mathrm{i}\pi(kx+\ell y)}\right)$$
$$-\pi^2\left(\sum_{k=-\infty}^{\infty}\sum_{\ell=-\infty}^{\infty}k\hat{a}_{k,\ell}\mathrm{e}^{\mathrm{i}\pi(kx+\ell y)}\right)\left(\sum_{k=-\infty}^{\infty}\sum_{\ell=-\infty}^{\infty}k\hat{u}_{k,\ell}\mathrm{e}^{\mathrm{i}\pi(kx+\ell y)}\right)$$
$$-\pi^2\left(\sum_{k=-\infty}^{\infty}\sum_{\ell=-\infty}^{\infty}\ell\hat{a}_{k,\ell}\mathrm{e}^{\mathrm{i}\pi(kx+\ell y)}\right)\left(\sum_{k=-\infty}^{\infty}\sum_{\ell=-\infty}^{\infty}\ell\hat{u}_{k,\ell}\mathrm{e}^{\mathrm{i}\pi(kx+\ell y)}\right)$$
$$=\sum_{k=-\infty}^{\infty}\sum_{\ell=-\infty}^{\infty}\hat{f}_{k,\ell}\mathrm{e}^{\mathrm{i}\pi(kx+\ell y)}$$

where

$$a(x,y)=\sum_{k=-\infty}^{\infty}\sum_{\ell=-\infty}^{\infty}\hat{a}_{k,\ell}\mathrm{e}^{\mathrm{i}\pi(kx+\ell y)}.$$

Next we replace products by convolutions – it is trivial to generalize (10.5) to the bivariate case by replacing one summation by two. Finally, we truncate the infinite-dimensional system to $-N/2+1\le k,\ell\le N/2$ and impose the normalization condition $\hat{u}_{0,0}=0$. All this results in a system of N^2-1 linear algebraic equations in the N^2-1 unknowns $u_{k,\ell}$, $k,\ell=-N/2+1,\ldots,N/2$, $(k,\ell)\ne(0,0)$. Typically, such a system is devoid of any useful structure and is not sparse. Yet its size is substantially smaller than anything we would have obtained, expecting similar precision, by the methods of Chapters 8 and 9.

This is, however, not the only means of constructing a spectral method for (10.20). Since $\mathcal{L}u=-\nabla\cdot(a\nabla u)$ is a positive-definite operator with respect to the standard Euclidean complex valued norm

$$\langle u,v\rangle=\int_{-1}^{1}\int_{-1}^{1}u(x,y)\overline{v(x,y)}\,\mathrm{d}x\,\mathrm{d}y,$$

we can use the finite element methods of Chapter 9 to construct a linear algebraic system, except that instead of the finite element basis (leading to a large, sparse matrix) we use the spectral basis $v_{k,\ell}(x,y)=\mathrm{e}^{\mathrm{i}\pi(kx+\ell y)}$, $-N/2+1\le k,\ell\le N/2$ (leading to small dense matrix). In particular, integrating by parts and using periodic boundary conditions, we have

$$\begin{aligned}\langle\mathcal{L}v_{k,\ell},v_{m,j}\rangle&=\int_{-1}^{1}\int_{-1}^{1}a(\nabla v_{k,\ell})\cdot(\nabla v_{m,j})\,\mathrm{d}x\,\mathrm{d}y\\&=\pi^2(km+\ell j)\int_{-1}^{1}\int_{-1}^{1}a(x,y)\mathrm{e}^{\mathrm{i}\pi[(k-m)x+(\ell-j)y]}\,\mathrm{d}x\,\mathrm{d}y\\&=4\pi^2(km+\ell j)\hat{a}_{k-m,\ell-j}.\end{aligned}$$

(Note that there is no need to resort to the formalism of bilinear forms, since our functions are smooth enough for the straightforward action of the operator $\mathcal{L}$.) In a similar way we obtain $\langle f,v_{k,\ell}\rangle=4\hat{f}_{k,\ell}$ (verify!). Therefore, using an FFT the Ritz equations (9.24) can be constructed fairly painlessly.

10.5 Chebyshev methods

And now to the bad news ... The efficacy of spectral methods depends completely on the analyticity and periodicity of the underlying problem, inclusive of boundary conditions and coefficients. Take away either and the rate of convergence drops to polynomial.[1] To obtain reasonable accuracy we then need a large number of variables and hence end up with a large matrix system but, unlike in the cases of finite differences or finite elements, the matrix is not sparse, so we have the worst of all worlds. Relatively few problems originating in real applications are genuinely periodic and this renders spectral methods of limited applicability: when they are good they are very very good but when they are bad, they are horrid.[2]

Once we wish spectral methods to be available in a more general setting, we need a framework that allows nonperiodic functions. This brings us to Chebyshev polynomials.

We let $T_n(x) = \cos(n \arccos x)$, $n \geq 0$; therefore

$$T_0(x) \equiv 1, \quad T_1(x) = x, \quad T_2(x) = 2x^2 - 1, \quad T_3(x) = 4x^3 - 3x, \quad \ldots$$

It is easy to verify (see Exercise 3.2) that each T_n is a polynomial of degree n: it is called the nth *Chebyshev polynomial* (of the first kind). Moreover, Chebyshev polynomials are orthogonal with respect to the weight function $(1-x^2)^{-1/2}$ in $(-1,1)$,

$$\int_{-1}^{1} T_m(x)T_n(x)\frac{\mathrm{d}x}{\sqrt{1-x^2}} = \begin{cases} \pi, & m=n=0, \\ \frac{1}{2}\pi, & m=n\geq 1, \\ 0, & m\neq n, \end{cases} \qquad m,n\in\mathbb{Z}, \tag{10.21}$$

and they obey the three-term recurrence relation

$$T_{n+1}(x) = 2xT_n(x) - T_{n-1}(x), \qquad n=1,2,\ldots$$

We consider the expansion of a general integrable function f in the orthogonal sequence $\{T_n\}_{n=0}^{\infty}$:

$$f(x) = \sum_{n=0}^{\infty} \breve{f}_n T_n(x). \tag{10.22}$$

Multiplying (10.22) by $T_m(x)(1-x^2)^{-1/2}$, integrating for $x \in (-1,1)$ and using the orthogonality conditions (10.21) results in

$$\breve{f}_0 = \frac{1}{\pi}\int_{-1}^{1} f(x)\frac{\mathrm{d}x}{\sqrt{1-x^2}}, \qquad \breve{f}_n = \frac{2}{\pi}\int_{-1}^{1} f(x)T_n(x)\frac{\mathrm{d}x}{\sqrt{1-x^2}}, \quad n=1,2,\ldots$$

[1] Not strictly true: if the data is C^∞ rather than analytic then spectral convergence can be recovered, a subject to which we will return before the end of this chapter. But in fact the difference between analytic and C^∞ data is mostly a matter of mathematical nicety, while piecewise-smooth data is fairly popular in applications.

[2] Applied mathematics and engineering departments abound in researchers forcing periodic conditions on their models, to render them amenable to spectral methods. This results in fast algorithms, nice pictures, but arguably only tenuous relevance to applications.

Letting $x = \cos\theta$, a simple change of variables confirms that

$$\int_{-1}^{1} f(x)T_n(x)\frac{\mathrm{d}x}{\sqrt{1-x^2}} = \int_0^{\pi} f(\cos\theta)\cos n\theta\,\mathrm{d}\theta = \frac{1}{2}\int_{-\pi}^{\pi} f(\cos\theta)\cos n\theta\,\mathrm{d}\theta.$$

Given that $\cos n\theta = \frac{1}{2}(\mathrm{e}^{\mathrm{i}n\theta} + \mathrm{e}^{-\mathrm{i}n\theta})$, the connection with Fourier expansions stands out. Specifically, letting $g(x) = f(\cos x)$ in place of f in (10.15), we have

$$\hat{g}_n = \frac{1}{2\pi}\int_{-\pi}^{\pi} g(\tau)\mathrm{e}^{-\mathrm{i}n\tau}\,\mathrm{d}\tau, \qquad n \in \mathbb{Z}.$$

Therefore

$$\int_{-1}^{1} f(x)T_n(x)\frac{\mathrm{d}x}{\sqrt{1-x^2}} = \frac{\pi}{2}(\hat{g}_{-n} + \hat{g}_n)$$

and we deduce that

$$\breve{f} = \begin{cases} \hat{g}_0, & n = 0, \\ \hat{g}_{-n} + \hat{g}_n, & n = 1, 2, \ldots. \end{cases} \tag{10.23}$$

The computation of the expansion (10.22) is therefore equivalent to the Fourier expansion of the function g. However, the latter is periodic with period 2π; therefore we can use a DFT to compute the $\breve{f}_n$ while enjoying all the benefits of periodic functions. In particular, if f can be extended analytically into an open neighbourhood of $[-1, 1]$ then the error decays at a spectral rate. Moreover, thanks to (10.23), Chebyshev coefficients $\breve{f}_n$ also decay spectrally fast for $n \gg 1$. Therefore we reap all the benefits of the rapid convergence of spectral methods without ever assuming that f is periodic!

Before we can apply Chebyshev expansions (10.22) in spectral-like methods for nonperiodic analytic functions, we must develop a toolbox for the algebraic and analytic manipulation of these expansions, along the lines of Section 10.2. Thus, let $\mathcal{B}$ denote the set of all analytic functions in $[-1, 1]$ that can be extended analytically into the complex plane. We identify each such function with its Chebyshev expansion. Like the set $\mathcal{A}$, we see that $\mathcal{B}$ is a linear space and is closed under multiplication. To derive an alternative to the convolution (10.5), we note that

$$\begin{aligned} T_m(x)T_n(x) &= \cos(m \arccos x)\cos(n \arccos x) \\ &= \tfrac{1}{2}\{\cos[(m-n)\arccos x] + \cos[(m+n)\arccos x]\} \\ &= \tfrac{1}{2}[T_{|m-n|}(x) + T_{m+n}(x)]. \end{aligned}$$

Therefore, after elementary algebra, we obtain

$$\begin{aligned} f(x)g(x) &= \sum_{m=0}^{\infty} \breve{f}_m T_m(x) \sum_{n=0}^{\infty} \breve{g}_n T_n(x) \\ &= \tfrac{1}{2}\sum_{m=0}^{\infty}\sum_{n=0}^{\infty} \breve{f}_m \breve{g}_n [T_{|m-n|}(x) + T_{m+n}(x)] \\ &= \tfrac{1}{2}\sum_{n=0}^{\infty}\sum_{m=0}^{\infty} \breve{f}_m (\breve{g}_{|m-n|} + \breve{g}_{m+n}) T_n(x). \end{aligned}$$

Finally, we need to express derivatives of functions in $\mathcal{B}$ as Chebyshev expansions. The analogous task was easy in $\mathcal{A}$, since $e^{i\pi nx}$ is an eigenfunction of the differential operator. In $\mathcal{B}$ this is somewhat more complicated. We note, however, that T_n' is a polynomial of degree $n-1$ and hence can be expressed in the basis $\{T_0, T_1, \ldots, T_{n-1}\}$. Moreover, each T_n is of the same parity as n (that is, T_{2m} is an even and T_{2m+1} an odd function); therefore T_n' is of opposite parity and the only surviving terms in the linear combination are $T_{n-1}, T_{n-3}, T_{n-5}, \ldots$

Lemma 10.3 *The derivatives of Chebyshev polynomials can be expressed explicitly as the linear combinations*

$$T_{2n}'(x) = 4n \sum_{\ell=0}^{n-1} T_{2l+1}(x), \tag{10.24}$$

$$T_{2n+1}'(x) = (2n+1)T_0(x) + 2(2n+1) \sum_{\ell=1}^{n} T_{2\ell}(x). \tag{10.25}$$

Proof We will prove only (10.24), since (10.25) follows by an identical argument. Thus,

$$\sin\theta \, T_{2n}'(\cos\theta) = 2n \sin 2n\theta$$

while, on telescoping series,

$$\begin{aligned} 4n \sin\theta \sum_{\ell=0}^{n-1} T_{2l+1}(\cos\theta) &= 4n \sum_{\ell=0}^{n-1} \sin\theta \cos(2\ell+1)\theta \\ &= 2n \sum_{\ell=0}^{n-1} [\sin(2\ell+2)\theta - \sin 2\ell\theta] = 2n \sin 2n\theta. \end{aligned}$$

This proves (10.24). ∎

The recursions (10.24) and (10.25) can be used to express the derivative of any term in $\mathcal{B}$ as a Chebyshev expansion. Moreover, they can be iterated in an obvious way to express arbitrarily high derivatives in this form.

Chebyshev methods can be assembled exactly like standard spectral methods, using the linearity of $\mathcal{B}$ and the product rules of Lemma 10.3. However, we must remember to translate correctly the computation of Chebyshev coefficients to the Fourier realm. In order to use the FFT, we sample the function g at N equidistant points in $[-\pi, \pi]$. Back in $[-1, 1]$, the world of the original function values, this means computing the function f at the *Chebyshev points* $\cos(2\pi k/N)$, $k = -N/2+1, \ldots, N/2$. (This, of course, can be translated linearly from $[-1, 1]$ into any other bounded interval.) Likewise, in two dimensions we need to sample our functions on the Chebyshev grid $(\cos(2\pi k/N), \cos(2\pi\ell/N))$, $k, \ell = -N/2+1, -N/2+2, \ldots, N/2$, which has the following form (for $N = 16$):

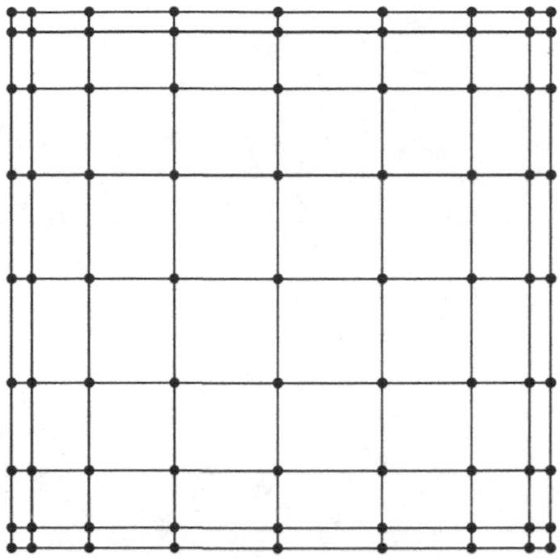

Thus, the computed function values are denser toward the edges. Insofar as the computation of elliptic problems, e.g. the Poisson equation and its generalization (10.20), are concerned, this is not problematic. However (and we comment further upon this below), sampling on the above grid can play havoc with numerical stability once we apply Chebyshev methods to initial value PDEs.

Comments and bibliography

There are many good texts on Fourier expansions, spanning the full range from the purely mathematical to the applied. For an introductory yet comprehensive exposition of the subject, painting on a broad canvass and conveying not just mathematical foundations but also the beauty and excitement of Fourier analysis, one can hardly do better than Körner (1988).

Within their own frame of reference – everything analytic, everything periodic – it is difficult to improve upon Fourier expansions and spectral methods. The question is, how much can we relax analyticity and periodicity while retaining the substantive advantages of Fourier expansions?

We can salvage spectral convergence once analyticity is replaced by the requirement that $f \in C^\infty(-1,1)$, in other words that $f^{(m)}(x)$ exists for all $x \in (-1,1)$ and $m = 0, 1, 2 \ldots$ For example, in Fig. 10.8 we plot $-(\log|\hat{f}_n|)/n^{0.44}$ for $n = 1, 2, \ldots, 100$ and the function $f(x) = \exp\left[-1/(1-x^2)\right]$. (Note that f is even, therefore $\hat{f}_{-n} = \hat{f}_n$ and it is enough to examine the Fourier coefficients for $n \geq 0$.) Note also that, while all derivatives of f exist in $(-1,1)$, this (periodic) function cannot be extended analytically because of essential singularities at ± 1. Yet, the figure indicates that, asymptotically, the scaled logarithm oscillates about a constant value. An easy calculation shows that asymptotically $|\hat{f}_n| \sim \mathcal{O}[\exp(-cn^\alpha)]$, where $c > 0$ and $\alpha \approx 0.44$. This is slower than the exponential decay of genuinely analytic functions, yet faster than $\mathcal{O}(n^{-p})$ for any integer p. Hence – and this is generally true for all periodic C^∞ functions – we have spectral convergence.

The difference between analytic and C^∞ functions is mostly of little relevance to real-life computing. Not so the difference between periodic and nonperiodic functions. Most differential equations are naturally equipped with Dirichlet or Neumann boundary conditions. Forcing periodicity on the solution is wrong since boundary conditions are not some sort of

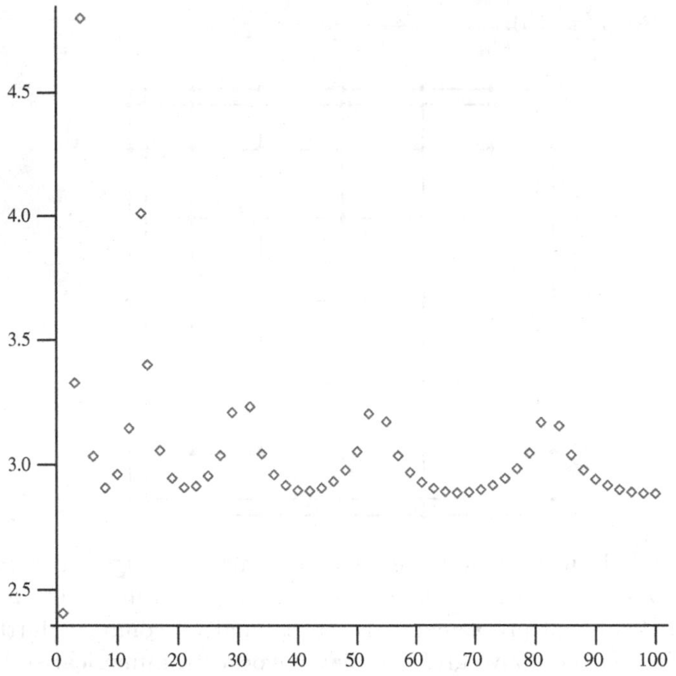

Figure 10.8 The quantity $-\log|\hat{f}_n|/n^{0.44}$ for $f(x) = \exp\left[-1/(1-x^2)\right]$ oscillates around a constant value, therefore illustrating the (roughly) $\mathcal{O}\left[\exp(-cn^{0.44})\right]$ decay of the Fourier coefficients.

optional extra: they are at the very heart of modelling nature by mathematics, so replacing them by periodic boundary conditions by fiat makes no sense.

Unfortunately, periodicity is necessary for spectral convergence. Even if f is analytic, Fourier series converge as $\mathcal{O}\bigl(N^{-1}\bigr)$ unless $f(-1) = f(+1)$; thus they are the equivalent of a first-order finite difference or finite element method. (Fourier series also converge when analyticity fails, subject to weaker conditions, but convergence might be even slower.) There are several techniques to speed up convergence. They (e.g. Gegenbauer filtering) are mostly outside the scope of our exposition, but we will mention one fairly elementary approach that, although coming short of inducing spectral convergence, speeds convergence up to $\mathcal{O}\bigl(N^{-p}\bigr)$ for $p \geq 2$: *polynomial subtraction.*

Suppose that an analytic function f is not periodic, yet $f(-1) = f(+1)$ (this is not contradictory since periodicity might fail with regard to its higher derivatives). Integrating by parts,

$$\hat{f}_n = -\frac{1}{\mathrm{i}\pi n}[f(1) - f(-1)] + \frac{1}{\mathrm{i}\pi n}\widehat{f'}_n = \frac{1}{\mathrm{i}\pi n}\widehat{f'}_n = \mathcal{O}\bigl(n^{-2}\bigr), \qquad |n| \gg 1,$$

since $\widehat{f'}_n = \mathcal{O}\bigl(n^{-1}\bigr)$. Thus, we gain one order: using the analysis of Section 10.1 we can show that the rate of convergence is $\mathcal{O}\bigl(N^{-2}\bigr)$.

In general, of course, $f(-1) \neq f(1)$, but we can force the values at the endpoints to be equal. Set $f(x) = \frac{1}{2}(1-x)f(-1) + \frac{1}{2}(1+x)f(+1) + g(x)$, where $g(x) = f(x) - \frac{1}{2}(1-x)f(-1) - \frac{1}{2}(1+x)f(+1)$. It is trivial to verify that $g(\pm 1) = 0$ and that if f is analytic then so is g.

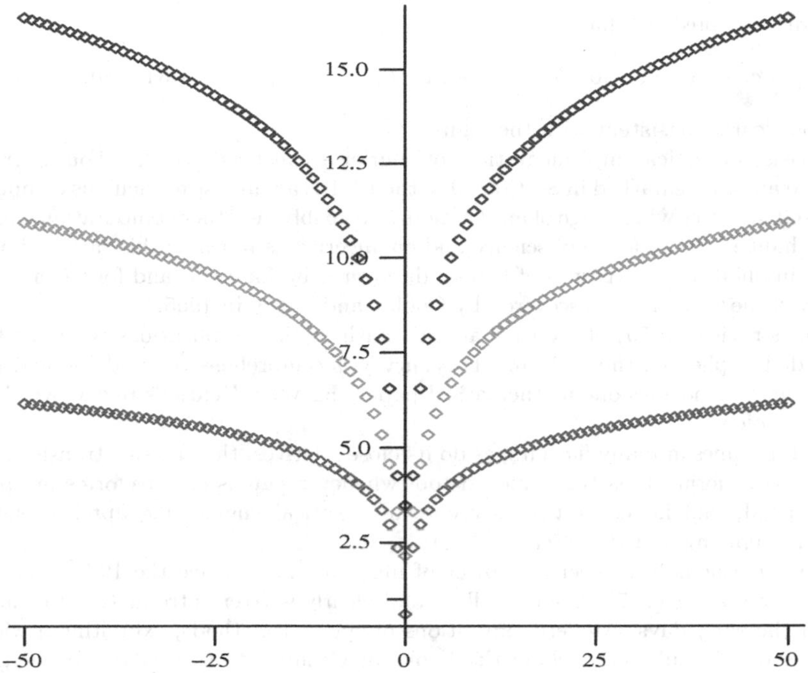

Figure 10.9 The logarithmic tree: the values of $\log |\hat{f}_n|$ (bottom branch), $\log |\hat{g}_n|$ (middle branch) and $\log |\hat{h}_n|$ for $f(x) = (2+x)^{-1}$.

The idea is now to represent f as a linear function plus the Fourier expansion of g:

$$f(x) = \tfrac{1}{2}(1-x)f(-1) + \tfrac{1}{2}(1+x)f(+1) + \sum_{n=-\infty}^{\infty} \hat{g}_n e^{i\pi n x}.$$

In principle there is nothing to stop us iterating this idea. Thus, setting

$$h(x) = f(x) - \tfrac{1}{4}(1-x)^2(2+x)f(-1) - \tfrac{1}{4}(1-x)^2(1+x)f'(-1) - \tfrac{1}{4}(1+x)^2(2-x)f(+1) + \tfrac{1}{4}(1+x)^2(1-x)f'(+1),$$

it is trivial to verify that $h(\pm 1), h'(\pm 1) = 0$ and consequently $\hat{h}_n = \mathcal{O}(n^{-3})$, $|n| \gg 1$. Setting

$$f(x) = \tfrac{1}{4}(1-x)^2(2+x)f(-1) + \tfrac{1}{4}(1-x)^2(1+x)f'(-1) + \tfrac{1}{4}(1+x)^2(2-x)f(+1) - \tfrac{1}{4}(1+x)^2(1-x)f'(+1) + \sum_{n=-\infty}^{\infty} \hat{h}_n e^{i\pi n x},$$

we thus have an $\mathcal{O}(N^{-3})$ rate of convergence. Convergence can be accelerated further in the same way. Of course, the *quid pro quo* is a rapid increase in complexity once we attempt to implement polynomial subtraction in tandem with spectral methods for PDEs. And, no matter how many levels of polynomial subtraction we employ, the convergence is never spectral. Figure 10.9 displays the logarithms of $|\hat{f}_n|$, $|\hat{g}_n|$ and $|\hat{h}_n|$ for the function $f(x) = (1+2x)^{-1}$.

Our analysis predicts that

$$\log|\hat{f}_n| \sim c_1 - \log|n|, \quad \log|\hat{g}_n| \sim c_2 - 2\log|n|, \quad \log|\hat{h}_n| \sim c_3 - 3\log|n|, \qquad |n| \gg 1,$$

and this is clearly consistent with the figure.

The basis of practical implementations of Fourier expansions is the fast Fourier transform. As we have already remarked in Section 10.3, the FFT is an almost miraculous computational device, used in a very wide range of applications. Arguably, no other computational algorithm has ever changed the practice of science and engineering as much as this rather simple trick – already implicit in the writings of Gauss, discovered by Lanczos (and forgotten) and, at a more opportune moment, rediscovered by Cooley and Tukey in 1965.

Henrici's review (1979) of the FFT and its mathematical applications is a must for every open-minded applied mathematician. This survey is comprehensive, readable and inspiring – if you plan to read just one mathematical paper this year, Henrici's review will be a very rewarding choice!

The FFT comes in many flavours, as do its close relatives, the fast sine transform and the fast cosine transform. It is the standard tool whenever signals or waveforms are processed or transmitted, and hence at the very core of electrical engineering applications: radio, television, telephony and the Internet.

Spectral methods have been a subject of much attention since the 1970s, and the text of Gottlieb & Orszag (1977) is an excellent and clearly written introduction to the state of the art in the early days. Modern expositions of spectral methods, exhibiting a wide range of outlooks on the subject, include the books of Canuto *et al.* (2006), Fornberg (1995), Hesthaven *et al.* (2007) and Trefethen (2000).

The menagerie of spectral methods is substantially greater than our very brief and elementary exposition would suggest (Canuto *et al.*, 2006). Let us mention briefly an important relative of spectral techniques, *pseudospectral methods* (Fornberg, 1995). Suppose that a function u is given on an equally spaced grid kh, $k = -M, \ldots, M$, where $h = 1/M$. (We do not assume that u is periodic.) Assume further that we wish to approximate u' at the grid points. This is the stuff of Chapter 8 and we already know how to do it using finite differences. This, however, is likely to lead to low-order methods.

As in Chapter 8, we wish to approximate $u'(mh)$ as a linear combination of the values of $u(kh)$ for $k = m - r, \ldots, m \ldots, m + s$, where $r, s \geq 0$. The highest-order approximation of this kind is (see Exercise 16.2)

$$u'(mh) \approx \frac{1}{h} \sum_{k=-r}^{s} \alpha_k u((m+k)h), \tag{10.26}$$

where

$$\alpha_k = \frac{(-1)^{k-1}}{k} \frac{r!s!}{(r+k)!(s-k)!}, \quad k \neq 0, \qquad \alpha_0 = -\sum_{\substack{j=-r \\ j\neq 0}}^{s} \frac{(-1)^{j-1}}{j} \frac{r!s!}{(r+j)!(s-j)!}.$$

It is possible to show that, for sufficiently smooth u, the error is $\mathcal{O}(h^{r+s}) = \mathcal{O}(M^{-r-s})$.

How to choose r and s? The logic of finite differences tells us to choose the same r and s everywhere except perhaps very near the edges, where it is no longer true that $r \leq M + k$, $s \leq M - k$. However, there is nothing to compel us to follow this finite difference logic; we could choose different r and s values at different grid points. In particular, note that at each $k \in \{-M, -M+1, \ldots, M\}$ we have $M + k$ points to our left and $M - k$ points to the right that can be legitimately employed in (10.26). We use *all* these points! In other words, we approximate u'_k as a linear combination of all the values of u_j on the grid, taking $r = r_k = M+k$, $s = s_k = M-k$. The result is a method whose error behaves like $\mathcal{O}(M^{-M})$, in other words it decays at spectral speed. True, the matrix is dense but we can choose a relatively small M value while obtaining high accuracy.

Sounds spectacular: spectral convergence without periodicity! Needless to say, we can repeat the trick for higher derivatives, construct numerical methods in this form... An experienced reader will rightly expect a catch! Indeed, the problem is that rapidly we are running into very large numbers. Expressing (10.26) as a matrix–vector product, $\boldsymbol{u}' \approx h^{-1}D\boldsymbol{u}$, say, it is possible to show that the norm of D grows very fast. Thus, for $M = 32$ we have $\|D\| \approx 2.18 \times 10^{17}$, while for $M = 64$ $\|D\| \approx 1.69 \times 10^{36}$ (we are using the Euclidean matrix norm). In general, $\|D\|$ increases exponentially fast in M.

We have just one means of salvaging exponential convergence of the pseudospectral approach while keeping the size of the differentiation matrices reasonably small: abandon the assumption of equally spaced grid points. We can think about the generation of a differentiation matrix as a two-step process: first interpolate u at the grid points by a polynomial (see A.2.2.1) and subsequently differentiate the polynomial in question at the grid points. Seen from this perspective, the size of $\|D\|$ is likely to be small once the interpolating polynomial approximates u well. It is known, though, that while equidistant points represent a very poor choice of interpolation points, an excellent choice is presented by the Chebyshev points $\cos(\pi k/M)$, $k = -M+1, \ldots, M$: it is possible to prove that in this case $\|D\| \approx \frac{9}{4}M^2$. For example, for $M = 32$ we obtain $\|D\| \approx 2.2801 \times 10^3$ and for $M = 64$ the norm is $\|D\| \approx 9.0619 \times 10^3$: the improvement is amazing. We are able to retain spectral convergence while keeping the matrices reasonably well conditioned. However, in this particular case we obtain a scheme equivalent to the Chebyshev method from Section 10.5. In more complicated settings, the Chebyshev and pseudospectral methods part company and lead to genuinely different algorithms.

In Chapters 15 and 16 we consider the solution of time-dependent PDEs, with an emphasis on finite differences methods. A major focus of attention in that setting is the stability (a fundamental concept that will be defined in due course). Spectral methods can be applied for time-dependent problems and, again, stability considerations are of central importance. Fourier-based methods do very well in this regard, but their applicability is restricted to periodic boundary conditions. Chebyshev-based and pseudospectral methods are more problematic but modern practice allows them to be used efficiently within this setting also (Fornberg, 1995; Hesthaven *et al.*, 2007).

Canuto, C., Hussaini, M.Y., Quarteroni, A. and Zang, T.A. (2006), *Spectral Methods. Fundamentals in Single Domains,* Springer Verlag, Berlin.

Fornberg, B. (1995), *A Practical Guide to Pseudospectral Methods,* Cambridge University Press, Cambridge.

Gottlieb, D. and Orszag, S.A. (1977), *Numerical Analysis of Spectral Methods: Theory and Applications,* SIAM, Philadelphia.

Henrici, P. (1979), Fast Fourier methods in computational complex analysis, *SIAM Review* **21,** 481–527.

Hesthaven, J.S., Gottlieb, S. and Gottlieb, D. (2007), *Spectral Methods for Time-Dependent Problems,* Cambridge University Press, Cambridge.

Körner, T.W. (1988), *Fourier Analysis,* Cambridge University Press, Cambridge.

Trefethen, L.N. (2000), *Spectral methods in MATLAB,* SIAM, Philadelphia.

Exercises

10.1 Given an analytic function f,

a prove that

$$\hat{f}_n = \frac{(-1)^{n-1}}{2\pi i n}[f(1) - f(-1)] + \frac{1}{\pi i n}\widehat{f'}_n, \qquad n \in \mathbb{Z} \setminus \{0\},$$

b deduce that for every $s = 1, 2, \ldots$ it is true for every $n \in \mathbb{Z} \setminus \{0\}$ that

$$\hat{f}_n = \frac{(-1)^{n-1}}{2} \sum_{m=0}^{s-1} \frac{1}{(\pi i n)^{m+1}} [f^{(m)}(1) - f^{(m)}(-1)] + \frac{1}{(\pi i n)^s} \widehat{f^{(s)}}_n.$$

10.2 Unless f is analytic, the rate of decay of its Fourier harmonics can be very slow, certainly slower than $\mathcal{O}(N^{-1})$. To explore this, let $f(x) = |x|^{-1/2}$.

a Prove that $\hat{f}_n = g(-n) + g(n)$, where $g(n) = \int_0^1 e^{i\pi n \tau^2}\, d\tau$.

b The *error function* is defined as the integral

$$\operatorname{erf} z = \frac{2}{\sqrt{\pi}} \int_0^z e^{-\tau^2}\, d\tau, \qquad z \in \mathbb{C}.$$

Show that its Fourier coefficients are

$$\hat{f}_n = \frac{\operatorname{erf}(\sqrt{i\pi n})}{2\sqrt{in}} + \frac{\operatorname{erf}(\sqrt{-i\pi n})}{2\sqrt{-in}}.$$

c Using without proof the asymptotic estimate $\operatorname{erf}(\sqrt{ix}) = 1 + \mathcal{O}(x^{-1})$ for $x \in \mathbb{R}$, $|x| \gg 1$, or otherwise, prove that

$$\hat{f}_n = \mathcal{O}\left(n^{-1/2}\right), \qquad |n| \gg 1.$$

(*It is possible to prove that this Fourier series converges to f, except at the origin. The proof is not easy.*)

10.3 Prove that Π_N satisfies all the axioms of a linear space (A.2.1.1). Find a basis of Π_N, thereby demonstrating that $\dim \Pi_N = N$.

10.4 Consider the solution of the two-point boundary value problem

$$(2 - \cos \pi x)u'' + u = 1, \quad -1 \le x \le 1, \qquad u(-1) = u(1),$$

using the spectral method.

a Plugging the Fourier expansion of u into this differential equation, show that the $\hat{u}_n$ obey a three-term recurrence relation.

b Computing $\hat{u}_0$ separately and using the fact that $\hat{u}_{-n} = \hat{u}_n$ (why?), prove that the computation of $\hat{u}_n$ for $-N/2+1 \leq n \leq N/2$ (assuming that $\hat{u}_n = 0$ outside this range of n) reduces to the solution of an $(N/2) \times (N/2)$ tridiagonal system of algebraic equations.

10.5 Let $a(x, y) = \cos \pi x + \cos \pi y$ and $f(x, y) = \sin \pi x + \sin \pi y$. Construct explicitly the linear algebraic system that needs to be computed once the equation $\nabla \cdot (a \nabla u) = f$, equipped with periodic boundary conditions, is solved for $-1 \leq x, y \leq 1$ by a spectral method.

10.6 Supposing that $\mathcal{B} \ni u = \sum_{n=0}^{\infty} \breve{u}_n T_n$, express u' in an explicit form as a Chebyshev expansion.

10.7 The two-point ODE $u'' + u = 1$, $u(-1) = u(1) = 0$, is solved by a Chebyshev method.

a Show that the odd coefficients are zero and that $u(x) = \sum_{n=0}^{\infty} \breve{u}_{2n} T_{2n}(x)$. Express the boundary conditions as a linear condition of the coefficients $\breve{u}_{2n}$.

b Express the differential equation as an infinite set of linear algebraic equations in the coefficients $\breve{u}_{2n}$.

c Discuss how to truncate the linear system and implement it as a proper, well-defined numerical method.

d Since $u(-1) = u(1)$, the solution is periodic. Yet we cannot expect a standard spectral method to converge at spectral speed. Why?

11

Gaussian elimination for sparse linear equations

11.1 Banded systems

Whether the objective is to solve the Poisson equation using finite differences, finite elements or a spectral method, the outcome of discretization is a set of linear algebraic equations, e.g. (8.16) or (9.7). The solution of such equations ultimately constitutes the lion's share of computational expenses. This is true not just with regard to the Poisson equation or even elliptic PDEs since, as will become apparent in Chapter 16, the practical computation of parabolic PDEs also requires the solution of linear algebraic systems.

The systems (8.16) and (9.7) share two important characteristics. Our first observation is that in practical situations such systems are likely to be very large. Thus, five-point equations in an 81×81 grid result in 6400 equations. Even this might sound large to the uninitiated but it is, actually, relatively modest compared to what is encountered on a daily basis in real-life situations. Consider the equations of motion of fluids or solids, for example. The universe is three-dimensional and typical GFD (geophysical fluid dynamics) codes employ 14 variables – three each for position and velocity, one each for density, pressure, temperature and, say, the concentrations of five chemical elements. (If you think that 14 variables is excessive, you might be interested to learn that in combustion theory, say, even this is regarded as rather modest.) Altogether, and unless some convenient symmetries allow us to simplify the task in hand, we are solving equations in a three-dimensional parallelepiped. Requiring 81 grid points in each spatial dimension spells $14 \times 80^3 = 7\,168\,000$ coupled linear equations!

The cost of computation using the familiar *Gaussian elimination* is $\mathcal{O}(d^3)$ for a $d \times d$ system, and this renders it useless for systems of size such as the above.[1] Even were we able to design a computer that can perform $(64\,000\,000)^3 \approx 2.6 \times 10^{23}$ operations, say, in a reasonable time, the outcome is likely to be useless because of an accumulation of roundoff error.[2]

[1] A brief remark about the $\mathcal{O}(\,)$ notation. Often, the meaning of '$f(x) = \mathcal{O}(x^\alpha)$ as $x \to x_0$' is that $\lim_{x\to x_0} x^{-\alpha} f(x)$ exists and is bounded. The $\mathcal{O}(\,)$ notation in this section can be formally defined in a similar manner, but it is perhaps more helpful to interpret $\mathcal{O}(d^3)$, say, in a more intuitive fashion: it means that a quantity equals roughly a constant times d^3.

[2] Of course, everybody knows that there are no such computers. Are they possible, however? Assuming serial computer architecture and considering that signals travel (at most) at the speed of light, the distance between the central processing unit and each random access memory cell should be at an atomic level.

Fortunately, linear systems originating in finite differences or finite elements have one redeeming grace: they are *sparse*.[3] In other words, each variable is coupled to just a small number of other variables (typically, neighbouring grid points or neighbouring vertices) and an overwhelming majority of elements in the matrix vanish. For example, in each row and column of a matrix originating in the five-point formula (8.16) at most four off-diagonal elements are nonzero. This abundance of zeros and the special structure of the matrix allow us to implement Gaussian elimination in a manner that brings systems with 80^2 equations into the realm of microcomputers and allows the sufficiently rapid solution of 7 168 000-variable systems on (admittedly, parallel) supercomputers.

The subject of our attention is the linear system

$$A\boldsymbol{x} = \boldsymbol{b}, \tag{11.1}$$

where the $d \times d$ real matrix A and the vector $\boldsymbol{b} \in \mathbb{R}^d$ are given. We assume that A is nonsingular and *well conditioned*; the latter means, roughly, that A is sufficiently far from being singular that its numerical solution by Gaussian elimination or its variants is always viable and does not require any special techniques such as pivoting (A.1.4.4). Elements of A, $\boldsymbol{x}$ and $\boldsymbol{b}$ will be denoted by $a_{k,\ell}$, x_k and b_ℓ respectively, $k, \ell = 1, 2, \ldots, d$. The size of d will play no further direct role, but it is always important to bear in mind that it motivates the whole discussion.

We say that A is a *banded* matrix of *bandwidth* s if $a_{k,\ell} = 0$ for every $k, \ell \in \{1, 2, \ldots, d\}$ such that $|k - \ell| > s$. Familiar examples are *tridiagonal* ($s = 1$) and *quindiagonal* ($s = 2$) matrices.

Recall that, subject to mild restrictions, a $d \times d$ matrix A can be factorized into the form

$$A = LU \tag{11.2}$$

where

$$L = \begin{bmatrix} 1 & 0 & \cdots & 0 \\ \ell_{2,1} & 1 & \ddots & \vdots \\ \vdots & \ddots & \ddots & 0 \\ \ell_{d,1} & \cdots & \ell_{d,d-1} & 1 \end{bmatrix} \quad \text{and} \quad U = \begin{bmatrix} u_{1,1} & u_{1,2} & \cdots & u_{1,d} \\ 0 & u_{2,2} & \ddots & \vdots \\ \vdots & \ddots & \ddots & u_{d-1,d} \\ 0 & \cdots & 0 & u_{d,d} \end{bmatrix}$$

Specifically, a back-of-the-envelope computation indicates that, were *all* the expense just in communication (at the speed of light!) and were the whole calculation to be completed in less than 24 hours on a serial computer – a reasonable requirement, e.g. in calculations originating in weather prediction – the average distance between the CPU and every memory cell should be roughly 10^{-7} millimetres, barely twice the radius of a hydrogen atom.

This, needless to say, is in the realm of fantasy. Even the bravest souls in the miniaturization business dare not contemplate realistic computers of this size and, anyway, quantum effects are bound to make an atomic-sized computer an uncertain (in Heisenberg's sense) proposition.

There is an important caveat to this emphatic statement: quantum computers are based upon different principles, which do not preclude this sort of mind-boggling speed. Yet, as things stand, quantum computers exist only in theory.

[3]Systems originating in spectral methods are dense, but much smaller: what is lost on the swings is regained on the roundabouts . . .

are lower triangular and upper triangular respectively (A.1.4.5). In general, it costs $\mathcal{O}(d^3)$ operations to calculate L and U. However, if A has bandwidth s then this can be significantly reduced.

To demonstrate that this is indeed the case (and, incidentally, to measure exactly the extent of the savings) we assume that $a_{1,1} \neq 0$, and we let

$$\boldsymbol{\ell} := \begin{bmatrix} 1 \\ a_{2,1}/a_{1,1} \\ a_{3,1}/a_{1,1} \\ \vdots \\ a_{d,1}/a_{1,1} \end{bmatrix}, \qquad \boldsymbol{u}^\top = \begin{bmatrix} a_{1,1} & a_{1,2} & \cdots & a_{1,d} \end{bmatrix}$$

and set $\tilde{A} := A - \boldsymbol{\ell}\boldsymbol{u}^\top$. Regardless of the bandwidth of A, the matrix $\tilde{A}$ has zeros along its first row and column, and we find that

$$L = \begin{bmatrix} \boldsymbol{\ell} & \begin{matrix} \boldsymbol{0}^\top \\ \hat{L} \end{matrix} \end{bmatrix}, \qquad U = \begin{bmatrix} & \boldsymbol{u}^\top & \\ \boldsymbol{0} & & \hat{U} \end{bmatrix},$$

where $\hat{A} = \hat{L}\hat{U}$, the matrix $\hat{A}$ having been obtained from $\tilde{A}$ by deleting the first row and column. Setting the first column of L and the first row of U to $\boldsymbol{\ell}$ and $\boldsymbol{u}^\top$ respectively, we therefore reduce the problem of LU-factorizing the $d \times d$ matrix A to that of an LU factorization of the $(d-1) \times (d-1)$ matrix $\hat{A}$. For a general matrix A it costs $\mathcal{O}(d^2)$ operations to evaluate $\hat{A}$, but the operation count is smaller if A is of bandwidth s. Since just $s+1$ top components of $\boldsymbol{\ell}$ and $s+1$ leftward components of $\boldsymbol{u}^\top$ are nonzero, we need to form just the top $(s+1) \times (s+1)$ minor of $\boldsymbol{\ell}\boldsymbol{u}^\top$. In other words, $\mathcal{O}(d^2)$ is replaced by $\mathcal{O}((s+1)^2)$.

Continuing by induction we obtain progressively smaller matrices and, after $d-1$ such steps, derive an *operation count* $\mathcal{O}((s+1)^2 d)$ for the LU factorization of a banded matrix. We assume, of course, that the *pivots* $a_{1,1}, \hat{a}_{1,1}, \ldots$ never vanish, otherwise the above procedure could not be completed, but mention in passing that substantial savings accrue even when there is a need for pivoting (see Exercise 11.1).

The matrices L and U share the bandwidth of A. This is a very important observation, since a common mistake is to regard the difficulty of solving (11.1) with very large A as being associated solely with the number of operations. *Storage* plays a crucial role as well! In place of d^2 storage 'units' required for a dense $d \times d$ matrix, a banded matrix requires only about $(2s+1)d$. Provided that $s \ll d$, this often makes as much difference as the reduction in the operation count. Since L and U also have bandwidth s and we obviously have no need to store known zeros, or for that matter known ones along the diagonal of L, we can reuse computer memory that has been devoted to the storing of A (in a sparse representation!) to store L and U instead.

Having obtained the LU factorization, we can solve (11.1) with relative ease by solving first $L\boldsymbol{y} = \boldsymbol{b}$ and then $U\boldsymbol{x} = \boldsymbol{y}$. On the face of it, this sounds like yet another mathematical nonsense – instead of solving one $d \times d$ linear system, we solve two! However, since both L and U are banded, this can be done considerably more cheaply. In general, for a dense matrix A the operation count is $\mathcal{O}(d^2)$, but this can be substantially reduced for banded matrices. Writing $L\boldsymbol{y} = \boldsymbol{b}$ in a form that pays heed to sparsity, we have

$$
\begin{aligned}
y_1 &= b_1,\\
\ell_{2,1}y_1 + y_2 &= b_2,\\
\ell_{3,1}y_1 + \ell_{3,2}y_2 + y_3 &= b_3,\\
&\vdots\\
\ell_{1,s}y_1 + \cdots + \ell_{s-1,s}y_{s-1} + y_s &= b_s,\\
\ell_{k-s,k}y_{k-s} + \cdots + \ell_{k-1,k}y_{k-1} + y_k &= b_k, \qquad k = s+1, s+2, \ldots, d,
\end{aligned}
$$

and hence $\mathcal{O}(sd)$ operations. A similar argument applies to $U\boldsymbol{x} = \boldsymbol{y}$.

Let us count the blessings of bandedness. Firstly, LU factorization 'costs' $\mathcal{O}(s^2 d)$ operations, rather than $\mathcal{O}(d^3)$. Secondly, the storage requirement is $\mathcal{O}((2s+1)d)$, compared with d^2 for a dense matrix. Finally, provided that we have already derived the factorization, the solution of (11.1) entails just $\mathcal{O}(sd)$ operations in place of $\mathcal{O}(d^2)$.

◇ **A few examples of banded matrices** The savings due to the exploitation of bandedness are at their most striking in the case of tridiagonal matrices. Then we need just $\mathcal{O}(d)$ operations for the factorization and a similar number for the solution of triangular systems, and just $4d-2$ real numbers (inclusive of the vector $\boldsymbol{x}$) need be stored at any one time. The implementation of banded LU factorization in the case $s = 1$ is sometimes known as the *Thomas algorithm*.

A more interesting case is presented by the five-point equations (8.16) in a square. To present them in the form (11.1) we need to rearrange, at least formally, the two-dimensional $m \times m$ grid from Fig. 8.4 into a vector in $\mathbb{R}^d$, $d = m^2$. Although there are $d!$ distinct ways of doing this, the most obvious is simply to append the columns of an array to each other, starting from the leftmost. Appropriately, this is known as *natural ordering*:

$$
\boldsymbol{x} = \begin{bmatrix} u_{1,1}\\ u_{2,1}\\ \vdots\\ u_{m,1}\\ u_{1,2}\\ \vdots\\ u_{m,2}\\ \vdots \end{bmatrix}, \qquad \tilde{\boldsymbol{b}} = (\Delta x)^2 \begin{bmatrix} f_{1,1}\\ f_{2,1}\\ \vdots\\ f_{m,1}\\ f_{1,2}\\ \vdots\\ f_{m,2}\\ \vdots \end{bmatrix}.
$$

The vector $\boldsymbol{b}$ is composed of $\tilde{\boldsymbol{b}}$ and the contribution of the boundary points is

$$
\begin{aligned}
b_1 &= \tilde{b}_1 - [u(\Delta x, 0) + u(0, \Delta x)], & b_2 &= \tilde{b}_2 - u(2\Delta x, 0), & &\ldots,\\
b_{m+1} &= \tilde{b}_{m+1} - u(0, 2\Delta x), & b_{m+2} &= \tilde{b}_{m+2}, & &\ldots
\end{aligned}
$$

It is convenient to represent the matrix A as composed of m blocks, each of size $m \times m$. For example, $m = 4$ results in the 16×16 matrix

$$\left[\begin{array}{cccc|cccc|cccc|cccc}
-4 & 1 & 0 & 0 & 1 & 0 & 0 & 0 & 0 & 0 & 0 & 0 & 0 & 0 & 0 & 0 \\
1 & -4 & 1 & 0 & 0 & 1 & 0 & 0 & 0 & 0 & 0 & 0 & 0 & 0 & 0 & 0 \\
0 & 1 & -4 & 1 & 0 & 0 & 1 & 0 & 0 & 0 & 0 & 0 & 0 & 0 & 0 & 0 \\
0 & 0 & 1 & -4 & 0 & 0 & 0 & 1 & 0 & 0 & 0 & 0 & 0 & 0 & 0 & 0 \\
\hline
1 & 0 & 0 & 0 & -4 & 1 & 0 & 0 & 1 & 0 & 0 & 0 & 0 & 0 & 0 & 0 \\
0 & 1 & 0 & 0 & 1 & -4 & 1 & 0 & 0 & 1 & 0 & 0 & 0 & 0 & 0 & 0 \\
0 & 0 & 1 & 0 & 0 & 1 & -4 & 1 & 0 & 0 & 1 & 0 & 0 & 0 & 0 & 0 \\
0 & 0 & 0 & 1 & 0 & 0 & 1 & -4 & 0 & 0 & 0 & 1 & 0 & 0 & 0 & 0 \\
\hline
0 & 0 & 0 & 0 & 1 & 0 & 0 & 0 & -4 & 1 & 0 & 0 & 1 & 0 & 0 & 0 \\
0 & 0 & 0 & 0 & 0 & 1 & 0 & 0 & 1 & -4 & 1 & 0 & 0 & 1 & 0 & 0 \\
0 & 0 & 0 & 0 & 0 & 0 & 1 & 0 & 0 & 1 & -4 & 1 & 0 & 0 & 1 & 0 \\
0 & 0 & 0 & 0 & 0 & 0 & 0 & 1 & 0 & 0 & 1 & -4 & 0 & 0 & 0 & 1 \\
\hline
0 & 0 & 0 & 0 & 0 & 0 & 0 & 0 & 1 & 0 & 0 & 0 & -4 & 1 & 0 & 0 \\
0 & 0 & 0 & 0 & 0 & 0 & 0 & 0 & 0 & 1 & 0 & 0 & 1 & -4 & 1 & 0 \\
0 & 0 & 0 & 0 & 0 & 0 & 0 & 0 & 0 & 0 & 1 & 0 & 0 & 1 & -4 & 1 \\
0 & 0 & 0 & 0 & 0 & 0 & 0 & 0 & 0 & 0 & 0 & 1 & 0 & 0 & 1 & -4
\end{array}\right]. \tag{11.3}$$

In general, it is easy to see that A has bandwidth $s = m$. In other words, we need just $\mathcal{O}(m^4)$ operations to LU-factorize it – a large number, yet significantly smaller than $\mathcal{O}(m^6)$, the operation count for 'dense' LU factorization.

Note that a matrix might possess a large number of zeros inside the band; (11.3) is a case in point. These zeros will, in all likelihood, be destroyed (or 'filled in') through LU factorization. The banded algorithm guarantees only that zeros *outside* the band are retained! ◇

Whenever a matrix originates in a one-dimensional arrangement of a multivariate grid, the exact nature of the ordering is likely to have a bearing on the bandwidth. Indeed, the secret of efficient implementation of banded LU factorization for matrices that originate in a planar (or higher-dimensional) grid is in finding a good arrangement of grid points. An example of a bad arrangement is provided by *red–black* ordering, which, as far as the matrix (11.3) is concerned, is

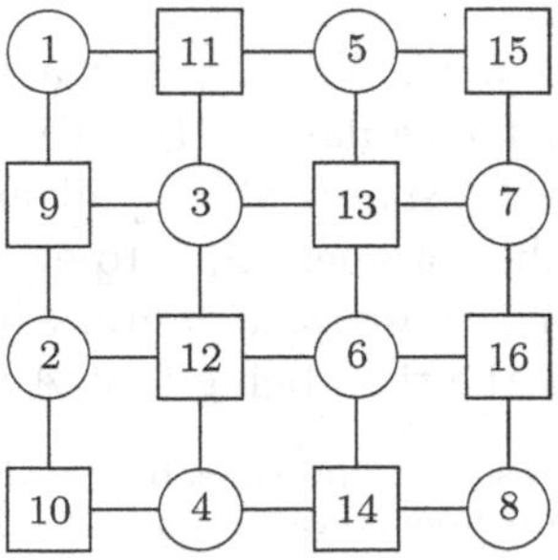

In other words, the grid is viewed as a chequerboard and black squares are selected

before the red ones. The outcome,

$$\left[\begin{array}{cccccccccccccccc}
\times & \circ & \circ & \circ & \circ & \circ & \circ & \circ & \times & \circ & \times & \circ & \circ & \circ & \circ & \circ \\
\circ & \times & \circ & \circ & \circ & \circ & \circ & \circ & \times & \times & \circ & \times & \circ & \circ & \circ & \circ \\
\circ & \circ & \times & \circ & \circ & \circ & \circ & \circ & \times & \circ & \times & \times & \times & \circ & \circ & \circ \\
\circ & \circ & \circ & \times & \circ & \circ & \circ & \circ & \circ & \times & \circ & \times & \circ & \times & \circ & \circ \\
\circ & \circ & \circ & \circ & \times & \circ & \circ & \circ & \circ & \circ & \times & \circ & \times & \circ & \times & \circ \\
\circ & \circ & \circ & \circ & \circ & \times & \circ & \circ & \circ & \circ & \circ & \times & \times & \times & \circ & \times \\
\circ & \circ & \circ & \circ & \circ & \circ & \times & \circ & \circ & \circ & \circ & \circ & \times & \circ & \times & \times \\
\circ & \circ & \circ & \circ & \circ & \circ & \circ & \times & \circ & \circ & \circ & \circ & \circ & \times & \circ & \times \\
\times & \times & \times & \circ & \circ & \circ & \circ & \circ & \times & \circ & \circ & \circ & \circ & \circ & \circ & \circ \\
\circ & \times & \circ & \times & \circ & \circ & \circ & \circ & \circ & \times & \circ & \circ & \circ & \circ & \circ & \circ \\
\times & \circ & \times & \circ & \times & \circ & \circ & \circ & \circ & \circ & \times & \circ & \circ & \circ & \circ & \circ \\
\circ & \times & \times & \times & \circ & \times & \circ & \circ & \circ & \circ & \circ & \times & \circ & \circ & \circ & \circ \\
\circ & \circ & \times & \circ & \times & \times & \times & \circ & \circ & \circ & \circ & \circ & \times & \circ & \circ & \circ \\
\circ & \circ & \circ & \times & \circ & \times & \circ & \times & \circ & \circ & \circ & \circ & \circ & \times & \circ & \circ \\
\circ & \circ & \circ & \circ & \times & \circ & \times & \circ & \circ & \circ & \circ & \circ & \circ & \circ & \times & \circ \\
\circ & \circ & \circ & \circ & \circ & \times & \times & \times & \circ & \circ & \circ & \circ & \circ & \circ & \circ & \times
\end{array}\right] \tag{11.4}$$

is of bandwidth $s = 10$ and is quite useless as far as sparse LU factorization is concerned. (Here and in the sequel we adopt the notation whereby '$\times$' and '$\circ$' stand for the nonzero and zero components respectively; after all, the exact numerical values of these components have no bearing on the underlying problem.) We mention in passing that, although red–black ordering is an exceedingly poor idea if the goal is to minimize the bandwith, it has certain virtues in the context of iterative methods (see Chapter 12).

In many situations it is relatively easy to find a configuration that results in a small bandwidth, but occasionally this might present quite a formidable problem. This is in particular the case with linear equations that originate from tessalations of two-dimensional sets into irregular triangles. There exist combinatorial algorithms that help to arrange arbitrary sets of equations into matrices having a 'good' bandwidth,[4] but they are outside the scope of this exposition.

11.2 Graphs of matrices and perfect Cholesky factorization

Let A be a symmetric $d \times d$ matrix. We say that a *Cholesky factorization* of A is $LL^\top$, where L is a $d \times d$ lower triangular matrix. (Note that the diagonal elements of L need not be ones.) Cholesky shares all the advantages of an LU factorization but requires only half the number of operations to evaluate and half the memory to store. Moreover, as long as A is positive definite, a Cholesky factorization always exists (A.1.4.6). We assume for the time being that A is indeed positive definite.

[4]A 'good' bandwidth is seldom the smallest possible, but this is frequently the case with combinatorial algorithms. 'Good' solutions are often relatively cheap to obtain, but finding the best solution is often much more expensive than solving the underlying equations with even the most inefficient ordering.

It follows at once from the proof of Theorem 8.4 that every matrix A obtained from the five-point formula and reasonable boundary schemes is negative definite. A similar statement is true in regard to matrices that arise when the finite element method is applied to the Poisson equation, provided that piecewise linear basis functions are used and that the geometry is sufficiently simple. Since we can solve $-A\boldsymbol{x} = -\boldsymbol{b}$ in place of $A\boldsymbol{x} = \boldsymbol{b}$, the stipulation of positive definiteness is not as contrived as it might perhaps seem at first glance.

We have already seen in Section 11.1 that the secret of good LU (and, for that matter, Cholesky) factorization is in the ordering of the equations and variables. In the case of a grid, this corresponds to ordering the grid points, but remember from the discussion in Chapter 8 that each such point corresponds to an equation and to a variable! Given a matrix A, we wish to find an ordering of equations and an ordering of variables such that the outcome is amenable to efficient factorization. Any rearrangement of equations (hence, of the rows of A) is equivalent to the product PA, where P is a $d \times d$ *permutation matrix* (A.1.2.5). Likewise, relabelling the variables is tantamount to rearranging the columns of A, hence to the product AQ, where Q is a permutation matrix. (Bearing in mind the purpose of the whole exercise, namely the solution of the linear system (11.1), we need to replace $\boldsymbol{b}$ by $P\boldsymbol{b}$ and $\boldsymbol{x}$ by $Q\boldsymbol{x}$; see Exercise 11.5.) The outcome, PAQ, retains symmetry and positive definiteness if $Q = P^\top$, hence we assume herewith that rows and columns are always reordered in unison. In practical terms, it means that if the (k,ℓ)th grid point corresponds to the jth equation, the variable $u_{k,\ell}$ becomes the jth unknown.

The matrix A is sparse and the purpose of a good Cholesky factorization is to retain as many zeros as possible. More formally, in any particular (symmetric) ordering of equations, we say that the *fill-in* is the number of pairs (i,j), $1 \le j < i \le d$, such that $a_{i,j} = 0$ and $\ell_{i,j} \ne 0$. Our goal is to devise an ordering that minimizes fill-in. In particular, we say that a Cholesky factorization of a specific matrix A is *perfect* if there exists an ordering that yields no fill-in whatsoever: every zero is retained.

An example of a perfect factorization is provided by a banded symmetric matrix, where we assume that no components vanish within the band. This is clear from the discussion in Section 11.1, which generalizes at once to a Cholesky factorization. Another example, at the other extreme, is a completely dense matrix (which, of course, is banded with a bandwidth of $s = d-1$, but to say this is to abuse the spirit, if not the letter, of the definition of bandedness).

A convenient way of analysing the sparsity patterns of symmetric matrices is afforded by *graph theory*. We have already mentioned graphs in a less formal setting, in the discussion at the end of Chapter 3.

Formally, a *graph* is the set $\mathbb{G} = \{\mathbb{V}, \mathbb{E}\}$, where $\mathbb{V} = \{1, 2, \ldots, d\}$ and $\mathbb{E} \subseteq \mathbb{V}^2$ consists of pairs of the form (i,j), $i < j$. The elements of $\mathbb{V}$ and $\mathbb{E}$ are said to be the *vertices* and *edges* of $\mathbb{G}$, respectively.

We say that $\mathbb{G}$ is the graph of a symmetric $d \times d$ matrix A if $(i,j) \in \mathbb{E}$ if and only if $a_{i,j} \ne 0$, $1 \le i < j \le d$. In other words, $\mathbb{G}$ displays the *sparsity pattern* of A, which we have presented in (11.4) using the symbols '$\circ$' and '$\times$'.

Although a graph can be represented by listing all the edges one by one, it is considerably more convenient to illustrate it pictorially in a self-explanatory manner.

Therefore we will now give a few examples of matrices (represented by their sparsity pattern) and their graphs.

tridiagonal: $\begin{bmatrix} \times & \times & \circ & \circ & \circ & \circ \\ \times & \times & \times & \circ & \circ & \circ \\ \circ & \times & \times & \times & \circ & \circ \\ \circ & \circ & \times & \times & \times & \circ \\ \circ & \circ & \circ & \times & \times & \times \\ \circ & \circ & \circ & \circ & \times & \times \end{bmatrix}$ $\Rightarrow$

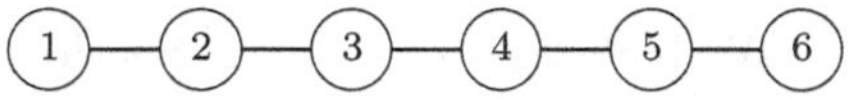

quindiagonal: $\begin{bmatrix} \times & \times & \times & \circ & \circ & \circ \\ \times & \times & \times & \times & \circ & \circ \\ \times & \times & \times & \times & \times & \circ \\ \circ & \times & \times & \times & \times & \times \\ \circ & \circ & \times & \times & \times & \times \\ \circ & \circ & \circ & \times & \times & \times \end{bmatrix}$ $\Rightarrow$

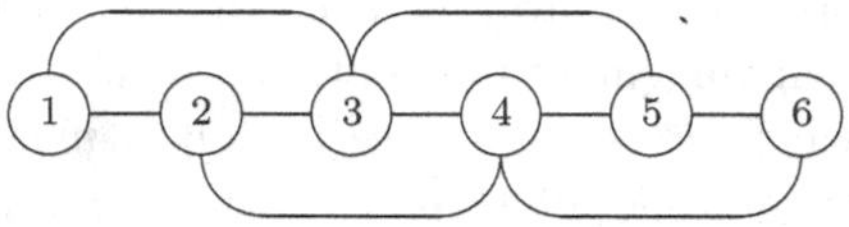

arrowhead: $\begin{bmatrix} \times & \times & \times & \times & \times & \times \\ \times & \times & \circ & \circ & \circ & \circ \\ \times & \circ & \times & \circ & \circ & \circ \\ \times & \circ & \circ & \times & \circ & \circ \\ \times & \circ & \circ & \circ & \times & \circ \\ \times & \circ & \circ & \circ & \circ & \times \end{bmatrix}$ $\Rightarrow$

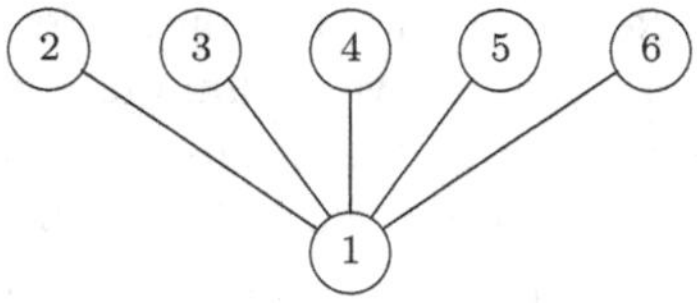

cyclic: $\begin{bmatrix} \times & \times & \circ & \circ & \circ & \times \\ \times & \times & \times & \circ & \circ & \circ \\ \circ & \times & \times & \times & \circ & \circ \\ \circ & \circ & \times & \times & \times & \circ \\ \circ & \circ & \circ & \times & \times & \times \\ \times & \circ & \circ & \circ & \times & \times \end{bmatrix}$ $\Rightarrow$

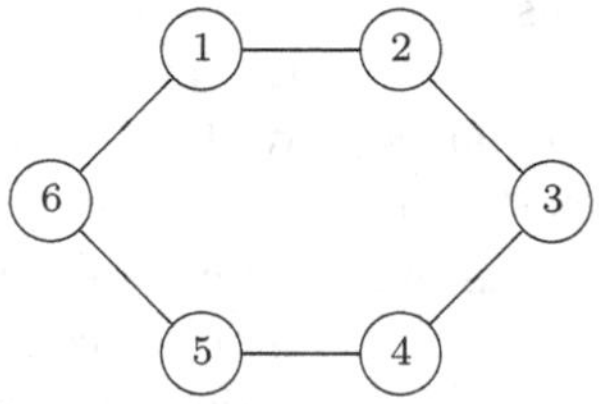

The graph of a matrix often reveals at a single glance its structure, which might not be evident from the sparsity pattern. Thus, consider the matrix

$$\begin{bmatrix} \times & \circ & \times & \circ & \times & \circ \\ \circ & \times & \circ & \times & \times & \circ \\ \times & \circ & \times & \circ & \circ & \times \\ \circ & \times & \circ & \times & \circ & \times \\ \times & \times & \circ & \circ & \times & \circ \\ \circ & \circ & \times & \times & \circ & \times \end{bmatrix}.$$

At a first glance, there is nothing to link it to any of the four matrices that we have just displayed, but its graph,

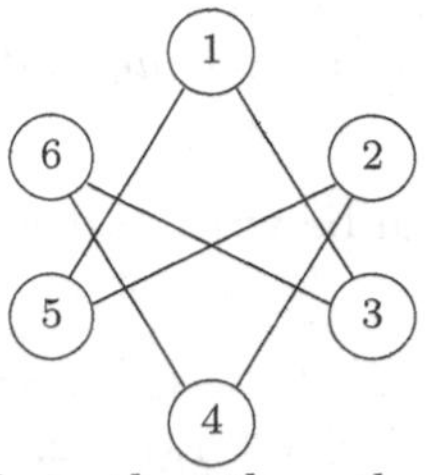

tells a different story – it is nothing other than the cyclic matrix in disguise! To see this, just relabel the vertices as follows

$$1 \to 1, \quad 2 \to 5, \quad 3 \to 2, \quad 4 \to 4, \quad 5 \to 6, \quad 6 \to 3.$$

This, of course, is equivalent to reordering (simultaneously) the equations and variables.

An ordered set of edges $\{(i_k, j_k)\}_{k=1}^{\nu} \subseteq \mathbb{E}$ is called a *path* joining the vertices α and β if $\alpha \in \{i_1, j_1\}$, $\beta \in \{i_\nu, j_\nu\}$ and for every $k = 1, 2, \ldots, \nu - 1$ the set $\{i_k, j_k\} \cap \{i_{k+1}, j_{k+1}\}$ contains exactly one member. It is a *simple path* if it does not visit any vertex more than once. We say that $\mathbb{G}$ is a *tree* if each two members of $\mathbb{V}$ are joined by a unique simple path. Both tridiagonal and arrowhead matrices correspond to trees, but this is not the case with either quindiagonal or cyclic matrices when $\nu \geq 3$.

Given a tree $\mathbb{G}$ and an arbitrary vertex $r \in \mathbb{V}$, the pair $\mathbb{T} = \langle \mathbb{G}, r \rangle$ is called a *rooted tree,* while r is said to be the *root.* Unlike an ordinary graph, $\mathbb{T}$ admits a natural partial ordering, which can best be explained by an analogy with a family tree. Thus, the root r is the *predecessor* of all the vertices in $\mathbb{V} \setminus \{r\}$ and these vertices are *successors* of r. Moreover, every $\alpha \in \mathbb{V} \setminus \{r\}$ is joined to v by a simple path and we designate each vertex along this path, except for r and α, as a *predecessor* of α and a *successor* of r. We say that the rooted tree $\mathbb{T}$ is *monotonically ordered* if each vertex is labelled before all its predecessors; in other words, we label the vertices from the top of the tree to the root. (As we have already said it, relabelling a graph is tantamount to permuting the rows and the columns of the underlying matrix.)

Every rooted tree can be monotonically ordered and, in general, such an ordering is not unique. We now give three monotone orderings of the same rooted tree:

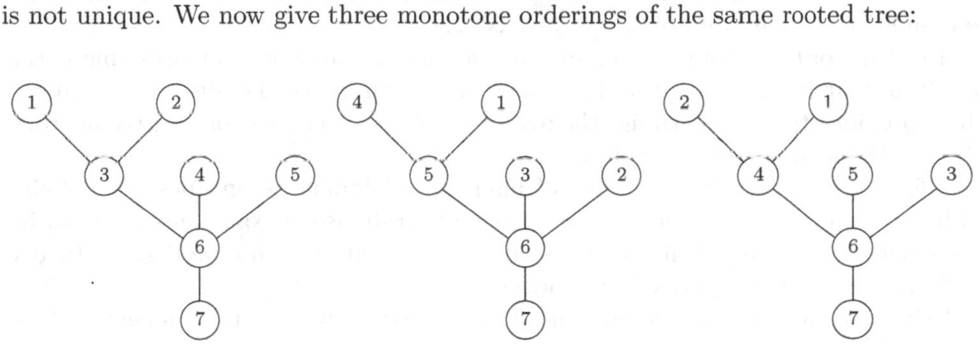

Theorem 11.1 *Let A be a symmetric matrix whose graph $\mathbb{G}$ is a tree. Choose a root $r \in \{1, 2, \ldots, d\}$ and assume that the rows and columns of A have been arranged so that $\mathbb{T} = \langle \mathbb{G}, r \rangle$ is monotonically ordered. Given that $A = LL^\top$ is a Cholesky factorization, it is true that*

$$\ell_{k,j} = \frac{a_{k,j}}{\ell_{j,j}}, \qquad k = j+1, j+2, \ldots, d, \quad j = 1, 2, \ldots, d-1. \tag{11.5}$$

Therefore $\ell_{k,j} = 0$ *whenever* $a_{k,j} = 0$ *and the matrix* A *can be Cholesky-factorized perfectly.*

Proof The coefficients of L can be written down explicitly (A.1.4.5). In particular,

$$\ell_{k,j} = \frac{1}{\ell_{j,j}}\left(a_{k,j} - \sum_{i=1}^{j-1} \ell_{k,i}\ell_{j,i}\right), \qquad k = j+1, j+2, \ldots, d, \quad j = 1, 2, \ldots, d-1. \tag{11.6}$$

It follows at once that the statement of the theorem is true with regard to the first column, since (11.6) yields $\ell_{k,1} = a_{k,1}/\ell_{1,1}$, $k = 2, 3, \ldots, d$. We continue by induction on j. Suppose thus that the theorem is true for $j = 1, 2, \ldots, q-1$, where $q \in \{2, 3, \ldots, d-1\}$. The rooted tree $\mathbb{T}$ is monotonically ordered, and this means that for every $i = 1, 2, \ldots, d-1$ there exists a *unique* vertex $\gamma_i \in \{i+1, i+2, \ldots, d\}$ such that $(i, \gamma_i) \in \mathbb{E}$.

Now, choose any $k \in \{q+1, q+2, \ldots, d\}$. If $\ell_{k,i} \neq 0$ for some $i \in \{1, 2, \ldots, q-1\}$ then, by the induction assumption, $a_{k,i} \neq 0$ also. This implies $(i, k) \in \mathbb{E}$, hence $k = \gamma_i$. We deduce that $q \neq \gamma_i$, therefore $(i, q) \notin \mathbb{E}$. Consequently $a_{i,q} = 0$ and, exploiting again the induction assumption, $\ell_{i,q} = 0$.

By an identical argument, if $\ell_{q,i} \neq 0$ for some $i \in \{1, 2, \ldots, q-1\}$ then $q = \gamma_i$, hence $k \neq \gamma_i$ for all $k = q+1, q+2, \ldots, d$, and this implies in turn that $\ell_{k,i} = 0$.

We let $j = q$ in (11.6). Since, as we have just proved, $\ell_{k,i}\ell_{q,i} = 0$ for all $i = 1, 2, \ldots, q-1$ and $k = i+1, i+2, \ldots, d$, the sum in (11.6) vanishes and we deduce that $\ell_{k,q} = a_{k,q}/\ell_{q,q}$, $k = q+1, q+2, \ldots, d$. This inductive proof of the theorem is thus complete. ■

An important observation is that the expense of Cholesky factorization of a matrix consistently with the conditions of Theorem 11.1 is proportional to the number of nonzero elements under the main diagonal. This is certainly true as far as $\ell_{k,j}$, $1 \le j < k \le d$, is concerned and it is possible to verify (see Exercise 11.8) that this is also the case for the calculation of $\ell_{1,1}, \ell_{2,2}, \ldots, \ell_{d,d}$.

Monotone ordering can lead to spectacular savings and a striking example is the arrowhead matrix. If we factorized it in a naive fashion we could easily cause total fill-in but, provided that we rearrange the rows and columns to correspond with monotone ordering, the factorization is perfect.

Unfortunately, very few matrices of interest are symmetric and positive definite and have a graph that is a tree. Perhaps the only truly useful example is provided by a (symmetric, positive definite) tridiagonal matrix – and we do not need graph theory to tell us that it can be perfectly factorized!

Positive definiteness is, however, not strictly necessary for our argument and we have used it only as an cast-iron guarantee that no pivots $\ell_{1,1}, \ell_{2,2}, \ldots, \ell_{d-1,d-1}$ ever vanish. Likewise, we can dispense – up to a point – with symmetry. All that matters is a symmetric sparsity structure, namely $a_{k,j} \neq 0$ if and only if $a_{j,k} \neq 0$ for all $d = 1, 2, \ldots, d$. We have LU factorization in place of Cholesky factorization, but a generalization of Theorem 11.1 presents no insurmountable difficulties.

The one truly restrictive assumption is that the graph is a tree and this renders Theorem 11.1 of little immediate interest in applications. There are three reasons why we nevertheless attend to it. Firstly, it provides the flavour of considerably more substantive results on matrices and graphs. Secondly, it is easy to generalize the theorem to *partitioned trees,* graphs where we have a tree-like structure, provided that instead of vertices we allow subsets of $\mathbb{V}$. An example illustrating this concept and its application to perfect factorization is presented in the comments below. Finally, the idea of using graphs to investigate the sparsity structure and factorization of matrices is such a beautiful example of lateral thinking in mathematics that its presentation can surely be justified on purely æsthetic grounds.

We complete this brief review of graph theory and the factorization of sparse matrices by remarking that, of course, there are many matrices whose graphs are not trees, yet which can be perfectly factorized. We have seen already that this is the case with a quindiagonal matrix and a less trivial example will be presented in the comments below. It is possible to characterize all graphs that correspond to matrices with a perfect (Cholesky or LU) factorization, but that requires considerably deeper graph theory.

Comments and bibliography

The solution of sparse algebraic systems is one of the main themes of modern scientific computing. Factorization that exploits sparsity is just one, and not necessarily the preferred, option and we shall examine alternative approaches – specifically, iterative methods and fast Poisson solvers – in Chapters 12–15.

References on sparse factorization (sometimes dubbed *direct solution,* to distinguish it from iterative methods) abound and we refer the reader to Duff *et al.* (1986); George & Liu (1981); Tewarson (1973). Likewise, many textbooks present graph theory and we single out Harary (1969) and Golumbic (1980). The latter includes an advanced treatment of graph-theoretical methods in sparse matrix factorization, including the characterization of all matrices that can be factorized perfectly.

It is natural to improve upon the concept of a banded matrix by allowing the bandwidth to vary. For example, consider the 8×8 matrix

$$\begin{bmatrix} \times & \times & \circ & \circ & \circ & \circ & \circ & \circ \\ \times & \times & \times & \times & \circ & \circ & \circ & \circ \\ \circ & \times & \times & \times & \circ & \circ & \circ & \circ \\ \circ & \times & \times & \times & \times & \times & \circ & \circ \\ \circ & \circ & \circ & \times & \times & \times & \circ & \circ \\ \circ & \circ & \circ & \times & \times & \times & \times & \circ \\ \circ & \circ & \circ & \circ & \circ & \times & \times & \times \\ \circ & \circ & \circ & \circ & \circ & \circ & \times & \times \end{bmatrix}$$

The portion of the matrix enclosed between the solid lines is called the *envelope.* It is easy to demonstrate that LU factorization can be performed in such a manner that fill-in cannot occur outside the envelope; see Exercise 11.2. It is evident even in this simple example that 'envelope factorization' might result in considerable savings over 'banded factorization'.

Sometimes it is easy to find a good banded structure or a good envelope of a given matrix by inspection – a procedure that might involve the rearrangement of equations and variables. In general, however, this is a formidable task, which needs to be accomplished by a combinatorial algorithm. An effective, yet relatively simple, such method is the *reverse*

Cuthill–McKee algorithm. We will not dwell further on this theme, which is explained well in George & Liu (1981).

Throughout our discussion of banded and envelope algorithms we have tacitly assumed that the underlying matrix A is symmetric or, at the very least, has a symmetric sparsity structure. As we have already commented, this is eminently sensible whenever we consider, for example, the equations that occur when the Poisson equation is approximated by the five-point formula. However, it is possible to extend much of the theory to arbitrary matrices; a good deal of the extension is trivial. For example, if there are nonzero elements in the set $\{a_{k,j} : k-2 \le j \le k+1\}$ then, assuming, as always, that the underlying factorization is well conditioned, $A = LU$, where $\ell_{k,j} = 0$ for $j \le k-3$ and $u_{k,j} = 0$ for $j \ge k+2$. However, the question of how to arrange elements of a nonsymmetric matrix so that it is amenable to this kind of treatment is substantially more formidable. Note that the correspondence between graphs and matrices assumes symmetry. Where the more advanced concepts of graph theory – specifically, directed graphs – have been applied to nonsymmetric sparsity structures, the results so far have met with only modest success.

To appreciate the power of graph theory in revealing the sparsity pattern of a symmetric matrix, thereby permitting its intelligent exploitation, consider the following example. At first glance, the matrix

$$\begin{bmatrix}
\times & \circ & \circ & \circ & \circ & \circ & \times & \circ & \circ & \circ & \circ & \circ & \circ \\
\circ & \times & \circ & \circ & \times & \circ & \circ & \circ & \times & \times & \circ & \circ & \times \\
\circ & \circ & \times & \circ & \circ & \times & \times & \circ & \circ & \circ & \times & \circ & \circ \\
\circ & \circ & \circ & \times & \circ & \circ & \circ & \circ & \circ & \times & \circ & \circ & \circ \\
\circ & \times & \circ & \circ & \times & \circ & \circ & \circ & \circ & \times & \circ & \times & \times \\
\circ & \circ & \times & \circ & \circ & \times & \circ & \circ & \circ & \circ & \circ & \circ & \times \\
\times & \circ & \times & \circ & \circ & \circ & \times & \circ & \circ & \circ & \times & \circ & \circ \\
\circ & \circ & \circ & \circ & \circ & \circ & \circ & \times & \circ & \circ & \times & \circ & \circ \\
\circ & \times & \circ & \circ & \circ & \circ & \circ & \circ & \times & \circ & \circ & \circ & \circ \\
\circ & \times & \circ & \times & \times & \circ & \circ & \circ & \circ & \times & \circ & \circ & \times \\
\circ & \circ & \times & \circ & \circ & \circ & \times & \times & \circ & \circ & \times & \circ & \circ \\
\circ & \circ & \circ & \circ & \times & \circ & \circ & \circ & \circ & \circ & \circ & \times & \circ \\
\circ & \times & \circ & \circ & \times & \times & \circ & \circ & \circ & \times & \circ & \circ & \times
\end{bmatrix}.$$

might appear as a completely unstructured mishmash of noughts and crosses, but we claim nonetheless that it can be perfectly factorized. This becomes more apparent upon an examination of its graph:

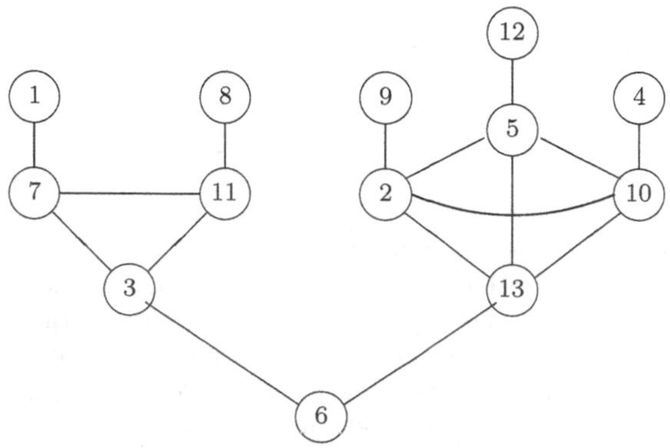

Although this is not a tree, a tree-like structure is apparent. In fact, it is a *partitioned tree,*

a 'super-graph' which can be represented as a tree of of 'super-vertices' that are themselves graphs.

The equations and unknowns are ordered so that the graph is traversed from top to bottom. This can be done in a variety of different ways and we herewith choose an ordering that keeps all vertices in each 'super-vertex' together; this is not really necessary but makes the exposition simpler. The outcome of this permutation is displayed in a block form, corresponding to the structure of the partitioned tree, as follows:

	1	4	8	9	12	7	11	3	2	5	10	13	6
1	×	○	○	○	○	×	○	○	○	○	○	○	○
4	○	×	○	○	○	○	○	○	○	○	×	○	○
8	○	○	×	○	○	○	×	○	○	○	○	○	○
9	○	○	○	×	○	○	○	○	×	○	○	○	○
12	○	○	○	○	×	○	○	○	○	×	○	○	○
7	×	○	○	○	○	×	×	×	○	○	○	○	○
11	○	○	×	○	○	×	×	×	○	○	○	○	○
3	○	○	○	○	○	×	×	×	○	○	○	○	×
2	○	○	○	×	○	○	○	○	×	×	×	×	○
5	○	○	○	○	×	○	○	○	×	×	×	×	○
10	○	×	○	○	○	○	○	○	×	×	×	×	○
13	○	○	○	○	○	○	○	○	×	×	×	×	×
6	○	○	○	○	○	○	○	×	○	○	○	×	×

It is now clear how to proceed. Firstly, factorize the vertices $1, 4, 8, 9$ and 12. This obviously causes no fill-in. If you cannot see it at once, consider the equivalent problem of Gaussian elimination. In the first column we need to eliminate just a single nonzero component and we do this by subtracting a multiple of row 1 from row 7. Likewise, we eliminate one component in the first row (remember that we need to maintain symmetry!). Neither operation causes fill-in and we proceed similarly with $4, 8, 9, 12$.

Having factorized the aforementioned five rows and columns, we are left with an 8×8 problem. We next factorize $7, 11, 3$, in this order; again, there is no fill-in. Finally, we factorize $2, 5, 10$ and, then, 13. The outcome is a perfect Cholesky factorization.

The last example, contrived as it might be, provides some insight into one of the most powerful techniques in the factorization of sparse matrices, the method of *partitioned trees*. Given a sparse matrix A with a graph $\mathbb{G} = \{\mathbb{V}, \mathbb{E}\}$, we will partition the set of vertices

$$\mathbb{V} = \bigcup_{i=1}^{s} \mathbb{V}_i \qquad \text{such that} \qquad \mathbb{V}_i \cap \mathbb{V}_j = \emptyset \qquad \text{for every} \qquad i, j = 1, 2, \ldots, s, \quad i \neq j.$$

Letting $\tilde{\mathbb{V}} := \{1, 2, \ldots, s\}$, we construct the set $\tilde{\mathbb{E}} \subseteq \tilde{\mathbb{V}} \times \tilde{\mathbb{V}}$ by assigning to it every pair (i, j) for which $i < j$ and there exist $\alpha \in \mathbb{V}_i$ and $\beta \in \mathbb{V}_j$ such that either $(\alpha, \beta) \in \mathbb{E}$ or $(\beta, \alpha) \in \mathbb{E}$. The set $\tilde{\mathbb{G}} := \{\tilde{\mathbb{V}}, \tilde{\mathbb{E}}\}$ is itself a graph. Suppose that we have partitioned $\mathbb{V}$ so that $\tilde{\mathbb{G}}$ is a tree. In that case we know from Theorem 11.1 that, by selecting a root in $\tilde{\mathbb{V}}$ and imposing monotone ordering on the partitioned tree, we can factorize the matrix without any fill-in taking place *between* partitioned vertices. There might well be fill-in inside each set $\mathbb{V}_i$, $i = 1, 2, \ldots, s$. However, provided that these sets are either small (as in the extreme case $s = d$, when each $\mathbb{V}_i$ is a singleton) or fairly dense (in our example, all the sets $\mathbb{V}_i$ are completely dense), the fill-in is likely to be modest.

Sometimes it is possible to find a good partitioned tree structure for a graph just by inspecting it, but there exist algorithms that produce good partitions automatically. Such methods are extremely unlikely to produce the best possible partition (in the sense of minimizing the fill-in), but it should be clear by now that, when it comes to combinatorial algorithms, the best is often the mortal enemy of the good.

We conclude these remarks with few sobering thoughts. Most numerical analysts regard direct factorization methods as *passé* and inferior to iterative algorithms. Bearing in mind the power of modern iterative schemes for linear equations – multigrid, preconditioned conjugate gradients, generalized minimal residuals (GMRes) etc. – this is probably a sensible approach. However, direct factorization has its place, not just in the obvious instances such as banded matrices; it can also, as we will note in Chapter 15, join forces with the (iterative) method of conjugate gradients to produce one of the most effective solvers of linear algebraic systems.

Duff, I.S., Erisman, A.M. and Reid, J.K. (1986), *Direct Methods for Sparse Matrices,* Oxford University Press, Oxford.

George, A. and Liu, J.W.-H. (1981), *Computer Solution of Large Sparse Positive Definite Systems,* Prentice–Hall, Englewood Cliffs, NJ.

Golumbic, M.C. (1980), *Algorithmic Graph Theory and Perfect Graphs,* Academic Press, New York.

Harary, F. (1969), *Graph Theory,* Addison–Wesley, Reading, MA.

Tewarson, R.P. (1973), *Sparse Matrices,* Academic Press, New York.

Exercises

11.1 Let A be a $d \times d$ nonsingular matrix with bandwidth $s \geq 1$ and suppose that Gaussian elimination with *column pivoting* is used to solve the system $A\boldsymbol{x} = \boldsymbol{b}$. This means that, before eliminating all nonzero compenents under the main diagonal in the jth column, where $j \in \{1, 2, \ldots, d-1\}$, we first find $|a_{k_j,j}| := \max_{i=j,j+1,\ldots,d} |a_{i,j}|$ and next exchange the jth and the k_jth rows of A, as well as the corresponding components of $\boldsymbol{b}$ (see A.1.4.4).

a Identify all the coefficients of the intermediate equations that might be filled in during this procedure.

b Prove that the operation count of LU factorization with column pivoting is $\mathcal{O}(s^2 d)$.

11.2 Let A be a $d \times d$ symmetric positive definite matrix and for every $j = 1, 2, \ldots, d-1$ define $k_j \in \{1, 2, \ldots, j\}$ as the least integer such that $a_{k_j,j} \neq 0$. We may assume without loss of generality that $k_1 \leq k_2 \leq \cdots \leq k_{d-1}$, since otherwise A can be brought into this form by row and column exchanges.

a Prove that the number of operations required to find a Cholesky factorization of A is

$$\mathcal{O}\left(\sum_{j=1}^{d-1} (j - k_j + 1)\right).$$

b Demonstrate that the result of an operation count for a Cholesky factorization of banded matrices is a special case of this formula.

11.3 Find the bandwidth of the $m^2 \times m^2$ matrix that is obtained from the nine-point equations (8.28) with natural ordering.

11.4 Suppose that a square is triangulated in the following fashion:

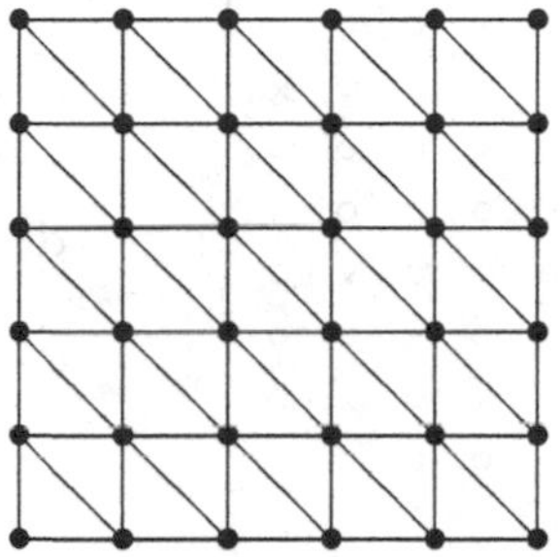

The Poisson equation in a square is discretized using the FEM and piecewise linear functions with this triangulation. Observe that every vertex can be identified with an unknown in the linear system (9.7).

a Arranging equations in an $m \times m$ grid in natural ordering, find the bandwidth of the $m^2 \times m^2$ matrix.

b What is the graph of this matrix?

11.5 Let P and Q be two $d \times d$ *permutation matrices* (A.1.2.5) and let $\tilde{A} := PAQ$, where the $d \times d$ matrix A is nonsingular. Prove that if $\tilde{\boldsymbol{x}} \in \mathbb{R}^d$ is the solution of $\tilde{A}\tilde{\boldsymbol{x}} = P\boldsymbol{b}$ then $\boldsymbol{x} = Q\tilde{\boldsymbol{x}}$ solves (11.1).

11.6 We say that a graph $\mathbb{G} = \{\mathbb{V}, \mathbb{E}\}$ is *connected* if any two vertices in $\mathbb{V}$ can be joined by a path of edges from $\mathbb{E}$; otherwise it is *disconnected.* Let A be a $d \times d$ symmetric matrix with graph $\mathbb{G}$. Prove that if $\mathbb{G}$ is disconnected then, after rearrangement of rows and columns, A can be written in the form

$$A = \begin{bmatrix} A_1 & O \\ O & A_2 \end{bmatrix},$$

where A_j is a matrix of size $d_j \times d_j$, $j = 1, 2$, and $d_1 + d_2 = d$.

11.7 Construct the graphs of the matrices with the following sparsity patterns:

$$
\mathbf{a}\quad \begin{bmatrix}
\times & \circ & \circ & \times & \circ & \circ & \circ & \circ \\
\circ & \times & \times & \circ & \times & \times & \times & \circ \\
\circ & \circ & \times & \circ & \times & \circ & \circ & \times \\
\times & \circ & \circ & \times & \circ & \circ & \times & \circ \\
\circ & \times & \times & \circ & \times & \circ & \circ & \circ \\
\circ & \times & \circ & \circ & \circ & \times & \circ & \times \\
\circ & \times & \circ & \times & \circ & \circ & \times & \circ \\
\circ & \circ & \times & \circ & \circ & \times & \circ & \times
\end{bmatrix}; \qquad
\mathbf{b}\quad \begin{bmatrix}
\times & \circ & \circ & \circ & \times & \times & \circ & \circ \\
\circ & \times & \circ & \circ & \circ & \circ & \circ & \times \\
\circ & \circ & \times & \times & \times & \circ & \times & \circ \\
\circ & \circ & \times & \times & \circ & \circ & \circ & \circ \\
\times & \circ & \times & \circ & \times & \circ & \circ & \times \\
\times & \circ & \circ & \circ & \circ & \times & \circ & \circ \\
\circ & \circ & \times & \circ & \circ & \circ & \times & \circ \\
\circ & \times & \circ & \circ & \times & \circ & \circ & \times
\end{bmatrix};
$$

$$
\mathbf{c}\quad \begin{bmatrix}
\times & \circ & \circ & \circ & \times & \circ & \circ & \circ \\
\circ & \times & \circ & \times & \times & \times & \circ & \times \\
\circ & \circ & \times & \circ & \times & \circ & \circ & \circ \\
\circ & \times & \circ & \times & \circ & \circ & \circ & \circ \\
\times & \times & \times & \circ & \times & \circ & \times & \circ \\
\circ & \times & \circ & \circ & \circ & \times & \circ & \circ \\
\circ & \circ & \circ & \circ & \times & \circ & \times & \circ \\
\circ & \times & \circ & \circ & \circ & \circ & \circ & \times
\end{bmatrix}; \qquad
\mathbf{d}\quad \begin{bmatrix}
\times & \circ & \times & \times & \circ & \circ & \circ & \circ \\
\circ & \times & \circ & \circ & \times & \times & \times & \times \\
\times & \circ & \times & \circ & \circ & \circ & \times & \circ \\
\times & \circ & \circ & \times & \circ & \circ & \circ & \times \\
\circ & \times & \circ & \circ & \times & \circ & \circ & \circ \\
\circ & \times & \circ & \circ & \circ & \times & \circ & \circ \\
\circ & \times & \times & \circ & \circ & \circ & \times & \circ \\
\circ & \times & \circ & \times & \circ & \circ & \circ & \times
\end{bmatrix}.
$$

Identify all trees. Suggest a monotone ordering for each tree.

11.8* Prove that a symmetric positive definite matrix with the sparsity pattern

$$
\begin{bmatrix}
\times & \circ & \circ & \circ & \circ & \times & \circ & \circ & \circ & \circ & \times & \circ & \circ & \circ & \circ & \circ \\
\circ & \times & \circ & \circ & \circ & \circ & \times & \circ & \circ & \circ & \circ & \times & \circ & \circ & \circ & \circ \\
\circ & \circ & \times & \circ & \circ & \circ & \circ & \times & \circ & \circ & \circ & \circ & \times & \circ & \circ & \circ \\
\circ & \circ & \circ & \times & \circ & \circ & \circ & \circ & \times & \circ & \circ & \circ & \circ & \times & \circ & \circ \\
\circ & \circ & \circ & \circ & \times & \circ & \circ & \circ & \circ & \times & \circ & \circ & \circ & \circ & \times & \circ \\
\times & \circ & \circ & \circ & \circ & \times & \circ & \circ & \circ & \circ & \times & \circ & \circ & \circ & \circ & \circ \\
\circ & \times & \circ & \circ & \circ & \circ & \times & \circ & \circ & \circ & \circ & \times & \circ & \circ & \circ & \circ \\
\circ & \circ & \times & \circ & \circ & \circ & \circ & \times & \circ & \circ & \circ & \circ & \times & \circ & \circ & \circ \\
\circ & \circ & \circ & \times & \circ & \circ & \circ & \circ & \times & \circ & \circ & \circ & \circ & \times & \circ & \circ \\
\circ & \circ & \circ & \circ & \times & \circ & \circ & \circ & \circ & \times & \circ & \circ & \circ & \circ & \times & \circ \\
\times & \circ & \circ & \circ & \circ & \times & \circ & \circ & \circ & \circ & \times & \circ & \circ & \circ & \circ & \times \\
\circ & \times & \circ & \circ & \circ & \circ & \times & \circ & \circ & \circ & \circ & \times & \circ & \circ & \circ & \times \\
\circ & \circ & \times & \circ & \circ & \circ & \circ & \times & \circ & \circ & \circ & \circ & \times & \circ & \circ & \times \\
\circ & \circ & \circ & \times & \circ & \circ & \circ & \circ & \times & \circ & \circ & \circ & \circ & \times & \circ & \times \\
\circ & \circ & \circ & \circ & \times & \circ & \circ & \circ & \circ & \times & \circ & \circ & \circ & \circ & \times & \times \\
\circ & \circ & \circ & \circ & \circ & \circ & \circ & \circ & \circ & \circ & \times & \times & \times & \times & \times & \times
\end{bmatrix}
$$

possesses a perfect Cholesky factorization.

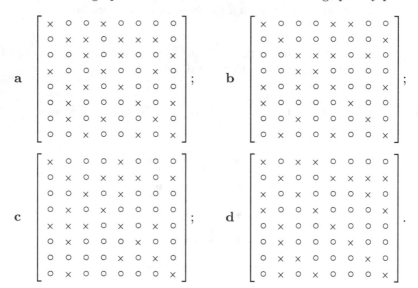

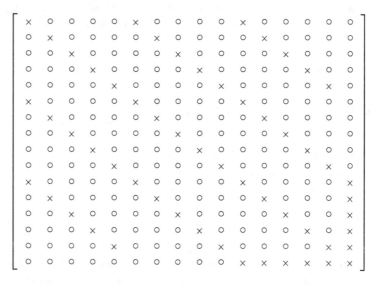

11.9* Theorem 11.1 states that the number of operations needed to form $\ell_{k,j}$, $1 \le j < k \le d$, is proportional to the number of sparse components underneath the diagonal of A. The purpose of this question is to prove a similar statement with regard to the diagonal terms $\ell_{k,k}$, $k = 1, 2, \ldots, d$. To this end one might use the explicit formula

$$\ell_{k,k}^2 = a_{k,k} - \sum_{i=1}^{k-1} \ell_{k,i}^2, \qquad k = 1, 2, \ldots, d,$$

and count the total number of nonzero terms $\ell_{k,i}^2$ in all the sums.

12

Classical iterative methods for sparse linear equations

12.1 Linear one-step stationary schemes

The theme of this chapter is iterative solution of the linear system

$$Ax = b, \tag{12.1}$$

where A is a $d \times d$ real nonsingular matrix and $\boldsymbol{b} \in \mathbb{R}^d$. The most general iterative method is a rule that for every $k = 0, 1, \ldots$ and $\boldsymbol{x}^{[0]}, \boldsymbol{x}^{[1]}, \ldots, \boldsymbol{x}^{[k]} \in \mathbb{R}^d$ generates a new vector $\boldsymbol{x}^{[k+1]} \in \mathbb{R}^d$. In other words, it is a family of functions $\{\boldsymbol{h}_k\}_{k=0}^{\infty}$ such that

$$\boldsymbol{h}_k : \overbrace{\mathbb{R}^d \times \mathbb{R}^d \times \cdots \times \mathbb{R}^d}^{k+1 \text{ times}} \to \mathbb{R}^d, \qquad k = 0, 1, \ldots,$$

and

$$\boldsymbol{x}^{[k+1]} = \boldsymbol{h}_k(\boldsymbol{x}^{[0]}, \boldsymbol{x}^{[1]}, \ldots, \boldsymbol{x}^{[k]}), \qquad k = 0, 1, \ldots \tag{12.2}$$

The most fundamental question with regard to the scheme (12.2) is about its convergence. Firstly, does it converge for every *starting value* $\boldsymbol{x}^{[0]} \in \mathbb{R}^d$?[1] Secondly, provided that it converges, is the limit bound to be the true solution of the linear system (12.1)?

Unless (12.2) always converges to the true solution the scheme is, obviously, unsuitable. However, not all convergent iterative methods are equally good. Our main consideration being to economize on computational cost, we must consider how fast convergence takes place and what is the expense of each iteration.

An iterative scheme (12.2) is said to be *linear* if each $\boldsymbol{h}_k$ is linear in all its arguments. It is *m-step* if $\boldsymbol{h}_k$ depends solely on $\boldsymbol{x}^{[k-m+1]}, \boldsymbol{x}^{[k-m+2]}, \ldots, \boldsymbol{x}^{[k]}$, $k = m-1, m, \ldots$ Finally, an m-step method is *stationary* if the function $\boldsymbol{h}_k$ does not vary with k for $k \geq m-1$. Each of these three concepts represents a considerable simplification, and it makes good sense to focus our effort on the most elementary model possible: a *linear one-step stationary* scheme. In that case (12.2) becomes

$$\boldsymbol{x}^{[k+1]} = H\boldsymbol{x}^{[k]} + \boldsymbol{v}, \tag{12.3}$$

[1]We are content when methods for *nonlinear* systems, e.g. functional iteration or the Newton–Raphson algorithm (cf. Chapter 7), converge for a suitably large set of starting values. When it comes to linear systems, however, we are more greedy!

where the $d \times d$ *iteration matrix* H and $\boldsymbol{v} \in \mathbb{R}^d$ are independent of k. (Of course, both H and $\boldsymbol{v}$ must depend on A and $\boldsymbol{b}$, otherwise convergence is impossible.)

Lemma 12.1 *Given an arbitrary linear system (12.1), a linear one-step stationary scheme (12.3) converges to a unique bounded limit $\hat{\boldsymbol{x}} \in \mathbb{R}^d$, regardless of the choice of starting value $\boldsymbol{x}^{[0]}$, if and only if $\rho(H) < 1$, where $\rho(\cdot)$ denotes the spectral radius (A.1.5.2). Provided that $\rho(H) < 1$, $\hat{\boldsymbol{x}}$ is the correct solution of the linear system (12.1) if and only if*

$$\boldsymbol{v} = (I - H)A^{-1}\boldsymbol{b}. \tag{12.4}$$

Proof Let us commence by assuming $\rho(H) < 1$. In this case we claim that

$$\lim_{k \to \infty} H^k = O. \tag{12.5}$$

To prove this statement, we make the simplifying assumption that H has a complete set of eigenvectors, hence that there exist a nonsingular $d \times d$ matrix V and a diagonal $d \times d$ matrix D such that $H = VDV^{-1}$ (A.1.5.3 and A.1.5.4). Hence $H^2 = (VDV^{-1}) \times (VDV^{-1}) = VD^2V^{-1}$, $H^3 = VD^3V^{-1}$ and, in general, it is trivial to prove by induction that $H^k = VD^kV^{-1}$, $k = 0, 1, 2, \ldots$ Therefore, passing to the limit,

$$\lim_{k \to \infty} H^k = V \left(\lim_{k \to \infty} D^k \right) V^{-1}.$$

The elements along the diagonal of D are the eigenvalues of H, hence $\rho(H) < 1$ implies $D^k \stackrel{k \to \infty}{\longrightarrow} O$ and we deduce (12.5).

If the set of eigenvectors is incomplete, (12.5) can be proved just as easily by using a Jordan factorization (see A.1.5.6 and Exercise 12.1).

Our next assertion is that

$$\boldsymbol{x}^{[k]} = H^k \boldsymbol{x}^{[0]} + (I - H)^{-1}(I - H^k)\boldsymbol{v}, \qquad k = 0, 1, 2, \ldots; \tag{12.6}$$

note that $\rho(H) < 1$ implies $1 \notin \sigma(H)$, where $\sigma(H)$ is the set of all eigenvalues (the *spectrum*) of H, therefore the inverse of $I - H$ exists.

The proof is by induction. It is obvious that (12.6) is true for $k = 0$. Hence, let us assume it for $k \geq 0$ and attempt its verification for $k + 1$. Using the definition (12.3) of the iterative scheme in tandem with the induction assumption (12.6), we readily obtain

$$\begin{aligned}
\boldsymbol{x}^{[k+1]} &= H\boldsymbol{x}^{[k]} + \boldsymbol{v} = H\left[H^k \boldsymbol{x}^{[0]} + (I - H)^{-1}(I - H^k)\boldsymbol{v}\right] + \boldsymbol{v} \\
&= H^{k+1}\boldsymbol{x}^{[0]} + \left[(I - H)^{-1}(H - H^{k+1}) + (I - H)^{-1}(I - H)\right]\boldsymbol{v} \\
&= H^{k+1}\boldsymbol{x}^{[0]} + (I - H)^{-1}(I - H^{k+1})\boldsymbol{v}
\end{aligned}$$

and the proof of (12.6) is complete.

Letting $k \to \infty$ in (12.6), (12.5) implies at once that the iterative process converges,

$$\lim_{k \to \infty} \boldsymbol{x}^{[k]} = \hat{\boldsymbol{x}} := (I - H)^{-1}\boldsymbol{v}. \tag{12.7}$$

We next consider the case $\rho(H) \geq 1$. Provided that $1 \notin \sigma(H)$, the matrix $I - H$ is invertible and $\hat{\boldsymbol{x}} = (I - H)^{-1}\boldsymbol{v}$ is the only possible bounded limit of the iterative scheme. For, suppose the existence of a bounded limit $\hat{\boldsymbol{y}}$. Then

$$\hat{\boldsymbol{y}} = \lim_{k\to\infty} \boldsymbol{x}^{[k+1]} = H \lim_{k\to\infty} \boldsymbol{x}^{[k]} + \boldsymbol{v} = H\hat{\boldsymbol{y}} + \boldsymbol{v}, \tag{12.8}$$

therefore $\hat{\boldsymbol{y}} = \hat{\boldsymbol{x}}$.

Even if $1 \in \sigma(H)$, it remains true that every possible limit $\hat{\boldsymbol{y}}$ must obey (12.8). To see this, let $\boldsymbol{w}$ be an eigenvector corresponding to the eigenvalue 1. Substitution into (12.8) verifies that $\hat{\boldsymbol{y}} + \boldsymbol{w}$ is also a solution. Hence either there is no limit or the limit is not unique and depends on the starting value; both cases are categorized as 'absence of convergence'. We thus assume that $1 \notin \sigma(H)$.

Choose $\lambda \in \sigma(H)$ such that $|\lambda| = \rho(H)$ and let $\boldsymbol{w}$ be a unit-length eigenvector corresponding to λ: $H\boldsymbol{w} = \lambda\boldsymbol{w}$ and $\|\boldsymbol{w}\| = 1$. ($\|\cdot\|$ denotes here – and elsewhere in this chapter – the Euclidean norm.)

We need to show that there always exists a starting value $\boldsymbol{x}^{[0]} \in \mathbb{R}^d$ for which the scheme (12.3) fails to converge.

Case 1 Let $\lambda \in \mathbb{R}$. Note that since λ is real, so is $\boldsymbol{w}$. We choose $\boldsymbol{x}^{[0]} = \boldsymbol{w} + \hat{\boldsymbol{x}}$ and claim that

$$\boldsymbol{x}^{[k]} = \lambda^k \boldsymbol{w} + \hat{\boldsymbol{x}}, \qquad k = 0, 1, \ldots \tag{12.9}$$

As we have already seen, (12.9) is true when $k = 0$. By induction,

$$\begin{aligned} \boldsymbol{x}^{[k+1]} &= H\left(\lambda^k \boldsymbol{w} + \hat{\boldsymbol{x}}\right) + \boldsymbol{v} = \lambda^k H\boldsymbol{w} + (I - H)^{-1}[H + (I - H)]\boldsymbol{v} \\ &= \lambda^{k+1}\boldsymbol{w} + \hat{\boldsymbol{x}}, \qquad k = 0, 1, \ldots, \end{aligned}$$

and we deduce (12.9).

Because $|\lambda| \geq 1$, (12.9) implies that

$$\|\boldsymbol{x}^{[k]} - \hat{\boldsymbol{x}}\| = |\lambda|^k \geq 1, \qquad k = 0, 1, \ldots$$

Therefore it is impossible for the sequence $\{\boldsymbol{x}^{[k]}\}_{k=0}^{\infty}$ to converge to $\hat{\boldsymbol{x}}$.

Case 2 Suppose that λ is complex. Therefore $\bar{\lambda}$ is also an eigenvalue of H (the bar denotes complex conjugation). Since $H\boldsymbol{w} = \lambda\boldsymbol{w}$, complex conjugation implies $H\bar{\boldsymbol{w}} = \bar{\lambda}\bar{\boldsymbol{w}}$, hence $\bar{\boldsymbol{w}}$ must be a unit-length eigenvector corresponding to the eigenvalue $\bar{\lambda}$. Furthermore $\bar{\lambda} \neq \lambda$, hence $\boldsymbol{w}$ and $\bar{\boldsymbol{w}}$ must be linearly independent otherwise they would correspond to the same eigenvalue. We define a function

$$g(z) := \|z\boldsymbol{w} + \bar{z}\bar{\boldsymbol{w}}\|, \qquad z \in \mathbb{C}.$$

It is trivial to verify that $g : \mathbb{C} \to \mathbb{R}$ is continuous, hence it attains its minimum in every closed, bounded subset of $\mathbb{C}$, in particular, in the unit circle. Therefore,

$$\inf_{-\pi\leq\theta\leq\pi} \|\mathrm{e}^{\mathrm{i}\theta}\boldsymbol{w} + \mathrm{e}^{-\mathrm{i}\theta}\bar{\boldsymbol{w}}\| = \min_{-\pi\leq\theta\leq\pi} \|\mathrm{e}^{\mathrm{i}\theta}\boldsymbol{w} + \mathrm{e}^{-\mathrm{i}\theta}\bar{\boldsymbol{w}}\| = \nu \geq 0,$$

say. Suppose that $\nu = 0$. Then there exists $\theta_0 \in [-\pi, \pi]$ such that

$$\|\mathrm{e}^{\mathrm{i}\theta_0}\boldsymbol{w} + \mathrm{e}^{-\mathrm{i}\theta_0}\bar{\boldsymbol{w}}\| = 0,$$

therefore $\bar{\boldsymbol{w}} = \mathrm{e}^{2\mathrm{i}\theta_0}\boldsymbol{w}$, in contradiction to the linear independence of $\boldsymbol{w}$ and $\bar{\boldsymbol{w}}$. Consequently $\nu > 0$.

The function g is homogeneous,

$$g(r\mathrm{e}^{\mathrm{i}\theta}) = r\|\mathrm{e}^{\mathrm{i}\theta}\boldsymbol{w} + \mathrm{e}^{-\mathrm{i}\theta}\bar{\boldsymbol{w}}\| = rg(\mathrm{e}^{\mathrm{i}\theta}), \qquad r > 0, \quad |\theta| \le \pi,$$

hence

$$g(z) = |z|\, g\!\left(\frac{z}{|z|}\right) \ge \nu|z|, \qquad z \in \mathbb{C} \setminus \{0\}. \tag{12.10}$$

We let $\boldsymbol{x}^{[0]} = \boldsymbol{w} + \bar{\boldsymbol{w}} + \hat{\boldsymbol{x}} \in \mathbb{R}^d$. An inductive argument identical to the proof of (12.9) affirms that

$$\boldsymbol{x}^{[k]} = \lambda^k\boldsymbol{w} + \bar{\lambda}^k\bar{\boldsymbol{w}} + \hat{\boldsymbol{x}}, \qquad k = 0, 1, \ldots$$

Therefore, substituting into the inequality (12.10),

$$\|\boldsymbol{x}^{[k]} - \hat{\boldsymbol{x}}\| = g(\lambda^k) \ge |\lambda|^k\nu \ge \nu > 0, \qquad k = 0, 1, \ldots$$

As in case 1, we obtain a sequence that is bounded away from its only possible limit, $\hat{\boldsymbol{x}}$; therefore it cannot converge.

To complete the proof, we need to demonstrate that (12.4) is true, but this is trivial: the exact solution of (12.1) being $\boldsymbol{x} = A^{-1}\boldsymbol{b}$, (12.4) follows by substitution into (12.8). ■

◇ **Incomplete LU factorization** Suppose that we can write the matrix A in the form $A = \tilde{A} - E$, the underlying assumption being that LU factorization of the nonsingular matrix $\tilde{A}$ can be evaluated with ease. For example, $\tilde{A}$ might be banded or (in the case of a symmetric sparsity structure) have a graph that is a tree; see Chapter 11. Moreover, we assume that E is small in comparison with $\tilde{A}$. Writing (12.1) in the form

$$\tilde{A}\boldsymbol{x} = E\boldsymbol{x} + \boldsymbol{b}$$

suggests the iterative scheme

$$\tilde{A}\boldsymbol{x}^{[k+1]} = E\boldsymbol{x}^{[k]} + \boldsymbol{b}, \qquad k = 0, 1, \ldots, \tag{12.11}$$

incomplete LU factorization (ILU). Its implementation requires just a single LU (or Cholesky, if $\tilde{A}$ is symmetric) factorization, which can be reused in each iteration.

To write (12.11) in the form (12.3), we let $H = -\tilde{A}^{-1}E$ and $\boldsymbol{v} = \tilde{A}^{-1}\boldsymbol{b}$. Therefore

$$\begin{aligned}(I - H)A^{-1}\boldsymbol{b} &= (I - \tilde{A}^{-1}E)(\tilde{A} - E)^{-1}\boldsymbol{b} = \tilde{A}^{-1}(\tilde{A} - E)(\tilde{A} - E)^{-1}\boldsymbol{b}\\ &= \tilde{A}^{-1}\boldsymbol{b} = \boldsymbol{v},\end{aligned}$$

consistently with (12.4). Note that this definition of H and $\boldsymbol{v}$ is purely formal. In reality we never compute them explicitly; we use (12.11) instead. ◇

The ILU iteration (12.11) is an example of a *regular splitting*. With greater generality, we make the splitting $A = P - N$, where P is a nonsingular matrix, and consider the iterative scheme

$$P\boldsymbol{x}^{[k+1]} = N\boldsymbol{x}^{[k]} + \boldsymbol{b}, \qquad k = 0, 1, \ldots \tag{12.12}$$

The underlying assumption is that a system having matrix P can be solved with ease, whether by LU factorization or by other means. Note that, formally, $H = P^{-1}N = P^{-1}(P - A) = I - P^{-1}A$ and $\boldsymbol{v} = P^{-1}\boldsymbol{b}$.

Theorem 12.2 *Suppose that both A and $P + P^\top - A$ are symmetric and positive definite. Then the method* (12.12) *converges.*

Proof Let $\lambda \in \mathbb{C}$ be an arbitrary eigenvalue of the iteration matrix H and suppose that $\boldsymbol{w}$ is a corresponding eigenvector. Recall that $H = I - P^{-1}A$, therefore

$$(I - P^{-1}A)\boldsymbol{w} = \lambda\boldsymbol{w}.$$

We multiply both sides by the matrix P, and this results in

$$(1 - \lambda)P\boldsymbol{w} = A\boldsymbol{w}. \tag{12.13}$$

Our first conclusion is that $\lambda \neq 1$, otherwise (12.13) implies $A\boldsymbol{w} = \boldsymbol{0}$, which contradicts our assumption that A, being positive definite, is nonsingular.

We deduce further from (12.13) that

$$\bar{\boldsymbol{w}}^\top A\boldsymbol{w} = (1 - \lambda)\,\bar{\boldsymbol{w}}^\top P\boldsymbol{w}. \tag{12.14}$$

However, A is symmetric, therefore $\bar{\boldsymbol{y}}^\top A\boldsymbol{y}$ is real for every $\boldsymbol{y} \in \mathbb{C}^d$. Therefore, taking conjugates in (12.14),

$$\bar{\boldsymbol{w}}^\top A\boldsymbol{w} = \overline{(1 - \lambda)\,\bar{\boldsymbol{w}}^\top P\boldsymbol{w}} = (1 - \bar{\lambda})\,\bar{\boldsymbol{w}}^\top P^\top\boldsymbol{w}.$$

This, together with (12.14) and $\lambda \neq 1$, implies the identity

$$\left(\frac{1}{1-\lambda} + \frac{1}{1-\bar{\lambda}} - 1\right)\bar{\boldsymbol{w}}^\top A\boldsymbol{w} = \bar{\boldsymbol{w}}^\top(P + P^\top - A)\boldsymbol{w}. \tag{12.15}$$

We note first that

$$\frac{1}{1-\lambda} + \frac{1}{1-\bar{\lambda}} - 1 = \frac{2 - 2\operatorname{Re}\lambda - |1-\lambda|^2}{|1-\lambda|^2} = \frac{1 - |\lambda|^2}{|1-\lambda|^2} \in \mathbb{R}.$$

Next, we let $\boldsymbol{w} = \boldsymbol{w}_{\mathrm{R}} + \mathrm{i}\boldsymbol{w}_{\mathrm{I}}$, where both $\boldsymbol{w}_{\mathrm{R}}$ and $\boldsymbol{w}_{\mathrm{I}}$ are real vectors, and we take the real part of (12.15). On the left-hand side we obtain

$$\operatorname{Re}\left[(\boldsymbol{w}_{\mathrm{R}} - \mathrm{i}\boldsymbol{w}_{\mathrm{I}})^\top A(\boldsymbol{w}_{\mathrm{R}} + \mathrm{i}\boldsymbol{w}_{\mathrm{I}})\right] = \boldsymbol{w}_{\mathrm{R}}^\top A\boldsymbol{w}_{\mathrm{R}} + \boldsymbol{w}_{\mathrm{I}}^\top A\boldsymbol{w}_{\mathrm{I}}$$

and a similar identity is true on the right-hand side with A replaced by $P + P^\top - A$. Therefore

$$\frac{1 - |\lambda|^2}{|1-\lambda|^2}\left(\boldsymbol{w}_{\mathrm{R}}^\top A\boldsymbol{w}_{\mathrm{R}} + \boldsymbol{w}_{\mathrm{I}}^\top A\boldsymbol{w}_{\mathrm{I}}\right) = \boldsymbol{w}_{\mathrm{R}}^\top(P+P^\top-A)\boldsymbol{w}_{\mathrm{R}} + \boldsymbol{w}_{\mathrm{I}}^\top(P+P^\top-A)\boldsymbol{w}_{\mathrm{I}}. \tag{12.16}$$

Recall that both A and $P+P^\top-A$ are positive definite. It is impossible for both $\boldsymbol{w}_{\mathrm{R}}$ and $\boldsymbol{w}_{\mathrm{I}}$ to vanish (since this would imply that $\boldsymbol{w}=\boldsymbol{0}$), therefore

$$\boldsymbol{w}_{\mathrm{R}}^\top A\boldsymbol{w}_{\mathrm{R}}+\boldsymbol{w}_{\mathrm{I}}^\top A\boldsymbol{w}_{\mathrm{I}},\ \ \boldsymbol{w}_{\mathrm{R}}^\top(P+P^\top-A)\boldsymbol{w}_{\mathrm{R}}+\boldsymbol{w}_{\mathrm{I}}^\top(P+P^\top-A)\boldsymbol{w}_{\mathrm{I}} > 0.$$

We therefore conclude from (12.16) that

$$\frac{1-|\lambda|^2}{|1-\lambda|^2}>0,$$

hence $|\lambda|<1$. This is true for every $\lambda\in\sigma(H)$, consequently $\rho(H)<1$ and we use Lemma 12.1 to argue that the iterative scheme (12.12) converges. ■

◇ **Tridiagonal matrices** A relatively simple demonstration of the power of Theorem 12.2 is provided by *tridiagonal* matrices. Thus, let us suppose that the $d\times d$ symmetric matrix

$$A=\begin{bmatrix}\alpha_1 & \beta_1 & 0 & \cdots & 0\\ \beta_1 & \alpha_2 & \beta_2 & \ddots & \vdots\\ 0 & \ddots & \ddots & \ddots & 0\\ \vdots & \ddots & \beta_{d-2} & \alpha_{d-1} & \beta_{d-1}\\ 0 & \cdots & 0 & \beta_{d-1} & \alpha_d\end{bmatrix}$$

is positive definite. Our claim is that the regular splittings

$$P=\begin{bmatrix}\alpha_1 & 0 & \cdots & 0\\ 0 & \alpha_2 & \ddots & \vdots\\ \vdots & \ddots & \ddots & 0\\ 0 & \cdots & 0 & \alpha_d\end{bmatrix},\qquad N=-\begin{bmatrix}0 & \beta_1 & 0 & \cdots & 0\\ \beta_1 & 0 & \beta_2 & \ddots & \vdots\\ 0 & \ddots & \ddots & \ddots & 0\\ \vdots & \ddots & \beta_{d-2} & 0 & \beta_{d-1}\\ 0 & \cdots & 0 & \beta_{d-1} & 0\end{bmatrix} \tag{12.17}$$

and

$$P=\begin{bmatrix}\alpha_1 & 0 & \cdots & \cdots & 0\\ \beta_1 & \alpha_2 & 0 & & \vdots\\ 0 & \ddots & \ddots & \ddots & \vdots\\ \vdots & \ddots & \beta_{d-2} & \alpha_{d-1} & 0\\ 0 & \cdots & 0 & \beta_{d-1} & \alpha_d\end{bmatrix},\qquad N=-\begin{bmatrix}0 & \beta_1 & \cdots & 0\\ 0 & 0 & \ddots & \vdots\\ \vdots & \ddots & \ddots & \beta_{d-1}\\ 0 & \cdots & 0 & 0\end{bmatrix} \tag{12.18}$$

– the *Jacobi splitting* and the *Gauss–Seidel splitting* respectively – result in convergent schemes (12.12).

Since A is positive definite, Theorem 12.2 and the positive definiteness of the matrix $Q:=P+P^\top-A$ imply convergence. For the splitting (12.17) we readily obtain

$$Q = \begin{bmatrix} \alpha_1 & -\beta_1 & 0 & \cdots & 0 \\ -\beta_1 & \alpha_2 & -\beta_2 & \ddots & \vdots \\ 0 & \ddots & \ddots & \ddots & 0 \\ \vdots & \ddots & -\beta_{d-2} & \alpha_{d-1} & -\beta_{d-1} \\ 0 & \cdots & 0 & -\beta_{d-1} & \alpha_d \end{bmatrix}.$$

Our claim is that Q is indeed positive definite, and this follows from the positive definiteness of A. Specifically, A is positive definite if and only if $\boldsymbol{x}^\top A\boldsymbol{x} > 0$ for all $\boldsymbol{x} \in \mathbb{R}^d$, $\boldsymbol{x} \neq \boldsymbol{0}$ (A.1.3.5). But

$$\boldsymbol{x}^\top A\boldsymbol{x} = \sum_{j=1}^{d} \alpha_j^2 x_j^2 + 2\sum_{j=1}^{d-1} \beta_j x_j x_{j+1} = \sum_{j=1}^{d} \alpha_j^2 y_j^2 - 2\sum_{j=1}^{d-1} \beta_j y_j y_{j+1} = \boldsymbol{y}^\top Q\boldsymbol{y},$$

where $y_j = (-1)^j x_j$, $j = 1,2,\ldots,d$. Therefore $\boldsymbol{y}^\top Q\boldsymbol{y} > 0$ for every $\boldsymbol{y} \in \mathbb{R}^d \setminus \{\boldsymbol{0}\}$ and we deduce that the matrix Q is indeed positive definite.

The proof for (12.18) is, if anything, even easier, since Q is simply the diagonal matrix

$$Q = \begin{bmatrix} \alpha_1 & 0 & \cdots & 0 \\ 0 & \alpha_2 & \ddots & \vdots \\ \vdots & \ddots & \ddots & 0 \\ 0 & \cdots & 0 & \alpha_d \end{bmatrix}.$$

Since A is positive definite and $\alpha_j = \boldsymbol{e}_j^\top A\boldsymbol{e}_j > 0$, $j = 1,2,\ldots,$ where $\boldsymbol{e}_j \in \mathbb{R}^d$ is the jth unit vector, $j = 1,2,\ldots,d$, it follows at once that Q also is positive definite.

Figure 12.1 displays the error in the solution of a $d \times d$ tridiagonal system with

$$\begin{aligned} &\alpha_1 = d, \qquad \alpha_j = 2j(d-j) + d, && j = 2,3,\ldots,d, \\ &\beta_j = -j(d-j), && j = 1,2,\ldots,d-1, \\ &b_j \equiv 1, \qquad x_j^{[0]} \equiv 0, && j = 1,2,\ldots,d. \end{aligned} \tag{12.19}$$

It is trivial to use the Geršgorin criterion (Lemma 8.3) to prove that the underlying matrix A is positive definite.

The system has been solved with both the Jacobi splitting (12.17) (upper row in the figure) and the Gauss–Seidel splitting (12.18) (lower row) for $d = 10, 20, 30$. Even superficial examination of the figure reveals a number of interesting features.

- Both Jacobi and Gauss–Seidel converge. This should come as no surprise since, as we have just proved, provided A is tridiagonal its positive definiteness is sufficient for both methods to converge.

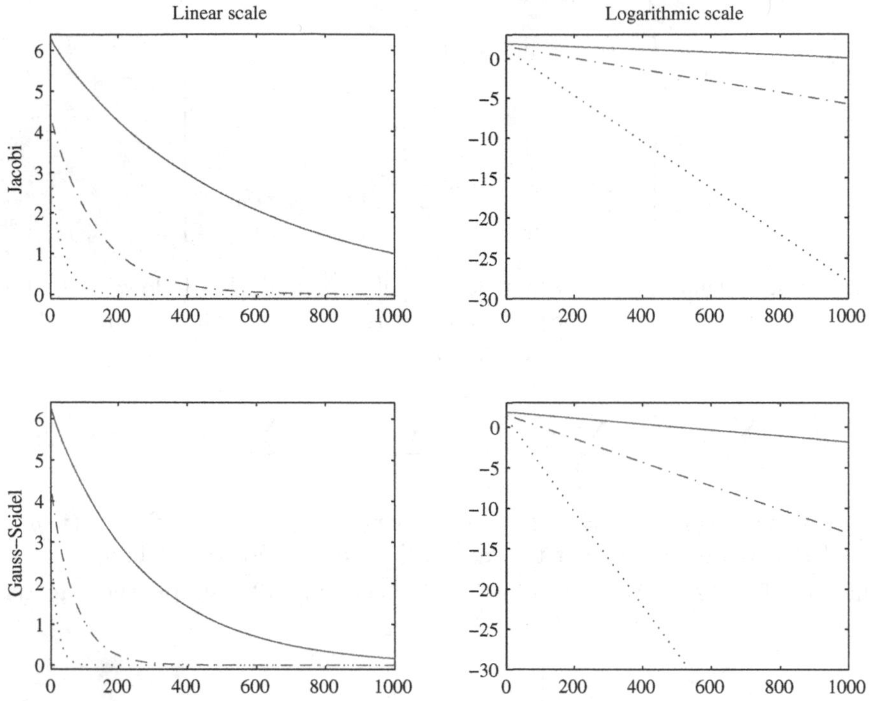

Figure 12.1 The error vs. the number of iterations in the Jacobi and Gauss–Seidel splittings for the system (12.19) with $d = 10$ (dotted line), $d = 20$ (broken-and-dotted line) and $d = 40$ (solid line). The first column displays the error (in the Euclidean norm) on a linear scale; the second column shows its logarithm.

- Convergence proceeds at a geometric speed; this is obvious from the second column, since the logarithm of the error is remarkably close to a linear function. This is not very surprising either since it is implicit in the proof of Lemma 12.1 (and made explicit in Exercise 12.2) that, at least asymptotically, the error decays like $[\rho(H)]^k$.
- The rate of convergence is slow and deteriorates markedly as d increases. This is a worrying feature since in practical computation we are interested in equations of considerably larger size than $d = 40$.
- Gauss–Seidel is better than Jacobi. Actually, careful examination of the rate of decay (which, obviously, is more transparent in the logarithmic scale) reveals that the error of Gauss–Seidel decays at twice the speed of Jacobi! In other words, we need just half the steps to attain the specified accuracy.

In the next section we will observe that the disappointing rate of decay of both methods, as well as the better performance of Gauss–Seidel, represent a fairly general state of affairs, rather than being just a feature of the linear system (12.19). ◇

12.2 Classical iterative methods

Let A be a real $d \times d$ matrix. We split it as shown below:

$$A \quad = \quad D \quad - \quad L_0 \quad - \quad U_0$$

Here the $d \times d$ matrices D, L_0 and U_0 are the diagonal, minus the strictly lower-triangular and minus the strictly upper-triangular portions of A, respectively. We assume that $a_{j,j} \neq 0$, $j = 1, 2, \ldots, d$. Therefore D is nonsingular and we let

$$L := D^{-1}L_0, \qquad U := D^{-1}U_0.$$

The *Jacobi* iteration is defined by setting in (12.3)

$$H = B := L + U, \qquad \boldsymbol{v} := D^{-1}\boldsymbol{b} \tag{12.20}$$

or, equivalently, considering a regular splitting (12.12) with $P = D$, $N = L_0 + U_0$. Likewise, we define the *Gauss–Seidel* iteration by specifying

$$H = \mathcal{L} := (I - L)^{-1}U, \qquad \boldsymbol{v} := (I - L)^{-1}D^{-1}\boldsymbol{b}, \tag{12.21}$$

and this is the same as the regular splitting $P = D - L_0$, $N = U_0$. Observe that (12.17) and (12.18) are nothing other than the Jacobi and Gauss–Seidel splittings, respectively, as applied to tridiagonal matrices.

The list of classical iterative schemes would be incomplete without mentioning the *successive over-relaxation* (SOR) scheme, which is defined by setting

$$H = \mathcal{L}_\omega := (I - \omega L)^{-1}[(1 - \omega)I + \omega U], \qquad \boldsymbol{v} = \omega(I - \omega L)^{-1}D^{-1}\boldsymbol{b}, \tag{12.22}$$

where $\omega \in [1, 2)$ is a parameter. Although this might not be obvious at a glance, the SOR scheme can be represented alternatively as a regular splitting with

$$P = \frac{1}{\omega}D - L_0, \qquad N = \left(\frac{1}{\omega} - 1\right)D + U_0 \tag{12.23}$$

(see Exercise 12.4).

Note that Gauss–Seidel is simply a special case of SOR, with $\omega = 1$. However, it makes good sense to single it out for special treatment.

All three methods (12.20)–(12.22) are consistent with (12.4), therefore Lemma 12.1 implies that if they converge, the limit is necessarily the true solution of the linear system.

The 'H–$\boldsymbol{v}$' notation is helpful within the framework of Lemma 12.1 but on the whole it is somewhat opaque. The three methods can be presented in a much simpler manner. Thus, writing the system (12.1) in the form

$$\sum_{j=1}^{d} a_{\ell,j}x_j = b_\ell, \qquad \ell = 1, 2, \ldots,$$

the Jacobi iteration reads

$$\sum_{j=1}^{\ell-1} a_{\ell,j}x_j^{[k]} + a_{\ell,\ell}x_\ell^{[k+1]} + \sum_{j=\ell+1}^{d} a_{\ell,j}x_j^{[k]} = b_\ell, \qquad \ell = 1, 2, \ldots, d, \quad k = 0, 1, \ldots$$

while the Gauss–Seidel scheme becomes

$$\sum_{j=1}^{\ell} a_{\ell,j}x_j^{[k+1]} + \sum_{j=\ell+1}^{d} a_{\ell,j}x_j^{[k]} = b_\ell, \qquad \ell = 1, 2, \ldots, d, \quad k = 0, 1, \ldots$$

In other words, the main difference between Jacobi and Gauss–Seidel is that in the first we always express each new component of $\boldsymbol{x}^{[k+1]}$ solely in terms of $\boldsymbol{x}^{[k]}$, while in the latter we use the elements of $\boldsymbol{x}^{[k+1]}$ whenever they are available. This is an important distinction as far as implementation is concerned. In each iteration (12.20) we need to store both $\boldsymbol{x}^{[k]}$ and $\boldsymbol{x}^{[k+1]}$ and this represents a major outlay in terms of computer storage – recall that $\boldsymbol{x}$ is a 'stretched' computational grid. (Of course, we do not store or even generate the matrix A if it originates in highly sparse finite difference or finite element equations. Instead, we need to know the 'rule' for constructing each linear equation, e.g. the five-point formula. If, however, A originates in a spectral method, we generate A and multiply it by vectors in the usual manner – but recall that for spectral methods the matrices are significantly smaller!) Clever programming and exploitation of the sparsity pattern can reduce the required amount of storage but this cannot ever compete with (12.21): in Gauss–Seidel we can throw away any ℓth component of $\boldsymbol{x}^{[k]}$ as soon as $\boldsymbol{x}_\ell^{[k+1]}$ has been generated, so both quantities can share the same storage.

The SOR iteration (12.22) can be also written in a similar form. Multiplying (12.23) by ω results in

$$\omega\sum_{j=1}^{\ell-1} a_{\ell,j}x_j^{[k+1]} + a_{\ell,\ell}x_\ell^{[k+1]} + (\omega-1)a_{\ell,\ell}x_\ell^{[k]} + \omega\sum_{j=\ell+1}^{d} a_{\ell,j}x_j^{[k]} = \omega b_\ell,$$
$$\ell = 1, 2, \ldots, d, \quad k = 0, 1, \ldots$$

Although precise estimates depend on the sparsity pattern of A, it is apparent that the cost of a single SOR iteration is not substantially larger than its counterpart for either Jacobi or Gauss–Seidel. Moreover, SOR shares with Gauss–Seidel the important virtue of requiring just a single copy of $\boldsymbol{x}$ to be stored at any one time.

The SOR iteration and its special case, the Gauss–Seidel method, share another feature. Their precise definition depends upon the ordering of the equations and the unknowns. As we have already seen in Chapter 11, the rearrangement of equations and unknowns is tantamount to acting on A with permutation matrices on the left and on the right respectively, and these two operations, in general, result in different iterative schemes. It is entirely possible that one of these arrangements converges, while the other fails to do so!

We have already observed in Section 12.1 that both Jacobi and Gauss–Seidel converge whenever A is a tridiagonal symmetric positive definite matrix and it is not difficult to verify that this is also the case with SOR for every $1 \le \omega < 2$ (cf. Exercise 12.5).

As far as convergence is concerned, Jacobi and Gauss–Seidel share similar behaviour for a wide range of linear systems. Thus, let A be *strictly diagonally dominant.* This means that

$$|a_{\ell,\ell}| \ge \sum_{\substack{j=1\\ j\neq\ell}}^{d} |a_{\ell,j}|, \qquad \ell = 1,2,\ldots,d \tag{12.24}$$

and the inequality is sharp for at least one $\ell \in \{1,2,\ldots,d\}$. (Some definitions require sharp inequality for *all* ℓ, but the present, weaker, definition is just perfect for our purposes.)

Theorem 12.3 *If the matrix A is irreducible and strictly diagonally dominant then both the Jacobi and Gauss–Seidel methods converge.*

Proof According to (12.20) and (12.24),

$$\sum_{j=1}^{d} |b_{\ell,j}| = \sum_{j=1,\ j\neq\ell}^{d} |b_{\ell,j}| = \frac{1}{|a_{\ell,\ell}|} \sum_{j=1,\ j\neq\ell}^{d} |a_{\ell,j}| \le 1, \qquad \ell = 1,2,\ldots,d$$

and the inequality is sharp for at least one ℓ. Therefore $\rho(B) < 1$ by the Geršgorin criterion (Lemma 8.3) and the Jacobi iteration converges.

The proof for Gauss–Seidel is slightly more complicated; essentially, we need to revisit the proof of Lemma 8.3 (i.e., Exercise 8.8) in a different framework. Choose $\lambda \in \sigma(\mathcal{L}_1)$ with an eigenvector $\boldsymbol{w}$. Therefore, multiplying $\mathcal{L}$ and $\boldsymbol{v}$ from (12.21) first by $I - L$ and then by D,

$$U_0\boldsymbol{w} = \lambda(D - L_0)\boldsymbol{w}.$$

It is convenient to rewrite this in the form

$$\lambda a_{\ell,\ell} w_\ell = \sum_{j=\ell+1}^{d} a_{\ell,j} w_j - \lambda \sum_{j=1}^{\ell-1} a_{\ell,j} w_j, \qquad \ell = 1,2,\ldots,d.$$

Therefore, by the triangle inequality,

$$|\lambda|\,|a_{\ell,\ell}|\,|w_\ell| \le \sum_{j=\ell+1}^{d} |a_{\ell,j}|\,|w_j| + |\lambda| \sum_{j=1}^{\ell-1} |a_{\ell,j}|\,|w_j| \tag{12.25}$$

$$\le \left(\sum_{j=\ell+1}^{d} |a_{\ell,j}| + |\lambda| \sum_{j=1}^{\ell-1} |a_{\ell,j}| \right) \max_{j=1,2,\ldots,d} |w_j|, \qquad j = 1,2,\ldots,d. \tag{12.26}$$

Let $\alpha \in \{1, 2, \ldots, d\}$ be such that

$$|w_\alpha| = \max_{j=1,2,\ldots,d} |w_j| > 0.$$

Substituting into (12.26), we have

$$|\lambda|\,|a_{\alpha,\alpha}| \le \sum_{j=\alpha+1}^{d} |a_{\alpha,j}| + |\lambda| \sum_{j=1}^{\alpha-1} |a_{\alpha,j}|.$$

Let us assume that $|\lambda| \ge 1$. We deduce that

$$|a_{\alpha,\alpha}| \le \sum_{\substack{j=1\\ j\neq\alpha}}^{d} |a_{\alpha,j}|,$$

and this can be consistent with (12.24) only if the weak inequality holds as an equality. Substitution in (12.25), in tandem with $|\lambda| \ge 1$, results in

$$|\lambda| \sum_{\substack{j=1\\ j\neq\alpha}}^{d} |a_{\alpha,j}|\,|w_\alpha| \le \sum_{j=\alpha+1}^{d} |a_{\alpha,j}|\,|w_j| + |\lambda| \sum_{j=1}^{\alpha-1} |a_{\alpha,j}|\,|w_j| \le |\lambda| \sum_{\substack{j=1\\ j\neq\alpha}}^{d} |a_{\alpha,j}|\,|w_j|.$$

This, however, can be consistent with $|w_\alpha| = \max |w_j|$ only if $|w_\ell| = |w_\alpha|$, $\ell = 1, 2, \ldots, d$. Therefore, every $\ell \in \{1, 2, \ldots, d\}$ can play the role of α and

$$|a_{\ell,\ell}| = \sum_{\substack{j=1\\ j\neq\ell}}^{d} |a_{\ell,j}|, \qquad \ell = 1, 2, \ldots, d,$$

in defiance of the definition of strict diagonal dominance. Therefore, having been led to a contradiction, our assumption that $|\lambda| \ge 1$ must be wrong. We deduce that $\rho(\mathcal{L}_1) < 1$, hence Lemma 12.1 implies convergence. ■

Another, less trivial, example where Jacobi and Gauss–Seidel converge in tandem is provided in the following theorem, which we state without proof.

Theorem 12.4 (The Stein–Rosenberg theorem) *Suppose that $a_{\ell,\ell} \neq 0$, $\ell = 1, 2, \ldots, d$, and that all the components of B are nonnegative. Then one of the floowing holds:*

$$\begin{array}{lllll} & \rho(\mathcal{L}_1) = \rho(B) = 0 & \text{or} & \rho(\mathcal{L}_1) < \rho(B) < 1 \\ \text{or} & \rho(\mathcal{L}_1) = \rho(B) = 1 & \text{or} & \rho(\mathcal{L}_1) > \rho(B) > 1. \end{array}$$

Hence, the Jacobi and Gauss–Seidel methods are either simultaneously convergent or simultaneously divergent. ■

◇ **An example of divergence** Lest there should be an impression that iterative methods are bound to converge or that they always share similar behaviour with regard to convergence, we give here a trivial counterexample,

$$A = \begin{bmatrix} 3 & 2 & 1 \\ 2 & 3 & 2 \\ 1 & 2 & 3 \end{bmatrix}.$$

The matrix is symmetric and positive definite; its eigenvalues are 2 and $\frac{1}{2}(7 \pm \sqrt{33}) > 0$. It is easy to verify, either directly or from Theorem 12.2 or Exercise 12.3, that $\rho(\mathcal{L}_1) < 1$ and the Gauss–Seidel method converges. However, $\rho(B) = \frac{1}{6}(1 + \sqrt{33}) > 1$ and the Jacobi method diverges.

An interesting variation on the last example is provided by the matrix

$$A = \begin{bmatrix} 3 & 1 & 2 \\ -1 & 3 & -2 \\ -2 & 2 & 3 \end{bmatrix}.$$

The spectrum of B is $\{0, \pm\mathrm{i}\}$, therefore the Jacobi method diverges marginally. Gauss–Seidel, however, proudly converges, since $\sigma(\mathcal{L}_1) = \left\{0, \frac{1}{54}(-23 \pm \sqrt{97})\right\} \in (0, 1)$.

Let us exchange the second and third rows and the second and third columns of A. The outcome is

$$\begin{bmatrix} 3 & 2 & 1 \\ -2 & 3 & 2 \\ -1 & -2 & 3 \end{bmatrix}.$$

The spectral radius of Jacobi remains intact, since it does not depend upon ordering. However, the eigenvalues of the new $\mathcal{L}_1$ are 0 and $\frac{1}{54}(-31 \pm \sqrt{1393})$, the spectral radius exceeds unity and the iteration diverges.

This demonstrates not just the sensitivity of (12.21) to ordering but also that the Jacobi iteration need not be the underachieving sibling of Gauss–Seidel; by replacing 3 with $3 + \varepsilon$, where $0 < \varepsilon \ll 1$, along the diagonal, we render Jacobi convergent (this is an immediate consequence of the Geršgorin criterion), while continuity of the eigenvalues of $\mathcal{L}_1$ as a function of ε means that Gauss–Seidel (in the second ordering) still diverges. ◇

To gain intuition with respect to the behaviour of classical iterative methods, let us first address ourselves in some detail to the matrix

$$A = \begin{bmatrix} -2 & 1 & 0 & \cdots & 0 \\ 1 & -2 & 1 & \ddots & \vdots \\ 0 & \ddots & \ddots & \ddots & 0 \\ \vdots & \ddots & 1 & -2 & 1 \\ 0 & \cdots & 0 & 1 & -2 \end{bmatrix}. \tag{12.27}$$

It is clear why such an A is relevant to our discussion: it is obtained from a central difference approximation to a second derivative.

A $d \times d$ matrix $T = (t_{k,\ell})_{k,\ell=1}^{d}$ is said to be *Toeplitz* if it is constant along all its diagonals, in other words, if there exist numbers $\tau_{-d+1}, \tau_{-d+2}, \ldots, \tau_0, \ldots, \tau_{d-1}$ such that

$$t_{k,\ell} = \tau_{k-\ell}, \qquad k, \ell = 1, 2, \ldots, d.$$

The matrix A, (12.27), is a Toeplitz matrix with

$$\tau_{-d+1} = \cdots = \tau_{-2} = 0, \quad \tau_{-1} = 1, \quad \tau_0 = -2, \quad \tau_1 = 1, \quad \tau_2 = \cdots = \tau_{d-1} = 0.$$

We say that a matrix is TST if it is Toeplitz, symmetric and tridiagonal. Therefore, T is TST if

$$\tau_j = 0, \quad |j| = 2, 3, \ldots, d-1, \qquad \tau_{-1} = \tau_1.$$

Matrices that are TST are important both for fast solution of the Poisson equation (see Chapter 14) and for the stability analysis of discretized PDEs of evolution (see Chapter 16) but, in the present context, we merely note that the matrix A, (12.27), is TST.

Lemma 12.5 *Let T be a $d \times d$ TST matrix and $\alpha := t_0$, $\beta := t_{-1} = t_1$. Then the eigenvalues of T are*

$$\lambda_j = \alpha + 2\beta \cos\left(\frac{\pi j}{d+1}\right), \qquad j = 1, 2, \ldots, d, \tag{12.28}$$

each with corresponding orthogonal eigenvector $\boldsymbol{q}_j$, where

$$q_{j,\ell} = \sqrt{\frac{2}{d+1}} \sin\left(\frac{\pi j \ell}{d+1}\right), \qquad j, \ell = 1, 2, \ldots, d. \tag{12.29}$$

Proof Although it is an easy matter to verify (12.28) and (12.29) directly from the definition of a TST matrix, we adopt a more roundabout approach since this will pay dividends later in this section.

We assume that $\beta \neq 0$, otherwise T reduces to a multiple of the identity matrix and the lemma is trivial. Let us suppose that λ is an eigenvalue of T with corresponding eigenvector $\boldsymbol{q}$. Letting $q_0 = q_{d+1} = 0$, we can write $A\boldsymbol{q} = \lambda\boldsymbol{q}$ in the form

$$\beta q_{\ell-1} + \alpha q_\ell + \beta q_{\ell+1} = \lambda q_\ell, \qquad \ell = 1, 2, \ldots, d$$

or, after a minor rearrangement,

$$\beta q_{\ell+1} + (\alpha - \lambda) q_\ell + \beta q_{\ell-1} = 0, \qquad \ell = 1, 2, \ldots, d.$$

This is a special case of a difference equation (4.19) and its general solution is

$$q_\ell = a\eta_+^\ell + b\eta_-^\ell, \qquad \ell = 0, 1, \ldots, d+1,$$

where $\eta_\pm$ are the zeros of the characteristic polynomial

$$\beta\eta^2 + (\alpha - \lambda)\eta + \beta = 0.$$

In other words,

$$\eta_{\pm} = \frac{1}{2\beta}\left\{\lambda - \alpha \pm \sqrt{(\lambda-\alpha)^2 - 4\beta^2}\right\}. \tag{12.30}$$

The constants a and b are determined by requiring $q_0 = q_{d+1} = 0$. The first condition yields $a + b = 0$, therefore $q_\ell = a(\eta_+^\ell - \eta_-^\ell)$ where $a \neq 0$ is arbitrary. To fulfil the second condition we need

$$\eta_+^{d+1} = \eta_-^{d+1}.$$

There are $d+1$ roots to this equation, namely

$$\eta_+ = \eta_- \exp\left(\frac{2\pi \mathrm{i} j}{d+1}\right), \qquad j = 0, 1, \ldots, d, \tag{12.31}$$

but we can discard at once the case $j = 0$, since it corresponds to $\eta_- = \eta_+$, hence to $q_\ell \equiv 0$.

We multiply (12.31) by $\exp[-\pi \mathrm{i} j/(d+1)]$ and substitute the values of $\eta_\pm$ from (12.30). Therefore

$$\left(\lambda - \alpha + \sqrt{(\lambda-\alpha)^2 - 4\beta^2}\right)\exp\left(\frac{-\pi \mathrm{i} j}{d+1}\right) = \left(\lambda - \alpha - \sqrt{(\lambda-\alpha)^2 - 4\beta^2}\right)\exp\left(\frac{\pi \mathrm{i} j}{d+1}\right)$$

for some $j \in \{1, 2, \ldots, d\}$. Rearrangement yields

$$\sqrt{(\lambda-\alpha)^2 - 4\beta^2}\cos\left(\frac{\pi j}{d+1}\right) = (\lambda - \alpha)\mathrm{i}\sin\left(\frac{\pi j}{d+1}\right)$$

and, squaring, we deduce

$$\left[(\lambda-\alpha)^2 - 4\beta^2\right]\cos^2\left(\frac{\pi j}{d+1}\right) = -(\lambda-\alpha)^2\sin^2\left(\frac{\pi j}{d+1}\right).$$

Therefore

$$(\lambda-\alpha)^2 = 4\beta^2\cos^2\left(\frac{\pi j}{d+1}\right)$$

and, taking the square root, we obtain

$$\lambda = \alpha \pm 2\beta\cos\left(\frac{\pi j}{d+1}\right).$$

Taking the plus sign we recover (12.28) with $\lambda = \lambda_j$, while the minus repeats $\lambda = \lambda_{d+1-j}$ and can be discarded. This concurs with the stipulated form of the eigenvalues.

Substituting (12.28) into (12.30), we readily obtain

$$\eta_\pm = \cos\left(\frac{\pi j}{d+1}\right) \pm \mathrm{i}\sin\left(\frac{\pi j}{d+1}\right) = \exp\left(\frac{\pm\pi \mathrm{i} j}{d+1}\right),$$

therefore

$$q_{j,\ell} = a(\eta_+^\ell - \eta_-^\ell) = 2a\mathrm{i}\sin\left(\frac{\pi j \ell}{d+1}\right), \qquad j, \ell = 1, 2, \ldots, d.$$

This will demonstrate that (12.29) is true if we can determine a value of a such that

$$\sum_{\ell=1}^{d} q_{j,\ell}^2 = 1$$

(note that symmetry implies that the eigenvectors are orthogonal, A.1.3.2). It is an easy exercise to show that

$$\sum_{\ell=1}^{d} \sin^2\left(\frac{\pi j \ell}{d+1}\right) = \tfrac{1}{2}(d+1), \qquad j = 1, 2, \ldots, d,$$

thereby providing a value of a that is consistent with (12.29). ■

Corollary *All $d \times d$ TST matrices commute with each other.*

Proof According to (12.29), all such matrices share the same set of eigenvectors, hence they commute (A.1.5.4). ■

Let us now return to the matrix A and to our discussion of classical iterative methods. It follows at once from (12.20) that the iteration matrix B is also a TST matrix, with $\alpha = 0$ and $\beta = \frac{1}{2}$. Therefore

$$\rho(B) = \cos\left(\frac{\pi}{d+1}\right) \approx 1 - \frac{\pi^2}{2d^2} < 1. \tag{12.32}$$

In other words, the Jacobi method converges; but we already know this from Section 12.1. However, (12.32) gives us an extra morsel of information, namely the speed of convergence. The news is not very good, unfortunately: the error is attenuated by $\mathcal{O}(d^{-2})$ in each iteration. In other words, if d is large, convergence up to any reasonable tolerance is very slow indeed.

Instead of debating Gauss–Seidel, we next leap all the way to the SOR scheme – after all, Gauss–Seidel is nothing other than SOR with $\omega = 1$. Although the matrix $\mathcal{L}_\omega$ is no longer Toeplitz, symmetric or tridiagonal, the method of proof of Lemma 12.5 is equally effective. Thus, let $\lambda \in \sigma(\mathcal{L}_\omega)$ and denote by $\boldsymbol{q}$ a corresponding eigenvector. It follows from (12.22) that

$$[(1-\omega)I + \omega U]\,\boldsymbol{q} = \lambda(I - \omega L)\boldsymbol{q}.$$

Therefore, letting $q_0 = q_{d+1} = 0$, we obtain

$$-2(1-\omega)q_\ell - \omega q_{\ell+1} = \lambda(\omega q_{\ell-1} - 2q_\ell), \qquad \ell = 1, 2, \ldots, d,$$

which we again rewrite as a difference equation,

$$\omega q_{\ell+1} - 2(\lambda - 1 + \omega)q_\ell + \omega\lambda q_{\ell-1} = 0, \qquad \ell = 1, 2, \ldots, d.$$

The solution is once more

$$q_\ell = a(\eta_+^\ell - \eta_-^\ell), \qquad \ell = 0, 1, \ldots, d+1,$$

except that (12.30) needs to be replaced by

$$\eta_{\pm} = \lambda - 1 + \omega \pm \sqrt{(\lambda - 1 + \omega)^2 - \omega^2 \lambda}. \tag{12.33}$$

We set $\eta_+^{d+1} = \eta_-^{d+1}$ and proceed as in the proof of Lemma 12.5. Substitution of the values of $\eta_{\pm}$ from (12.33) results in

$$(\lambda - 1 + \omega)^2 = \omega^2 \kappa \lambda, \tag{12.34}$$

where

$$\kappa = \cos^2\left(\frac{\pi \ell}{d+1}\right)$$

for some $\ell \in \{1, 2, \ldots, d\}$.

In the special case of Gauss–Seidel, (12.34) yields

$$\lambda^2 = \omega^2 \kappa \lambda$$

and we deduce that

$$\begin{aligned} &0, \cos^2\left(\frac{\pi}{d+1}\right), \cos^2\left(\frac{2\pi}{d+1}\right), \ldots, \cos^2\left(\frac{d\pi}{d+1}\right) \\ &\subseteq \sigma(\mathcal{L}_1) \subseteq \{0\} \cup \left\{\cos^2\left(\frac{\pi \ell}{d+1}\right) : \ell = 1, 2, \ldots, d\right\}. \end{aligned}$$

In particular,

$$\rho(\mathcal{L}_1) = \cos^2\left(\frac{\pi}{d+1}\right) \approx 1 - \frac{\pi^2}{d^2}. \tag{12.35}$$

Comparison with (12.32) demonstrates that, as far as the specific matrix A is concerned, Gauss–Seidel converges at exactly *twice* the rate of Jacobi. In other words, each iteration of Gauss–Seidel is, at least asymptotically, as effective as two iterations of Jacobi! Recall that Gauss–Seidel also has important advantages over Jacobi in terms of storage, while the number of operations in each iteration is identical in the two schemes. Thus, remarkably, it appears that Gauss–Seidel wins on every score. There is, however, an important reason why the Jacobi iteration is of interest and we address ourselves to this theme later in this section. At present, we wish to debate the convergence of SOR for different values of $\omega \in [1, 2)$. Note that our goal is not merely to check the convergence and assess its speed. The whole point of using SOR with an optimal value of ω, rather than Gauss–Seidel, rests in the exploitation of the parameter to accelerate convergence. We already know from (12.35) that a particular choice of ω is associated with convergence; now we seek to improve upon this result by identifying ω_{opt}, the optimal value of ω.

We distinguish between the following cases.

Case 1 $\kappa\omega^2 \le 4(\omega - 1)$, hence the roots of (12.34) form a complex conjugate pair. It is easy, substituting the explicit value of κ, to verify that this is indeed the case when

$$2\,\frac{1 - |\sin[\pi\ell/(d+1)]|}{\cos^2[\pi\ell/(d+1)]} \le \omega \le 2\,\frac{1 + |\sin[\pi\ell/(d+1)]|}{\cos^2[\pi\ell/(d+1)]}.$$

Moreover,

$$\frac{1+|\sin[\pi\ell/(d+1)]|}{\cos^2[\pi\ell/(d+1)]} \geq 1$$

and we restrict our attention to $\omega \leq 2$. Therefore case 1 corresponds to

$$\tilde{\omega} := 2\,\frac{1-|\sin[\pi\ell/(d+1)]|}{\cos^2[\pi\ell/(d+1)]} \leq \omega < 2. \tag{12.36}$$

The two solutions of (12.34) are

$$\lambda = 1-\omega+\tfrac{1}{2}\kappa\omega^2 \pm \sqrt{\left(1-\omega+\tfrac{1}{2}\kappa\omega^2\right)^2-(1-\omega)^2}; \tag{12.37}$$

consequently

$$|\lambda|^2 = \left(1-\omega+\tfrac{1}{2}\kappa\omega^2\right)^2 + \left[(1-\omega)^2-\left(1-\omega+\tfrac{1}{2}\kappa\omega^2\right)^2\right] = (\omega-1)^2$$

and we obtain

$$|\lambda| = \omega - 1. \tag{12.38}$$

Case 2 $\kappa\omega^2 \geq 4(\omega-1)$ and both zeros of (12.34) are real. Differentiating (12.34) with respect to ω yields

$$2(\lambda-1+\omega)(\lambda_\omega+1) = 2\kappa\omega\lambda + \kappa\omega^2\lambda_\omega,$$

where $\lambda_\omega = \mathrm{d}\lambda/\mathrm{d}\omega$. Therefore λ_ω may vanish only for

$$\lambda = \frac{1-\omega}{1-\kappa\omega}.$$

Substitution into (12.34) results in $\kappa(1-\omega) = 1-\kappa\omega$, hence in $\kappa = 1$ – but this is impossible, because $\kappa = \cos^2[\pi\ell/(d+1)] \in (0,1)$. Therefore $\lambda_\omega \neq 0$ in

$$1 < \omega \leq \tilde{\omega}$$

(cf. (12.36)) and the zeros of (12.34) are monotone in this interval. Since they are continuous functions of ω, it follows that, to track the zero of largest magnitude, it is enough to restrict attention to the endpoints 1 and $\tilde{\omega}$.

At $\omega = 1$ we are back to the Gauss–Seidel case and the spectral radius is given by (12.35). At the other endpoint (which is the meeting point of cases 1 and 2) we have $\kappa\omega^2 - 4\omega + 4 = 0$, hence

$$\omega = \frac{2\left(1-\sqrt{1-\kappa}\right)}{\kappa}.$$

We have taken the minus sign rather than the plus sign in front of the square root, otherwise $\omega \notin [1,2)$. Therefore, by (12.38),

$$|\lambda| = \rho(\kappa) := 2\frac{1-\sqrt{1-\kappa}}{\kappa} - 1.$$

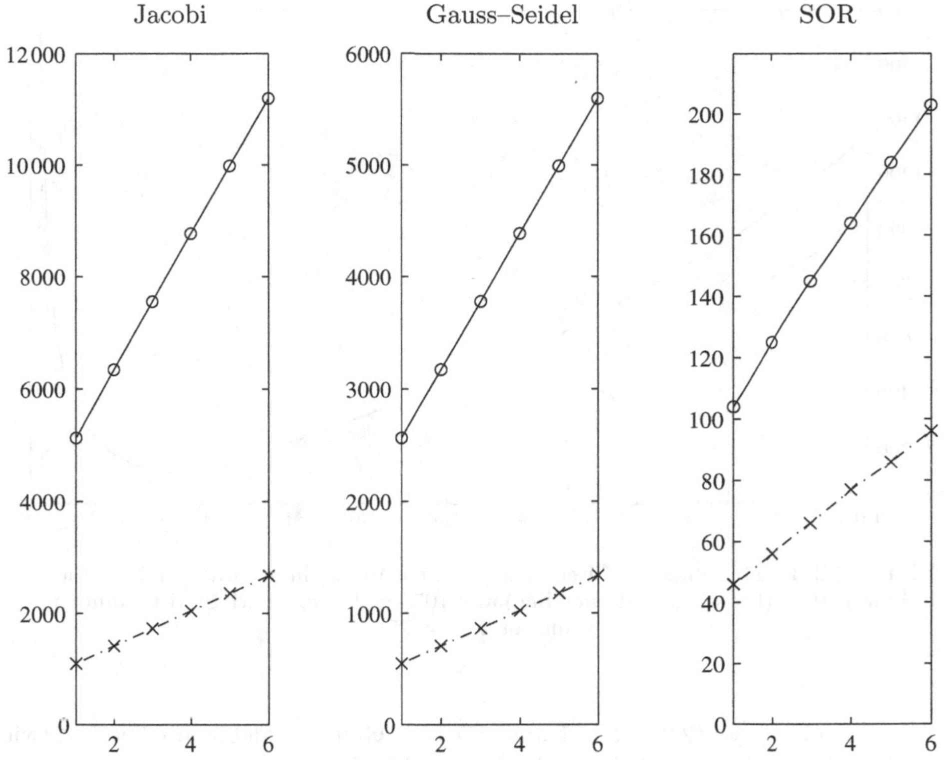

Figure 12.2 The number of iterations in the solution of $A\boldsymbol{x} = \boldsymbol{b}$, where A is given by (12.27), $\boldsymbol{x}^{[0]} = \boldsymbol{0}$ and $\boldsymbol{b} = \boldsymbol{1}$, required to reach accuracy up to a given tolerance. The x-axis displays the number of correct significant digits, while the number of iterations can be read from the y-axis. The broken-and-dotted line corresponds to $d = 25$ and the solid line to $d = 50$.

The above is true for every $\kappa = \cos^2[\pi\ell/(d+1)]$, $\ell = 1, 2, \ldots, d$. It is, however, elementary to verify that ρ increases strictly monotonically as a function of κ. Thus, the maximum is attained for the largest value of κ, i.e. when $\ell = 1$, and we deduce that

$$\omega_{\text{opt}} = 2\frac{1 - \sin[\pi/(d+1)]}{\cos^2[\pi/(d+1)]} \tag{12.39}$$

and

$$\rho(\mathcal{L}_{\omega_{\text{opt}}}) = \omega_{\text{opt}} - 1 = \left\{\frac{1 - \sin[\pi/(d+1)]}{\cos[\pi/(d+1)]}\right\}^2 \approx 1 - \frac{2\pi}{d}. \tag{12.40}$$

Casting our eyes at the expressions (12.32), (12.35) and (12.40), for the spectral radii of Jacobi, Gauss–Seidel and SOR (with ω_{opt}) respectively, it is difficult not to notice the drastic improvement inherent in the SOR scheme. This observation is vividly demonstrated in Fig. 12.2, where the three methods are employed to solve

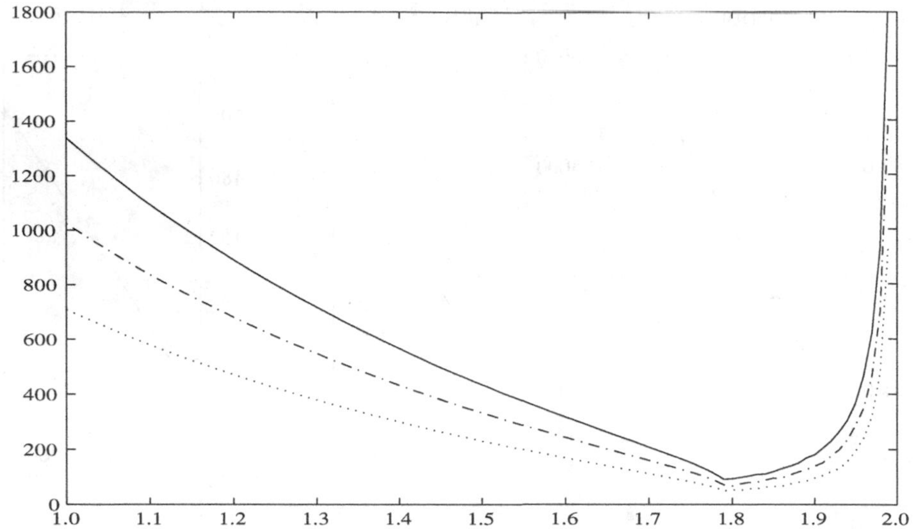

Figure 12.3 The number of iterations required to attain accuracy of 10^{-2} (dotted line), 10^{-4} (broken-and-dashed line) and 10^{-6} (solid line) with SOR for different values of ω ($d = 25$).

$A\boldsymbol{x} = \boldsymbol{b}$; A is given by (12.27), $\boldsymbol{b} = \mathbf{1}$ and $\boldsymbol{x}^{[0]} = \mathbf{0}$. Gauss–Seidel is, predictably, twice as efficient as Jacobi but SOR leaves both far behind!

Figure 12.3 displays the number of iterations required to approach the solution of $A\boldsymbol{x} = \boldsymbol{b}$ (for $d = 25$, with the same $\boldsymbol{b}$ and $\boldsymbol{x}^{[0]}$ as in Fig. 12.2) to within a given accuracy. The sensitivity of the rate of convergence to the value of ω is striking.

Although our analysis of the performance of classical iteration schemes has been restricted to a very special matrix A, (12.27), its conclusions are relevant in a substantially wider framework. It is easy, for example, to generalize it to an arbitrary TST matrix, provided the underlying Jacobi iteration converges (in the terminology of Lemma 12.5 this corresponds to $2|\beta| \le |\alpha|$; cf. Exercise 12.6).

In the next section we present an outline of a more general theory. This answers the question of convergence and its rate for classical iterative methods, and identifies the optimal SOR parameter, for an extensive family of matrices occurring in the numerical solution of partial differential equations.

12.3 Convergence of successive over-relaxation

In this section we address ourselves to matrices that possess a specific sparsity pattern. For brevity and in order to steer clear of complicated algebra, we have omitted several proofs, and the section should be regarded as no more than a potted outline of general SOR theory.

In Chapter 11 we have had already an opportunity to observe that the terminology of graph theory confers a useful means to express sparsity patterns. As we wish at present to discuss matrices that neither are symmetric nor conform with symmetric sparsity patterns, we say that $\mathbb{G} = \{\mathbb{V}, \mathbb{E}\}$ is the *digraph* (an abbreviation for *directed graph*) of a $d \times d$ matrix A if $\mathbb{V} = \{1, 2, \ldots, d\}$ and $(m, \ell) \in \mathbb{E}$ for $m, \ell \in \{1, 2, \ldots, d\}$, $m \neq \ell$, if and only if $a_{m,\ell} \neq 0$.

To distinguish between $(\ell, m) \in \mathbb{E}$ and $(m, \ell) \in \mathbb{E}$, a pictorial representation of a digraph uses arrows to denote the 'direction'. Thus, (ℓ, m) is represented by a curve joining the ℓth and the mth vertex, with an arrow pointing at the latter. The restriction $\ell \neq m$ is often dropped in a definition of a digraph. We insist on it in the present context because it simplifies the notation to some extent.

Given a $d \times d$ matrix A, we say that $\boldsymbol{j} \in \mathbb{Z}^d$ is an *ordering vector* if $|j_\ell - j_m| = 1$ for every $(\ell, m) \in \mathbb{E}$. Moreover, $\boldsymbol{j}$ is a *compatible ordering vector* if, in addition,

$$\begin{aligned} \ell \geq m+1 \quad &\Longrightarrow \quad j_\ell - j_m = +1, \\ \ell \leq m-1 \quad &\Longrightarrow \quad j_\ell - j_m = -1. \end{aligned}$$

Lemma 12.6 *If the matrix A has an ordering vector then there exists a permutation matrix P such that the matrix $\tilde{A} := PAP^{-1}$ has a compatible ordering vector.*

Proof Each similarity transformation by a permutation matrix is merely a relabelling of variables; this theme underlies the discussion in Section 11.2.[2] Therefore the graph of $\tilde{A}$ is simply $\tilde{\mathbb{G}} = \{\boldsymbol{\pi}(\mathbb{V}), \boldsymbol{\pi}(\mathbb{E})\}$, where $\boldsymbol{\pi}$ is the corresponding permutation of $\{1, 2, \ldots, d\}$; in other words, $\boldsymbol{\pi}(\mathbb{V}) = \{\pi(1), \pi(2), \ldots, \pi(d)\}$ and $(\pi(\ell), \pi(m)) \in \boldsymbol{\pi}(\mathbb{E})$ if and only if $(\ell, m) \in \mathbb{E}$.

Let $\boldsymbol{j}$ be the ordering vector whose existence has been stipulated in the statement of the lemma and set $i_\ell := j_{\pi^{-1}(\ell)}$, $\ell = 1, 2, \ldots, d$, where $\boldsymbol{\pi}^{-1}$ is the inverse permutation of $\boldsymbol{\pi}$. It is easy to verify that $\boldsymbol{i}$ is an ordering vector of $\tilde{A}$ since $(\ell, m) \in \boldsymbol{\pi}(\mathbb{E})$ implies $(\pi^{-1}(\ell), \pi^{-1}(m)) \in \mathbb{E}$. Therefore, by the definition of an ordering vector, $|j_{\pi^{-1}(\ell)} - j_{\pi^{-1}(m)}| = 1$. Recalling the way we have constructed the vector $\boldsymbol{i}$, it follows that $|i_\ell - i_m| = 1$ and that $\boldsymbol{i}$ is indeed an ordering vector of $\tilde{A}$.

Let us choose a permutation $\boldsymbol{\pi}$ such that $i_1 \leq i_2 \leq \cdots \leq i_d$; this, of course, corresponds to $j_{\pi^{-1}(1)} \leq j_{\pi^{-1}(2)} \leq \cdots \leq j_{\pi^{-1}(d)}$ and can always be done. Given $(\ell, m) \in \pi(\mathbb{E})$, $\ell \geq m+1$, we then obtain $i_\ell - i_m \geq 0$; therefore, $\boldsymbol{i}$ being an ordering vector, $i_\ell - i_m = +1$. Likewise $(\ell, m) \in \pi(\mathbb{E})$, $\ell \leq m-1$, implies that $i_\ell - i_m = -1$. We thus deduce that $\boldsymbol{i}$ is a compatible ordering vector of $\tilde{A}$. ■

◇ **Ordering vectors** Consider a matrix A with the following sparsity pattern:

$$\begin{bmatrix} \times & \circ & \times & \circ & \times \\ \circ & \times & \times & \circ & \circ \\ \times & \times & \times & \circ & \circ \\ \circ & \circ & \circ & \times & \times \\ \times & \circ & \circ & \times & \times \end{bmatrix}.$$

Here $\mathbb{E} = \{(1,3), (3,1), (1,5), (5,1), (2,3), (3,2), (4,5), (5,4)\}$, and it is easy to

[2] It is trivial to prove that P^{-1} is also a permutation matrix and that it reverses the action of P.

verify that

$$\boldsymbol{j} = \begin{bmatrix} 2 \\ 2 \\ 1 \\ 2 \\ 3 \end{bmatrix}$$

is an ordering vector. However, it is not a compatible ordering vector since, for example, $(3,1) \in \mathbb{E}$ and $j_3 - j_1 = -1$.

To construct a compatible ordering vector $\boldsymbol{i}$ for a permutation $\boldsymbol{\pi}$, we require $j_{\pi^{-1}(1)} \le j_{\pi^{-1}(2)} \le j_{\pi^{-1}(3)} \le j_{\pi^{-1}(4)} \le j_{\pi^{-1}(5)}$. This will be the case if we let, for example, $\pi^{-1}(1) = 3$, $\pi^{-1}(2) = 1$, $\pi^{-1}(3) = 2$, $\pi^{-1}(4) = 4$ and $\pi^{-1}(5) = 5$, in other words,

$$\boldsymbol{\pi} = \begin{bmatrix} 2 \\ 3 \\ 1 \\ 4 \\ 5 \end{bmatrix}, \qquad P = \begin{bmatrix} 0 & 0 & 1 & 0 & 0 \\ 1 & 0 & 0 & 0 & 0 \\ 0 & 1 & 0 & 0 & 0 \\ 0 & 0 & 0 & 1 & 0 \\ 0 & 0 & 0 & 0 & 1 \end{bmatrix} \qquad \text{and} \qquad \tilde{A} = \begin{bmatrix} \times & \times & \times & \circ & \circ \\ \times & \times & \circ & \circ & \times \\ \times & \circ & \times & \circ & \circ \\ \circ & \circ & \circ & \times & \times \\ \circ & \times & \circ & \times & \times \end{bmatrix}.$$

Incidentally, $\boldsymbol{i}$ is not the only possible compatible ordering vector. To demonstrate this and, incidentally, that a compatible ordering vector need not be unique, we render the digraph of A pictorially,

and observe that A is itself a permutation of a tridiagonal matrix. It is easy to identify a compatible ordering vector for any tridiagonal matrix with nonvanishing off-diagonal elements – a task that is relegated to Exercise 12.8 – hence producing yet another compatible ordering vector for a permutation of A. ◇

The existence of a compatible ordering vector confers on a matrix several interesting properties which are of great relevance to the behaviour of classical iterative methods.

Lemma 12.7 *If the matrix A has a compatible ordering vector then the function*

$$g(s,t) := \det\left(tL_0 + \frac{1}{t}U_0 - sD\right), \qquad s \in \mathbb{R}, \quad t \in \mathbb{R} \setminus \{0\}$$

is independent of t.

Proof Since A and $H(s,t) := tL_0 + (1/t)U_0 - sD$ share the same sparsity pattern for all $t \neq 0$, it follows from the definition that every compatible ordering vector of A is also a compatible ordering vector of $H(s,t)$. In particular, we deduce that the matrix $H(s,t)$ possesses a compatible ordering vector.

By the definition of a determinant,

$$g(s,t) = \sum_{\pi \in \Pi_d} (-1)^{|\pi|} \prod_{i=1}^{d} h_{i,\pi(i)}(s,t),$$

where $H(s,t) = (h_{j,\ell}(s,t))_{j,\ell=1}^{d}$, Π_d is the set of all permutations of $\{1,2,\ldots,d\}$ and $|\pi|$ is the sign of $\pi \in \Pi_d$. Since

$$h_{j,\ell}(s,t) = -t^{\sigma_{j-\ell}} s^{1-|\sigma_{j-\ell}|} a_{j,\ell}, \qquad j,\ell = 1,2,\ldots,d,$$

where

$$\sigma_m = \begin{cases} +1, & m > 0, \\ -1, & m < 0, \\ 0, & m = 0, \end{cases}$$

we deduce that

$$g(s,t) = (-1)^d \sum_{\pi \in \Pi_d} (-1)^{|\pi|} t^{d_{\mathrm{L}}(\pi) - d_{\mathrm{U}}(\pi)} s^{d - d_{\mathrm{L}}(\pi) - d_{\mathrm{U}}(\pi)} \prod_{i=1}^{d} a_{i,\pi(i)}, \tag{12.41}$$

where $d_{\mathrm{L}}(\pi)$ and $d_{\mathrm{U}}(\pi)$ denote the number of elements $\ell \in \{1,2,\ldots,d\}$ such that $\ell > \pi(\ell)$ and $\ell < \pi(\ell)$ respectively.

Let $\boldsymbol{j}$ be the compatible ordering vector of $H(s,t)$, whose existence we have already deduced from the statement of the lemma, and choose an arbitrary $\pi \in \Pi_d$ such that

$$a_{1,\pi(1)}, a_{2,\pi(2)}, \ldots, a_{d,\pi(d)} \neq 0.$$

It follows from the definition that

$$d_{\mathrm{L}}(\pi) = \sum_{\substack{\ell=1 \\ \pi(\ell)<\ell}}^{d} [j_\ell - j_{\pi(\ell)}], \qquad d_{\mathrm{U}}(\pi) = \sum_{\substack{\ell=1 \\ \pi(\ell)>\ell}}^{d} [j_{\pi(\ell)} - j_\ell],$$

hence

$$d_{\mathrm{L}}(\pi) - d_{\mathrm{U}}(\pi) = \sum_{\substack{\ell=1 \\ \pi(\ell)\neq\ell}}^{d} [j_\ell - j_{\pi(\ell)}] = \sum_{\ell=1}^{d} [j_\ell - j_{\pi(\ell)}] = \sum_{\ell=1}^{d} j_\ell - \sum_{\ell=1}^{d} j_{\pi(\ell)}.$$

Recall, however, that π is a permutation of $\{1,2,\ldots,d\}$; therefore

$$\sum_{\ell=1}^{d} j_{\pi(\ell)} = \sum_{\ell=1}^{d} j_\ell$$

and $d_{\mathrm{L}}(\pi) - d_{\mathrm{U}}(\pi) = 0$ for every $\pi \in \Pi_d$ such that $a_{\ell,\pi(\ell)} \neq 0$, $\ell = 1,2,\ldots,d$. Therefore

$$t^{d_{\mathrm{L}}(\pi) - d_{\mathrm{U}}(\pi)} \prod_{i=1}^{d} a_{i,\pi(i)} = \prod_{i=1}^{d} a_{i,\pi(i)}, \qquad \pi \in \Pi_d,$$

and it follows from (12.41) that $g(s,t)$ is indeed independent of $t \in \mathbb{R} \setminus \{0\}$. ■

Theorem 12.8 *Suppose that the matrix A has a compatible ordering vector and let $\mu \in \mathbb{C}$ be an eigenvalue of the matrix B, the iteration matrix of the Jacobi method. Then also*

(i) $-\mu \in \sigma(B)$ *and the multiplicities of $+\mu$ and $-\mu$ (as eigenvalues of B) are identical;*

(ii) *given any $\omega \in (0,2)$, every $\lambda \in \mathbb{C}$ that obeys the equation*

$$\lambda + \omega - 1 = \omega\mu\lambda^{1/2} \tag{12.42}$$

belongs to $\sigma(\mathcal{L}_\omega)$;[3]

(iii) *for every $\lambda \in \sigma(\mathcal{L}_\omega)$, $\omega \in (0,2)$, there exists $\mu \in \sigma(B)$ such that the equation (12.42) holds.*

Proof According to Lemma 12.7, the presence of a compatible ordering vector of A implies that

$$\det(L_0 + U_0 - \mu D) = g(\mu, 1) = g(\mu, -1) = \det(-L_0 - U_0 - \mu D).$$

Moreover, $\det(-C) = (-1)^d \det C$ for any $d \times d$ matrix C and we thus deduce that

$$\det(L_0 + U_0 - \mu D) = (-1)^d \det(L_0 + U_0 + \mu D). \tag{12.43}$$

By the definition of an eigenvalue, $\mu \in \sigma(B)$ if and only if $\det(B - \mu I) = 0$ (see A.1.5.1). But, according to (12.20) and (12.43),

$$\begin{aligned}\det(B - \mu I) &= \det\left[D^{-1}(L_0 + U_0) - \mu I\right] = \det[D^{-1}(L_0 + U_0 - \mu D)] \\ &= \frac{1}{\det D}\det(L_0 + U_0 - \mu D) = \frac{(-1)^d}{\det D}\det(L_0 + U_0 + \mu D) \\ &= (-1)^d \det(B + \mu I).\end{aligned}$$

This proves (i).

The matrix $I - \omega L$ is lower triangular with ones across the diagonal, therefore $\det(I - \omega L) \equiv 1$. Hence, it follows from the definition (12.22) of $\mathcal{L}_\omega$ that

$$\begin{aligned}\det(\mathcal{L}_\omega - \lambda I) &= \det\left\{(I - \omega L)^{-1}[\omega U + (1-\omega)I] - \lambda I\right\} \\ &= \frac{1}{\det(I - \omega L)}\det[\omega U + \omega\lambda L - (\lambda + \omega - 1)I] \\ &= \det[\omega U + \omega\lambda L - (\lambda + \omega - 1)I].\end{aligned} \tag{12.44}$$

[3]The SOR iteration was defined in (12.22) for $\omega \in [1,2)$, while now we allow $\omega \in (0,2)$. This should cause no difficulty whatsoever.

Suppose that $\lambda = 0$ lies in $\sigma(\mathcal{L}_\omega)$. Then (12.44) implies that $\det[\omega U - (\omega - 1)I] = 0$. Recall that U is strictly upper triangular, therefore

$$\det[\omega U - (\omega - 1)I] = (1 - \omega)^d$$

and we deduce $\omega = 1$. It is trivial to check that $(\lambda, \omega) = (0, 1)$ obeys the equation (12.42). Conversely, if $\lambda = 0$ satisfies (12.42) then we immediately deduce that $\omega = 1$ and, by (12.44), $0 \in \sigma(\mathcal{L}_1)$. Therefore, (ii) and (iii) are true in the special case $\lambda = 0$.

To complete the proof of the theorem, we need to discuss the case $\lambda \neq 0$. According to (12.44),

$$\frac{1}{\omega^d \lambda^{d/2}} \det(\mathcal{L}_\omega - \lambda I) = \det\left(\lambda^{1/2} L + \lambda^{-1/2} U - \frac{\lambda + \omega - 1}{\omega \lambda^{1/2}} I\right)$$

and, again using Lemma 12.7,

$$\frac{1}{\omega^d \lambda^{d/2}} \det(\mathcal{L}_\omega - \lambda I) = \det\left(L + U - \frac{\lambda + \omega - 1}{\omega \lambda^{1/2}} I\right) = \det\left(B - \frac{\lambda + \omega - 1}{\omega \lambda^{1/2}} I\right). \tag{12.45}$$

Let $\mu \in \sigma(B)$ and suppose that λ obeys the equation (12.42). Then

$$\mu = \frac{\lambda + \omega - 1}{\omega \lambda^{1/2}}$$

and substitution in (12.45) proves that $\det(\mathcal{L}_\omega - \lambda I) = 0$, hence $\lambda \in \sigma(\mathcal{L}_\omega)$. However, if $\lambda \in \sigma(\mathcal{L}_\omega)$, $\lambda \neq 0$, then, according to (12.45), $(\lambda + \omega - 1)/(\omega \lambda^{1/2}) \in \sigma(B)$. Consequently, there exists $\mu \in \sigma(B)$ such that (12.42) holds; thus the proof of the theorem is complete. ■

Corollary *Let A be a tridiagonal matrix and suppose that $a_{j,\ell} \neq 0$, $|j - \ell| \leq 1$, $j, \ell = 1, 2, \ldots$ Then $\rho(\mathcal{L}_1) = [\rho(B)]^2$.*

Proof We have already mentioned that every tridiagonal matrix with a nonvanishing off-diagonal has a compatible ordering vector, a statement whose proof was consigned to Exercise 12.8. Therefore (12.42) holds and, ω being unity, reduces to

$$\lambda = \mu \lambda^{1/2}.$$

In other words, either $\lambda = 0$ or $\lambda = \mu^2$. If $\lambda = 0$ for all $\lambda \in \sigma(\mathcal{L}_1)$ then part (iii) of the theorem implies that all the eigenvalues of B vanish as well. Hence $\rho(\mathcal{L}_1) = [\rho(B)]^2 = 0$. Otherwise, there exists $\lambda \neq 0$ in $\sigma(\mathcal{L}_1)$ and, since $\lambda = \mu^2$, parts (ii) and (iii) of the theorem imply that $\rho(\mathcal{L}_1) \geq [\rho(B)]^2$ and $\rho(\mathcal{L}_1) \leq [\rho(B)]^2$ respectively. This proves the corollary. ■

The statement of the corollary should not come as a surprise, since we have already observed behaviour consistent with $\rho(\mathcal{L}_1) = [\rho(B)]^2$ in Fig. 12.1 and proved it in Section 12.2 for a specific TST matrix.

The importance of Theorem 12.8 ranges well beyond a comparison of the Jacobi and Gauss–Seidel schemes. It comes into its own when applied to SOR and its convergence.

Theorem 12.9 *Let A be a $d \times d$ matrix. If $\rho(\mathcal{L}_\omega) < 1$ and the SOR iteration* (12.22) *converges then necessarily $\omega \in (0, 2)$. Moreover, if A has a compatible ordering vector* *and all the eigenvalues of B are real then the iteration converges for every $\omega \in (0,2)$ if and only if $\rho(B) < 1$ and the Jacobi method converges for the same matrix.*

Proof Let $\sigma(\mathcal{L}_\omega) = \{\lambda_1, \lambda_2, \ldots, \lambda_d\}$, therefore

$$\det \mathcal{L}_\omega = \prod_{\ell=1}^{d} \lambda_\ell. \tag{12.46}$$

Using a previous argument, see (12.44), we obtain

$$\det \mathcal{L}_\omega = \det[\omega U - (\omega - 1)I] = (1-\omega)^d$$

and substitution in (12.46) leads to the inequality

$$\rho(\mathcal{L}_\omega) = \max_{\ell=1,2,\ldots,d} |\lambda_\ell| \geq \left| \prod_{\ell=1}^{d} \lambda_\ell \right|^{1/d} = |1-\omega|.$$

Therefore $\rho(\mathcal{L}_\omega) < 1$ is inconsistent with either $\omega \leq 0$ or $\omega \geq 2$ and the first statement of the theorem is true.

We next suppose that A possesses a compatible ordering vector. Thus, according to Theorem 12.8, for every $\lambda \in \sigma(\mathcal{L}_\omega)$ there exists $\mu \in \sigma(B)$ such that $p(\lambda^{1/2}) = 0$, where $p(z) := z^2 - \omega\mu z + (\omega - 1)$ (equivalently, λ is a solution of (12.42)).

Recall the *Cohn–Schur criterion* (Lemma 4.9): Both zeros of the quadratic $\alpha w^2 + \beta w + \gamma$, $\alpha \neq 0$, reside in the closed complex unit disc if and only if $|\alpha|^2 \geq |\gamma|^2$ and $(|\alpha|^2 - |\gamma|^2)^2 \geq |\alpha\bar{\beta} - \beta\bar{\gamma}|^2$. Similarly, it is possible to prove that both zeros of this quadratic reside in the *open* unit disc if and only if

$$|\alpha|^2 > |\gamma|^2 \qquad \text{and} \qquad (|\alpha|^2 - |\gamma|^2)^2 > |\alpha\bar{\beta} - \beta\bar{\gamma}|^2.$$

Letting $\alpha = 1$, $\beta = -\omega\mu$, $\gamma = \omega - 1$ and bearing in mind that $\mu \in \mathbb{R}$, these two conditions become $(\omega - 1)^2 < 1$ (which is the same as $\omega \in (0,2)$) and $\mu^2 < 1$. Therefore, provided $\rho(B) < 1$, it is true that $|\mu| < 1$ for all $\mu \in \sigma(B)$, therefore $|\lambda|^{1/2} < 1$ for all $\lambda \in \sigma(\mathcal{L}_\omega)$ and $\rho(\mathcal{L}_\omega) < 1$ for all $\omega \in (0,2)$. Likewise, if $\rho(\mathcal{L}_\omega) < 1$ then part (iii) of Theorem 12.8 implies that all the eigenvalues of B reside in the open unit disc. ■

The condition that all the zeros of B are real is satisfied in the important special case where B is symmetric. If it fails, the second statement of the theorem need not be true; see Exercise 12.10, where the reader can prove that $\rho(B) < 1$ and $\rho(\mathcal{L}_\omega) > 1$ for $\omega \in (1,2)$ for the matrix

$$A = \begin{bmatrix} 2 & 1 & 0 & \cdots & 0 \\ -1 & 2 & 1 & \ddots & \vdots \\ 0 & \ddots & \ddots & \ddots & 0 \\ \vdots & \ddots & -1 & 2 & 1 \\ 0 & \cdots & 0 & -1 & 2 \end{bmatrix},$$

provided that d is sufficiently large.

Within the conditions of Theorem 12.9 there is not much to choose between our three iterative procedures regarding convergence: either they all converge or they all fail on that score. The picture changes when we take the speed of convergence into account. Thus, the corollary to Theorem 12.8 affirms that Gauss–Seidel is asymptotically *twice as good* as Jacobi. Bearing in mind that Gauss–Seidel is but a special case of SOR, we thus expect to improve the rate of convergence further by choosing a superior value of ω.

Theorem 12.10 *Suppose that A possesses a compatible ordering vector, that $\sigma(B) \subset \mathbb{R}$ and that $\tilde{\mu} := \rho(B) < 1$. Then*

$$\rho(\mathcal{L}_{\omega_{\rm opt}}) < \rho(\mathcal{L}_\omega), \qquad \omega \in (0,2) \setminus \{\omega_{\rm opt}\},$$

where

$$\omega_{\rm opt} := \frac{2}{1+\sqrt{1-\tilde{\mu}^2}} = 1 + \left(\frac{\tilde{\mu}}{1+\sqrt{1-\tilde{\mu}^2}}\right)^2 \in (1,2). \tag{12.47}$$

Proof Although it might not be immediately obvious, the proof is but an elaboration of the detailed example from Section 12.2. Having already done all the hard work, we can allow ourselves to proceed at an accelerated pace.

Solving the quadratic (12.42) yields

$$\lambda = \tfrac{1}{4}\left[\omega\mu \pm \sqrt{(\omega\mu)^2 - 4(\omega-1)}\right]^2.$$

According to Theorem 12.8, both roots reside in $\sigma(\mathcal{L}_\omega)$.

Since μ is real, the term inside the square root is nonpositive when

$$\tilde{\omega} := \frac{2(1-\sqrt{1-\mu^2})}{\mu^2} \le \omega < 2.$$

In this case $\lambda \in \mathbb{C} \setminus \mathbb{R}$ and it is trivial to verify that $|\lambda| = \omega - 1$.

In the remaining portion of the range of ω both roots λ are positive and the larger one equals $\frac{1}{4}[f(\omega, |\mu|)]^2$, where

$$f(\omega, t) := \omega t + \sqrt{(\omega t)^2 - 4(\omega - 1)}, \qquad \omega \in (0, \tilde{\omega}], \quad t \in [0,1).$$

It is an easy matter to ascertain that for any fixed $\omega \in (0, \tilde{\omega}]$ the function $f(\omega, \cdot)$ increases strictly monotonically for $t \in [0,1)$. Likewise, $\tilde{\omega}$ increases strictly monotonically as a function of $\mu \in [0,1)$. Therefore the spectral radius of $\mathcal{L}_\omega$ in the range $\omega \in \left(0, 2(1-\sqrt{1-\tilde{\mu}^2})/\tilde{\mu}^2\right)$ is $\frac{1}{4}f(\omega, \tilde{\mu})$. Note that the endpoint of the interval is

$$2\frac{1-\sqrt{1-\tilde{\mu}^2}}{\tilde{\mu}^2} = \frac{2}{1+\sqrt{1-\tilde{\mu}^2}} = \omega_{\rm opt},$$

as given in (12.47). The function $f(\cdot, t)$ decreases strictly monotonically in $\omega \in (0, \omega_{\rm opt})$ for any fixed $t \in [0,1)$, thereby reaching its minimum at $\omega = \omega_{\rm opt}$.

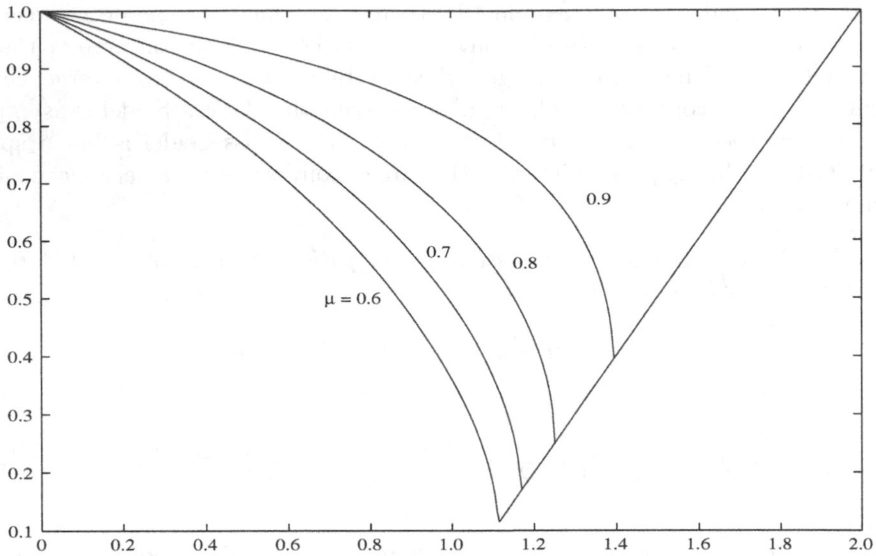

Figure 12.4 The graph of $\rho(\mathcal{L}_\omega)$ for different values of $\tilde{\mu}$.

As far as the interval $[\omega_{\rm opt}, 2)$ is concerned, the modulus of each λ corresponding to $\mu \in \sigma(B)$, $|\mu| = \tilde{\mu}$, equals $\omega - 1$ and is at least as large as the magnitude of any other eigenvalues of $\mathcal{L}_\omega$. We thus deduce that

$$\rho(\mathcal{L}_\omega) = \begin{cases} \frac{1}{4}\left[\omega\tilde{\mu} + \sqrt{(\omega\tilde{\mu})^2 - 4(\omega-1)}\right]^2, & \omega \in (0, \omega_{\rm opt}], \\ \omega - 1, & \omega \in [\omega_{\rm opt}, 2) \end{cases} \tag{12.48}$$

and, for all $\omega \in (0,2)$, $\omega \neq \omega_{\rm opt}$, it is true that

$$\rho(\mathcal{L}_\omega) > \rho(\mathcal{L}_{\omega_{\rm opt}}) = \left(\frac{\tilde{\mu}}{1 + \sqrt{1 - \tilde{\mu}^2}}\right)^2.$$

■

Figure 12.4 displays $\rho(\mathcal{L}_\omega)$ for different values of $\tilde{\mu}$ for matrices that are consistent with the conditions of the theorem. As apparent from (12.48), each curve is composed of two smooth portions, joining at $\omega_{\rm opt}$.

In practical computation the value of $\tilde{\mu}$ is frequently estimated rather than derived in an explicit form. An important observation from Fig. 12.4 is that it is always a sound policy to overestimate (rather than underestimate) $\omega_{\rm opt}$, since the curve has a larger slope to the left of the optimal value and so overestimation is punished less severely.

The figure can be also employed as an illustration of the method of proof. Thus, instead of visualizing each individual curve as corresponding to a different matrix, think of them as plots of $|\lambda|$, where $\lambda \in \sigma(\mathcal{L}_\omega)$, as a function of ω. The spectral radius for any given value of ω is provided by the top curve – and it can be observed in Fig. 12.4 that this top curve is associated with $\tilde{\mu}$.

Neither Theorem 12.8 nor Theorem 12.9 requires the knowledge of a compatible ordering vector – it is enough that such a vector exists. Unfortunately, it is not a trivial matter to verify directly from the definition whether a given matrix possesses a compatible ordering vector.

We have established in Lemma 12.6 that, provided A has an ordering vector, there exists a rearrangement of its rows and columns that possesses a compatible ordering vector. There is more to the lemma than meets the eye, since its proof is constructive and can be used as a numerical algorithm in a most straightforward manner. In other words, it is enough to find an ordering vector (provided that it exists) and the algorithm from the proof of Lemma 12.6 takes care of compatibility!

A $d \times d$ matrix with a digraph $\mathbb{G} = \{\mathbb{V}, \mathbb{E}\}$ is said to possess *property* A if there exists a partition $\mathbb{V} = \mathbb{S}_1 \cup \mathbb{S}_2$, where $\mathbb{S}_1 \cap \mathbb{S}_2 = \emptyset$, such that for every $(j, \ell) \in \mathbb{E}$ either $j \in \mathbb{S}_1$ and $\ell \in \mathbb{S}_2$ or $\ell \in \mathbb{S}_1$ and $j \in \mathbb{S}_2$.

◇ **Property A** As often in matters involving sparsity patterns, pictorial representation conveys more information than many a formal definition.

Let us consider, for example, a 6×6 matrix with a symmetric sparsity structure, as follows:

$$\begin{bmatrix} \times & \times & \circ & \times & \circ & \circ \\ \times & \times & \times & \circ & \times & \circ \\ \circ & \times & \times & \times & \circ & \times \\ \times & \circ & \times & \times & \circ & \circ \\ \circ & \times & \circ & \circ & \times & \times \\ \circ & \circ & \times & \circ & \times & \times \end{bmatrix}$$

We claim that $\mathbb{S}_1 = \{1, 3, 5\}$, $\mathbb{S}_2 = \{2, 4, 6\}$ is a partition consistent with property A. To confirm this, write the digraph $\mathbb{G}$ in the following fashion:

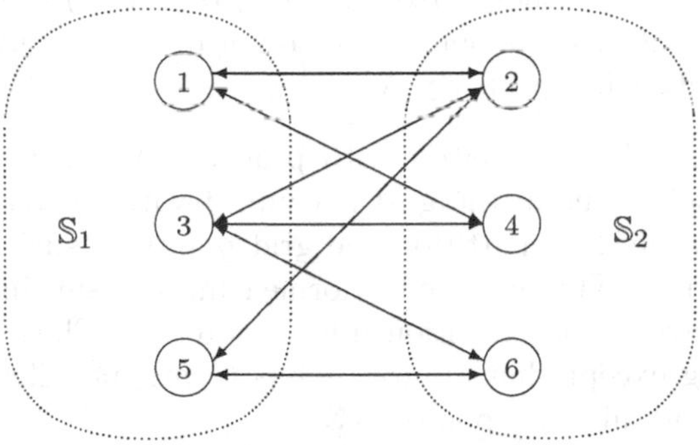

(In the interests of clarity, we have replaced each pair of arrows pointing in opposite directions by a single, 'two-sided' arrow.) Evidently, no edges join vertices in the same set, be it $\mathbb{S}_1$ or $\mathbb{S}_2$, and this is precisely the meaning of property A.

An alternative interpretation of property A comes to light when we rearrange the matrix in such a way that rows and columns corresponding to $\mathbb{S}_1$ precede those of $\mathbb{S}_2$. In our case, we permute rows and columns in the order $1,3,5,2,4,6$ and the resultant sparsity pattern is then

$$\left[\begin{array}{ccc|ccc} \times & \circ & \circ & \times & \times & \circ \\ \circ & \times & \circ & \times & \times & \times \\ \circ & \circ & \times & \times & \circ & \times \\ \hline \times & \times & \times & \times & \circ & \circ \\ \times & \times & \circ & \circ & \times & \circ \\ \circ & \times & \times & \circ & \circ & \times \end{array}\right].$$

In other words, the partitioned sparsity pattern has two diagonal blocks along the main diagonal. $\diamond$

The importance of property A is encapsulated in the following result.

Lemma 12.11 *A matrix possesses property* A *if and only if it has an ordering vector.*

Proof Suppose first that a $d \times d$ matrix has property A and set

$$j_\ell = \begin{cases} 1, & \ell \in \mathbb{S}_1, \\ 2, & \ell \in \mathbb{S}_2, \end{cases} \qquad \ell = 1, 2, \ldots, d.$$

For any $(\ell, m) \in \mathbb{E}$ it is true that ℓ and m belong to different sets, therefore $j_\ell - j_m = \pm 1$ and we deduce that $\boldsymbol{j}$ is an ordering vector.

To establish the proof in the opposite direction assume that the matrix has an ordering vector $\boldsymbol{j}$ and let

$$\mathbb{S}_1 := \{\ell \in \mathbb{V} : j_\ell \text{ is odd}\}, \qquad \mathbb{S}_2 := \{\ell \in \mathbb{V} : j_\ell \text{ is even}\}.$$

Clearly, $\mathbb{S}_1 \cup \mathbb{S}_2 = \mathbb{V}$ and $\mathbb{S}_1 \cap \mathbb{S}_2 = \emptyset$, therefore $\{\mathbb{S}_1, \mathbb{S}_2\}$ is indeed a partition of $\mathbb{V}$. For any $\ell, m \in \mathbb{V}$ such that $(\ell, m) \in \mathbb{E}$ it follows from the definition of an ordering vector that $j_\ell - j_m = \pm 1$. In other words, the integers j_ℓ and j_m are of different parity, hence it follows from our construction that ℓ and m belong to different partition sets. Consequently, the matrix has property A. $\blacksquare$

An important example of a matrix with property A follows from the five-point equations (8.16). Each point in the grid is coupled with its vertical and horizontal neighbours, hence we need to partition the grid points in such a way that $\mathbb{S}_1$ and $\mathbb{S}_2$ separate neighbours. This can be performed most easily in terms of *red–black* ordering, which we have already mentioned in Chapter 11. Thus, we traverse the grid as in natural ordering except that all grid points (ℓ, m) such that $\ell + m$ is odd, say, are consigned to $\mathbb{S}_1$ and all other points to $\mathbb{S}_2$.

An example, corresponding to a five-point formula in a 4×4 square, is presented in (11.4). Of course, the real purpose of the exercise is not simply to verify property A or, equivalently, to prove that an ordering vector exists. Rather, our goal is to identify a permutation that yields a compatible ordering vector. As we have already mentioned, this can be performed by the method of proof of Lemma 12.6. However, in the present circumstances we can single out such a vector directly *for the natural ordering*. For example, as far as (11.4) is concerned we associate with every grid point (which, of course, corresponds to an equation and a variable in the linear system) an integer as follows:

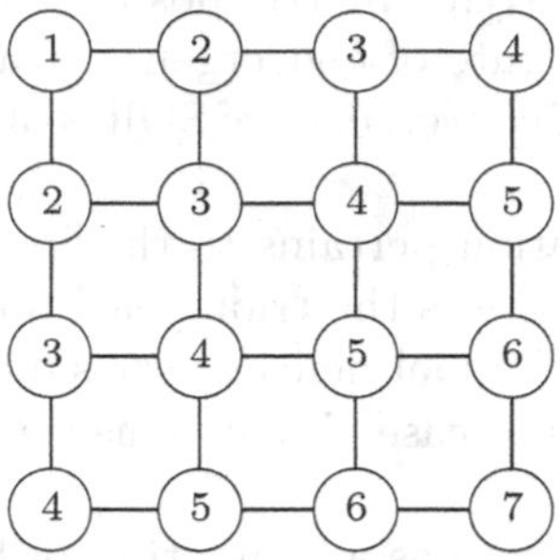

As can be easily verified, natural ordering results in a compatible ordering vector. All this can be easily generalized to rectangular grids of arbitrary size (see Exercise 12.11).

The exploitation of red–black ordering in the search for property A is not restricted to rectangular grids. Thus, consider the L-shaped grid

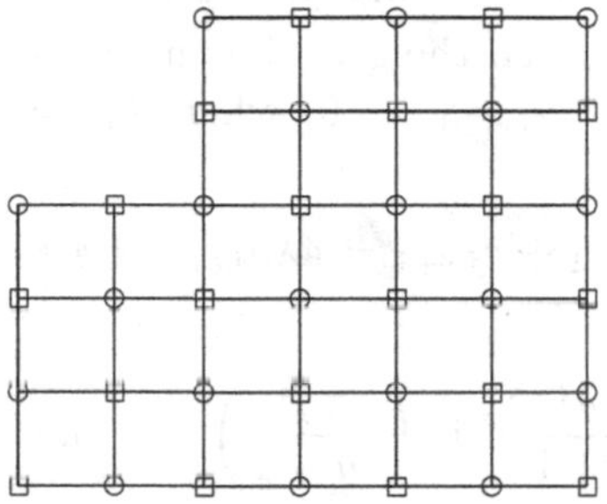

(12.49)

where '□' and '○' denote vertices in $\mathbb{S}_1$ and $\mathbb{S}_2$, respectively. Note that (12.49) serves a dual purpose: it is both the depiction of the computational grid and the graph of a matrix. It is quite clear that here this underlying matrix has property A. The task of finding explicitly a compatible ordering vector is relegated to Exercise 12.12.

12.4 The Poisson equation

Figure 12.5 displays the error attained by four different iterative methods, when applied to the Poisson equation (8.33) on a 16×16 grid. The first row depicts the line relaxation method (a variant of the incomplete LU factorization (12.11) – read on for details), the second corresponds to the Jacobi iteration (12.20), next comes the Gauss–Seidel method (12.21) and, finally, the bottom row displays the error in the successive over-relaxation (SOR) method (12.22) with optimal choice of the parameter ω. Each column corresponds to a different number of iterations, specifically 50, 100 and 150, except that there is little point in displaying the error for ≥ 100 iterations for SOR since, remarkably, the error after 50 iterations is already close to machine accuracy![4]

Our first observation is that, evidently, all four methods converge. This is hardly a surprise in the case of Jacobi, Gauss–Seidel and SOR since we have already noted in the last section that the underlying matrix possesses property A. The latter feature explains also the very different rate of convergence: Gauss–Seidel converges twice as fast as Jacobi while the speed of convergence of SOR is of a different order of magnitude altogether.

Another interesting observation pertains to the line relaxation method, a version of ILU from Section 12.1, where $\tilde{A}$ is the tridiagonal portion of A. Fig. 12.5 suggests that line relaxation and Gauss–Seidel deliver very similar performances and we will prove later that this is indeed the case. We commence our discussion, however, with classical iterative methods.

Because the underlying matrix has a compatible ordering vector, as noted in Section 12.3, we need to determine $\tilde{\mu} = \rho(B)$; and, by virtue of Theorems 12.8 and 12.10, $\tilde{\mu}$ determines completely both $\omega_{\rm opt}$ and the rates of convergence of Gauss–Seidel and SOR.

Let $V = (v_{j,\ell})_{j,\ell=1}^{m}$ be an eigenvector of the matrix B from (12.20) and let λ be the corresponding eigenvalue. We assume that the matrix A originates in the five-point formula (7.16) in a $m \times m$ square. Formally, V is a matrix; to obtain a genuine vector $\boldsymbol{v} \in \mathbb{R}^{m^2}$ we would need to stretch the grid, but in fact this will not be necessary.

Setting $v_{0,\ell}$, $v_{m+1,\ell}$, $v_{k,0}$, $v_{k,m+1} := 0$, where $k, \ell = 1, 2, \ldots, m$, we can express $A\boldsymbol{v} = \lambda \boldsymbol{v}$ in the form

$$v_{j-1,\ell} + v_{j+1,\ell} + v_{j,\ell-1} + v_{j,\ell+1} = 4\lambda v_{j,\ell}, \qquad j, \ell = 1, 2, \ldots, m. \tag{12.50}$$

Our claim is that

$$v_{j,\ell} = \sin\left(\frac{\pi p j}{m+1}\right) \sin\left(\frac{\pi q \ell}{m+1}\right), \qquad j, \ell = 1, 2, \ldots, m,$$

where p and q are arbitrary integers in $\{1, 2, \ldots, m\}$. If this is true then

$$\begin{aligned} v_{j-1,\ell} + v_{j+1,\ell} &= \left\{\sin\left[\frac{\pi p(j-1)}{m+1}\right] + \sin\left[\frac{\pi p(j+1)}{m+1}\right]\right\} \sin\left(\frac{\pi q \ell}{m+1}\right) \\ &= 2 v_{j,\ell} \cos\left(\frac{\pi p}{m+1}\right), \\ v_{j,\ell-1} + v_{j,\ell+1} &= \sin\left(\frac{\pi p j}{m+1}\right) \left\{\sin\left[\frac{\pi q(\ell-1)}{m+1}\right] + \sin\left[\frac{\pi q(\ell+1)}{m+1}\right]\right\} \\ &= 2 v_{k,\ell} \cos\left(\frac{\pi q}{m+1}\right), \end{aligned}$$

and substitution into (12.50) confirms that

$$\lambda = \lambda_{p,q} = \frac{1}{2}\left[\cos\left(\frac{\pi p}{m+1}\right) + \cos\left(\frac{\pi q}{m+1}\right)\right]$$

[4] To avoid any misunderstanding, at this point we emphasize that by 'error' we mean departure from the solution of the corresponding five-point equations (8.16) *not* departure from the exact solution of the Poisson equation.

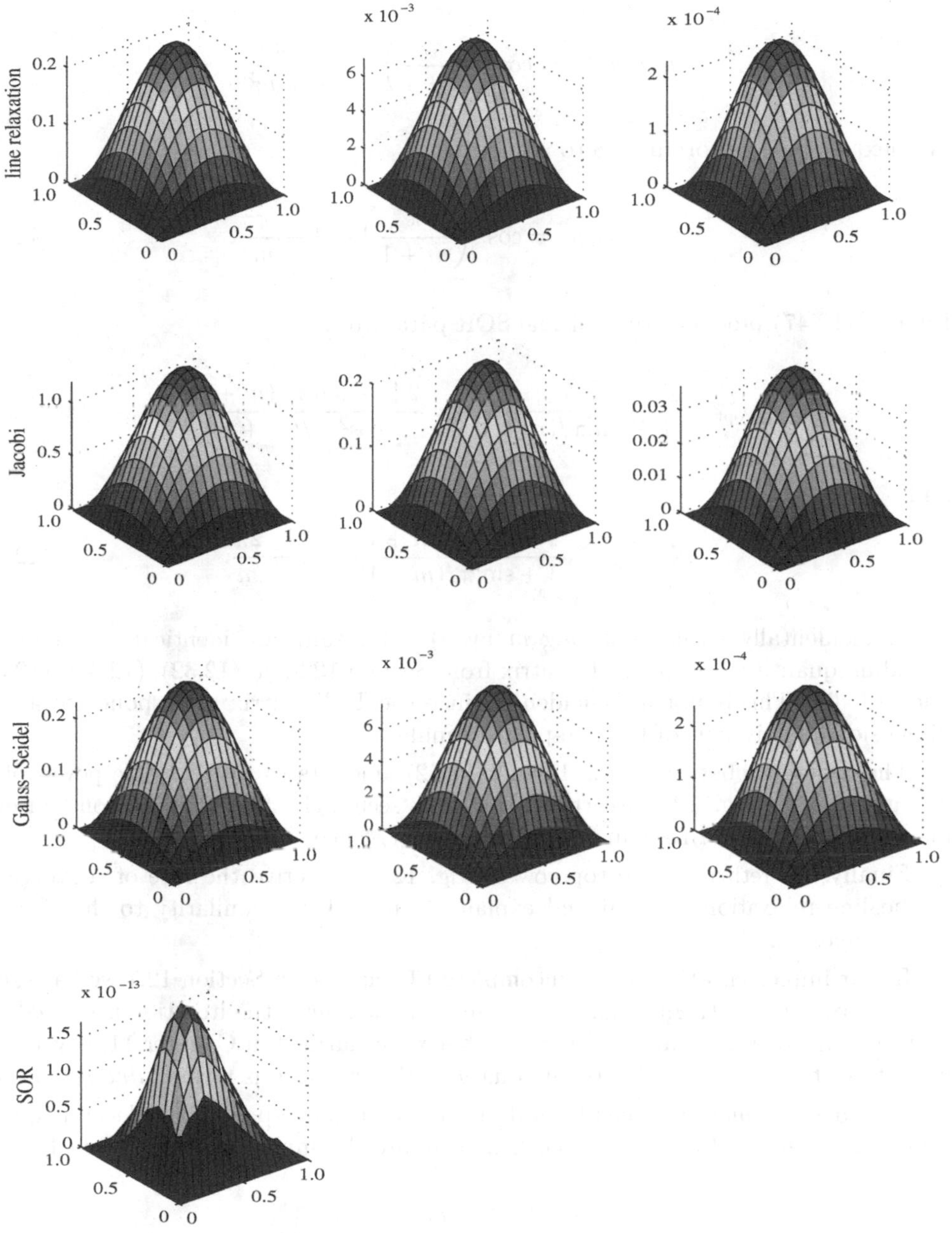

Figure 12.5 The error in the line relaxation, Jacobi, Gauss–Seidel and SOR (with ω_{opt}) methods for the Poisson equation (8.33) for $m = 16$ after 100, 200 and 300 iterations. Note the differences in scale.

is an eigenvalue of A for every $p, q = 1, 2, \ldots, m$. This procedure yields all m^2 eigenvalues of A and we therefore deduce that

$$\tilde{\mu} = \rho(B) = \cos\left(\frac{\pi}{m+1}\right) \approx 1 - \frac{\pi^2}{2m^2}.$$

We next employ Theorem 12.8 to argue that

$$\rho(\mathcal{L}_1) = \tilde{\mu}^2 = \cos^2\left(\frac{\pi}{m+1}\right) \approx 1 - \frac{\pi^2}{m^2}. \tag{12.51}$$

Finally, (12.47) produces the optimal SOR parameter,

$$\omega_{\text{opt}} = \frac{2}{1 + \sin[\pi/(m+1)]} = \frac{2\{1 - \sin[\pi/(m+1)]\}}{\cos^2[\pi/(m+1)]},$$

and

$$\rho(\mathcal{L}_{\omega_{\text{opt}}}) = \frac{1 - \sin[\pi/(m+1)]}{1 + \sin[\pi/(m+1)]} \approx 1 - \frac{2\pi}{m}. \tag{12.52}$$

Note, incidentally, that (replacing m by d) our results are identical to the corresponding quantites for the TST matrix from Section 12.2; cf. (12.32), (12.35), (12.39) and (12.40). This is not a coincidence, since the TST matrix corresponds to a one-dimensional equivalent of the five-point formula.

The difference between (12.51) and (12.52) amounts to just a single power of m but glancing at Fig. 12.5 ascertains that this seemingly minor distinction causes a most striking improvement in the speed of convergence.

Finally, we return to the top row of Fig. 12.5, to derive the rate of convergence of the line relaxation method and explain its remarkable similarity to that for the Gauss–Seidel method.

In our implementation of the incomplete LU method in Section 12.1 we have split the matrix A into a tridiagonal portion and a remainder – the iteration is carried out on the tridiagonal part and, for reasons that were clarified in Chapter 11, is very low in cost. In the context of five-point equations this splitting is termed *line relaxation.*

Provided the matrix A has been derived from the five-point formula in a square, we can write it in a block form that has been already implied in (11.3), namely

$$A = \begin{bmatrix} C & I & O & \cdots & O \\ I & C & I & \ddots & \vdots \\ O & \ddots & \ddots & \ddots & O \\ \vdots & \ddots & I & C & I \\ O & \cdots & O & I & C \end{bmatrix}, \tag{12.53}$$

where I and O are the $m \times m$ identity and zero matrices respectively, and

$$C = \begin{bmatrix} -4 & 1 & 0 & \cdots & 0 \\ 1 & -4 & 1 & \ddots & \vdots \\ 0 & \ddots & \ddots & \ddots & 0 \\ \vdots & \ddots & 1 & -4 & 1 \\ 0 & \ddots & 0 & 1 & -4 \end{bmatrix}.$$

In other words, A is block-TST and each block is itself a TST matrix.

Line relaxation (12.11) splits A into a tridiagonal part $\tilde{A}$ and a remainder $-E$. The matrix C being itself tridiagonal, we deduce that $\tilde{A}$ is block-diagonal, with C's along the main diagonal, while E consists of the off-diagonal blocks. Let λ and $\boldsymbol{v}$ be an eigenvalue and a corresponding eigenvector of the iteration matrix $\tilde{A}^{-1}E$. Therefore

$$E\boldsymbol{v} = \lambda \tilde{A} \boldsymbol{v}$$

and, rendering as before the vector $\boldsymbol{v} \in \mathbb{R}^{m^2}$ as an $m \times m$ matrix V, we obtain

$$v_{j,\ell-1} + v_{j,\ell+1} + \lambda(v_{j-1,\ell} - 4v_{j,\ell} + v_{j+1,\ell}) = 0, \qquad j,\ell = 1,2,\ldots,d. \tag{12.54}$$

As before, we have assumed zero 'boundary values': $v_{j,0}, v_{j,m+1}, v_{0,\ell}, v_{m+1,\ell} = 0$, $j,\ell = 1,2,\ldots,d$.

Our claim (which, with the benefit of experience, was hardly surprising) is that

$$v_{j,\ell} = \sin\left(\frac{\pi p j}{m+1}\right) \sin\left(\frac{\pi q \ell}{m+1}\right), \qquad j,\ell = 1,2,\ldots,m,$$

for some $p, q \in \{1,2,\ldots,m\}$. Since

$$\sin\left[\frac{\pi p(j-1)}{m+1}\right] - 4\sin\left(\frac{\pi p j}{m+1}\right) + \sin\left[\frac{\pi p(j+1)}{m+1}\right] = 2\left[\cos\left(\frac{\pi p}{m+1}\right) - 2\right] \sin\left(\frac{\pi p j}{m+1}\right),$$

$$\sin\left[\frac{\pi q(\ell-1)}{m+1}\right] + \sin\left[\frac{\pi q(\ell+1)}{m+1}\right] = 2\cos\left(\frac{\pi q}{m+1}\right) \sin\left(\frac{\pi q \ell}{m+1}\right),$$

substitution in (12.54) results in

$$\lambda = \lambda_{p,q} = \frac{-\cos[\pi q/(m+1)]}{2 - \cos[\pi p/(m+1)]}.$$

Letting p, q range across $\{1,2,\ldots,m\}$, we recover all m^2 eigenvalues and, in particular, determine the spectral radius of the iteration matrix:

$$\rho(\tilde{A}^{-1}E) = \frac{\cos[\pi/(m+1)]}{2 - \cos[\pi/(m+1)]} \approx 1 - \frac{\pi^2}{m^2}. \tag{12.55}$$

Comparison of (12.55) with (12.51) verifies our observation from Fig. 12.5 that line relaxation and Gauss–Seidel have very similar rates of convergence. As a matter of fact, it is easy to prove that Gauss–Seidel is marginally better, since

$$\cos^2\varphi < \frac{\cos\varphi}{2 - \cos\varphi} < \cos^2\varphi + \frac{\varphi^2}{2}, \qquad 0 < \varphi < \frac{\pi}{2}.$$

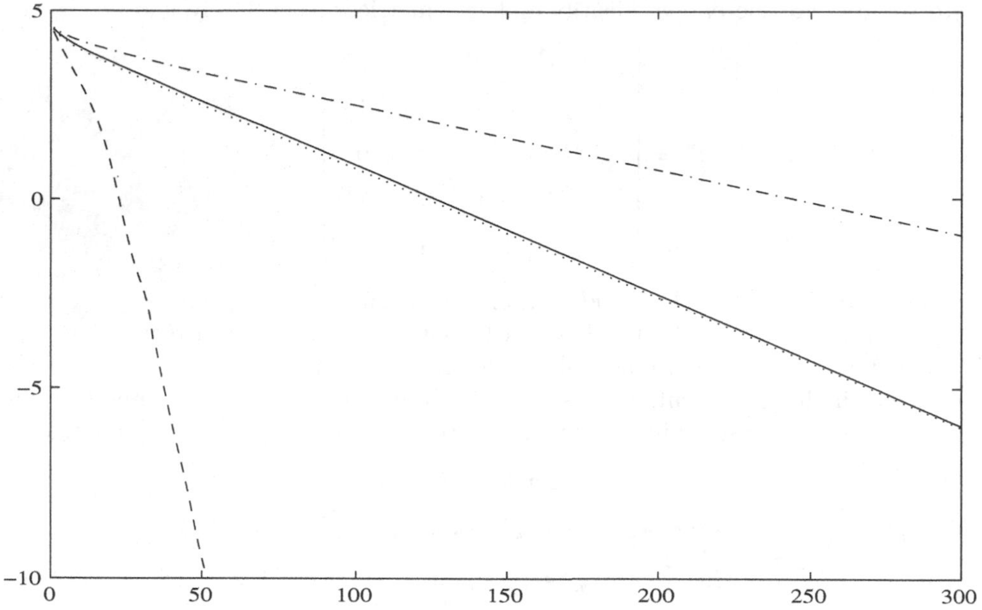

Figure 12.6 The logarithm of the error in the Euclidean norm, $\log \|\boldsymbol{x}^{[k]} - \hat{\boldsymbol{x}}\|$, for Jacobi (broken-and-dotted line), Gauss–Seidel (dotted line), SOR (broken line) and line relaxation (solid line) for $m = 16$ in the first 300 iterations. The starting vector is $\boldsymbol{x}^{[0]} = \mathbf{0}$.

Fig 12.6 displays (on a logarithmic scale) the decay of the Euclidean norm of the error after a given number of iterations – thus, the information in each three-dimensional surface from Fig. 12.5 is reduced to a single number. Having analysed and understood the four methods in some detail, we can again observe and compare their features. There is however, an interesting new detail in Fig. 12.6. In principle, the size of the spectral radius determines the speed of convergence only in an asymptotic sense and there is nothing in our analysis to tell how soon – or how late – the asymptotic regime occurs. However, we can observe in Fig. 12.6 (and, for that matter, though for a different equation, in Fig. 12.1) that the onset of asymptotic behaviour is pretty rapid for a general starting vector $\boldsymbol{x}^{[0]}$.

Comments and bibliography

Numerical mathematics is an old art and its history is replete with the names of intellectual giants – Newton, Euler, Lagrange, Legendre, Gauss, Jacobi... It is fair, however, to observe that the most significant milestone in its long journey has been the invention of the electronic computer. Numerical linear algebra, including iterative methods for sparse linear systems, is a case in point. A major research effort in the 1950s – the dawn of the computer era – led to an enhanced understanding of classical iterative methods and forms the cornerstone of our exposition.

A large number of textbooks and monographs deal with the theme of this chapter and we single out the books of Axelsson (1994), Varga (1962) and Young (1971); see also Hageman & Young (1981). Readers who are at home with the terminology of functional analysis will also enjoy the concise monograph of Nevanlinna (1993).

The theory of SOR can be developed significantly beyond the material of Section 12.3. This involves much beautiful and intricate mathematics but is, arguably, of mainly theoretical interest, for reasons that will become clearer in Chapter 13 – in a nutshell, we describe there how to accelerate Gauss–Seidel iteration in a manner that is much more powerful than SOR.

Classical iterative methods are neither the only nor, indeed, the best means to solve sparse linear systems by iteration, and we wish to single out three other approaches.

Recall the line relaxation method (12.11). We split the matrix A, which originated from the five-point discretization of the Poisson equation in a square, into $\tilde{A} - E$, where $\tilde{A}$ is its tridiagonal part, subsequently iterating $\tilde{A}\boldsymbol{x}^{[k+1]} = E\boldsymbol{x}^{[k]} + \boldsymbol{b}$. Suppose, however, that instead of ordering the grid by columns (the natural ordering), we do so by rows and apply line relaxation to the new system. This is just as logical – or illogical – as employing a column-wise ordering but leads to a different iterative scheme (note that the matrix A stays intact but both $\boldsymbol{x}$ and $\boldsymbol{b}$ are permuted as a consequence of the row-wise rearrangement). Which variant should we adopt, ordering by column or by row? A natural approach is to alternate. In other words, we write $A = A_x + A_y$, where A_x and A_y originate in central differencing in the x- and y- directions respectively. Column-wise line relaxation (12.11) reads

$$(A_y - 2I)\boldsymbol{x}^{[k+1]} = -(A_x + 2I)\boldsymbol{x}^{[k]} + \boldsymbol{b}, \qquad k = 0, 1, \ldots,$$

while a row-wise rearrangement results in

$$(A_x - 2I)\boldsymbol{x}^{[k+1]} = -(A_y + 2I)\boldsymbol{x}^{[k]} + \boldsymbol{b}, \qquad k = 0, 1, \ldots$$

For greater generality, we choose parameters $\alpha_0, \alpha_1, \ldots$ and iterate

$$\left.\begin{aligned}(A_x - \alpha_{2k}I)\boldsymbol{x}^{[2k+1]} &= -(A_y + \alpha_{2k})\boldsymbol{x}^{[2k]} + \boldsymbol{b},\\ (A_y - \alpha_{2k+1}I)\boldsymbol{x}^{[2k+2]} &= -(A_x + \alpha_{2k+1})\boldsymbol{x}^{[2k+1]} + \boldsymbol{b},\end{aligned}\right\} \qquad k = 0, 1, \ldots$$

This is the *alternate directions implicit* (ADI) method (Wachspress, 1966). As often in numerical analysis, the devil is in the parameter. However, it is known how to choose $\{\alpha_k\}$ so as to accelerate ADI a great deal. The outcome, at least in certain cases, e.g. the Poisson equation in a square, is an iterative method that clearly outperforms SOR. This, however, falls outside the scope of the present volume.

Another example of iterative nonstationary schemes are the *Krylov subspace methods,* in particular the method of conjugate gradients. They are of such fundamental importance and wide-ranging applicability that they deserve a chapter all of their own in our book; see Chapter 14.

Axelsson, O. (1994), *Iterative Solution Methods,* Cambridge University Press, Cambridge.

Hageman, L.A. and Young, D.M. (1981), *Applied Iterative Methods,* Academic Press, New York.

Nevanlinna, O. (1993), *Convergence of Iterations for Linear Equations,* Birkhäuser, Basel.

Varga, R.S. (1962), *Matrix Iterative Analysis,* Prentice–Hall, Englewood Cliffs, NJ.

Wachspress, E.L. (1966), *Iterative Solution of Elliptic Systems, and Applications to the Neutron Diffusion Equations of Reactor Physics,* Prentice–Hall, Englewood Cliffs, NJ.

Young, D.M. (1971), *Iterative Solution of Large Linear Systems,* Academic Press, New York.

Exercises

12.1 Let $\lambda \in \mathbb{C}$ be such that $|\lambda| < 1$ and define

$$J = \begin{bmatrix} \lambda & 1 & 0 & \cdots & 0 \\ 0 & \lambda & \ddots & \ddots & \vdots \\ \vdots & \ddots & \ddots & 1 & 0 \\ \vdots & & \ddots & \lambda & 1 \\ 0 & \cdots & \cdots & 0 & \lambda \end{bmatrix}.$$

Prove that $\lim_{k\to\infty} J^k = O$.

12.2 Suppose that the $d \times d$ matrix H has a full set of eigenvectors, i.e., that there exist d linearly independent vectors $\boldsymbol{w}_\ell \in \mathbb{C}^d$ and numbers $\lambda_1, \lambda_2, \ldots, \lambda_d \in \mathbb{C}$ such that $H\boldsymbol{w}_\ell = \lambda_\ell \boldsymbol{w}_\ell$, $\ell = 1, 2, \ldots, d$. We consider the iterative scheme (12.3). Let $\boldsymbol{r}^{[k]} := (I - H)\boldsymbol{x}^{[k]} - \boldsymbol{v}$ be the residual in the kth iteration, and suppose that

$$\boldsymbol{r}^{[0]} = \sum_{\ell=1}^{d} \alpha_\ell \boldsymbol{w}_\ell$$

(such $\alpha_1, \alpha_2, \ldots, \alpha_d$ always exist – why?). Prove that

$$\boldsymbol{r}^{[k]} = \sum_{\ell=1}^{d} \alpha_\ell \lambda_\ell^k \boldsymbol{w}_\ell, \qquad k = 0, 1, \ldots$$

Outline an alternative proof of Lemma 12.1 using this representation.

12.3 Prove that the Gauss–Seidel iteration converges whenever the matrix A is symmetric and positive definite.

12.4 Show that the SOR method is a regular splitting (12.12) with

$$P = \omega^{-1}D - L_0, \qquad N = (\omega^{-1} - 1)D + U_0.$$

12.5 Let A be a symmetric tridiagonal positive definite matrix. Prove that the SOR method converges for this matrix and for $0 < \omega < 2$.

12.6 Let A be a TST matrix such that $a_{1,1} = \alpha$ and $a_{1,2} = \beta$. Show that the Jacobi iteration converges if $2|\beta| < |\alpha|$. Moreover, prove that if convergence is required for *all* $d \geq 1$ then this inequality is necessary as well as sufficient.

12.7 Demonstrate that

$$\sum_{\ell=1}^{d} \sin^2\left(\frac{\pi j \ell}{d+1}\right) = \tfrac{1}{2}(d+1), \qquad j = 1, 2, \ldots, d,$$

thereby verifying (12.29).

12.8 Let

$$A = \begin{bmatrix} \alpha_1 & \beta_1 & 0 & \cdots & 0 \\ \gamma_1 & \alpha_2 & \beta_2 & \ddots & \vdots \\ 0 & \ddots & \ddots & \ddots & 0 \\ \vdots & \ddots & \gamma_{d-2} & \alpha_{d-1} & \beta_{d-1} \\ 0 & \cdots & 0 & \gamma_{d-1} & \alpha_d \end{bmatrix},$$

where $\beta_\ell, \gamma_\ell \neq 0$, $\ell = 1, 2, \ldots, d-1$. Prove that $\boldsymbol{j}$, where $j_\ell = \ell$, $\ell = 1, 2, \ldots, d$, is a compatible ordering vector of A.

12.9 Find an ordering vector for a matrix with the following sparsity pattern:

$$\begin{bmatrix} \times & \times & \circ & \times & \circ & \circ & \circ & \circ & \circ & \times \\ \times & \times & \times & \circ & \circ & \circ & \circ & \circ & \times & \circ \\ \circ & \times & \times & \times & \circ & \times & \circ & \circ & \circ & \circ \\ \times & \circ & \times & \times & \times & \circ & \circ & \circ & \circ & \circ \\ \circ & \circ & \circ & \times & \times & \times & \circ & \times & \circ & \circ \\ \circ & \circ & \times & \times & \times & \times & \times & \circ & \circ & \circ \\ \circ & \circ & \circ & \circ & \circ & \times & \times & \times & \circ & \times \\ \circ & \circ & \circ & \circ & \times & \circ & \times & \times & \times & \circ \\ \circ & \times & \circ & \circ & \circ & \circ & \circ & \times & \times & \times \\ \times & \circ & \circ & \circ & \circ & \circ & \times & \circ & \times & \times \end{bmatrix}.$$

12.10* We consider the $d \times d$ tridiagonal Toeplitz matrix

$$A = \begin{bmatrix} 2 & 1 & 0 & \cdots & 0 \\ -1 & 2 & 1 & \ddots & \vdots \\ 0 & \ddots & \ddots & \ddots & 0 \\ \vdots & \ddots & -1 & 2 & 1 \\ 0 & \cdots & 0 & -1 & 2 \end{bmatrix}.$$

a Prove that A possesses a compatible ordering vector.

b Find explicitly all the eigenvalues of the Jacobi iteration matrix B. (*Hint: Solve explicitly the difference equation obeyed by the components of an eigenvector of B.*) Conclude that this iterative scheme diverges.

c Using Theorem 12.8, or otherwise, show that $\rho(\mathcal{L}_\omega) > 1$ and that the SOR iteration diverges for all choices of $\omega \in (0, 2)$.

12.11 Let A be an $m^2 \times m^2$ matrix that originates in the implementation of the five-point formula in a square $m \times m$ grid. For every grid point (r, s) we let

$$j_{(r,s)} := m + s - r, \qquad r, s, = 1, 2, \ldots, m,$$

and we construct a vector $\boldsymbol{j}$ by assembling the components in the same order as that used in the matrix A. Prove that $\boldsymbol{j}$ is an ordering vector of A and identify a permutation for which $\boldsymbol{j}$ is a compatible ordering vector.

12.12 Find a compatible ordering vector for the L-shaped grid (12.49).

13

Multigrid techniques

13.1 In lieu of a justification . . .

How good is the Gauss–Seidel iteration (12.21) at solving the five-point equations on an $m \times m$ grid? On the face of it, posing this question just after we have completed a whole chapter devoted to iterative methods is neither necessary nor appropriate. According to (12.51), the spectral radius of the iteration matrix is $\cos^2[\pi/(m+1)] \approx 1 - \pi^2 m^{-2}$ and inspection of the third row of Fig. 12.5 will convince us that this presents a fair estimate of the behaviour of the scheme. Yet, by its very nature, the spectral radius displays the *asymptotic* attenuation rate of the error and it is entirely legitimate to query how well (or badly) Gauss–Seidel performs before the onset of its asymptotic regime.

Figure 13.1 displays the logarithm of the Euclidean norm of the residual for $m = 10, 20, 40, 80$; we remind the reader that, given the equation

$$Ax = b \tag{13.1}$$

and a sequence of iterations $\{x^{[k]}\}_{i=0}^{\infty}$, the residual is defined as $r^{[k]} = Ax^{[k]} - b$, $k \geq 0$.[1] The emerging picture is startling: the norm drops dramatically in the first few iterations! Only after a while does the rate of attenuation approach the linear curve predicted by the spectral radius of $\mathcal{L}_1$. Moreover, this phenomenon – unlike the asymptotic rate of decay of $\ln \|r^{[k]}\|$ – appears to be fairly independent of the magnitude of m.

A similar lesson can be drawn from Fig. 13.2, where we have displayed in detail the residuals for the first six even numbers of iterations for $m = 20$. Evidently, a great deal of the error disappears very fast indeed, while after about ten iterations nothing much changes and each further iteration removes roughly $\cos^2(\pi/21) \approx 0.9778$ of the remaining residual.

The explanation of this phenomenon is quite interesting, as far as the understanding of Gauss–Seidel is concerned. More importantly, it provides a clue about how to accelerate iterative schemes for linear algebraic equations that originate in finite difference and finite element discretizations.

We hasten to confess that limitations of space and of the degree of mathematical sophistication that we allow ourselves in this book preclude us from providing a compre-

[1]There exists an intimate connection between $r^{[k]}$ and the error $\varepsilon^{[k]} = x^{[k]} - \hat{x}$, where $\hat{x}$ is the solution of (13.1) – see Exercise 13.1.

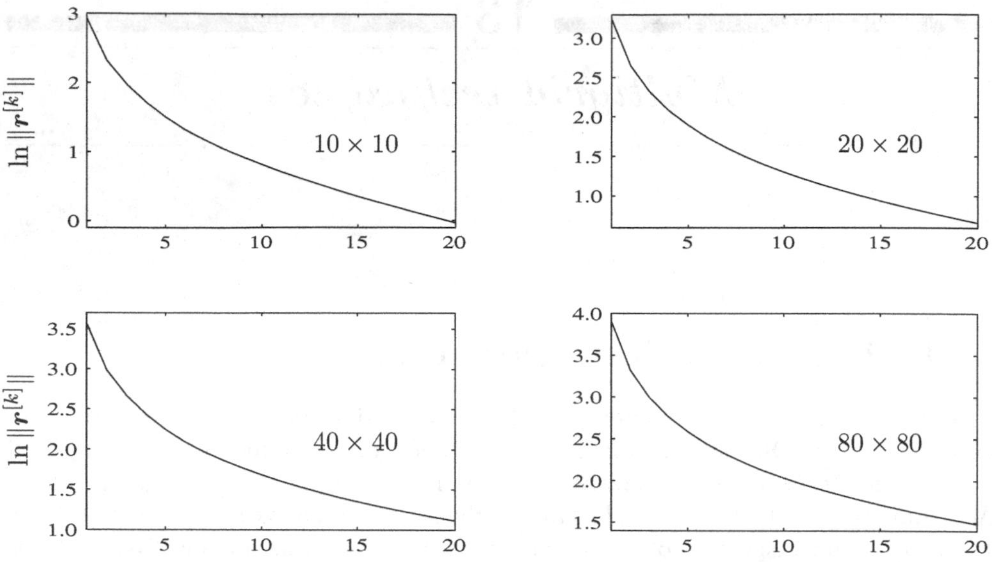

Figure 13.1 The logarithm of the norm of the residual in the first 20 Gauss–Seidel iterations for the five-point discretization of the Poisson equation (8.33).

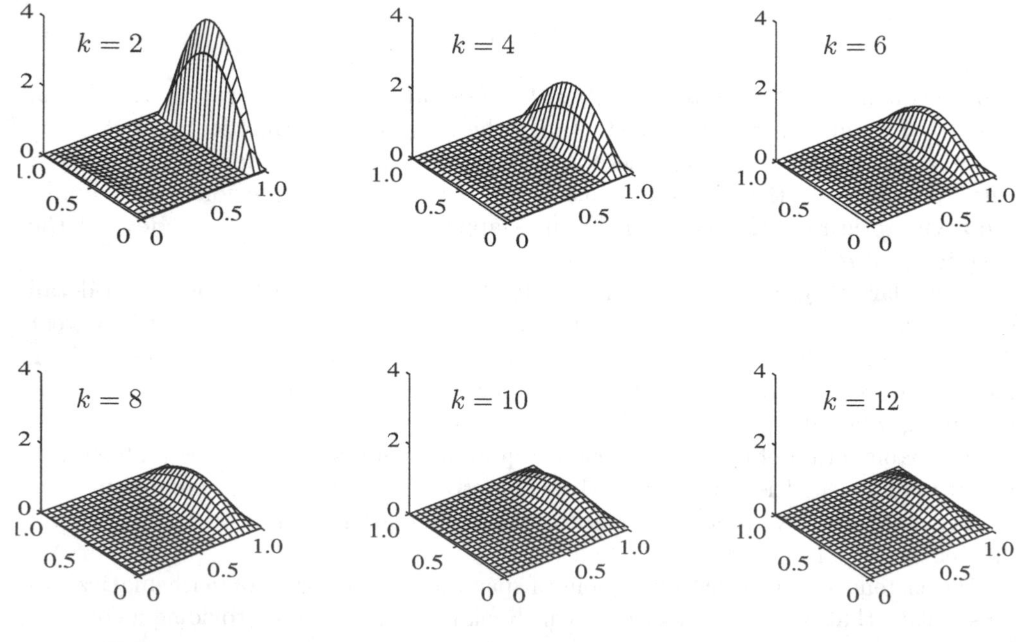

Figure 13.2 The residual after various small numbers k of Gauss–Seidel iterations for the five-point discretization of the Poisson equation (8.33) with $m = 20$.

hensive explanation of the aforementioned phenomenon. Instead, we plan to indulge in a great deal of mathematical hand-waving, our excuse being that it conveys the spirit, if not the letter, of a complete analysis and sets us on the right path to exploit the phenomenon in presenting superior iterative schemes.

Let us subtract the exact five-point equations,

$$u_{j-1,\ell} + u_{j,\ell-1} + u_{j+1,\ell} + u_{j,\ell+1} - 4u_{j,\ell} = (\Delta x)^2 f_{j,\ell}, \qquad j,\ell = 1,2,\ldots,m,$$

from the Gauss–Seidel scheme

$$u_{j-1,\ell}^{[k+1]} + u_{j,\ell-1}^{[k+1]} + u_{j+1,\ell}^{[k]} + u_{j,\ell+1}^{[k]} - 4u_{j,\ell}^{[k+1]} = (\Delta x)^2 f_{j,\ell}, \qquad j,\ell = 1,2,\ldots,m.$$

The outcome is

$$\varepsilon_{j-1,\ell}^{[k+1]} + \varepsilon_{j,\ell-1}^{[k+1]} + \varepsilon_{j+1,\ell}^{[k]} + \varepsilon_{j,\ell+1}^{[k]} - 4\varepsilon_{j,\ell}^{[k+1]} = 0, \qquad j,\ell = 1,2,\ldots,m, \tag{13.2}$$

where $\varepsilon_{j,\ell}^{[k]} := u_{j,\ell}^{[k]} - u_{j,\ell}$ is the error after k iterations at the (j,ℓ)th grid point. Since we assume Dirichlet boundary conditions, $u_{j,\ell}$ and $u_{j,\ell}^{[k]}$ are identical at all boundary grid points, therefore $\varepsilon_{j,\ell}^{[k]} = 0$ there.

Let

$$p^{[k]}(\theta,\psi) = \sum_{j=1}^{m}\sum_{\ell=1}^{m} \varepsilon_{j,\ell}^{[k]} \mathrm{e}^{\mathrm{i}(j\theta+\ell\psi)}, \qquad 0 \le \theta,\psi \le 2\pi,$$

be a *bivariate Fourier transform* of the sequence $\{\varepsilon_{j,\ell}^{[k]}\}_{j,\ell=1}^{m}$. We have already considered Fourier transforms in a more formal setting in Section 12.3 and subsequently we will employ them (to entirely different ends) in Chapters 15–17. In the present section our treatment of Fourier transfroms is therefore perfunctory and we will hint at proofs rather than providing any degree of detail.

We measure the magnitude of $p^{[k]}$ in the Euclidean norm[2]

$$|||g||| = \left[\frac{1}{4\pi^2}\int_{-\pi}^{\pi}\int_{-\pi}^{\pi} |g(\theta,\psi)|^2 \,\mathrm{d}\theta\,\mathrm{d}\psi\right]^{1/2}.$$

Therefore

$$\begin{aligned}
|||p^{[k]}|||^2 &= \frac{1}{4\pi^2}\int_{-\pi}^{\pi}\int_{-\pi}^{\pi} \left|\sum_{j=1}^{m}\sum_{\ell=1}^{m} \varepsilon_{j,\ell}^{[k]} \mathrm{e}^{\mathrm{i}(j\theta+\ell\psi)}\right|^2 \mathrm{d}\theta\,\mathrm{d}\psi \\
&= \frac{1}{4\pi^2}\sum_{j_1=1}^{m}\sum_{j_2=1}^{m}\sum_{\ell_1=1}^{m}\sum_{\ell_2=1}^{m} \varepsilon_{j_1,\ell_1}^{[k]}\bar{\varepsilon}_{j_2,\ell_2}^{[k]} \int_{-\pi}^{\pi} \mathrm{e}^{\mathrm{i}(j_1-j_2)\theta}\,\mathrm{d}\theta \int_{-\pi}^{\pi} \mathrm{e}^{\mathrm{i}(\ell_1-\ell_2)\psi}\,\mathrm{d}\psi \\
&= \sum_{j=1}^{m}\sum_{\ell=1}^{m} |\varepsilon_{j,\ell}^{[k]}|^2 = \|\varepsilon^{[k]}\|^2,
\end{aligned}$$

[2]Here for convenience we use triple verticals to indicate the Euclidean norm.

where

$$\|\boldsymbol{y}\| = \left(\sum_{j=1}^{m} \sum_{\ell=1}^{m} |y_{j,\ell}|^2 \right)^{1/2}$$

is the standard Euclidean norm on vectors in $\mathbb{C}^{m^2}$ (which, for convenience, we arrange in $m \times m$ arrays. Here $\|\cdot\|$ should not be mistaken for the Euclidean *matrix* norm; see A.1.3.4). We have outlined a proof of a remarkable identity, which is at the root of many applications of Fourier transforms: provided that we measure both vectors and their transforms in the corresponding Euclidean norms, their magnitudes are the same.[3] Recalling our goal, to measure the rate of decay of the residuals, we deduce that monitoring $\|\!|p^{[k]}\|\!|$ or $\|\boldsymbol{\varepsilon}^{[k]}\|$ is equivalent.

We multiply (13.2) by $\mathrm{e}^{\mathrm{i}(j\theta+\ell\psi)}$ and sum for $j, \ell = 1, 2, \ldots, d$. Since

$$\sum_{j=1}^{m}\sum_{\ell=1}^{m} \varepsilon_{j-1,\ell}^{[k+1]} \mathrm{e}^{\mathrm{i}(j\theta+\ell\psi)} = \sum_{j=0}^{m-1}\sum_{\ell=1}^{m} \varepsilon_{j,\ell}^{[k+1]} \mathrm{e}^{\mathrm{i}((j+1)\theta+\ell\psi)} = \mathrm{e}^{\mathrm{i}\theta} p^{[k]}(\theta,\psi) - \mathrm{e}^{\mathrm{i}(m+1)\theta} \sum_{\ell=1}^{m} \varepsilon_{m,\ell}^{\mathrm{i}\ell\psi},$$

applying similar algebra to the other terms in (13.2) we obtain the identity

$$\begin{aligned}(4 - \mathrm{e}^{\mathrm{i}\theta} - \mathrm{e}^{\mathrm{i}\psi}) p^{[k+1]}(\theta,\psi) &= (\mathrm{e}^{-\mathrm{i}\theta} + \mathrm{e}^{-\mathrm{i}\psi}) p^{[k]}(\theta,\psi) \\ &\quad - \left\{ \mathrm{e}^{\mathrm{i}(m+1)\theta} \sum_{\ell=1}^{m} \varepsilon_{m,\ell}^{[k+1]} \mathrm{e}^{\mathrm{i}\ell\psi} + \mathrm{e}^{\mathrm{i}(m+1)\psi} \sum_{j=1}^{m} \varepsilon_{j,m}^{[k+1]} \mathrm{e}^{\mathrm{i}j\theta} \right. \\ &\quad \left. + \sum_{\ell=1}^{m} \varepsilon_{1,\ell}^{[k]} \mathrm{e}^{\mathrm{i}\ell\psi} + \sum_{j=1}^{m} \varepsilon_{j,1}^{[k]} \mathrm{e}^{\mathrm{i}j\theta} \right\}.\end{aligned}$$

This is a moment when we commit a mathematical crime and assume that the term in the curly brackets is so small in comparison with $p^{[k]}$ and $p^{[k+1]}$ that it can be harmlessly neglected. Our half-hearted excuse is that this term sums over m components, whereas $p^{[k]}$, say, sums over m^2, and that if boundary conditions were periodic rather than Dirichlet, it would have disappeared altogether. However, the true justification, as for most other crimes, is that it pays.

The main idea now is to consider how fast the Gauss–Seidel iteration attentuates each individual *wavenumber* (θ, ψ). In other words, we are interested in the *local attenuation factor*

$$\rho^{[k]}(\theta,\psi) := \left| \frac{p^{[k+1]}(\theta,\psi)}{p^{[k]}(\theta,\psi)} \right|, \qquad |\theta|, |\psi| \le \pi.$$

Having agreed that

$$(4 - \mathrm{e}^{\mathrm{i}\theta} - \mathrm{e}^{\mathrm{i}\psi}) p^{[k+1]}(\theta,\psi) \approx (\mathrm{e}^{-\mathrm{i}\theta} + \mathrm{e}^{-\mathrm{i}\psi}) p^{[k]}(\theta,\psi), \qquad |\theta|, |\psi| \le \pi,$$

we can make the following estimate

$$\rho^{[k]}(\theta,\psi) \approx \tilde{\rho}(\theta,\psi) := \left| \frac{\mathrm{e}^{\mathrm{i}\theta} + \mathrm{e}^{\mathrm{i}\psi}}{4 - \mathrm{e}^{\mathrm{i}\theta} - \mathrm{e}^{\mathrm{i}\psi}} \right|, \qquad |\theta|, |\psi| \le \pi. \tag{13.3}$$

[3] Lemma 16.9 provides a more formal statement of this important result, as well as a complete proof in a single dimension. The generalization to bivariate – indeed, multivariate – Fourier transforms is straightforward.

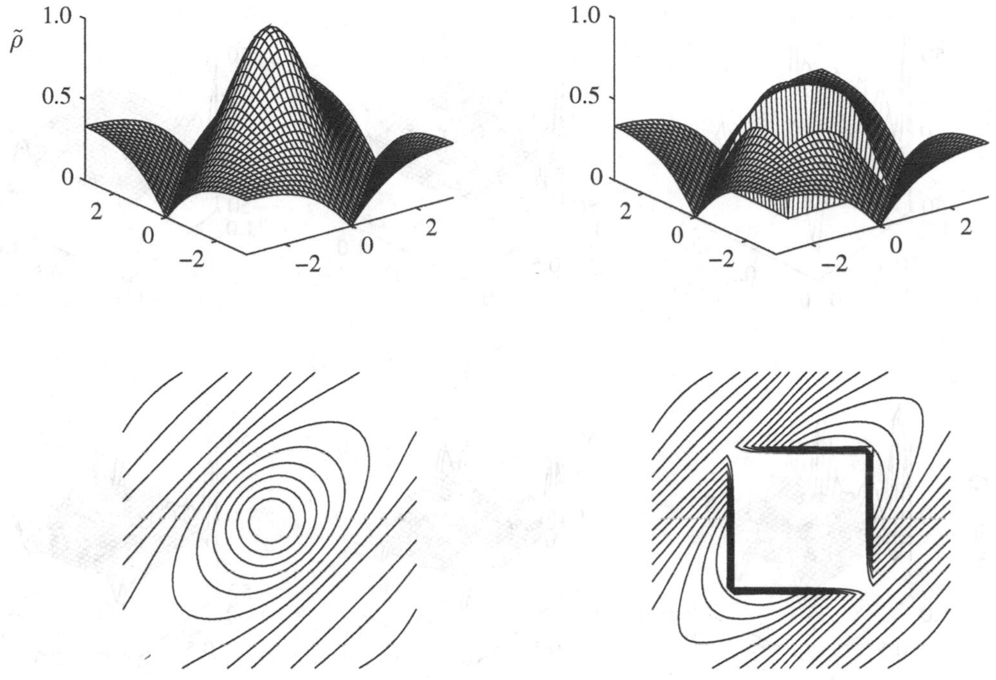

Figure 13.3 The function $\tilde\rho$ as a three-dimensional surface and as a contour plot. The left-hand column displays the whole square $[-\pi,\pi]\times[-\pi,\pi]$, while the right-hand column displays only the set $\mathbb{O}_0$.

Note that the function $\tilde\rho$ is independent of k. On the left in Fig. 13.3 the whole square $[-\pi,\pi]\times[-\pi,\pi]$ is displayed and there are no surprises there; thus, $\tilde\rho(\theta,\psi)\le 1$ for all $|\theta|,|\psi|\le\pi$.

Considerably more interesting is the right-hand column of Fig. 13.3, where the function $\tilde\rho$ is displayed just in the set

$$\mathbb{O}_0 := \left\{(\theta,\psi) : \tfrac{1}{2}\pi \le \max\{|\theta|,|\psi|\} \le \pi\right\}$$

of *oscillatory* wavenumbers. A remarkable feature emerges: provided only such wavenumbers are considered, the function $\tilde\rho$ peaks at the value $\frac{1}{2}$.

Forearmed with this observation, we formally evaluate the maximum of $\tilde\rho$ within the set $\mathbb{O}_0$. It is not difficult to verify that

$$\max_{(\theta,\psi)\in\mathbb{O}_0} \tilde\rho(\theta,\psi) = \tilde\rho\left(\tfrac{\pi}{2},\tan^{-1}\tfrac{3}{4}\right) = \tfrac{1}{2}.$$

This confirms our observation: as soon as we disregard non-oscillatory wavenumbers, *the amplitude of the error is halved in each iteration!* This at last explains the phenomenon that we observe in Figs. 13.1 and 13.2. To start with, the error is typically a linear combination of many wavenumbers, oscillatory as well as non-oscillatory. A Gauss–Seidel iteration attenuates the oscillatory components much faster, and this means that the contribution of the latter is, to all practical purposes,

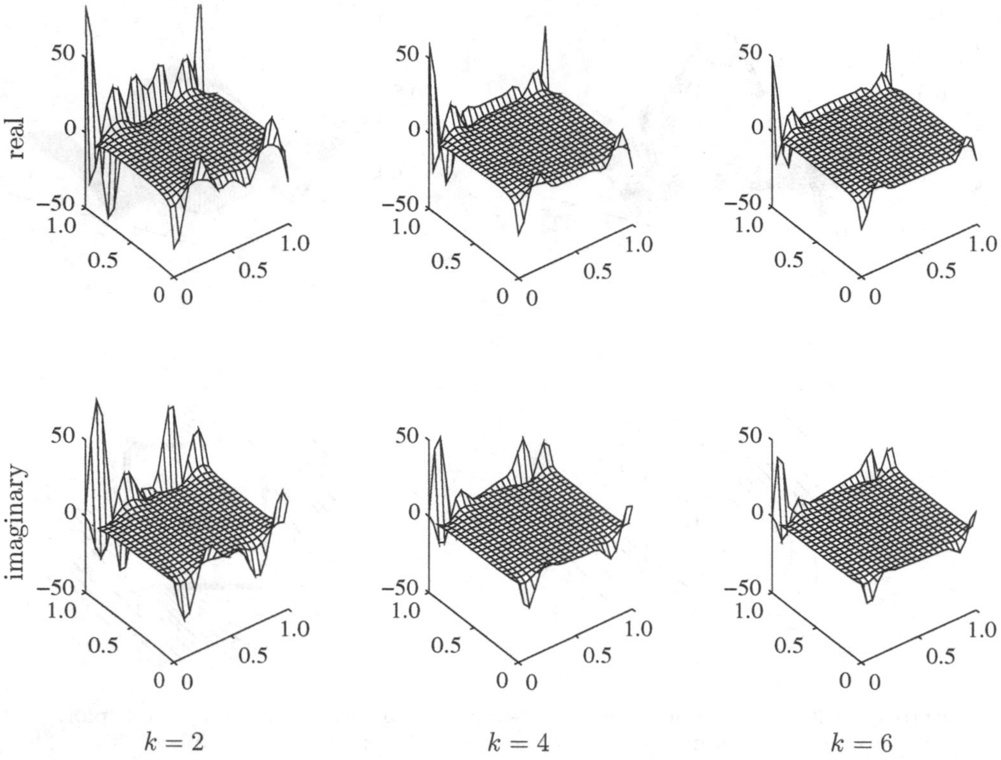

Figure 13.4 Real and imaginary components of $\varepsilon^{[k]}$ for $m = 20$ and $k = 2, 4, 6$. Note the rapid elimination rate of the highly oscillatory components.

completely eliminated after only a few iterations. The non-oscillatory terms, however, are left and they account for the sedate and plodding rate of attenuation in the asymptotic regime.

To rephrase this state of affairs, the Gauss–Seidel scheme is a *smoother:* its effect after a few iterations is to filter out high frequencies from the 'signal'. This becomes apparent in Fig. 13.4, where the real and imaginary parts of $\varepsilon^{[k]}$ are displayed for a 20×20 grid and $k = 2, 4, 6$. This is perhaps the place to emphasize that not all iterative methods from Chapter 12 are smoothers; far from it. In Exercise 13.2, for example, it is demonstrated that this attribute is absent from the Jacobi method.

Had this been all there were to it, the smoothing behaviour of Gauss–Seidel would be not much more than a mathematical curiosity. Suppose that we wish to solve the five-point equations to a given tolerance $\delta > 0$. The fast attenuation of the highly oscillatory components does not advance perceptibly the instant when $\|\varepsilon^{[k]}\| < \delta$ (or, in a realistic computer program, $\|\boldsymbol{r}^{[k]}\| < \delta$); it is the straggling non-oscillatory wavenumbers that dictate the rate of convergence. Figure 12.5 does not lie: Gauss–Seidel, in complete agreement with the theory of Chapter 12, will perform just twice as well as Jacobi (which, according to Exercise 13.2, is not a smoother). Fortunately, there is much more to the innocent phrase 'highly oscillatory components', and this

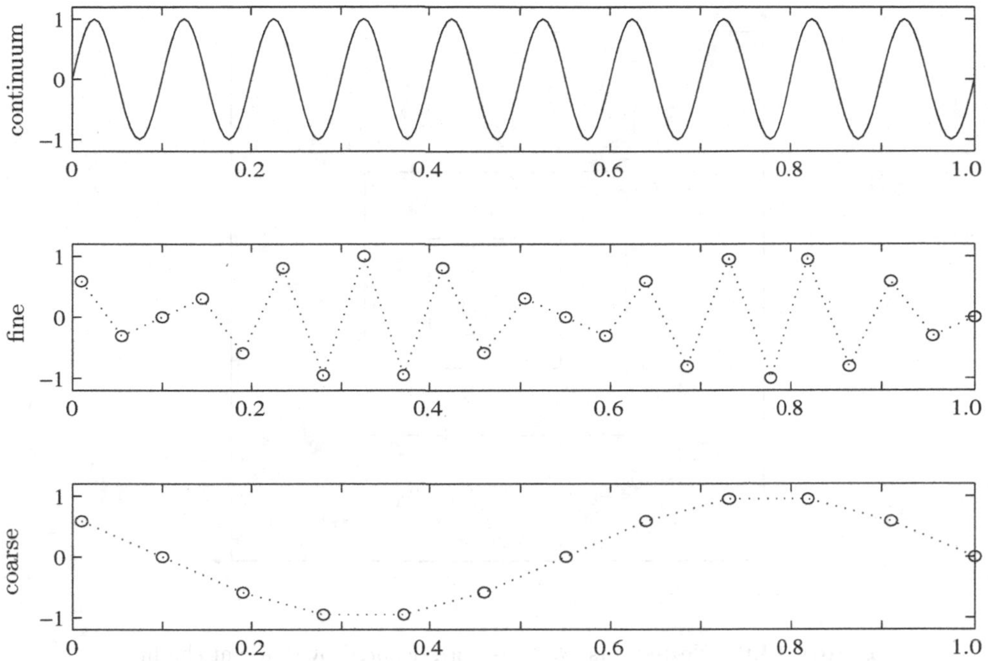

Figure 13.5 Now you see it, now you don't...: A highly oscillatory component and its restrictions to a fine and to a coarse grid.

forms our final clue about how to accelerate the Gauss–Seidel iteration.

Let us ponder for a moment the meaning of 'highly oscillatory components'. A grid – any grid – is a set of peepholes to the continuum, say $[0,1] \times [0,1]$. The continuum supports all possible frequencies and wavenumbers, but this is not the case with a grid. Suppose that the frequency is so high that a wave oscillates more than once between grid points – *this high oscillation will be invisible on the grid!* More precisely, observing the continuum through the narrow slits of the grid, we will, in all probability, register the wave as non-oscillatory. An example is presented in Fig. 13.5 where, for simplicity, we have confined ourselves to a single dimension. The top graph displays the highly oscillatory wave $\sin 20\pi x$, $x \in [0,1]$. In the middle graph the signal has been sampled at 23 equidistant points, and this renders faithfully the oscillatory nature of the sinusoidal wave. However, in the bottom graph we have thrown away every second point. The new graph, with 12 points, completely misses the high frequency!

The concept of a 'high oscillation' is, thus, a feature of a specific grid. This means that on grids of *different* spacing the Gauss–Seidel iteration attenuates *different* wavenumbers rapidly. Suppose that we coarsen a grid by taking out every second point, the outcome being a new square grid in $[0,1] \times [0,1]$ but with Δx replaced by $2\Delta x$. The range of the former high frequencies $\mathbb{O}_0$ is no longer visible on the coarse grid. Instead, the new grid has its own range of high frequencies, on which Gauss–Seidel performs well – as far as the fine grid is concerned, these correspond to the

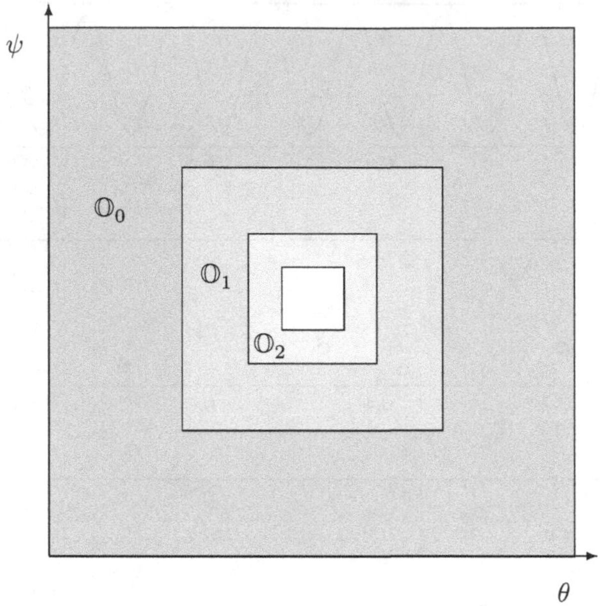

Figure 13.6 Nested sets $\mathbb{O}_s \subset [-\pi, \pi]$, denoted by different shading.

wavenumbers

$$\mathbb{O}_1 := \left\{(\theta, \psi) \,:\, \tfrac{1}{4}\pi \le \max\{|\theta|, |\psi|\} \le \tfrac{1}{2}\pi\right\}.$$

Needless to say, there is no need to stop with just a single coarsening. In general, we can cover the whole range of frequencies by a *hierarchy* of grids, embedded into each other, whose (grid-specific) high frequencies correspond, as far as the fine grid is concerned, to the sets

$$\mathbb{O}_s := \left\{(\theta, \psi) \,:\, 2^{-s-1}\pi \le \max\{|\theta|, |\psi|\} \le 2^{-s}\pi\right\}, \qquad s = 1, 2, \ldots, \lfloor \log_2(m+1) \rfloor.$$

The sets $\mathbb{O}_s$ nest inside each other (see Fig. 13.6) and their totality is the whole of $[-\pi, \pi] \times [-\pi, \pi]$. In the next section we describe a computational technique that sweeps across the sets $\mathbb{O}_s$, damping the highly oscillatory terms and using Gauss–Seidel in its 'fast' mode throughout the entire iterative process.

13.2 The basic multigrid technique

Let us suppose for simplicity that $m = 2^s - 1$ and let us embed our grid (and from here on we designate it as the *finest* grid) in a hierarchy of successively coarser grids, as indicated in Fig. 13.7.

The main idea behind the *multigrid* technique is to travel up and down the grid hierarchy, using Gauss–Seidel iterations to dampen the (locally) highly oscillating components of the error. *Coarsening* means that we are descending down the hierarchy to a coarser grid (in other words, getting rid of every other point), while *refinement*

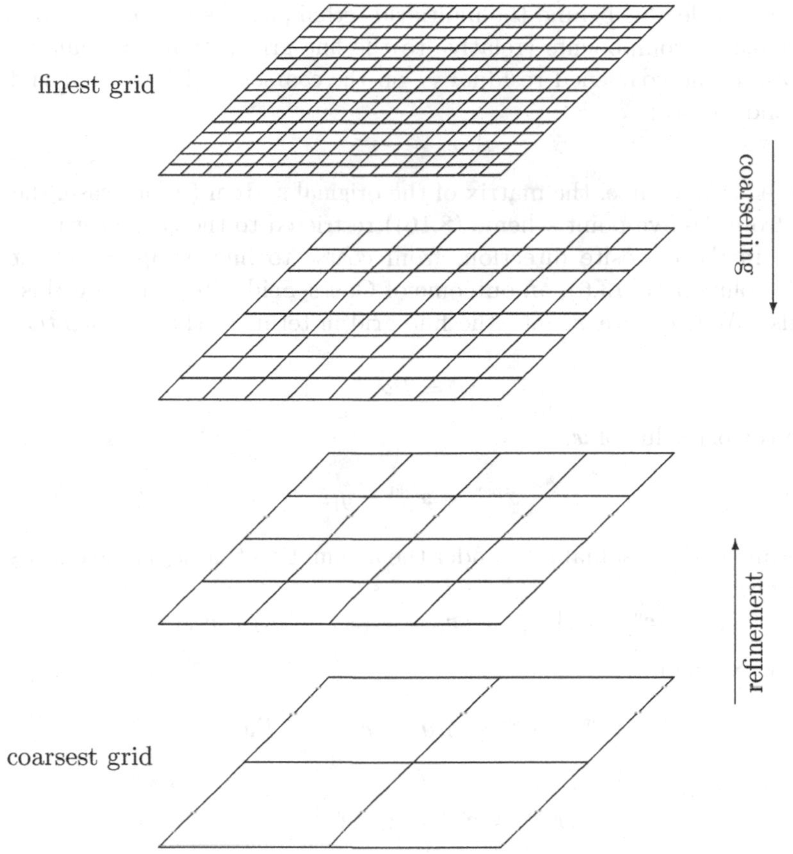

Figure 13.7 Nested grids, from the finest to the coarsest.

is the exact opposite, ascending from a coarser to a finer grid. Our goal is to solve the five-point equations *on the finest grid* – the coarser grids are just a means to that end.

In order to describe a multigrid algorithm we need to explain exactly how each coarsening or refinement step is performed, as well as to specify the exact strategy of how to start, when to coarsen, when to refine and when to terminate the entire procedure.

To describe refinement and coarsening it is enough to assume just two grids, one fine and one coarse. Suppose that we are solving the equation

$$A_{\mathrm{f}}\boldsymbol{x}_{\mathrm{f}} = \boldsymbol{v}_{\mathrm{f}} \tag{13.4}$$

on the fine grid. Having performed a few Gauss–Seidel iterations, so as to smooth the high frequencies, we let $\boldsymbol{r}_{\mathrm{f}} := A_{\mathrm{f}}\boldsymbol{x}_{\mathrm{f}} - \boldsymbol{v}_{\mathrm{f}}$ be the residual. This residual needs to be translated into the coarser grid. This is done by means of a *restriction matrix* R such that

$$\boldsymbol{r}_{\mathrm{c}} = R\boldsymbol{r}_{\mathrm{f}}. \tag{13.5}$$

Remember the whole idea behind the multigrid technique: the vector $\boldsymbol{r}_{\rm f}$ is constructed from low-frequency components (relative to the fine grid). Hence it makes sense to go on smoothing the coarsened residual $\boldsymbol{r}_{\rm c}$ on the coarser grid.[4] To that end we set $\boldsymbol{v}_{\rm c} := -\boldsymbol{r}_{\rm c}$, and so solve

$$A_{\rm c}\boldsymbol{x}_{\rm c} = -\boldsymbol{r}_{\rm c}. \tag{13.6}$$

The matrix $A_{\rm c}$ is, of course, the matrix of the original system (in our case, the matrix originating from the five-point scheme (8.16)) restricted to the coarser grid.

To move in the opposite direction, from coarse to fine, suppose that $\boldsymbol{x}_{\rm c}$ is an approximate solution of (13.6), an outcome of Gauss–Seidel iterations on this and yet coarser grids. We translate $\boldsymbol{x}_{\rm c}$ into the fine grid in terms of the *prolongation matrix* P, where

$$\boldsymbol{y}_{\rm f} = P\boldsymbol{x}_{\rm c} \tag{13.7}$$

and update the old value of $\boldsymbol{x}_{\rm f}$,

$$\boldsymbol{x}_{\rm f}^{\rm new} = \boldsymbol{x}_{\rm f}^{\rm old} + \boldsymbol{y}_{\rm f}. \tag{13.8}$$

Let us evaluate the residual $\boldsymbol{r}_{\rm f}^{\rm new}$ under the assumption that $\boldsymbol{x}_{\rm c}$ is the exact solution of (13.6). Since

$$\boldsymbol{r}_{\rm f}^{\rm new} = A_{\rm f}\boldsymbol{x}_{\rm f}^{\rm new} - \boldsymbol{v}_{\rm f} = A_{\rm f}(\boldsymbol{x}_{\rm f}^{\rm old} + \boldsymbol{y}_{\rm f}) - \boldsymbol{v}_{\rm f},$$

(13.7) and (13.8) yield

$$\boldsymbol{r}_{\rm f}^{\rm new} = \boldsymbol{r}_{\rm f}^{\rm old} + A_{\rm f}\boldsymbol{y}_{\rm f} = \boldsymbol{r}_{\rm f}^{\rm old} + A_{\rm f}P\boldsymbol{x}_{\rm c}.$$

Therefore, by (13.6),

$$\boldsymbol{r}_{\rm f}^{\rm new} = \boldsymbol{r}_{\rm f}^{\rm old} - A_{\rm f}PA_{\rm c}^{-1}\boldsymbol{r}_{\rm c}.$$

Finally, invoking (13.5), we deduce that

$$\boldsymbol{r}_{\rm f}^{\rm new} = (I - A_{\rm f}PA_{\rm c}^{-1}R)\,\boldsymbol{r}_{\rm f}^{\rm old}. \tag{13.9}$$

Thus, the sole contribution to the new residual comes from replacing the fine grid by a coarser one. Similar reasoning is valid even if $\boldsymbol{x}_{\rm c}$ is an approximate solution of (13.6), provided that some bandwidths of wavenumbers have been eliminated in the course of the iteration. Moreover, suppose that (other) bandwidths of wavenumbers have been already filtered out of the residual $\boldsymbol{r}_{\rm f}^{\rm old}$. Upon the update (13.8), the contribution of both bandwidths is restricted to the minor ill effects of the restriction and prolongation matrices.

Both the restriction and prolongation matrices are rectangular, but it is a very poor idea to execute them naively as matrix products. The proper procedure is to describe their effect on individual components of the grid, since this provides a convenient and cheap algorithm as well as clarifying what are we trying to do in mathematical terms. Let $\boldsymbol{w}_{\rm f} = P\boldsymbol{w}_{\rm c}$, where $\boldsymbol{w}_{\rm c} = (w_{j,\ell}^{\rm c})_{j,\ell=1}^{m}$ (a subscript has just been promoted to a

[4]To be exact, we have advanced an argument to justify this assertion for the error, rather than the residual. However, it is clear from Exercise 13.1 that the two assertions are equivalent. Of course, the residual, unlike the error, has an important virtue: we can calculate it without knowing the exact solution of the linear system. . .

superscript, for notational convenience) and $\boldsymbol{w}_{\mathrm{f}} = (w^{\mathrm{f}}_{j,\ell})^{2m+1}_{j,\ell=1}$. The simplest way of restricting a grid is *injection,*

$$w^{\mathrm{c}}_{j,\ell} = w^{\mathrm{f}}_{2j,2\ell}, \qquad j,\ell = 1,2,\ldots,m, \tag{13.10}$$

but a popular alternative is *full weighting*

$$\begin{aligned} w^{\mathrm{c}}_{j,\ell} = \tfrac{1}{4}w^{\mathrm{f}}_{2j,2\ell} &+ \tfrac{1}{8}(w^{\mathrm{f}}_{2j-1,2\ell} + w^{\mathrm{f}}_{2j,2\ell-1} + w^{\mathrm{f}}_{2j+1,2\ell} + w^{\mathrm{f}}_{2j,2\ell+1}) + \tfrac{1}{16}(w^{\mathrm{f}}_{2j-1,2\ell-1} \\ &+ w^{\mathrm{f}}_{2j+1,2\ell-1} + w^{\mathrm{f}}_{2j-1,2\ell+1} + w^{\mathrm{f}}_{2j+1,2\ell+1}), \qquad j,\ell = 1,2,\ldots,m. \end{aligned} \tag{13.11}$$

The latter can be rendered as a computational stencil (see Section 8.2) in the form

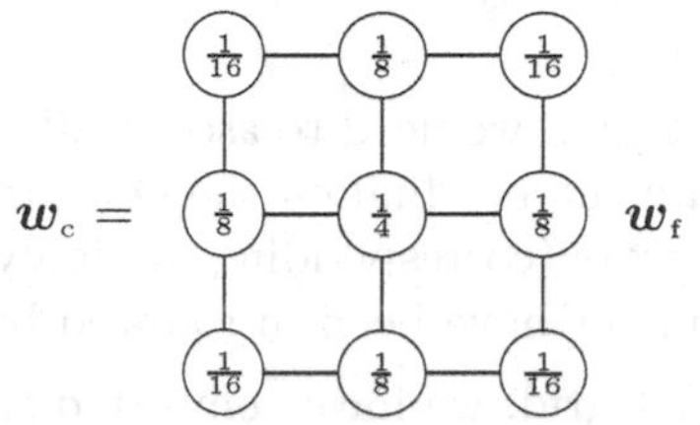

Why bother with (13.11), given the availability of the more natural injection (13.10)? One reason is that in the latter case $R = \frac{1}{4}P^{\top}$ for the prolongation P that we are just about to introduce in (13.12) (the factor $\frac{1}{4}$ originates in the fourfold decrease in the number of grid points in coarsening), and this has important theoretical and practical advantages. Another is that in this manner *all* points from the finer grid contribute equally.

There is just one sensible way of prolonging a grid: *linear interpolation.* The exact equations are

$$\begin{aligned} w^{\mathrm{f}}_{2j-1,2\ell-1} &= w^{\mathrm{c}}_{j,\ell}, && j,\ell = 1,2,\ldots,m; \\ w^{\mathrm{f}}_{2j-1,2\ell} &= \tfrac{1}{2}(w^{\mathrm{c}}_{j,\ell} + w^{\mathrm{c}}_{j,\ell+1}), && j = 1,2,\ldots,m-1, \quad \ell = 1,2,\ldots,m; \\ w^{\mathrm{f}}_{2j,2\ell-1} &= \tfrac{1}{2}(w^{\mathrm{c}}_{j,\ell} + w^{\mathrm{c}}_{j+1,\ell}), && j = 1,2,\ldots,m, \quad \ell = 1,2,\ldots,m-1; \\ w^{\mathrm{f}}_{2j,2\ell} &= \tfrac{1}{4}(w^{\mathrm{c}}_{j,\ell} + w^{\mathrm{c}}_{j,\ell+1} \\ &\qquad + w^{\mathrm{c}}_{j+1,\ell} + w^{\mathrm{c}}_{j+1,\ell+1}), && j,\ell = 1,1,\ldots,m-1. \end{aligned} \tag{13.12}$$

The values of $\boldsymbol{w}_{\mathrm{f}}$ along the boundary are, of course, zero; recall that we are dealing with residuals!

Having learnt how to travel across the hierarchy of nested grids, we now need to specify an itinerary. There are many distinct multigrid strategies and here we mention just the simplest (and most popular), the *V-cycle*

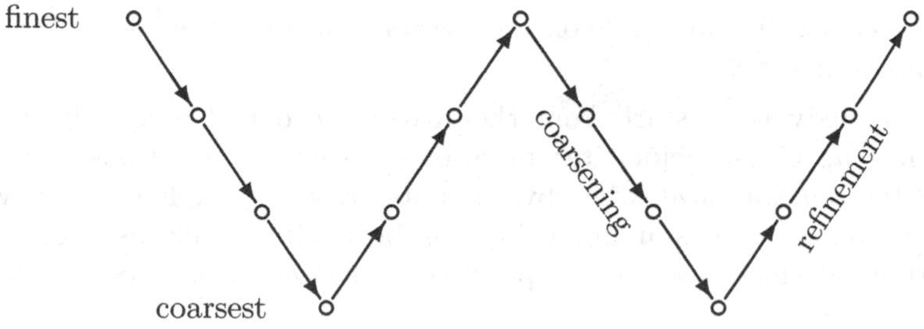

The whole procedure commences and ends at the finest grid. To start with, we stipulate an initial condition, let $\boldsymbol{v}_{\mathrm{f}} = \boldsymbol{b}$ (the original right-hand side of the linear system (13.1)) and iterate a small number of times – n_{r}, say – with Gauss–Seidel. Subsequently we evaluate the residual $\boldsymbol{r}_{\mathrm{f}}$, restrict it to the coarser grid, perform n_{r} further Gauss–Seidel iterations, evaluate the residual, again restrict and so on, until we reach the coarsest grid, with just a single grid point, which we solve exactly. (In principle, it is possible to stop this procedure earlier, deciding that a 15×15 system, say, can be solved directly without further coarsening.) Having reached this stage, we have successively damped the influence of error components in the entire range of wavenumbers supported by the finest grid, except that a small amount of error might have been added by restriction.

When we reach the coarsest grid, we need to ascend all the way back to the finest. In each step we prolong, update the residual on the new grid and perform n_{p} Gauss–Seidel iterations to eliminate errors (corresponding to highly oscillatory wavenumbers on the grid in question) that might have been introduced by past prolongations.

Having returned to the finest grid, we have completed the V-cycle. It is now, and only now, that we check for convergence, by measuring the size of the residual vector. Provided that the error is below the required tolerance, the iteration is terminated; otherwise the V-cycle is repeated. This completes the description of the multigrid algorithm in its simplest manifestation.

13.3 The full multigrid technique

An obvious Achilles heel of all iterative methods is the choice of the starting vector $\boldsymbol{x}^{[0]}$. Although the penalty for a wrong choice is not as drastic as in methods for nonlinear algebraic equations (see Chapter 7), it is nonetheless likely to increase the cost a great deal. By the same token, an astute choice of $\boldsymbol{x}^{[0]}$ is bound to lead to considerable savings.

So far, throughout Chapters 12 and 13, we have assumed that $\boldsymbol{x}^{[0]} = \boldsymbol{0}$, a choice which is likely to be as good or as bad as many others for most iterative methods.

The logic of the multigrid approach – working in unison on a whole hierarchy of embedded grids – can be complemented by a superior choice of starting value. Why not use an approximate solution from a coarser grid as the starting value on the finest? Of course, at the beginning of the iteration, exactly when the starting value is required, we have no solution available on the coarser grid, since the V-cycle iteration commences from the finest.

The obvious remedy is to start from the coarsest grid and ascend by prolongation, performing n_{p} Gauss–Seidel iterations on each grid. This leads to a technique known as the *full multigrid,* whereby, upon its arrival at the finest grid (where the V-cycles commence), the starting value has been already cleansed of a substantial proportion of smooth error components. The self-explanatory pattern is

illustrated by the graph

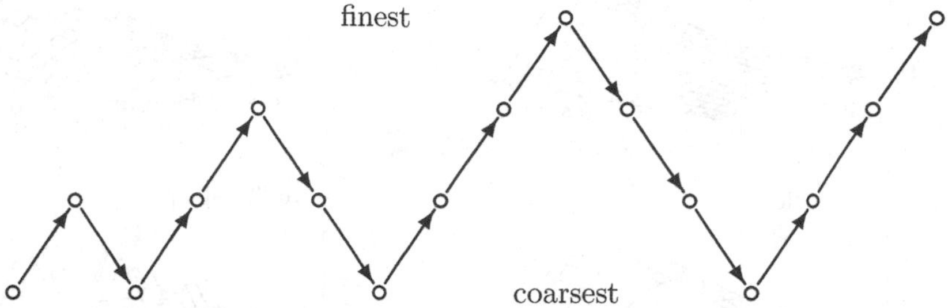

The speed-up in convergence of the full multigrid technique, as will be evidenced in the results of Section 13.4, is spectacular.

The full multigrid combines two ideas: the first is the multigrid concept of using Gauss–Seidel, say, to smooth the highly oscillatory components by progressing from fine to coarse grids; the second is *nested iteration.* The latter uses estimates from a coarse grid as a starting value for an iteration on a fine grid. In principle, nested iteration can be used whenever an iterative scheme is applied in a grid, without necessarily any reference to multigrid. An example is provided by the solution of nonlinear algebraic equations by means of functional iteration or Newton–Raphson (see Chapter 7). However, it comes into its own in conjunction with multigrid.

13.4 Poisson by multigrid

This chapter is short on theory and, to remedy the situation, we have made it long on computational results.

Since Chapter 8 we have used a particular Poisson equation, the problem (8.33), as a yardstick to measure the behaviour of numerical methods, and we will continue this practice here. The finest grid used is always 63×63 (that is, with $\Delta x = \frac{1}{64}$. We measure the performance of the methods by the size of the error at the end of each V-cycle (disregarding, in the case of full multigrid, all but the 'complete V-cycles', from the finest to the coarsest grid and back again). It is likely that, in practical error estimation, the residual rather than the error is calculated. This might lead to different numbers but it will give the same qualitative picture.

We have tested three different choices of the pair (n_r, n_p) for both the 'regular' multigrid from Section 13.2 and the full multigrid technique, Section 13.3. The results are displayed in Figs. 13.8 and 13.9.

Each figure displays three detailed iteration strategies: (*a*) $n_r = 1$, $n_p = 1$; (*b*) $n_r = 2$, $n_p = 1$; and (*c*) $n_r = 3$, $n_p = 2$. We have not printed the outcome of seven V-cycles (four in Fig. 13.9), since the error is so small that it is likely to be a roundoff artefact.

To assess the cost of a single V-cycle, we disregard the expense of restriction and prolongation, counting just the number of smoothing (i.e., Gauss–Seidel) iterations. The latter are performed on grids of vastly different sizes, but this can be easily incorporated into our estimate by observing that the cost of Gauss–Seidel is linear in

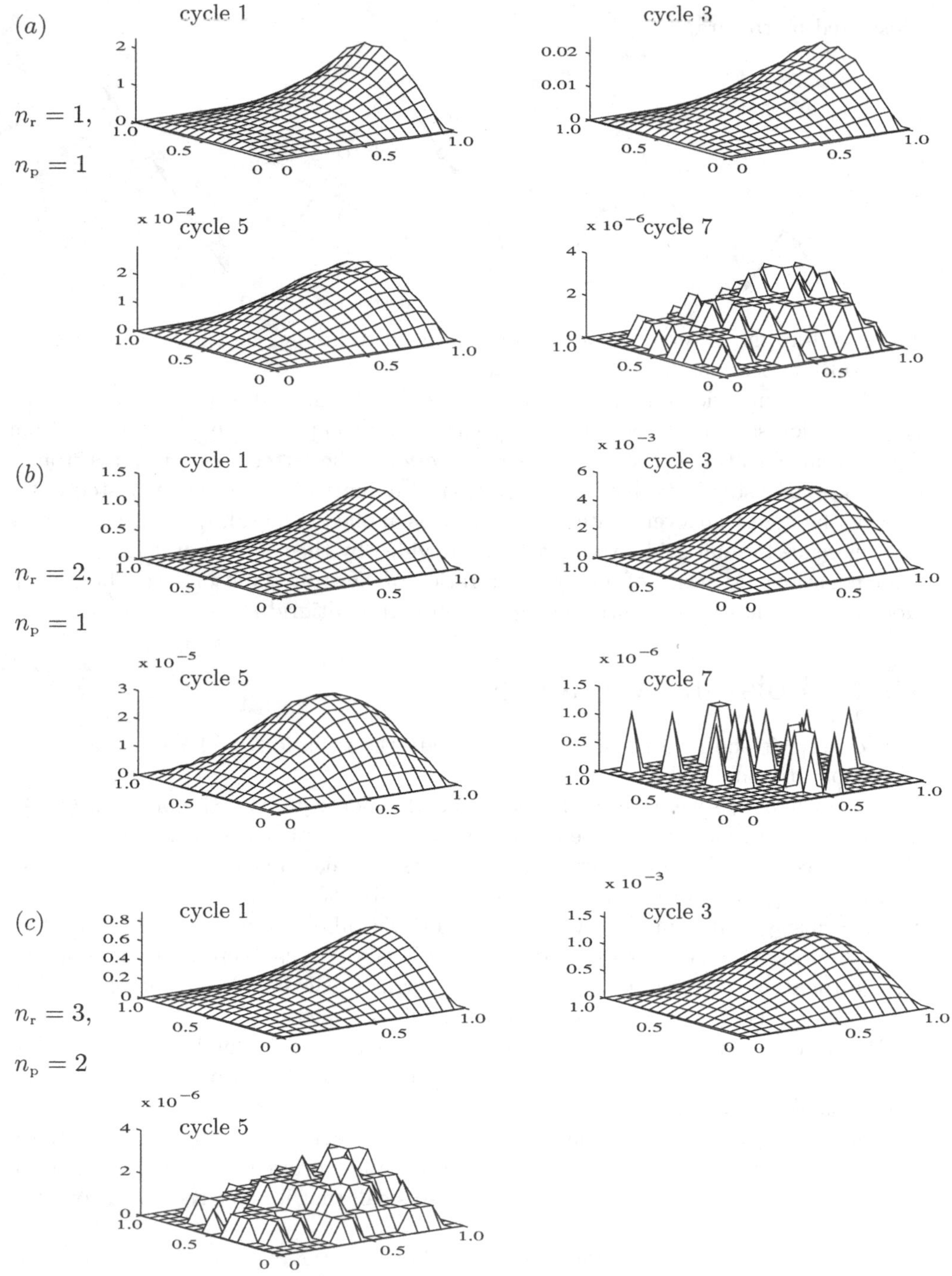

Figure 13.8 The V-cycle multigrid method for the Poisson equation (8.33).

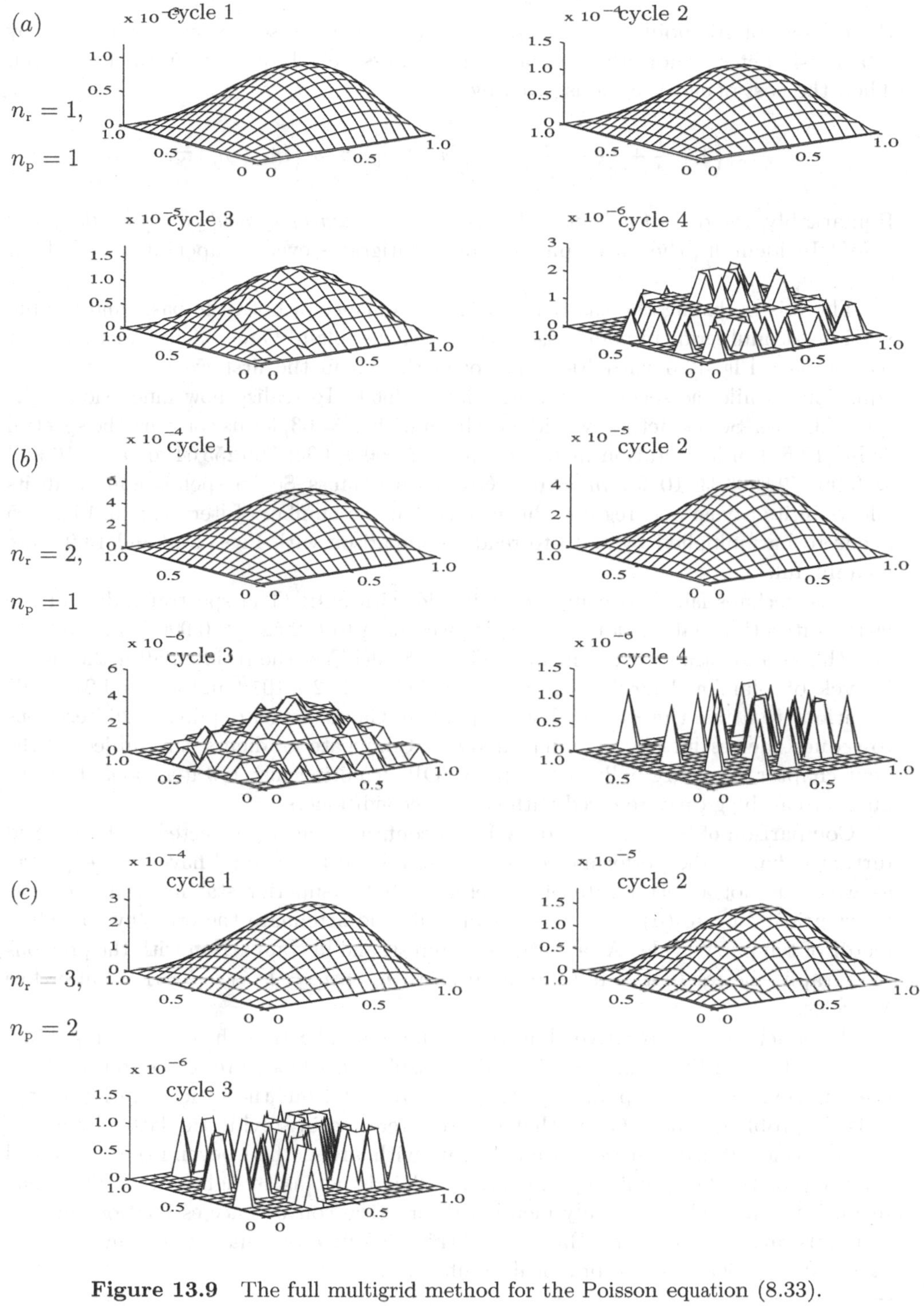

Figure 13.9 The full multigrid method for the Poisson equation (8.33).

the number of grid points, hence a single coarsening decreases its operations count by a factor 4. Let ϖ denote the cost of a single Gauss–Seidel iteration on the finest grid. Then the cost of one V-cycle is given by

$$\left(1+\frac{1}{4}+\frac{1}{4^2}+\frac{1}{4^3}+\cdots\right)(n_{\rm r}+n_{\rm p})\,\varpi\approx\tfrac{4}{3}(n_{\rm r}+n_{\rm p})\,\varpi.$$

Remarkably, *the cost of a V-cycle is linear in the number of grid points on the finest grid!*[5] Incidentally, the initial phase of full multigrid is even cheaper: it 'costs' about $\frac{4}{9}(n_{\rm r}+n_{\rm p})\,\varpi$.

There is no hiding the vastly superior performance of both the basic and the full versions of multigrid, in comparison with, say, the 'plain' Gauss–Seidel method. Let us compare Fig. 12.5 with Fig. 13.8, even though in the first we have $15^2 = 225$ equations, while the second comprises $63^2 = 3969$. To realize how much slower the 'plain' Gauss–Seidel method would have been with $m = 63$, let us compare the spectral radii (12.51) of its iteration matrices: we find $\approx 0.961\,939\,766\,255\,64$ for $m = 16$ and $\approx 0.997\,592\,363\,336\,10$ for $m = 63$. Given that Gauss–Seidel spends almost all its efforts in its asymptotic regime, this means that the number of iterations in Fig. 12.5 needs to be multiplied by ≈ 16 to render comparison with Figs. 13.8 and 13.9 more meaningful.

It is perhaps fairer to compare multigrid with SOR. The spectral radius of the latter's iteration matrix (for $m = 63$) is, according to (12.52), $\approx 0.906\,454\,701\,582\,76$, and this is a great improvement upon Gauss–Seidel. Yet the residual after the eighth V-cycle of 'plain' multigrid (with $n_{\rm r} = n_{\rm p} = 1$) is $\approx 2.62 \times 10^{-5}$ and we need 243 SOR iterations to attain this value. (By comparison, Gauss–Seidel requires 6526 iterations to reduce the residual by a similar amount. Conjugate gradients, the subject of the next chapter, are marginally better than SOR, requiring just 179 iterations, but this number can be greatly reduced with good preconditioners.)

Comparison of Figs. 13.8 and 13.9 also confirms that, as expected, full multigrid further enhances the performance. The reason – and this should have been expected as well – is not a better rate of convergence but a superior starting value (on the finest grid): in case (*a*) both versions of multigrid attenuate the error by roughly a factor of ten per V-cycle. As a matter of interest, and in comparison with the previous paragraph, the residual of full multigrid (with $n_{\rm r} = n_{\rm p} = 1$) is $\approx 5.8510^{-6}$ after five V-cycles.

We conclude this 'iterative olympics' with a reminder that the errors in Figs. 13.8 and 13.9 (and in Figs. 12.5 and 12.7 also) display the departure of the iterates from the solution of the five-point equations (8.16), not from the exact solution of the Poisson problem (8.33). Given that we are interested in solving the latter by means of the former, it makes little sense to iterate with any method beyond the theoretical accuracy of the five-point approximation. This is not as straightforward as it may seem, since, as we have already mentioned, practical convergence estimation employs residuals rather than errors. Having said this, seeking a residual lower than 10^{-5} (for $m = 63$), is probably of no practical significance.

[5] Our assumption is that all the calculations are performed in a serial, as distinct from a parallel, computer architecture. Otherwise the results are likely to be even more spectacular.

Comments and bibliography

The idea of using a hierarchy of grids has been around for a while, mainly in the context of nested iteration, but the first modern treatment of the multigrid technique was presented by Brandt (1977).

There exist a number of good introductory texts on multigrid techniques, e.g. Briggs (1987); Hackbusch (1985) and Wesseling (1992). Convergence and complexity (in the present context complexity means the estimation of computational cost) are addressed in the book of Bramble (1993) and in a survey by Yserentant (1993). It is important to emphasize that, although multigrid techniques can be introduced and explained in an elementary fashion, their convergence analysis is fairly challenging from a mathematical point of view. The reason is that the multigrid is an example of a *multiscale* phenomenon, which coexists along a hierarchy of different scales. Such phenomena occur in applications (the ingredients of a physical model often involve different orders of magnitude in both space and time) and are playing an increasingly greater role in modern scientific computation.

Our treatment of multigrid has centred on just one version, the V-cycle, and we mention in passing that other strategies are perfectly viable and often preferable, e.g. the W-cycle:

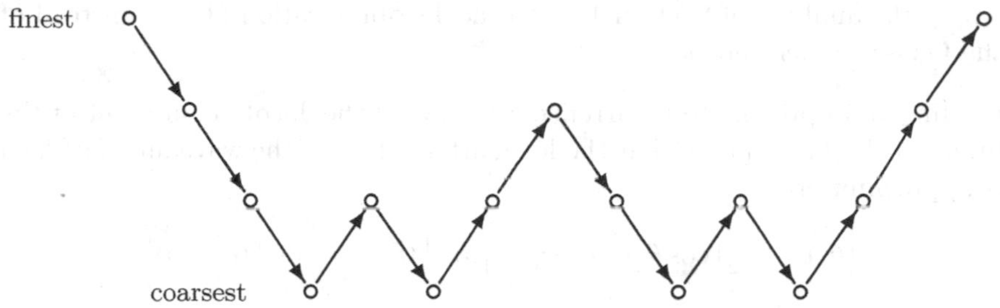

The number of different strategies and implementations of multigrid is a source of major preoccupation to professionals, although it might be at times slightly baffling to other numerical analysts and to users of computational algorithms.

Gauss–Seidel is not the only smoother, although neither the Jacobi iteration nor SOR (with $\omega \neq 1$) possess this welcome property (see Exercise 13.2). An example of a smoother is provided by a version of the incomplete LU factorization (not the Jacobi line relaxation from Section 12.1, though; see Exercise 13.3). Another example is *Jacobi over-relaxation* (JOR), an iterative scheme that is to Jacobi what SOR is to Gauss–Seidel, with a particular parameter value.

Multigrid methods would be of little use were their applicability restricted to the five-point equations in a square. Indeed, possibly the greatest virtue of multigrid is its versatility. Provided linear equations are specified in one grid and we can embed this into a hierarchy of nested grids of progressive coarseness, multigrid confers an advantage over a single-grid implementation of iterative methods. Indeed, we use the word 'grid' in a loose sense, since multigrid is, if anything, even more useful for finite elements than for finite differences!

Bramble, J.H. (1993), *Multigrid Methods,* Longman, Harlow, Essex.

Brandt, A. (1977), Multi-level adaptive solutions to boundary-value problems, *Mathematics of Computation* **31,** 333–390.

Briggs, W.L. (1987), *A Multigrid Tutorial,* SIAM, Philadelphia.

Hackbusch, W. (1985), *Multi-Grid Methods and Applications,* Springer-Verlag, Berlin.

Wesseling, P. (1992), *An Introduction to Multigrid Methods,* Wiley, Chichester.

Yserentant, H. (1993), Old and new convergence proofs for multigrid methods, *Acta Numerica* **2**, 285–326.

Exercises

13.1 Let $\hat{\boldsymbol{x}}$ be the solution of (13.1), $\boldsymbol{\varepsilon}^{[k]} := \boldsymbol{x}^{[k]} - \hat{\boldsymbol{x}}$ and $\boldsymbol{r}^{[k]} := A\boldsymbol{x}^{[b]} - \boldsymbol{b}$. Show that $\boldsymbol{r}^{[k]} = A\boldsymbol{\varepsilon}^{[k]}$. Further supposing that A is symmetric and that its eigenvalues reside in the interval $[\lambda_-, \lambda_+]$, prove that the inequality

$$\min\{|\lambda_-|, |\lambda_+|\}\|\boldsymbol{\varepsilon}^{[k]}\| \leq \|\boldsymbol{r}^{[k]}\| \leq \max\{|\lambda_-|, |\lambda_+|\}\|\boldsymbol{\varepsilon}^{[k]}\|$$

holds in the Euclidean norm.

13.2 Apply the analysis of Section 13.1 to the Jacobi iteration (12.20) instead of the Gauss–Seidel iteration.

a Finding an approximate recurrence relation for the Jacobi equivalent of the function $p^{[k]}(\theta, \psi)$, prove that the local attenuation of the wavenumber (θ, ψ) is approximately

$$\tilde{\rho}(\theta, \psi) = \tfrac{1}{2}|\cos\theta + \cos\psi| = \left|\cos\tfrac{1}{2}(\theta + \psi)\cos\tfrac{1}{2}(\theta - \psi)\right|.$$

b Show that the best upper bound on $\tilde{\rho}$ in $\mathbb{O}_0$ is unity and hence that the Jacobi iteration does not smooth highly oscillatory terms.

13.3 Using the same method as in the last exercise, show that the line relaxation method from Section 12.4 is not a good smoother. You should prove that

$$\tilde{\rho}(\theta, \psi) = \frac{|\cos\psi|}{2 - \cos\theta}$$

and that it can attain unity in the set $\mathbb{O}_0$.

13.4 Assuming that $w^{\mathrm{f}}_{j,\ell} = g(j, \ell)$, where g is a linear function of both its arguments, and that $\boldsymbol{w}^{\mathrm{c}}$ has been obtained by the fully weighted restriction (13.11), prove that $w^{\mathrm{c}}_{j,\ell} = g(2j, 2\ell)$.

14

Conjugate gradients

14.1 Steepest, but slow, descent

Our approach to iterative methods in Chapter 12 was based, at least implicitly, on dynamical systems. The solution of the linear system

$$A\boldsymbol{x} = \boldsymbol{b}, \tag{14.1}$$

where A is a $d \times d$ real nonsingular matrix and $\boldsymbol{b} \in \mathbb{R}^d$, was formulated as an *iterated map*

$$\boldsymbol{x}^{[k+1]} = \boldsymbol{h}(\boldsymbol{x}^{[k]}), \qquad k = 0, 1, 2, \ldots, \tag{14.2}$$

where $\boldsymbol{h} : \mathbb{R}^d \to \mathbb{R}^d$. The convergence of this recursive procedure was a consequence of basic features of the map $\boldsymbol{h}$: its contractivity (in the spirit of Section 7.1) and fixed points. Indeed, much of the effort required to design, analyse and understand methods of this kind is a reflection of the tension between mathematical attributes of the map $\boldsymbol{h}$, which ensure convergence to the right limit, and numerical desiderata that each iteration should be cheap and that convergence should occur rapidly.

The basic pattern of one-step stationary iteration (14.2) can be generalized by the inclusion of past values of $\boldsymbol{x}^{[k]}$ or by allowing $\boldsymbol{h}$ to vary. In this chapter we intend to adopt a different point of departure altogether and view the problem from the standpoint of the *theory of optimization.* The main underlying idea is to restate (14.1) as the minimization of some function $f : \mathbb{R}^d \to \mathbb{R}$ and apply an optimization algorithm.

Let us assume for the time being that the matrix A in (14.1) is symmetric and positive definite.

Lemma 14.1 *The unique minimum of the function*

$$f(\boldsymbol{x}) = \tfrac{1}{2}\boldsymbol{x}^\top A\boldsymbol{x} - \boldsymbol{b}^\top \boldsymbol{x}, \qquad \boldsymbol{x} \in \mathbb{R}^d, \tag{14.3}$$

is the solution of the linear system (14.1).

Proof We note that $\nabla f(\boldsymbol{x}) = A\boldsymbol{x} - \boldsymbol{b}$, therefore (14.3) has a unique stationary point $\boldsymbol{x}$ which is the solution of (14.1). Moreover $\nabla^2 f(\boldsymbol{x}) = A$ is positive definite, therefore $\boldsymbol{x}$ is indeed a minimum of f. ■

We are concerned with iterative algorithms of the following general form. We pick a starting vector $\boldsymbol{x}^{[0]} \in \mathbb{R}^d$. For any $k = 0, 1, \ldots$ the calculation stops if the residual

$\|\nabla f(\boldsymbol{x}^{[k]})\| = \|A\boldsymbol{x}^{[k]} - \boldsymbol{b}\|$ is sufficiently small. (We are using here the usual Euclidean norm.) Otherwise, we seek a *search direction* $\boldsymbol{d}^{[k]} \in \mathbb{R}^d \setminus \{\boldsymbol{0}\}$ that satisfies the *descent condition*

$$\left.\frac{\mathrm{d}f(\boldsymbol{x}^{[k]} + \omega\boldsymbol{d}^{[k]})}{\mathrm{d}\omega}\right|_{\omega=0} = \nabla f(\boldsymbol{x}^{[k]})^\top \boldsymbol{d}^{[k]} < 0. \tag{14.4}$$

In other words,

$$f(\boldsymbol{x}^{[k]} + \omega\boldsymbol{d}^{[k]}) = f(\boldsymbol{x}^{[k]}) + \omega\nabla f(\boldsymbol{x}^{[k]})^\top \boldsymbol{d}^{[k]} + \mathcal{O}(\omega^2)$$

implies that $f(\boldsymbol{x}^{[k]} + \omega\boldsymbol{d}^{[k]}) < f(\boldsymbol{x}^{[k]})$ for a sufficiently small step $\omega > 0$.

The obvious way forward is to choose such a 'sufficiently small $\omega > 0$' and let $\boldsymbol{x}^{[k+1]} = \boldsymbol{x}^{[k]} + \omega\boldsymbol{d}^{[k]}$. This will create a monotonically decreasing sequence of non-negative values $f(\boldsymbol{x}^{[k]})$ and, according to an elementary theorem of calculus, such a sequence descends to a limit. However, we can do better and choose the *best* value of ω. Note that

$$f(\boldsymbol{x}^{[k]} + \omega\boldsymbol{d}^{[k]}) = f(\boldsymbol{x}^{[k]}) + \omega\nabla f(\boldsymbol{x}^{[k]})^\top \boldsymbol{d}^{[k]} + \tfrac{1}{2}\omega^2 \boldsymbol{d}^{[k]\top} A\boldsymbol{d}^{[k]}$$

is a quadratic function. Therefore, we can easily find the value of ω that minimizes $f(\boldsymbol{x}^{[k]} + \omega\boldsymbol{d}^{[k]})$ by setting its derivative to zero. Letting $\boldsymbol{g}^{[k]} = \nabla f(\boldsymbol{x}^{[k]})$, we thus have

$$\omega^{[k]} = -\frac{\boldsymbol{d}^{[k]\top}\boldsymbol{g}^{[k]}}{\boldsymbol{d}^{[k]\top} A\boldsymbol{d}^{[k]}}. \tag{14.5}$$

(Observe that $\boldsymbol{d}^{[k]\top} A\boldsymbol{d}^{[k]} > 0$ for $\boldsymbol{d}^{[k]} \neq \boldsymbol{0}$, because A is positive definite and we are indeed at a minimum.) In other words,

$$\boldsymbol{x}^{[k+1]} = \boldsymbol{x}^{[k]} + \omega^{[k]}\boldsymbol{d}^{[k]} = \boldsymbol{x}^{[k]} - \frac{\boldsymbol{d}^{[k]\top}\boldsymbol{g}^{[k]}}{\boldsymbol{d}^{[k]\top} A\boldsymbol{d}^{[k]}}\boldsymbol{d}^{[k]}. \tag{14.6}$$

The description of our method is not complete without a means of choosing 'good' directions $\boldsymbol{d}^{[k]}$ which ensure that the target function f decays rapidly in each iteration.

The obvious idea is to choose the search direction $\boldsymbol{d}^{[k]}$ for which the function f decays the fastest at $\boldsymbol{x}^{[k]}$. Since the gradient there is $\boldsymbol{g}^{[k]} \neq \boldsymbol{0}$ (if the gradient vanishes we are already at the minimum and our labour is over!), we can take $\boldsymbol{d}^{[k]} = -\boldsymbol{g}^{[k]}$. This is known as *the steepest descent method.*[1]

Although this choice of steepest descent is natural, it leads to a method with unacceptably slow convergence. As an example, we take a 20×20 TST matrix A such that $a_{k,k} = 2$ and $a_{k,k+1} = a_{k+1,k} = -1$ (it follows at once from Lemma 12.5 that this matrix is positive definite) and $\boldsymbol{b} \in \mathbb{R}^{20}$ with $b_k = \cos[(k-1)\pi/19]$, $k = 1, \ldots, 20$. (There is special significance in this particular matrix, but it will be revealed only in Section 14.3.) The upper plot in Fig. 14.1 displays the first 100 values of $f(\boldsymbol{x}^{[k]})$ and, on the face of it, all is fine: the values decrease monotonically and clearly tend to

[1] This must not be confused with the identically named, but totally different, method of steepest descent in the theory of asymptotic expansions of highly oscillatory integrals.

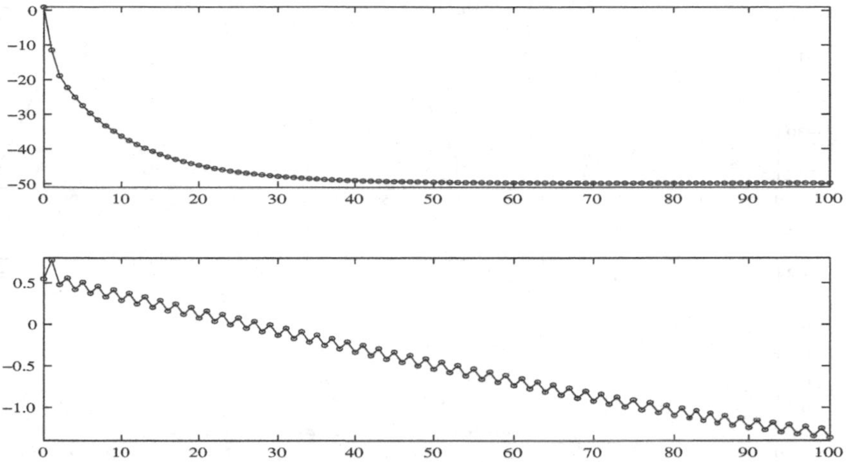

Figure 14.1 The values of $f(\boldsymbol{x}^{[k]})$ (upper plot) and of the logarithm of the residual norm $\log_{10}\|A\boldsymbol{x}^{[k]}-\boldsymbol{b}\|$ (lower plot), for the steepest descent method.

a limit. Unfortunately, this is only half the story and the lower plot is rather more disheartening. It exhibits the behaviour of $\log_{10}\|A\boldsymbol{x}^{[k]}-\boldsymbol{b}\|$. The norm of the residual $\|A\boldsymbol{x}^{[k]}-\boldsymbol{b}\|$ evidently does not decay monotonically (and there is absolutely no reason why should it, since in our choice of $\omega^{[k]}$ we have arranged for monotone decay of $f(\boldsymbol{x}^{[k]})$, not of the residual) but evidently it does decay on average at an exponential rate. Even so, the speed is excruciatingly slow and the graph demonstrates that after 100 iterations we cannot expect even two significant digits of accuracy.

The reason for this sluggish performance of the steepest descent method is that if the ratio of the greatest and smallest eigenvalues of A is large then the level sets of the function f are exceedingly elongated hyperellipsoids with steep faces. Instead of travelling down to the bottom of a hyperellipsoid, the iterates bounce ping-pong-like across the valley. This is demonstrated vividly in Fig. 14.2, where we have taken $d=2$ and

$$A=\begin{bmatrix}100 & 1\\ 1 & 1\end{bmatrix},\qquad \boldsymbol{b}=\begin{bmatrix}20\\ 0\end{bmatrix};$$

note that the ratio of the eigenvalues of A is large. The upper plot describes the sequence $(x_1^{[k]}-x_1^{[k-1]})/(x_2^{[k]}-x_2^{[k-1]})$, which evidently bounces up and down: very similar directions are repeated in this back-and-forth journey. (This figure does not describe the size of a step, only its direction.) Indeed, it is evident from the lower plot that the distances $\|\boldsymbol{x}^{[k]}-\boldsymbol{x}^{[k-1]}\|$ do decrease after a while and tend to zero, albeit not very rapidly. It is, however, the zig-zag pattern of directions that makes the method so ineffective.

It is possible to prove that the method of steepest descent converges, but this is of little comfort. It should be apparent by now, having studied Chapter 12, that we seek

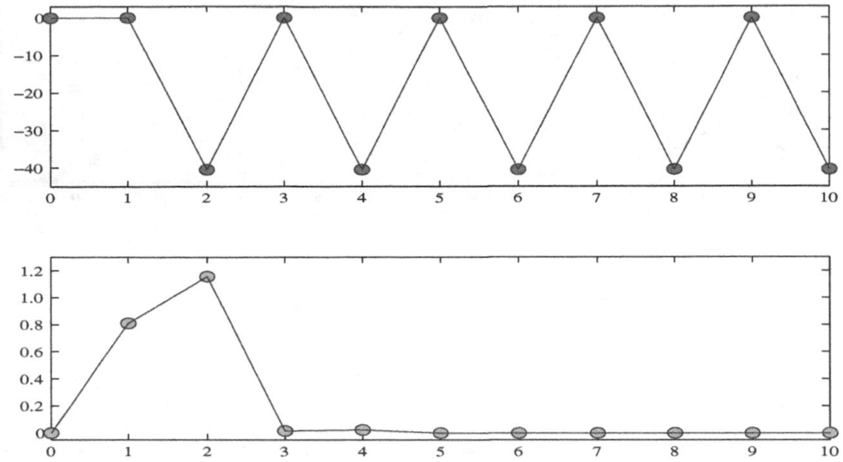

Figure 14.2 The zig-zag pattern of directions for steepest descent: the sequences $(x_1^{[k]} - x_1^{[k-1]})/(x_2^{[k]} - x_2^{[k-1]})$ (upper plot) and $\|\boldsymbol{x}^{[k]} - \boldsymbol{x}^{[k-1]}\|$ (lower plot).

rapid convergence. There is little point in designing or describing a new approach, unless it can compete with other leading methods.

The problem is not, we hasten to say, with the general idea of minimizing the function f. Moreover, the approach of choosing a descent direction $\boldsymbol{d}^{[k]}$, employing a line search to pick the optimal $\omega^{[k]}$ and updating the iteration according to (14.6) is perfectly sound. The problem lies in our intuitive and rash choice of the steepest direction $\boldsymbol{d}^{[k]} = -\boldsymbol{g}^{[k]}$. It should be clear by this stage in this book that the right criteria in the choice of computational methods are overwhelmingly global. What looks locally good is often globally disastrous!

14.2 The method of conjugate gradients

To eliminate the root cause of the sluggishness in the steepest descent method we are compelled to use directions that, rather than repeating themselves, are set well apart.

Recalling that the matrix A is positive definite, we say that the vectors $\boldsymbol{u}, \boldsymbol{v} \in \mathbb{R}^d$ are *conjugate* with respect to A if they are nonzero and satisfy $\boldsymbol{u}^\top A \boldsymbol{v} = 0$.[2]

Multiplying $\boldsymbol{x}^{[k+1]} = \boldsymbol{x}^{[k]} + \omega^{[k]}\boldsymbol{d}^{[k]}$ from the left by the matrix A and subtracting $\boldsymbol{b}$ from both sides, we have

$$A\boldsymbol{x}^{[k+1]} - \boldsymbol{b} = A\boldsymbol{x}^{[k]} - \boldsymbol{b} + \omega^{[k]} A\boldsymbol{d}^{[k]}.$$

Since $\boldsymbol{g}^{[k]} = \boldsymbol{\nabla} f(\boldsymbol{x}^{[k]}) = A\boldsymbol{x}^{[k]} - \boldsymbol{b}$, we thus deduce that

$$\boldsymbol{g}^{[k+1]} = \boldsymbol{g}^{[k]} + \omega^{[k]} A\boldsymbol{d}^{[k]}. \tag{14.7}$$

[2]Conjugacy is a generalization of the more familiar concept of orthogonality (A.1.3.2).

Lemma 14.2 *Let us suppose that $\boldsymbol{d}^{[k]}$ is conjugate to any vector $\boldsymbol{a} \in \mathbb{R}^d$ which is orthogonal to $\boldsymbol{g}^{[k]}$ (in other words, such that $\boldsymbol{a}^\top \boldsymbol{g}^{[k]} = 0$). Then $\boldsymbol{g}^{[k+1]}$ is orthogonal to $\boldsymbol{a}$.*

Proof We multiply (14.7) by $\boldsymbol{a}^\top$ from the left. The lemma follows because $\boldsymbol{a}^\top \boldsymbol{g}^{[k]} = 0$ (orthogonality) and $\boldsymbol{a}^\top A \boldsymbol{d}^{[k]} = 0$ (conjugacy). ■

The main idea of the *method of conjugate gradients (CG)* is to select search directions $\boldsymbol{d}^{[k]}$ which are conjugate to each other,

$$\boldsymbol{d}^{[k]\top} A \boldsymbol{d}^{[\ell]} = 0, \qquad k, \ell = 0, 1, \ldots, \quad k \neq \ell. \tag{14.8}$$

Specifically, we commence the iterative procedure with the steepest descent direction, $\boldsymbol{d}^{[0]} = -\boldsymbol{g}^{[0]}$, while choosing

$$\boldsymbol{d}^{[k+1]} = -\boldsymbol{g}^{[k+1]} + \beta^{[k]} \boldsymbol{d}^{[k]}, \qquad \text{where} \quad \beta^{[k]} = \frac{\boldsymbol{g}^{[k+1]\top} A \boldsymbol{d}^{[k]}}{\boldsymbol{d}^{[k]\top} A \boldsymbol{d}^{[k]}}, \qquad k = 0, 1, \ldots \tag{14.9}$$

We note that

$$\boldsymbol{d}^{[k+1]\top} A \boldsymbol{d}^{[k]} = (-\boldsymbol{g}^{[k+1]} + \beta^{[k]} \boldsymbol{d}^{[k]})^\top A \boldsymbol{d}^{[k]} = -\boldsymbol{g}^{[k+1]\top} A \boldsymbol{d}^{[k]} + \beta^{[k]} \boldsymbol{d}^{[k]\top} A \boldsymbol{d}^{[k]} = 0,$$

because of (14.9). Therefore $\boldsymbol{d}^{[k+1]}$ is conjugate to $\boldsymbol{d}^{[k]}$. We will soon prove that the substantially stronger statement (14.8) is true. First, however, we argue that the direction defined in (14.9) obeys the descent conditon (14.4).

Using (14.7) and substituting the value of $\omega^{[k]}$ from (14.5), we have

$$\boldsymbol{d}^{[k]\top} \boldsymbol{g}^{[k+1]} = \boldsymbol{d}^{[k]\top} (\boldsymbol{g}^{[k]} + \omega^{[k]} A \boldsymbol{d}^{[k]}) = \boldsymbol{d}^{[k]\top} \boldsymbol{g}^{[k]} + \omega^{[k]} \boldsymbol{d}^{[k]\top} A \boldsymbol{d}^{[k]} = 0. \tag{14.10}$$

The definition (14.9) of the new search direction, in tandem with (14.10), implies that

$$\boldsymbol{d}^{[k+1]\top} \boldsymbol{g}^{[k+1]} = (-\boldsymbol{g}^{[k+1]} + \beta^{[k]} \boldsymbol{d}^{[k]})^\top \boldsymbol{g}^{[k+1]} = -\|\boldsymbol{g}^{[k+1]}\|^2 < 0$$

(recall that $\boldsymbol{g}^{[k+1]} \neq \boldsymbol{0}$, otherwise we are already at the minimum and the iterative process terminates) and that $\boldsymbol{d}^{[k+1]}$ is indeed a descent direction.

We wish to prove that (14.8) holds and the directions are conjugate. This will be done as part of a larger technical theorem, exploring a number of important features of the CG algorithm. We commence by defining for each $k = 0, 1, \ldots$ the linear spaces

$$\mathcal{D}_k = \mathrm{Sp}\left\{\boldsymbol{d}^{[0]}, \boldsymbol{d}^{[1]}, \ldots, \boldsymbol{d}^{[k]}\right\}, \qquad \mathcal{G}_k = \mathrm{Sp}\left\{\boldsymbol{g}^{[0]}, \boldsymbol{g}^{[1]}, \ldots, \boldsymbol{g}^{[k]}\right\},$$

where the *span* Sp of a set of vectors in $\mathbb{R}^d$ was defined in Section 9.1.

Theorem 14.3 *The following assertions are true for all $k = 1, 2, \ldots$*

Assertion 1. *The linear spaces $\mathcal{D}_{k-1}$ and $\mathcal{G}_{k-1}$ are the same.*

Assertion 2. *The direction* $\boldsymbol{d}^{[k-1]}$ *is conjugate to* $\boldsymbol{d}^{[j]}$ *for* $k \geq 2$ *and* $j = 0, 1, \ldots, k-2$.

Assertion 3. *The gradients satisfy the orthogonality condition* $\boldsymbol{g}^{[j]\top}\boldsymbol{g}^{[k]} = 0$, $j = 0, 1, \ldots, k-1$.

Proof All three assertions are trivial for $k = 1$: the first follows from $\boldsymbol{d}^{[0]} = -\boldsymbol{g}^{[0]}$, the second is immediate and the last comes from $\boldsymbol{g}^{[0]} = -\boldsymbol{d}^{[0]}$ by letting $k = 0$ in (14.10).

We continue by induction. Suppose that the assertions of the theorem are true for k. The first assertion is easy. Since $\boldsymbol{g}^{[k]} \in \mathcal{G}_k$ and, by induction, $\boldsymbol{d}^{[k-1]} \in \mathcal{D}_{k-1} = \mathcal{G}_{k-1} \subset \mathcal{G}_k$, it follows from (14.9) that

$$\boldsymbol{d}^{[k]} = -\boldsymbol{g}^{[k]} + \beta^{[k-1]}\boldsymbol{d}^{[k-1]} \in \mathcal{G}_k,$$

therefore $\mathcal{D}_k \subseteq \mathcal{G}_k$. Likewise, since $\boldsymbol{d}^{[k-1]}, \boldsymbol{d}^{[k]} \in \mathcal{D}_k$, we again deduce from (14.9) that

$$\boldsymbol{g}^{[k]} = \beta^{[k-1]}\boldsymbol{d}^{[k-1]} - \boldsymbol{d}^{[k]} \in \mathcal{D}_k,$$

therefore $\mathcal{G}_k \subseteq \mathcal{D}_k$. Consequently $\mathcal{D}_k = \mathcal{G}_k$ and the first assertion of the theorem is true for $k+1$.

We turn our attention to the second assertion and note that we have already shown that $\boldsymbol{d}^{[k]\top}A\boldsymbol{d}^{[k-1]} = 0$. Therefore, to advance the inductive argument we need to show that $\boldsymbol{d}^{[k]\top}A\boldsymbol{d}^{[j]} = 0$ for $j = 0, 1, \ldots, k-2$. (If $k = 1$ then there is nothing to show!) According to (14.9), this is equivalent to

$$-\boldsymbol{g}^{[k]\top}A\boldsymbol{d}^{[j]} + \beta^{[k-1]}\boldsymbol{d}^{[k-1]\top}A\boldsymbol{d}^{[j]} = 0, \qquad j = 0, 1, \ldots, k-2,$$

but according to the induction assumption $\boldsymbol{d}^{[k-1]\top}A\boldsymbol{d}^{[j]} = 0$ within this range. Therefore it is enough to demonstrate that $\boldsymbol{g}^{[k]\top}A\boldsymbol{d}^{[j]} = 0$ for $j = 0, 1, \ldots, k-2$.

It follows from (14.7), replacing k by j, that $\boldsymbol{g}^{[j+1]} - \boldsymbol{g}^{[j]} = \omega^{[j]}A\boldsymbol{d}^{[j]}$. Therefore

$$\omega^{[j]}\boldsymbol{g}^{[k]\top}A\boldsymbol{d}^{[j]} = \omega^{[j]}\boldsymbol{g}^{[k]\top}(\boldsymbol{g}^{[j+1]} - \boldsymbol{g}^{[j]}) = 0$$

for $j = 0, 1, \ldots, k-2$, because the third assertion and the inductive argument mean that $\boldsymbol{g}^{[k]\top}\boldsymbol{g}^{[j]} = 0$ for $j = 0, 1, \ldots, k-1$. Since $\omega^{[j]} \neq 0$ (actually, $\omega^{[j]} > 0$, because $\boldsymbol{d}^{[j]\top}\boldsymbol{g}^{[j]} < 0$ and $\boldsymbol{d}^{[j]\top}A\boldsymbol{d}^{[j]} > 0$, cf. (14.5)), it follows that $\boldsymbol{g}^{[k]\top}A\boldsymbol{d}^{[j]} = 0$, $j = 0, 1, \ldots, k-2$ and we have proved that the second assertion is valid for $k+1$.

All that remains is to prove that we can advance the third assertion from k to $k+1$, i.e. that $\boldsymbol{g}^{[j]\top}\boldsymbol{g}^{[k+1]} = 0$, $j = 0, 1, \ldots, k$. However, we have already proved that $\mathcal{G}_k = \mathcal{D}_k$, therefore this is equivalent to $\boldsymbol{d}^{[j]\top}\boldsymbol{g}^{[k+1]} = 0$, $j = 0, 1, \ldots, k$. Furthermore, (14.10) implies that the latter is true for $j = k$, therefore we need to check just the range $j = 0, 1, \ldots, k-1$.

According to the induction assumption applied to the third assertion, it is true that $\boldsymbol{d}^{[j]\top}\boldsymbol{g}^{[k]} = 0$, $j = 0, 1, \ldots, k-1$. Therefore

$$\boldsymbol{d}^{[j]\top}\boldsymbol{g}^{[k+1]} = 0 \quad \Longleftrightarrow \quad \boldsymbol{d}^{[j]\top}(\boldsymbol{g}^{[k+1]} - \boldsymbol{g}^{[k]}) = 0, \qquad j = 0, 1, \ldots, k-1,$$

and it is the claim on the right that we now prove. Because of (14.7), this statement is identical to

$$\omega^{[k]} \boldsymbol{d}^{[j]\top} A \boldsymbol{d}^{[k]} = 0, \qquad j = 0, 1, \ldots, k-1,$$

which follows at once from the conjugacy of $\boldsymbol{d}^{[k]}$ and $\boldsymbol{d}^{[j]}$, $j = 0, 1, \ldots, k-1$, the second assertion of the theorem, which we have already proved.

This completes the inductive step and the proof of the theorem. ■

Note the clever way in which the second and third assertions are intertwined in the proof of the theorem: we need the third assertion to advance the induction for the second, but this is repaid by the second assertion, which is required to prove the third.

Corollary *Once the CG method is applied in exact arithmetic, it terminates in at most d steps.*

Proof Because of the third assertion of the theorem, the sequence $\{\boldsymbol{g}^{[0]}, \boldsymbol{g}^{[1]}, \ldots\}$ consists of mutually orthogonal nonzero vectors unless $\boldsymbol{g}^{[r]} = \boldsymbol{0}$ for some $r \geq 0$, whence the method terminates. Since there cannot be more than d mutually orthogonal nonzero vectors in $\mathbb{R}^d$, we deduce that the iterative procedure terminates and, in addition, $r \leq d$. ■

Real computers work in finite-precision arithmetic. Once d is large, as it inevitably is in the problems of concern in this book, roundoff errors accumulate and cause gradual deterioration in the orthogonality of the $\boldsymbol{g}^{[k]}$. Thus the method does not necessarily terminate in at most d (or any finite number of) steps. Even so, its convergence represents a vast improvement upon the method of steepest descent.

◇ **A simple example of conjugate gradients** We now revisit the linear system from Section 14.1 that demonstrated the sluggishness of the method of steepest descent. Thus, A is a 20×20 TST matrix with $a_{k,k} = 2$ and $a_{k+1,k} = a_{k,k+1} = -1$, while $\boldsymbol{b} \in \mathbb{R}^{20}$ is defined by $b_k = \cos(k-1)\pi/19$.

Figure 14.3 depicts the values of $f(x^{[k]})$ and, in the lower plot, the decimal logarithm of the norm of the residual, the same information that we have already reported in Fig. 14.1 for the method of steepest descent (except that now we stop after just 20 iterations). The difference could not be greater! The logarithm of the norm (which roughly corresponds to the number of exact decimal digits in the solution) decreases gently for a while and then, in the ninth iteration, drops suddenly down to the least value allowed by machine accuracy. Not much happens afterwards: the iterative procedure delivers all it can in nine iterations.

Note another interesting point. The corollary to Theorem 14.3 stated that in exact arithmetic we need at least $d = 20$ steps, but in reality we have reached the exact solution (up to machine precision) in half that number of iterations. This is not an accident of fate but a structural feature of the CG method, which we will exploit to good effect in the next section. ◇

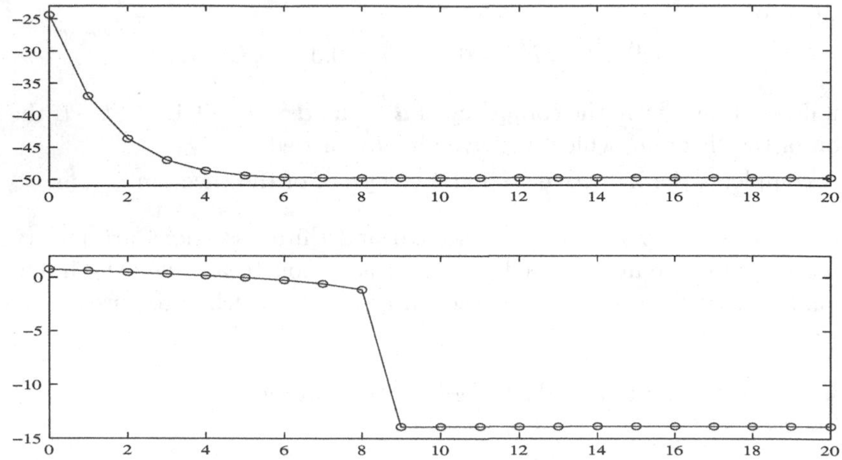

Figure 14.3 The values of $f(\boldsymbol{x}^{[k]})$ (upper plot) and of the logarithm of the residual norm, $\log_{10}\|A\boldsymbol{x}^{[k]}-\boldsymbol{b}\|$ (lower plot), for the conjugate gradients method.

The time has come to gather all the strands together and present the CG algorithm in a convenient form. To this end we let $\boldsymbol{r}^{[k]} = -\boldsymbol{g}^{[k]} = \boldsymbol{b} - A\boldsymbol{x}^{[k]}$, $k = 0, 1, \ldots$, be the residual. Putting together (14.7) and (14.9), we have

$$\beta^{[k]} = \frac{\boldsymbol{g}^{[k+1]\top} A\boldsymbol{d}^{[k]}}{\boldsymbol{d}^{[k]\top} A\boldsymbol{d}^{[k]}} = \frac{\boldsymbol{g}^{[k+1]\top}(\boldsymbol{g}^{[k+1]} - \boldsymbol{g}^{[k]})}{\boldsymbol{d}^{[k]\top}(\boldsymbol{g}^{[k+1]} - \boldsymbol{g}^{[k]})}.$$

However, by Theorem 14.3 the $\boldsymbol{g}^{[k]}$ are orthogonal, therefore

$$\boldsymbol{g}^{[k+1]\top}(\boldsymbol{g}^{[k+1]} - \boldsymbol{g}^{[k]}) = \|\boldsymbol{g}^{[k+1]}\|^2 = \|\boldsymbol{r}^{[k+1]}\|^2.$$

Moreover, by (14.10) we have $\boldsymbol{d}^{[k]\top}\boldsymbol{g}^{[k+1]} = \boldsymbol{d}^{[k-1]\top}\boldsymbol{g}^{[k]} = 0$. Therefore it follows from (14.9) that

$$\boldsymbol{d}^{[k]\top}(\boldsymbol{g}^{[k+1]} - \boldsymbol{g}^{[k]}) = -\boldsymbol{d}^{[k]\top}\boldsymbol{g}^{[k]} = -(-\boldsymbol{g}^{[k]} + \beta^{[k-1]}\boldsymbol{d}^{[k-1]})^\top\boldsymbol{g}^{[k]} = \|\boldsymbol{g}^{[k]}\|^2 = \|\boldsymbol{r}^{[k]}\|^2.$$

We deduce the somewhat neater form

$$\beta^{[k]} = \frac{\|\boldsymbol{r}^{[k+1]}\|^2}{\|\boldsymbol{r}^{[k]}\|^2}.$$

The standard form of the CG algorithm The 'plain vanilla' conjugate gradients method consists of the following steps.

Step 1. Set $\boldsymbol{x}^{[0]} = \boldsymbol{0} \in \mathbb{R}^d$, $\boldsymbol{r}^{[0]} = \boldsymbol{b}$ and $\boldsymbol{d}^{[0]} = \boldsymbol{r}^{[0]}$. Let $k = 0$.

Step 2. Stop when $\|\boldsymbol{r}^{[k]}\|$ is acceptably small.

Step 3. If $k \geq 1$ (i.e., except for the initial step) set $\beta^{[k-1]} = \|\boldsymbol{r}^{[k]}\|^2 / \|\boldsymbol{r}^{[k-1]}\|^2$ and $\boldsymbol{d}^{[k]} = \boldsymbol{r}^{[k]} + \beta^{[k-1]}\boldsymbol{d}^{[k-1]}$.

Step 4. Calculate the matrix–vector product $\boldsymbol{v}^{[k]} = A\boldsymbol{d}^{[k]}$, subsequently letting $\omega^{[k]} = \|\boldsymbol{r}^{[k]}\|^2 / \left(\boldsymbol{d}^{[k]\top}\boldsymbol{v}^{[k]}\right)$.

Step 5. Form the new iteration $\boldsymbol{x}^{[k+1]} = \boldsymbol{x}^{[k]} + \omega^{[k]}\boldsymbol{d}^{[k]}$ and the new residual $\boldsymbol{r}^{[k+1]} = \boldsymbol{r}^{[k]} - \omega^{[k]}\boldsymbol{v}^{[k]}$.

Step 6. Increase k by one and go back to step 2.

Perhaps the most remarkable feature of the CG algorithm is not apparent at first glance: the *only* way the matrix A enters into the calculation (and the only computationally significant part of the algorithm) is in the formation of the auxiliary vector $\boldsymbol{v}^{[k]}$ in step 4. This has two important implications.

Firstly, often we do not need even to form the matrix A explicitly in order to execute the matrix–vector product. It is enough to have a constructive rule to formulate it! Thus, if A originates in the five-point formula (8.16) then the rule in forming $\boldsymbol{v} = A\boldsymbol{d}$ is 'for every (k, ℓ) on the grid add the component d_i value of $\boldsymbol{d}$ corresponding to the vertical and horizontal neighbours of the point and subtract four times the d_i corresponding to the grid point'. This use of a 'multiplication rule' rather than direct matrix mutiplication has been already evident in the iterative methods of Chapter 12; it allows a drastic reduction in cost. Thus, an $m \times m$ grid results in $d = m^2$ equations and naive matrix–vector multiplication would require $\mathcal{O}(m^4)$ operations, whereas using the above rule results in $\mathcal{O}(m^2)$ operations. It is precisely this sort of reasoning that converts computational methods from ugly ducklings to fully fledged swans.

The second implication is that we can often lift the restrictive condition that A is symmetric and positive definite. Suppose thus that we wish to solve the linear system $B\boldsymbol{x} = \boldsymbol{c}$, where the $d \times d$ matrix B is nonsingular. We convert it to the form (14.1) by letting $A = B^\top B$ and $\boldsymbol{b} = B^\top \boldsymbol{c}$: note that A is indeed symmetric and positive definite. Of course, in practical applications, and bearing in mind the previous paragraph, we never actually form the matrix A, a fairly costly procedure. Instead, to calculate $\boldsymbol{v}^{[k]}$ we first use the 'multiplication rule' to form $\boldsymbol{u} = B\boldsymbol{d}^{[k]}$ and next employ the transpose of that rule to evaluate $\boldsymbol{v}^{[k]} = B^\top \boldsymbol{u}$.

14.3 Krylov subspaces and preconditioners

There is more to Fig. 14.3 than meets the eye. The rapid drop in error, down to machine accuracy, after just nine iterations is not accidental; it is implicit in our choice of the matrix A and the vector $\boldsymbol{b}$. The right terminology in which to express this behaviour and harness it to accelerate the CG method is the formalism of Krylov subspaces.

Given a $d \times d$ matrix A (which need be neither symmetric nor positive definite), a vector $\boldsymbol{v} \in \mathbb{R}^d \setminus \{\boldsymbol{0}\}$ and a natural number m, we call the linear space

$$\mathcal{K}_m(A, \boldsymbol{v}) = \mathrm{Sp}\{A^j \boldsymbol{v} : j = 0, 1, \ldots, m-1\}$$

the *mth Krylov subspace* of $\mathbb{R}^d$. It is trivial to verify that $\mathcal{K}_m(A, \boldsymbol{v})$ is indeed a linear space.

Lemma 14.4 *Let ℓ_m be the dimension of the Krylov subspace $\mathcal{K}_m(A, \boldsymbol{v})$. The sequence $\{\ell_m\}_{m=0,1,\ldots}$ increases monotonically. Moreover, there exists a natural number s with the following property: for every $m = 1, 2, \ldots, s$ it is true that $\ell_m = m$, while $\ell_m = s$ for all $m \geq s$.*

Suppose further that $\boldsymbol{v} = \sum_{i=1}^{r} c_i \boldsymbol{w}_i$, where $\boldsymbol{w}_1, \boldsymbol{w}_2, \ldots, \boldsymbol{w}_r$ are eigenvectors of A corresponding to distinct eigenvalues and $c_1, c_2, \ldots, c_r \neq 0$. Then $s = r$.

Proof Since it follows from the definition of Krylov subspaces that $\mathcal{K}_m(A, \boldsymbol{v}) \subseteq \mathcal{K}_{m+1}(A, \boldsymbol{v})$, we deduce that $\ell_m \leq \ell_{m+1}$, $m = 0, 1, \ldots$: we indeed have a monotonically increasing sequence. Moreover, $\ell_m \leq d$, because $\mathcal{K}_m(A, \boldsymbol{v}) \subseteq \mathbb{R}^d$, while $\boldsymbol{v} \neq \boldsymbol{0}$ implies that $\ell_1 = 1$. Finally, since $\mathcal{K}_m(A, \boldsymbol{v})$ is spanned by m vectors, necessarily $\ell_m \leq m$. To sum up,

$$1 = \ell_1 \leq \ell_2 \leq \ell_3 \leq \cdots \leq \ell_m \leq \min\{m, d\}, \qquad m \geq 3.$$

Let s be the greatest integer such that $\ell_s = s$ and note that $s \geq 1$. Since $\ell_m \leq m$, we deduce that $\ell_m \leq m - 1$ for $m \geq s + 1$, in particular $\ell_{s+1} \leq s$. However, by the definition of s, it is true that $s = \ell_s \leq \ell_{s+1}$. Therefore $\ell_{s+1} = \ell_s$ and we deduce that $\mathcal{K}_{s+1}(A, \boldsymbol{v}) = \mathcal{K}_s(A, \boldsymbol{v})$. This means that $A^s \boldsymbol{v} \in \mathcal{K}_s(A, \boldsymbol{v})$, hence that there exist $\alpha_0, \alpha_1, \ldots, \alpha_{s-1}$ such that $A^s \boldsymbol{v} = \sum_{i=0}^{s-1} \alpha_i A^i \boldsymbol{v}$. Multiplying both sides by A^j for any $j = 0, 1, \ldots,$ we have

$$A^{s+j} \boldsymbol{v} = \sum_{i=0}^{s-1} \alpha_i A^{i+j} \boldsymbol{v}.$$

Therefore, if $A^j \boldsymbol{v}, A^{j+1} \boldsymbol{v}, \ldots, A^{j+s-1} \boldsymbol{v} \in \mathcal{K}_s(A, \boldsymbol{v})$ then necessarily also $A^{j+s} \boldsymbol{v} \in \mathcal{K}_s(A, \boldsymbol{v})$. Since, as we have just seen, this is true for $j = 0$, it follows by induction that $A^j \boldsymbol{v} \in \mathcal{K}_s(A, \boldsymbol{v})$ for all $j = 0, 1, \ldots,$ hence that $\mathcal{K}_m(A, \boldsymbol{v}) = \mathcal{K}_s(A, \boldsymbol{v})$ and $\ell_m = \ell_s$ for all $m \geq k$.

To complete the proof, we assume that $\boldsymbol{v}$ can be written as a linear combination of $\boldsymbol{w}_1, \ldots, \boldsymbol{w}_r$, eigenvectors of A corresponding to distinct eigenvalues $\lambda_1, \ldots, \lambda_r$ respectively,

$$\boldsymbol{v} = \sum_{i=1}^{r} c_i \boldsymbol{w}_i,$$

where the coefficients $c_1, \ldots, c_r$ are all nonzero. Therefore $A^j \boldsymbol{v} = \sum_{i=1}^{r} c_i \lambda_i^j \boldsymbol{w}_i$, $j = 0, 1, \ldots,$ and we conclude that

$$\mathcal{K}_s(A, \boldsymbol{v}) = \operatorname{Sp}\{\boldsymbol{v}, A\boldsymbol{v}, \ldots, A^{s-1}\boldsymbol{v}\} \subseteq \operatorname{Sp}\{\boldsymbol{w}_1, \boldsymbol{w}_2, \ldots, \boldsymbol{w}_r\}.$$

Eigenvectors corresponding to distinct eigenvalues are linearly independent and we thus deduce that $s \leq r$.

Assume next that $s < r$. Then, by the definition of s, it is necessarily true that $\ell_r = \ell_s = s$ and this means that the r vectors $A^j \boldsymbol{v}$, $j = 0, 1, \ldots, r - 1$, are linearly dependent: there exist scalars $\beta_0, \beta_1, \ldots, \beta_{r-1}$, not all zero, such that $\sum_{j=1}^{r-1} \beta_j A^j \boldsymbol{v} = \boldsymbol{0}$

Therefore

$$\mathbf{0}=\sum_{j=0}^{r-1}\beta_jA^j\boldsymbol{v}=\sum_{j=0}^{r-1}\beta_jA^j\sum_{i=1}^{r}c_i\boldsymbol{w}_i=\sum_{j=0}^{r-1}\beta_j\sum_{i=1}^{r}c_i\lambda_i^j\boldsymbol{w}_i=\sum_{i=1}^{r}c_i\left(\sum_{j=0}^{r-1}\beta_j\lambda_i^j\right)\boldsymbol{w}_i.$$

Since eigenvectors are linearly independent and $c_1, c_2, \ldots, c_r \neq 0$, we thus deduce that

$$p(\lambda_i)=0, \quad i=1,2,\ldots,r, \qquad \text{where} \qquad p(z)=\sum_{j=0}^{r-1}\beta_jz^j.$$

Now, p is a polynomial of degree $r-1$ and it is not identically zero. But we have just proved that it vanishes at the r distinct points $\lambda_1, \lambda_2, \ldots, \lambda_r$. This is a contradiction, following from our assumption that $s < r$. Therefore this assumption must be false, $r = s$ and the proof is complete. ■

Many methods in linear algebra can be phrased in the terminology of Krylov subspaces and this often leads to their better understanding – and, once we understand methods, we can often improve them!

Theorem 14.5 *Each residual $\boldsymbol{r}^{[m]}$ generated by the method of conjugate gradients belongs to the Krylov subspace $\mathcal{K}_{m+1}(A, \boldsymbol{b})$, $m = 0, 1, \ldots$*

Proof The first three residuals are explicitly

$$\begin{aligned}
\boldsymbol{r}^{[0]} &= \boldsymbol{b} \in \mathcal{K}_1(A,\boldsymbol{b}),\\
\boldsymbol{r}^{[1]} &= \boldsymbol{r}^{[0]} - \omega^{[0]}A\boldsymbol{d}^{[0]} = (I-\omega^{[0]}A)\boldsymbol{b} \in \mathcal{K}_2(A,\boldsymbol{b}),\\
\boldsymbol{r}^{[2]} &= \boldsymbol{r}^{[1]} - \omega^{[1]}A\boldsymbol{d}^{[1]} = \boldsymbol{r}^{1} - \omega^{[1]}A(\boldsymbol{r}^{[1]}+\beta^{[0]}\boldsymbol{b})\\
&= [(I-\omega^{[1]}A)(I-\omega^{[0]}A)-\omega^{[1]}\beta^{[0]}A]\boldsymbol{b} \in \mathcal{K}_3(A,\boldsymbol{b}).
\end{aligned}$$

Thus, the claim of the theorem is true for $m = 0, 1, 2$ and we note that the first assertion of Theorem 14.3 now implies also that $\boldsymbol{d}^{[m]} \in \mathcal{K}_{m+1}(A, \boldsymbol{b})$ for $m = 0, 1, 2$.

We continue by induction. Assume that $\boldsymbol{r}^{[j]}, \boldsymbol{d}^{[j]} \in \mathcal{K}_{j+1}(A, \boldsymbol{b})$ for $j \leq m$. Since $\boldsymbol{r}^{[m+1]} = \boldsymbol{r}^{[m]} - \omega^{[m]}A\boldsymbol{d}^{[m]}$, it follows from the definition of Krylov subspaces that $\boldsymbol{r}^{[m+1]} \in \mathcal{K}_{m+2}(A, \boldsymbol{b})$. Hence, according to the first assertion of Theorem 14.3, the same is true for $\boldsymbol{d}^{[m+1]}$ and our proof is complete. ■

Corollary *The CG method in exact arithmetic terminates in at most ℓ_d steps, where ℓ_d is the dimension of $\mathcal{K}_d(A, \boldsymbol{b})$.*

Proof According to the third assertion of Theorem 14.3 the residuals $\boldsymbol{r}^{[m]}$ are orthogonal to each other. Therefore the number of nonzero residuals is bounded by the dimension of $\mathcal{K}_d(A, \boldsymbol{b})$. ■

The difference between the corollary to Theorem 14.3 (convergence in at most d iterations) and the corollary to Theorem 14.5 (convergence in at most ℓ_d iterations) is the key to improving upon conjugate gradients. If only we can make ℓ_d significantly smaller than d, we can expect the method to perform significantly better.

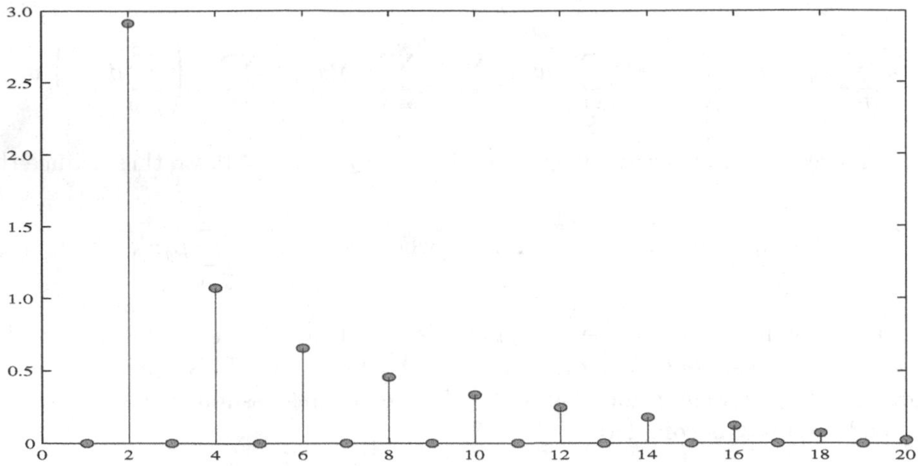

Figure 14.4 The coefficients $|u_k|$ for the example from Figs 14.1 and 14.3.

Sometimes ℓ_d is small by good fortune. In the example that we have already considered in Figs 14.1 and 14.3, the right-hand side $\boldsymbol{b}$ can be expressed as a linear combination of just ten eigenvectors. Thus let $A = WDW^{-1}$, where W is the matrix of the eigenvectors of A (which is orthogonal, since A is symmetric) and the diagonal matrix D comprises of its eigenvalues (A.1.5.4). Then

$$\boldsymbol{b} = \sum_{k=1}^{20} u_k \boldsymbol{w}_k, \qquad \text{where} \qquad \boldsymbol{u} = W^{-1}\boldsymbol{b} = W^\top \boldsymbol{b}.$$

Figure 14.4 displays the quantities $|u_k|$ and it is evident that $u_{2k+1} = 0$ for $k = 0, 1, \ldots, 9$. Therefore, resorting to the notation of Lemma 14.4, we can express $\boldsymbol{v} = \boldsymbol{b}$ as a linear combination of just ten eigenvectors $c_k = u_{2k}$ and the dimension of $\mathcal{K}_{20}(A, \boldsymbol{b})$ is just $\ell_{20} = 10$. This explains the lower plot in Fig. 14.3.

In general, we can hardly expect the matrix A and the vector $\boldsymbol{b}$ to be in such a perfect relationship: serendipity can take us only so far! It is a general rule in life and numerical analysis that, to be lucky, we must make our own luck.

Consider the problems

$$B\boldsymbol{z} = \boldsymbol{g} \qquad \text{and} \qquad B^\top \boldsymbol{z} = \boldsymbol{g}, \tag{14.11}$$

where $\boldsymbol{g} \in \mathbb{R}^d$ is arbitrary while the $d \times d$ matrix B is nonsingular. Assume further that either of the systems (14.11) can be solved very easily and cheaply: for example, B might be tridiagonal or banded. The idea is to incorporate repeated solution of systems of the form (14.1) into the CG method, to accelerate it. This procedure is known as *preconditioning* and the outcome is the method of *preconditioned conjugate gradients (PCG)*.

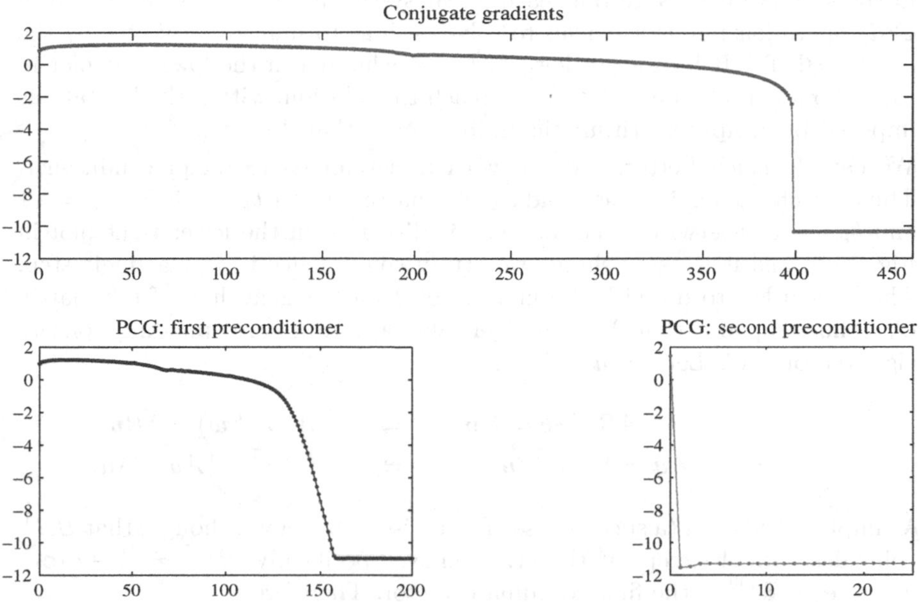

Figure 14.5 The logarithms of the residuals $\log_{10} \|\boldsymbol{r}^{[k]}\|$ for 'plain' conjugate gradients and for two PCG methods, applied to a 400×400 TST matrix.

We set $\boldsymbol{h} = B^{-1}\boldsymbol{b}$, $\boldsymbol{y} = B^{\top}\boldsymbol{x}$ and $C = B^{-1}AB^{-\top}$. Then

$$A\boldsymbol{x} = \boldsymbol{b} \quad \Rightarrow \quad AB^{-\top}\boldsymbol{y} = B\boldsymbol{h} \quad \Rightarrow \quad C\boldsymbol{y} = \boldsymbol{h}.$$

In place of A and $\boldsymbol{b}$, we apply the CG method with C and $\boldsymbol{h}$. Note that we need to change just two ingredients of the CG algorithm. In step 1 we calculate $\boldsymbol{r}^{[0]} = \boldsymbol{h}$ by solving the linear system $B\boldsymbol{h} = \boldsymbol{b}$, while in step 4 we compute $\boldsymbol{v}^{[k]} = C\boldsymbol{d}^{[k]}$ in two stages: firstly we evaluate $\boldsymbol{u} \in \mathbb{R}^d$ such that $B^{\top}\boldsymbol{u} = \boldsymbol{d}^{[k]}$ and subsequently find $\boldsymbol{v}^{[k]}$ by solving the linear system $B\boldsymbol{v}^{[k]} = A\boldsymbol{u}$. All these calculations involve the solution of linear systems of the form (14.11), which we have assumed is easy. In addition we need to add a final step, to recover $\boldsymbol{x} = B^{-\top}\boldsymbol{y}$.

◇ **The TST matrix, again ...** We consider again a TST matrix A with 2's along the main diagonal and −1's in the first off-diagonal, except that we now let $d = 400$. The vector $\boldsymbol{b}$ is defined by $b_k = 1/\sqrt{k}$, $k = 1, 2, \ldots, 400$.

The upper plot in Fig. 14.5 depicts the size (on a logarithmic scale) of the residual for the 'plain vanilla' CG method. As predicted by our theory, the iteration lumbers along for 400 steps, decaying fairly gently, and then in a single step the error drops down to eleven significant digits – as much as computer arithmetic will allow.

How to precondition our system? Our first shot is to choose B as the lower-triangular portion of A, i.e. a matrix with 2's along the diagonal and −2's in the subdiagonal. Note that each linear system in (14.11) can be solved in $\mathcal{O}(d)$ operations: it is as cheap to solve each as to multiply a vector by the matrix B! The behaviour of $\log_{10}\|\boldsymbol{r}^{[k]}\|$ is exhibited in the lower left plot in Fig. 14.5 and it can be seen that we reach the solution, within the limitations imposed by computer arithmetic, in little more than 150 steps.[3]

We can do much better, though, with a cleverer choice of preconditioner. Thus, we choose again B as a bidiagonal matrix but let $b_{k,k} = 1$, $b_{k+1,k} = -1$ and $b_{k,\ell} = 0$ otherwise. The outcome is displayed in the lower right plot in Fig. 14.5, and it is astonishing: we attain convergence in just a single step! The reason has to do with the number of distinct eigenvalues of the matrix C. Thus, suppose that λ is an eigenvalue and $\boldsymbol{w}$ a corresponding nonzero eigenvector of C. Letting $\boldsymbol{u} = B^{-\top}\boldsymbol{v}$,

$$\begin{aligned} & B^{-1}AB^{-\top}\boldsymbol{w} = \lambda\boldsymbol{w} && \Rightarrow && A(B^{-\top}\boldsymbol{w}) = \lambda B\boldsymbol{w} \\ \Rightarrow\quad & A\boldsymbol{u} = \lambda(BB^{\top})\boldsymbol{u} && \Rightarrow && (BB^{\top})^{-1}A\boldsymbol{u} = \lambda\boldsymbol{u}. \end{aligned}$$

A simple calculation (a special case of Exercise 14.6) shows, though, that $BB^{\top}$ coincides with A except at the $(1,1)$ entry. Specifically, $BB^{\top} = A - \boldsymbol{e}_1\boldsymbol{e}_1^{\top}$, where $\boldsymbol{e}_1 \in \mathbb{R}^{400}$ is the first coordinate vector. Therefore

$$F := (BB^{\top})^{-1}A = I - (BB^{\top})^{-1}\boldsymbol{e}_1\boldsymbol{e}_1^{\top} = I - \boldsymbol{\gamma}\boldsymbol{e}_1^{\top},$$

where $\boldsymbol{\gamma} = (BB^{\top})^{-1}\boldsymbol{e}_1$, a rank-1 perturbation of the identity matrix. It is now a trivial exercise to verify that all the eigenvalues of F, except for one, are equal to unity. (The remaining eigenvalue is $1 - \gamma_1$.) But the eigenvalues of F and C coincide, and so we deduce that the matrix C has just two distinct eigenvalues. Therefore, by Lemma 14.4, the dimension of $\mathcal{K}_m(C, \boldsymbol{h})$ is at most 2 and convergence in a single step follows from the corollary to Theorem 14.5. ◇

Our example looks, and indeed is, too good to be true. (Anyway, we do not need iterative methods to solve tridiagonal systems, the direct method of Chapter 11 will do!) In general, even the cleverest preconditioner cannot reduce the number of iterations down to one or two. Our example is artificial, yet it emphasizes the potential benefits that follow from a good choice of preconditioner.

How in general should we choose a good preconditioner? The purpose being to reduce the maximal dimension of $\mathcal{K}_m(B^{-1}AB^{-\top}, B^{-1}\boldsymbol{b})$, we note that for every $j = 0, 1, \ldots$ it is true that

$$(B^{-1}AB^{-\top})^j B^{-1} = B^{-1}(AB^{-\top}B^{-1})^j = B^{-1}(AS^{-1})^{-1},$$

where $S = BB^{\top}$. Therefore

$$\boldsymbol{y} \in \mathcal{K}_m(AS^{-1}, \boldsymbol{b}) \qquad \Leftrightarrow \qquad B^{-1}\boldsymbol{y} \in \mathcal{K}_m(B^{-1}AB^{-\top}, B^{-1}\boldsymbol{b})$$

[3] Note that, unlike in the case of plain conjugate gradients, here computer arithmetic has a very minor effect on accuracy. The reason is simply that the entire procedure requires less computation, hence generates less roundoff error.

and, since B is nonsingular, we deduce that the dimensions of $\mathcal{K}_m(B^{-1}AB^{-\top}, B^{-1}\boldsymbol{b})$ and $\mathcal{K}_m(AS^{-1}, \boldsymbol{b})$ are the same.

A popular technique is to choose a symmetric positive definite matrix S such that $\|A - S\|$ is small and S can be Cholesky-factorized easily. Yet, an insistence on small $\|A - S\|$ might be misleading, since the dimension of $\mathcal{K}_m(AS^{-1}, \boldsymbol{b})$ does not change when S is replaced by aS for any $a > 0$.

An obvious choice of preconditioner, which we have used already in the above example, is to take B as the lower triangular part of A. Another option is to choose S as a banded portion of A (provided that it is positive definite) and use the approach of Section 11.1 to factorize it into the form $S = BB^\top$, where B is lower triangular. This, of course, means that the systems (14.11) can be solved rapidly, as is necessary for preconditioning.

A more sophisticated approach adopts the graph-theoretical elimination methods of Section 11.2. Suppose that the graph corresponding to the matrix A is not a tree but that we can convert it to a tree by setting to zero a small number of entries. We obtain in this manner a matrix S (of course, we need to check that it is positive definite) which can be subjected to perfect Gaussian elimination while being very close to the original matrix A.

An alternative to preconditioners based upon direct methods is to mix conjugate gradients and classical iterative methods. For example, we could use a preconditioner that consists of a number of Jacobi (or Gauss–Seidel, or SOR) iterations, or (if we really feel sophisticated and brave) a multigrid preconditioner.

14.4 Poisson by conjugate gradients

The CG method was applied to the Poisson problem (8.33) on a 20×20 grid, hence with $d = 400$. The results are reported at the top of Fig. 14.6. We display there $\log_{10} \|\boldsymbol{r}^{[k]}\|$, the accuracy (in decimal digits) of the residual. Evidently, the residual decreases at a fairly even pace for about 85 iterations, during which time the iterative procedure converges within the limitations of finite computer arithmetic. Recall that $d = 400$ implies convergence in *at most* 400 steps, but the situation in Fig. 14.6 is typical: convergence occurs more rapidly than in the worst-case scenario.

How should we precondition the matrix A originating in the five-point formula? One possibility is to choose S as the tridiagonal portion of A. (Note that since we are choosing S we need to verify that it is positive definite, something which is trivial in this case. Had we started by choosing *any* nonsingular B, the positive definiteness of $S = BB^\top$ would have been assured.) To obtain B we Cholesky-factorize S, a procedure which can be accomplished very rapidly (see Section 11.1). Moreover, B is a bidiagonal lower triangular matrix and both systems (14.11) can be solved with great ease. The result features in the second graph of Fig. 14.6 and is only marginally better than plain CG, not really worth the effort.

An alternative to the tridiagonal preconditioner is to take B as equal to the lower triangular part of A, a choice that allows for rapid solution of both systems (14.11) by back substitution. The outcome is displayed in the bottom graph of Fig. 14.6 and we can see that the number of iterations is cut by more than a factor of 2, while each

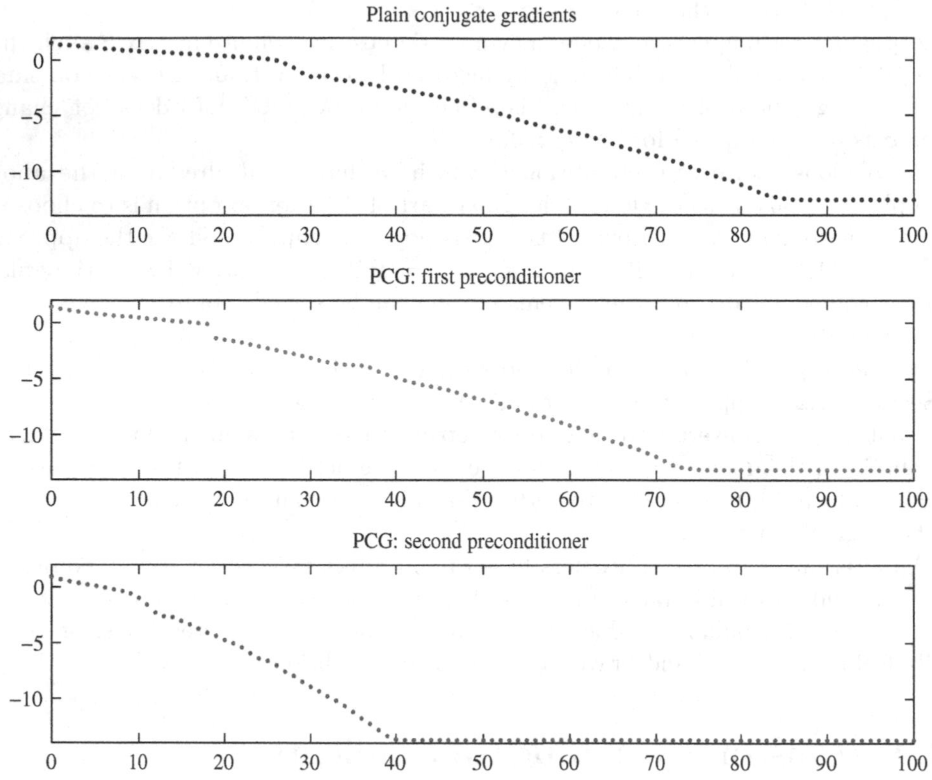

Figure 14.6 The logarithm of the residual for the CG method and two different PGC methods, applied to the Poisson equation (8.33) on a 20×20 grid.

iteration is of similar cost to that of the 'plain vanilla' CG method.

All three methods converge within the confines of finite computer arithmetic, but even here the second preconditioner beats the competition and delivers roughly 14 significant digits. The reason is clear: the less computation we have, the smaller the accumulation of roundoff error.

Why is the second preconditioner so much better than the first? A useful approach is to examine the eigenvalues of the matrix, whether A (for the plain GC method) or C. A good preconditioner typically 'squashes' eigenvalues and renders small the ratio of the largest and the smallest (denoted by $\kappa(A)$ or $\kappa(C)$, respectively, and known as the *spectral condition number*). (Remember that all eigenvalues are positive, since the matrix in question is positive definite.) Figure 14.7 displays histograms of the eigenvalues of the relevant matrix, A or C. The eigenvalues of A fit snugly into the interval $[0, 8]$ and, since the least eigenvalue is fairly small, the spectral conditioning number is large, $\kappa(A) \approx 178.06$. Matters are somewhat better for the first preconditioner, yet the presence of small eigenvalues renders the spectral condition number

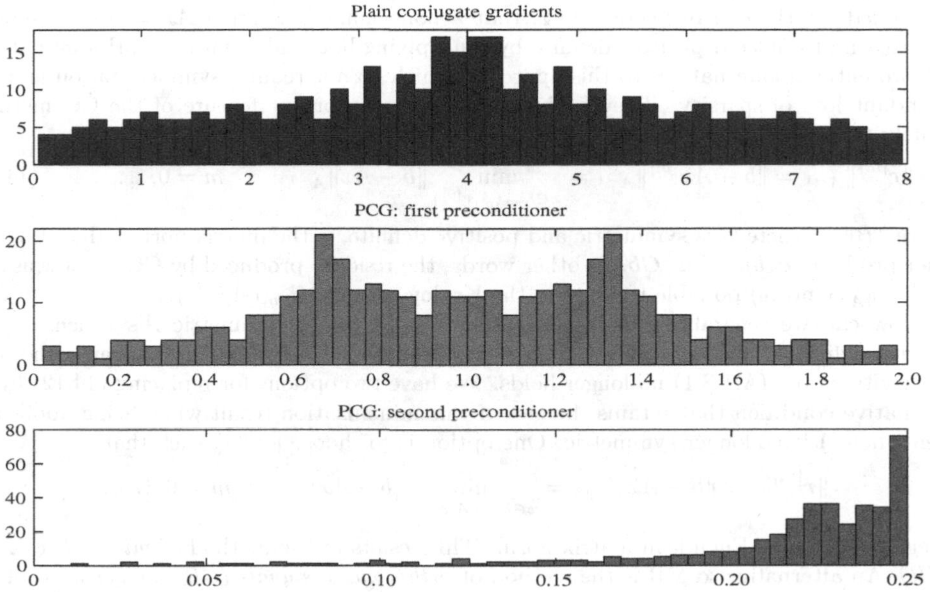

Figure 14.7 Histograms of the eigenvalues of the underlying matrices for the CG method and two different PGC methods, applied to the Poisson equation (8.33) on a 20×20 grid.

large, $\kappa(C) \approx 89.53$. For the second preconditioner, however, we have $\kappa(C) \approx 23.11$, a much smaller number. This does not necessarily prove that the second preconditioner is superior; nevertheless, minimizing the spectral condition number is a very convenient rule of thumb.

The second preconditioner is by no means the best we can do (just changing $b_{k,k}$ to $\frac{5}{2}$ decreases the number of iterations to 30), but then the role of this discussion was not to describe the state of the art in the application of conjugate gradients to the Poisson equation but to highlight the importance of good preconditioning.

Comments and bibliography

Much of the narrative of this chapter, in particular Theorem 14.3 and Lemma 14.4, is based on lecture notes for our Cambridge numerical analysis course, originally compiled by my colleague and friend Michael J.D. Powell. There are many alternative proofs of conjugacy and of the essential features of a Krylov subspace but, to my mind, Mike Powell's approach is the most beautiful and it is pleasure to share it, with due acknowledgement, with readers outside Cambridge.

Many iterative methods lend themselves to the formalism of Krylov subspaces. Thus, the kth iterate of the standard one-step stationary scheme $\boldsymbol{x}^{[k+1]} = H\boldsymbol{x}^{[k]} + \boldsymbol{v}$ from Section 12.1, with starting value $\boldsymbol{x}^{[0]} = \mathbf{0}$, lives in $\mathcal{K}_k(H, \boldsymbol{v})$.

More interesting in the context of conjugate gradients are two of its generalizations to nonsymmetric matrices, both couched in the language of Krylov subspaces. We have already mentioned, at the end of Section 14.2, that a nonsymmetric system $A\boldsymbol{x} = \boldsymbol{b}$ can be symmetrized and rendered positive definite by multiplying both sides with $A^\top$. However, there are two enticing alternatives to this approach which do not require symmetrization and the attendant loss of sparsity. They both generalize an important feature of the CG method, namely that, assuming $\boldsymbol{x}^{[0]} = \boldsymbol{0}$,

$$\|\boldsymbol{r}^{[m]}\|_{A^{-1}} = \|\boldsymbol{b} - A\boldsymbol{x}^{[m]}\|_{A^{-1}} = \min_{\boldsymbol{x}\in\mathcal{K}_m(A,\boldsymbol{r}^{[0]})} \|\boldsymbol{b} - A\boldsymbol{x}\|_{A^{-1}}, \qquad m = 0, 1, \ldots, \tag{14.12}$$

where $\|B\|_G$, where G is symmetric and positive definite, is the matrix norm induced by the inner product $\langle \boldsymbol{a}, \boldsymbol{b}\rangle_G = \boldsymbol{a}^\top G\boldsymbol{b}$. In other words, the residual produced by CG is the least (in the $\|\cdot\|_{A^{-1}}$ norm) possible residual in the Krylov subspace $\mathcal{K}_m(A, \boldsymbol{r}^{[0]})$.

How can we generalize this to nonsymmetric A's (or to symmetric A's which are not positive definite)? In such cases $\langle\,\cdot\,,\,\cdot\,\rangle_{A^{-1}}$ is no longer an inner product, because the non-negativity axiom (A.1.3.1) no longer holds. We have two options for replacing (14.12) by an alternative condition that retains the gist of this minimization result while being applicable even when A is no longer symmetric. One option is to choose $\|\boldsymbol{x}^{[m]}\|$ such that

$$\|\boldsymbol{r}^{[m]}\|_2 = \|\boldsymbol{b} - A\boldsymbol{x}^{[m]}\|_2 = \min_{\boldsymbol{x}\in\mathcal{K}_m(A,\boldsymbol{r}^{[0]})} \|\boldsymbol{b} - A\boldsymbol{x}\|_2, \qquad m = 0, 1, \ldots,$$

where $\|\cdot\|_2$ is the Euclidean matrix norm. This results in the method of *minimal residuals (MR)*. An alternative to MR is the method of *orthogonal residuals (OR)*, in the spirit of the Galerkin methods from Chapter 9. Thus, we seek $\boldsymbol{x}^{[m]} \in \mathcal{K}_m(A, \boldsymbol{b})$ such that

$$\boldsymbol{a}^\top \boldsymbol{r}^{[m]} = 0, \qquad \boldsymbol{a} \in \mathcal{K}_m(A, \boldsymbol{b}).$$

There exist many algorithmic implementations of both the MR and OR approaches and, needless to say, there has been a great deal of work on their preconditioners. Good references are Axelsson (1994), Golub & Van Loan (1996) and Greenbaum (1997), but perhaps the most readable, brief and gentle introduction to the subject is Freud *et al.* (1992). Here we outline because of its importance the famed GMRes (generalized minimal residuals) method, which implements the MR approach. Our point of departure is the *Arnoldi iteration,* which is defined by the following algorithmic steps.

Step 1. Choose $\boldsymbol{x}^{[0]} \in \mathbb{R}^d$, compute $\boldsymbol{r}^{[0]}$ and, assuming that it is nonzero (otherwise we terminate!), let $m = 1$ and $\boldsymbol{v}^{[0]} = \boldsymbol{r}^{[0]}/|\boldsymbol{r}^{[0]}\|_2$.

Step 2. For every $k = 0, 1, \ldots, m-1$ compute $h_{k,m} = \boldsymbol{v}^{[k]\top} A\boldsymbol{v}^{[m]}$.

Step 3. Set $\tilde{\boldsymbol{v}}^{[m]} = A\boldsymbol{v}^{[m-1]} - \sum_{k=0}^{m-1} h_{k,m-1}\boldsymbol{v}^{[k]}$ and let $h_{m,m-1} = \|\tilde{\boldsymbol{v}}^{[m]}\|_2$.

Step 4. If $h_{m,m-1} = 0$ (or is of suitably small magnitude) then stop. Otherwise let $\boldsymbol{v}^{[m]} = h_{m,m-1}^{-1}\tilde{\boldsymbol{v}}^{[m]}$, subsequently stepping m up by one, and go to step 2.

Now, once we have $\boldsymbol{v}^{[0]}, \ldots, \boldsymbol{v}^{[m-1]}$, we let

$$\boldsymbol{x}^{[m]} = \boldsymbol{x}^{[0]} + \sum_{j=0}^{m-1} \alpha_j \boldsymbol{v}^{[j]},$$

where the vector $\boldsymbol{\alpha} \in \mathbb{R}^m$ minimizes $\|\boldsymbol{\phi}^{[m]} - H^{[m]}\boldsymbol{\alpha}\|_2$ with

$$\boldsymbol{\phi}^{[m]} = \|\boldsymbol{r}^{[0]}\|_2 \boldsymbol{e}_{m+1}, \qquad H^{[m]} = \begin{bmatrix} h_{0,0} & \cdots & h_{0,m-2} & h_{0,m-1} \\ h_{1,0} & \ddots & & h_{1,m-1} \\ \vdots & \ddots & \ddots & \vdots \\ 0 & \cdots & h_{m-1,m-2} & h_{m-1,m-1} \\ 0 & \cdots & 0 & h_{m,m-1} \end{bmatrix}.$$

Note that the $(m+1) \times m$ matrix $H^{[m]}$ is of rank m.

The menagerie of different iterative methods that can be expressed in a Krylov subspace formalism is very extensive indeed. We do not propose to dwell further on the alphabet soup of BCG, BI-CGSTAB, CGNE, CGNR, CGS, GCG, GMRes, MINRES, MR, OR, QMR, SYMMBK, SYMMLQ and TFQMR (only a partial list) as well as the more humanely named Arnoldi, Chebyshev and Lanczos methods, or on their diverse preconditioners.

Axelsson, O. (1994), *Iterative Solution Methods,* Cambridge University Press, Cambridge.

Freund, R.W., Golub, G.H. and Nachtigal, N.M. (1992), Iterative solution of linear systems, *Acta Numerica* **1,** 57–100.

Golub, G.H. and Van Loan, C.F. (1996), *Matrix Computations* (3rd edn), Johns Hopkins Press, Baltimore.

Greenbaum, A. (1997), *Iterative Methods for Solving Linear Systems,* SIAM, Philadelphia.

Exercises

14.1 We consider the one-step stationary method $M\boldsymbol{x}^{[k+1]} = N\boldsymbol{x}^{[k]} + \boldsymbol{b}$, where $M - N = A$ and the matrix M is nonsingular.

a Prove that $\boldsymbol{x}^{[k]} - \boldsymbol{x} = H^k \boldsymbol{e}^{[0]}$, where $H = M^{-1}N$ is the iteration matrix, $\boldsymbol{e}^{[0]} = \boldsymbol{x}^{[0]} - \boldsymbol{x}$ and $\boldsymbol{x}$ is the exact solution of the linear system.

b Given $m \geq 1$, we form a new candidate solution $\boldsymbol{y}^{[m]}$ by the linear combination

$$\boldsymbol{y}^{[m]} = \sum_{k=0}^{m} \nu_k \boldsymbol{x}^{[k]}, \qquad \text{where} \qquad \sum_{k=0}^{m} \nu_k = 1.$$

Prove that $\boldsymbol{y}^{[m]} - \boldsymbol{x} = \sum_{k=0}^{m} \nu_k H^k \boldsymbol{e}^{[0]}$, and thus deduce that

$$\|\boldsymbol{y}^{[m]} - \boldsymbol{x}\|_2 \leq \|p_m(H)\|_2 \|\boldsymbol{e}^{[0]}\|_2, \qquad \text{where} \qquad p(z) = \sum_{k=0}^{m} \nu_k z^k.$$

c Suppose that it is known that all the eigenvalues of H are real and reside in the interval $[\alpha, \beta]$ and that the matrix has a full set of eigenvectors. Prove that

$$\|\boldsymbol{y}^{[m]} - \boldsymbol{x}\|_2 \leq \|V\|_2 \|V^{-1}\|_2 \|\boldsymbol{e}^{[0]}\|_2 \max_{x \in [\alpha, \beta]} |p_m(x)|,$$

where V is the matrix of the eigenvectors of H.

d We now use our freedom of choice of the parameters $\nu_0, \ldots, \nu_m$, hence of the polynomial p_m such that $p_m(1) = 1$ (why this condition?), to minimize $|p(x)|$ for $x \in [\alpha, \beta]$. To this end prove that the Chebyshev polynomial T_m (see Exercise 3.2) satisfies the inequality $|T_m(x)| \leq 1$, $x \in [-1, 1]$. (Since

$T_n(1) = 1$, this inequality cannot be improved by any other polynomial q such that $\max_{x\in[-1,1]} |q(x)| = 1$.) Deduce that the best choice of p_m is

$$p_m(x) = \frac{T_m(2(x-\alpha)/(\beta-\alpha)-1)}{T_m(2(1-\alpha)/(\beta-\alpha)-1)}.$$

e Show that this algorithm can be formulated in a Krylov subspace formalism.

(*This is the famed Chebyshev iterative method. Note, however, that naive implementation of this iterative procedure is problematic.*)

14.2 Apply the plain conjugate gradient method to the linear system

$$\begin{bmatrix} 1 & 0 & 0 \\ 0 & 2 & 0 \\ 0 & 0 & 3 \end{bmatrix} \boldsymbol{x} = \begin{bmatrix} 1 \\ 1 \\ 1 \end{bmatrix},$$

starting as usual with $\boldsymbol{x}^{[0]} = \boldsymbol{0}$. Verify that the residuals $\boldsymbol{r}^{[0]}, \boldsymbol{r}^{[1]}$ and $\boldsymbol{r}^{[2]}$ are mutually orthogonal, that the search directions $\boldsymbol{d}^{[0]}$, $\boldsymbol{d}^{[1]}$ and $\boldsymbol{d}^{[2]}$ are mutually conjugate and that $\boldsymbol{x}^{[3]}$ satisfies the linear system.

14.3 Let the plain conjugate gradient method be applied when A is positive definite. Express $\boldsymbol{d}^{[k]}$ in terms of $\boldsymbol{r}^{[j]}$ and $\beta^{[j]}$, $j = 0, 1, \ldots, k-1$. Then, commencing with the formula $\boldsymbol{x}^{[k+1]} = \sum_{j=0}^{k} \omega^{[j]} \boldsymbol{d}^{[j]}$, from $\omega^{[j]} > 0$ and with Theorem 14.3, deduce in a few lines that the sequence $\{\|\boldsymbol{x}^{[j]}\| : j = 0, 1, \ldots, k+1\}$ increases monotonically.

14.4 The polynomial $p(x) = x^m + \sum_{l=0}^{m-1} c_l x^l$ is the *minimal polynomial* of the $d \times d$ matrix A if it is the polynomial of lowest degree that satisfies $p(A) = O$. Note that $m \le d$ holds because of the Cayley–Hamilton theorem from linear algebra.

a Give an example of a 3×3 symmetric matrix with a quadratic minimal polynomial.

b Prove that (in exact arithmetic) the conjugate gradient method requires at most m iterations to calculate the exact solution of $A\boldsymbol{v} = \boldsymbol{b}$, where m is the degree of the minimal polynomial of A.

14.5 Let $A = I + B$ be a symmetric positive definite matrix and suppose that the rank of B is s. Prove that the CG algorithm converges in at most s steps.

14.6 Let A be a $d \times d$ TST matrix with $a_{k,k} = \alpha$ and $a_{k,k+1} = a_{k+1,k} = \beta$.

a Verify that $\alpha \ge 2|\beta| > 0$ implies that the matrix is positive definite.

b Now we precondition the CG method for $A\boldsymbol{x} = \boldsymbol{b}$ with the Toeplitz lower-triangular bidiagonal matrix B,

$$b_{k,\ell} = \begin{cases} \gamma, & k = \ell, \\ \delta, & k = \ell + 1, \\ 0, & \text{otherwise.} \end{cases}$$

Determine real numbers γ and δ such that $BB^\top$ differs from A in just the $(1,1)$ coordinate.

c Prove that with this choice of γ and δ the PCG method converges in a single step.

14.7 Find the spectral condition number $\kappa(A)$ when the matrix A corresponds to the five-point method in an $m \times m$ grid.

14.8 Let

$$A = \begin{bmatrix} A_1 & A_2 \\ A_2^\top & A_3 \end{bmatrix}, \qquad S = \begin{bmatrix} A_1 & O \\ O & A_2 \end{bmatrix},$$

where A_1, A_3 are symmetric $d \times d$ matrices and the rank of the $d \times d$ matrix A_2 is $r \le d-1$. We further stipulate that the $(2d) \times (2d)$ matrix A is positive definite.

a Let $A_1 = L_1 L_1^T$, $A_3 = L_3 L_3^\top$ be Cholesky factorizations and assume that the preconditioner B is the lower-triangular Cholesky factor of S (hence $BB^\top = S$). Prove that

$$C = B^{-1} A B^{-\top} = \begin{bmatrix} I & F \\ F^\top & I \end{bmatrix}, \qquad \text{where} \qquad F = L_1^{-1} A_2 L_3^{-T}.$$

b Supposing that the eigenvalues of C are $\lambda_1, \ldots, \lambda_{2d}$, while the eigenvalues of $FF^\top$ are $\mu_1, \ldots, \mu_d \ge 0$, prove that, without loss of generality,

$$\lambda_k = 1 - \sqrt{\mu_k}, \quad \lambda_{d+k} = 1 + \sqrt{\mu_k}, \qquad k = 1, 2, \ldots, d.$$

c Prove that the rank of $FF^\top$ is r, thereby deducing that C has at most $2r+1$ distinct eigenvalues. What does this tell you about the number of steps before the PCG method terminates in exact arithmetic?

15

Fast Poisson solvers

15.1 TST matrices and the Hockney method

This chapter is concerned with yet another approach to the solution of the linear equations that occur when the Poisson equation is discretized by finite differences. This approach is an alternative to the direct methods of Chapter 11 and to the iterative schemes of Chapters 12–14. We intend to present two techniques for the very fast approximation of $\nabla^2 u = f$, one in a rectangle and the other in a disc. These techniques share two features. Firstly, they originate in numerical solution of the Poisson equation – hence their sobriquet, *fast Poisson solvers*. Secondly, the secret of their efficacy rests in a clever use of the *fast Fourier transform* (FFT).

In the present section we assume that the Poisson equation (8.13) with Dirichlet boundary conditions (8.14) is solved in a rectangle with either the five-point formula (8.15) or the nine-point formula (8.28) (or, for that matter, the modified nine-point formula (8.32) – the matrix of the linear system does not depend on whether the nine-point method has been modified). In either case we assume that the linear equations have been assembled in *natural ordering*.

Suppose that the grid is $m_1 \times m_2$. The linear system $A\boldsymbol{x} = \boldsymbol{b}$ can be written in the block-TST form (recall from Section 12.2 that 'TST' stands for 'Toeplitz, symmetric and tridiagonal')

$$\begin{bmatrix} S & T & O & \cdots & O \\ T & S & T & \ddots & \vdots \\ O & \ddots & \ddots & \ddots & O \\ \vdots & \ddots & T & S & T \\ O & \cdots & O & T & S \end{bmatrix} \begin{bmatrix} \boldsymbol{x}_1 \\ \boldsymbol{x}_2 \\ \vdots \\ \boldsymbol{x}_{m_2} \end{bmatrix} = \begin{bmatrix} \boldsymbol{b}_1 \\ \boldsymbol{b}_2 \\ \vdots \\ \boldsymbol{b}_{m_2} \end{bmatrix}, \tag{15.1}$$

where $\boldsymbol{x}_\ell$ and $\boldsymbol{b}_\ell$ correspond to the variables and to the portion of $\boldsymbol{b}$ along the ℓth column of the grid, respectively:

$$\boldsymbol{x}_\ell = \begin{bmatrix} u_{1,\ell} \\ u_{2,\ell} \\ \vdots \\ u_{m_1,\ell} \end{bmatrix}, \qquad \boldsymbol{b}_\ell = \begin{bmatrix} b_{1,\ell} \\ b_{2,\ell} \\ \vdots \\ b_{m_1,\ell} \end{bmatrix}, \qquad \ell = 1, 2, \ldots, m_2.$$

Both S and T are themselves $m_1 \times m_1$ TST matrices:

$$S = \begin{bmatrix} -4 & 1 & 0 & \cdots & 0 \\ 1 & -4 & 1 & \ddots & \vdots \\ 0 & \ddots & \ddots & \ddots & 0 \\ \vdots & \ddots & 1 & -4 & 1 \\ 0 & \cdots & 0 & 1 & -4 \end{bmatrix} \qquad \text{and} \qquad T = \begin{bmatrix} 1 & 0 & \cdots & 0 \\ 0 & 1 & \ddots & \vdots \\ \vdots & \ddots & \ddots & 0 \\ 0 & \cdots & 0 & 1 \end{bmatrix}$$

for the five-point formula and

$$S = \begin{bmatrix} -\frac{10}{3} & \frac{2}{3} & 0 & \cdots & 0 \\ \frac{2}{3} & -\frac{10}{3} & \frac{2}{3} & \ddots & \vdots \\ 0 & \ddots & \ddots & \ddots & 0 \\ \vdots & \ddots & \frac{2}{3} & -\frac{10}{3} & \frac{2}{3} \\ 0 & \cdots & 0 & \frac{2}{3} & -\frac{10}{3} \end{bmatrix} \qquad \text{and} \qquad T = \begin{bmatrix} \frac{2}{3} & \frac{1}{6} & 0 & \cdots & 0 \\ \frac{1}{6} & \frac{2}{3} & \frac{1}{6} & \ddots & \vdots \\ 0 & \ddots & \ddots & \ddots & 0 \\ \vdots & \ddots & \frac{1}{6} & \frac{2}{3} & \frac{1}{6} \\ 0 & \cdots & 0 & \frac{1}{6} & \frac{2}{3} \end{bmatrix}$$

in the case of the nine-point formula.

We rewrite (15.1) in the form

$$T\boldsymbol{x}_{\ell-1} + S\boldsymbol{x}_\ell + T\boldsymbol{x}_{\ell+1} = \boldsymbol{b}_\ell, \qquad \ell = 1, 2, \ldots, m_2, \tag{15.2}$$

where $\boldsymbol{x}_0, \boldsymbol{x}_{m_2+1} := \boldsymbol{0} \in \mathbb{R}^{m_1}$, and recall from Lemma 12.5 that the eigenvalues and eigenvectors of TST matrices are known and that all TST matrices of similar dimension share the same eigenvectors. In particular, according to (12.29) we have

$$S = QD_SQ, \qquad T = QD_TQ, \tag{15.3}$$

where

$$q_{j,\ell} = \sqrt{\frac{2}{m_1+1}} \sin\left(\frac{\pi j \ell}{m_1+1}\right), \qquad j, \ell = 1, 2, \ldots, m_1.$$

(Note that Q is both orthogonal and symmetric; thus, for example, $S = QD_SQ^{-1} = QD_SQ^\top = QD_SQ$. Such a matrix is called an *orthogonal involution.*) Both D_S and D_T are $m_1 \times m_1$ diagonal matrices whose diagonal components consist of the eigenvalues of S and T respectively, $\lambda_1^{(S)}, \lambda_2^{(S)}, \ldots, \lambda_{m_1}^{(S)}$ and $\lambda_1^{(T)}, \lambda_2^{(T)}, \ldots, \lambda_{m_1}^{(T)}$, say; cf. (12.28). We substitute (15.3) into (15.2) and multiply with $Q = Q^{-1}$ from the left. The outcome is

$$D_T\boldsymbol{y}_{\ell-1} + D_S\boldsymbol{y}_\ell + D_T\boldsymbol{y}_{\ell+1} = \boldsymbol{c}_\ell, \qquad \ell = 1, 2, \ldots, m_2, \tag{15.4}$$

where

$$\boldsymbol{y}_\ell := Q\boldsymbol{x}_\ell, \quad \boldsymbol{c}_\ell := Q\boldsymbol{b}_\ell, \qquad \ell = 1, 2, \ldots, m_2.$$

The crucial difference between (15.2) and (15.4) is that in the latter we have diagonal, rather than TST, matrices. To exploit this, we recall that $\boldsymbol{x}$ and $\boldsymbol{b}$ (and, indeed, the matrix A) have been obtained from a natural ordering of a rectangular grid by columns. Let us now reorder the $\boldsymbol{y}_\ell$ and the $\boldsymbol{c}_\ell$ *by rows.* Thus,

$$\tilde{\boldsymbol{y}}_j := \begin{bmatrix} y_{j,1} \\ y_{j,2} \\ \vdots \\ y_{j,m_2} \end{bmatrix}, \quad \tilde{\boldsymbol{c}}_j := \begin{bmatrix} c_{j,1} \\ c_{j,2} \\ \vdots \\ c_{j,m_2} \end{bmatrix}, \qquad j = 1, 2, \ldots, m_1.$$

To derive linear equations that are satisfied by the $\tilde{\boldsymbol{y}}_j$, let us consider the first equation in each of the m_2 blocks in (15.4),

$$\begin{aligned} \lambda_1^{(S)} y_{1,1} + \lambda_1^{(T)} y_{1,2} &= c_{1,1}, \\ \lambda_1^{(T)} y_{1,\ell-1} + \lambda_1^{(S)} y_{1,\ell} + \lambda_1^{(T)} y_{1,\ell+1} &= c_{1,\ell}, \qquad \ell = 2, 3, \ldots, m_2 - 1, \\ \lambda_1^{(T)} y_{1,m_2-1} + \lambda_1^{(S)} y_{1,m_2} &= c_{1,m_2}, \end{aligned}$$

or, in a matrix notation,

$$\begin{bmatrix} \lambda_1^{(S)} & \lambda_1^{(T)} & 0 & \cdots & 0 \\ \lambda_1^{(T)} & \lambda_1^{(S)} & \lambda_1^{(T)} & \ddots & \vdots \\ 0 & \ddots & \ddots & \ddots & 0 \\ \vdots & \ddots & \lambda_1^{(T)} & \lambda_1^{(S)} & \lambda_1^{(T)} \\ 0 & \cdots & 0 & \lambda_1^{(T)} & \lambda_1^{(S)} \end{bmatrix} \tilde{\boldsymbol{y}}_1 = \tilde{\boldsymbol{c}}_1.$$

Likewise, collecting together the jth equation from each block in (15.4) results in

$$\Gamma_j \tilde{\boldsymbol{y}}_j = \tilde{\boldsymbol{c}}_j, \qquad j = 1, 2, \ldots, m_1, \tag{15.5}$$

where

$$\Gamma_j := \begin{bmatrix} \lambda_j^{(S)} & \lambda_j^{(T)} & 0 & \cdots & 0 \\ \lambda_j^{(T)} & \lambda_j^{(S)} & \lambda_j^{(T)} & \ddots & \vdots \\ 0 & \ddots & \ddots & \ddots & 0 \\ \vdots & \ddots & \lambda_j^{(T)} & \lambda_j^{(S)} & \lambda_j^{(T)} \\ 0 & \cdots & 0 & \lambda_j^{(T)} & \lambda_j^{(S)} \end{bmatrix}, \qquad j = 1, 2, \ldots, m_2.$$

Hence, *switching from column-wise to row-wise ordering uncouples the* $(m_1 m_2) \times (m_1 m_2)$ *system* (15.4) *into* m_2 *systems, each of dimension* $m_1 \times m_1$.

The matrices Γ_j are also TST; hence their eigenvalues and eigenvectors are known and, in principle, can be used to solve (15.5). This, however, is a bad idea, since it is considerably easier to compute these linear systems by banded LU factorization (see Chapter 11). This costs just $\mathcal{O}(m_1)$ operations per system, altogether $\mathcal{O}(m_1 m_2)$ operations.

◇ **Counting operations** Measuring the cost of numerical calculation is a highly uncertain business. At the dawn of the computer era (not such a long time ago!) it was usual to count multiplications, since they were significantly more expensive than additions or subtractions (divisions were even more expensive, avoided if at all possible). As often in the computer business, technological developments have rendered this point of view obsolete. In modern processors, operations like multiplication – and, for that matter, exponentiation, square-rooting, the evaluation of logarithms and of trigonometric functions – are built into the hardware and can be performed exceedingly fast. Thus, a more up-to-date measure of computational cost is a *flop,* an abbreviation for *floating point operation.* A single flop is considered the equivalent of a `FORTRAN` statement

```
A(I) = B(I,J) * C(J) + D(J)
```

or alternative statements in `Pascal`, `C++` or any other high-level computer language. Note that a flop involves a product, an addition and a number of calculations of indices – none of these operations is free and the above statement, whose form is familiar even to the novice scientific programmer, combines them in a useful manner. The reader might have also encountered flops as a yardstick of computer speed, in which case flops (kiloflops, megaflops, gigaflops, teraflops and perhaps, one day, petaflops) are measured per second.

Even flops, though, are increasingly uncertain as measures of computational cost, because modern computers – or, at least, the large computers used for *macho* applications of scientific computing – typically involve parallel architectures having different configurations. This means that the cost of computation varies between computers and no single number can provide a complete comparison. Matters are complicated further by the fact that calculation is not the only time-consuming task of parallel multi-processor computers. The other is communication and message-passing among the different processors.

To make a long story short – and to avoid excessive departures from the main theme of this book – we thereafter provide only the order of magnitude (indicated by the $\mathcal{O}(\,\cdot\,)$ notation) of the cost of an algorithm. This can be easily converted to a 'multiplication count' or to flops, but, for all intents and purposes, the order of magnitude is illustrative enough. All our counts are in serial architecture – parallel processing is likely to change everything! Having said this, we hasten to add that, regardless of the underlying architecture, most good methods remain good and most bad methods remain bad. It is still better to use sparse LU factorization, say, for banded matrices, multigrid still outperforms Gauss–Seidel and fast Poisson solvers are still ... fast. ◇

Having solved the tridiagonal systems, we end up with the vectors $\tilde{\boldsymbol{y}}_1, \tilde{\boldsymbol{y}}_2, \ldots, \tilde{\boldsymbol{y}}_{m_1}$, which we 'translate' back to $\boldsymbol{y}_1, \boldsymbol{y}_2, \ldots, \boldsymbol{y}_{m_2}$ by rearranging rows to columns. Note that this rearrangement is free of any computational cost since, in reality, we are (or, at least, should) hold the information in the form of a $m_1 \times m_2$ array, corresponding to the computational grid. Column-wise or row-wise natural orderings are purely notational devices! Finally, we reverse the effects of the multiplication by Q and let $\boldsymbol{x}_\ell = Q\boldsymbol{y}_\ell$, $\ell = 1, 2, \ldots, m_2$ (recall that $Q = Q^{-1}$).

Let us review briefly the stages in this, the *Hockney method,* estimating their computational cost.

(1) At the outset, we have the vectors $\boldsymbol{b}_1, \boldsymbol{b}_2, \ldots, \boldsymbol{b}_{m_2} \in \mathbb{R}^{m_1}$ and we form the products $\boldsymbol{c}_\ell = Q\boldsymbol{b}_\ell$, $\ell = 1, 2, \ldots, m_2$. This costs $\mathcal{O}(m_1^2 m_2)$ operations.

(2) We rearrange columns into rows, i.e., $\boldsymbol{c}_\ell \in \mathbb{R}^{m_1}$, $\ell = 1, 2, \ldots, m_2$, into $\tilde{\boldsymbol{c}}_j \in \mathbb{R}^{m_2}$, $j = 1, 2, \ldots, m_1$. This is purely a change in notation and is free of any computational cost.

(3) The tridiagonal systems $\Gamma_j \tilde{\boldsymbol{y}}_j = \tilde{\boldsymbol{c}}_j$, $j = 1, 2, \ldots, m_1$, are solved by banded LU factorization, and the cost of this procedure is $\mathcal{O}(m_1 m_2)$.

(4) We next rearrange rows into columns, i.e., $\tilde{\boldsymbol{y}}_j \in \mathbb{R}^{m_2}$, $j = 1, 2, \ldots, m_1$ into $\boldsymbol{y}_\ell \in \mathbb{R}^{m_1}$, $\ell = 1, 2, \ldots, m_2$. Again, this costs nothing.

(5) Finally, we find the solution of the discretized Poisson equation by forming the products $\boldsymbol{x}_\ell = Q\boldsymbol{y}_\ell$, $\ell = 1, 2, \ldots, m_2$, at the cost of $\mathcal{O}(m_1^2 m_2)$ operations.

Provided both m_1 and m_2 are large, matrix multiplications dominate the computational cost. Our first, trivial observation is that, the expense being $\mathcal{O}(m_1^2 m_2)$, it is a good policy to choose $m_1 \leq m_2$ (because of symmetry, we can always rotate the rectangle, in other words proceed from row to columns and to rows again). However, a considerably more important observation is that the special form of the matrix Q can be exploited to make products of the form $\boldsymbol{s} = Q\boldsymbol{p}$, say, substantially cheaper than $\mathcal{O}(m_1^2)$. Because of (12.29), we have

$$s_\ell = c\sum_{j=1}^{m_1} p_j \sin\left(\frac{\pi j \ell}{m_1+1}\right) = c\,\mathrm{Im}\left[\sum_{j=0}^{m_1} p_j \exp\left(\frac{\pi \mathrm{i} j \ell}{m_1+1}\right)\right], \qquad \ell = 1, 2, \ldots, m_1, \tag{15.6}$$

where $c = [2/(m_1+1)]^{1/2}$ is a multiplicative constant.

And this is the very point when the Hockney method becomes a powerful computational tool, rather than a matter of mathematical curiosity: *the sum on the right of* (15.6) *is a discrete Fourier transform!* Therefore, using the FFT (see Section 10.3), it can be computed in $\mathcal{O}(m_1 \log_2 m_1)$ operations. The cost of each of steps 1 and 5, which dominates the Hockney method, drops from $\mathcal{O}(m_1^2 m_2)$ to $\mathcal{O}(m_1 m_2 \log_2 m_1)$.

The DFT and the FFT were introduced in Chapter 10 in the context of Fourier expansions and the computation of their coefficients. There are no overt Fourier expansions here! It is the special form of the eigenvectors of TST matrices that renders them amenable to a technique which we introduced earlier in a very different context.

An important remark is that in this section we have not used the fact that we are solving the Poisson equation! The crucial feature of the underlying linear system is that it is block-TST and each of the blocks is itself a TST matrix. There is nothing to prevent us from using the same approach for other equations that possess this structure, regardless of their origin. Moreover, it is an easy matter to extend this approach to, say, Poisson equations in three variables with Dirichlet conditions along the boundary of a parallelepiped: the matrix partitions into blocks of block-TST matrices and each such block-TST matrix is itself composed of TST matrices.

15.2 Fast Poisson solver in a disc

Let us suppose that the Poisson equation $\nabla^2 u = g_0$ is given in the open unit disc

$$\mathbb{D} = \{(x,y) \in \mathbb{R}^2 \,:\, x^2 + y^2 < 1\},$$

together with Dirichlet boundary conditions along the unit circle,

$$u(\cos\theta, \sin\theta) = \phi(\theta), \qquad 0 \le \theta \le 2\pi, \tag{15.7}$$

where $\phi(0) = \phi(2\pi)$. It is convenient to translate the equation from Cartesian to polar coordinates. Thus, we let

$$\left.\begin{aligned} v(r,\theta) &= u(r\cos\theta, r\sin\theta), \\ g(r,\theta) &= g_0(r\cos\theta, r\sin\theta), \end{aligned}\right. \qquad 0 < r < 1, \quad 0 \le \theta \le 2\pi.$$

The form of ∇^2 in polar coordinates readily gives us

$$\frac{\partial^2 v}{\partial r^2} + \frac{1}{r}\frac{\partial v}{\partial r} + \frac{1}{r^2}\frac{\partial^2 v}{\partial \theta^2} = g, \qquad 0 < r < 1, \quad 0 \le \theta \le 2\pi. \tag{15.8}$$

The boundary conditions, however, are more delicate. Switching from Cartesian to polar means, in essence, that the disc $\mathbb{D}$ is replaced by the square

$$\widetilde{\mathbb{D}} = \{(r,\theta) \,:\, 0 < r < 1,\ 0 \le \theta \le 2\pi\}.$$

Unlike $\mathbb{D}$, which has just one boundary 'segment' – its circumference – the set $\widetilde{\mathbb{D}}$ boasts four portions of boundary and we need to allocate appropriate conditions at all of them.

The segment $r = 1$, $0 \le \theta \le 2\pi$, is the easiest, being the destination of the original boundary condition (15.7). Hence, we set

$$v(1,\theta) = \phi(\theta), \qquad 0 \le \theta \le 2\pi. \tag{15.9}$$

Next in order of difficulty are the line segments $0 < r < 1$, $\theta = 0$, and $0 < r < 1$, $\theta = 2\pi$. They both correspond to the same segment, namely $0 < x < 1$, $y = 0$, in the original disc $\mathbb{D}$. The value of u on this segment is, of course, unknown, but, at the very least, we know that it is the same whether we assign it to $\theta = 0$ or $\theta = 2\pi$. In other words, we have the *periodic* boundary condition

$$v(r,0) = v(r,2\pi), \qquad 0 < r < 1. \tag{15.10}$$

Finally, we pay attention to the remaining portion of $\partial\widetilde{\mathbb{D}}$, namely $r = 0$, $0 \le \theta \le 2\pi$. This whole line corresponds to just a single point in $\mathbb{D}$, namely the origin $x = y = 0$

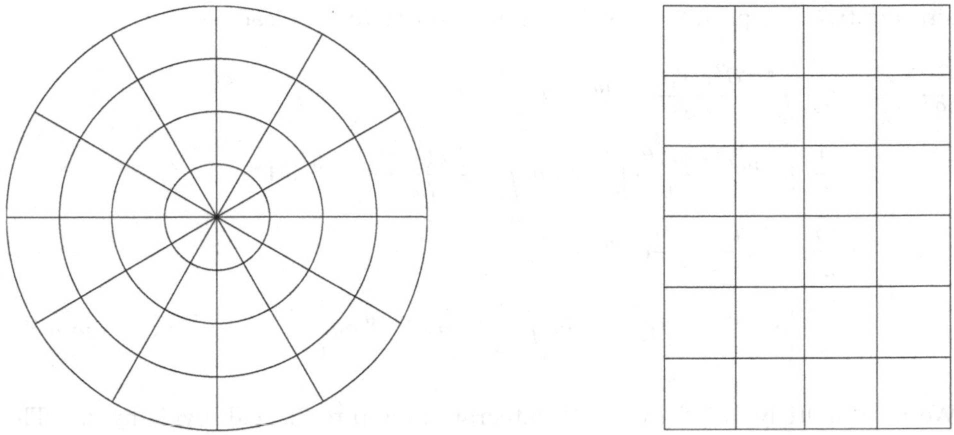

Figure 15.1 Computational grids (x, y) in the disc $\mathbb{D}$ and (r, θ) in the square $\widetilde{\mathbb{D}}$, associated by the Cartesian-to-polar transformation.

(see Fig. 15.1). Therefore, v is constant along that line or, to express it in a more manageable form, we obtain the *Neumann* boundary condition

$$\frac{\partial}{\partial\theta}v(0,\theta) = 0, \qquad 0 \le \theta \le 2\pi. \tag{15.11}$$

We can approximate the solution of (15.8) with boundary conditions (15.9)–(15.11) by inscribing a square grid into $\widetilde{\mathbb{D}}$, approximating $\partial v/\partial r$, $\partial^2 v/\partial r^2$ and $\partial^2 v/\partial\theta^2$ by central differences, say, at the grid points and taking adequate care of the boundary conditions. The outcome is certainly preferable by far to imposing a square grid on the original disc $\mathbb{D}$, a procedure that leads to excessively unpleasant equations at the near-boundary grid points. Having solved the Poisson equation in $\widetilde{\mathbb{D}}$, we can map the outcome to the disc, i.e., to the concentric grid of Fig. 15.1. This, however, is not the fast solver that is the goal of this section. To calculate (15.8) considerably faster, we again resort to FFTs.

We have already defined in Chapter 10 the *Fourier transform* (10.15) of an arbitrary complex-valued integrable periodic function g in $\mathbb{R}$. Because of the periodic boundary condition (15.10) we can Fourier-transform $v(r, \cdot)$, and this results in the sequence

$$\hat{v}_m(r) = \frac{1}{2\pi}\int_0^{2\pi} v(r,\theta)\mathrm{e}^{-\mathrm{i}m\theta}\,\mathrm{d}\theta, \qquad m \in \mathbb{Z}.$$

Our goal is to convert the PDE (15.8) into an infinite set of ODEs that are satisfied by the *Fourier coefficients* $\{v_m\}_{m=-\infty}^{\infty}$. It is easy to deduce from (10.15) that

$$\widehat{\left(\frac{\partial v}{\partial r}\right)}_m = v_m'(r) \qquad \text{and} \qquad \widehat{\left(\frac{\partial^2 v}{\partial r^2}\right)}_m = v_m''(r), \qquad m \in \mathbb{Z},$$

but the second derivative with respect to θ is less trivial – fortunately not by much! Integrating twice by parts and exploiting periodicity readily leads to

$$\begin{aligned}\left(\widehat{\frac{\partial^2 v}{\partial \theta^2}}\right)_m &= \frac{1}{2\pi}\int_0^{2\pi} \frac{\partial^2 v(r,\theta)}{\partial \theta^2} \mathrm{e}^{-\mathrm{i}m\theta}\,\mathrm{d}\theta \\ &= \frac{1}{2\pi}\left[\mathrm{e}^{-\mathrm{i}m\theta}\frac{\partial v(r,\theta)}{\partial\theta}\Big|_0^{2\pi} + \mathrm{i}m\int_0^{2\pi}\frac{\partial v(r,\theta)}{\partial\theta}\mathrm{e}^{-\mathrm{i}m\theta}\,\mathrm{d}\theta\right] \\ &= \frac{\mathrm{i}m}{2\pi}\int_0^{2\pi}\frac{\partial v(r,\theta)}{\partial\theta}\mathrm{e}^{-\mathrm{i}m\theta}\,\mathrm{d}\theta \\ &= \frac{\mathrm{i}m}{2\pi}\left[\mathrm{e}^{-\mathrm{i}m\theta}v(r,\theta)\Big|_0^{2\pi} + \mathrm{i}m\int_0^{2\pi} v(r,\theta)\mathrm{e}^{-\mathrm{i}m\theta}\,\mathrm{d}\theta\right] = -m^2\hat{v}_m(r), \qquad m \in \mathbb{Z}.\end{aligned}$$

We now multiply (15.8) by $\mathrm{e}^{-\mathrm{i}m\theta}$, integrate from 0 to 2π and divide by 2π. The outcome is the ODE

$$\hat{v}_m'' + \frac{1}{r}\hat{v}_m' - \frac{m^2}{r^2}\hat{v}_m = \hat{g}_m, \tag{15.12}$$

which is obeyed by each Fourier coefficient $\hat{v}_m$, $m \in \mathbb{Z}$, for $0 < r < 1$. The right-hand side is the mth Fourier coefficient of the inhomogeneous term g.

What are the boundary conditions associated with the differential equation (15.12)? Firstly, the periodic conditions (15.10), having played their role in validating the use of the Fourier transform, disappear in tandem with the variable θ. Secondly, the right-hand condition (15.9) translates at once into

$$\hat{v}_m(1) = \hat{\phi}_m, \qquad m \in \mathbb{Z}. \tag{15.13}$$

Finally, we must render the natural condition (15.11) in the language of Fourier coefficients. This presents more of a challenge.

We commence by using the inverse Fourier transform formula (10.13), which allows us to synthesize v from its coefficients:

$$v(r,\theta) = \sum_{m=-\infty}^{\infty} \hat{v}_m(r)\mathrm{e}^{\mathrm{i}m\theta}, \qquad 0 \le r \le 1, \qquad 0 \le \theta \le 2\pi.$$

Next, we differentiate with respect to θ and set $r = 0$. Comparison with (15.11) yields

$$\mathrm{i}\sum_{m=-\infty}^{\infty} m\hat{v}_m(0)\mathrm{e}^{\mathrm{i}m\theta} \equiv 0, \qquad 0 \le \theta \le 2\pi.$$

The trigonometric functions $\mathrm{e}^{\mathrm{i}m\theta}$, $m \in \mathbb{Z}$, are linearly independent. Therefore, whenever their linear combination vanishes identically, all its components must be zero.[1] The outcome is the boundary condition

$$\hat{v}_m(0) = 0, \qquad m \in \mathbb{Z} \setminus \{0\}. \tag{15.14}$$

[1] The argument is *slightly* more complicated, since the standard theorem about linear independence refers to a finite number of components. For reasons of brevity, we take for granted the generalization to linear combinations having an infinite number of components.

This leaves us with just a single missing item of information, namely the boundary value for the zeroth harmonic. The Cartesian-to-polar transformation,

$$v(r,\theta) = u(r\cos\theta, r\sin\theta)$$

implies

$$\frac{\partial v(0,\theta)}{\partial r} = \cos\theta\,\frac{\partial u(0,0)}{\partial x} + \sin\theta\,\frac{\partial u(0,0)}{\partial y}.$$

Therefore

$$\hat{v}'(0) = \frac{1}{2\pi}\int_0^{2\pi}\frac{\partial v(0,\theta)}{\partial r}\,\mathrm{d}\theta = \frac{1}{2\pi}\int_0^{2\pi}\cos\theta\,\mathrm{d}\theta\frac{\partial u(0,0)}{\partial x} + \frac{1}{2\pi}\int_0^{2\pi}\sin\theta\,\mathrm{d}\theta\frac{\partial u(0,0)}{\partial y} = 0,$$

since

$$\int_0^{2\pi}\cos\theta\,\mathrm{d}\theta,\ \int_0^{2\pi}\sin\theta\,\mathrm{d}\theta = 0.$$

We thus obtain the missing boundary condition,

$$\hat{v}'(0) = 0. \tag{15.15}$$

The crucial fact about the ODEs (15.12) (with initial conditions (15.13)–(15.15)) is that the Fourier transform *uncouples* the harmonics – the equation for each $\hat{v}_m$ is a two-point boundary value problem, whose solution is independent of all other Fourier coeffcients!

The two-point boundary problem (15.12) can be solved easily by finite differences (see Exercise 15.4 for an alternative that uses the finite element method and, if really daring, design a Chebyshev spectral method for this problem). Thus, we choose a positive integer d and cover the interval $[0,1]$ with d subintervals of length $\Delta r = 1/d$. Adopting the usual notation from Chapters 1–7, $\hat{v}_{m,k}$ denotes an approximation to $\hat{v}_m(k\Delta r)$. Employing the simplest central difference approximation,

$$\hat{v}'_m(k\Delta r) \approx \frac{1}{2\Delta r}\left(\hat{v}_{m,k+1} - \hat{v}_{m,k-1}\right),$$
$$\hat{v}''_m(k\Delta r) \approx \frac{1}{(\Delta r)^2}\left(\hat{v}_{m,k+1} - 2\hat{v}_{m,k} + \hat{v}_{m,k-1}\right),$$

the ODE (15.12) leads to the difference equation

$$\frac{1}{(\Delta r)^2}\left(\hat{v}_{m,k+1} - 2\hat{v}_{m,k} + \hat{v}_{m,k-1}\right) + \frac{2}{2k(\Delta r)^2}\left(\hat{v}_{m,k+1} - \hat{v}_{m,k-1}\right) - \frac{m^2}{k^2(\Delta r)^2}\hat{v}_{m,k} = \hat{g}_{m,k},$$

where $\hat{g}_{m,k} = \hat{g}_m(k\Delta r)$ and k ranges in $\{1,2,\ldots,d-1\}$. Rearranging terms, we arrive at

$$\left(1 - \frac{1}{2k}\right)\hat{v}_{m,k-1} - \left(2 + \frac{m^2}{k^2}\right)\hat{v}_{m,k} + \left(1 + \frac{1}{2k}\right)\hat{v}_{m,k+1} = (\Delta r)^2\hat{g}_{m,k},$$
$$k = 1,2,\ldots,d-1. \tag{15.16}$$

We complement (15.16) with boundary values. Thus, (15.13) yields

$$\hat{v}_{m,d} = \hat{\phi}_m, \qquad m \in \mathbb{Z},$$

and (15.14) results in

$$\hat{v}_{m,0} = 0, \qquad m \in \mathbb{Z} \setminus \{0\}.$$

Finally, we use forward differences to approximate (15.15): for example,

$$-\tfrac{5}{4}\hat{v}_{0,0} + \tfrac{3}{2}\hat{v}_{0,1} - \tfrac{1}{4}\hat{v}_{0,2} = 0.$$

The outcome is a tridiagonal linear system for every $m \neq 0$ and an almost tridiagonal system for $m = 0$, with just one 'rogue' element outside a three-diagonal band.[2] Such systems can be solved with minimal effort by sparse LU factorization, see Chapter 11.

The practical computation of (15.16) should be confined, needless to say, to a finite subset of $m \in \mathbb{Z}$, for example $-m^* + 1 \leq m \leq m^*$. We have already observed in Section 10.2 that as long as g and ϕ are analytic their periodicity implies the spectral decay of Fourier coefficients. Therefore the error decays roughly like $\mathcal{O}(\mathrm{e}^{-cm^*})$ for some $c > 0$. We can use small values of m^*, and this translates into small linear algebraic systems.

Let us turn our attention to the nuts and bolts of a fast Poisson solver in the disc $\mathbb{D}$. We commence by choosing a positive integer n and let $m^* = 2^{n-1}$. Next we approximate $\{\hat{g}_m\}_{m=-m^*+1}^{m^*}$ and $\{\hat{\phi}_m\}_{m=-m^*+1}^{m^*}$ with FFTs, a task that carries a computational price tag of $\mathcal{O}(m^* \log_2 m^*)$ operations. As we commented in Section 10.3, the error in such a procedure is very small and, provided that g and ϕ are sufficiently well behaved, it decays at spectral speed as a function of m^*.

Having calculated the two transforms and chosen a positive integer d, we proceed to solve the linear systems (15.16) for the relevant range of m by employing sparse LU factorization. The total cost of this procedure is $\mathcal{O}(dm^*)$ operations.

Finally, having evaluated $\hat{v}_{m,k}$ for $-m^* + 1 \leq m \leq m^*$ and $k = 1, 2, \ldots, d-1$, we employ $d - 1$ inverse FFTs to produce values of v on a $d \times (2m^*)$ square grid, or, alternatively, on a concentric grid in $\mathbb{D}$ (see Fig. 15.1). Specifically, we use Fourier coefficients to reconstruct the function, in line with the formula (10.13):

$$u\left(k\Delta r \cos(\pi\ell/m^*), k\Delta r \sin(\pi\ell/m^*)\right) = v(k\Delta x, \omega_{2m^*}^{\ell}) = \sum_{m=-m^*+1}^{m^*} \hat{v}_{m,k}\omega_{2m^*}^{\ell m}$$

for $k = 1, 2, \ldots, d-1$ and $m = 0, 1, \ldots, 2m^* - 1$.[3] Here $\omega_r = \exp(2\pi\mathrm{i}/r)$ is the rth primitive root of unity (cf. Section 10.2). This is the most computationally intense part of the algorithm and the cost totals $\mathcal{O}(dm^* \log_2 m^*)$. It is comparable, though, with the expense of the Hockney method from Section 15.1 (which, of course, acts in a different geometry) and very modest indeed in comparison with other computational alternatives.

[2] Of course, we have $d + 1$ unknowns and a matching number of equations when $m = 0$.

[3] We do not need to perform an inverse FFT for $d = 0$ since, obviously, $u(0,0) = v(0,\theta) = \hat{v}_{0,0}$.

Why is the Poisson problem in a disc so important as to deserve a section all its own? According to the *conformal mapping theorem*, given any simply connected open set $\mathbb{B} \subset \mathbb{C}$ with a sufficiently smooth boundary, there exists an analytic univalent (that is, one-to-one) function χ that maps $\mathbb{B}$ onto the complex unit disc and $\partial\mathbb{B}$ onto the unit circle. (There are many such functions but to attain uniqueness it is enough, for example, to require that an arbitrary $z_0 \in \mathbb{B}$ is mapped into the origin with a positive derivative.) Such a function χ is called a *conformal mapping* of $\mathbb{B}$ on the complex unit disc.

Suppose that the Poisson equation

$$\nabla^2 w = f \tag{15.17}$$

is given for all $(x, y) \in \mathbb{B}^\star$, the natural projection of $\mathbb{B}$ on $\mathbb{R}^2$, where

$$(x, y) \in \mathbb{B}^\star \qquad \text{if and only if} \qquad x + \mathrm{i}y \in \mathbb{B}.$$

We accompany (15.17) with Dirichlet boundary conditions $w = \psi$ across $\partial\mathbb{B}^\star$.

Provided that a conformal mapping χ from $\mathbb{B}$ onto the complex unit disc is known, it is possible to translate the problem of numerically solving (15.17) into the unit disc. Being one-to-one, χ possesses an inverse $\eta = \chi^{-1}$.

We let

$$u(x, y) = w\left(\operatorname{Re}\eta(x + \mathrm{i}y), \operatorname{Im}\eta(x + \mathrm{i}y)\right), \qquad (x, y) \in \operatorname{cl}\mathbb{D}.$$

Therefore

$$\frac{\partial^2 u}{\partial x^2} = \left(\frac{\partial^2 \operatorname{Re}\eta}{\partial x^2}\right)\frac{\partial w}{\partial x} + \left(\frac{\partial^2 \operatorname{Im}\eta}{\partial x^2}\right)\frac{\partial w}{\partial y} + \left(\frac{\partial \operatorname{Re}\eta}{\partial x}\right)^2 \frac{\partial^2 w}{\partial x^2} + \left(\frac{\partial \operatorname{Im}\eta}{\partial x}\right)^2 \frac{\partial^2 w}{\partial y^2},$$

$$\frac{\partial^2 u}{\partial y^2} = \left(\frac{\partial^2 \operatorname{Re}\eta}{\partial y^2}\right)\frac{\partial w}{\partial x} + \left(\frac{\partial^2 \operatorname{Im}\eta}{\partial y^2}\right)\frac{\partial w}{\partial y} + \left(\frac{\partial \operatorname{Re}\eta}{\partial y}\right)^2 \frac{\partial^2 w}{\partial x^2} + \left(\frac{\partial \operatorname{Im}\eta}{\partial y}\right)^2 \frac{\partial^2 w}{\partial y^2}$$

and we deduce that

$$\begin{aligned}\nabla^2 u = {} & \left[\left(\frac{\partial \operatorname{Re}\eta}{\partial x}\right)^2 + \left(\frac{\partial \operatorname{Re}\eta}{\partial y}\right)^2\right]\frac{\partial^2 w}{\partial x^2} + \left[\left(\frac{\partial \operatorname{Im}\eta}{\partial x}\right)^2 + \left(\frac{\partial \operatorname{Im}\eta}{\partial y}\right)^2\right]\frac{\partial^2 w}{\partial y^2} \\ & + (\nabla^2 \operatorname{Re}\eta)\frac{\partial w}{\partial x} + (\nabla^2 \operatorname{Im}\eta)\frac{\partial w}{\partial y}.\end{aligned} \tag{15.18}$$

The function η is the inverse of a univalent analytic function, hence it is itself analytic and obeys the *Cauchy–Riemann* equations

$$\frac{\partial \operatorname{Re}\eta}{\partial x} = \frac{\partial \operatorname{Im}\eta}{\partial y}, \qquad \frac{\partial \operatorname{Re}\eta}{\partial y} = -\frac{\partial \operatorname{Im}\eta}{\partial x}, \qquad (x, y) \in \operatorname{cl}\mathbb{D}.$$

We conclude that

$$\frac{\partial^2 \operatorname{Re}\eta}{\partial y^2} = \frac{\partial}{\partial y}\left(-\frac{\partial \operatorname{Im}\eta}{\partial x}\right) = -\frac{\partial}{\partial x}\left(\frac{\partial \operatorname{Im}\eta}{\partial y}\right) = -\frac{\partial^2 \operatorname{Re}\eta}{\partial x^2},$$

consequently $\nabla^2 \operatorname{Re}\eta = 0$. Likewise $\nabla^2 \operatorname{Im}\eta = 0$ and we deduce the familiar theorem that the real and imaginary parts of an analytic function are *harmonic* (i.e., they obey the Laplace equation).

Another outcome of the Cauchy–Riemann equations is that

$$\left(\frac{\partial \operatorname{Re}\eta}{\partial x}\right)^2 + \left(\frac{\partial \operatorname{Re}\eta}{\partial y}\right)^2 = \left(\frac{\partial \operatorname{Im}\eta}{\partial x}\right)^2 + \left(\frac{\partial \operatorname{Im}\eta}{\partial y}\right)^2 := \kappa(x,y),$$

say. Therefore, substitution in (15.18), in tandem with the Poisson equation (15.17), yields

$$\begin{aligned}\nabla^2 u &= \kappa(x,y)\,\nabla^2 w\left(\operatorname{Re}\eta(x+\mathrm{i}y), \operatorname{Im}\eta(x+\mathrm{i}y)\right)\\ &= \kappa(x,y) f\left(\operatorname{Re}\eta(x+\mathrm{i}y), \operatorname{Im}\eta(x+\mathrm{i}y)\right) := g_0(x,y), \qquad (x,y)\in\mathbb{D}\end{aligned}$$

and we are back to the Poisson equation in a unit disc! The boundary condition is $u(x,y) = \psi(\operatorname{Re}\eta(x+\mathrm{i}y), \operatorname{Im}\eta(x+\mathrm{i}y))$, $(x,y)\in\partial\mathbb{D}$.

Even if χ is unknown, all is not lost since there are very effective numerical methods for its approximation. Their efficacy is based – again – on a clever use of the FFT technique. Moreover, approximate maps χ and η are typically expressible as DFTs, and this means that, having solved the equation in a unit disc, we return to $\mathbb{B}^\star$ with an FFT ... The outcome is a numerical solution on a curved grid, the image of the concentric grid under the function η (see Fig. 15.2).

Comments and bibliography

Golub's survey (1971) and Pickering's monograph (1986) are probably the most comprehensive surveys of fast Poisson solvers, although some methods appear also in Henrici's survey (1979) and elsewhere.

The name 'Poisson solver' is frequently a misnomer. While the rationale behind Section 15.2 is intimately linked with the Laplace operator, this is not the case with the Hockney method of Section 15.1 and, indeed, with many other 'fast Poisson solvers'. A more appropriate name, in line with the comments that conclude Section 15.1, would be 'fast block-TST solvers'. See Exercise 15.2 for an application of a fast block-TST solver to the Helmholtz equation.

We conclude these remarks with a brief survey of a fast Poisson (or, again, a fast block-TST) solver that does not employ the FFT: *cyclic odd–even reduction and factorization.*

The starting point for our discussion is the block-TST equations (15.2), where both S and T are $m_1 \times m_1$ matrices. Neither S nor T need be TST, but we assume that they commute (an assumption which, of course, is certainly true when they are TST). We assume that $m_2 = 2^n$ for some $n \geq 1$. For every $\ell = 1, 2, \ldots, 2^{n-1}$ we multiply the $(2\ell-1)$th equation by T, the (2ℓ)th equation by S and the $(2\ell+1)$th equation by T. Therefore

$$\begin{aligned} T^2\boldsymbol{x}_{2\ell-2} + TS\boldsymbol{x}_{2\ell-1} + T^2\boldsymbol{x}_{2\ell} \qquad\qquad\qquad\qquad\qquad &= T\boldsymbol{b}_{2\ell-1},\\ -\;ST\boldsymbol{x}_{2\ell-1} - S^2\boldsymbol{x}_{2\ell} - ST\boldsymbol{x}_{2\ell+1} \qquad\qquad &= -S\boldsymbol{b}_{2\ell},\\ T^2\boldsymbol{x}_{2\ell} + TS\boldsymbol{x}_{2\ell+1} + T^2\boldsymbol{x}_{2\ell+2} &= T\boldsymbol{b}_{2\ell+1}\end{aligned}$$

and summation, in tandem with $ST = TS$, results in

$$T^2\boldsymbol{x}_{2(\ell-1)} + (2T^2 - S^2)\boldsymbol{x}_{2\ell} + T^2\boldsymbol{x}_{2(\ell+1)} = T(\boldsymbol{b}_{2\ell-1} + \boldsymbol{b}_{2\ell+1}) - S\boldsymbol{b}_{2\ell}, \qquad \ell = 1, 2, \ldots, 2^{n-1}. \tag{15.19}$$

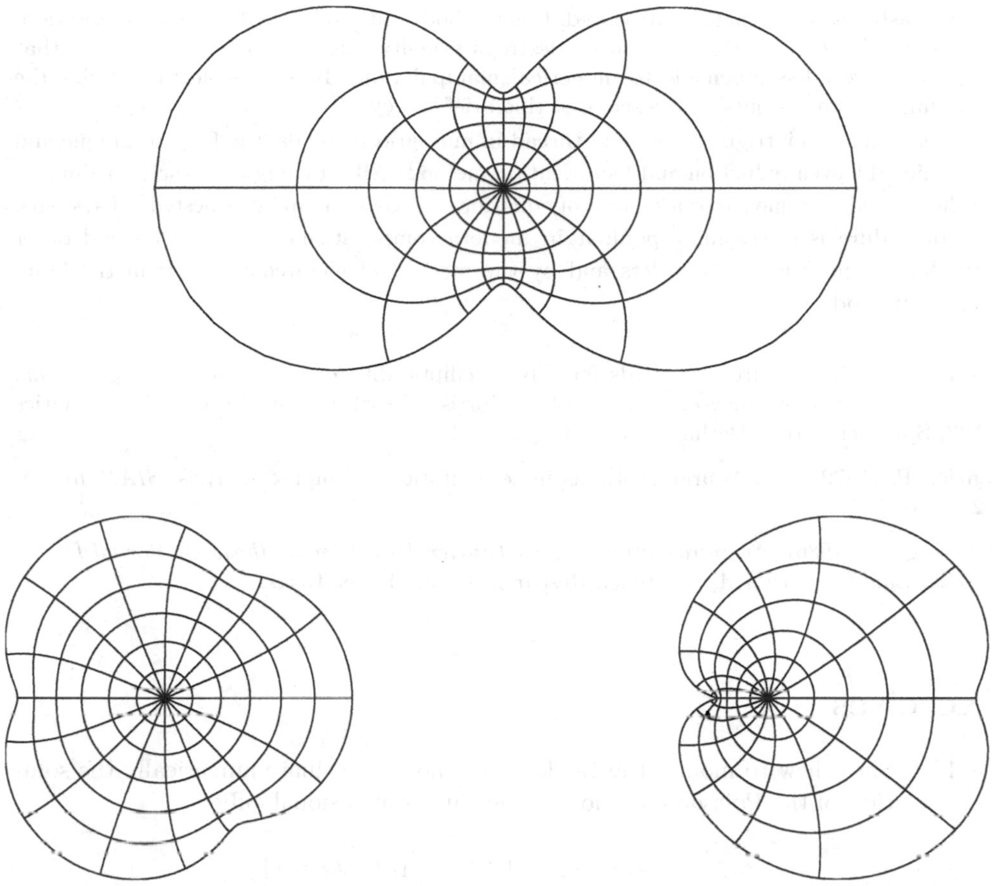

Figure 15.2 Conformal mappings and induced grids for three subsets of $\mathbb{R}^2$. The corresponding mapppings are $z/\left(\frac{3}{2}-z^2\right)$, $z(4+z^2)^{1/2}$ and $(2+z^3)^{1/2}/\left(\frac{7}{4}+z\right)^{3/2}$.

The linear system (15.19) is also block-TST and it possesses exactly half the number of blocks of (15.2). Moreover, T^2 and $2T^2 - S^2$ commute. Provided that the solution of (15.19) is known, we can easily recover the missing components by solving the $m_1 \times m_1$ linear systems

$$S\boldsymbol{x}_{2\ell-1} = \boldsymbol{b}_{2\ell-1} - T(\boldsymbol{x}_{2\ell-2} + \boldsymbol{x}_{2\ell}), \qquad \ell = 1, 2, \ldots, 2^{n-1}.$$

Our choice of $m_2 = 2^n$ already gives the game away – we continue by repeating this procedure again and again, each time reducing the size of the system. Thus, we let

$$\begin{aligned} S^{[0]} &:= S, & S^{[r+1]} &:= 2(T^{[r]})^2 - (S^{[r]})^2, \\ T^{[0]} &:= T, & T^{[r+1]} &:= (T^{[r]})^2, \\ \boldsymbol{b}_\ell^{[0]} &:= \boldsymbol{b}, & \boldsymbol{b}^{[r+1]} &:= T^{[r]}(\boldsymbol{b}_{\ell-2^r}^{[r]} + \boldsymbol{b}_{\ell+2^r}^{[r]}) - S^{[r]}\boldsymbol{b}_\ell^{[r]} \end{aligned}$$

and recover the missing components by iterating backwards:

$$S^{[r-1]}\boldsymbol{x}_{j2^r-2^{r-1}} = \boldsymbol{b}_{j2^r-2^{r-1}}^{[r-1]} - T^{[r-1]}(\boldsymbol{x}_{j2^r} + \boldsymbol{x}_{(j-1)2^r}), \qquad j = 1, 2, \ldots, 2^{n-r-1}.$$

We hasten to warn that, as presented, this method is ill conditioned since each 'iteration' $S^{[r]} \to S^{[r+1]}$, $T^{[r]} \to T^{[r+1]}$ not only destroys sparsity but also produces matrices that are progressively less amenable to numerical manipulation. It is possible to stabilize the algorithm, but this is outside the scope of this brief survey.

There exists an intriguing common thread in multigrid methods, the FFT technique and the cyclic odd–even reduction and factorization method. All these organize their 'medium' – whether a grid, a sequence or a system of equations – into a hierarchy of nested subsystems. This procedure is increasingly popular in modern computational mathematics and other examples are provided by wavelets and by the method of hierarchical bases in the finite element method.

Golub, G.H. (1971), Direct methods for solving elliptic difference equations, in *Symposium on the Theory of Numerical Analysis* (J.L. Morris, editor), Lecture Notes in Mathematics 193, Springer-Verlag, Berlin.

Henrici, P. (1979), Fast Fourier methods in computational complex analysis, *SIAM Review* **21,** 481–527.

Pickering, M. (1986), *An Introduction to Fast Fourier Transform Methods for Partial Differential Equations, with Applications,* Research Studies Press, Herts.

Exercises

15.1 Show how to modify the Hockney method to evaluate numerically the solution of the Poisson equation in the three-dimensional cube

$$\{(x_1, x_2, x_3) \,:\, 0 \le x_1, x_2, x_3 \le 1\}$$

with Dirichlet boundary conditions.

15.2 The Helmholtz equation

$$\nabla^2 u + \lambda u = g$$

is given in a rectangle in $\mathbb{R}^2$, accompanied by Dirichlet boundary conditions. Here λ cannot be an eigenvalue of the operator $-\nabla^2$ (cf. Section 8.2), because if it were then in general a solution would not exist. Amend the Hockney method so that it can provide fast numerical solution of this equation.

15.3* An alternative to solving (15.12) by the finite difference equations (15.16) is the finite element method.

a Find a variational problem whose Euler–Lagrange equation is (15.12).

b Formulate explicitly the Ritz equations.

c Discuss a choice of finite element functions that is likely to produce an accuracy similar to the finite difference method (15.16).

15.4 The Poisson integral formula

$$v(r,\theta) = \frac{1}{\pi}\int_0^{2\pi}\left[\frac{1-r^2}{1-2r\cos(\theta-\tau)+r^2}\right]g(\tau)\,\mathrm{d}\tau$$

confers an alternative to the natural boundary condition (15.15).

a Find explicitly the value of $v(0,\theta)$.

b Deduce the value of $\hat{v}_0(0)$. (*Hint: Express* $v(0,\theta)$ *as a linear combination of Fourier coefficients, in line with* (10.13).)

15.5 Amend the fast Poisson solver from Section 15.2 to approximate the solution of $\nabla^2 u = g_0$ in the unit disc, but with (15.7) replaced by the Neumann boundary condition

$$\frac{\partial u(\cos\theta,\sin\theta)}{\partial x}\cos\theta + \frac{\partial u(\cos\theta,\sin\theta)}{\partial y}\sin\theta = \phi(\theta), \qquad 0 \le \theta \le 2\pi,$$

where $\phi(0) = \phi(2\pi)$ and

$$\int_0^{2\pi}\phi(\theta)\,\mathrm{d}\theta = 0.$$

15.6 Describe a fast Poisson solver for the Poisson equation $\nabla^2 u = g_0$ with Dirichlet boundary conditions in the annulus

$$\{(x,y) \,:\, \rho < x^2+y^2 < 1\},$$

where $\rho \in (0,1)$ is given.

15.7* Generalize the fast Poisson solver from Section 15.2 from the unit disc to the three-dimensional cylinder

$$\{(x_1,x_2,x_3) \,:\, x_1^2+x_2^2 < 1,\; 0 < x_3 < 1\}.$$

You may assume Dirichlet boundary conditions.

PART III

Partial differential equations of evolution

16

The diffusion equation

16.1 A simple numerical method

It is often useful to classify partial differential equations into two kinds: *steady-state* equations, where all the variables are spatial, and *evolutionary* equations, which combine differentiation with respect to space and to time. We have already seen some examples of steady-state equations, namely the Poisson equation and the biharmonic equation. Typically, equations of this type describe physical phenomena whose behaviour depends on the minimization of some quantity, e.g. potential energy, and they are ubiquitous in mechanics and elasticity theory.[1] Evolutionary equations, however, model systems that undergo change as a function of time and they are important *inter alia* in the description of wave phenomena, thermodynamics, diffusive processes and population dynamics.

It is usual in the theory of PDEs to distinguish between *elliptic, parabolic* and *hyperbolic* equations. We do not wish to pursue here this formalism – or even provide the requisite definitions – except to remark that elliptic equations are of the steady-state type whilst both parabolic and hyperbolic PDEs are evolutionary. A brief explanation of this distinction rests in the different kind of characteristic curves admitted by the three types of equations.

Evolutionary differential equations are, in a sense, reminiscent of ODEs. Indeed, one can view ODEs as evolutionary equations without space variables. We will see in what follows that there are many similarities between the numerical treatment of ODEs and of evolutionary PDEs and that, in fact, one of the most effective means to compute the latter is by approximate conversion to an ODE system. However, this similarity is deceptive. The numerical solution of evolutionary PDEs requires us to discretize both in time and in space and, in a successful algorithm, these two procedures are not independent. The concepts underlying the numerical analysis of PDEs of evolution might sound familiar but they are often surprisingly more intricate and subtle than the comparable concepts from Chapters 1–3.

Our first example of an evolutionary equation is the *diffusion equation,*

$$\frac{\partial u}{\partial t} = \frac{\partial^2 u}{\partial x^2}, \qquad 0 \le x \le 1, \quad t \ge 0, \tag{16.1}$$

also known as the *heat conduction equation.* The function $u = u(x,t)$ is accompanied

[1] This minimization procedure can be often rendered in the language of the theory of variations, and this provides a bridge to the material of Chapter 9.

by two kinds of 'side condition', namely an *initial condition*

$$u(x,0) = g(x), \qquad 0 \le x \le 1, \tag{16.2}$$

and the *boundary conditions*

$$u(0,t) = \varphi_0(t), \quad u(1,t) = \varphi_1(t), \qquad t \ge 0 \tag{16.3}$$

(of course, $g(0) = \varphi_0(0)$ and $g(1) = \varphi_1(0)$). As its name implies, (16.1) models diffusive phenomena, e.g. in thermodynamics, epidemiology, financial mathematics and image processing.[2]

The equation (16.1) is the simplest form of a diffusion equation and it can be generalized in several ways:

- by allowing more *spatial variables,* giving
$$\frac{\partial u}{\partial t} = \nabla^2 u, \tag{16.4}$$
where $u = u(x,y,t)$, say;
- by adding to (16.1) a *forcing term* f, giving
$$\frac{\partial u}{\partial t} = \frac{\partial^2 u}{\partial x^2} + f, \tag{16.5}$$
where $f = f(x,t)$;
- by allowing a variable *diffusion coefficient* a, giving
$$\frac{\partial u}{\partial t} = \frac{\partial}{\partial x}\left[a(x)\frac{\partial u}{\partial x}\right], \tag{16.6}$$
where $a = a(x)$ is a differentiable function such that $0 < a(x) < \infty$ for all $x \in [0,1]$;
- by letting x range in an arbitrary interval of $\mathbb{R}$. The most important special case is the *Cauchy problem,* where $-\infty < x < \infty$ and the boundary conditions (16.3) are replaced by the requirement that the solution $u(\,\cdot\,,t)$ is square integrable for all t, i.e.,
$$\int_{-\infty}^{\infty} [u(x,t)]^2 \, \mathrm{d}x < \infty, \qquad t \ge 0.$$

Needless to say, we can combine several such generalizations.

We commence from the most elementary framework but will address ourselves hereafter to various generalizations. Our intention being to approximate (16.1) by finite differences, we choose a positive integer d and inscribe into the strip

$$\{(x,t) \,:\, x \in [0,1],\; t \ge 0\}$$

[2] This ability to look beyond the obvious and discover similar structural patterns across different physical and societal phenomena – in this instance, a flow of 'stuff' across a medium from high-concentration to low-concentration areas – is exactly what makes mathematics into such a powerful tool in the mission to understand the world.

a rectangular grid

$$\{(\ell\Delta x, n\Delta t), \quad \ell = 0, 1, \ldots, d+1, \; n \geq 0\},$$

where $\Delta x = 1/(d+1)$. The approximation of $u(\ell\Delta x, n\Delta t)$ is denoted by u_ℓ^n. Observe that in the latter, n is a superscript not a power – we employ this notation to establish a firm and clear distinction between space and time, a central *leitmotif* in the numerical analysis of evolutionary equations.

Let us replace the second spatial derivative and the first temporal derivative respectively by the central difference

$$\frac{\partial^2 u(x,t)}{\partial x^2} \approx \frac{1}{(\Delta x)^2}[u(x-\Delta x,t) - 2u(x,t) + u(x+\Delta x,t)] + \mathcal{O}\big((\Delta x)^2\big), \qquad \Delta x \to 0,$$

and the forward difference

$$\frac{\partial u(x,t)}{\partial t} = \frac{1}{\Delta t}[u(x,t+\Delta t) - u(x,t)] + \mathcal{O}((\Delta t)), \qquad \Delta t \to 0.$$

Substitution into (16.1) and multiplication by Δt results in the *Euler method*

$$u_\ell^{n+1} = u_\ell^n + \mu(u_{\ell-1}^n - 2u_\ell^n + u_{\ell+1}^n), \qquad \ell = 1, 2, \ldots, d, \quad n = 0, 1, \ldots, \tag{16.7}$$

where the ratio

$$\mu = \frac{\Delta t}{(\Delta x)^2}$$

is important enough to be given a name all of its own, the *Courant number*.

To launch the recursive procedure (16.7) we use the initial condition (16.2), setting

$$u_\ell^0 = g(\ell\Delta x), \qquad \ell = 1, 2, \ldots, d.$$

Note that the calculation of (16.7) for $\ell = 1$ and $\ell = d$ requires us to substitute boundary values from (16.3), namely $u_0^n = \varphi_0(n\Delta t)$ and $u_{d+1}^n = \varphi_1(n\Delta t)$ respectively.

How accurate is the method (16.7)? In line with our definiton of the order of a numerical scheme for ODEs, we observe that

$$\frac{u(x,t+\Delta t) - u(x,t)}{\Delta t} - \frac{u(x-\Delta x,t) - 2u(x,t) + u(x+\Delta x,t)}{(\Delta x)^2} = \mathcal{O}\big((\Delta x)^2, \Delta t\big) \tag{16.8}$$

for $\Delta x, \Delta t \to 0$. Let us assume that Δx and Δt approach zero in such a manner that μ stays constant – it will be seen later that this assumption makes perfect sense! Therefore $\Delta t = \mu(\Delta x)^2$ and (16.8) becomes

$$\frac{u(x,t+\Delta t) - u(x,t)}{\Delta t} - \frac{u(x-\Delta x,t) - 2u(x,t) + u(x+\Delta x,t)}{(\Delta x)^2} = \mathcal{O}\big((\Delta x)^2\big)$$

for $\Delta x \to 0$. We say that the Euler method (16.7) is of *order* 2.[3]

[3]There is some room for confusion here, since for ODE methods an error of $\mathcal{O}\big(h^{p+1}\big)$ means an error of order p. The reason for the present definition of the order will be made clear in the proof of Theorem 16.1.

The concept of order is important in studying how well a finite difference scheme models a continuous differential equation but – as was the case with ODEs in Chapter 2 – our main concern is *convergence*, not order. We say that (16.7) (or, for that matter, any other finite difference method) is convergent if, given any $t^* > 0$, it is true that

$$\lim_{\Delta x \to 0} \left[\lim_{\ell \to x/\Delta x} \left(\lim_{n \to t/\Delta t} u_\ell^n \right) \right] = u(x,t) \qquad \text{for all} \qquad x \in [0,1], \quad t \in [0, t^*].$$

As before, $\mu = \Delta t/(\Delta x)^2$ is kept constant.

Theorem 16.1 *If* $\mu \le \frac{1}{2}$ *then the method* (16.7) *is convergent.*

Proof Let $t^* > 0$ be an arbitrary constant and define

$$e_\ell^n := u_\ell^n - u(\ell \Delta x, n \Delta t), \qquad \ell = 0, 1, \ldots, d+1, \quad n = 0, 1, \ldots, n_{\Delta t},$$

where $n_{\Delta t} = \lfloor t^*/\Delta t \rfloor = \lfloor t^*/(\mu(\Delta x)^2) \rfloor$ is the right-hand endpoint of the range of n. The definition of convergence can be expressed in the terminology of the variables e_ℓ^n as

$$\lim_{\Delta x \to 0} \left[\max_{\ell = 0,1,\ldots,d+1} \left(\max_{n = 0,1,\ldots,n_{\Delta t}} |e_\ell^n| \right) \right] = 0.$$

Letting

$$\eta_n := \max_{\ell = 0,1,\ldots,d+1} |e_\ell^n|, \qquad n = 0, 1, \ldots, n_{\Delta t},$$

we rewrite this as

$$\lim_{\Delta x \to 0} \left(\max_{n = 0,1,\ldots,n_{\Delta t}} \eta_n \right) = 0. \tag{16.9}$$

Since

$$\begin{aligned} u_\ell^{n+1} &= u_\ell^n + \mu(u_{\ell-1}^n - 2u_\ell^n + u_{\ell+1}^n), \\ \tilde{u}_\ell^{n+1} &= \tilde{u}_\ell^n + \mu(\tilde{u}_{\ell-1}^n - 2\tilde{u}_\ell^n + \tilde{u}_{\ell+1}^n) + \mathcal{O}\big((\Delta x)^4\big), \\ &\qquad \ell = 0, 1, \ldots, d+1, \quad n = 0, 1, \ldots, n_{\Delta t} - 1, \end{aligned}$$

where $\tilde{u}_\ell^n = u(\ell \Delta x, n \Delta t)$, subtraction results in

$$\begin{aligned} e_\ell^{n+1} &= e_\ell^n + \mu(e_{\ell-1}^n - 2e_\ell^n + u_{\ell+1}^n) + \mathcal{O}\big((\Delta x)^4\big), \\ &\qquad \ell = 0, 1, \ldots, d+1, \quad n = 0, 1, \ldots, n_{\Delta t} - 1. \end{aligned}$$

In the same way as in the proof of Theorem 1.1, we may now deduce that, provided u is sufficiently smooth (as it will be, provided that the initial and boundary conditions are ssufficiently smooth; but we choose not to elaborate this point further), there exists a constant $c > 0$, independent of Δx, such that, for every $\ell = 0, 1, \ldots, d+1$,

$$\begin{aligned} |e_\ell^{n+1} - e_\ell^n - \mu(e_{\ell-1}^n - 2e_\ell^n + e_{\ell+1}^n)| &\le c(\Delta x)^4, \\ &\qquad \ell = 0, 1, \ldots, d+1, \quad n = 0, 1, \ldots, n_{\Delta t} - 1. \end{aligned}$$

Therefore, by the triangle inequality and the definition of η^n,

$$\begin{aligned}|e_\ell^{n+1}| &\le \left|e_\ell^n + \mu(e_{\ell-1}^n - 2e_\ell^n + e_{\ell+1}^n)\right| + c(\Delta x)^4\\ &\le \mu|e_{\ell-1}^n| + |1-2\mu|\,|e_\ell^n| + \mu|e_{\ell+1}^n| + c(\Delta x)^4\\ &\le (2\mu + |1-2\mu|)\eta^n + c(\Delta x)^4, \qquad n = 0,1,\ldots,n_{\Delta t}-1.\end{aligned}$$

Because $\mu \le \frac{1}{2}$, we may deduce that

$$\eta^{n+1} = \max_{\ell=0,1,\ldots,d+1} |e_\ell^{n+1}| \le \eta^n + c(\Delta x)^4, \qquad n = 0,1,\ldots,n_{\Delta t}-1.$$

Thus, by induction

$$\eta^{n+1} \le \eta^n + c(\Delta x)^4 \le \eta^{n-1} + 2c(\Delta x)^4 \le \eta^{n-2} + 3c(\Delta x)^4 \le \cdots$$

and we conclude that

$$\eta^n \le \eta^0 + nc(\Delta x)^4, \qquad n = 0,1,\ldots,n_{\Delta t}.$$

Since $\eta^0 = 0$ (because the initial conditions at the grid points match for the exact and the discretized equation) and $n(\Delta x)^2 = n\Delta t/\mu \le t^*/\mu$, we deduce that

$$\eta^n \le \frac{ct^*}{\mu}(\Delta x)^2, \qquad n = 0,1,\ldots,n_{\Delta t}.$$

Therefore $\lim_{\Delta x\to 0} \eta^n = 0$ for all n, and comparison with (16.9) completes the proof of convergence. ■

Note that the error in η^n in the proof of the theorem behaves like $\mathcal{O}\big((\Delta x)^2\big)$. This justifies the statement that the method (16.7) is second order.

◇ **A numerical example** Let us consider the diffusion equation (16.1) with the initial and boundary conditions

$$\begin{aligned} g(x) &= \sin\tfrac{1}{2}\pi x + \tfrac{1}{2}\sin 2\pi x, \qquad 0 \le x \le 1, \\ \varphi_0(t) &\equiv 0, \quad \varphi_1(t) = \mathrm{e}^{-\pi^2 t/4}, \qquad t \ge 0, \end{aligned} \tag{16.10}$$

respectively. Its exact solution is, incidentally,

$$u(x,t) = \mathrm{e}^{-\pi^2 t/4}\sin\tfrac{1}{2}\pi x + \tfrac{1}{2}\mathrm{e}^{-4\pi^2 t}\sin 2\pi x, \qquad 0 \le x \le 1, \quad t \ge 0.$$

Fig. 16.1 displays the error in the solution of this equation by the Euler method (16.7) with two choices of μ, one at the edge of the interval $(0, \frac{1}{2}]$ and one outside. It is evident that, while for $\mu = \frac{1}{2}$ the solution looks right, for the second choice of the Courant number it soon deteriorates into complete nonsense.

A different aspect of the solution is highlighted in Fig. 16.2, where μ is kept constant (and within a 'safe' range), while the size of the spatial grid is doubled. The error can be observed to be roughly divided by 4 with each doubling of d, a behaviour entirely consistent with our statement that (16.7) is a second-order method. ◇

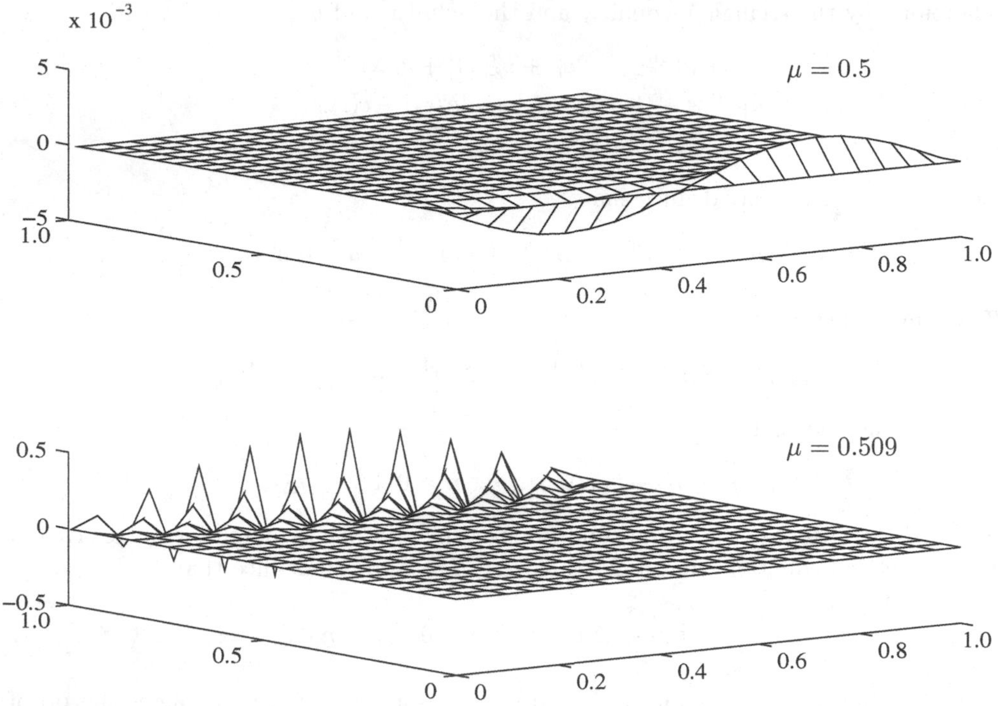

Figure 16.1 The error in the solution of the diffusion equation (16.10) by the Euler method (16.7) with $d = 20$ and two choices of μ, 0.5 and 0.509.

Two important remarks are in order. Firstly, unless a method converges it should not be used – the situation is similar to the numerical analysis of ODEs, with one important exception, as follows. An ODE method is either convergent or not, whereas a method for evolutionary PDEs (e.g. for the diffusion equation (16.1)) possesses a parameter μ and it is entirely possible that it converges only for some values of μ. (We will see later examples of methods that converge for all $\mu > 0$ and, in Exercise 16.13, a method which diverges for all $\mu > 0$.)

Secondly, 'keeping μ constant' means in practice that each time we refine Δx, we need to amend Δt so that the quotient $\mu = \Delta t/(\Delta x)^2$ remains constant. This implies that Δt is likely to be considerably smaller than Δx; for example, $d = 20$ and $\mu = \frac{1}{2}$ yields $\Delta x = \frac{1}{20}$ and $\Delta t = \frac{1}{800}$, leading to a very large computational cost.[4] For example, the lower right-hand surface in Fig. 16.2 was produced with

$$\Delta x = \frac{1}{160} \qquad \text{and} \qquad \Delta t = \frac{1}{64\,000}.$$

Much of the effort associated with designing and analysing numerical methods for the diffusion equation is invested in circumventing such restrictions and attaining

[4]To be fair, we have not yet proved that $\mu > \frac{1}{2}$ is bound to lead to a loss of convergence, although Fig. 16.1 certainly seems to indicate that this is likely. The proof of the necessity of $\mu \le \frac{1}{2}$ is deferred to Section 16.5.

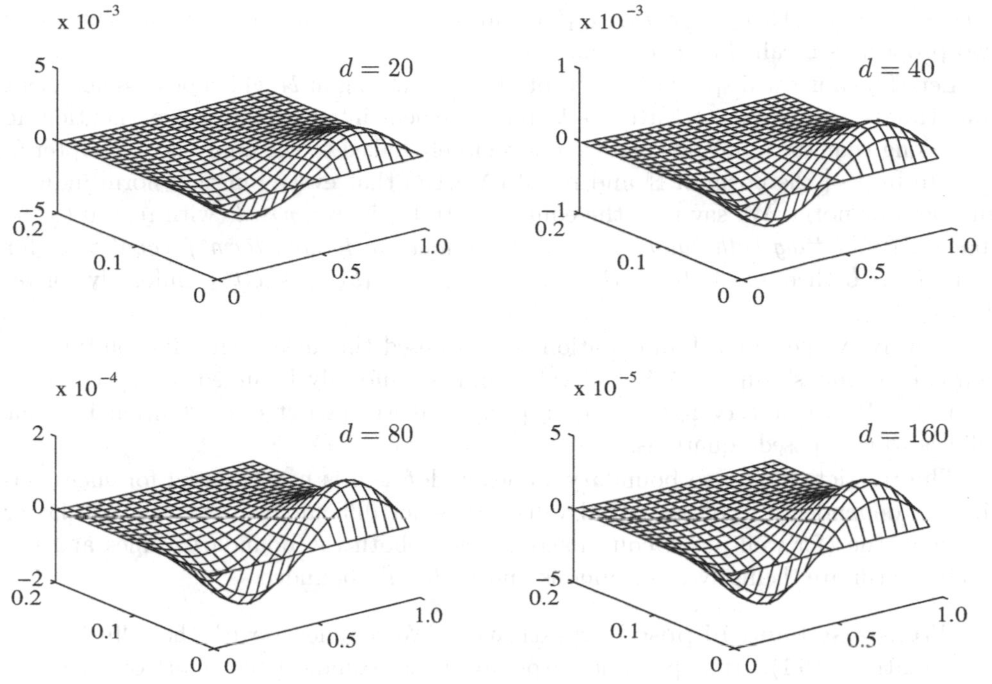

Figure 16.2 The numerical error in the solution of the diffusion equation (16.10) by the Euler method (16.7) with $d = 20, 40, 80, 160$ and $\mu = \frac{2}{5}$.

convergence in regimes of Δx and Δt that are of a more comparable size.

16.2 Order, stability and convergence

In the present section we wish to discuss the numerical solution by finite differences of a general linear PDE of evolution,

$$\frac{\partial u}{\partial t} = \mathcal{L}u + f, \qquad \boldsymbol{x} \in \mathcal{U}, \quad t \geq 0, \tag{16.11}$$

where $\mathcal{U} \subset \mathbb{R}^s$, $u = u(\boldsymbol{x}, t)$, $f = f(t, \boldsymbol{x})$ and $\mathcal{L}$ is a *linear differential operator*,

$$\mathcal{L} = \sum_{i_1+i_2+\cdots+i_s \leq r} a_{i_1,i_2,\ldots,i_s} \frac{\partial^{i_1+i_2+\cdots+i_s}}{\partial x_1^{i_1} \partial x_2^{i_2} \cdots \partial x_s^{i_s}}.$$

We assume that the equation (16.11) is, as usual, provided with an initial value $u(\boldsymbol{x}, 0) = g(\boldsymbol{x})$, $\boldsymbol{x} \in \mathcal{U}$, as well as appropriate boundary conditions.

We express the solution of (16.11) in the form $u(\boldsymbol{x}, t) = \mathcal{E}(t)g(\boldsymbol{x})$, where $\mathcal{E}$ is the *evolution operator* of $\mathcal{L}$. In other words, $\mathcal{E}(t)$ takes an initial value and maps it to the solution at time t. As an aside, note that $\mathcal{E}(0) = \mathcal{I}$, the identity operator, and that

$\mathcal{E}(t_1+t_2) = \mathcal{E}(t_1)\mathcal{E}(t_2) = \mathcal{E}(t_2)\mathcal{E}(t_1)$ for all $t_1, t_2 \geq 0$. An operator with the latter two properties is called a *semigroup*.

Let $\mathbb{H}$ be a normed space (A.2.1.4) of functions acting in $\mathcal{U}$ which possess sufficient smoothness – we prefer to leave the latter statement intentionally vague, mentioning in passing that the requisite smoothness is closely linked to the analysis in Chapter 9. Denote by $\|\cdot\|$ the norm of $\mathbb{H}$ and recall (A.2.1.8) that every function norm induces an operator norm. We say that the equation (16.11) is *well posed* (with regard to the space $\mathbb{H}$) if, *letting both boundary conditions and the forcing term f equal zero,* for every $t^* > 0$ there exists $0 < c(t^*) < \infty$ such that $\|\mathcal{E}(t)\| \leq c(t^*)$ uniformly for all $0 \leq t \leq t^*$.

Intuitively speaking, if an equation is well posed this means that its solution depends continuously upon its initial value and is uniformly bounded in any compact interval. This is a very important property and we restrict our attention in what follows to well-posed equations.

The restriction to zero boundary values and $f \equiv 0$ is not essential for such continuous dependence upon an initial value. It is not difficult to prove that the latter remains true (for well-posed equations) provided both the boundary values and the forcing term are themselves continuous and uniformly bounded.

◇ **Well-posed and ill-posed equations** We commence with the diffusion equation (16.1). It is possible to prove by the standard technique of separation of variables that, provided the initial condition g possesses a Fourier expansion,

$$g(x) = \sum_{m=1}^{\infty} \alpha_m \sin \pi m x, \qquad 0 \leq x \leq 1,$$

the solution of (16.1) can be written explicitly in the form

$$u(x,t) = \sum_{m=1}^{\infty} \alpha_m \mathrm{e}^{-\pi^2 m^2 t} \sin \pi m x, \qquad 0 \leq x \leq 1, \quad t \geq 0. \tag{16.12}$$

Note that u does indeed obey zero boundary conditions.

Suppose that $\|\cdot\|$ is the familiar Euclidean norm,

$$\|f\| = \left\{ \int_0^1 [f(x)]^2 \, \mathrm{d}x \right\}^{1/2}.$$

Then, according to (16.12) (and allowing ourselves to exchange the order of summation and integration)

$$\begin{aligned}\|\mathcal{E}(t)g\|^2 &= \int_0^1 [u(x,t)]^2 \, \mathrm{d}x = \int_0^1 \left(\sum_{m=1}^{\infty} \alpha_m \mathrm{e}^{-\pi^2 m^2 t} \sin \pi m x \right)^2 \mathrm{d}x \\ &= \sum_{m=1}^{\infty} \sum_{j=1}^{\infty} \alpha_m \alpha_j \mathrm{e}^{-\pi^2 (m^2+j^2) t} \int_0^1 \sin \pi m x \sin \pi j x \, \mathrm{d}x.\end{aligned}$$

But

$$\int_0^1 \sin \pi m x \sin \pi j x \, dx = \begin{cases} \frac{1}{2}, & m = j, \\ 0, & \text{otherwise;} \end{cases}$$

consequently

$$\|\mathcal{E}(t)g\|^2 = \tfrac{1}{2} \sum_{m=1}^{\infty} \alpha_m^2 e^{-2\pi^2 m^2 t} \le \tfrac{1}{2} \sum_{m=1}^{\infty} \alpha_m^2 = \|g\|^2.$$

Therefore $\|\mathcal{E}(t)\| \le 1$ for every $t \ge 0$ and we deduce that (16.1) is well posed.

Another example of a well-posed equation is provided by the *advection equation*

$$\frac{\partial u}{\partial t} = \frac{\partial u}{\partial x},$$

which, for simplicity, we define for all $x \in \mathbb{R}$; therefore, there is no need to specify boundary conditions although, of course, we still need to define u at $t = 0$. We will encounter this equation time and again in Chapter 17.

The exact solution of the advection equation is a unilateral shift, $u(x, t) = g(x + t)$ (verify this!). Therefore, employing again the Euclidean norm, we have

$$\|\mathcal{E}g\|^2 = \int_{-\infty}^{\infty} [g(x + t)]^2 \, dx = \int_{-\infty}^{\infty} [g(x)]^2 \, dx = \|g\|^2,$$

and the equation is well posed.

For an example of an ill-posed problem we resort to the 'reversed-time' diffusion equation

$$\frac{\partial u}{\partial t} = -\frac{\partial^2 u}{\partial x^2}.$$

Its solution, obtained by separation of variables, is almost identical to (16.12), except that we need to replace the decaying exponential by an increasing one. Therefore

$$\|\mathcal{E}(t) \sin \pi m x\| = e^{\pi^2 m^2 t} \|\sin \pi m x\|, \qquad m = 1, 2, \ldots,$$

and it is easy to ascertain that $\|\mathcal{E}\|$ is unbounded.

That the 'reversed-time' diffusion equation is ill posed is intimately linked to one of the main principles of thermodynamics, namely that it is impossible to tell the thermal history of an object from its present temperature distribution. ◇

There are, basically, two avenues toward the design of finite difference schemes for the PDE (16.11). Firstly, we can replace the derivatives with respect to each of the variables $t, x_1, x_2, \ldots, x_s$, by finite differences. The outcome is a linear recurrence relation that allows us to advance from $t = n\Delta t$ to $t = (n+1)\Delta t$; the method (16.7) is a case in point. Arranging all the components at the time level $n\Delta t$ in a vector $\boldsymbol{u}_{\Delta x}^n$, we can write a general *full discretization* (FD) of (16.11) in the form

$$\boldsymbol{u}_{\Delta x}^{n+1} = A_{\Delta x} \boldsymbol{u}_{\Delta x}^n + \boldsymbol{k}_{\Delta x}^n, \qquad n = 0, 1, \ldots, \tag{16.13}$$

where the vector $\boldsymbol{k}_{\Delta x}^n$ contains the contributions of the forcing term f and the influence of the boundary values. The elements of the matrix $A_{\Delta x}$ and of the vector $\boldsymbol{k}_{\Delta x}^n$ may depend upon Δx and the *Courant number* $\mu = \Delta t/(\Delta x)^r$ (recall that r is the largest order of differentiation in $\mathcal{L}$).

◇ **The Euler method as FD** The Euler method (16.7) can be written in the form (16.13) with

$$A_{\Delta x} = \begin{bmatrix} 1-2\mu & \mu & 0 & \cdots & 0 \\ \mu & 1-2\mu & \mu & \ddots & \vdots \\ 0 & \ddots & \ddots & \ddots & 0 \\ \vdots & \ddots & \mu & 1-2\mu & \mu \\ 0 & \cdots & 0 & \mu & 1-2\mu \end{bmatrix}, \qquad (16.14)$$

and $\boldsymbol{k}_{\Delta x}^n \equiv \mathbf{0}$. It might appear that neither $A_{\Delta x}$ nor $\boldsymbol{k}_{\Delta x}^n$ depend upon Δx and that we have followed the bad habit of excessive mathematical nitpicking. Not so! Δx expresses itself via the dimension of the space, since $(d+1)\Delta x = 1$. ◇

Let us denote the exact solution of (16.11) at the time level $n\Delta t$, arranged into a similar vector, by $\tilde{\boldsymbol{u}}_{\Delta x}^n$. We say that the FD method (16.13) is of *order* p if, for every initial condition,

$$\tilde{\boldsymbol{u}}_{\Delta x}^{n+1} - A_{\Delta x}\tilde{\boldsymbol{u}}_{\Delta x}^n - \boldsymbol{k}_{\Delta x}^n = \mathcal{O}\big((\Delta x)^{p+r}\big), \qquad \Delta x \to 0, \qquad (16.15)$$

for all $n \geq 0$ and if there exists at least one initial condition g for which the $\mathcal{O}((\Delta x)^{p+r})$ term on the right-hand side does not vanish. The reason why the exponent $p+r$, rather than p, features in the definition, is nontrivial and it will be justified in the proof of Lemma 16.2.

We equip the underlying linear space with the Euclidean norm

$$\|\boldsymbol{g}_{\Delta x}\|_{\Delta x} = \Big(\Delta x \sum |g_j|^2\Big)^{1/2},$$

where the sum ranges over all the grid points.

◇ **Why the factor Δx?** Before we progress further, this is the place to comment on the presence of the mysterious factor $(\Delta x)^{1/2}$ in our definition of the vector norm. Recall that in the present context vectors approximate functions and suppose that $g_{\Delta x,\ell} = g(\ell\Delta x)$, $\ell = 1, 2, \ldots, d$, where $(d+1)\Delta x = 1$. Provided that g is square integrable and letting Δx tend to zero in the Riemann sum, it follows from elementary calculus that

$$\lim_{\Delta x \to 0} \|\boldsymbol{g}_{\Delta x}\|_{\Delta x} = \left\{\int_0^1 [g(x)]^2\,\mathrm{d}x\right\}^{1/2} = |\!|\!| g |\!|\!|.$$

Thus, scaling by $(\Delta x)^{1/2}$ provides for continuous passage from a vector to a function norm. ◇

A method is *convergent* if for every initial condition and all $t^* > 0$ it is true that

$$\lim_{\Delta x \to 0} \left(\max_{n=0,1,\ldots,\lfloor t^*/\Delta t \rfloor} \|\boldsymbol{u}^n_{\Delta x} - \tilde{\boldsymbol{u}}^n_{\Delta x}\|_{\Delta x} \right) = 0.$$

We stipulate that the Courant number is constant as $\Delta x, \Delta t \to 0$, hence $\Delta t = \mu(\Delta x)^r$.

Lemma 16.2 *Let $\|A_{\Delta x}\|_{\Delta x} \le 1$ and suppose that the order condition* (16.15) *holds. Then, for every $t^* > 0$, there exists $c = c(t^*) > 0$ such that*

$$\|\boldsymbol{u}^n_{\Delta x} - \tilde{\boldsymbol{u}}^n_{\Delta x}\|_{\Delta x} \le c(\Delta x)^p, \qquad n = 0, 1, \ldots, \lfloor t^*/\Delta t \rfloor,$$

for all sufficiently small $\Delta x > 0$.

Proof We subtract (16.15) from (16.13), therefore

$$\boldsymbol{e}^{n+1}_{\Delta x} = A_{\Delta x}\boldsymbol{e}^n_{\Delta x} + \mathcal{O}\big((\Delta x)^{p+r}\big), \qquad \Delta x \to 0,$$

where $\boldsymbol{e}^n_{\Delta x} := \boldsymbol{u}^n_{\Delta x} - \tilde{\boldsymbol{u}}^n_{\Delta x}$, $n \ge 0$. The errors $\boldsymbol{e}^n_{\Delta x}$ obey zero initial conditions as well as zero boundary conditions. Hence, and provided that $\Delta x > 0$ is small and the solution of the differential equation is sufficiently smooth (the latter condition depends solely on the requisite smoothness of initial and boundary conditions), there exists $c = c(t^*)$ such that

$$\|\boldsymbol{e}^{n+1}_{\Delta x} - A_{\Delta x}\boldsymbol{e}^n_{\Delta x}\|_{\Delta x} \le c(\Delta x)^{p+r}, \qquad n = 0, 1, \ldots, n_{\Delta t} - 1,$$

where, as in the proof of Theorem 16.1, $n_{\Delta t} := \lfloor t^*/\Delta t \rfloor$. We deduce that

$$\|\boldsymbol{e}^{n+1}_{\Delta x}\|_{\Delta x} \le \|A_{\Delta x}\|_{\Delta x} \times \|\boldsymbol{e}^n_{\Delta x}\|_{\Delta x} + c(\Delta x)^{p+r}, \qquad n = 0, 1, \ldots, n_{\Delta t} - 1;$$

induction, in tandem with $\boldsymbol{e}^0_{\Delta x} = \boldsymbol{0}_{\Delta x}$, then readily yields

$$\|\boldsymbol{e}^n_{\Delta x}\|_{\Delta x} \le c\big(1 + \|A_{\Delta x}\|_{\Delta x} + \cdots + \|A_{\Delta x}\|^{n-1}_{\Delta x}\big)\,(\Delta x)^{p+r}, \qquad n = 0, 1, \ldots, n_{\Delta t}.$$

Since $\|A_{\Delta x}\|_{\Delta x} \le 1$, this leads to

$$\|\boldsymbol{e}^n_{\Delta x}\|_{\Delta x} \le cn(\Delta x)^{p+r} \le c\frac{t^*}{\Delta t}(\Delta x)^{p+r} = \frac{ct^*}{\mu}(\Delta x)^p, \qquad n = 0, 1, \ldots, n_{\Delta t},$$

and the proof of the lemma is complete. ■

We mention in passing that, with little extra effort, the condition $\|A_{\Delta x}\|_{\Delta x} \le 1$ in the above lemma can be replaced by the weaker condition that the underlying method is stable – although, of course, we have not yet said what is meant by stability! This is a good moment to introduce this important concept.

Let us suppose that $f \equiv 0$ and that the boundary conditions are continuous and uniformly bounded. Reiterating that $\Delta t = \mu(\Delta x)^r$, we say that (16.13) is *stable* if for every t^* there exists a constant $c(t^*) > 0$ such that

$$\|\boldsymbol{u}^n_{\Delta x}\|_{\Delta x} \le c(t^*), \qquad n = 0, 1, \ldots, \lfloor t^*/\Delta t \rfloor, \quad \Delta x \to 0. \tag{16.16}$$

Suppose further that the boundary values vanish. In that case $\boldsymbol{k}^n_{\Delta x} \equiv \mathbf{0}_{\Delta x}$, the solution of (16.13) is $\boldsymbol{u}^n_{\Delta x} = A^n_{\Delta x}\boldsymbol{u}^0_{\Delta x}$, $n = 0, 1, \ldots$, and (16.16) becomes equivalent to

$$\lim_{\Delta x \to 0} \left(\max_{n=0,1,\ldots,\lfloor t^*/\Delta t \rfloor} \|A^n_{\Delta x}\|_{\Delta x} \right) \le c(t^*). \tag{16.17}$$

Needless to say, (16.16) and (16.17) each require that the Courant number be kept constant.

An important feature of both order and stability is that they are not attributes of any single numerical scheme (16.13) but of the totality of such schemes as $\Delta x \to 0$. This distinction, to which we will return in Chapter 17, is crucial to the understanding of stability.

Before we make use of this concept, it is only fair to warn the reader that *there is no connection between the present concept of stability and the notion of A-stability from Chapter 4.* Mathematics is replete with diverse concepts bearing the identical sobriquet 'stability' and a careful mathematician should always verify whether a casual reference to 'stability' has to do with stable ultrafilters in logic, with stable fluid flow, stable dynamical systems or, perhaps, with (16.17).

The purpose of our definition of stability is the following theorem. Without much exaggeration, it can be singled out as the lynchpin of the modern numerical theory of evolutionary PDEs.

Theorem 16.3 (The Lax equivalence theorem) *Provided that the linear evolutionary PDE* (16.11) *is well posed, the fully discretized numerical method* (16.13) *is convergent if and only if it is stable and of order* $p \ge 1$. ■

The last theorem plays a similar role to the Dahlquist equivalence theorem (Theorem 2.2) in the theory of multistep numerical methods for ODEs. Thus, on the one hand the concept of convergence might be the central goal of our analysis but it is, in general, difficult to verify from first principles – Theorem 16.1 is almost the exception that proves the rule! On the other hand, it is easy to derive the order and, as we will see in Sections 16.4 and 16.5 and Chapter 17, a number of powerful techniques are available to determine whether a given method (16.13) is stable. Exactly as in Theorem 2.2, we replace an awkward analytic requirement by more manageable algebraic conditions.

Discretizing all the derivatives in line with the recursion (16.13) is not the only possible – or, indeed, useful – approach to the numerical solution of evolutionary PDEs. An alternative technique follows by subjecting only the spatial derivatives to finite difference discretization. This procedure, which we term *semi-discretization* (SD), converts a PDE into a system of coupled ODEs. Using a similar notation to that for FD schemes, in particular 'stretching' grid points into a long vector, we write an archetypical SD method in the form

$$\boldsymbol{v}'_{\Delta x} = P_{\Delta x}\boldsymbol{v}_{\Delta x} + \boldsymbol{h}_{\Delta x}(t), \qquad t \ge 0, \tag{16.18}$$

where $\boldsymbol{h}_{\Delta x}$ consists of the contributions of the forcing term and the boundary values. Note that the use of a prime to denote a derivative is unambiguous: since we have replaced all spatial derivatives by differences, only the temporal derivative is left.

Needless to say, having derived (16.18) we next solve the ODEs, putting to use the theory of Chapters 1–7. Although the outcome is a full discretization – at the end of the day, both spatial and temporal variables are discretized – it is, in general, simpler to derive an SD scheme first and then apply to it the considerable apparatus of numerical ODE methods. Moreover, instead of using finite differences to discretize in space, there is nothing to prevent us from employing finite elements (via the Galerkin approach), spectral methods or other means, e.g. boundary element methods. Only limitations of space (no pun intended) prevent us from debating these issues further.

The method (16.18) is occasionally termed the *method of lines*, mainly in the more traditional numerical literature, to reflect the fact that each component of $\boldsymbol{v}_{\Delta x}$ describes a variable along the line $t \geq 0$. To prevent confusion and for assorted æsthetic reasons we will not use this name.

The concepts of order, convergence and stability can be generalized easily to the SD framework. Denoting by $\tilde{\boldsymbol{v}}_{\Delta x}(t)$ the vector of exact solutions of (16.11) at the (spatial) grid points, we say that the method (16.18) is of *order* p if for all initial conditions it is true that

$$\tilde{\boldsymbol{v}}'_{\Delta x}(t) - P_{\Delta x}\tilde{\boldsymbol{v}}_{\Delta x}(t) - \boldsymbol{h}_{\Delta x}(t) = \mathcal{O}((\Delta x)^p), \qquad \Delta x \to 0, \quad t \geq 0, \tag{16.19}$$

and if the error is precisely $\mathcal{O}((\Delta x)^p)$ for some initial condition. It is *convergent* if for every initial condition and all $t^* > 0$ it is true that

$$\lim_{\Delta x \to 0} \left(\max_{t \in [0,t^*]} \|\boldsymbol{v}_{\Delta x}(t) - \tilde{\boldsymbol{v}}_{\Delta x}(t)\|_{\Delta x} \right) = 0.$$

The semi-discretized method is *stable* if, whenever $f \equiv 0$ and the boundary values are uniformly bounded, for every $t^* > 0$ there exists a constant $c(t^*) > 0$ such that

$$\|\boldsymbol{v}_{\Delta x}(t)\|_{\Delta x} \leq c(t^*), \qquad t \in [0, t^*]. \tag{16.20}$$

Now suppose that the boundary values vanish, in which case $\boldsymbol{h}_{\Delta x} \equiv \boldsymbol{0}_{\Delta x}$ and the solution of (16.18) is

$$\boldsymbol{v}_{\Delta x}(t) = \mathrm{e}^{tP_{\Delta x}}\boldsymbol{v}_{\Delta x}(0), \qquad t \geq 0.$$

We recall that the exponential of an arbitrary square matrix B is defined by means of the Taylor series

$$\mathrm{e}^B = \sum_{k=0}^{\infty} \frac{1}{k!} B^k,$$

which always converges (see Exercise 16.4). Therefore, (16.20) becomes equivalent to

$$\lim_{\Delta x \to 0} \left(\max_{t \in [0,t^*]} \|\mathrm{e}^{tP_{\Delta x}}\|_{\Delta x} \right) \leq c(t^*). \tag{16.21}$$

Theorem 16.4 (The Lax equivalence theorem for SD schemes) *Provided that the linear evolutionary PDE* (16.11) *is well posed, the semi-discretized numerical method* (16.18) *is convergent if and only if it is stable and of order* $p \geq 1$. ■

Approximating a PDE by an ODE is, needless to say, only half the computational job and the effect of the best semi-discretization can be undone by an inappropriate choice of ODE solver for the equations (16.18). We will return to this issue later.

The two equivalence theorems establish a firm bedrock and a starting point for a proper theory of discretized PDEs of evolution. It is easy to discretize a PDE and to produce numbers, but only methods that adhere to the conditions of these theorems allow us to regard such numbers with a modicum of trust.

16.3 Numerical schemes for the diffusion equation

We have already seen one method for the equation (16.1), namely the Euler scheme (16.7). In the present section we follow a more systematic route toward the design of numerical methods – semi-discretized and fully discretized alike – for the diffusion equation (16.1) and some of its generalizations.

Let

$$v'_\ell = \frac{1}{(\Delta x)^2} \sum_{k=-\alpha}^{\beta} a_k v_{\ell+k}, \qquad \ell = 1, 2, \ldots, d, \tag{16.22}$$

be a general SD scheme for (16.1). Note, incidentally, that, unless $\alpha, \beta \le 1$, we need somehow to provide additional information in order to implement (16.22). For example, if $\alpha = 2$ we could require a value for v_{-1} (which is not provided by the boundary conditions (16.3)). Alternatively, we need to replace the first equation in (16.22) by a different scheme. This procedure is akin to boundary effects for the Poisson equation (see Chapter 8) and, more remotely, to the requirement for additional starting values to launch a multistep method (see Chapter 2).

Now set

$$a(z) := \sum_{k=-\alpha}^{\beta} a_k z^k, \qquad z \in \mathbb{C}.$$

Theorem 16.5 *The SD method* (16.22) *is of order p if and only if there exists a constant $c \ne 0$ such that*

$$a(z) = (\ln z)^2 + c(z-1)^{p+2} + \mathcal{O}\left(|z-1|^{p+3}\right), \qquad z \to 1. \tag{16.23}$$

Proof We employ the terminology of finite difference operators from Section 8.1, except that we add to each operator a subscript that denotes the variable. For example, $\mathcal{D}_t = \mathrm{d}/\mathrm{d}t$, whereas $\mathcal{E}_x$ stands for the shift operator along the x-axis. Letting

$$\tilde{v}_\ell = u(\ell\Delta x, \cdot), \qquad \ell = 0, 1, \ldots, d+1,$$

we can thus write the error in the form

$$\tilde{e}_\ell := \tilde{v}'_\ell - \frac{1}{(\Delta x)^2} \sum_{k=-\alpha}^{\beta} a_k \tilde{v}_{\ell+k} = \left[\mathcal{D}_t - \frac{1}{(\Delta x)^2} a(\mathcal{E}_x)\right] \tilde{v}_\ell, \qquad \ell = 1, 2, \ldots, d.$$

Recall that the function $\tilde{v}_\ell$ is the solution of the diffusion equation (16.1) at $x = \ell\Delta x$. In other words,

$$\mathcal{D}_t \tilde{v}_\ell = \frac{\partial u(\ell\Delta x, t)}{\partial t} = \frac{\partial^2 u(\ell\Delta x, t)}{\partial x^2} = \mathcal{D}_x^2 \tilde{v}_\ell, \qquad \ell = 1, 2, \ldots, d.$$

Consequently,

$$\tilde{e}_\ell = \left[\mathcal{D}_x^2 - \frac{1}{(\Delta x)^2} a(\mathcal{E}_x)\right] \tilde{v}_\ell, \qquad \ell = 1, 2, \ldots, d.$$

According to (8.2), however, it is true that

$$\mathcal{D}_x = \frac{1}{\Delta x} \ln \mathcal{E}_x,$$

allowing us to deduce that

$$\tilde{\boldsymbol{e}} = \frac{1}{(\Delta x)^2} \left[(\ln \mathcal{E}_x)^2 - a(\mathcal{E}_x)\right] \tilde{\boldsymbol{v}}, \tag{16.24}$$

where $\tilde{\boldsymbol{e}} = [\, \tilde{e}_1 \quad \tilde{e}_2 \quad \cdots \quad \tilde{e}_d \,]^\top$.

Since, formally, $\mathcal{E}_x = \mathcal{I} + \mathcal{O}(\Delta x)$, it follows that (16.23) is equivalent to

$$[(\ln \mathcal{E}_x)^2 - a(\mathcal{E}_x)]\tilde{\boldsymbol{v}} = c(\Delta x)^{p+2}\mathcal{D}_x^{p+2}\tilde{\boldsymbol{v}} + \mathcal{O}\big((\Delta x)^{p+3}\big), \qquad \Delta x \to 0,$$

provided that the solution u of (16.1) is sufficiently smooth. In particular, substitution into (16.24) gives

$$\tilde{\boldsymbol{e}} = c(\Delta x)^p \mathcal{D}_x^{p+2}\tilde{\boldsymbol{v}} + \mathcal{O}\big((\Delta x)^{p+1}\big), \qquad \Delta x \to 0.$$

It now follows from (16.19) that the SD scheme (16.22) is indeed of order p. ■

◇ **Examples of SD methods** Our first example is

$$v_\ell' = \frac{1}{(\Delta x)^2}(v_{\ell-1} - 2v_\ell + v_{\ell+1}), \qquad \ell = 1, 2, \ldots, d. \tag{16.25}$$

In this case $a(z) = z^{-1} - 2 + z$ and, to derive the order, we let $z = \mathrm{e}^{\mathrm{i}\theta}$; hence

$$a(\mathrm{e}^{\mathrm{i}\theta}) = \mathrm{e}^{-\mathrm{i}\theta} - 2 + \mathrm{e}^{\mathrm{i}\theta} = -4\sin^2 \tfrac{1}{2}\theta = -\theta^2 + \tfrac{1}{12}\theta^4 + \cdots, \qquad \theta \to 0,$$

while $(\ln \mathrm{e}^{\mathrm{i}\theta})^2 = (\mathrm{i}\theta)^2 = -\theta^2$. Therefore (16.25) is of order 2.

Bearing in mind that $a(\mathcal{E}_x)$ is nothing other than a finite difference approximation of $(\Delta x \mathcal{D}_x)^2$, the form of (16.25) is not very surprising, once we write it in the language of finite difference operators:

$$v_\ell' = \frac{1}{(\Delta x)^2}\Delta_{0,x}^2 v_\ell, \qquad \ell = 1, 2, \ldots, d. \tag{16.26}$$

Likewise, we can use (8.8) as a starting point for the SD scheme

$$\begin{aligned} v_\ell' &= \frac{1}{(\Delta x)^2}\left(\Delta_{0,x}^2 - \tfrac{1}{12}\Delta_{0,x}^4\right) v_\ell \qquad\qquad (16.27)\\ &= -\tfrac{1}{12}v_{\ell-2} + \tfrac{4}{3}v_{\ell-1} - \tfrac{5}{2}v_\ell + \tfrac{4}{3}v_{\ell+1} - \tfrac{1}{12}v_{\ell+2}, \qquad \ell = 1, 2, \ldots, d, \end{aligned}$$

where, needless to say, at $\ell = 1$ and $\ell = d$ a special 'fix' might be required to cover for the missing values. It is left to the reader in Exercise 16.5 to verify that (16.27) is of order 4.

Both (16.25) and (16.27) were constructed using central differences and their coefficients display an obvious spatial symmetry. We will see in Section 16.4 that this state of affairs confers an important advantage. Other things being equal, we prefer such schemes and this is the rule for equation (16.1). In Chapter 17, though, while investigating different equations we will encounter a situation where 'other things' are not equal. ◇

The method (16.22) can be easily amended to cater for (16.4), the diffusion equation in several space variables, and it can withstand the addition of a forcing term. Examples are the counterpart of (16.25) in a square,

$$v'_{k,\ell} = \frac{1}{(\Delta x)^2}(v_{k-1,\ell} + v_{k,\ell-1} + v_{k+1,\ell} + v_{k,\ell+1} - 4v_{k,\ell}), \qquad k,\ell = 1,2,\ldots,d \quad (16.28)$$

(unsurprisingly, the terms on the right-hand side are the five-point discretization of the Laplacian ∇^2), and an SD scheme for (16.5), the diffusion equation with a forcing term,

$$v'_\ell = \frac{1}{(\Delta x)^2}(v_{\ell-1} - 2v_\ell + v_{\ell+1}) + f_\ell(t), \qquad \ell = 1,2,\ldots,d. \quad (16.29)$$

Both (16.28) and (16.29) are second-order discretizations.

Extending (16.22) to the case of a variable diffusion coefficient, e.g. to equation (16.6), is equally easy if done correctly. We extend (16.25) by replacing (16.26) with

$$v'_\ell = \frac{1}{(\Delta x)^2}\Delta_{0,x}\left(a_\ell \Delta_{0,x} v_\ell\right), \qquad \ell = 1,2,\ldots,d,$$

where $a_\gamma = a(\kappa\Delta x)$, $\kappa \in [0, d+1]$. The outcome is

$$\begin{aligned} v'_\ell &= \frac{1}{(\Delta x)^2}\Delta_{0,x}[a_\ell(v_{\ell+1/2} - v_{\ell-1/2})] \\ &= \frac{1}{(\Delta x)^2}[a_{\ell-1/2}v_{\ell-1} - (a_{\ell-1/2} + a_{\ell+1/2})v_\ell + a_{\ell+1/2}v_{\ell+1}], \qquad \ell = 1,2,\ldots,d, \end{aligned} \quad (16.30)$$

and it involves solely the values of v on the grid. Again, it is easy to prove that, subject to the requisite smoothness of a, the method is second order.

The derivation of FD schemes can proceed along two distinct avenues, which we explore in the case of the 'plain-vanilla' diffusion equation (16.1). Firstly, we may combine the SD scheme (16.22) with an ODE solver. Three ODE methods are of sufficient interest in this context to merit special mention.

- The Euler method (1.4), that is

$$\boldsymbol{y}_{n+1} = \boldsymbol{y}_n + \Delta t \boldsymbol{f}(n\Delta t, \boldsymbol{y}_n),$$

yields the similarly named *Euler* scheme

$$u_\ell^{n+1} = u_\ell^n + \mu \sum_{k=-\alpha}^{\beta} a_k u_{\ell+k}^n, \qquad \ell = 1,2,\ldots,d, \quad n \geq 0. \quad (16.31)$$

- An application of the trapezoidal rule (1.9),

$$\boldsymbol{y}_{n+1}=\boldsymbol{y}_n+\tfrac{1}{2}\Delta t[\boldsymbol{f}(n\Delta t,\boldsymbol{y}_n)+\boldsymbol{f}((n+1)\Delta t,\boldsymbol{y}_{n+1})]$$

results, after minor manipulation, in the *Crank–Nicolson* scheme

$$u_\ell^{n+1}-\tfrac{1}{2}\mu\sum_{k=-\alpha}^{\beta}a_k u_{\ell+k}^{n+1}=u_\ell^n+\tfrac{1}{2}\mu\sum_{k=-\alpha}^{\beta}a_k u_{\ell+k}^n,\qquad \ell=1,2,\ldots,d,\quad n\geq 0. \tag{16.32}$$

Unlike (16.31), the Crank-Nicolson method is *implicit* – to advance the recursion by a single step, we need to solve a system of linear equations.

- The explicit midpoint rule

$$\boldsymbol{y}_{n+2}=\boldsymbol{y}_n+2\Delta t\boldsymbol{f}((n+1)\Delta t,\boldsymbol{y}_{n+1})$$

(see Exercise 2.5), in tandem with (16.22), yields the *leapfrog* method

$$u_\ell^{n+2}=2\mu\sum_{k=-\alpha}^{\beta}a_k u_{\ell+k}^{n+1}+u_\ell^n,\qquad \ell=1,2,\ldots,d,\quad n\geq 1. \tag{16.33}$$

The leapfrog scheme is *multistep* (specifically, two-step). This is not very surprising, given that the explicit midpoint method itself requires two steps.

Suppose that the SD scheme is of order p_1 and the ODE solver is of order p_2. Hence, the contribution of the semi-discretization to the error is $\Delta t\,\mathcal{O}((\Delta x)^{p_1})=\mathcal{O}\big((\Delta x)^{p_1+2}\big)$, while the ODE solver adds $\mathcal{O}\big((\Delta t)^{p_2+1}\big)=\mathcal{O}\big((\Delta x)^{2p_2+2}\big)$. Altogether, according to (16.15), the order of the FD method is thus

$$p=\min\{p_1,2p_2\} \tag{16.34}$$

(see also Exercise 16.6).

◇ **FD from SD** Let us marry the SD scheme (16.25) with the ODE solvers (16.29)–(16.31). In the first instance this yields the Euler method (16.7). Since, according to Theorem 16.5, $p_1=2$ and since, of course, $p_2=1$, we deduce from (16.34) that the order is 2 – a result that is implicit in the proof of Theorem 16.1.

Putting (16.25) into (16.32) yields the Crank–Nicolson scheme

$$-\tfrac{1}{2}\mu u_{\ell-1}^{n+1}+(1+\mu)u_\ell^{n+1}-\tfrac{1}{2}\mu u_{\ell+1}^{n+1}=\tfrac{1}{2}\mu u_{\ell-1}^n+(1-\mu)u_\ell^n+\tfrac{1}{2}\mu u_{\ell+1}^n. \tag{16.35}$$

Since $p_1=2$ (the trapezoidal rule is second order, see Section 1.3) and $p_2=2$, we have order 2. The superior order of the trapezoidal rule has not helped in improving the order of Crank–Nicolson beyond that of Euler's method (16.7). Bearing in mind that (16.35) is, as well as everything else, implicit, it is fair to query why should we bother with it in the first place. The one-word answer, which will be discussed at length in Sections 16.4 and 16.5, is its stability.

The explicit midpoint rule is also of order 2, and so is the order of the leapfrog scheme

$$u_\ell^{n+2} = 2\mu(u_{\ell+1}^{n+1} - 2u_\ell^{n+1} + u_{\ell-1}^{n+1}) + u_\ell^n. \tag{16.36}$$

Similar reasoning can be applied to more general versions of the diffusion equation and to the SD schemes (16.26)–(16.28). ◇

An alternative technique in designing FD schemes follows similar logic to Theorems 2.1 and 16.5, identifying the order of a method with the order of approximation to a certain function. In line with (16.22), we write a general FD scheme for the diffusion equation (16.1) in the form

$$\sum_{k=-\gamma}^{\delta} b_k(\mu)u_{\ell+k}^{n+1} = \sum_{k=-\alpha}^{\beta} c_k(\mu)u_{\ell+k}^n, \qquad \ell = 1,2,\ldots,d, \quad n \geq 0, \tag{16.37}$$

where, as before, a different type of discretization might be required near the boundary if $\max\{\alpha,\beta,\gamma,\delta\} \geq 2$. We assume that the identity

$$\sum_{k=-\gamma}^{\delta} b_k(\mu) \equiv 1 \tag{16.38}$$

holds and that $b_{-\gamma}, b_\delta, c_{-\alpha}, c_\beta \not\equiv 0$. Otherwise the coefficients $b_k(\mu)$ and $c_k(\mu)$ are, for the time being, arbitrary. If $\gamma = \delta = 0$ then (16.37) is explicit, otherwise the method is implicit.

We set

$$\tilde{a}(z,\mu) := \frac{\sum_{k=-\alpha}^{\beta} c_k(\mu)z^k}{\sum_{k=-\gamma}^{\delta} b_k(\mu)z^k}, \qquad z \in \mathbb{C}, \quad \mu > 0.$$

Theorem 16.6 *The method* (16.37) *is of order* p *if and only if*

$$\tilde{a}(z,\mu) = \mathrm{e}^{\mu(\ln z)^2} + c(\mu)(z-1)^{p+2} + \mathcal{O}\left(|z-1|^{p+3}\right), \qquad z \to 1, \tag{16.39}$$

where $c \not\equiv 0$.

Proof The argument is similar to the proof of Theorem 16.5, hence we will just present its outline. Thus, applying (16.37) to the exact solution, we obtain

$$\begin{aligned} \tilde{e}_\ell^n &= \sum_{k=-\gamma}^{\delta} b_k(\mu)\tilde{u}_{\ell+k}^{n+1} - \sum_{k=-\alpha}^{\beta} c_k(\mu)\tilde{u}_{\ell+k}^n \\ &= \left[\mathcal{E}_t \sum_{k=-\gamma}^{\delta} b_k(\mu)\mathcal{E}_x^k - \sum_{k=-\alpha}^{\beta} c_k(\mu)\mathcal{E}_x^k\right]\tilde{u}_\ell^n. \end{aligned}$$

We deduce from the differential equation (16.1) and the finite difference calculus in Section 8.1 that

$$\mathcal{E}_t = \mathrm{e}^{(\Delta t)\mathcal{D}_t} = \mathrm{e}^{\mu(\Delta x \mathcal{D}_x)^2} = \mathrm{e}^{\mu(\ln \mathcal{E}_x)^2},$$

and this renders $\tilde{e}_\ell^n$ in the language of $\mathcal{E}_x$:

$$\tilde{e}_\ell^n = \left[e^{\mu(\ln \mathcal{E}_x)^2} \sum_{k=-\gamma}^{\delta} b_k(\mu)\mathcal{E}_x^k - \sum_{k=-\alpha}^{\beta} c_k(\mu)\mathcal{E}_x^k \right] \tilde{u}_\ell^n.$$

Next, we conclude from (16.39), from $\mathcal{E}_x = \mathcal{I} + \mathcal{O}(\Delta x)$ and from the normalization condition (16.38) that

$$e^{\mu(\ln \mathcal{E}_x)^2} \sum_{k=-\gamma}^{\delta} b_k(\mu)\mathcal{E}_x^k - \sum_{k=-\alpha}^{\beta} c_k(\mu)\mathcal{E}_x^k = \mathcal{O}\big((\Delta x)^{p+2}\big)$$

and comparison with (16.15) completes the proof. ■

◇ **FD from the function $\tilde{a}$** We commence by revisiting methods that have already been presented in this chapter. As we saw earlier in this section, it is a useful practice to let $z = e^{i\theta}$, so that $z \to 1$ is replaced by $\theta \to 0$.

In the case of the Euler method (16.7) we have

$$\tilde{a}(z, \mu) = 1 + \mu(z^{-1} - 2 + z);$$

therefore

$$\begin{aligned}\tilde{a}(e^{i\theta}) &= 1 - 4\mu \sin^2 \tfrac{1}{2}\theta = 1 - \mu\theta^2 + \tfrac{1}{12}\mu\theta^4 + \mathcal{O}(\theta^6) \\ &= e^{-\mu\theta^2} + \mathcal{O}(\theta^4), \qquad \theta \to 0,\end{aligned}$$

and we deduce order 2 from (16.39).

For the Crank–Nicolson method (16.35) we have

$$\tilde{a}(z, \mu) = \frac{1 + \frac{1}{2}\mu(z^{-1} - 2 + z)}{1 - \frac{1}{2}\mu(z^{-1} - 2 + z)}$$

(note that (16.38) is satisfied), hence

$$\begin{aligned}\tilde{a}(e^{i\theta}) &= \frac{1 - 2\mu \sin^2 \frac{1}{2}\theta}{1 + 2\mu \sin^2 \frac{1}{2}\theta} = 1 - \mu\theta^2 + \left(\tfrac{1}{3}\mu + \tfrac{1}{4}\mu^2\right)\theta^4 + \mathcal{O}(\theta^6) \\ &= e^{-\mu\theta^2} + \mathcal{O}(\theta^4), \qquad \theta \to 0.\end{aligned}$$

Again, we obtain order 2.

The leapfrog method (16.36) does not fit into the framework of Theorem 16.6, but it is not difficult to derive order conditions along the lines of (16.39) for two-step methods, a task left to the reader. ◇

Using the approach of Theorems 16.5 and 16.6, it is possible to express order conditions as a problem in approximation in the two-dimensional case also. This is not so, however, for a variable diffusion coefficient; the quickest practical route toward FD schemes for (16.6) lies in combining the SD method (16.30) with, say, the trapezoidal rule.

16.4 Stability analysis I: Eigenvalue techniques

Throughout this section we will restrict our attention, mainly for the sake of simplicity and brevity, to the case of zero boundary conditions. Therefore, the relevant stability requirements are (16.17) and (16.21) for FD and SD schemes respectively.

A real square matrix B is *normal* if $BB^\top = B^\top B$ (A.1.2.5). Important special cases are symmetric and skew-symmetric matrices.

Two properties of normal matrices are relevant to the material of this section. Firstly, every $d \times d$ normal matrix B possesses a complete set of unitary eigenvectors; in other words, the eigenvectors of B span a d-dimensional linear space and $\bar{\boldsymbol{w}}_j^\top \boldsymbol{w}_\ell = 0$ for any two distinct eigenvectors $\boldsymbol{w}_j, \boldsymbol{w}_\ell \in \mathbb{C}^d$ (A.1.5.3). Secondly, all normal matrices B obey the identity $\|B\| = \rho(B)$, where $\|\cdot\|$ is the Euclidean norm and ρ is the spectral radius. The proof is easy and we leave it to the reader (see Exercise 16.10).

We denote the usual Euclidean inner product by $\langle \cdot, \cdot \rangle$, hence

$$\langle \boldsymbol{x}, \boldsymbol{y} \rangle = \boldsymbol{x}^\top \boldsymbol{y}, \qquad \boldsymbol{x}, \boldsymbol{y} \in \mathbb{R}^d. \tag{16.40}$$

Theorem 16.7 *Let us suppose that the matrix $A_{\Delta x}$ is normal for every sufficiently small $\Delta x > 0$ and that there exists $\nu \geq 0$ such that*

$$\rho(A_{\Delta x}) \leq \mathrm{e}^{\nu \Delta t}, \qquad \Delta x \to 0. \tag{16.41}$$

Then the FD method (16.13) is stable.[5]

Proof We choose an arbitrary $t^* > 0$ and, as before, let $n_{\Delta t} := t^*/\Delta t$. Hence it is true for every vector $\boldsymbol{w}_{\Delta x} \neq \boldsymbol{0}_{\Delta x}$ that

$$\begin{aligned} \|A_{\Delta x}^n \boldsymbol{w}_{\Delta x}\|_{\Delta x}^2 &= \langle A_{\Delta x}^n \boldsymbol{w}_{\Delta x},\, A_{\Delta x}^n \boldsymbol{w}_{\Delta x} \rangle_{\Delta x} \\ &= \langle \boldsymbol{w}_{\Delta x},\, (A_{\Delta x}^n)^\top A_{\Delta x}^n \boldsymbol{w}_{\Delta x} \rangle_{\Delta x}, \qquad n = 0, 1, \ldots, n_{\Delta t}. \end{aligned}$$

Note that we have used here the identity $\langle B\boldsymbol{x}, \boldsymbol{y} \rangle = \langle \boldsymbol{x}, B^\top \boldsymbol{y} \rangle$, which follows at once from (16.40).

It is trivial to verify by induction, using the normalcy of $A_{\Delta x}$, that

$$(A_{\Delta x}^n)^\top A_{\Delta x}^n = (A_{\Delta x}^\top A_{\Delta x})^n, \qquad n = 0, 1, \ldots, n_{\Delta t}.$$

Therefore, by the triangle inequality (A.1.3.3) and the definition of a matrix norm (A.1.3.4), we have

$$\begin{aligned} \|A_{\Delta x}^n \boldsymbol{w}_{\Delta x}\|_{\Delta x}^2 &= \langle \boldsymbol{w}_{\Delta x},\, (A_{\Delta x}^\top A_{\Delta x})^n \boldsymbol{w}_{\Delta x} \rangle_{\Delta x} \\ &\leq \|\boldsymbol{w}_{\Delta x}\|_{\Delta x} \times \|(A_{\Delta x}^\top A_{\Delta x})^n \boldsymbol{w}_{\Delta x}\|_{\Delta x} \\ &\leq \|\boldsymbol{w}_{\Delta x}\|_{\Delta x}^2 \times \|(A_{\Delta x}^\top A_{\Delta x})^n\|_{\Delta x} \\ &\leq \|\boldsymbol{w}_{\Delta x}\|_{\Delta x}^2 \times \|A_{\Delta x}\|_{\Delta x}^{2n} \end{aligned}$$

[5] We recall that $\Delta t \to 0$ as $\Delta x \to 0$ and that the Courant number remains constant.

for $n = 0, 1, \ldots, n_{\Delta t}$. Recalling that $A_{\Delta x}$ is normal, hence that its norm and spectral radius coincide, we deduce from (16.41) the inequality

$$\frac{\|A^n_{\Delta x}\boldsymbol{w}_{\Delta x}\|_{\Delta x}}{\|\boldsymbol{w}_{\Delta x}\|_{\Delta x}} \le [\rho(A_{\Delta x})]^n \le \mathrm{e}^{\nu n\Delta t} \le \mathrm{e}^{\nu t^*}, \qquad n = 0, 1, \ldots, n_{\Delta t}. \tag{16.42}$$

The crucial observation about (16.42) is that it holds *uniformly* for $\Delta x \to 0$. Since by the definition of a matrix norm

$$\|A^n_{\Delta x}\|_{\Delta x} = \max_{\boldsymbol{w}_{\Delta x}\neq \boldsymbol{0}_{\Delta x}} \frac{\|A^n_{\Delta x}\boldsymbol{w}_{\Delta x}\|_{\Delta x}}{\|\boldsymbol{w}_{\Delta x}\|_{\Delta x}},$$

it follows that (16.17) is satisfied by $c(t^*) = \mathrm{e}^{\nu t^*}$ and the method (16.13) is stable. ■

It is of interest to consider an alternative proof of the theorem, which highlights the role of normalcy and clarifies why, in its absence, the condition (16.41) may not be sufficient for stability. Suppose, thus, that $A_{\Delta x}$ has a complete set of eigenvectors but is not necessarily normal. We can factorize $A_{\Delta x}$ as $V_{\Delta x}D_{\Delta x}V^{-1}_{\Delta x}$, where $V_{\Delta x}$ is the matrix of the eigenvectors, while $D_{\Delta x}$ is a diagonal matrix of eigenvalues (A.1.5.4). It follows that, for every $n = 0, 1, \ldots, n_{\Delta t}$,

$$\begin{aligned}\|A^n_{\Delta x}\| &= \|(V_{\Delta x}D_{\Delta x}V^{-1}_{\Delta x})^n\|_{\Delta x} = \|V_{\Delta x}D^n_{\Delta x}V^{-1}_{\Delta x}\|_{\Delta x}\\ &\le \|V_{\Delta x}\|_{\Delta x} \times \|D^n_{\Delta x}\|_{\Delta x} \times \|V^{-1}_{\Delta x}\|_{\Delta x}.\end{aligned}$$

The matrix $D_{\Delta x}$ is diagonal and its diagonal components, $d_{j,j}$, say, are the eigenvalues of $A_{\Delta x}$. Therefore

$$\|D^n_{\Delta x}\|_{\Delta x} = \max_j |d^n_{j,j}| = (\max_j |d_{j,j}|)^n = [\rho(A_{\Delta x})]^n$$

and we deduce that

$$\|A^n_{\Delta x}\| \le \kappa_{\Delta x}[\rho(A_{\Delta x})]^n, \tag{16.43}$$

where

$$\kappa_{\Delta x} := \|V_{\Delta x}\|_{\Delta x} \times \|V^{-1}_{\Delta x}\|_{\Delta x}$$

is the *spectral condition number* of the matrix $V_{\Delta x}$.

On the face of it, we could have continued from (16.43) in a manner similar to the proof of Theorem 16.7, thereby proving the inequality

$$\|A^n_{\Delta x}\|_{\Delta x} \le \kappa_{\Delta x}\mathrm{e}^{\nu t^*}, \qquad n = 0, 1, \ldots, n_{\Delta t}.$$

This looks deceptively like a proof of stability without assuming normalcy in the process. The snag, of course, is in the number $\kappa_{\Delta x}$: as Δx tends to zero, it is entirely possible that $\kappa_{\Delta x}$ becomes infinite! However, if $A_{\Delta x}$ is normal then its eigenvectors are orthogonal, therefore $\|V_{\Delta x}\|_{\Delta x}, \|V^{-1}_{\Delta x}\|_{\Delta x} \equiv 1$ for all Δx (A.1.3.4) and we can indeed use (16.43) to construct an alternative proof of the theorem.

Using the same approach as in Theorem 16.7, we can prove a stability condition for SD schemes with normal matrices.

Theorem 16.8 *Let the matrix $P_{\Delta x}$ be normal for every sufficiently small $\Delta x > 0$. If there exists $\eta \in \mathbb{R}$ such that*

$$\operatorname{Re}\lambda \le \eta \qquad \textit{for every} \qquad \lambda \in \sigma(P_{\Delta x}) \qquad \textit{and} \qquad \Delta x \to 0 \tag{16.44}$$

then the SD method (16.18) *is stable.*

Proof Let $t^* > 0$ be given. Because of the normalcy of $P_{\Delta x}$, it follows along similar lines to the proof of Theorem 16.7 that

$$\begin{aligned}
\|e^{tP_{\Delta x}}\boldsymbol{w}_{\Delta x}\|_{\Delta x}^2 &= \langle e^{tP_{\Delta x}}\boldsymbol{w}_{\Delta x}, e^{tP_{\Delta x}}\boldsymbol{w}_{\Delta x}\rangle_{\Delta x} = \left\langle \boldsymbol{w}_{\Delta x}, (e^{tP_{\Delta x}})^\top e^{tP_{\Delta x}}\boldsymbol{w}_{\Delta x}\right\rangle_{\Delta x} \\
&= \left\langle \boldsymbol{w}_{\Delta x}, e^{tP_{\Delta x}^\top} e^{tP_{\Delta x}}\boldsymbol{w}_{\Delta x}\right\rangle_{\Delta x} = \left\langle \boldsymbol{w}_{\Delta x}, e^{t(P_{\Delta x}^\top + P_{\Delta x})}\boldsymbol{w}_{\Delta x}\right\rangle_{\Delta x} \\
&\le \|\boldsymbol{w}_{\Delta x}\|_{\Delta x}^2 \times \|e^{t(P_{\Delta x}^\top + P_{\Delta x})}\|_{\Delta x} = \|\boldsymbol{w}_{\Delta x}\|_{\Delta x}^2\, \rho(e^{t(P_{\Delta x}^\top + P_{\Delta x})}) \\
&= \|\boldsymbol{w}_{\Delta x}\|_{\Delta x}^2 \max\left\{e^{2t\operatorname{Re}\lambda} : \lambda \in \sigma(P_{\Delta x})\right\} \le \|\boldsymbol{w}_{\Delta x}\|_{\Delta x}^2 e^{2\eta t^*}, \quad t \in [0, t^*].
\end{aligned}$$

We leave it to the reader to verify that for all normal matrices B and for $t \ge 0$ it is true that

$$(e^{tB})^\top = e^{tB^\top} \qquad e^{tB^\top} e^{tB} = e^{t(B^\top + B)}$$

(note that the second identity might fail unless B is normal!) and that

$$\sigma(e^{t(B^\top + B)}) = \{e^{2t\operatorname{Re}\lambda},\ \lambda \in \sigma(P_{\Delta x})\}.$$

We deduce stability from the definition (16.21). ■

The *spectral abscissa* of a square matrix B is the real number

$$\tilde{\alpha}(B) := \max\{\operatorname{Re}\lambda : \lambda \in \sigma(B)\}.$$

We can rephrase Theorem 16.8 by requiring $\tilde{\alpha}(P_{\Delta x}) \le \eta$ for all $\Delta x \to 0$.

The great virtue of Theorems 16.7 and 16.8 is that they reduce the task of determining stability to that of locating the eigenvalues of a normal matrix. Even better, to establish stability it is often sufficient to bound the spectral radius or the spectral abscissa. According to a broad principle mentioned in Chapter 2, we replace the analytic – and difficult – stability conditions (16.17) and (16.21) by algebraic requirements.

◇ **Eigenvalues and stability of methods for the diffusion equation** The matrix associated with the SD method (16.25) is (in the natural ordering of grid points, from left to right) TST and, according to Lemma 12.5, its eigenvalues are $-4\sin^2[\pi\ell/(d+1)]$, $\ell = 1, 2, \ldots, d$. Therefore

$$\tilde{\alpha}(P_{\Delta x}) = -4\sin^2(\pi\Delta x) \le 0, \qquad \Delta x > 0$$

(recall that $(d+1)\Delta x = 1$) and the method is stable.

Next we consider Euler's FD scheme (16.7). The matrix $A_{\Delta x}$ is again TST and its eigenvalues are $1-4\mu\sin^2[\pi\ell/(d+1)]$, $\ell=1,2,\ldots,d$. Therefore

$$\rho(A_{\Delta x}) \equiv |1-4\mu|, \qquad \Delta x > 0.$$

Consequently, (16.41) is satisfied by $\nu=0$ for $\mu\le\frac{1}{2}$, whereas no ν will do for $\mu>\frac{1}{2}$. This, in tandem with the Lax equivalence theorem (Theorem 16.3) and our observation from Section 16.3 that (16.7) is a second-order method, provides a brief alternative proof of Theorem 16.1.

The next candidate for our attention is the Crank–Nicolson scheme (16.35), which we also render in a vector form. Disregarding for a moment our assumption that the forcing term and boundary contributions vanish, we have

$$A_{\Delta x}^{[+]}\boldsymbol{u}_{\Delta x}^{n+1} = A_{\Delta x}^{[-]}\boldsymbol{u}_{\Delta x}^{n} + \tilde{\boldsymbol{k}}_{\Delta x}^{n}, \qquad n \ge 0,$$

where the matrices $A_{\Delta x}^{[\pm]}$ are TST while the vector $\tilde{\boldsymbol{k}}_{\Delta x}^{n}$ contains the contribution of both forcing and boundary terms. Therefore $A_{\Delta x} = {A_{\Delta x}^{[+]}}^{-1} A_{\Delta x}^{[-]}$ and $\boldsymbol{k}_{\Delta x}^{n} = {A_{\Delta x}^{[+]}}^{-1}\tilde{\boldsymbol{k}}_{\Delta x}^{n}$. (We insist on the presence of the forcing terms $\tilde{\boldsymbol{k}}_{\Delta x}^{n}$ before eliminating them for the sake of stability analysis, since this procedure illustrates how to construct the form (16.13) for general implicit FD schemes.)

According to Lemma 12.5, TST matrices of the same dimension share the same set of eigenvectors. Moreover, these eigenvactors span the whole space, consequently $A_{\Delta x}^{\pm} = V_{\Delta x} D_{\Delta x}^{[\pm]} V_{\Delta x}^{-1}$, where $V_{\Delta x}$ is the matrix of the eigenvectors and $D_{\Delta x}^{[\pm]}$ are diagonal. Therefore

$$A_{\Delta x} = V_{\Delta x} {D_{\Delta x}^{[+]}}^{-1} D_{\Delta x}^{[-]} V_{\Delta x}^{-1}$$

and the eigenvalues of the quotient matrix of two TST matrices – itself, in general, not TST – are the quotients of the eigenvalues of $A_{\Delta x}^{[\pm]}$. Employing again Lemma 12.5, we write down explicitly the eigenvalues of the latter,

$$\sigma(A_{\Delta x}^{[\pm]}) = \left\{1 \pm 2\mu\sin^2\left(\frac{\pi\ell}{2(d+1)}\right) \,:\, \ell=1,2,\ldots,d\right\},$$

hence

$$\sigma(A_{\Delta x}) = \left\{\frac{1-2\mu\sin^2\{\pi\ell/[2(d+1)]\}}{1+2\mu\sin^2\{\pi\ell/[2(d+1)]\}} \,:\, \ell=1,2,\ldots,d\right\}$$

and we deduce that

$$\rho(A_{\Delta x}) = \frac{|1-2\mu\sin^2(\pi\Delta x/2)|}{1+2\mu\sin^2(\pi\Delta x/2)} \le 1.$$

Therefore the Crank–Nicolson scheme is stable *for all* $\mu > 0$.

All three aforementioned examples make use of TST matrices, but this technique is, unfortunately, of limited scope. Consider, for example, the SD scheme (16.30) for the variable diffusion coefficient PDE (16.6). Writing this in the form (16.18), we obtain for $(\Delta x)^2 P_{\Delta x}$ the following matrix:

$$\begin{bmatrix} -a_{-1/2}-a_{1/2} & a_{1/2} & 0 & \cdots & 0 \\ a_{1/2} & -a_{1/2}-a_{3/2} & a_{3/2} & \ddots & \vdots \\ 0 & \ddots & \ddots & \ddots & 0 \\ \vdots & \ddots & a_{d-3/2} & -a_{d-3/2}-a_{d-1/2} & a_{d-1/2} \\ 0 & \cdots & 0 & a_{d-1/2} & -a_{d-1/2}-a_{d+1/2} \end{bmatrix}.$$

Clearly, in general $P_{\Delta x}$ is not a Toeplitz matrix and so we are not allowed to use Lemma 12.5. However, it is symmetric, hence normal, and we are within the conditions of Theorem 16.8.

Although we cannot find the eigenvalues of $P_{\Delta x}$, we can exploit the Geršgorin criterion (Lemma 8.3) to derive enough information about their location to prove stability. Since $a(x) > 0$, $x \in [0, 1]$, it follows at once that all the Geršgorin discs $\mathbb{S}_i$, $i = 1, 2, \ldots, d$, lie in the closed complex left half-plane. Therefore $\tilde{\alpha}(P_{\Delta x}) \leq 0$, hence we have stability. ◇

16.5 Stability analysis II: Fourier techniques

We commence this section by assuming that we are solving an evolutionary PDE given (in a single spatial dimension) for all $x \in \mathbb{R}$ and that in place of boundary conditions, say, (16.3) we impose the requirement that the function $u(\,\cdot\,, t)$ is square-integrable in $\mathbb{R}$ for all $t \geq 0$. As we mentioned in Section 16.1, this is known as the *Cauchy* problem.

The technique of the present section is valid whenever a Cauchy problem for an arbitrary linear PDE of evolution with constant coefficients is solved by a method – either SD or FD – that employs an identical formula at each grid point. For simplicity, however, we restrict ourselves here to the diffusion equation and to the general SD and FD schemes (16.22) and (16.37) respectively (except that the range of ℓ now extends across all $\mathbb{Z}$). The reader should bear in mind, however, that special properties of the diffusion equation – except in the narrowest technical sense, e.g. with regard to the power of Δx in (16.22) and (16.37) – are never used in our exposition. This makes for entirely straightforward generalization.

The definition of stability depends on the underlying norm and throughout this section we consider exclusively the Euclidean norm over bi-infinite sequences. The set $\ell[\mathbb{Z}]$ is the linear space of all complex sequences, indexed over the integers, that are bounded in the Euclidean vector norm. In other words,

$$\boldsymbol{w} = \{w_m\}_{m=-\infty}^{\infty} \in \ell_2[\mathbb{Z}] \qquad \text{if and only if} \qquad \|\boldsymbol{w}\| := \left(\sum_{m=-\infty}^{\infty} |w_m|^2\right)^{1/2} < \infty.$$

Note that throughout the present section we omit the factor $(\Delta x)^{1/2}$ in the definition of the Euclidean norm, mainly to unclutter the notation and to bring it into line with the standard terminology of Fourier analysis. We also allow ourselves the liberating convention of dropping the subscripts $_{\Delta x}$ in our formulae, the reason being that Δx no longer expresses the reciprocal of the number of grid points – which is infinite for all Δx. The only influence of Δx on the underlying equations is expressed in the multiplier $(\Delta x)^{-2}$ for SD equations and – most importantly – in the spacing of the grid along which we are sampling the initial condition g.

We let $L[0, 2\pi]$ denote the set of all complex, square-integrable functions in $[0, 2\pi]$, equipped with the Euclidean function norm:

$$w \in L[0, 2\pi] \qquad \text{if and only if} \qquad \|w\| = \left[\frac{1}{2\pi}\int_0^{2\pi} |w(\theta)|^2 \,\mathrm{d}\theta\right]^{1/2} < \infty.$$

Solutions of either (16.22) or (16.37) live in $\ell[\mathbb{Z}]$ (remember that the index ranges across all integers!), consequently we phrase their stability in terms of the norm in that space. As it turns out, however, it is considerably more convenient to investigate stability in $L[0, 2\pi]$. The opportunity to abandon $\ell[\mathbb{Z}]$ in favour of $L[0, 2\pi]$ is conferred by the *Fourier transform*. We have already encountered a similar concept in Chapters 10, 13 and 15 in a different context. For our present purpose, we choose a definition different from that in Chapter 10, letting

$$\hat{w}(\theta) = \sum_{m=-\infty}^{\infty} w_m \mathrm{e}^{-\mathrm{i}m\theta}, \qquad \boldsymbol{w} = (w_m)_{m=-\infty}^{\infty} \subset \ell[\mathbb{Z}]. \tag{16.45}$$

Lemma 16.9 *The mapping* (16.45) *takes* $\ell[\mathbb{Z}]$ *onto* $L[0, 2\pi]$. *It is an* isomorphism *(i.e., a one-to-one mapping) and its inverse is given by*

$$w_m = \frac{1}{2\pi}\int_0^{2\pi} \hat{w}(\theta)\mathrm{e}^{\mathrm{i}m\theta}\,\mathrm{d}\theta, \qquad m \in \mathbb{Z}, \quad \hat{w} \in L[0, 2\pi]. \tag{16.46}$$

Moreover, (16.45) *is an* isometry:

$$\|\hat{w}\| = \|\boldsymbol{w}\|, \qquad \boldsymbol{w} \in \ell[\mathbb{Z}]. \tag{16.47}$$

Proof We combine the proof that $\hat{w} \in L[0, 2\pi]$ (hence, that (16.45) indeed takes $\ell[\mathbb{Z}]$ to $L[0, 2\pi]$) with the proof of (16.47), by evaluating the norm of $\hat{w}$:

$$\begin{aligned}\|\hat{w}\|^2 &= \frac{1}{2\pi}\int_0^{2\pi} \sum_{m=-\infty}^{\infty}\sum_{j=-\infty}^{\infty} w_m \bar{w}_j \mathrm{e}^{\mathrm{i}(j-m)\theta}\,\mathrm{d}\theta \\ &= \frac{1}{2\pi}\sum_{m=-\infty}^{\infty}\sum_{j=-\infty}^{\infty} w_m \bar{w}_j \int_0^{2\pi} \mathrm{e}^{\mathrm{i}(j-m)\theta}\,\mathrm{d}\theta = \sum_{m=-\infty}^{\infty} |w_m|^2 = \|\boldsymbol{w}\|^2.\end{aligned}$$

Note our use of the identity

$$\frac{1}{2\pi}\int_0^{2\pi} \mathrm{e}^{\mathrm{i}k\theta}\,\mathrm{d}\theta = \begin{cases} 1, & k = 0, \\ 0, & \text{otherwise}, \end{cases} \qquad k \in \mathbb{Z}.$$

The argument required to prove that the mapping $\boldsymbol{w} \mapsto \hat{w}$ is an isomorphism onto $L[0, 2\pi]$ and that its inverse is given by (16.46) is an almost exact replica of the proof of Lemma 10.2. ■

We will call $\hat{w}$ the *Fourier transform* of $\boldsymbol{w}$. This is at variance with the terminology of Chapter 10 – by rights, we should call $\hat{w}$ the *inverse* Fourier transform of $\boldsymbol{w}$. The present usage, however, has the advantage of brevity.

The isomorphic isometry of the Fourier transform is perhaps the main reason for its importance in a wide range of applications. A Euclidean norm typically measures the energy of physical systems and a major consequence of (16.47) is that, while travelling back and forth between $\ell[\mathbb{Z}]$ and $L[0, 2\pi]$ by means of the Fourier transform and its inverse, the energy stays intact.

We commence our analysis with the SD scheme (16.22), recalling that the index ranges across all $\ell \in \mathbb{Z}$. We multiply the equation by $\mathrm{e}^{-\mathrm{i}\ell\theta}$ and sum over ℓ; the outcome is

$$\begin{aligned}
\frac{\partial \hat{v}(\theta, t)}{\partial t} &= \sum_{\ell=-\infty}^{\infty} v_\ell' \mathrm{e}^{-\mathrm{i}\ell\theta} = \frac{1}{(\Delta x)^2} \sum_{\ell=-\infty}^{\infty} \sum_{k=-\alpha}^{\beta} a_k v_{\ell+k} \mathrm{e}^{-\mathrm{i}\ell\theta} \\
&= \frac{1}{(\Delta x)^2} \sum_{k=-\alpha}^{\beta} a_k \sum_{\ell=-\infty}^{\infty} v_{\ell+k} \mathrm{e}^{-\mathrm{i}\ell\theta} = \frac{1}{(\Delta x)^2} \sum_{k=-\alpha}^{\beta} a_k \sum_{\ell=-\infty}^{\infty} v_\ell \mathrm{e}^{-\mathrm{i}(\ell-k)\theta} \\
&= \frac{1}{(\Delta x)^2} \sum_{k=-\alpha}^{\beta} a_k \mathrm{e}^{\mathrm{i}k\theta} \sum_{\ell=-\infty}^{\infty} v_\ell \mathrm{e}^{-\mathrm{i}\ell\theta} = \frac{a(\mathrm{e}^{\mathrm{i}\theta})}{(\Delta x)^2} \hat{v}(\theta, t),
\end{aligned}$$

where the function $a(\,\cdot\,)$ was defined in Section 16.3. The crucial step in the above argument is the shift of the index from ℓ to $\ell - k$ without changing the endpoints of the summation, a trick that explains why we require that ℓ should range across all the integers.

We have just proved that the Fourier transform $\hat{v} = \hat{v}(\theta, t)$ obeys, as a function of t, the linear ODE

$$\frac{\partial \hat{v}}{\partial t} = \frac{a(\mathrm{e}^{\mathrm{i}\theta})}{(\Delta x)^2} \hat{v}, \qquad t \geq 0, \quad \theta \in [0, 2\pi].$$

with initial condition $\hat{v}(\theta, 0) = \hat{g}$, where $g_m = u(m\Delta x, 0)$, $m \in \mathbb{Z}$, is the projection on the grid of the initial condition of the PDE. The solution of the ODE can be written down explicitly:

$$\hat{v}(\theta, t) = \hat{g}(\theta) \exp\left[\frac{a(\mathrm{e}^{\mathrm{i}\theta})t}{(\Delta x)^2}\right], \qquad t \geq 0, \quad \theta \in [0, 2\pi]. \tag{16.48}$$

Suppose that

$$\operatorname{Re} a(\mathrm{e}^{\mathrm{i}\theta}) \leq 0, \qquad \theta \in [0, 2\pi]. \tag{16.49}$$

In that case it follows from (16.48) that

$$|||\hat{v}|||^2 = \frac{1}{2\pi} \int_0^{2\pi} |\hat{g}(\theta)|^2 \exp\left[\frac{2 \operatorname{Re} a(\mathrm{e}^{\mathrm{i}\theta})}{(\Delta x)^2}\right] \mathrm{d}\theta \leq \frac{1}{2\pi} \int_0^{2\pi} |\hat{g}(\theta)|^2 \,\mathrm{d}\theta = |||\hat{g}|||^2,$$

Therefore, according to (16.47),

$$\|\boldsymbol{v}(t)\| \leq \|\boldsymbol{v}(0)\|$$

for all possible initial conditions $\boldsymbol{v} \in \ell[\mathbb{Z}]$. We thus conclude that, according to (16.20), the method is stable.

Next, we consider the case when the condition (16.49) is violated, in other words, when there exists $\theta_0 \in [0, 2\pi]$ such that $\operatorname{Re} a(\mathrm{e}^{\mathrm{i}\theta_0}) > 0$. Since $a(\mathrm{e}^{\mathrm{i}\theta})$ is continuous in θ, there exist $\varepsilon > 0$ and $0 \leq \theta_- < \theta_+ < 2\pi$ such that

$$\operatorname{Re} a(\mathrm{e}^{\mathrm{i}\theta}) > \varepsilon, \qquad \theta \in [\theta_-, \theta_+].$$

We choose an initial condition g such that $\hat{g}$ is a characteristic function of the interval $[\theta_-, \theta_+]$:

$$\hat{g}(\theta) = \begin{cases} 1, & \theta \in [\theta_-, \theta_+], \\ 0 & \text{otherwise} \end{cases}$$

(it is possible to identify easily a square-integrable initial condition g with the above $\hat{g}$). It follows from (16.48) that

$$\begin{aligned} \|\hat{v}\|^2 &= \frac{1}{2\pi} \int_{\theta_-}^{\theta_+} \exp\left[\frac{2 \operatorname{Re} a(\mathrm{e}^{\mathrm{i}\theta}) t}{(\Delta x)^2}\right] \mathrm{d}\theta \geq \frac{1}{2\pi} \int_{\theta_-}^{\theta_+} \exp\left[\frac{2\varepsilon t}{(\Delta x)^2}\right] \mathrm{d}\theta \\ &= \frac{\theta_+ - \theta_-}{2\pi} \exp\left[\frac{2\varepsilon t}{(\Delta x)^2}\right]. \end{aligned}$$

Therefore $\|\hat{v}\|$ cannot be uniformly bounded for $t \in [0, t^*]$ (regardless of the size of $t^* > 0$) as $\Delta x \to 0$. We will again exploit isometry to argue that (16.22) is unstable.

Theorem 16.10 *The SD method* (16.22)*, when applied to a Cauchy problem, is stable if and only if the inequality* (16.49) *is obeyed.* ■

Fourier analysis can be applied with similarly telling effect to the FD scheme (16.37) – again, with $\ell \in \mathbb{Z}$. The argument is almost identical, hence we present it with greater brevity.

Theorem 16.11 *The FD method* (16.37)*, when applied to a Cauchy problem, is stable for a specific value of the Courant number* μ *if and only if*

$$|\tilde{a}(\mathrm{e}^{\mathrm{i}\theta}, \mu)| \leq 1, \qquad \theta \in [0, 2\pi], \tag{16.50}$$

where

$$\tilde{a}(z, \mu) = \frac{\sum_{k=-\alpha}^{\beta} c_k(\mu) z^k}{\sum_{k=-\gamma}^{\delta} b_k(\mu) z^k}, \qquad z \in \mathbb{C}.$$

Proof We multiply both sides of (16.37) by $\mathrm{e}^{-\mathrm{i}\ell\theta}$ and sum over $\ell \in \mathbb{Z}$. The outcome is the recursive relationship

$$\hat{u}^{n+1} = \tilde{a}(\mathrm{e}^{\mathrm{i}\theta}, \mu) \hat{u}^n, \qquad n \geq 0,$$

between the Fourier transforms in adjacent time levels. Iterating this recurrence results in the explicit formula

$$\hat{u}^n = \left[\tilde{a}(e^{i\theta}, \mu)\right]^n \hat{u}^0, \qquad n \ge 0,$$

where, of course, $\hat{u}^0 = \hat{g}$. Therefore

$$\|\boldsymbol{u}^n\|^2 = \|\hat{u}^n\|^2 = \frac{1}{2\pi}\int_0^{2\pi} \left[\tilde{a}(e^{i\theta})\right]^n \hat{u}^0(\theta)\,d\theta, \qquad n \ge 0. \tag{16.51}$$

If (16.50) is satisfied we deduce from (16.51) that

$$\|\boldsymbol{u}^n\| \le \|\boldsymbol{u}^0\|, \qquad n \ge 0.$$

Stability follows from (16.16) by virtue of isometry.

The course of action when (16.50) fails is identical to our analysis of SD methods. We have $\varepsilon > 0$ such that $|\tilde{a}(e^{i\theta}, \mu)| \ge 1 + \varepsilon$ for all $\theta \in [\theta_-, \theta_+]$. Picking $\hat{u}^0 = \hat{g}$ as the characteristic function of the interval $[\theta_-, \theta_+]$, we exploit (16.51) to argue that

$$\|\boldsymbol{u}^n\|^2 = \|\hat{u}^n\|^2 = \frac{1}{2\pi}\int_0^{2\pi} \left|\tilde{a}(e^{i\theta}, \mu)\right|^{2n} g(\theta)\,d\theta \ge \frac{\theta_+ - \theta_-}{2\pi}(1+\varepsilon)^n, \qquad n \ge 0.$$

This concludes the proof of instability. ■

◇ **The Fourier technique in practice** It is trivial to use Theorem 16.10 to prove that the SD method (16.25) is stable, since $a(e^{i\theta}) = -4\sin^2\frac{1}{2}\theta$. Let us attempt a more ambitious goal, the fourth-order SD scheme (16.27). We have

$$\begin{aligned} a(e^{i\theta}) &= -\tfrac{1}{12}e^{-2i\theta} + \tfrac{4}{3}e^{-i\theta} - \tfrac{5}{2} + \tfrac{4}{3}e^{i\theta} - \tfrac{1}{12}e^{2i\theta} \\ &= -\tfrac{7}{3} + \tfrac{8}{3}\cos\theta - \tfrac{1}{3}\cos^2\theta = -\tfrac{1}{3}(1-\cos\theta)(7-\cos\theta) \le 0 \end{aligned}$$

for all $\theta \in [0, 2\pi]$, hence stability.

Whenever the Fourier technique can be put to work, results are easily obtained and this is also true with regard to FD schemes. The Euler method (16.7) yields

$$\tilde{a}(e^{i\theta}, \mu) = 1 - 4\mu\sin^2\tfrac{1}{2}\theta, \qquad \theta \in [0, 2\pi],$$

and it is trivial to deduce that (16.50) implies stability if and only if $\mu \le \frac{1}{2}$. Likewise, for the Crank–Nicolson method we have

$$\tilde{a}(e^{i\theta}, \mu) = \frac{1 - 2\mu\sin^2\frac{1}{2}\theta}{1 + 2\mu\sin^2\frac{1}{2}\theta} \in [-1, 1], \qquad \theta \in [0, 2\pi],$$

and hence stability for all $\mu > 0$.

Let us set ourselves a fairer challenge. Solving the SD scheme (16.25) with the Adams–Bashforth method (2.6) results in the second-order two-step scheme

$$u_\ell^{n+2} = u_\ell^{n+1} + \tfrac{3}{2}\mu(u_{\ell-1}^{n+1} - 2u_\ell^{n+1} + u_{\ell+1}^{n+1}) - \tfrac{1}{2}(u_{\ell-1}^n - 2u_\ell^n + u_{\ell+1}^n). \tag{16.52}$$

It is not difficult to extend the Fourier technique to multistep methods. We multiply (16.52) by $\mathrm{e}^{-\mathrm{i}\ell\theta}$ and sum for all $\ell \in \mathbb{Z}$; the outcome is the three-term recurrence relation

$$\hat{u}^{n+2} - \left(1 - 6\mu \sin^2 \tfrac{1}{2}\theta\right) \hat{u}^{n+1} - 2\mu \left(\sin^2 \tfrac{1}{2}\theta\right) \hat{u}^n = 0, \qquad n \geq 0. \qquad (16.53)$$

The general solution of the difference equation (16.53) is

$$\hat{u}^n = q_-(\theta)[\omega_-(\theta)]^n + q_+(\theta)[\omega_+(\theta)]^n, \qquad n = 0, 1, \ldots,$$

where $\omega_\pm$ are zeros of the characteristic equation

$$\omega^2 - \left(1 - 6\mu \sin^2 \tfrac{1}{2}\theta\right) \omega - 2\mu \sin^2 \tfrac{1}{2}\theta = 0$$

and $q_\pm$ depend on the starting values (see Section 4.4 for the solution of comparable difference equations). As before, the condition for stability is uniform boundedness of the set $\{|\!|\!|\hat{u}^n|\!|\!|\}_{n=0,1\ldots}$, since this implies that the vectors $\{\|\boldsymbol{u}^n\|\}_{n=0,1,\ldots}$ are uniformly bounded. Evidently, the Fourier transforms are uniformly bounded for all $q_\pm$ if and only if $|\omega_\pm(\theta)| \leq 1$ for all $\theta \in [0, 2\pi]$ and, whenever $|\omega_\pm(\theta)| = 1$ for some θ, the two zeros are distinct – in other words, the root condition all over again!

We use Lemma 4.9 to verify the root condition and this, after some trivial yet tedious algebra, results in the stability condition $\mu \leq \frac{2}{5}$.

Another example of a two-step method, the leapfrog scheme (16.36), features in Exercise 16.13. ◇

The scope of the Fourier technique can be generalized in several directions. The easiest is from one to several spatial dimensions and this requires a multivariate counterpart of the Fourier transform (16.45).

More interesting is a relaxation of the ban on boundary conditions in finite time – after all, most physical objects subjected to mathematical modelling possess finite size! In Chapter 17 we will mention briefly periodic boundary conditions, which lend themselves to the same treatment as the Cauchy problem. Here, though, we address ourselves to the Dirichlet boundary conditions (16.3), which are more characteristic of parabolic equations. Without going into any proofs we simply state that, provided that an SD or an FD method uses just one point from the right and one from the left (in other words, $\max\{\alpha, \beta, \gamma, \delta\} = 1$), the scope of the Fourier technique extends to finite intervals. Thus, the outcome of the Fourier analysis for the SD (16.25), the Euler method (16.7), the Crank–Nicolson scheme (16.35) and, indeed, the Adams–Bashforth two-step FD (16.52) extends *in toto* to Dirichlet boundary conditions, but this is not the case with the fourth-order SD scheme (16.27). There, everything depends on our treatment of the 'missing' values near the boundary. This is an important subject – admittedly more important in the context of hyperbolic differential equations, the theme of Chapter 17 – which requires a great deal of advanced mathematics and is well outside the scope of this book.

This section would not be complete without the mention of a remarkable connection, which might have already caught the eye of a vigilant reader. Let us consider a simple example, the 'basic' SD (16.25). The Fourier condition for stability is

$$\operatorname{Re} a(\mathrm{e}^{\mathrm{i}\theta}) = -4\sin^2 \tfrac{1}{2}\theta \le 0, \qquad \theta \in [0, 2\pi],$$

while the eigenvalue condition is nothing other than

$$\operatorname{Re} a(\omega_{d+1}^{\ell}) = -4\sin^2\left(\frac{\pi\ell}{d+1}\right) \le 0, \qquad \ell = 1, 2, \ldots, d,$$

where $\omega_d = \exp[2\mathrm{i}\pi/(d+1)]$ is the dth root of unity. A similar connection exists for Euler's method and Crank–Nicolson. Before we get carried away, we need to clarify that this coincidence is restricted, at least in the context of the Cauchy problem, mostly to methods that are constructed from TST matrices.

16.6 Splitting

Even the stablest and the most heavily analysed method must be, ultimately, run on a computer. This can be even more expensive for PDEs of evolution than for the Poisson equation; in a sense, using an implicit method for (16.1) in two spatial dimensions, say, and with a forcing term is equivalent to solving a Poisson equation in every time step. Needless to say, by this stage we know full well that effective solution of the diffusion equation calls for implicit schemes; otherwise, we would need to advance with such a miniscule step Δt as to render the whole procedure unrealistically expensive.

The emphasis on two (or more) space dimensions is important, since in one dimension the algebraic equations originating in the Crank–Nicolson scheme, say, are fairly small and tridiagonal (cf. (16.32)) and can be easily solved with banded LU factorization from Chapter 11.[6]

We restrict our analysis to the diffusion equation

$$\frac{\partial u}{\partial t} = \nabla(a\nabla u), \qquad 0 \le x, y \le 1, \tag{16.54}$$

where the diffusion coefficient $a = a(x, y)$ is bounded and positive in $[0,1] \times [0,1]$.

The starting point of our discussion is an extension of the SD equations (16.29) to two dimensions,

$$\begin{aligned} v'_{k,\ell} = \frac{1}{(\Delta x)^2} \Big[& a_{k-1/2,\ell}v_{k-1,\ell} + a_{k,\ell-1/2}v_{k,\ell-1} + a_{k+1/2,\ell}v_{k+1,\ell} + a_{k,\ell+1/2}v_{k,\ell+1} \\ & - (a_{k-1/2,\ell} + a_{k,\ell-1/2} + a_{k+1/2,\ell} + a_{k,\ell+1/2})v_{k,\ell}\Big] + h_{k,\ell}, \\ & k, \ell = 1, \ldots, d, \end{aligned}$$

where $h_{k,\ell}$ includes the contribution of the boundary values. (We could have also added a forcing term without changing the equation materially.) We commence by

[6]Even in two space dimensions we can obtain small – although dense – algebraic systems using spectral methods. If they are too small for your liking, try three dimensions instead.

assuming that $h_{k,\ell} = 0$ for all $k, \ell = 1, 2, \ldots, d$ and, employing natural ordering, write the method in a vector form,

$$\boldsymbol{v}' = \frac{1}{(\Delta x)}(B_x + B_y)\boldsymbol{v}, \quad t \geq 0, \qquad \boldsymbol{v}(0) \text{ given.} \tag{16.55}$$

Here B_x and B_y are $d^2 \times d^2$ matrices that contain the contribution of the differentiation in the x- and y- variables respectively. In other words, B_y is a block-diagonal matrix and its diagonal is constructed from the tridiagonal $d \times d$ matrices:

$$\begin{bmatrix} -(b_{1/2}+b_{3/2}) & b_{3/2} & 0 & \cdots & 0 \\ b_{3/2} & -(b_{3/2}+b_{5/2}) & b_{5/2} & \ddots & \vdots \\ 0 & \ddots & \ddots & \ddots & 0 \\ \vdots & \ddots & b_{d-3/2} & -(b_{d-3/2}+b_{d-1/2}) & b_{d-1/2} \\ 0 & \cdots & 0 & b_{d-1/2} & -(b_{d-1/2}+b_{d+1/2}) \end{bmatrix},$$

where $b_\ell = a_{k,\ell}$ and $k = 1, 2, \ldots, d$. The matrix B_x contains all the remaining terms. A crucial observation is that its sparsity pattern is also block-diagonal, with tridiagonal blocks, *provided that the grid is ordered by rows rather than by columns.*

Letting $\boldsymbol{v}^n := \boldsymbol{v}(n\Delta t)$, $n \geq 0$, the solution of (16.55) can be written explicitly as

$$\boldsymbol{v}^{n+1} = e^{\mu(B_x+B_y)}\boldsymbol{v}^n, \qquad n \geq 0. \tag{16.56}$$

It might be remembered that the exponential of a matrix has been already defined in Section 16.2 (see also Exercise 16.4). To solve (16.56) we can discretize the exponential by means of the *Padé approximation* $\hat{r}_{1/1}$ (Theorem 4.5). The outcome,

$$\boldsymbol{u}^{n+1} = \hat{r}_{1/1}(\mu(B_x+B_y))\boldsymbol{u}^n = \left[I - \tfrac{1}{2}\mu(B_x + B_y)\right]^{-1}\left[I + \tfrac{1}{2}\mu(B_x + B_y)\right]\boldsymbol{u}^n, \qquad n \geq 0,$$

is nothing other than the Crank–Nicolson method (in two dimensions) in disguise. Advancing the solution by a single time step is tantamount to solving a linear algebraic system by use of the matrix $I - \frac{1}{2}\mu(B_x + B_y)$, a task which can be quite expensive, even with the fast methods of Chapters 13 and 14, when repeated for a large number of steps.

An exponential, however, is a very special function. In particular, we are all aware of the identity $e^{z_1+z_2} = e^{z_1}e^{z_2}$, where $z_1, z_2 \in \mathbb{C}$. Were this identity true for matrices, so that

$$e^{t(Q+S)} = e^{tQ}e^{tS}, \qquad t \geq 0, \tag{16.57}$$

for all square matrices Q and S of equal dimension, we could replace Crank–Nicolson by

$$\begin{aligned} \boldsymbol{u}^{n+1} &= \hat{r}_{1/1}(\mu B_x)\hat{r}_{1/1}(\mu B_y)\boldsymbol{u}^n \qquad\qquad (16.58) \\ &= \left(I - \tfrac{1}{2}\mu B_x\right)^{-1}\left(I + \tfrac{1}{2}\mu B_x\right)\left(I - \tfrac{1}{2}\mu B_y\right)^{-1}\left(I + \tfrac{1}{2}\mu B_y\right)\boldsymbol{u}^n, \qquad n \geq 0. \end{aligned}$$

The implementation of (16.58) would have a great advantage over the unadulterated form of Crank–Nicolson. We would need to solve two linear systems to advance one step, but the second matrix, $I-\frac{1}{2}\mu B_y$, is tridiagonal whilst the first, $I-\frac{1}{2}\mu B_x$, can be converted into a tridiagonal form by reordering the grid by rows. Hence, (16.58) could be solved by sparse LU factorization in $\mathcal{O}(d^2)$ operations!

Unfortunately, in general the identity (16.57) is false. Thus, let $[Q,S] := QS - SQ$ be the *commutator* of Q and S. Since

$$\begin{aligned} e^{tQ}e^{tS} - e^{t(Q+S)} &= \left(I + tQ + \tfrac{1}{2}t^2Q^2 + \cdots\right)\left(I + tS + \tfrac{1}{2}t^2S^2 + \cdots\right) \\ &\quad - \left\{I + t(Q+S) + \tfrac{1}{2}t^2(Q+S)^2 + \cdots\right\} = \tfrac{1}{2}t^2[S,Q] + \mathcal{O}(t^3), \end{aligned} \tag{16.59}$$

we deduce that (16.57) cannot be true unless Q and S commute. If $a \equiv 1$ and (16.54) reduces to (16.4) then $[B_x, B_y] = O$ (see Exercise 16.14) and we are fully justified in using (16.58), but this will not be the case when the diffusion coefficient μ is allowed to vary.

As with every good policy, the rule that mathematical injunctions must always be followed has its exceptions. For instance, were we to disregard for a moment the breakdown in commutativity and use (16.58) with a variable diffusion coefficient

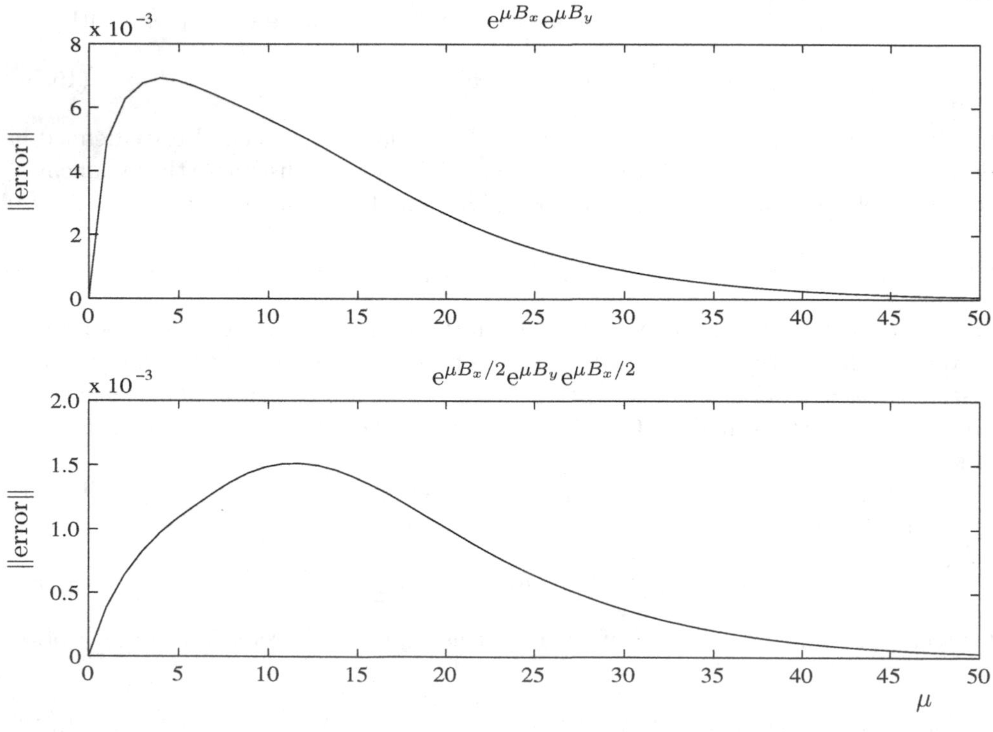

Figure 16.3 The norm of the error in approximating $\exp \mu(B_x + B_y)$ by the 'naive' splitting $e^{\mu B_x}e^{\mu B_y}$ and by the Strang splitting $e^{\mu B_x/2}e^{\mu B_y}e^{\mu B_x/2}$ for $a(x,y) = 1 + \frac{1}{4}(x-y)$ and $d = 10$.

a, the difference would hardly register. The reason is explained by Fig. 16.3, where we plot (in the upper graph) the error $\|\mathrm{e}^{\mu B_x}\mathrm{e}^{\mu B_y} - \mathrm{e}^{\mu(B_x+B_y)}\|$ for a specific variable diffusion coefficient. Evidently, the loss of commutativity does not necessarily cause a damaging loss of accuracy. Part of the reason is that, according to (16.59), $\mathrm{e}^{\mu B_x}\mathrm{e}^{\mu B_y} - \mathrm{e}^{\mu(B_x+B_y)} = \mathcal{O}(\mu^2)$; but this hardly explains the phenomenon, since we are interested in large values of μ. Another clue is that, provided B_x, B_y and $B_x + B_y$ have their eigenvalues in the left half-plane, all the exponents vanish for $\mu \to \infty$ (cf. Section 4.1). More justification is provided in Exercise 16.16. In any case, the error in the *splitting*

$$\mathrm{e}^{\mu B_x}\mathrm{e}^{\mu B_y} \approx \mathrm{e}^{\mu(B_x+B_y)}$$

is sufficiently small to justify the use of (16.58) even in the absence of commutativity.

Even better is the *Strang splitting*

$$\mathrm{e}^{\mu B_x/2}\mathrm{e}^{\mu B_y}\mathrm{e}^{\mu B_x/2} \approx \mathrm{e}^{\mu(B_x+B_y)}.$$

It can be observed from Fig. 16.3 that it produces a smaller error – Exercise 16.15 is devoted to proving that this is $\mathcal{O}(\mu^3)$.

In general, splitting presents an affordable alternative to the 'full' Crank–Nicolson and the technique can be generalized to other PDEs of evolution and diverse computational schemes.

We have assumed zero boundary conditions in our exposition, but this is not strictly necessary. Let us add to (16.54) nonzero boundary conditions and, possibly, a forcing term. Thus, in place of (16.55) we now have

$$\boldsymbol{v}' = \frac{1}{(\Delta x)^2}(B_x + B_y)\boldsymbol{v} + \boldsymbol{h}(t), \quad t \geq 0, \qquad \boldsymbol{v}(0) \text{ given},$$

an equation whose explicit solution is

$$\boldsymbol{v}^{n+1} = \mathrm{e}^{\mu(B_x+B_y)}\boldsymbol{v}^n + \Delta t \int_0^1 \mathrm{e}^{(1-\tau)\mu(B_x+B_y)}\boldsymbol{h}((n+\tau)\Delta t)\,\mathrm{d}\tau, \qquad n \geq 0.$$

We replace the integral using the trapezoidal rule – a procedure whose error is within the same order of magnitude as that of the original SD scheme. The outcome is

$$\begin{aligned}\tilde{\boldsymbol{v}}^{n+1} &= \mathrm{e}^{\mu(B_x+B_y)}\tilde{\boldsymbol{v}}^n + \tfrac{1}{2}\Delta t\left[\mathrm{e}^{\mu(B_x+B_y)}\boldsymbol{h}(n\Delta t) + \boldsymbol{h}((n+1)\Delta t)\right]\\ &= \mathrm{e}^{\mu(B_x+B_y)}\left[\tilde{\boldsymbol{v}}^n + \tfrac{1}{2}\Delta t\,\boldsymbol{h}(n\Delta t)\right] + \tfrac{1}{2}\Delta t\,\boldsymbol{h}((n+1)\Delta), \qquad n \geq 0,\end{aligned}$$

and we form an FD scheme by splitting the exponential and approximating it with the $\hat{r}_{1/1}$ Padé approximation.

Comments and bibliography

Numerical theory for PDEs of evolution is sometimes presented in a deceptively simple way. On the face of it, nothing could be more straightforward: discretize all spatial derivatives by finite differences and apply a reputable ODE solver, without paying heed to that fact that, actually, one is attempting to solve a PDE. This nonsense has, unfortunately, taken root in many textbooks and lecture courses, which, not to mince words, propagate shoddy mathematics and poor numerical practice. Reputable literature is surprisingly scarce, considering the importance and the depth of the subject. The main source and a good reference to much of the advanced theory is the monograph of Richtmyer & Morton (1967). Other surveys of finite differences that get stability and convergence right are Gekeler (1984), Hirsch (1988) and Mitchell & Griffiths (1980), while the slim volume of Gottlieb & Orszag (1977), whose main theme is entirely different, contains a great deal of useful material applicable to the stability of finite difference schemes for evolutionary PDEs.

Both the eigenvalue approach and the Fourier technique are often – and confusingly – termed in the literature 'the von Neumann method'. While paying due homage to John von Neumann, who originated both techniques, we prefer a more descriptive and less ambiguous terminology.

Modern stability theory ranges far and wide beyond the exposition of this chapter. A useful technique, the *energy method,* will be introduced in Chapter 17. Perhaps the most significant effort in generalizing the framework of stability theory has been devoted to boundary conditions in the Fourier technique. This is perhaps more significant in the context of hyperbolic equations, the theme of Chapter 17; our only remark here is that these are *very* deep mathematical waters. The original reference, not for the faint-hearted, is Gustaffson *et al.* (1972).

Another interesting elaboration on the theme of stability is connected with the *Kreiss matrix theorem,* its generalizations and applications (Gottlieb & Orszag, 1977; van Dorsselaer *et al.*, 1993).

The splitting algorithms of Section 16.6 are a popular means of solving multivariate PDEs of evolution and they have much in common with the composition methods from Section 5.4. An alternative, first pioneered by electrical engineers and subsequently adopted and enhanced by numerical analysts, is *waveform relaxation.* Like splitting, it is concerned with effective solution of the ODEs that occur in the course of semi-discretization. Let us suppose that an SD method can be written in the form

$$\boldsymbol{v}' = \frac{1}{(\Delta x)^2} P\boldsymbol{v} + \boldsymbol{h}(t), \quad t \geq 0, \qquad \boldsymbol{v}(0) = \boldsymbol{v}_0,$$

and that we can express P as a sum of two matrices, $P = Q + S$, say, such that it is easy to solve linear ODE systems with the matrix Q; for example, Q might be diagonal (*Jacobi* waveform relaxation) or lower triangular (*Gauss–Seidel* waveform relaxation). We replace the ODE system by the recursion

$$\left(\boldsymbol{v}^{[k+1]}\right)' = \frac{1}{(\Delta x)^2} Q\boldsymbol{v}^{[k+1]} + \frac{1}{(\Delta x)^2} S\boldsymbol{v}^{[k]} + \boldsymbol{h}(t), \quad t \geq 0, \qquad \boldsymbol{v}^{[k+1]} = \boldsymbol{v}_0, \qquad k = 0, 1, \ldots, \tag{16.60}$$

where $\boldsymbol{v}^{[k+1]}(0) \equiv \boldsymbol{v}_0$. In each kth iteration we apply a standard ODE solver, e.g. a multistep method, to (16.60) until the procedure converges to our satisfaction.[7] This idea might appear to be very strange indeed – to replace a single ODE by an infinite (in principle) system of such equations. However, solving the original, unamended, ODE by conversion into an algebraic system and employing an iterative procedure from Chapters 12–14 also replaces a single equation by an infinite recursion ...

There exists a respectable theory that predicts the rate of convergence of (16.60) with k, similar in spirit to the convergence theory from Sections 12.2 and 12.3. An important

[7]This brief description does little justice to a complicated procedure. For example, for 'intermediate' values of k there is no need to solve the implicit equations with high precision, and this leads to substantial savings (Vandewalle, 1993).

advantage of waveform relaxation is that it can easily be programmed in a way that takes full advantage of parallel computer architectures, and it can also be combined with multigrid techniques (Vandewalle, 1993).

Gekeler, E. (1984), *Discretization Methods for Stable Initial Value Problems,* Springer-Verlag, Berlin.

Gottlieb, D. and Orszag, S.A. (1977), *Numerical Analysis of Spectral Methods: Theory and Applications,* SIAM, Philadelphia,

Gustafsson, B., Kreiss, H.-O. and Sundström, A. (1972), Stability theory of difference approximations for mixed initial boundary value problems, *Mathematics of Computation* **26,** 649–686.

Hirsch, C. (1988), *Numerical Computation of Internal and External Flows, Vol. I: Fundamentals of Numerical Discretization,* Wiley, Chichester.

Mitchell, A.R. and Griffiths, D.F. (1980), *The Finite Difference Method in Partial Differential Equations,* Wiley, London.

Richtmyer, R.D. and Morton, K.W. (1967), *Difference Methods for Initial-Value Problems,* Interscience, New York.

Vandewalle, S. (1993), *Parallel Multigrid Waveform Relaxation for Parabolic Problems,* B.G. Teubner, Stuttgart.

van Dorsselaer, J.L.M., Kraaijevanger, J.F.B.M. and Spijker, M.N. (1993), Linear stability analysis in the numerical solution of initial value problems, *Acta Numerica* **2,** 199–237.

Exercises

16.1 Extend the method of proof of Theorem 16.1 to furnish a direct proof that the Crank–Nicolson method (16.32) converges.

16.2 Let

$$u_\ell^{n+1} = u_\ell^n + \mu(u_{\ell-1}^n - 2u_\ell^n + u_{\ell+1}^n) - \tfrac{1}{2}b\mu\Delta x(u_{\ell+1}^n - u_{\ell-1}^n)$$

be an FD scheme for the *convection–diffusion* equation

$$\frac{\partial u}{\partial t} = \frac{\partial^2 u}{\partial x^2} - b\frac{\partial u}{\partial x}, \qquad 0 \le x \le 1, \quad t \ge 0,$$

where $b > 0$ is given. Prove from first principles that the method converges. (*You can take for granted that the convection–diffusion equation is well posed.*)

16.3 Let

$$c(t) := \|u(\,\cdot\,, t)\| = \left\{\int_0^1 [u(x,t)]^2\,\mathrm{d}x\right\}^{1/2}, \qquad t \ge 0,$$

be the Euclidean norm of the exact solution of the diffusion equation (16.1) with zero boundary conditions.

a Prove that $c'(t) \le 0$, $t \ge 0$, hence $c(t) \le c(0)$, $t \ge 0$, thereby deducing an alternative proof that (16.1) is well posed.

b Let $\boldsymbol{u}^n = (u_\ell^n)_{\ell=0}^{d+1}$ be the Crank–Nicolson solution (16.32) and define

$$c^n := \|\boldsymbol{u}^n\|_{\Delta x} = \left(\Delta x \sum_{\ell=0}^{d+1} |u_\ell^n|^2\right)^{1/2}, \qquad n \ge 0,$$

as the discretized counterpart of the function c. Demonstrate that

$$(c^{n+1})^2 = (c^n)^2 - \frac{\Delta t}{2\Delta x} \sum_{\ell=1}^{d} \left(u_\ell^{n+1} + u_\ell^n - u_{\ell-1}^{n+1} - u_{\ell-1}^n\right)^2.$$

Consequently $c_n \le c_0$, $n \ge 0$, and this furnishes yet another proof that the Crank–Nicolson method is stable. (*This is an example of the* energy method, *which we will encounter again in Chapter 17.*)

16.4 The exponential of a $d \times d$ matrix B is defined by the Taylor series

$$e^B = \sum_{k=0}^{\infty} \frac{1}{k!} B^k.$$

a Prove that the series converges and that

$$\|e^B\| \le e^{\|B\|}.$$

(*This particular result does not depend on the choice of a norm and you should be able to prove it directly from the definition of the induced matrix norm in A.1.3.3.*)

b Suppose that $B = VDV^{-1}$, where V is nonsingular. Prove that

$$e^{tB} = Ve^{tD}V^{-1}, \qquad t \ge 0.$$

Deduce that, provided B has distinct eigenvalues $\lambda_1, \lambda_2, \ldots, \lambda_d$, there exist $d \times d$ matrices $E_1, E_2, \ldots, E_d$ such that

$$e^{tB} = \sum_{m=1}^{d} e^{t\lambda_m} E_m, \qquad t \ge 0.$$

c Prove that the solution of the linear ODE system

$$\boldsymbol{y}' = B\boldsymbol{y}, \quad t \ge t_0, \qquad \boldsymbol{y}(t_0) = \boldsymbol{y}_0,$$

is

$$\boldsymbol{y}(t) = e^{(t-t_0)B}\boldsymbol{y}_0, \qquad t \ge 0.$$

d Generalize the result from **c**, proving that the explicit solution of

$$\boldsymbol{y}' = B\boldsymbol{y} + \boldsymbol{p}(t), \quad t \ge t_0, \qquad \boldsymbol{y}(t_0) = \boldsymbol{y}_0,$$

is

$$\boldsymbol{y}(t) = \mathrm{e}^{(t-t_0)B}\boldsymbol{y}_0 + \int_{t_0}^{t} \mathrm{e}^{(t-\tau)B}\boldsymbol{p}(\tau)\,\mathrm{d}\tau, \qquad t \ge t_0.$$

e Let $\|\cdot\|$ be the Euclidean norm and let B be a normal matrix. Prove that $\|\mathrm{e}^{tB}\| \le \mathrm{e}^{t\tilde{\alpha}(B)}$, $t \ge 0$, where $\tilde{\alpha}(\cdot)$, the *spectral abscissa,* was defined in Section 16.4.

16.5 Prove that the SD method (16.27) is of order 4.

16.6 Suppose that an SD scheme of order p_1 for the PDE (16.11) is computed with an ODE solver of order p_2, and that this results in an FD method (possibly multistep). Show that this method is of order $\min\{p_1, rp_2\}$.

16.7 The diffusion equation (16.1) is solved by the fully discretized scheme

$$u_\ell^{n+1} - \tfrac{1}{2}(\mu-\zeta)\left(u_{\ell-1}^{n+1} - 2u_\ell^{n+1} + u_{\ell+1}^{n+1}\right) = u_\ell^n + \tfrac{1}{2}(\mu+\zeta)\left(u_{\ell-1}^n - 2u_\ell^n + u_{\ell+1}^n\right), \tag{16.61}$$

where ζ is a given constant. Prove that (16.61) is a second-order method for all $\zeta \ne \frac{1}{6}$, while for the choice $\zeta = \frac{1}{6}$ (the *Crandall method*) it is of order 4.

16.8 Determine the order of the SD method

$$v_\ell' = \frac{1}{(\Delta x)^2}\left(\tfrac{11}{12}v_{\ell-1} - \tfrac{5}{3}v_\ell + \tfrac{1}{2}v_{\ell+1} + \tfrac{1}{3}v_{\ell+2} - \tfrac{1}{12}v_{\ell+3}\right)$$

for the diffusion equation (16.1). Is it stable? (*Hint: Express the function* $\mathrm{Re}\, a(\mathrm{e}^{\mathrm{i}\theta})$ *as a cubic polynomial in* $\cos\theta$.)

16.9 The SD scheme (16.30) for the diffusion equation with a variable coefficient (16.6) is solved by means of the Euler method.

a Write down the fully discretized equations.

b Prove that the FD method is stable, provided that $\mu \le 1/(2a_{\min})$, where $a_{\min} = \min\{a(x) : 0 \le x \le 1\} > 0$.

16.10 Let B be a $d \times d$ *normal* matrix and let $\boldsymbol{y} \in \mathbb{C}^d$ be an arbitrary vector such that $\|\boldsymbol{y}\| = 1$ (in the Euclidean norm).

a Prove that there exist numbers $\alpha_1, \alpha_2, \ldots, \alpha_d$ such that $\boldsymbol{y} = \sum_{k=1}^{d} \alpha_k \boldsymbol{w}_k$, where $\boldsymbol{w}_1, \boldsymbol{w}_2, \ldots, \boldsymbol{w}_d$ are the eigenvectors of B. Express $\|\boldsymbol{y}\|^2$ explicitly in terms of α_k, $k = 1, 2, \ldots, d$.

b Let $\lambda_1, \lambda_2, \ldots, \lambda_d$ be the eigenvalues of B, $B\boldsymbol{w}_k = \lambda_k \boldsymbol{w}_k$, $k = 1, 2, \ldots, d$. Prove that

$$\|B\boldsymbol{y}\|^2 = \sum_{k=1}^{d} |\alpha_k \lambda_k|^2.$$

c Deduce that $\|B\| = \rho(B)$.

16.11 Apply the Fourier stability technique to the FD scheme

$$u_\ell^{n+1} = \tfrac{1}{2}(2 - 5\mu + 6\mu^2)u_\ell^n + \tfrac{2}{3}\mu(2 - 3\mu)(u_{\ell-1}^n + u_{\ell+1}^n) - \tfrac{1}{12}\mu(1 - 6\mu)(u_{\ell-2}^n + u_{\ell+2}^n), \qquad \ell \in \mathbb{Z}.$$

You should find that stability occurs if and only if $0 \le \mu \le \frac{2}{3}$. (*We have not specified which equation – if any – the scheme is supposed to solve, but this, of course, has no bearing on the question of stability.*)

16.12 Investigate the stability of the FD scheme (16.61) (see Exercise 16.7) for different values of ζ using both the eigenvalue and the Fourier technique.

16.13* Prove that the leapfrog scheme (16.33) for the diffusion equation is unstable for every choice of $\mu > 0$. (*An experienced student of mathematics will not be surprised to hear that this was the first-ever discretization method for the diffusion equation to be published in the scientific literature. Sadly, it is still occasionally used by the unwary – those who forget the history of mathematics are condemned to repeat its mistakes…*)

16.14* Prove that the matrices B_x and B_y from Section 16.6 commute when $a \equiv$ constant. (*Hint: Employ the techniques from Section 12.1 to factorize these matrices and demonstrate that they share the same eigenvalues.*)

16.15 Prove that

$$e^{tQ/2}e^{tS}e^{tQ/2} = e^{t(Q+S)} + \mathcal{O}(t^3), \qquad t \to 0,$$

for any $d \times d$ matrices Q and S, thereby establishing the order of the *Strang splitting*.

16.16* Let $E(t) := e^{tQ}e^{tS}$, $t \ge 0$, where Q and S are $d \times d$ matrices.

a Prove that

$$E' = (Q + S)E + [e^{tQ}, S]e^{tS}, \qquad t \ge 0.$$

b Using the explicit formula from Exercise 16.4d – or otherwise – show that

$$E(t) = e^{t(Q+S)} + \int_0^t e^{(t-\tau)(Q+S)}[e^{\tau Q}, S]e^{\tau S}\,d\tau, \qquad t \ge 0.$$

c Let Q, S and $Q + S$ be symmetric negative definite matrices. Prove that

$$\|e^{tQ}e^{tS} - e^{t(Q+S)}\| \le 2\|S\| \int_0^t \exp\{(t-\tau)\tilde{\alpha}(Q + S) + \tau[\tilde{\alpha}(Q) + \tilde{\alpha}(S)]\}\,d\tau$$

for $t \ge 0$, where $\tilde{\alpha}(\,\cdot\,)$, the spectral abscissa, was defined in Section 16.4. (*Hint: Use the estimate from Exercise 16.4e.*)

17

Hyperbolic equations

17.1 Why the advection equation?

Much of the discussion in this chapter is centred upon the *advection equation*

$$\frac{\partial u}{\partial t} + \frac{\partial u}{\partial x} = 0, \qquad 0 \le x \le 1, \quad t \ge 0, \tag{17.1}$$

which is specified in tandem with an initial value

$$u(x, 0) = g(x), \qquad 0 \le x \le 1, \tag{17.2}$$

as well as the boundary condition

$$u(0, t) = \varphi_0(t), \qquad t \ge 0, \tag{17.3}$$

where $g(0) = \varphi_0(0)$.

The first and most natural question pertaining to any mathematical construct should not be 'how?' (the knee-jerk reaction of many a trained mathematical mind) but 'why?'. This is a particularly pointed remark with regard to equation (17.1), whose exact solution is both well-known and trivial:

$$u(x,t) = \begin{cases} g(x-t), & t \le x, \\ \varphi_0(t-x), & x \le t. \end{cases} \tag{17.4}$$

Note that (17.4) can be verified at once by direct differentiation and that it makes clear why the single boundary condition (17.3) is sufficient.

There are three reasons why numerical study of the advection equation (17.1) is of interest. Firstly, by its very simplicity, it affords an insight into a multitude of computational phenomena that are specific to *hyperbolic PDEs*. It is a fitting counterpart of the linear ODE $y' = \lambda y$, which was so fruitful in Chapter 4 in elucidating the behaviour of ODE solvers for stiff equations. Secondly, various generalizations of (17.1) lead to PDEs that are crucial in many applications for which in practice we require numerical solutions: for example the advection equation with a variable coefficient,

$$\frac{\partial u}{\partial t} + \tau(x)\frac{\partial u}{\partial x} = 0, \qquad 0 \le x \le 1, \quad t \ge 0, \tag{17.5}$$

the advection equation in two dimensions,

$$\frac{\partial u}{\partial t} + \frac{\partial u}{\partial x} + \frac{\partial u}{\partial y} = 0, \qquad 0 \le x, y \le 1, \quad t \ge 0, \tag{17.6}$$

and the *wave equation*

$$\frac{\partial^2 u}{\partial t^2} = \frac{\partial^2 u}{\partial x^2}, \qquad -1 \le x \le 1, \quad t \ge 0. \tag{17.7}$$

Equations (17.5)–(17.7) need to be equipped with appropriate initial and boundary conditions and we will address this problem later. Here we just mention the connection that takes us from (17.1) to (17.7). Consider two coupled advection equations, specifically

$$\begin{aligned} \frac{\partial u}{\partial t} + \frac{\partial v}{\partial x} &= 0, \\ \frac{\partial v}{\partial t} + \frac{\partial u}{\partial x} &= 0, \end{aligned} \qquad 0 \le x \le 1, \quad t \ge 0. \tag{17.8}$$

It follows that

$$\frac{\partial^2 u}{\partial t^2} = \frac{\partial}{\partial t}\left(\frac{\partial u}{\partial t}\right) = \frac{\partial}{\partial t}\left(-\frac{\partial v}{\partial x}\right) = -\frac{\partial}{\partial x}\left(\frac{\partial v}{\partial t}\right) = -\frac{\partial}{\partial x}\left(-\frac{\partial u}{\partial x}\right) = \frac{\partial^2 u}{\partial x^2}$$

and so u obeys the wave equation. The system (17.8) is a special case of the vector advection equation

$$\frac{\partial \boldsymbol{u}}{\partial t} + A\frac{\partial \boldsymbol{u}}{\partial x} = \boldsymbol{0}, \qquad 0 \le x \le 1, \quad t \ge 0, \tag{17.9}$$

where the matrix A is diagonalizable and has only real eigenvalues.

The third, and perhaps the most interesting, reason why the humble advection equation is so important leads us into the realm of the nonlinear hyperbolic equations that are pervasive in wave theory and in quantum mechanics, e.g. the *Burgers equation*

$$\frac{\partial u}{\partial t} + \frac{1}{2}\frac{\partial u^2}{\partial x} = 0, \qquad -\infty < x < \infty, \quad t \ge 0 \tag{17.10}$$

and the *Korteweg–de-Vries* equation

$$\frac{\partial u}{\partial t} + \kappa\frac{\partial u}{\partial x} + \frac{3\kappa}{4\eta}\frac{\partial u^2}{\partial x} + \frac{\kappa\eta^2}{6}\frac{\partial^3 u}{\partial x^3} = 0, \qquad -\infty < x < \infty, \quad t \ge 0, \tag{17.11}$$

whose name is usually abbreviated to KdV. Both display a wealth of nonlinear phenomena of a kind that we have not encountered previously in this volume.

Figure 17.1 displays the evolution of the solution of the Burgers equation (17.10) in the interval $0 \le x \le 2\pi$ with periodic boundary condition $u(2\pi, t) = u(0, t)$, $t \ge 0$. The initial condition is $g(x) = \frac{5}{2} + \sin x$, $0 \le x \le 2\pi$ and, as t increases from the origin, $g(x)$ is transported with unit speed to the right – as we can expect from the original advection equation – and simultaneously evolves into a function with an increasingly sharper profile which, after a while, looks (and is!) discontinuous. The same picture emerges even more vividly from Fig. 17.2, where six 'snapshots' of the solution are

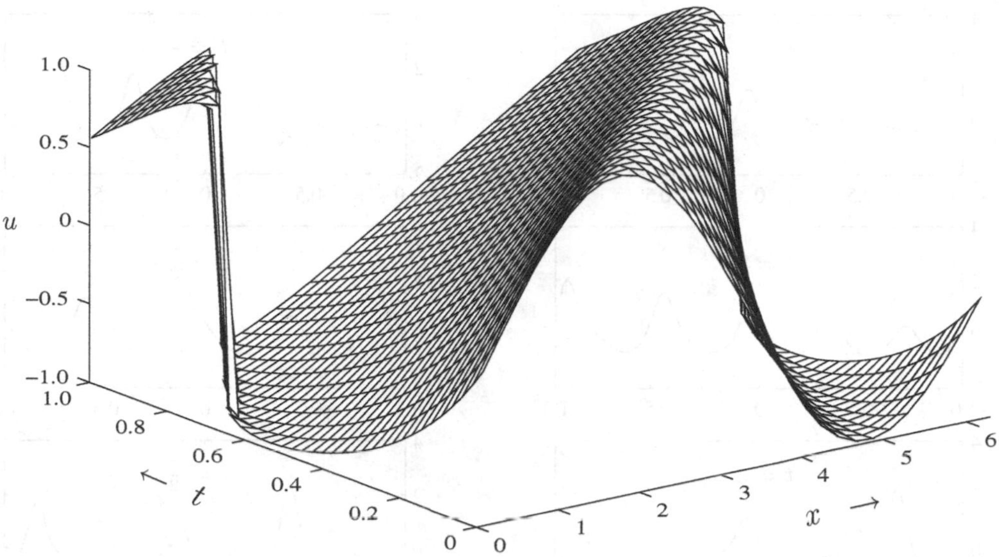

Figure 17.1 The solution of the Burgers equation (17.10) with periodic boundary conditions and $u(x,0) = \frac{5}{2} + \sin x$, $x \in [0, 2\pi)$.

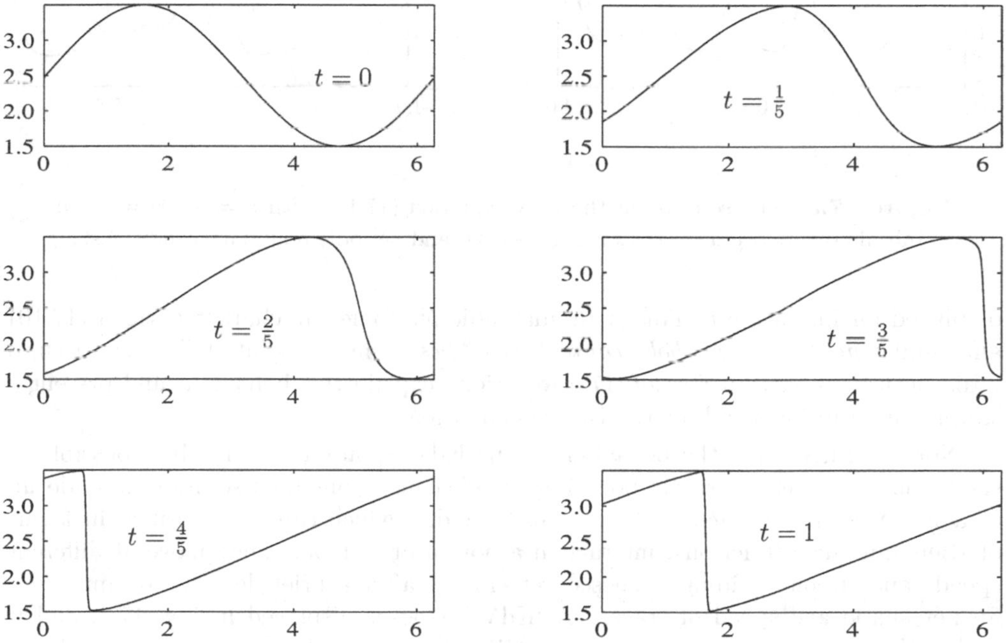

Figure 17.2 The solution of the Burgers equation from Fig. 17.1 at times $t = i/5$, $i = 1, 2, \ldots, 5$.

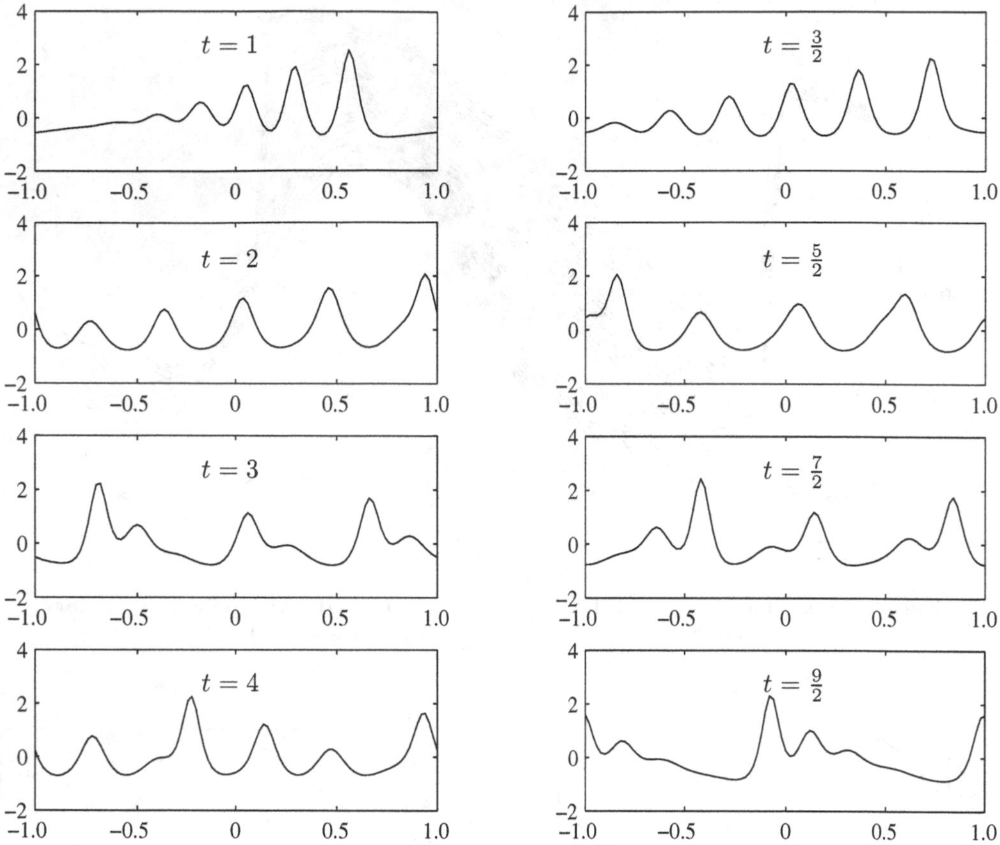

Figure 17.3 The solution of the KdV equation (17.11) with $\kappa = \frac{1}{10}$, $\eta = \frac{1}{5}$, an initial condition $g(x) = \cos \pi x$, $x \in [-1, 1)$, and periodic boundary conditions.

displayed for increasing t. This is a remarkable phenomenon, characteristic of (17.10) and similar *nonlinear hyperbolic conservation laws*: a smooth solution degenerates into a discontinuous one. In Section 17.5 we briefly explain this behaviour and present a simple numerical method for the Burgers equation.

Not less intricate is the behaviour of the KdV equation (17.11). It is possible to show that, for every periodic boundary condition, a nontrivial solution is made up of a finite number of *active modes*. Such modes, which can be described in terms of Riemann theta functions, interact in a nonlinear fashion. They move at different speeds and, upon colliding, coalesce yet emerge after a brief delay to resume their former shape and speed of travel. A 'KdV movie' is displayed in Fig. 17.3, and it makes this concept more concrete. We will not pursue further the interesting theme of modelling KdV and other equations with such *soliton* solutions.

Although – hopefully – we have argued to the satisfaction of even the most discerning reader why numerical schemes for the humble advection equation might be of interest, the task of motivating the present chapter is not yet complete. For, have we not just spent a whole chapter deliberating in some detail how to discretize evolutionary PDEs by finite differences and discussing questions of stability and implementation? According to this comforting point of view, we just need to employ finite difference operators to construct a numerical method, evaluate an eigenvalue or two to prove stability ... and the task of computing the solution of (17.1) will be complete. Nothing could be further from the truth!

To convince a sceptical reader (and all good readers ought to be sceptical!) that hyperbolic equations require a subtly different approach, we prove a theorem. Its statement might sound at first quite incredible – as, of course, it is. Having carefully studied Chapter 16, the reader should be adequately equipped to verify – or reject – the veracity of the following statement.

'Theorem' $1 = 2$.

Proof We construct the simplest possible genuine finite difference method for (17.1) by replacing the time derivative by forward differences and the space derivative by backward differences. The outcome is the *Euler scheme*

$$u_\ell^{n+1} = u_\ell^n - \mu(u_\ell^n - u_{\ell-1}^n) = \mu u_{\ell-1}^n + (1-\mu)u_\ell^n, \qquad \ell = 1, 2, \ldots, d, \quad n \geq 0, \quad (17.12)$$

where

$$\mu = \frac{\Delta t}{\Delta x}$$

is the Courant number. Assuming for even greater simplicity that the boundary value φ_0 is identically zero, we pose the question 'What is the set of all numbers μ that bring about stability?'

We address this problem by two different techniques, based on eigenvalue analysis (Section 16.4) and on Fourier transforms (Section 16.5) respectively. Firstly, we write (17.12) in the vector form

$$\boldsymbol{u}^{n+1} = A\boldsymbol{u}^n, \qquad n \geq 0, \qquad \text{where} \qquad A = \begin{bmatrix} 1-\mu & 0 & \cdots & \cdots & 0 \\ \mu & 1-\mu & \ddots & & \vdots \\ 0 & \ddots & \ddots & \ddots & \vdots \\ \vdots & \ddots & \mu & 1-\mu & 0 \\ 0 & \cdots & 0 & \mu & 1-\mu \end{bmatrix}.$$

Since A is lower triangular, its eigenvalues all equal $1-\mu$. Requiring $|1-\mu| \leq 1$ for stability, we thus deduce that

$$\text{stability} \iff \mu \in (0, 2]. \qquad (17.13)$$

Next we turn our attention to the Fourier approach. Multiplying (17.12) by $e^{-i\ell\theta}$ and summing over ℓ we easily deduce that the Fourier transform obeys the recurrence

$$\hat{u}^{n+1} = [\mu e^{-i\theta} + (1-\mu)]\hat{u}^n, \qquad \theta \in [0, 2\pi], \quad n \geq 0.$$

Therefore, the Fourier stability condition is

$$|1-\mu(1-\mathrm{e}^{-\mathrm{i}\theta})| \le 1, \qquad \theta \in [0, 2\pi].$$

Straightforward algebra renders this in the form

$$1-|1-\mu(1-\mathrm{e}^{-\mathrm{i}\theta})|^2 = 4\mu(1-\mu)\sin^2 \tfrac{1}{2}\theta \ge 0, \qquad \theta \in [0, 2\pi],$$

hence $\mu(1-\mu) \ge 0$ and we conclude that

$$\text{stability} \quad \Longleftrightarrow \quad \mu \in (0, 1]. \tag{17.14}$$

Comparison of (17.13) with (17.14) proves the assertion of the theorem. ■

Before we get carried away by the last theorem, it is fair to give the game away and confess that it is, after all, just a prank. It is a prank with a point, though; more precisely, three points. Firstly, its 'proof' is entirely consistent with several books of numerical analysis. Secondly, it is the author's experience that a fair number of professional numerical analysts fail to spot exactly what is wrong. Thirdly, although a rebuttal of the proof should be apparent after a careful reading of Chapter 16, it affords us an opportunity to emphasize the very different ground rules that apply for hyperbolic PDEs and their discretizations.

Which part of the proof is wrong? In principle, both, except that the second part can be amended with relative ease while the first rests upon on a blunder, pure and simple.

The Fourier stability technique from Section 16.5 is based on the assumption that we are analysing a Cauchy problem: the range of x is the whole real axis and there are no boundary conditions except for the requirement that the solution is square integrable. In the proof of the theorem, however, we have stipulated that (17.1) holds in $[0, 1]$ with zero boundary conditions at $x = 0$. This can be easily amended and we can convert this equation into a Cauchy problem without changing the nonzero portion of the solution of (17.12). To that end let us define

$$u_\ell^n = 0, \qquad \ell \in \mathbb{Z} \setminus \{1, 2, \ldots, d\}, \tag{17.15}$$

and let the index ℓ in (17.12) range in $\mathbb{Z}$ rather than just $\{1, 2, \ldots, d\}$. We denote the new solution sequence by $\breve{\boldsymbol{u}}^n$ and claim that $\|\breve{\boldsymbol{u}}^n\| \ge \|\boldsymbol{u}^n\|$, $n \ge 0$. This is obvious from the following diagram, describing the flow of information in the scheme (17.12). This diagram also clarifies why only a left-hand side boundary condition is required to implement this method. We denote by '•' a point that belongs to the original grid and by '○' any value that we have set to zero, whether as a consequence of letting $\varphi_0 = 0$ or of (17.15) or of the FD scheme. Finally, we denote by '⊙' any value of u_ℓ^n for $\ell \ge d+1$ that is rendered nonzero by (17.12).

The arrows denote the flow of information, which is from each u_ℓ^n to u_ℓ^{n+1} and to $u_{\ell+1}^n$ – and we can see at once that padding $\boldsymbol{u}^0$ with zeros does not introduce any changes in the solution for $\ell = 1, 2, \ldots, d$. Therefore, $\|\breve{\boldsymbol{u}}^n\| \geq \|\boldsymbol{u}^n\|$ for all $n \geq 0$ and we have thereby deduced the stability of the original problem by Fourier analysis.

Before advancing further, we should perhaps use this opportunity to comment that often it is more natural to solve (17.1) in $x \in [0, \infty)$ (or, more specifically, in $x \in [0, 1+t)$, $t \geq 0$), since typically it is of interest to follow the wave-like phenomena modelled by hyperbolic PDEs throughout their evolution and at all their destinations. Thus, it is $\breve{\boldsymbol{u}}^n$, rather than $\boldsymbol{u}^n$, that should be measured for the purposes of stability analysis, in which case (17.14) is valid, as an 'if and only if' statement, by virtue of Theorem 16.11.

Unlike (17.14), the stability condition (17.13) is false. Recall from Section 16.4 that using eigenvalues and spectral radii to prove stability is justified only if the underlying matrix A is *normal,* which is not so in the present case. It might seem that this clear injunction should be enough to deter anybody from 'proving' stability by eigenvalues. Unfortunately, most students of numerical analysis are weaned on the diffusion equation (where all reasonable finite difference schemes are symmetric) and then given a brief treatment of the wave equation (where, as we will see in Section 17.4, all reasonable finite difference schemes are skew-symmetric). Sooner or later the limited scope of eigenvalue analysis is likely to be forgotten ...

It is easy to convince ourselves that the present matrix A is not normal by verifying that $A^\top A \neq AA^\top$. It is of interest, however, to go back to the theme of Section 16.4 and see exactly what goes wrong with this matrix. The purpose of stability analysis is to deduce uniform bounds on *norms,* while eigenvalue analysis delivers *spectral radii.* Had A been normal, it would have been true that $\rho(A) = \|A\|$ and, in greater generality, $\rho(A^n) = \|A^n\|$. Let us estimate the norm of $A = A_{\Delta x}$ for (17.12), demonstrating that it is consistent with the Fourier estimate rather than the eigenvalue estimate.

In greater generality, we let

$$S_d = \begin{bmatrix} s & 0 & \cdots & \cdots & 0 \\ q & s & \ddots & & \vdots \\ 0 & \ddots & \ddots & \ddots & \vdots \\ \vdots & \ddots & q & s & 0 \\ 0 & \cdots & 0 & q & s \end{bmatrix}$$

be a *bidiagonal* $d \times d$ matrix and assume that $s, q \neq 0$. (Letting $s = 1 - \mu$ and $q = \mu$ recovers the matrix A.) To evaluate $\|S_d\|$ (in the usual Euclidean norm) we recall from A.1.5.2 that $\|B\| = [\rho(B^\top B)]^{1/2}$ for any real square matrix B. Let us thus form the product

$$S_d^\top S_d = \begin{bmatrix} s^2+q^2 & sq & 0 & \cdots & 0 \\ sq & s^2+q^2 & sq & \ddots & \vdots \\ 0 & \ddots & \ddots & \ddots & 0 \\ \vdots & \ddots & sq & s^2+q^2 & sq \\ 0 & \cdots & 0 & sq & s^2 \end{bmatrix}.$$

This is almost a TST matrix – just a single rogue element prevents us from applying Lemma 12.5 to determine its norm! Instead, we take a more roundabout approach. Firstly, it readily follows from the Geršgorin criterion (Lemma 8.3) that

$$\|S_d\|^2 = \rho(S_d^\top S_d) \le \max\{s^2 + |sq|,\ s^2 + q^2 + 2|sq|\} = (|s| + |q|)^2. \tag{17.16}$$

Secondly, set $w_{d,\ell} := (\operatorname{sgn} s/q)^{\ell-1}$, $\ell = 1, 2, \ldots, d$ and $\boldsymbol{w}_d = (w_{d,\ell})_{\ell=1}^{d}$. Since

$$S_d^\top S_d \boldsymbol{w}_d = \begin{bmatrix} s^2 + |sq| \\ (|s| + |q|)^2 \\ \vdots \\ (|s| + |q|)^2 \\ s^2 + |sq| + q^2 \end{bmatrix} \boldsymbol{w}_d = (|s| + |q|)^2 \boldsymbol{w}_d - \begin{bmatrix} w_{d,1}|sq| \\ 0 \\ \vdots \\ 0 \\ q^2 + (2 - w_{d,1})|sq| \end{bmatrix},$$

it follows from the definition of a matrix norm (A.1.3.4) that

$$\|S_d\|^2 = \|S_d^\top S_d\| = \max_{\boldsymbol{y} \ne \boldsymbol{0}} \frac{\|S_d^\top S_d \boldsymbol{y}\|}{\|\boldsymbol{y}\|} \ge \frac{\|S_d^\top S_d \boldsymbol{w}_d\|}{\|\boldsymbol{w}_d\|} = (|s| + |q|)^2 + \mathcal{O}\left(d^{-1/2}\right), \qquad d \to \infty.$$

Comparison with (17.16) demonstrates that

$$\|S_d\| = |s| + |q| + \mathcal{O}\left(d^{-1/2}\right), \qquad d \to \infty$$

which, returning to the matrix A, becomes

$$\|A\| = |1 - \mu| + \mu + \mathcal{O}\left((\Delta x)^{1/2}\right), \qquad \Delta x \to 0.$$

Hence $|1 - \mu| + \mu \le 1$ is sufficient for stability, as we have already deduced by Fourier analysis.

Hopefully, we have made the case that the numerical solution of hyperbolic equations deserves further elaboration and effort.

17.2 Finite differences for the advection equation

We are concerned with semi-discretizations of the form

$$v_\ell' + \frac{1}{\Delta x} \sum_{k=-\alpha}^{\beta} a_k v_{k+\ell} = 0 \tag{17.17}$$

and with fully discretized schemes

$$\sum_{k=-\gamma}^{\delta} b_k(\mu) u_{\ell+k}^{n+1} = \sum_{k=-\alpha}^{\beta} c_k(\mu) u_{\ell+k}^{n}, \qquad n \ge 0, \tag{17.18}$$

where

$$\sum_{k=-\gamma}^{\delta} b_k(\mu) \equiv 1$$

(cf. (16.22) and (16.37) respectively), when applied to the advection equation (17.1).

To address the question of stability, we will need to augment (17.1) by boundary conditions; we plan to devote most of our attention to the Cauchy problem and to *periodic* boundary conditions. For the time being we focus on the *orders* of (17.17) and of (17.18), a task for which it is not yet necessary to specify the exact range of ℓ.

Theorem 17.1 *The SD method* (17.17) *is of order p if and only if*

$$a(z) := \sum_{k=-\alpha}^{\beta} a_k z^k = \ln z + c(z-1)^{p+1} + \mathcal{O}\left(|z-1|^{p+2}\right), \qquad z \to 1, \tag{17.19}$$

where $c \neq 0$*, while the FD scheme* (17.18) *is of order p for Courant number* $\mu = \Delta t/\Delta x$ *if and only if there exists* $c(\mu) \neq 0$ *such that*

$$\tilde{a}(z,\mu) := \frac{\sum_{k=-\alpha}^{\beta} c_k(\mu) z^k}{\sum_{k=-\gamma}^{\delta} b_k(\mu) z^k} = z^{-\mu} + c(\mu)(z-1)^{p+1} + \mathcal{O}\left(|z-1|^{p+2}\right), \qquad z \to 1. \tag{17.20}$$

Proof Our analysis is similar to that in Section 16.3 but, if anything, easier. Letting $\tilde{v}_\ell(t) := u(\ell\Delta x, t)$ and $\tilde{u}_\ell^n := u(\ell\Delta x, n\Delta t)$ stand for the exact solution at the grid points, we have

$$\tilde{v}_\ell' + \frac{1}{\Delta x}\sum_{k=-\alpha}^{\beta} a_k \tilde{v}_{k+\ell} = \left(\mathcal{D}_t + \frac{1}{\Delta x}\sum_{k=-\alpha}^{\beta} a_k \mathcal{E}_x^k\right)\tilde{v}_\ell.$$

As far as the exact solution of the advection equation is concerned, we have $\mathcal{D}_t = -\mathcal{D}_x$ and, by (8.1), $\mathcal{D}_x = (\Delta x)^{-1}\ln\mathcal{E}_x$. Therefore

$$\tilde{v}_\ell' + \frac{1}{\Delta x}\sum_{k=-\alpha}^{\beta} a_k \tilde{v}_{k+\ell} = -\frac{1}{\Delta x}[\ln\mathcal{E}_x - a(\mathcal{E}_x)]v_\ell$$

and we deduce, using a method similar to that in the proof of Theorem 16.5, that (17.19) is necessary and sufficient for order p.

The order condition for FD schemes is based on the same argument and its derivation proceeds along the lines of Theorem 16.6. Thus, briefly,

$$\sum_{k=-\gamma}^{\delta} b_k(\mu)\tilde{u}_{\ell+k}^{n+1} - \sum_{k=-\alpha}^{\beta} c_k(\mu)\tilde{u}_{\ell+k}^{n} = \left[\mathcal{E}_t \sum_{k=-\gamma}^{\delta} b_k(\mu)\mathcal{E}_x^k - \sum_{k=-\alpha}^{\beta} c_k(\mu)\mathcal{E}_x^k\right]\tilde{u}_\ell$$

while, by (17.1) and Section 8.1,

$$\mathcal{E}_t = e^{(\Delta t)\mathcal{D}_t} = e^{-(\Delta t)\mathcal{D}_x} = e^{-\mu(\Delta x)\mathcal{D}_x} = e^{-\mu\ln\mathcal{E}_x} = \mathcal{E}_x^{-\mu}.$$

Hence

$$\sum_{k=-\gamma}^{\delta} b_k(\mu)\tilde{u}_{\ell+k}^{n+1} - \sum_{k=-\alpha}^{\beta} c_k(\mu)\tilde{u}_{\ell+k}^{n} = \left[\mathcal{E}_x^{-\mu} \sum_{k=-\gamma}^{\delta} b_k(\mu)\mathcal{E}_x^k - \sum_{k=-\alpha}^{\beta} c_k(\mu)\mathcal{E}_x^k\right]\tilde{u}_\ell.$$

This and the normalization of the denominator of $\tilde{a}$ are now used to complete the proof that the pth-order condition is indeed (17.20). ■

◇ **Examples of methods and their order** It is possible to show that, given any $\alpha, \beta \geq 0$, $\alpha + \beta \geq 1$, there exists for the advection equation a unique SD method of order $\alpha+\beta$ and that no other method may attain this bound. The coefficients of such a method are not difficult to derive explicitly, a task that is deferred to Exercise 17.2. Here we present four such schemes for future consideration. In each case we specify the function a. The schemes are as follows:

$$\alpha = 1, \quad \beta = 0, \qquad a(z) = -z^{-1} + 1; \tag{17.21}$$

$$\alpha = 0, \quad \beta = 1, \qquad a(z) = z - 1; \tag{17.22}$$

$$\alpha = 1, \quad \beta = 1, \qquad a(z) = -\tfrac{1}{2}z^{-1} + \tfrac{1}{2}z; \tag{17.23}$$

$$\alpha = 3, \quad \beta = 1, \qquad a(z) = -\tfrac{1}{12}z^{-3} + \tfrac{1}{2}z^{-2} - \tfrac{3}{2}z^{-1} + \tfrac{5}{6} + \tfrac{1}{4}z. \tag{17.24}$$

To demonstrate the power of Theorem 17.1 we address ourselves to the most complicated method above, (17.24), verifying that its order is indeed 4. Letting $z = \mathrm{e}^{\mathrm{i}\theta}$, (17.19) becomes equivalent to

$$a(\mathrm{e}^{\mathrm{i}\theta}) = \mathrm{i}\theta + \mathcal{O}(\theta^{p+1}), \qquad \theta \to 0.$$

For (17.24) we have

$$\begin{aligned} a(\mathrm{e}^{\mathrm{i}\theta}) &= -\tfrac{1}{12}\mathrm{e}^{-3\mathrm{i}\theta} + \tfrac{1}{2}\mathrm{e}^{-2\mathrm{i}\theta} - \tfrac{3}{2}\mathrm{e}^{-\mathrm{i}\theta} + \tfrac{5}{6} + \tfrac{1}{4}\mathrm{e}^{\mathrm{i}\theta} \\ &= -\tfrac{1}{12}\left(1 - 3\mathrm{i}\theta - \tfrac{9}{2}\theta^2 + \tfrac{9}{2}\mathrm{i}\theta^3 + \tfrac{27}{8}\theta^4\right) + \tfrac{1}{2}\left(1 - 2\mathrm{i}\theta - 2\theta^2 + \tfrac{4}{3}\mathrm{i}\theta^3 + \tfrac{2}{3}\theta^4\right) \\ &\quad - \tfrac{3}{2}\left(1 - \mathrm{i}\theta - \tfrac{1}{2}\theta^2 + \tfrac{1}{6}\mathrm{i}\theta^3 + \tfrac{1}{24}\theta^4\right) + \tfrac{5}{6} \\ &\quad + \tfrac{1}{4}\left(1 + \mathrm{i}\theta - \tfrac{1}{2}\theta^2 - \tfrac{1}{6}\theta^3 + \tfrac{1}{24}\theta^4\right) + \mathcal{O}(\theta^5) \\ &= \mathrm{i}\theta + \mathcal{O}(\theta^5), \qquad \theta \to 0, \end{aligned}$$

hence the method is of order 4. It is substantially easier to check that the orders of both (17.21) and (17.22) are 1 and that (17.23) is a second-order scheme.

As was the case with the diffusion equation in Chapter 16, the easiest technique in the design of FD schemes is the combination of an SD method with an ODE solver (typically, of at least the same order, cf. Exercise 16.6). Thus, pairing (17.21) with the Euler method (1.4) results in the FD scheme (17.12), which we have already encountered in Section 17.1 in somewhat strange circumstances. The marriage of (1.4) and (17.22) yields

$$u_\ell^{n+1} = u_\ell^n - \mu(u_{\ell+1}^n - u_\ell^n) = (1+\mu)u_\ell^n - \mu u_{\ell+1}^n, \qquad n \geq 0, \tag{17.25}$$

a method that looks very similar to (17.12) – but, as we will see later, is quite different.

The SD scheme (17.23) is of order 2 and we consider two popular schemes that are obtained when it is combined with second-order ODE schemes. Our first example is the *Crank–Nicolson method*

$$-\tfrac{1}{4}\mu u_{\ell-1}^{n+1} + u_\ell^{n+1} + \tfrac{1}{4}\mu u_{\ell+1}^{n+1} = \tfrac{1}{4}\mu u_{\ell-1}^n + u_\ell^n - \tfrac{1}{4}\mu u_{\ell+1}^n, \qquad n \geq 0, \tag{17.26}$$

which originates in the trapezoidal rule. Although we can deduce directly from Exercise 16.6 that it is of order 2, we can also prove it by using (17.20). To that end, we again exploit the substitution $z = \mathrm{e}^{\mathrm{i}\theta}$. Since

$$\begin{aligned}\tilde{a}(\mathrm{e}^{\mathrm{i}\theta},\mu) &= \frac{\frac{1}{4}\mu\mathrm{e}^{-\mathrm{i}\theta}+1-\frac{1}{4}\mu\mathrm{e}^{\mathrm{i}\theta}}{-\frac{1}{4}\mu\mathrm{e}^{-\mathrm{i}\theta}+1+\frac{1}{4}\mu\mathrm{e}^{\mathrm{i}\theta}} = \frac{1-\frac{1}{2}\mathrm{i}\mu\sin\theta}{1+\frac{1}{2}\mathrm{i}\mu\sin\theta}\\ &= \left(1-\tfrac{1}{2}\mathrm{i}\mu\sin\theta\right)\left(1-\tfrac{1}{2}\mathrm{i}\mu\sin\theta-\tfrac{1}{4}\mu^2\sin^2\theta+\tfrac{1}{8}\mathrm{i}\mu^3\sin^3\theta+\cdots\right)\\ &= 1-\mathrm{i}\mu\theta-\tfrac{1}{2}\mu^2\theta^2+\tfrac{1}{12}\mathrm{i}\mu(2+3\mu^2)\theta^3+\mathcal{O}(\theta^4), \qquad \theta\to 0,\end{aligned}$$

and

$$\mathrm{e}^{-\mathrm{i}\mu\theta} = 1-\mathrm{i}\mu\theta-\tfrac{1}{2}\mu^2\theta^2+\tfrac{1}{6}\mathrm{i}\mu^3\theta^3+\mathcal{O}(\theta^4), \qquad \theta\to 0,$$

we deduce that

$$\tilde{a}(\mathrm{e}^{\mathrm{i}\theta},\mu) = \mathrm{e}^{-\mathrm{i}\mu\theta}+\tfrac{1}{12}\mathrm{i}\mu(2+\mu^2)\theta^3+\mathcal{O}(\theta^4), \qquad \theta\to 0.$$

Thus, Crank–Nicolson is a second-order scheme.

Another popular scheme originates when (17.23) is combined with the explicit midpoint rule from Exercise 2.5. The outcome, the *leapfrog method,* uses two steps:

$$u_\ell^{n+2} = \mu(u_{\ell-1}^{n+1}-u_{\ell+1}^{n+1})+u_\ell^n, \qquad n\ge 0 \tag{17.27}$$

(cf. (16.33)). Although we have addressed ourselves in Theorem 17.1 to one-step FD methods, a generalization to two steps is easy: since

$$\mathrm{e}^{-2\mathrm{i}\mu\theta}-\mu(\mathrm{e}^{-\mathrm{i}\theta}-\mathrm{e}^{\mathrm{i}\theta})\mathrm{e}^{-\mathrm{i}\mu\theta}-1 = -\tfrac{1}{3}\mathrm{i}\mu(1-\mu^2)\theta^3+\mathcal{O}(\theta^4), \qquad \theta\to 0,$$

this method is also of order 2. Note that the error constant $c(\mu) = -\frac{1}{3}\mathrm{i}\mu(1-\mu^2)$ vanishes at $\mu = 1$ and so the method is of superior order for this value of the Courant number μ. An explanation of this phenomenon is the theme of Exercise 17.3.

Not all interesting FD methods can be derived easily from semi-discretized schemes; an example is the *angled derivative method*

$$u_\ell^{n+2} = (1-2\mu)(u_\ell^{n+1}-u_{\ell-1}^{n+1})+u_{\ell-1}^n, \qquad n\ge 0. \tag{17.28}$$

It can be proved that the method is of order 2, a task that we relegate to Exercise 17.4. Exercise 17.4 also includes the order analysis of the *Lax–Wendroff scheme*

$$u_\ell^{n+1} = \tfrac{1}{2}\mu(1+\mu)u_{\ell-1}^n+(1-\mu^2)u_\ell^n-\tfrac{1}{2}\mu(1-\mu)u_{\ell+1}^n, \qquad n\ge 0. \tag{17.29}$$

◇

Proceeding next to the stability analysis of the advection equation (17.1) and paying heed to the lesson of the 'theorem' from Section 17.1, we choose not to use eigenvalue techniques.[1] Our standard tool in the remainder of this section is Fourier analysis.

[1] Eigenvalues retain a marginal role, since some methods yield normal matrices; see Exercise 17.5.

It is of little surprise, thus, that we commence by considering the Cauchy problem, where the initial condition is given on the whole real line. Not much change is needed in the theory of Section 16.5 – as far as stability analysis is concerned, the exact identity of the PDE is irrelevant! The one obvious exception is that, since the space derivative in (17.1) is on the left-hand side, the inequality in the stability condition for SD schemes needs to be reversed. Without further ado, we thus formulate an equivalent of Theorems 16.10 and 16.11 appropriate to the current discussion.

Theorem 17.2 *The SD method* (17.17) *is stable (for a Cauchy problem) if and only if*

$$\mathrm{Re}\, a(\mathrm{e}^{\mathrm{i}\theta}) \geq 0, \qquad \theta \in [0, 2\pi], \tag{17.30}$$

where the function a is defined in (17.19). *Likewise, the FD method* (17.18) *is stable (for a Cauchy problem) for a given Courant number $\mu \in \mathbb{R}$ if*

$$|\tilde{a}(\mathrm{e}^{\mathrm{i}\theta}, \mu)| \leq 1, \qquad \theta \in [0, 2\pi]; \tag{17.31}$$

the function $\tilde{a}$ is defined in (17.20). ∎

It is easy to observe that (17.21) and (17.23) are stable, while (17.22) is not. This is an important point, intimately connected to the lack of symmetry in the advection equation. Since the exact solution is a unilateral shift, each value is transported to the right at a constant speed. Hence, numerical methods have a privileged direction and it is popular to choose schemes – whether SD or FD – that employ more points to the left than to the right of the current point, a practice known under the name of *upwinding*. Both (17.21) and (17.24) are upwind schemes, (17.23) is symmetric and the downwind scheme (17.22) seeks information at the wrong venue.

Being upwind is not a guarantee of stability, but it certainly helps. Thus, (17.24) is stable, since

$$\begin{aligned}
\mathrm{Re}\, a(\mathrm{e}^{\mathrm{i}\theta}) &= \mathrm{Re}\,\left(-\tfrac{1}{12}\mathrm{e}^{-3\mathrm{i}\theta} + \tfrac{1}{2}\mathrm{e}^{-2\mathrm{i}\theta} - \tfrac{3}{2}\mathrm{e}^{-\mathrm{i}\theta} + \tfrac{5}{6} + \tfrac{1}{4}\mathrm{e}^{\mathrm{i}\theta}\right) \\
&= -\tfrac{1}{12}\cos 3\theta + \tfrac{1}{2}\cos 2\theta - \tfrac{5}{4}\cos\theta + \tfrac{5}{6} \\
&= -\tfrac{1}{12}(4\cos^3\theta - 3\cos\theta) + \tfrac{1}{2}(2\cos^2\theta - 1) - \tfrac{5}{4}\cos\theta + \tfrac{5}{6} \\
&= -\tfrac{1}{3}\cos^3\theta + \cos^2\theta - \cos\theta + \tfrac{1}{3} \\
&= \tfrac{1}{3}(1-\cos\theta)^2 \geq 0, \qquad \theta \in [0, 2\pi].
\end{aligned}$$

Fully discretized schemes lend themselves to Fourier analysis just as easily. We have already seen that the Euler method (17.12) is stable for $\mu \in (0, 1]$. The outlook is less promising with regard to the downwind scheme (17.25) and, indeed,

$$|\tilde{a}(\mathrm{e}^{\mathrm{i}\theta}, \mu)|^2 = |1 + \mu - \mu\mathrm{e}^{\mathrm{i}\theta}|^2 = 1 + 4\mu(1+\mu)\sin^2\tfrac{1}{2}\theta, \qquad \theta \in [0, 2\pi],$$

exceeds unity for every $\theta \neq 0$ or 2π and $\mu > 0$. Before we discard this method, however, let us pause for a while and recall the vector equation (17.9). Suppose that $A = VDV^{-1}$, where D is diagonal. The elements along the diagonal of D are real since, as we have already mentioned, $\sigma(A) \subset \mathbb{R}$, but they might be negative or positive (in particular, in the important case of the wave equation (17.8), one is positive and the other negative). Letting $\boldsymbol{w}(x,t) := V^{-1}\boldsymbol{u}(x,t)$, equation (17.9) factorizes into

$$\frac{\partial \boldsymbol{w}}{\partial t} + D\frac{\partial \boldsymbol{w}}{\partial x} = \boldsymbol{0}, \qquad t \geq 0,$$

and hence into

$$\frac{\partial w_k}{\partial t} + \lambda_k \frac{\partial w_k}{\partial x} = 0, \qquad t \geq 0, \quad k = 1, 2, \ldots, m, \tag{17.32}$$

where m is the dimension of $\boldsymbol{u}$ and $\lambda_1, \lambda_2, \ldots, \lambda_m$ are the eigenvalues of A (and form the diagonal of D). A similar transformation can be applied to a numerical method, replacing $\boldsymbol{u}^n$ by, say, $\boldsymbol{w}^n$, and it is obvious that the two solution sequences are uniformly bounded (or otherwise) in norm for the same values of μ. Let us suppose that $\mathcal{M} \subseteq \mathbb{R}$ is the set of all numbers (positive, negative or zero) such that $\mu \in \mathcal{M}$ implies that an FD method is stable for equation (17.1). If we wish to apply this FD scheme to (17.9), it follows from (17.32) that we require

$$\lambda_k \mu \in \mathcal{M}, \qquad k = 1, 2, \ldots, m. \tag{17.33}$$

We recognize a situation, familiar from Section 4.2, in which the interval $\mathcal{M}$ plays a role similar to the linear stability domain of an ODE solver. In most cases of interest, $\mathcal{M}$ is a closed interval, which we denote by $[\mu_-, \mu_+]$.

Provided all eigenvalues are positive, (17.33) merely rescales μ by $\rho(A)$. If they are all negative, the method (17.25) becomes stable (for appropriate values of μ), while (17.12) loses its stability. More interesting, though, is the situation, as in (17.8), when some eigenvalues are positive and others negative since then, unless $\mu_- < 0$ and $0 < \mu_+$, no value of μ may coexist with stability. Both (17.12) and (17.21) fail in this situation, but this is not the case with Crank–Nicolson, since

$$|\tilde{a}(\mathrm{e}^{\mathrm{i}\theta}, \mu)|^2 = \left|\frac{1 - \frac{1}{2}\mathrm{i}\sin\theta}{1 + \frac{1}{2}\mathrm{i}\sin\theta}\right| \equiv 1, \qquad \theta \in [0, 2\pi];$$

hence we have stability for all $\mu \in (-\infty, \infty)$! Another example is the Lax–Wendroff scheme, whose explicit form confers important advantages in comparison with Crank–Nicolson. Since

$$\begin{aligned} |\tilde{a}(\mathrm{e}^{\mathrm{i}\theta}, \mu)|^2 &= \left|\tfrac{1}{2}\mu(1+\mu)\mathrm{e}^{-\mathrm{i}\theta} + (1-\mu^2) - \tfrac{1}{2}\mu(1-\mu)\mathrm{e}^{\mathrm{i}\theta}\right|^2 \\ &= |1 - \mu(1-\cos\theta) - \mathrm{i}\mu\sin\theta|^2 \\ &= 1 - 4\mu^2(1-\mu^2)\sin^4\tfrac{1}{2}\theta, \qquad \theta \in [0, 2\pi], \end{aligned}$$

we obtain $\mu_- = -1$, $\mu_+ = 1$.

Periodic boundary conditions, our next theme, are important in the context of the wave-like phenomena that are typically described by hyperbolic PDEs. Thus, let us complement the advection equation (17.1) with, say, the boundary condition

$$u(0, t) = u(1, t), \qquad t \geq 0. \tag{17.34}$$

The exact solution is no longer (17.4) but is instead periodic in t: the initial condition is transported to the right with unit speed but, as soon as it disappears through $x = 1$, it reappears from the other end; hence $u(x, 1) = g(x)$, $0 \le x \le 1$.

To emphasize the difference between Dirichlet and periodic boundary conditions we write the Lax–Wendroff scheme (17.29) in a matrix form, $\boldsymbol{u}^{n+1} = A\boldsymbol{u}^n$, say. Assuming (zero) Dirichlet conditions, we have

$$A = A^{[\mathrm{D}]} := \begin{bmatrix} 1-\mu^2 & \frac{1}{2}\mu(\mu-1) & 0 & \cdots & 0 \\ \frac{1}{2}\mu(1+\mu) & 1-\mu^2 & \frac{1}{2}\mu(\mu-1) & \ddots & \vdots \\ 0 & \ddots & \ddots & \ddots & 0 \\ \vdots & \ddots & \frac{1}{2}\mu(1+\mu) & 1-\mu^2 & \frac{1}{2}\mu(\mu-1) \\ 0 & \cdots & 0 & \frac{1}{2}\mu(1+\mu) & 1-\mu^2 \end{bmatrix},$$

while periodic boundary conditions yield

$$A = A^{[\mathrm{p}]} := \begin{bmatrix} 1-\mu^2 & \frac{1}{2}\mu(\mu-1) & 0 & \cdots & 0 & \frac{1}{2}\mu(1+\mu) \\ \frac{1}{2}\mu(1+\mu) & 1-\mu^2 & \frac{1}{2}\mu(\mu-1) & \ddots & \vdots & 0 \\ 0 & \frac{1}{2}\mu(1+\mu) & \ddots & \ddots & 0 & \vdots \\ \vdots & 0 & \ddots & \ddots & \frac{1}{2}\mu(\mu-1) & 0 \\ 0 & \vdots & \ddots & \frac{1}{2}\mu(1+\mu) & 1-\mu^2 & \frac{1}{2}\mu(\mu-1) \\ \frac{1}{2}\mu(\mu-1) & 0 & \cdots & 0 & \frac{1}{2}\mu(1+\mu) & 1-\mu^2 \end{bmatrix}.$$

The reason for the discrepancies in the top right-hand and lower left-hand corners is that, in the presence of periodic boundary conditions, each time we need a value from outside the set $\{0, 1, \ldots, d-1\}$ at one end, we borrow it from the other end.[2]

The difference between $A^{[\mathrm{D}]}$ and $A^{[\mathrm{p}]}$ does seem minor – just two entries in what are likely to be very large matrices. However, as we will see soon, these two matrices could hardly be more dissimilar in their properties. In particular, while stability analysis with Dirichlet boundary conditions, a subject to which we will have returned briefly by the end of this section, is very intricate, periodic boundary conditions surrender their secrets much more easily. In fact, we have the unexpected comfort that both eigenvalue and Fourier analysis are absolutely straightforward in the periodic case!

The matrix $A^{[\mathrm{p}]}$ is a special case of a *circulant* – the latter being a $d \times d$ matrix C whose jth row, $j = 2, 3, \ldots, d$, is a 'right-rotated' $(j-1)$th row,

$$C = C(\boldsymbol{\kappa}) = \begin{bmatrix} \kappa_0 & \kappa_1 & \kappa_2 & \cdots & \kappa_{d-1} \\ \kappa_{d-1} & \kappa_0 & \kappa_1 & \cdots & \kappa_{d-2} \\ \kappa_{d-2} & \kappa_{d-1} & \kappa_0 & \cdots & \kappa_{d-3} \\ \vdots & \vdots & \vdots & & \vdots \\ \kappa_1 & \kappa_2 & \kappa_3 & \cdots & \kappa_0 \end{bmatrix}; \tag{17.35}$$

[2]We have just tacitly adopted the convention that the unknowns in a periodic problem are the points with spatial coordinates $\ell\Delta x$, $\ell = 0, 1, \ldots, d-1$, where $\Delta x = 1/d$. This makes for a somewhat less unwieldy notation.

specifically, $\kappa_0 = 1-\mu^2$, $\kappa_1 = \frac{1}{2}\mu(\mu-1)$, $\kappa_2 = \cdots = \kappa_{d-2} = 0$ and $\kappa_{d-1} = \frac{1}{2}\mu(1+\mu)$.

Lemma 17.3 *The eigenvalues of $C(\boldsymbol{\kappa})$ are $\kappa(\omega_d^j)$, $j = 0, 1, \ldots, d-1$, where*

$$\kappa(z) := \sum_{\ell=0}^{d-1} \kappa_\ell z^\ell, \qquad z \in \mathbb{C}$$

and $\omega_d = \exp(2\pi \mathrm{i}/d)$ is the dth primitive root of unity. To each $\lambda_j = \kappa(\omega_d^j)$ there corresponds the eigenvector

$$\boldsymbol{w}_j = \begin{bmatrix} 1 \\ \omega_d^j \\ \omega_d^{2j} \\ \vdots \\ \omega_d^{(d-1)j} \end{bmatrix}, \qquad j = 0, 1, \ldots, d-1.$$

Proof We show directly that $C(\boldsymbol{\kappa})\boldsymbol{w}_j = \lambda_j \boldsymbol{w}_j$ for all $j = 0, 1, \ldots, d-1$. To that end we observe that in (17.35) the mth row of $C(\boldsymbol{\kappa})$ is

$$\begin{bmatrix} \kappa_{d-m} & \kappa_{d-m+1} & \cdots & \kappa_{d-1} & \kappa_0 & \kappa_1 & \cdots & \kappa_{d-m-1} \end{bmatrix},$$

hence the mth component of $C(\boldsymbol{\kappa})\boldsymbol{w}_j$ is

$$\sum_{\ell=0}^{d-1} c_{m,\ell} w_{j,\ell} = \sum_{\ell=0}^{m-1} \kappa_{d-m+\ell}\omega_d^{j\ell} + \sum_{\ell=m}^{d-1} \kappa_{\ell-m}\omega_d^{j\ell}.$$

Let us replace the summation indices on the right by $\ell' = d-m+\ell$, $\ell = 1, 2, \ldots, m-1$ and $\ell' = \ell - m$, $\ell = m, m+1, \ldots, d-1$ respectively. Since $\omega_d^d = 1$, the outcome (dropping the prime from the index ℓ') is

$$\begin{aligned} \sum_{\ell=0}^{d-1} c_{m,\ell} w_{j,\ell} &= \sum_{\ell=d-m}^{d-1} \kappa_\ell \omega_d^{j(\ell-d+m)} + \sum_{\ell=0}^{d-1-m} \kappa_\ell \omega_d^{j(\ell+m)} \\ &= \left(\sum_{\ell=0}^{d-1} \kappa_\ell \omega_d^{j\ell} \right) \omega_d^{jm} = \lambda_j w_{j,m}, \qquad m = 0, 1, \ldots, d-1. \end{aligned}$$

We conclude that the $\boldsymbol{w}_j$ are indeed eigenvectors corresponding to the eigenvalues $\kappa(\omega_d^j)$, $j = 0, 1, \ldots, d-1$, respectively. ■

The lemma has several interesting consequences. For example, since the matrix of the eigenvectors is *exactly* the inverse discrete Fourier transform (10.13), the theory of Section 10.3 demonstrates that multiplying an arbitrary $d \times d$ circulant by a vector can be executed very fast by FFT. More interestingly from our point of view, the eigenvectors of $C(\boldsymbol{\kappa})$ do not depend on $\boldsymbol{\kappa}$ at all: all $d \times d$ circulants share the same eigenvectors, hence *all such matrices commute.*

The matrix of eigenvectors,

$$[\boldsymbol{w}_0 \quad \boldsymbol{w}_1 \quad \cdots \quad \boldsymbol{w}_{d-1}],$$

is *unitary* since, trivially,

$$\langle \boldsymbol{w}_j, \boldsymbol{w}_\ell \rangle = \bar{\boldsymbol{w}}_j^\top \boldsymbol{w}_\ell = 0, \qquad j, \ell = 0, 1, \ldots, d-1, \quad j \neq \ell.$$

Therefore every circulant is *normal* (A.1.2.5). An alternative proof is left to Exercise 17.9. As we already know from Theorems 16.7 and 16.8, the stability of finite difference schemes with normal matrices can be completely specified in terms of the eigenvalues of the latter. Since these eigenvalues were fully described in Lemma 17.3, stability analysis becomes almost as easy as painting by numbers. Thus, for Lax–Wendroff,

$$\begin{aligned}
&\kappa_0 = 1 - \mu^2, \qquad \kappa_1 = \tfrac{1}{2}\mu(\mu - 1), \qquad \kappa_{d-1} = \tfrac{1}{2}\mu(\mu + 1)\\
\Rightarrow\quad &\lambda_j = (1 - \mu^2) + \tfrac{1}{2}\mu(\mu - 1)\omega_d^j + \tfrac{1}{2}\mu(\mu + 1)\omega_d^{(d-1)j}\\
&\quad = (1 - \mu^2) + \tfrac{1}{2}\mu(\mu - 1)\exp(2\pi \mathrm{i} j/d) + \tfrac{1}{2}\mu(\mu + 1)\exp(-2\pi \mathrm{i} j/d)\\
&\quad = \tilde{a}(\exp(2\pi \mathrm{i} j/d), \mu), \qquad j = 0, \ldots, d-1,
\end{aligned}$$

where $\tilde{a}$ has been already encountered in the context of both order and Fourier stability analysis.

There is nothing special about the Lax–Wendroff scheme; it is the presence of periodic boundary conditions that makes the difference. The identity

$$\lambda_j = \tilde{a}(\exp(2\pi \mathrm{i} j/d), \mu), \qquad j = 0, 1, \ldots, d-1,$$

is valid for all FD methods (17.18). Letting $d \to \infty$, we can now use Theorem 16.7 (or a similar analysis, in tandem with Theorem 16.8, in the case of SD schemes) to extend the scope of Theorem 17.2 to the realm of periodic boundary conditions.

Theorem 17.4 *Let us assume the periodic boundary conditions* (17.34). *The SD method* (17.17) *is stable subject to the inequality* (17.30), *and the FD scheme* (17.18) *is stable subject to the inequality* (17.31). ■

An alternative route to Theorem 17.4 proceeds via Fourier analysis with a *discrete Fourier transform (DFT)*. It is identical in both content and consequences; as far as circulants are concerned, the main difference between Fourier and eigenvalue analysis is just a matter of terminology.

The stability analysis for a Dirichlet boundary problem is considerably more complicated and the conditions of Theorem 17.2, say, are necessary but often far from sufficient. The Euler method (17.12) is a double exception. Firstly, the Fourier conditions are both necessary and sufficient for stability. Secondly, the statement in the previous sentence can be proved by elementary means (cf. Section 17.1). In general, even if (17.30) or (17.31) are sufficient to attain stability, the proof is far from elementary.

Let us consider the solution of (17.1) with the initial condition $g(x) = \sin 8\pi x$, $x \in [0,1]$, and the Dirichlet boundary condition $\varphi_0(t) = -\sin 8\pi t$, $t \geq 0$. The exact solution, according to (17.4), is $u(x,t) = \sin 8\pi(x-t)$, $x \in [0,1]$, $t \geq 0$.

To illustrate the difficulty of stability analysis in the presence of the Dirichlet boundary conditions, we now solve this equation with the leapfrog method (17.27), evaluating the first step with the Lax–Wendroff scheme (17.29). However, in attempting to implement the leapfrog method, it is soon realized that a vital item of data is missing: since there is no boundary condition at $x = 1$, (17.27) cannot be executed for $\ell = d$. So, let us simply substitute

$$u_d^{n+1} = 0, \qquad n \geq 0, \tag{17.36}$$

which seems a safe bet – what could be more stable than zero?! Figure 17.4 displays the solution for $d = 40$ and $d = 80$ in the interval $t \in [0,6]$ and it is quite apparent that it looks nothing like the expected sinusoidal curve. Worse, the solution deteriorates when the grid is refined, a hallmark of instability.

The mechanism that causes instability and deterioration of the solution is indeed the rogue 'boundary scheme' (17.36). This perhaps becomes more evident upon an examination of Fig. 17.5, which displays snapshots of $\boldsymbol{u}^n$ for time intervals of equal length. The oscillatory overlay on the (correct) sinusoidal curve at $t = 0.5$ gives the game away: the substitution (17.36) allows for increasing oscillations that enter the interval $[0,1]$ at $x = 1$ and travel leftwards, *in the wrong direction.*

This amazing sensitivity to the choice of just one point (whose value does not influence at all the *exact* solution in $[0,1)$) is further emphasized in Fig. 17.6, where the very same leapfrog has been used to solve an identical equation, except that, in place of (17.36), we have used

$$u_d^{n+1} = u_{d-1}^n, \qquad n \geq 0. \tag{17.37}$$

Like Wordsworth's 'Daffodils', the sinusoidal curves of Fig. 17.6 are 'stretch'd in a never-ending line', perfectly aligned and stable. This is already apparent from a cursory inspection of the solution, while the four 'snapshots' are fully consistent with the numerical error of $\mathcal{O}\big((\Delta x)^{-2}\big)$ that is to be expected from a second-order convergent scheme.

The general rules governing stability in the presence of boundaries are far too complicated for an introductory text; they require sophisticated mathematical machinery. Our simple example demonstrates that adding boundaries is a genuine issue, not simply a matter of mathematical nitpicking, and that a wrong choice of a 'boundary fix' might well corrupt a stable scheme.

17.3 The energy method

Both the eigenvalue technique and Fourier analysis are, as should have been amply demonstrated, of limited scope. Sooner or later – sooner if we set our mind on solving nonlinear PDEs – we are bound to come across a numerical scheme that defies both methods. The one means left is the recourse of the desperate, the *energy method.* The

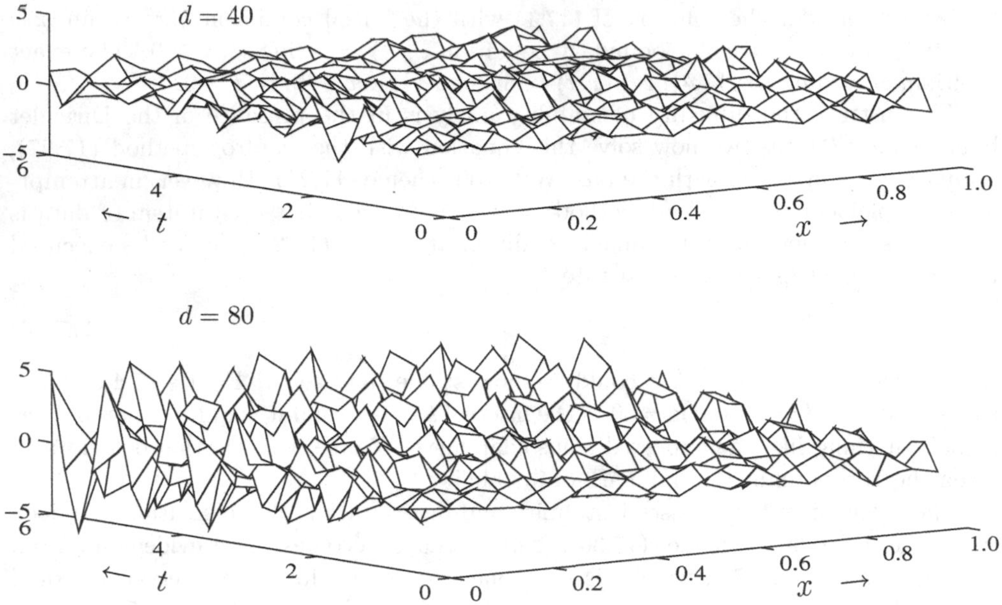

Figure 17.4 A leapfrog solution of (17.1) with Dirichlet boundary conditions.

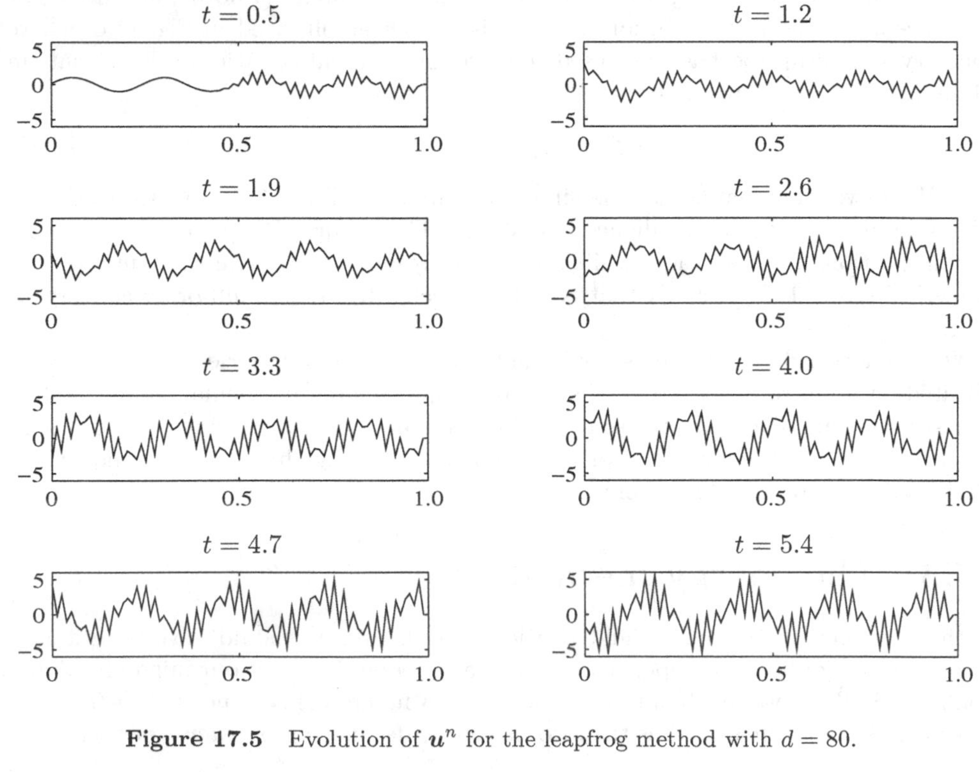

Figure 17.5 Evolution of $\boldsymbol{u}^n$ for the leapfrog method with $d = 80$.

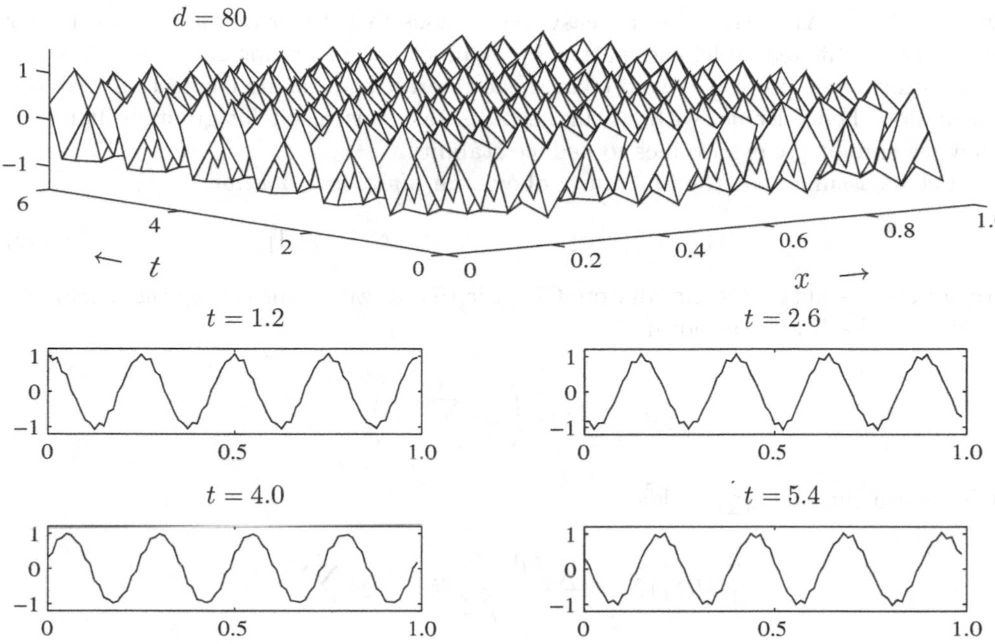

Figure 17.6 A leapfrog solution of (17.1) with the stable *artificial boundary condition* (17.37).

truth of the matter is that, far from being a coherent technique, the energy method is essentially a brute force approach toward proving the stability conditions (16.16) or (16.20) by direct manipulation of the underlying scheme.

We demonstrate the energy method by a single example, namely numerical solution of the variable-coefficient advection equation (17.5), with zero boundary condition $\varphi_0 = 0$, by the SD scheme

$$v'_\ell = \frac{\tau_\ell}{2\Delta x}(v^n_{\ell-1} - v^n_{\ell+1}), \qquad \ell = 1, 2, \ldots, d, \quad t \geq 0, \tag{17.38}$$

where $\tau_\ell := \tau(\ell\Delta x)$, $\ell = 1, 2, \ldots, d$. Being a generalization of (17.23), we note that this method is of order 2, but our current goal is to investigate its stability.

Fourier analysis is out of the question; the whole point about this technique is that it requires exactly the same difference scheme at every ℓ, and a variable function τ_ℓ renders this impossible. It takes more effort to demonstrate that the eigenvalue technique is not up to the task either. The matrix of the SD system (17.38) is

$$P_{\Delta x} = \frac{1}{2\Delta x}\begin{bmatrix} 0 & -\tau_1 & 0 & \cdots & 0 \\ \tau_2 & 0 & -\tau_2 & \ddots & \vdots \\ 0 & \ddots & \ddots & \ddots & 0 \\ \vdots & \ddots & \tau_{d-1} & 0 & -\tau_{d-1} \\ 0 & \cdots & 0 & \tau_d & 0 \end{bmatrix},$$

where $(d+1)\Delta x = 1$. It is an easy yet tedious task to prove that, subject to τ being twice differentiable, the matrix $P_{\Delta x}$ cannot be normal for $\Delta x \to 0$ unless τ is a constant. The proof is devoid of intrinsic interest and has no connection with our discussion; hence let us, without further ado, take this result for granted. It means that we cannot use eigenvalues to deduce stability.

Let us assume that the function τ obeys the *Lipschitz condition*

$$|\tau(x) - \tau(y)| \le \lambda |x - y|, \qquad x, y \in [0,1], \tag{17.39}$$

for some constant $\lambda \ge 0$. Recall from Chapter 16 that we are measuring the magnitude of $\boldsymbol{v}_{\Delta x}$ in the Euclidean norm

$$\|\boldsymbol{w}_{\Delta x}\|_{\Delta x} = \left(\Delta x \sum_{\ell=1}^{d} w_\ell^2\right)^{1/2}.$$

Differentiating $\|\boldsymbol{v}_{\Delta x}\|_{\Delta x}^2$ yields

$$\frac{\mathrm{d}}{\mathrm{d}t}\|\boldsymbol{v}_{\Delta x}\|_{\Delta x}^2 = \Delta x \frac{\mathrm{d}}{\mathrm{d}t}\sum_{\ell=1}^{d} v_\ell^2 = 2\Delta x \sum_{\ell=1}^{d} v_\ell v_\ell'.$$

Substituting the value of v_ℓ from the SD equations (17.38) and changing the order of summation, we obtain

$$\begin{aligned}\frac{\mathrm{d}}{\mathrm{d}t}\|\boldsymbol{v}_{\Delta x}\|_{\Delta x}^2 &= \sum_{\ell=1}^{d} \tau_\ell v_\ell (v_{\ell-1} - v_{\ell+1}) = \sum_{\ell=0}^{d-1} \tau_{\ell+1} v_\ell v_{\ell+1} - \sum_{\ell=1}^{d} \tau_\ell v_\ell v_{\ell+1} \\ &= \sum_{\ell=1}^{d} (\tau_{\ell+1} - \tau_\ell) v_\ell v_{\ell+1}.\end{aligned}$$

Note that we have used the zero boundary condition $v_0 = 0$. Observe next that the Lipschitz condition (17.39) implies

$$|\tau_{\ell+1} - \tau_\ell| = |\tau((\ell+1)\Delta x) - \tau(\ell\Delta x)| \le \lambda \Delta x;$$

therefore

$$\frac{\mathrm{d}}{\mathrm{d}t}\|\boldsymbol{v}_{\Delta x}\|_{\Delta x}^2 \le \left|\sum_{\ell=1}^{d} (\tau_{\ell+1} - \tau_\ell) v_\ell v_{\ell+1}\right| \le \sum_{\ell=1}^{d} |\tau_{\ell+1} - \tau_\ell|\, |v_\ell v_{\ell+1}| \le \lambda \Delta x \sum_{\ell=1}^{d} |v_\ell v_{\ell+1}|.$$

Finally, we resort to the *Cauchy–Schwarz inequality* (A.1.3.1) to deduce that

$$\frac{\mathrm{d}}{\mathrm{d}t}\|\boldsymbol{v}_{\Delta x}\|_{\Delta x}^2 \le \lambda \Delta x \left(\sum_{\ell=1}^{d} v_\ell^2\right)^{1/2} \left(\sum_{\ell=1}^{d} v_{\ell+1}^2\right)^{1/2} \le \lambda \|\boldsymbol{v}_{\Delta x}\|_{\Delta x}^2. \tag{17.40}$$

It follows at once from (17.40) that

$$\|\boldsymbol{v}_{\Delta x}(t)\|_{\Delta x}^2 \le \mathrm{e}^{\lambda t}\|\boldsymbol{v}_{\Delta x}(0)\|_{\Delta x}^2, \qquad t \in [0, t^*].$$

Since

$$\lim_{\Delta x \to 0} \|\boldsymbol{v}_{\Delta x}(0)\|_{\Delta x} = \|g\| < \infty,$$

where g is the initial condition (recall our remark on Riemann sums in Section 16.2!), it is possible to bound $\|\boldsymbol{v}_{\Delta x}(0)\|_{\Delta x} \le c$, say, uniformly for sufficiently small Δx, thereby deducing the inequality

$$\|\boldsymbol{v}_{\Delta x}(t)\|_{\Delta x}^2 \le c^2 \mathrm{e}^{\lambda t^*}, \qquad t \in [0, t^*].$$

This is precisely what is required for (16.20), the definition of stability for SD schemes, and we thus deduce that the scheme (17.38) is stable.

17.4 The wave equation

As we have already noted in Section 17.1, the wave equation can be expressed as a system of advection equations (17.8). At least in principle, this enables us to exploit the theory of Sections 17.2 and 17.3 to produce finite difference schemes for the wave equation. Unfortunately, we soon encounter two practical snags. Firstly, the wave equation is equipped with two initial conditions, namely

$$u(x,0) = g_0(x), \qquad \frac{\partial u(x,0)}{\partial t} = g_1(x), \qquad 0 \le x \le 1, \tag{17.41}$$

and, typically, two Dirichlet boundary conditions,

$$u(0,t) = \varphi_0(t), \quad u(1,t) = \varphi_1(t), \qquad t \ge 0 \tag{17.42}$$

(we will return later to the matter of boundary conditions). Rendering (17.41) and (17.42) in the terminology of a vector advection equation (17.9) makes for strange-looking conditions that needlessly complicate the exposition. The second difficulty comes to light as soon as we attempt to generalize the SD method (17.23), say, to cater for the system (17.9). On the face of it, nothing could be easier: just replace $(\Delta x)^{-1}$ by $(\Delta x)^{-1}A$, thereby obtaining

$$\boldsymbol{v}'_\ell + \frac{1}{2\Delta x} A(\boldsymbol{v}_{\ell+1} - \boldsymbol{v}_{\ell-1}) = \boldsymbol{0}. \tag{17.43}$$

This is entirely reasonable so far as a general matrix A is concerned. However, choosing

$$A = \begin{bmatrix} 0 & 1 \\ 1 & 0 \end{bmatrix}$$

converts (17.43) into

$$\begin{aligned} v'_{1,\ell} &= -\frac{1}{2\Delta x}(v_{2,\ell+1} - v_{2,\ell-1}), \\ v'_{2,\ell} &= -\frac{1}{2\Delta x}(v_{1,\ell+1} - v_{1,\ell-1}). \end{aligned}$$

According to Section 17.1, $v_\ell := v_{1,\ell}$ approximates the solution of the wave equation. Further differentiation helps us to eliminate the second coordinate,

$$v_\ell'' = -\frac{1}{2\Delta x}(v_{2,\ell+1}' - v_{2,\ell-1}') = -\frac{1}{2\Delta x}\left[-\frac{1}{2\Delta x}(v_{\ell+2} - v_\ell) + \frac{1}{2\Delta x}(v_\ell - v_{\ell-2})\right],$$

and results in the SD scheme

$$v_\ell'' = \frac{1}{4(\Delta x)^2}(v_{\ell-2} - 2v_\ell + v_{\ell+2}). \tag{17.44}$$

Although (17.44) is a second-order scheme, it makes very little sense. There is absolutely no good reason to make v_ℓ'' depend on $v_{\ell\pm2}$ rather than on $v_{\ell\pm1}$ and we have at least one powerful incentive for the latter course – it is likely to make the numerical error significantly smaller. A simple trick can sort this out: replace (17.43) by the formal scheme

$$\boldsymbol{v}_\ell' + \frac{1}{\Delta x}A(\boldsymbol{v}_{\ell+1/2} - \boldsymbol{v}_{\ell-1/2}).$$

In general this is nonsensical, but for the present matrix A the outcome is the SD scheme

$$v_\ell'' = \frac{1}{(\Delta x)^2}(v_{\ell-1} - 2v_\ell + v_{\ell+1}), \tag{17.45}$$

which is of exactly the right form. An alternative route leading to (17.45) is to discretize the second spatial derivative using central differences, exactly as we did in Chapters 8 and 15.

Choosing to follow the path of analytic expansion, along the lines of Sections 16.3 and 17.2, we consider the general SD method

$$v_\ell'' = \frac{1}{(\Delta x)^2}\sum_{k=-\alpha}^{\beta} a_k v_{\ell+k}. \tag{17.46}$$

The only difference from (16.22) and (17.7) is that this scheme possesses a second time derivative, hence being of the right form to satisfy both the initial conditions (17.41).

Letting $\tilde{v}_\ell(t) := u(\ell\Delta x, t)$, $t \geq 0$, and engaging without any further ado in the already familiar calculus of finite difference operators, we deduce from (17.8) that

$$\tilde{v}_\ell'' - \frac{1}{(\Delta x)^2}\sum_{k=-\alpha}^{\beta} a_k \tilde{v}_{\ell+k} = \frac{1}{(\Delta x)^2}\left[(\ln \mathcal{E}_x)^2 - \sum_{k=-\alpha}^{\beta} a_k \mathcal{E}_x^k\right]\tilde{v}_\ell.$$

Therefore (17.46) is of order p if and only if

$$a(z) := \sum_{k=-\alpha}^{\beta} a_k z^k = (\ln z)^2 + c(z-1)^{p+2} + \mathcal{O}\left(|z-1|^{p+3}\right), \qquad z \to 1, \tag{17.47}$$

for some $c \neq 0$.

It is an easy matter to verify that both (17.44) and (17.45) are of order 2. A little more effort is required to demonstrate that the SD scheme

$$v_\ell'' = \frac{1}{(\Delta x)^2}\left(-\tfrac{1}{12}v_{\ell-2} + \tfrac{4}{3}v_{\ell-1} - \tfrac{5}{2}v_\ell + \tfrac{4}{3}v_{\ell+1} - \tfrac{1}{12}v_{\ell+2}\right)$$

is fourth order.

Our next step consists of discretizing the ODE system (17.46) and it affords us an opportunity to discuss the numerical solution of second-order ODEs with greater generality. Consider thus the equations

$$\boldsymbol{z}'' = \boldsymbol{f}(t,\boldsymbol{z}), \quad t \ge t_0, \qquad \boldsymbol{z}(t_0) = \boldsymbol{z}_0, \quad \boldsymbol{z}'(t_0) = \boldsymbol{z}_0', \tag{17.48}$$

where $\boldsymbol{f}$ is a given function. Note that we can easily cast the semi-discretized scheme (17.46) in this form.

The easiest way of solving (17.48) numerically is to convert it into a first-order system having twice the number of variables. It can be verified at once that, subject to the substitution $\boldsymbol{y}_1(t) := \boldsymbol{z}(t)$, $\boldsymbol{y}_2(t) := \boldsymbol{z}'(t)$, $t \ge t_0$, (17.48) is equivalent to the ODE system

$$\begin{aligned} \boldsymbol{y}_1' &= \boldsymbol{y}_2, \\ \boldsymbol{y}_2' &= \boldsymbol{f}(t,\boldsymbol{y}_1), \end{aligned} \qquad t \ge t_0, \tag{17.49}$$

with the initial condition

$$\boldsymbol{y}_1(t_0) = \boldsymbol{z}_0, \qquad \boldsymbol{y}_2(t_0) = \boldsymbol{z}_0'.$$

On the face of it, we may choose any ODE scheme from Chapters 1–3 and apply it to (17.49). The outcome can be surprising ...

Suppose, thus, that (17.49) is solved with Euler's method (1.4), hence, in the notation of Chapters 1–7, we have

$$\begin{aligned} \boldsymbol{y}_{1,n+1} &= \boldsymbol{y}_{1,n} + h\boldsymbol{y}_{2,n}, \\ \boldsymbol{y}_{2,n+1} &= \boldsymbol{y}_{2,n} + h\boldsymbol{f}(t_n,\boldsymbol{y}_{1,n}), \end{aligned} \qquad n \ge 0.$$

According to the first equation,

$$\boldsymbol{y}_{2,n} = \frac{1}{h}(\boldsymbol{y}_{1,n+1} - \boldsymbol{y}_{1,n}).$$

Substituting this twice (once with n and once with $n+1$) into the second equation allows us to eliminate $\boldsymbol{y}_{2,n}$ altogether:

$$\frac{1}{h}(\boldsymbol{y}_{1,n+2} - \boldsymbol{y}_{1,n+1}) = \frac{1}{h}(\boldsymbol{y}_{1,n+1} - \boldsymbol{y}_{1,n}) + h\boldsymbol{f}(t_n,\boldsymbol{y}_{1,n}).$$

The outcome is the two-step explicit method

$$\boldsymbol{z}_{n+2} - 2\boldsymbol{z}_{n+1} + \boldsymbol{z}_n = h^2\boldsymbol{f}(t_n,\boldsymbol{z}_n), \qquad n \ge 0. \tag{17.50}$$

A considerably cleverer approach is to solve the first set of equations in (17.49) with the backward Euler method (1.15), while retaining the usual Euler method for the second set. This yields

$$\begin{aligned} \boldsymbol{y}_{1,n+1} &= \boldsymbol{y}_{1,n} + h\boldsymbol{y}_{2,n+1}, \\ \boldsymbol{y}_{2,n+1} &= \boldsymbol{y}_{2,n} + h\boldsymbol{f}(t_n, \boldsymbol{y}_{1,n}), \end{aligned} \qquad n \ge 0.$$

Substitution of

$$\boldsymbol{y}_{2,n+1} = \frac{1}{h}(\boldsymbol{y}_{1,n+1} - \boldsymbol{y}_{1,n})$$

in the second equation and a shift in the index results in the *Störmer* method

$$\boldsymbol{z}_{n+2} - 2\boldsymbol{z}_{n+1} + \boldsymbol{z}_n = h^2 \boldsymbol{f}(t_{n+1}, \boldsymbol{z}_{n+1}), \qquad n \ge 0, \qquad (17.51)$$

which we encountered in a different context in (5.26). Although we have used the backward Euler method in its construction, the Störmer method is explicit.

A numerical method for the ODE (17.49) is of *order p* if substitution of the exact solution results in a perturbation of $\mathcal{O}(h^{p+2})$. As can be expected, the method (17.50) is of order 1 – after all, it is nothing other than the Euler scheme. However, as far as (17.51) is concerned, symmetry and the subtle interplay between the forward and backward Euler methods mean that its order is increased and its performance improved significantly in comparison with what we might have naively expected. Let $\tilde{\boldsymbol{z}}_n = \boldsymbol{z}(t_n)$, $n \ge 0$, be the exact solution of (17.48). Substitution of this into (17.51), expansion about the point t_{n+1} and the differential equation (17.48) yield

$$\begin{aligned} &\tilde{\boldsymbol{z}}_{n+2} - 2\tilde{\boldsymbol{z}}_{n+1} + \tilde{\boldsymbol{z}}_n - h^2 \boldsymbol{f}(t_{n+1}, \tilde{\boldsymbol{z}}_{n+1}) \\ &= \left[\tilde{\boldsymbol{z}}_{n+1} + h\tilde{\boldsymbol{z}}'_{n+1} + \tfrac{1}{2}h^2\tilde{\boldsymbol{z}}''_{n+1} + \tfrac{1}{6}h^3\tilde{\boldsymbol{z}}'''_{n+1} + \mathcal{O}(h^4)\right] - 2\tilde{\boldsymbol{z}}_{n+1} \\ &\quad + \left[\tilde{\boldsymbol{z}}_{n+1} - h\tilde{\boldsymbol{z}}'_{n+1} + \tfrac{1}{2}h^2\tilde{\boldsymbol{z}}''_{n+1} - \tfrac{1}{6}h^3\tilde{\boldsymbol{z}}'''_{n+1} + \mathcal{O}(h^4)\right] - h^2\tilde{\boldsymbol{z}}''_{n+1} \\ &= \mathcal{O}(h^4), \end{aligned}$$

and thus we see that the Störmer method is of order 2. Both the methods (17.50) and (17.51) are two-step and explicit. Their implementation is likely to entail a very similar expense. Yet, (17.51) is of order 2, while (17.50) is just first-order – yet another example of a free lunch in numerical mathematics.[3]

Applying Störmer's method (17.51) to the semi-discretized scheme (17.45) results in the following two-step FD recursion, the *leapfrog* method,

$$u_\ell^{n+2} - 2u_\ell^{n+1} + u_\ell^n = \mu^2(u_{\ell-1}^{n+1} - 2u_\ell^{n+1} + u_{\ell+1}^{n+1}), \qquad n \ge 0, \qquad (17.52)$$

where $\mu = \Delta t/\Delta x$.

Being composed of a second-order space discretization and a second-order approximation in time, (17.52) is itself a second-order method. To analyse its stability we commence by assuming a Cauchy problem and proceeding as in our investigation of the Adams–Bashforth-like method (16.52) in Section 16.5. Relocating to Fourier space, straightforward manipulation results in

$$\hat{u}^{n+2} - 2(1 - 2\mu^2 \sin^2 \tfrac{1}{2}\theta)\hat{u}^{n+1} + \hat{u}^n = 0, \qquad \theta \in [0, 2\pi].$$

[3]To add insult to injury, (17.50) leads to an *unstable* FD scheme for all $\mu > 0$ (see Exercise 17.12).

All solutions of this three-term recurrence are uniformly bounded (and the underlying leapfrog method is stable) if and only if the zeros of the quadratic

$$\omega^2 - 2(1 - 2\mu^2 \sin^2 \tfrac{1}{2}\theta)\omega + 1 = 0$$

both reside in the closed unit disc for all $\theta \in [0, 2\pi]$. Although we could now use Lemma 8.3, it is perhaps easier to write the zeros explicitly,

$$\omega_\pm = 1 - 2\mu \sin^2 \tfrac{1}{2}\theta \pm 2\mathrm{i}\mu \sin \tfrac{1}{2}\theta \left(1 - \mu^2 \sin^2 \tfrac{1}{2}\theta\right)^{1/2}.$$

Provided $0 < \mu \le 1$, both ω_+ and ω_- are of unit modulus, while if μ exceeds unity then so does the magnitude of one of the zeros. We deduce that the leapfrog method is stable for all $0 < \mu \le 1$ insofar as the Cauchy problem is concerned. However, as in the Euler method for the advection equation in Section 17.1, we are allowed to infer from Cauchy to Dirichlet. For, suppose that the wave equation (17.7) is given for $0 \le x \le 1$ with zero boundary conditions $\varphi_0, \varphi_1 \equiv 0$. Since the leapfrog scheme (17.52) is explicit and each u_ℓ^{n+2} is coupled to just the nearest neighbours on the spatial grid, it follows that, as long as $u_0^n, u_{d+1}^n \equiv 0$, $n \ge 0$, we can assign arbitrary values to u_ℓ^n for $\ell \le -1$ and $\ell \ge d+1$ without any influence upon $u_1^n, u_2^n, \ldots, u_d^n$. In particular, we can embed a Dirichlet problem into a Cauchy one by padding with zeros, without amending the Euclidean norm of $\boldsymbol{u}^n$. Thus we have stability in a Dirichlet setting for $0 < \mu \le 1$.

To launch the first iteration of the leapfrog method we first need to derive the vector $\boldsymbol{u}^1$ by other means. Recall that both u and $\partial u/\partial t$ are specified along $x = 0$, and this can be exploited in the derivation of a second-order approximation at $t = \Delta t$. Let $\breve{u}_\ell^0 = g_1(\ell\Delta x)$, $\ell = 1, 2, \ldots, d$. Expanding about $(\ell\Delta x, 0)$ in a Taylor series, we have

$$u(\ell\Delta x, \Delta t) = u(\ell\Delta x, 0) + \Delta t \frac{\partial u(\ell\Delta x, 0)}{\partial t} + \tfrac{1}{2}(\Delta t)^2 \frac{\partial^2 u(\ell\Delta x, 0)}{\partial t^2} + \mathcal{O}\left((\Delta t)^3\right).$$

Substituting initial values and the second derivative from the SD scheme (17.45) results in

$$u_\ell^1 = u_\ell^0 + (\Delta t)\breve{u}_\ell^0 + \tfrac{1}{2}\mu^2(u_{\ell-1}^0 - 2u_\ell^0 + u_{\ell+1}^0), \qquad \ell = 1, 2, \ldots, d, \tag{17.53}$$

a scheme whose order is consistent with the leapfrog method (17.52).

The Dirichlet boundary conditions (17.42) are not the only interesting means for determining the solution of the wave equation. It is a well-known peculiarity of hyperbolic differential equations that, for every $t_0 \ge 0$, an initial condition along an interval $[x_-, x_+]$, say, determines uniquely the solution for all (x, t) in a set $\mathbb{D}_{x_-,x_+}^{t_0} \subset \mathbb{R} \times [t_0, \infty)$, the *domain of dependence.* For example, it is easy to deduce from (17.4) that the domain of dependence of the advection equation is the parallelogram

$$\mathbb{D}_{x_-,x_+}^{t_0} = \{(x, t) : t \ge t_0,\ x_- + t - t_0 \le x \le x_+ + t - t_0\}.$$

The domain of dependence of the wave equation, also known as the *Monge cone,* is the triangle

$$\mathbb{D}_{x_-,x_+}^{t_0} = \left\{(x, t) : t_0 \le t \le t_0 + \tfrac{1}{2}(x_+ - x_-),\ x_- + t - t_0 \le x \le x_+ - t + t_0\right\}.$$

In other words, provided that we specify $u(x,t_0)$ and $\partial u(x,t_0)/\partial t$ in $[x_-,x_+]$, the solution of (17.6) can be uniquely determined in the Monge cone without any need for boundary conditions.

This dependence of hyperbolic PDEs on local data has two important implications in their numerical analysis. Firstly, consider an explicit FD scheme of the form

$$u_\ell^{n+1} = \sum_{k=-\alpha}^{\beta} c_k(\mu) u_{\ell+k}^n, \qquad n \ge 0, \tag{17.54}$$

where $c_{-\alpha}(\mu), c_\beta(\mu) \neq 0$. Remember that each u_j^m approximates the solution at $(j\Delta x, m\mu\Delta x)$ and suppose that for sufficiently many ℓ and n in the region of interest it is true that

$$\Big(\ell\Delta x, (n+1)\mu\Delta x\Big) \not\in \mathbb{D}^{n\mu\Delta x}_{(\ell-\alpha)\Delta x,(\ell+\beta)\Delta x}.$$

As $\Delta x \to 0$, the points that we wish to determine stay persistently outside their domain of dependence, hence the numerical solution cannot converge there. In other words, a necessary condition for the stability of explicit FD schemes (and, for that matter, SD schemes) for hyperbolics is that μ should be small enough that the new point 'almost always' fits into the domain of dependence of the 'footprint'.[4] This is the celebrated *Courant–Friedrichs–Lewy* condition, usually known under its acronym, the CFL condition.

As an example, let us consider again the method (17.12) for the advection equation. Comparing the flow of information (see the diagrams soon after (17.15)) with the shape of the domain of dependence, we obtain

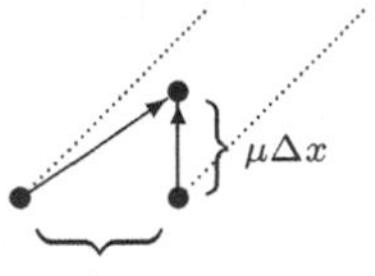

where the domain of dependence is enclosed between the parallel dotted lines. We deduce at once that stability requires $\mu \le 1$, which, of course, we already know. The CFL condition, however, is more powerful than that! Thus suppose that, in an explicit method for the advection equation u_ℓ^{n+s} depends on u_ℓ^{n+i} and $u_{\ell+1}^{n+i}$ for $i = 0, 1, \ldots, s-1$. No matter how we choose the coefficients, the above geometrical argument demonstrates at once that stability is inconsistent with $\mu > 1$.

In the case of the wave equation and the leapfrog method (17.52), the diagram is as follows:

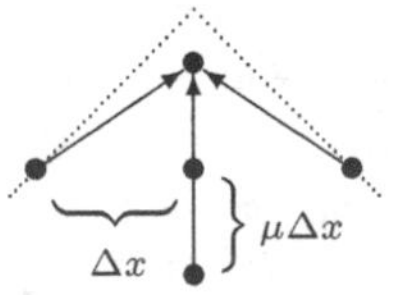

[4]We do not propose to elaborate on the meaning of 'almost always' here. It is enough to remark that for both the advection equation and the wave equation the condition is either obeyed for all ℓ and n or violated for all ℓ and n – a clear enough distinction.

and, again, we need $\mu \le 1$ otherwise the method overruns the Monge cone.

Suppose that the wave equation is specified with the initial conditions (17.41) but without boundary conditions. Since

$$\mathbb{D}^0_{0,1} = \left\{(x,t) : 0 \le t \le \tfrac{1}{2},\ t \le x \le 1-t\right\},$$

it makes sense to use the leapfrog method (17.52), in tandem with the starting method (17.53), to derive the solution there. Each new point depends on its immediate neighbours at the previous time level, and this, together with the value of $\mu \in (0,1]$, restricts the portion of $\mathbb{D}^0_{0,1}$ that can be reached with the leapfrog method. This is illustrated in the following diagram, for $\mu = \frac{3}{5}$:

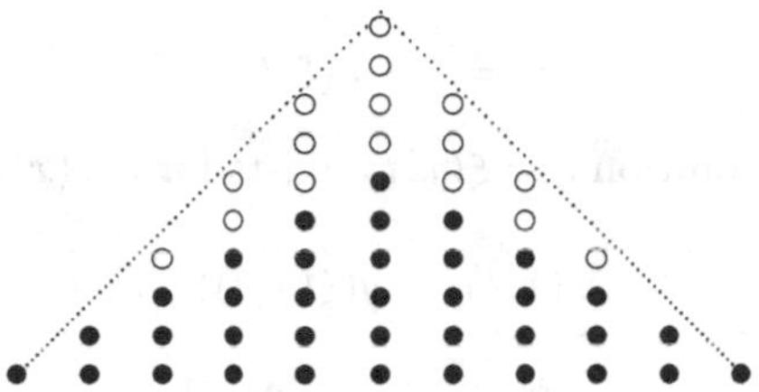

Here, the reachable points are shown as solid and we can observe that they leave a portion of the Monge cone out of reach of the numerical method.

17.5 The Burgers equation

The Burgers equation (17.10) is the simplest *nonlinear hyperbolic conservation law* and its generalization gives us the *Euler equations* of inviscid compressible fluid dynamics. We have already seen in Figs 17.1 and 17.2 that its solution displays strange behaviour and might generate discontinuities from an arbitrarily smooth initial condition. Before we take up the challenge of numerically solving it, let us first devote some attention to the analytic properties of the solution.

◇ **Why analysis?** The assembling of analytic information before even considering computation is a hallmark of good numerical analysis, while the ugly instinct of 'discretizing everything in sight and throwing it on the nearest computer' is the worst kind of advice. Partial differential equations are complicated constructs and, as soon as nonlinearities are allowed, they may exhibit a great variety of difficult phenomena. Before we even pose the question of how well a numerical algorithm is performing, we need to formulate more precisely what 'well' means!

In reality, it is a two-way traffic between analysis and computation since, when it comes to truly complicated equations, the best way for a pure mathematician to guess what should be proved is by using numerical experimentation. Recent advances in the understanding of nonlinear behaviour in PDEs have been not the work of narrow specialists in hermetically sealed intellectual compartments but the outcome of collaboration between mathematical analysts, computational experts, applied mathematicians and even experimenters in their laboratories. There is little doubt that future advances will increasingly depend on a greater permeability of discipline boundaries. ◇

Throughout this section we assume a Cauchy problem,

$$u(x,0) = g(x), \quad -\infty < x < \infty, \qquad \text{where} \qquad \int_{-\infty}^{\infty} [g(x)]^2 \, \mathrm{d}x < \infty$$

and g is differentiable, since this allows us to disregard boundaries and simplify the notation. As a matter of fact, introducing Dirichlet boundary conditions leaves most of our conclusions unchanged.

Let us choose $(x,t) \in \mathbb{R} \times [0,\infty)$ and consider the algebraic equation

$$x = \xi + g(\xi)t. \tag{17.55}$$

Supposing that a unique solution $\xi = \xi(x,t)$ exists for all (x,t) in a suitable subset of $\mathbb{R} \times [0,\infty)$, we set there

$$w(x,t) := g(\xi(x,t)).$$

Therefore

$$\begin{aligned} \frac{\partial w}{\partial x} &= g'(\xi)\frac{\partial \xi}{\partial x}, \\ \frac{\partial w}{\partial t} &= g'(\xi)\frac{\partial \xi}{\partial t} \end{aligned} \tag{17.56}$$

(it can be easily proved that, provided the solution of (17.55) is unique, the function ξ is differentiable with respect to x and t). The partial derivatives of ξ can be readily obtained by differentiating (17.55):

$$\begin{aligned} 1 &= \frac{\partial \xi}{\partial x} + g'(\xi)t\frac{\partial \xi}{\partial x} \qquad &\Rightarrow \qquad \frac{\partial \xi}{\partial x} &= \frac{1}{1+g'(\xi)t}, \\ 0 &= \frac{\partial \xi}{\partial t} + g'(\xi)t\frac{\partial \xi}{\partial t} + g(\xi) \qquad &\Rightarrow \qquad \frac{\partial \xi}{\partial t} &= -\frac{g(\xi)}{1+g'(\xi)t}. \end{aligned}$$

Substituting in (17.56) gives

$$\frac{\partial w}{\partial t} + \frac{1}{2}\frac{\partial w^2}{\partial x} = \left[\frac{\partial \xi}{\partial t} + g(\xi)\frac{\partial \xi}{\partial x}\right] g'(\xi) = \left[-\frac{g(\xi)}{1+g'(\xi)t} + \frac{g(\xi)}{1+g'(\xi)t}\right] g'(\xi) = 0,$$

thus proving that the function w obeys the Burgers equation.

Since the solution of (17.55) for $t = 0$ is $\xi = x$, it follows that $w(x,0) = g(x)$. In other words, the function w obeys both the correct equation and the required initial conditions, hence within its domain of definition it coincides with u. (We have just used – without a proof – the uniqueness of the solution of (17.10). In fact, and as we are about to see, the solution need not be unique for general (x,t), but it is so within the domain of definition of w.)

Let us examine our conclusion that $u(x,t) = g(\xi(x,t))$ in a new light. We choose an arbitrary $\xi \in \mathbb{R}$. It follows from (17.55) that for every $0 \le t < t_\xi$, say, it is true that $u(\xi + g(\xi)t, t) = g(\xi)$. In other words, at least for a short while, the solution of the Burgers equation is constant along a straight line.

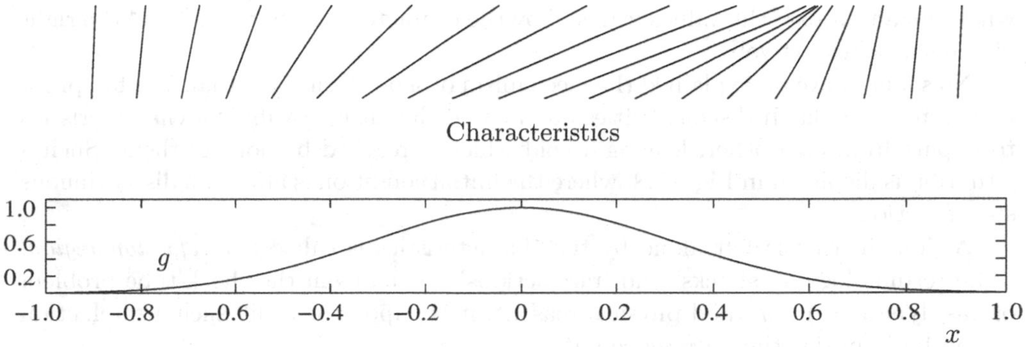

Figure 17.7 Characteristics for $g(x) = \mathrm{e}^{-5x^2}$ and shock formation.

We have already seen an example of similar behaviour in the case of the advection equation, since, according to (17.4), its solution is constant along straight lines of slope +1. The crucial difference is that for the Burgers equation the slopes of the straight lines depend on an initial condition g and, in general, vary with ξ. This immediately creates a problem: what if the straight lines collide? At such a point of collision, which might occur for an arbitrarily small $t \geq 0$, we cannot assign an unambiguous value to the solution – it has a discontinuity there. Figure 17.7 displays the straight lines (under their proper name, *characteristics*) for the function $g(x) = \mathrm{e}^{-5x^2}$. The formation of a discontinuity is apparent and it is easy to ascertain (cf. Exercise 17.16) that it starts to develop from the very beginning, at the point $x = 0$.

Another illustration of how discontinuities develop in the solution of the Burgers equation is seen in Figs. 17.1 and 17.2.

A discontinuity that originates in a collision of characteristics is called a *shock* and its position is determined by the requirement that characteristics must always flow *into* a shock and never emerge from it. Representing the position of a shock at time t by $\eta(t)$, say, it is not difficult to derive from elementary geometric considerations the *Rankine–Hugoniot* condition

$$\eta'(t) = \tfrac{1}{2}(u_{\mathrm{L}} + u_{\mathrm{R}}), \qquad (17.57)$$

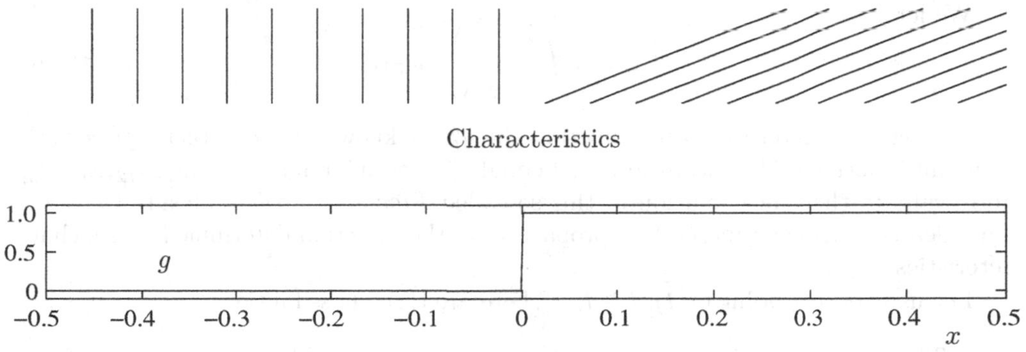

Figure 17.8 Characteristics for a step function and formation of a rarefaction region.

where u_{L} and u_{R} are the values 'carried' by the characteristics to the left and the right of the shock respectively.

No sooner have we explained the mechanism of shock formation than another problem comes to light. If discontinuities are allowed then it is possible for characteristics to depart from each other, leaving a void which is reached by none of them. Such a situation is displayed in Fig. 17.8, where the initial condition is already a discontinuous step function.

A domain that is left alone by the characteristics is called a *rarefaction region.* (The terminology of 'shocks' and 'rarefactions' originates in the shock-tube problem of gas dynamics.) For valid physical reasons, it is important to fill such a rarefaction region by imposing the *entropy condition*

$$\frac{1}{2}\frac{\partial u^2}{\partial t}+\frac{1}{3}\frac{\partial u^3}{\partial x}\le 0. \tag{17.58}$$

The origin of (17.58) is the Burgers equation with *artificial viscosity,*

$$\frac{\partial u}{\partial t}+\frac{1}{2}\frac{\partial u^2}{\partial x}=\nu\frac{\partial^2 u}{\partial x^2},$$

where $\nu>0$ is small. The addition of the parabolic term $\nu\partial^2u/\partial x^2$ causes dissipation and the solution is smooth, with neither shocks nor rarefaction regions. Letting $\nu\to 0$, it is possible to derive the inequality (17.58). More importantly, it is possible to prove that, subject to the Rankine–Hugoniot condition (17.57) and the entropy condition (17.58), the Burgers equation possesses a unique solution.

There are many numerical schemes for solving the Burgers equation but we restrict our exposition to perhaps the simplest algorithm that takes on board the special structure of (17.10) – the *Godunov* method.

The main idea behind this approach is to approximate locally the solution by a piecewise-constant function. Since, as we will see soon, the Godunov method amends the step size, we can no longer assume that Δt is constant. Instead, we denote by Δt_n the step that takes us from the nth to the $(n+1)$th time level, $n\ge 0$, and let $u_\ell^n\approx u(\ell\Delta x,t_n)$, where $t_0=0$ and $t_{n+1}=t_n+\Delta t_n$, $n\ge 0$.

We let

$$u_\ell^0=\frac{1}{\Delta x}\int_{(\ell-1/2)\Delta x}^{(\ell+1/2)\Delta x} g(x)\,\mathrm{d}x \tag{17.59}$$

for all ℓ-values of interest. Supposing that the u_ℓ^n are known, we construct a piecewise-constant function $w^{[n]}(\cdot,t_n)$ by letting it equal u_ℓ^n in each interval $\mathbb{I}_\ell:=(x_{\ell-1/2},x_{\ell+1/2}]$ and evaluate the exact solution of this so-called *Riemann problem* ahead of $t=t_n$. The idea is to let each interval $\mathbb{I}_\ell$ 'propagate' in the direction determined by its characteristics.

Let us choose a point (x,t), $t\ge t_n$. There are three possibilities.

(1) There exists a unique ℓ such that the point is reached by a characteristic from $\mathbb{I}_\ell$. Since characteristics propagate constant values, the solution of the Riemann problem at this point is u_ℓ^n.

(2) There exists a unique ℓ such that the point is reached by characteristics from the intervals $\mathbb{I}_\ell$ and $\mathbb{I}_{\ell+1}$. In this case, as the two intervals 'propagate' in time, they are separated by a shock. It is trivial to verify from (17.57) that the shock advances along a straight line that commences at $(\ell+\frac{1}{2})\Delta x$ and whose slope is the average of the slopes in $\mathbb{I}_\ell$ and $\mathbb{I}_{\ell+1}$ – in other words, it is $\frac{1}{2}(u_\ell^n + u_{\ell+1}^n)$. Let us denote this line by ρ_ℓ.[5] The value at (x,t) is u_ℓ^n if $x < \rho_\ell(t)$ and $u_{\ell+1}^n$ if $x > \rho_\ell(t)$. (We disregard the case when $x = \rho_\ell(t)$ and the point resides on the shock, since it makes absolutely no difference to the algorithm.)

(3) Characteristics from more than two intervals reach the point (x,t). In this case we cannot assign a value to the point.

Simple geometrical considerations demonstrate that case (3), which we must avoid, occurs (for some x) for $t > \tilde{t}$, where $\tilde{t} > t_n$ is the least solution of the equation $\rho_\ell(t) = \rho_{\ell+1}(t)$ for some ℓ. This becomes obvious upon an examination of Fig. 17.9.

Let us consider the vertical lines rising from the points $(\ell+\frac{1}{2})\Delta x$. Unless the original solution is identically zero, sooner or later one such line is bound to hit one of the segments ρ_j. We let $\breve{t}$ be the time at which the first such encounter takes place, choose $t_{n+1} \in (t_n, \breve{t}\,]$ and set $\Delta t_n = t_{n+1} - t_n$. Since $t_{n+1} \in (t_n, \tilde{t}\,]$ (see Fig. 17.9), cases (1) and (2) can be used to construct a unique solution $w^{[n]}(x,t)$ for all $t_n \le t \le t_{n+1}$. We choose the u_ℓ^{n+1} as averages of $w^{[n]}(\,\cdot\,, t_{n+1})$ along the intervals $\mathbb{I}_\ell$,

$$u_\ell^{n+1} = \frac{1}{\Delta x}\int_{(\ell-1/2)\Delta x}^{(\ell+1/2)\Delta x} w^{[n]}(x, t_{n+1})\,\mathrm{d}x. \tag{17.60}$$

Our description of the Godunov method is complete, except for an important remark: the integral in (17.60) can be calculated with great ease. Disregarding shocks, the function $w^{[n]}$ obeys the Burgers equation for $t \in [t_n, t_{n+1}]$. Therefore, integrating in t,

$$\frac{\partial w^{[n]}}{\partial t} + \frac{1}{2}\frac{\partial [w^{[n]}]^2}{\partial x} = 0 \qquad \Rightarrow \qquad w^{[n]}(x, t_{n+1}) = w^{[n]}(x, t_n) - \frac{1}{2}\int_{t_n}^{t_{n+1}} \frac{\partial [w^{[n]}(x,t)]^2}{\partial x}\,\mathrm{d}t.$$

Substitution into (17.60) results in

$$u_\ell^{n+1} = \frac{1}{\Delta x}\int_{(\ell-1/2)\Delta x}^{(\ell+1/2)\Delta x} \left\{ w^{[n]}(x, t_n) - \frac{1}{2}\int_{t_n}^{t_{n+1}} \frac{\partial [w^{[n]}(x,t)]^2}{\partial x}\,\mathrm{d}t \right\} \mathrm{d}x.$$

Since the u_ℓ^n have been obtained by an averaging procedure as given in (17.60) (this is the whole purpose of (17.59)), we have, after changing of the order of integration,

$$\begin{aligned} u_\ell^{n+1} &= u_\ell^n - \frac{1}{2\Delta x}\int_{t_n}^{t_{n+1}} \int_{(\ell-1/2)\Delta x}^{(\ell+1/2)\Delta x} \frac{\partial [w^{[n]}(x,t)]^2}{\partial x}\,\mathrm{d}x\,\mathrm{d}t \\ &= u_\ell^n - \frac{1}{2\Delta x}\int_{t_n}^{t_{n+1}} \left\{ \left[w^{[n]}\left((\ell+\tfrac{1}{2})\Delta x, t\right)\right]^2 - \left[w^{[n]}\left((\ell-\tfrac{1}{2})\Delta x, t\right)\right]^2 \right\} \mathrm{d}t. \end{aligned}$$

[5] Not every ρ_ℓ is a shock, but this makes no difference to the method.

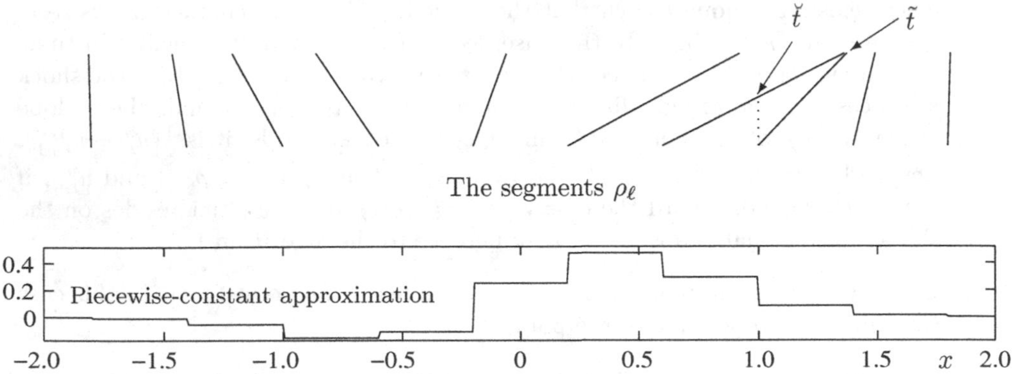

Figure 17.9 The graph shows a piecewise-constant approximation; the upper diagram shows the line segments ρ_ℓ and the first vertical line to collide with ρ_ℓ (dotted).

Let us now recall our definition of t_{n+1}. No vertical line segments $((\ell + \frac{1}{2})\Delta x, t)$, $t \in [t_n, t_{n+1}]$, may cross the discontinuities ρ_j, therefore the value of $w^{[n]}$ across each such segment is constant – equalling either u_ℓ^n or $u_{\ell+1}^n$ (depending on the slope of ρ_ℓ: if it points rightwards it is u_ℓ^n, otherwise $u_{\ell+1}^n$). Let us denote this value by $\chi_{\ell+1/2}$; then

$$u_\ell^{n+1} = u_\ell^n - \tfrac{1}{2}\mu_n(\chi_{\ell+1/2}^2 - \chi_{\ell-1/2}^2), \tag{17.61}$$

where

$$\mu_n := \frac{\Delta t_n}{\Delta x}.$$

The Godunov method is a first-order approximation to the solution of the Burgers equation, since the only error that we have incurred comes from replacing the values along each step by piecewise-constant approximants. It satisfies the Rankine–Hugoniot condition and it is possible to prove that it is stable. However, more work, outside the scope of this exposition, is required to ensure that the entropy condition (17.58) is obeyed as well.

It is possible to generalize the Godunov method to more complicated nonlinear hyperbolic conservation laws

$$\frac{\partial u}{\partial t} + \frac{\partial f(u)}{\partial x} = 0,$$

where f is a general differentiable function, as well as to systems of such equations. In each case we obtain a recursion of the form (17.61), except that the definition of the *flux* $\chi_{\ell+1/2}$ needs to be amended and is slightly more complicated. In the special case $f(u) = u$ we are back to the advection equation and the Godunov method (17.61) becomes the familiar scheme (17.12) with $0 < \mu \le 1$.

Comments and bibliography

Fluid and gas dynamics, relativity theory, quantum mechanics, aerodynamics – this is just a partial list of subjects that need hyperbolic PDEs to describe their mathematical foundations.

Such equations – the Euler equations of inviscid compressible flow, the Schrödinger equation of wave mechanics, Einstein's equations of general relativity etc. – are nonlinear and generally multivariate and multidimensional, and their numerical solution presents a formidable challenge. This perhaps explains the major effort that has gone into the computation of hyperbolic PDEs in the last few decades. A bibliographical journey through the hyperbolic landscape might commence with texts on their theory, mainly in a nonlinear setting – thus Drazin & Johnson (1988) on solitons, Lax (1973) and LeVeque (1992) on conservation laws and Whitham (1974) for a general treatment of wave theory. The next destination might be the classic volume of Richtmyer & Morton (1967), still the best all-round volume on the foundations of the numerical treatment of evolutionary PDEs, followed by more specialized sources, LeVeque (1992), Morton & Sonar (2007) or Hirsch (1988) on numerical conservation laws. Finally, there is an abundance of texts on themes that bear some relevance to the subject matter: Gustaffson *et al.* (1972) on the influence of boundary conditions on stability; Trefethen (1992) on an alternative treatment, by means of pseudospectra, of numerical stability in the absence of normalcy; Iserles & Nørsett (1991) on how to derive optimal schemes for the advection equation, a task that bears a striking similarity to some of the themes from Chapter 4; and Davis (1979) on circulant matrices.

As soon as we concern ourselves with computational wave mechanics (which, in a way, is exactly what the numerical solution of hyperbolic PDEs is all about), there are additional considerations besides order and stability. In Fig. 17.10 we display a numerical solution of the advection equation (17.1) with initial condition

$$g(x) = e^{-100(x-1/2)^2} \sin 20\pi x, \qquad -\infty \le x \le \infty. \tag{17.62}$$

The function g is a *wave packet* – a highly oscillatory wave modulated by a sharply decaying exponential so that, for all intents and purposes, it vanishes outside a small support. The exact solution of (17.1) and (17.62) at time t is the function g, unilaterally translated rightwards by a distance t. In Fig. 17.10 we can observe what happens to the wave packet under the influence of discretization by three *stable* FD schemes. Firstly, the leapfrog scheme evidently moves the wave packet at the wrong speed, distorting it in the process. The energy of the packet – that is, the Euclidean norm of the solution – stays constant, as it does in the exact solution: the leapfrog is a *conservative* method. However, the energy is transported at an altogether wrong speed. This wrong speed of propagation depends on the wavenumber: the higher the oscillation (in comparison with Δx – recall from Chapter 13 that frequencies larger than $\pi/\Delta x$ are 'invisible' on a grid scale), the more false the reading and, for sufficiently high frequencies, a wave can be transported in the wrong direction altogether. Of course, we can always decrease Δx so as to render the frequency of any particular wave small on the grid scale, although this, obviously, increases the cost of computation. Unfortunately, if the initial condition is discontinuous then its Fourier transform (i.e., its decomposition as a linear combination of periodic 'waves') contains all frequencies that are 'visible' in a grid and this cannot be changed by decreasing Δx; see Fig. 17.11.

The behaviour of the Lax–Wendroff method as shown in Fig. 17.10 poses another difficulty. Not only does the wave packet lag somewhat; the main problem is that it has almost disappeared! Its energy has decreased by about a factor 3 and this is unacceptable. Lax–Wendroff is *dissipative,* rather than conservative. The dissipation is governed by the size of Δx and it disappears as $\Delta x \to 0$, yet it might be highly problematic in some applications.

Unlike either leapfrog or Lax–Wendroff, the angled derivative method (17.28) displays the correct qualitative behaviour: virtually no dissipation; little dispersion; high frequencies are transported at roughly the right speed (Trefethen, 1982).

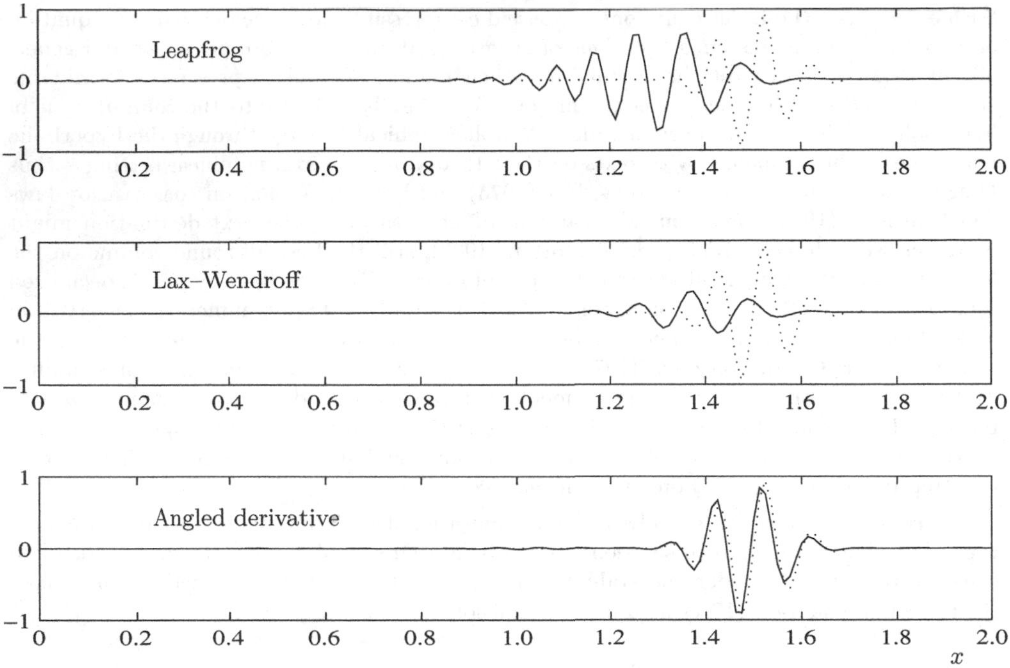

Figure 17.10 Numerical wave propagation by three FD schemes. The dotted line presents the position of the exact solution at time $t = 1$. We have used $\Delta x = \frac{1}{80}$ and $\mu = \frac{2}{3}$.

A similar picture emerges from Fig. 17.11, where we have displayed the evolution of the piecewise-constant function

$$g(x) = \begin{cases} 1, & \frac{1}{4} \leq x < \frac{3}{4}, \\ 0, & \text{otherwise.} \end{cases}$$

Leapfrog emerges the worst, both degrading the shock front and transporting some waves too slowly or, even worse, in the wrong direction. Lax–Wendroff is much better: although it also smoothes the sharp shock front, the 'missing mass' is simply dissipated, rather than reappearing in the wrong place. The angled derivative method displays the sharpest profile, but the *quid pro quo* is spurious oscillations at high wavenumbers. This dead heat between Lax–Wendroff and angled derivative emphasizes that no method is perfect and different methods often possess contrasting advantages and disadvantages.

By this stage, the reader should be well aware why the correct propagation of shock fronts is so important. Fig. 17.11 reaffirms a principle that underlies much of the discussion of hyperbolic PDEs: methods should follow characteristics.

In the particular context of conservation laws, 'following characteristics' means upwinding. This is easy for the advection equation but becomes a more formidable task when the characteristics change direction, e.g. for the Burgers equation (17.10). Seen in this light, the Godunov method from Section 17.5 is all about the local determination of the upwind

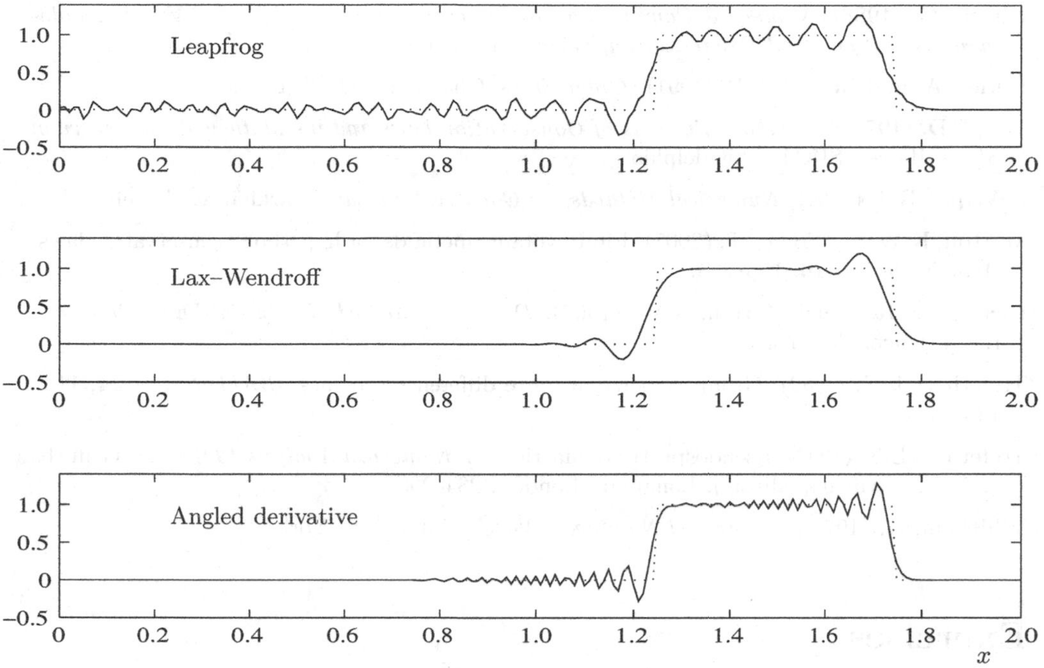

Figure 17.11 Numerical shock propagation by three FD schemes. The dotted line presents the position of the exact solution at time $t = 1$. We have used $\Delta x = \frac{1}{100}$ and $\mu = \frac{2}{3}$.

direction. Another popular choice of an upwinding technique is the use of *Engquist–Osher switches*

$$f_-(y) := [\min\{y,0\}]^2, \quad f_+(y) := [\max\{y,0\}]^2, \qquad y \in \mathbb{R},$$

to form the SD scheme

$$u'_\ell + \frac{1}{\Delta x}[\Delta_+ f_-(u_\ell) + \Delta_- f_+(u_\ell)] = 0.$$

If $u_{\ell-1}, u_\ell$ and $u_{\ell+1}$ are all positive and the characteristics propagate rightwards then we have $\Delta_+ f_-(u_\ell) = 0$ and $\Delta_- f_+(u_\ell) = [u_\ell]^2 - [u_{\ell-1}]^2$, while if all three values are negative then $\Delta_+ f_-(u_\ell) = [u_{\ell+1}]^2 - [u_\ell]^2$ and $\Delta_- f_+(u_\ell) = 0$. Again, the scheme determines an upwind direction on a local basis.

Numerical methods for nonlinear conservation laws are among the great success stories of modern numerical analysis. They come in many shapes and sizes – finite differences, finite elements, particle methods, finite volume methods, spectral methods, ... , but all good schemes have in common elements of upwinding and attention to shock propagation.

Davis, P.J. (1979), *Circulant Matrices,* Wiley, New York.

Drazin, P.G. and Johnson, R.S. (1988), *Solitons: An Introduction,* Cambridge University Press, Cambridge.

Gustafsson, B., Kreiss, H.-O. and Sundström, A. (1972), Stability theory of difference approximations for mixed initial boundary value problems, *Mathematics of Computation* **26,** 649–686.

Hirsch, C. (1988), *Numerical Computation of Internal and External Flows, Vol. 1: Fundamentals of Numerical Discretization,* Wiley, Chichester.

Iserles, A. and Nørsett, S.P. (1991), *Order Stars,* Chapman & Hall, London.

Lax, P.D. (1973), *Hyperbolic Systems of Conservation Laws and the Mathematical Theory of Shock Waves,* SIAM, Philadelphia.

LeVeque, R.J. (1992), *Numerical Methods for Conservation Laws,* Birkhäuser, Basel.

Morton, K.W. and Sonar, T. (2007), Finite volume methods for hyperbolic conservation laws, *Acta Numerica* **16**, 155–238.

Richtmyer, R.D. and Morton, K.W. (1967), *Difference Methods for Initial-Value Problems,* Interscience, New York.

Trefethen, L.N. (1982), Group velocity in finite difference schemes, *SIAM Review* **24**, 113–136.

Trefethen, L.N. (1992), Pseudospectra of matrices, in *Numerical Analysis 1991* (D.F. Griffiths and G.A. Watson, editors), Longman, London, 234–266.

Whitham, G. (1974), *Linear and Nonlinear Waves,* Wiley, New York.

Exercises

17.1 In Section 17.1 we proved that the Fourier stability analysis of the Euler method (17.12) for the advection equation can be translated from the real axis (where it is a Cauchy problem) to the interval $[0,1]$ and a zero boundary condition (17.3). Can we use the same technique of proof to analyse the stability of the Euler method (16.7) for the diffusion equation (16.1) with zero Dirichlet boundary conditions (16.3)?

17.2* Let α and β be nonnegative integers, $\alpha+\beta \geq 1$, and set

$$p_{\alpha,\beta}(z) = \sum_{k=-\alpha,\ k\neq 0}^{\beta} \breve{a}_k z^k + q_{\alpha,\beta}, \qquad z \in \mathbb{C},$$

where

$$\breve{a}_k = \frac{(-1)^{k-1}}{k} \frac{\alpha!\beta!}{(\alpha+k)!(\beta-k)!}, \qquad k = -\alpha, -\alpha+1, \ldots, \beta, \quad k \neq 0,$$

and the constant $q_{\alpha,\beta}$ is such that $p_{\alpha,\beta}(1) = 0$.

a Evaluate $p'_{\alpha,\beta}$, proving that

$$p'_{\alpha,\beta}(z) = \frac{1}{z} + \mathcal{O}\left(|z-1|^{\alpha+\beta}\right), \qquad z \to 1.$$

b Integrate the last expression, thereby demonstrating that

$$p_{\alpha,\beta}(z) = \ln z + \mathcal{O}\left(|z-1|^{\alpha+\beta+1}\right), \qquad z \to 1.$$

c Determine the order of the SD scheme

$$u'_\ell + \frac{1}{\Delta x}\left(\sum_{k=-\alpha}^{-1} \breve{a}_k u_{\ell+k} + q_{\alpha,\beta} u_\ell + \sum_{k=1}^{\beta} \breve{a}_k u_{\ell+k}\right) = 0$$

for the advection equation (17.1).

17.3 Show that the leapfrog method (17.27) recovers the exact solution of the advection equation when the Courant number μ equals unity. (*It should be noted that this is of little or no relevance to the solution of the system* (17.9) *or to nonlinear equations.*)

17.4 Analyse the order of the following FD methods for the advection equation:

a the angled derivative scheme (17.28);

b the Lax–Wendroff scheme (17.29);

c the *Lax–Friedrichs* scheme

$$u_\ell^{n+1} = \tfrac{1}{2}(1+\mu)u_{\ell-1}^n + \tfrac{1}{2}(1-\mu)u_{\ell+1}^n, \qquad n \geq 0.$$

17.5 Carefully justifying your arguments, use the eigenvalue technique from Section 16.4 to prove that the SD scheme (17.23) is stable.

17.6 Find a third-order upwind SD method (17.17) with $\alpha = 3$, $\beta = 0$ and prove that it is unstable for the Cauchy problem.

17.7 Determine the range of Courant numbers μ for which

a the leapfrog scheme (17.27) and

b the angled derivative scheme (17.28)

are stable for the Cauchy problem. Use the method of proof from Section 17.1 to show that the result for the angled derivative scheme can be generalized from the Cauchy problem to a zero Dirichlet boundary condition.

17.8 Find the order of the *box method*

$$(1-\mu)u_{\ell-1}^{n+1} + (1+\mu)u_\ell^{n+1} = (1+\mu)u_{\ell-1}^n + (1-\mu)u_\ell^n, \qquad n \geq 0.$$

Determine the range of Courant numbers μ for which the method is stable.

17.9 Show that the transpose of a circulant matrix is itself a circulant and use this to prove that every circulant matrix is normal.

17.10 Find the order of the method

$$u_{j,\ell}^{n+1} = \tfrac{1}{2}\mu(\mu-1)u_{j+1,\ell+1}^n + (1-\mu^2)u_{j,\ell}^n + \tfrac{1}{2}\mu(\mu+1)u_{j-1,\ell-1}^n, \qquad n \geq 0,$$

for the bivariate advection equation (17.6). Here $\Delta x = \Delta y$, $\Delta t = \mu\Delta x$ and $u_{j,\ell}^n$ approximates $u(j\Delta x, \ell\Delta x, n\Delta t)$.

17.11 Determine the order of the *Numerov* method

$$z_{n+2} - 2z_{n+1} + z_n = \tfrac{1}{12}h^2 \left[\boldsymbol{f}(t_{n+2}, z_{n+2}) \right.$$
$$\left. + 10\boldsymbol{f}(t_{n+1}, z_{n+1}) + \boldsymbol{f}(t_n, z_n) \right], \qquad n \geq 0,$$

for the second-order ODE system (17.48).

17.12 Application of the method (17.50) to the second-order ODE system (17.45) results in a two-step FD scheme for the wave equation. Prove that this method is unstable for all $\mu > 0$ by two alternative techniques:

a by Fourier analysis;

b directly from the CFL condition.

17.13 Determine the order of the scheme

$$u_\ell^{n+2} - u_\ell^{n+1} + u_\ell^n = \tfrac{1}{12}\mu^2(-u_{\ell-2}^{n+1} + 16u_{\ell-1}^{n+1}$$
$$- 30u_\ell^{n+1} + 16u_{\ell+1}^{n+1} - u_{\ell+2}^{n+1}), \qquad n \geq 0,$$

for the solution of the wave equation (17.8) and find the range of Courant numbers μ for which the method is stable for the Cauchy problem.

17.14 Let $\tilde{u}_\ell^n = u(\ell\Delta x, n\Delta x)$, $\ell = 1, 2, \ldots, d+1$, $n \geq 0$, where $(d+1)\Delta x = 1$ and u is the solution of the wave equation (17.8) with the initial conditions (17.41) and the boundary conditions (17.42).

a Prove the identity

$$\tilde{u}_\ell^{n+2} - \tilde{u}_\ell^n = \tilde{u}_{\ell+1}^{n+1} - \tilde{u}_{\ell-1}^{n+1}, \qquad \ell = 1, 2, \ldots, d, \quad n \geq 0.$$

(*Hint: You might use without proof the d'Alembert solution of the wave equation*

$$u(x, t) = g(x - t) + h(x + t)$$

for all $0 \leq x \leq 1$ *and* $t \geq 0$.)

b Suppose that the wave equation is solved using the FD method (17.52) with Courant number $\mu = 1$ and write $e_\ell^n := u_\ell^n - \tilde{u}_\ell^n$, $\ell = 0, 1, \ldots, d+1$, $n \geq 0$. Prove by induction that

$$e_\ell^n = \sum_{j=0}^{n-1} (-1)^j e_{\ell+n-2j-1}^1, \qquad \ell = 1, 2, \ldots, d, \quad n \geq 1,$$

where we let $e_\ell^1 = 0$ for $\ell \notin \{1, 2, \ldots, d\}$.

c The leapfrog method cannot be used to obtain u_ℓ^1 so, instead, we use the scheme (17.53). Assuming that it is known that $|e_\ell^1| \leq \varepsilon$, $\ell = 1, 2, \ldots, d$, prove that

$$|e_\ell^n| \leq \min\{n, \lfloor \tfrac{1}{2}(d+1) \rfloor\}\varepsilon, \qquad n \geq 0. \tag{17.63}$$

(*Naive considerations provide a geometrically increasing upper bound on the error, but* (17.63) *demonstrates that it is much too large and that the increase in the error is at most linear.*)

17.15 Prove that it is impossible for any explicit FD method (17.54) for the advection equation to be convergent for $\mu > \alpha$.

17.16 Let g be a continuous function and suppose that $x_0 \in \mathbb{R}$ is its strict (local) maximum. The Burgers equation (17.10) is solved with the initial condition $u(x,0) = g(x)$, $x \in \mathbb{R}$. Prove that a shock propagates from the point x_0.

17.17 Show that the entropy condition is satisfied by a solution of the Burgers equation as an equality in any portion of $\mathbb{R} \times [0, \infty)$ that is neither a shock nor a rarefaction region. (*Hint: Multiply the equation by u.*)

17.18* Amend the Godunov method from Section 17.5 to solve the advection equation (17.1) rather than the Burgers equation. Prove that the result is the Euler scheme (17.12) with μ automatically restricted to the stable range $(0, 1]$.

Appendix

Bluffer's guide to useful mathematics

HEALTH WARNING

This is not a review of undergraduate mathematics *or a distillation of the wisdom of many lecture courses into a few pages. Certainly, nobody should use it to understand new material. Mathematics is not learnt from crib-sheets and brief compendia but by careful study of definitions, theorems and – most importantly, perhaps – proofs, by elucidating the intuition behind ideas and grasping the interconnectedness between what might seem disparate concepts at first glance. There are no shortcuts and no cherry-tasting knowledge capsules to help you along your path…*

A conscious attempt has been made throughout the volume not to take for granted any knowledge that an advanced mathematics undergraduate is unlikely to possess. If we need it, we explain it. However, every book has to start from somewhere.

Unless you have a basic knowledge of the first two years of university or college mathematics, this appendix will not help you and, indeed, this is the wrong book for you. However, it is not unusual for students to attend a lecture course, study material, absorb it, pass an exam with flying colours – and yet, a year or two later, a concept is perhaps not entirely forgotten but resides so deep in the recesses of memory that it cannot be used here and now. In these circumstances a virtuous reader consults another textbook or perhaps her lecture notes. A less virtuous reader usually means to do so – not just yet – but in the meantime plunges ahead with a decreased level of comprehension. This appendix has been written in recognition of the poverty and scarcity of virtue.

While trying to read a mathematical textbook, nothing can be worse than gradually losing the thread, progressively understanding less and less. This can happen either because the reader fails to understand the actual material – and the fault may well rest with the author – or when she encounters unfamilar mathematical constructs.

If in this volume you occasionally come across a mathematical concept and, for the life of you, simply cannot recall exactly what it means (or perhaps are not sure of the finer details of its definition), do glance in this appendix – you might find it here! However, if these glances become a habit, rather than an exception, perhaps you had better use a proper textbook!

There are two sections to this appendix, one on linear algebra and the second on analysis. Neither is complete – they both endeavour to answer possible queries arising from this book, rather than providing a potted summary of a subject.

There is nothing on basic calculus. Unless you are familiar with calculus then, I am afraid, you are trying to dance the samba before you can walk.

A.1 Linear algebra

A.1.1 Vector spaces

A.1.1.1 A *vector space* or a *linear space* $\mathbb{V}$ over the field $\mathbb{F}$ (which in our case will be either $\mathbb{R}$, the reals, or $\mathbb{C}$, the complex numbers) is a set of elements closed with respect to addition, i.e.,

$$\boldsymbol{x}_1, \boldsymbol{x}_2 \in \mathbb{V} \qquad \text{implies} \qquad \boldsymbol{x}_1 + \boldsymbol{x}_2 \in \mathbb{V},$$

and multiplication by a *scalar,* i.e.,

$$\alpha \in \mathbb{F},\ \boldsymbol{x} \in \mathbb{V} \qquad \text{implies} \qquad \alpha\boldsymbol{x} \in \mathbb{V}.$$

Addition obeys the axioms of an abelian group: $\boldsymbol{x}_1 + \boldsymbol{x}_2 = \boldsymbol{x}_2 + \boldsymbol{x}_1$ for all $\boldsymbol{x}_1, \boldsymbol{x}_2 \in \mathbb{V}$ (commutativity); $(\boldsymbol{x}_1 + \boldsymbol{x}_2) + \boldsymbol{x}_3 = \boldsymbol{x}_1 + (\boldsymbol{x}_2 + \boldsymbol{x}_3)$, $\boldsymbol{x}_1, \boldsymbol{x}_2, \boldsymbol{x}_3 \in \mathbb{V}$ (associativity); there exists an element $\mathbf{0} \in \mathbb{V}$ (the zero element) such that $\boldsymbol{x} + \mathbf{0} = \boldsymbol{x}$, $\boldsymbol{x} \in \mathbb{V}$; and for every $\boldsymbol{x}_1 \in \mathbb{V}$ there exists an element $\boldsymbol{x}_2 \in \mathbb{V}$ (the inverse) such that $\boldsymbol{x}_1 + \boldsymbol{x}_2 = \mathbf{0}$ (we write $\boldsymbol{x}_2 = -\boldsymbol{x}_1$). Multiplication by a scalar is also commutative: $\alpha(\beta\boldsymbol{x}) = (\alpha\beta)\boldsymbol{x}$ for all $\alpha, \beta \in \mathbb{F}$, $\boldsymbol{x} \in \mathbb{V}$, and multiplication by the unit element of $\mathbb{F}$ leaves $\boldsymbol{x} \in \mathbb{V}$ intact, $1\boldsymbol{x} = \boldsymbol{x}$. Moreover, addition and multiplication by a scalar are linked by the distributive laws $\alpha(\boldsymbol{x}_1 + \boldsymbol{x}_2) = \alpha\boldsymbol{x}_1 + \alpha\boldsymbol{x}_2$ and $(\alpha + \beta)\boldsymbol{x}_1 = \alpha\boldsymbol{x}_1 + \beta\boldsymbol{x}_1$, $\alpha, \beta \in \mathbb{F}$, $\boldsymbol{x}_1, \boldsymbol{x}_2 \in \mathbb{V}$.

The elements of $\mathbb{V}$ are sometimes called *vectors.*

If $\mathbb{V}_1 \subseteq \mathbb{V}_2$, where both $\mathbb{V}_1$ and $\mathbb{V}_2$ are vector spaces, we say that $\mathbb{V}_1$ is a *subspace* of $\mathbb{V}_2$.

A.1.1.2 The vectors $\boldsymbol{x}_1, \boldsymbol{x}_2, \ldots, \boldsymbol{x}_m \in \mathbb{V}$ are *linearly independent* if

$$\exists\, \alpha_1, \alpha_2, \ldots, \alpha_m \in \mathbb{F} \quad \text{such that} \quad \sum_{\ell=1}^{m} \alpha_\ell \boldsymbol{x}_\ell = \mathbf{0} \qquad \Longrightarrow \qquad \alpha_1, \alpha_2, \ldots, \alpha_m = 0.$$

A vector space $\mathbb{V}$ is of *dimension* $\dim \mathbb{V} = d$ if there exist d linearly independent elements $\boldsymbol{y}_1, \boldsymbol{y}_2, \ldots, \boldsymbol{y}_d \in \mathbb{V}$ such that for every $\boldsymbol{x} \in \mathbb{V}$ we can find scalars $\beta_1, \beta_2, \ldots, \beta_d \in \mathbb{F}$ for which

$$\boldsymbol{x} = \sum_{\ell=1}^{d} \beta_\ell \boldsymbol{y}_\ell.$$

The set $\{\boldsymbol{y}_1, \boldsymbol{y}_2, \ldots, \boldsymbol{y}_d\} \subset \mathbb{V}$ is then said to be a *basis* of $\mathbb{V}$. In other words, all the elements of $\mathbb{V}$ can be expressed by forming linear combinations of its basis elements.

A.1.1.3 The vector space $\mathbb{R}^d$ (over $\mathbb{F} = \mathbb{R}$, the reals) consists of all real d-tuples

$$\boldsymbol{x} = \begin{bmatrix} x_1 \\ x_2 \\ \vdots \\ x_d \end{bmatrix},$$

with addition and multiplication by a scalar defined by

$$\begin{bmatrix} x_1 \\ x_2 \\ \vdots \\ x_d \end{bmatrix} + \begin{bmatrix} y_1 \\ y_2 \\ \vdots \\ y_d \end{bmatrix} = \begin{bmatrix} x_1 + y_1 \\ x_2 + y_2 \\ \vdots \\ x_d + y_d \end{bmatrix} \quad \text{and} \quad \alpha \begin{bmatrix} x_1 \\ x_2 \\ \vdots \\ x_d \end{bmatrix} = \begin{bmatrix} \alpha x_1 \\ \alpha x_2 \\ \vdots \\ \alpha x_d \end{bmatrix}$$

respectively. It is of dimension d with a *canonical* basis

$$\boldsymbol{e}_1 = \begin{bmatrix} 1 \\ 0 \\ 0 \\ \vdots \\ 0 \end{bmatrix}, \quad \boldsymbol{e}_2 = \begin{bmatrix} 0 \\ 1 \\ 0 \\ \vdots \\ 0 \end{bmatrix}, \quad \ldots, \quad \boldsymbol{e}_d = \begin{bmatrix} 0 \\ 0 \\ \vdots \\ 0 \\ 1 \end{bmatrix}$$

of *unit vectors.*

Likewise, letting $\mathbb{F} = \mathbb{C}$ we obtain the vector space $\mathbb{C}^d$ of all complex d-tuples, with similarly defined operations of addition and multiplication by a scalar. It is again of dimension d and possesses an identical basis $\{\boldsymbol{e}_1, \boldsymbol{e}_2, \ldots, \boldsymbol{e}_d\}$ of unit vectors.

In what follows, unless explicitly stated, we restrict our review to $\mathbb{R}^d$. Transplantation to $\mathbb{C}^d$ is straightforward.

A.1.2 Matrices

A.1.2.1 A *matrix* A is an $d \times n$ array of real numbers,

$$A = \begin{bmatrix} a_{1,1} & a_{1,2} & \cdots & a_{1,n} \\ a_{2,1} & a_{2,2} & \cdots & a_{2,n} \\ \vdots & \vdots & & \vdots \\ a_{d,1} & a_{d,2} & \cdots & a_{d,n} \end{bmatrix}.$$

It is said to have d *rows* and n *columns.* The addition of two $d \times n$ matrices is defined by

$$\begin{bmatrix} a_{1,1} & a_{1,2} & \cdots & a_{1,n} \\ a_{2,1} & a_{2,2} & \cdots & a_{2,n} \\ \vdots & \vdots & & \vdots \\ a_{d,1} & a_{d,2} & \cdots & a_{d,n} \end{bmatrix} + \begin{bmatrix} b_{1,1} & b_{1,2} & \cdots & b_{1,n} \\ b_{2,1} & b_{2,2} & \cdots & b_{2,n} \\ \vdots & \vdots & & \vdots \\ b_{d,1} & b_{d,2} & \cdots & b_{d,n} \end{bmatrix} = \begin{bmatrix} a_{1,1} + b_{1,1} & a_{1,2} + b_{1,2} & \cdots & a_{1,n} + b_{1,n} \\ a_{2,1} + b_{2,1} & a_{2,2} + b_{2,2} & \cdots & a_{2,n} + b_{2,n} \\ \vdots & \vdots & & \vdots \\ a_{d,1} + b_{d,1} & a_{d,2} + b_{d,2} & \cdots & a_{d,n} + b_{d,n} \end{bmatrix}$$

and multiplication by a scalar is defined by

$$\alpha \begin{bmatrix} a_{1,1} & a_{1,2} & \cdots & a_{1,n} \\ a_{2,1} & a_{2,2} & \cdots & a_{2,n} \\ \vdots & \vdots & & \vdots \\ a_{d,1} & a_{d,2} & \cdots & a_{d,n} \end{bmatrix} = \begin{bmatrix} \alpha a_{1,1} & \alpha a_{1,2} & \cdots & \alpha a_{1,n} \\ \alpha a_{2,1} & \alpha a_{2,2} & \cdots & \alpha a_{2,n} \\ \vdots & \vdots & & \vdots \\ \alpha a_{d,1} & \alpha a_{d,2} & \cdots & \alpha a_{d,n} \end{bmatrix}.$$

Given an $m \times d$ matrix A and an $d \times n$ matrix B, the product $C = AB$ is the $m \times n$ matrix

$$C = \begin{bmatrix} c_{1,1} & c_{1,2} & \cdots & c_{1,n} \\ c_{2,1} & c_{2,2} & \cdots & c_{2,n} \\ \vdots & \vdots & & \vdots \\ c_{m,1} & c_{m,2} & \cdots & c_{m,n} \end{bmatrix},$$

where

$$c_{i,j} = \sum_{\ell=1}^{d} a_{i,\ell} b_{\ell,j}, \qquad i = 1, 2, \ldots, m, \quad j = 1, 2, \ldots, n.$$

Any $\boldsymbol{x} \in \mathbb{R}^d$ is itself an $d \times 1$ matrix. Hence, the *matrix–vector product* $\boldsymbol{y} = A\boldsymbol{x}$, where A is $m \times d$, is an element of $\mathbb{R}^m$ such that

$$y_i = \sum_{\ell=1}^{d} a_{i,\ell} x_\ell, \qquad i = 1, 2, \ldots, m.$$

In other words, any $m \times d$ matrix A is a *linear transformation* that maps $\mathbb{R}^d$ to $\mathbb{R}^m$.

A.1.2.2 The *identity matrix* is the $d \times d$ matrix

$$I = \begin{bmatrix} 1 & 0 & \cdots & 0 \\ 0 & 1 & \ddots & \vdots \\ \vdots & \ddots & \ddots & 0 \\ 0 & \cdots & 0 & 1 \end{bmatrix}.$$

It is true that $IA = A$ and $BI = B$ for any $d \times n$ matrix A and $m \times d$ matrix B respectively.

A.1.2.3 A matrix is *square* if it is $d \times d$ for some $d \geq 1$. The *determinant* of a square matrix A can be defined by induction. If $d = 1$, so that A is simply a real number, $\det A = A$. Otherwise

$$\det A = \sum_{j=1}^{d} (-1)^{d+j} a_{d,j} \det A_j,$$

where $A_1, A_2, \ldots, A_d$ are $(d-1) \times (d-1)$ matrices given by

$$A_j = \begin{bmatrix} a_{1,1} & \cdots & a_{1,j-1} & a_{1,j+1} & \cdots & a_{1,d} \\ a_{2,1} & \cdots & a_{2,j-1} & a_{2,j+1} & \cdots & a_{2,d} \\ \vdots & & \vdots & \vdots & & \vdots \\ a_{d-1,1} & \cdots & a_{d-1,j-1} & a_{d-1,j+1} & \cdots & a_{d-1,d} \end{bmatrix}, \qquad j = 1, 2, \ldots, d.$$

An alternative means of defining $\det A$ is as as follows:

$$\det A = \sum_{\sigma} (-1)^{|\sigma|} \prod_{j=1}^{d} a_{j,\sigma(j)},$$

where the summation is carried across all $d!$ *permutations* $\boldsymbol{\sigma}$ of the numbers $1, 2, \ldots, d$. The *parity* $|\boldsymbol{\sigma}|$ of the permutation $\boldsymbol{\sigma}$ is the minimal number of two-term exchanges that are needed to convert it to the unit permutation $\boldsymbol{i} = (1, 2, \ldots, d)$.

Provided that $\det A \neq 0$, the $d \times d$ matrix A possesses a unique *inverse* A^{-1}; this is a $d \times d$ matrix such that $A^{-1}A = AA^{-1} = I$. An explicit definition of A^{-1} is

$$A^{-1} = \frac{\operatorname{adj} A}{\det A},$$

where the (i, j)th component $b_{i,j}$ of the $d \times d$ *adjugate matrix* adj A is

$$b_{i,j} = (-1)^{i+j} \det \begin{bmatrix} a_{1,1} & \cdots & a_{1,j-1} & a_{1,j+1} & \cdots & a_{1,d} \\ \vdots & & \vdots & \vdots & & \vdots \\ a_{i-1,1} & \cdots & a_{i-1,j-1} & a_{i-1,j+1} & \cdots & a_{i-1,d} \\ a_{i+1,1} & \cdots & a_{i+1,j-1} & a_{i+1,j+1} & \cdots & a_{i+1,d} \\ \vdots & & \vdots & \vdots & & \vdots \\ a_{d,1} & \cdots & a_{d,j-1} & a_{d,j+1} & \cdots & a_{d,d} \end{bmatrix}, \qquad i, j = 1, 2, \ldots, d.$$

A matrix A such that $\det A \neq 0$ is *nonsingular;* otherwise it is *singular.*

A.1.2.4 The *transpose* $A^\top$ of a $d \times n$ matrix A is an $n \times d$ matrix such that

$$A^\top = \begin{bmatrix} a_{1,1} & a_{2,1} & \cdots & a_{n,1} \\ a_{1,2} & a_{2,2} & \cdots & a_{n,2} \\ \vdots & \vdots & & \vdots \\ a_{1,d} & a_{2,d} & \cdots & a_{n,d} \end{bmatrix}.$$

A.1.2.5 A square $d \times d$ matrix A is

- *diagonal* if $a_{j,\ell} = 0$ for every $j \neq \ell$, $j, \ell = 1, 2, \ldots, d$.
- *symmetric* if $A^\top = A$;
- *Hermitian* or *self-adjoint* if A is complex and $\bar{A}^\top = A$;
- *skew-symmetric* (or *anti-symmetric*) if $A^\top = -A$;
- *skew-Hermitian* if A is complex and $\bar{A}^\top = -A$;
- *orthogonal* if $A^\top A = I$, the identity matrix, which is equivalent to

$$\sum_{\ell=1}^{d} a_{\ell,i} a_{\ell,j} = \begin{cases} 1, & i = j, \\ 0, & i \neq j, \end{cases} \qquad i, j = 1, 2, \ldots, d.$$

  Note that in this case $A^{-1} = A^\top$ and that $A^\top$ is also orthogonal;
- *unitary* if A is complex and $\bar{A}^\top A = I$;

- a *permutation matrix* if all its elements are either 0 or 1 and there is exactly one 1 in each row and column. The matrix–vector product $A\boldsymbol{x}$ permutes the elements of $\boldsymbol{x} \in \mathbb{R}^d$, while $A^\top \boldsymbol{y}$ causes an inverse permutation – therefore $A^\top = A^{-1}$ and A is orthogonal;
- *tridiagonal* if $a_{i,j} = 0$ for every $|i-j| \geq 2$, in other words

$$A = \begin{bmatrix} a_{1,1} & a_{1,2} & 0 & \cdots & 0 \\ a_{2,1} & a_{2,2} & a_{2,3} & \ddots & \vdots \\ 0 & \ddots & \ddots & \ddots & 0 \\ \vdots & \ddots & a_{d-1,d-2} & a_{d-1,d-1} & a_{d-1,d} \\ 0 & \cdots & 0 & a_{d,d-1} & a_{d,d} \end{bmatrix};$$

- a *Vandermonde matrix* if

$$A = \begin{bmatrix} 1 & \xi_1 & \xi_1^2 & \cdots & \xi_1^{d-1} \\ 1 & \xi_2 & \xi_2^2 & \cdots & \xi_2^{d-1} \\ \vdots & \vdots & \vdots & & \vdots \\ 1 & \xi_d & \xi_d^2 & \cdots & \xi_d^{d-1} \end{bmatrix}$$

  for some $\xi_1, \xi_2, \ldots, \xi_d \in \mathbb{C}$. It is true in this case that

$$\det A = \prod_{i=2}^{d} \prod_{j=1}^{i-1} (\xi_i - \xi_j),$$

  hence a Vandermonde matrix is nonsingular if and only if $\xi_1, \xi_2, \ldots, \xi_d$ are distinct numbers;
- *reducible* if $\{1, 2, \ldots, d\} = I_1 \cup I_2$ such that $I_1 \cap I_2 = \emptyset$ and $a_{i,j} = 0$ for all $i \in I_1$, $j \in I_2$. Otherwise it is *irreducible*;
- *normal* if $A^\top A = AA^\top$. Symmetric, skew-symmetric and orthogonal matrices are all normal.

A.1.3 Inner products and norms

A.1.3.1 An *inner product*, also known as a *scalar product*, is a function $\langle \cdot, \cdot \rangle : \mathbb{R}^d \times \mathbb{R}^d \to \mathbb{R}$ with the following properties.

(1) *Nonnegativity:* $\langle \boldsymbol{x}, \boldsymbol{x} \rangle \geq 0$ for every $\boldsymbol{x} \in \mathbb{R}^d$ and $\langle \boldsymbol{x}, \boldsymbol{x} \rangle = 0$ if and only if $\boldsymbol{x} = \boldsymbol{0}$, the *zero vector*.

(2) *Linearity:* $\langle \alpha\boldsymbol{x} + \beta\boldsymbol{y}, \boldsymbol{z} \rangle = \alpha\langle \boldsymbol{x}, \boldsymbol{z} \rangle + \beta\langle \boldsymbol{y}, \boldsymbol{z} \rangle$ for all $\alpha, \beta \in \mathbb{R}$ and $\boldsymbol{x}, \boldsymbol{y}, \boldsymbol{z} \in \mathbb{R}^d$.

(3) *Symmetry:* $\langle \boldsymbol{x}, \boldsymbol{y} \rangle = \langle \boldsymbol{y}, \boldsymbol{x} \rangle$ for all $\boldsymbol{x}, \boldsymbol{y} \in \mathbb{R}^d$.

(4) *The Cauchy–Schwarz inequality:* For every $\boldsymbol{x}, \boldsymbol{y} \in \mathbb{R}^d$ it is true that

$$|\langle \boldsymbol{x}, \boldsymbol{y} \rangle| \leq [\langle \boldsymbol{x}, \boldsymbol{x} \rangle]^{1/2} [\langle \boldsymbol{y}, \boldsymbol{y} \rangle]^{1/2}.$$

A particular example is the *Euclidean* (or ℓ_2) inner product

$$\langle \boldsymbol{x}, \boldsymbol{y} \rangle = \boldsymbol{x}^\top \boldsymbol{y}, \qquad \boldsymbol{x}, \boldsymbol{y} \in \mathbb{R}^d.$$

In the case of the complex-valued vector space $\mathbb{C}^d$, the symmetry axiom of the inner product needs to be replaced by $\langle \boldsymbol{x}, \boldsymbol{y} \rangle = \overline{\langle \boldsymbol{y}, \boldsymbol{x} \rangle}$, $\boldsymbol{x}, \boldsymbol{y} \in \mathbb{C}^d$, where the bar denotes conjugation, while the complex Euclidean inner product is

$$\langle \boldsymbol{x}, \boldsymbol{y} \rangle = \bar{\boldsymbol{x}}^\top \boldsymbol{y}, \qquad \boldsymbol{x}, \boldsymbol{y} \in \mathbb{C}^d.$$

A.1.3.2 Two vectors $\boldsymbol{x}, \boldsymbol{y} \in \mathbb{R}^d$ such that $\langle \boldsymbol{x}, \boldsymbol{y} \rangle = 0$ are said to be *orthogonal.* A basis $\{\boldsymbol{y}_1, \boldsymbol{y}_2, \ldots, \boldsymbol{y}_d\}$ constructed from orthogonal vectors is called an *orthogonal basis,* or, if $\langle \boldsymbol{y}_\ell, \boldsymbol{y}_\ell \rangle = 1$, $\ell = 1, 2, \ldots, d$, an *orthonormal basis.* The canonical basis $\{\boldsymbol{e}_1, \boldsymbol{e}_2, \ldots, \boldsymbol{e}_d\}$ is orthonormal with respect to the Euclidean norm.

A.1.3.3 A vector *norm* is a function $\|\cdot\| : \mathbb{R}^d \to \mathbb{R}$ that obeys the following axioms.

(1) *Nonnegativity:* $\|\boldsymbol{x}\| \geq 0$ for every $\boldsymbol{x} \in \mathbb{R}^d$ and $\|\boldsymbol{x}\| = 0$ if and only if $\boldsymbol{x} = \boldsymbol{0}$.

(2) *Rescaling:* $\|\alpha \boldsymbol{x}\| = |\alpha| \|\boldsymbol{x}\|$ for every $\alpha \in \mathbb{R}$ and $\boldsymbol{x} \in \mathbb{R}^d$.

(3) *The triangle inequality:* $\|\boldsymbol{x} + \boldsymbol{y}\| \leq \|\boldsymbol{x}\| + \|\boldsymbol{y}\|$ for every $\boldsymbol{x}, \boldsymbol{y} \in \mathbb{R}^d$.

Any inner product $\langle \cdot, \cdot \rangle$ induces a norm $\|\boldsymbol{x}\| = [\langle \boldsymbol{x}, \boldsymbol{x} \rangle]^{1/2}$, $\boldsymbol{x} \in \mathbb{R}^d$. In particular, the Euclidean inner product induces the ℓ_2 norm, also known as the *Euclidean* norm, the *least squares* or the *energy* norm, $\|\boldsymbol{x}\| = (\boldsymbol{x}^\top \boldsymbol{x})^{1/2}$, $\boldsymbol{x} \in \mathbb{R}^d$ (or $\|\boldsymbol{x}\| = (\bar{\boldsymbol{x}}^\top \boldsymbol{x})^{1/2}$, $\boldsymbol{x} \in \mathbb{C}^d$).

Not every vector norm is induced by an inner product. Well-known and useful examples are the ℓ_1 norm (also known as the *Manhattan* norm)

$$\|\boldsymbol{x}\| = \sum_{\ell=1}^{d} |x_\ell|, \qquad \boldsymbol{x} \in \mathbb{R}^d,$$

and the ℓ_∞ norm (also known as the *Chebyshev* norm, the *uniform* norm, the *max* norm or the *sup* norm),

$$\|\boldsymbol{x}\| = \max_{\ell=1,2,\ldots,d} |x_\ell|, \qquad \boldsymbol{x} \in \mathbb{R}^d.$$

A.1.3.4 Every vector norm $\|\cdot\|$ acting on $\mathbb{R}^d$ can be extended to a norm on $d \times d$ matrices, the *induced matrix norm,* by letting

$$\|A\| = \max_{\boldsymbol{x} \in \mathbb{R}^d,\, \boldsymbol{x} \neq \boldsymbol{0}} \frac{\|A\boldsymbol{x}\|}{\|\boldsymbol{x}\|} = \max_{\boldsymbol{x} \in \mathbb{R}^d,\, \|\boldsymbol{x}\|=1} \|A\boldsymbol{x}\|.$$

It is always true that

$$\|A\boldsymbol{x}\| \leq \|A\| \times \|\boldsymbol{x}\|, \qquad \boldsymbol{x} \in \mathbb{R}^d$$

and

$$\|AB\| \leq \|A\| \times \|B\|,$$

where both A and B are $d \times d$ matrices.

The Euclidean norm of an orthogonal matrix always equals unity.

A.1.3.5 A $d \times d$ symmetric matrix A is said to be *positive definite* if

$$\langle A\boldsymbol{x}, \boldsymbol{x}\rangle > 0 \qquad \boldsymbol{x} \in \mathbb{R}^d \setminus \{\boldsymbol{0}\}$$

and *negative definite* if the above inequality is reversed. It is *positive semidefinite* or *negative semidefinite* if

$$\langle A\boldsymbol{x}, \boldsymbol{x}\rangle \geq 0 \qquad \text{or} \qquad \langle A\boldsymbol{x}, \boldsymbol{x}\rangle \leq 0$$

for all $\boldsymbol{x} \in \mathbb{R}^d$, respectively.

A.1.4 Linear systems

A.1.4.1 The *linear system*

$$\begin{aligned} a_{1,1}x_1 + a_{1,2}x_2 + \cdots + a_{1,d}x_d &= b_1, \\ a_{2,1}x_1 + a_{2,2}x_2 + \cdots + a_{2,d}x_d &= b_2, \\ &\vdots \\ a_{d,1}x_1 + a_{d,2}x_2 + \cdots + a_{d,d}x_d &= b_d, \end{aligned}$$

is written in vector notation as $A\boldsymbol{x} = \boldsymbol{b}$. It possesses a unique solution $\boldsymbol{x} = A^{-1}\boldsymbol{b}$ if and only if A is nonsingular.

A.1.4.2 If a square matrix A is singular then there exists a nonzero solution to the *homogeneous* linear system $A\boldsymbol{x} = \boldsymbol{0}$. The *kernel* of A, denoted by $\ker A$, is the set of all $\boldsymbol{x} \in \mathbb{R}^d$ such that $A\boldsymbol{x} = \boldsymbol{0}$. If A is nonsingular then $\ker A = \{\boldsymbol{0}\}$, otherwise $\ker A$ is a subspace of $\mathbb{R}^d$ of dimension ≥ 1.

Recall that a $d \times d$ matrix A is a linear transformation mapping $\mathbb{R}^d$ into itself. If A is nonsingular ($\Leftrightarrow \det A \neq 0 \Leftrightarrow \ker A = \{\boldsymbol{0}\}$) then this mapping is an *isomorphism* (in other words, it has a well-defined and unique inverse linear transformation A^{-1}, acting on the image $A\mathbb{R}^d := \{A\boldsymbol{x} : \boldsymbol{x} \in \mathbb{R}^d\}$ and mapping it back to $\mathbb{R}^d$) *on* $\mathbb{R}^d$ (i.e., the image $A\mathbb{R}^d$ is all $\mathbb{R}^d$). However, if A is singular ($\Leftrightarrow \det A = 0 \Leftrightarrow \dim\ker A \geq 1$) then $A\mathbb{R}^d$ is a proper vector subspace of $\mathbb{R}^d$ and $\dim(A\mathbb{R}^d) = d - \dim\ker A \leq d - 1$.

A.1.4.3 The practical solution of linear systems such as the above can be performed by *Gaussian elimination*. We commence by subtracting from the ℓth equation, $\ell = 2, 3, \ldots, d$, the product of the first equation by the real number $a_{\ell,1}/a_{1,1}$. This does not change the solution $\boldsymbol{x}$ of the linear system, while setting zeros in the first column, except in the first equation, and replacing the system by

$$\begin{aligned} a_{1,1}x_1 + a_{1,2}x_2 + \cdots + a_{1,d}x_d &= b_1, \\ \tilde{a}_{2,2}x_2 + \cdots + \tilde{a}_{2,d} &= \tilde{b}_2, \\ &\vdots \\ \tilde{a}_{d,2}x_2 + \cdots + \tilde{a}_{d,d} &= \tilde{b}_d, \end{aligned}$$

where

$$\tilde{a}_{\ell,j} = a_{\ell,j} - \frac{a_{1,1}}{a_{\ell,1}} a_{1,j}, \quad j = 2,3,\ldots,d, \qquad \tilde{b}_\ell = b_\ell - \frac{a_{1,1}}{a_{\ell,1}} b_1, \qquad \ell = 2,3,\ldots,d.$$

The unknown x_1 does not feature in equations $2, 3, \ldots, d$ because it has been *eliminated;* the latter thereby constitute a set of $d - 1$ equations in $d - 1$ unknowns. Continuing this process by induction results in an *upper triangular* linear system:

$$\begin{aligned} a_{1,1}^{(1)} x_1 + a_{1,2}^{(1)} x_2 + a_{1,3}^{(1)} x_3 + \cdots + a_{1,d}^{(1)} x_d &= b_1^{(1)}, \\ a_{2,2}^{(2)} x_2 + a_{2,3}^{(2)} x_3 + \cdots + a_{2,d}^{(2)} x_d &= b_2^{(2)}, \\ a_{3,3}^{(3)} x_3 + \cdots + a_{3,d}^{(3)} x_d &= b_3^{(3)} \\ &\ \vdots \\ a_{d,d}^{(d)} x_d &= b_d^{(d)}, \end{aligned}$$

where $a_{1,j}^{(1)} = a_{1,j}$, $a_{2,j}^{(2)} = \tilde{a}_{2,j}$, $a_{3,j}^{(3)} = a_{3,j}^{(2)} - \left(a_{2,2}^{(2)}/a_{3,2}^{(2)}\right) a_{2,j}^{(2)}$ etc.

The upper triangular system is solved successively from the bottom:

$$\begin{aligned} x_d &= \frac{1}{a_{d,d}^{(d)}} b_d^{(d)}, \\ x_{d-1} &= \frac{1}{a_{d-1,d-1}^{(d-1)}} \left[b_{d-1}^{(d-1)} - a_{d-1,d}^{(d)} x_d \right], \\ &\ \vdots \\ x_1 &= \frac{1}{a_{1,1}^{(1)}} \left[b_1^{(1)} - \sum_{j=2}^{d} a_{1,j}^{(1)} x_j \right]. \end{aligned}$$

The whole procedure depends on the *pivots* $a_{\ell,\ell}^{(\ell)}$ being nonzero, otherwise it cannot be carried out in this fashion. It is perfectly possible for a pivot to vanish even if A is nonsingular and (with few important exceptions) it is impractical to determine whether a pivot vanishes by inspecting the elements of A. Moreover, if some pivots are exceedingly small (in modulus), even if none vanishes, large rounding errors can be introduced by computer arithmetic, thereby destroying the precision of Gaussian elimination and rendering it unusable.

A.1.4.4 A standard means of preventing pivots becoming small is *column pivoting.* Instead of eliminating the ℓth equation at the ℓth stage, we first search for $m \in \{\ell, \ell+1, \ldots, d\}$ such that $|a_{m,\ell}^{(\ell)}| \geq |a_{i,\ell}^{(\ell)}|$ for all $i = \ell, \ell+1, \ldots, d$, interchange the ℓth and the ith equations and only then eliminate.

A.1.4.5 An alternative formulation of Gaussian elimination is by means of the *LU factorization* $A = LU$, where the $d \times d$ matrices L and U are *lower* and *upper* *triangular,* respectively, and all diagonal elements of L equal unity,

$$L = \begin{bmatrix} 1 & 0 & \cdots & \cdots & 0 \\ \ell_{2,1} & 1 & 0 & & \vdots \\ \vdots & \ddots & \ddots & \ddots & \vdots \\ \ell_{d-1,1} & \cdots & \ell_{d-1,d-2} & 1 & 0 \\ \ell_{d,1} & \cdots & \ell_{d,d-2} & \ell_{d,d-1} & 1 \end{bmatrix} \quad \text{and} \quad U = \begin{bmatrix} u_{1,1} & u_{1,2} & \cdots & \cdots & u_{1,d} \\ 0 & u_{2,2} & u_{2,3} & & \vdots \\ \vdots & \ddots & \ddots & \ddots & \vdots \\ 0 & \cdots & 0 & u_{d-1,d-1} & u_{d-1,d} \\ 0 & \cdots & \cdots & 0 & u_{d,d} \end{bmatrix}.$$

Provided that an LU factorization of A is known, we replace the linear system $A\boldsymbol{x} = \boldsymbol{b}$ by the two systems $L\boldsymbol{y} = \boldsymbol{b}$ and $U\boldsymbol{x} = \boldsymbol{y}$. The first is lower triangular, hence

$$\begin{aligned} y_1 &= b_1, \\ y_2 &= b_2 - \ell_{2,1} y_1, \\ &\vdots \\ y_d &= b_d - \sum_{j=1}^{d-1} \ell_{d,j} y_j, \end{aligned}$$

while the second is upper triangular – and we have already seen in A.1.4.3 how to solve upper triangular linear systems.

The practical evaluation of an LU factorization can be done explicitly, by letting, consecutively for $k = 1, 2, \ldots, d$,

$$\begin{aligned} u_{k,j} &= a_{k,j} - \sum_{i=1}^{j-1} \ell_{k,i} u_{i,j}, & j &= k, k+1, \ldots, d, \\ \ell_{j,k} &= \frac{1}{u_{k,k}} \left(a_{j,k} - \sum_{i=1}^{k-1} \ell_{j,i} u_{i,k} \right), & j &= k+1, k+2, \ldots, d. \end{aligned}$$

As always, empty sums are nil and empty number ranges are disregarded.

LU factorization, like Gaussian elimination, is prone to failure due to small values of the pivots $u_{1,1}, u_{2,2}, \ldots, u_{d,d}$ and the remedy – column pivoting – is identical. After all, LU factorization is but a recasting of Gaussian elimination into a form that is more convenient for various applications.

A.1.4.6 A positive definite matrix A can be factorized into the product $A = LL^\top$, where L is lower triangular with $\ell_{j,j} > 0$, $j = 1, 2, \ldots, d$; this is the *Cholesky factorization.* It can be evaluated explicitly via

$$\begin{aligned} \ell_{k,j} &= \frac{1}{\ell_{j,j}} \left(a_{k,j} - \sum_{i=1}^{j-1} \ell_{k,i} \ell_{j,i} \right), \qquad j = 1, 2, \ldots, k-1, \\ \ell_{k,k} &= \left(a_{k,k} - \sum_{i=1}^{k-1} \ell_{k,i}^2 \right)^{1/2} \end{aligned}$$

for $k = 1, 2, \ldots, d$.

Having evaluated a Cholesky factorization, the linear system $A\boldsymbol{x} = \boldsymbol{b}$ can be solved by a sequential evaluation of two tridiagonal systems, firstly computing $L\boldsymbol{y} = \boldsymbol{b}$ and then $L^\top \boldsymbol{x} = \boldsymbol{y}$.

An advantage of Cholesky factorization is that it requires half the storage and half the computational cost of an LU factorization.

A.1.5 Eigenvalues and eigenvectors

A.1.5.1 We say that $\lambda \in \mathbb{C}$ is an *eigenvalue* of the $d \times d$ matrix A if $\det(A - \lambda I) = 0$. The set of all eigenvalues of a square matrix A is called the *spectrum* and denoted by $\sigma(A)$.

Each $d \times d$ matrix has exactly d eigenvalues. All the eigenvalues of a symmetric matrix are real, all the eigenvalues of a skew-symmetric matrix are pure imaginary and, in general, the eigenvalues of a real matrix are either real or form complex conjugate pairs.

If all the eigenvalues of a symmetric matrix are positive then it is positive definite. A similarly worded statement extends to negative, semipositive and seminegative matrices.

We say that an eigenvalue is of *algebraic multiplicity* $r \geq 1$ if it is a zero of multiplicity r of the *characteristic polynomial* $p(z) = \det(A - zI)$; in other words, if

$$p(\lambda) = \frac{\mathrm{d}p(\lambda)}{\mathrm{d}z} = \cdots = \frac{\mathrm{d}^{r-1}p(\lambda)}{\mathrm{d}z^{r-1}} = 0, \qquad \frac{\mathrm{d}^r p(\lambda)}{\mathrm{d}z^r} \neq 0.$$

An eigenvalue of algebraic multiplicity 1 is said to be *distinct*.

A.1.5.2 The *spectral radius* of A is a nonnegative number

$$\rho(A) = \max\{|\lambda| \,:\, \lambda \in \sigma(A)\}.$$

It always obeys the inequality

$$\rho(A) \leq \|A\|$$

(where $\|\cdot\|$ is the Euclidean norm) but $\rho(A) = \|A\|$ for a normal matrix A. It is possible to express the Euclidean norm of a general square matrix A in the form

$$\|A\| = [\rho(A^\top A)]^{1/2}.$$

A.1.5.3 If $\lambda \in \sigma(A)$ it follows that $\dim \ker(A - \lambda I) \geq 1$, therefore there are nonzero vectors in the *eigenspace* $\ker(A - \lambda I)$. Each such vector is called an *eigenvector* of A corresponding to the eigenvalue λ. An alternative formulation is that $\boldsymbol{v} \in \mathbb{R}^d \setminus \{\boldsymbol{0}\}$ is an eigenvector of A, corresponding to $\lambda \in \sigma(A)$, if $A\boldsymbol{v} = \lambda\boldsymbol{v}$. Note that even if A is real, its eigenvectors – like its eigenvalues – may be complex.

The *geometric multiplicity* of λ is the dimension of its eigenspace and it is always true that

$$1 \leq \text{geometric multiplicity} \leq \text{algebraic multiplicity}.$$

If the geometric and algebraic multiplicities are equal for all its eigenvalues, A is said to have a *complete set of eigenvectors.* Since different eigenspaces are linearly independent and the sum of algebraic multiplicities is always d, a matrix possessing a complete set of eigenvectors provides a basis of $\mathbb{R}^d$ formed by its eigenvectors – specifically, the union over all bases of its eigenspaces.

If all the eigenvalues of A are distinct then it has a complete set of eigenvectors. A normal matrix also shares this feature and, moreover, it always has an orthogonal basis of eigenvectors.

A.1.5.4 If a $d \times d$ matrix A has a complete set of eigenvectors then it possesses the *spectral factorization*

$$A = VDV^{-1}.$$

Here D is a diagonal matrix and $d_{\ell,\ell} = \lambda_\ell$, $\sigma(A) = \{\lambda_1, \lambda_2, \ldots, \lambda_d\}$, the ℓth column of the $d \times d$ matrix V is an eigenvector in the eigenspace of λ_ℓ and the columns of V are selected so that $\det V \neq 0$, in other words so that the columns form a basis of $\mathbb{R}^d$. This is possible according to A.1.5.3.

If A is normal, it is possible to normalize its eigenvectors (specifically, by letting them be of unit Euclidean norm) so that V is an orthogonal matrix.

Let two $d \times d$ matrices A and B share a complete set of eigenvectors; then $A = VD_AV^{-1}$, $B = VD_BV^{-1}$, say. Since diagonal matrices always commute,

$$\begin{aligned} AB &= (VD_AV^{-1})(VD_BV^{-1}) = (VD_A)(V^{-1}V)(D_BV^{-1}) = V(D_AD_B)V^{-1} \\ &= V(D_BD_A)V^{-1} = (VD_B)(V^{-1}V)(B_AV^{-1}) = (VD_BV^{-1})(VD_AV^{-1}) = BA \end{aligned}$$

and the matrices A and B also commute.

A.1.5.5 Let

$$f(z) = \sum_{k=0}^{\infty} f_k z^k, \qquad z \in \mathbb{C},$$

be an arbitrary power series that converges for all $|z| \le \rho(A)$, where A is a $d \times d$ matrix. The matrix function

$$f(A) := \sum_{k=0}^{\infty} f_k A^k$$

then converges. Moreover, if $\lambda \in \sigma(A)$ and $\boldsymbol{v}$ is in the eigenspace of λ then $f(A)\boldsymbol{v} = f(\lambda)\boldsymbol{v}$. In particular,

$$\sigma(f(A)) = \{f(\lambda_j) \,:\, \lambda_j \in \sigma(A), \quad j = 1, 2, \ldots, d\}.$$

If A has a spectral factorization $A = VDV^{-1}$ then $f(A)$ factorizes as follows:

$$f(A) = Vf(D)V^{-1}.$$

A.1.5.6 Every $d \times d$ matrix A possesses a *Jordan factorization*

$$A = W\Lambda W^{-1},$$

where $\det W \neq 0$ and the $d \times d$ matrix Λ can be written in block form as

$$\Lambda = \begin{bmatrix} \Lambda_1 & O & \cdots & O \\ O & \Lambda_2 & \ddots & \vdots \\ \vdots & \ddots & \ddots & O \\ O & \cdots & O & \Lambda_s \end{bmatrix}.$$

Here $\lambda_1, \lambda_2, \ldots, \lambda_s \in \sigma(A)$ and the kth *Jordan block* is

$$\Lambda_k = \begin{bmatrix} \lambda_k & 1 & 0 & \cdots & 0 \\ 0 & \lambda_k & 1 & \ddots & \vdots \\ \vdots & \ddots & \ddots & \ddots & 0 \\ \vdots & & \ddots & \lambda_k & 1 \\ 0 & \cdots & \cdots & 0 & \lambda_k \end{bmatrix}, \qquad k = 1, 2, \ldots, s.$$

Bibliography

Halmos, P.R. (1958), *Finite-Dimensional Vector Spaces,* van Nostrand–Reinhold, Princeton, NJ.

Lang, S. (1987), *Introduction to Linear Algebra* (2nd edn), Springer-Verlag, New York.

Strang, G. (1987), *Linear Algebra and its Applications* (3rd edn), Harcourt Brace Jovanovich, San Diego, CA.

A.2 Analysis

A.2.1 Introduction to functional analysis

A.2.1.1 A *linear space* is an arbitrary collection of objects closed under addition and multiplication by a scalar. In other words, $\mathbb{V}$ is a linear space over the field $\mathbb{F}$ (*a scalar field*) if there exist operations $+ : \mathbb{V} \times \mathbb{V} \to \mathbb{V}$ and $\cdot : \mathbb{F} \times \mathbb{V} \to \mathbb{V}$, consistent with the following axioms (for clarity we denote the elements of $\mathbb{V}$ as vectors):

(1) *Commutativity:* $\boldsymbol{x} + \boldsymbol{y} = \boldsymbol{y} + \boldsymbol{x}$ for every $\boldsymbol{x}, \boldsymbol{y} \in \mathbb{V}$.

(2) *Existence of zero:* There exists a unique element $\mathbf{0} \in \mathbb{V}$ such that $\boldsymbol{x} + \mathbf{0} = \boldsymbol{x}$ for every $\boldsymbol{x} \in \mathbb{V}$.

(3) *Existence of inverse:* For every $\boldsymbol{x} \in \mathbb{V}$ there exists a unique element $-\boldsymbol{x} \in \mathbb{V}$ such that $\boldsymbol{x} + (-\boldsymbol{x}) = \mathbf{0}$.

(4) *Associativity:* $(\boldsymbol{x} + \boldsymbol{y}) + \boldsymbol{z} = \boldsymbol{x} + (\boldsymbol{y} + \boldsymbol{z})$ for every $\boldsymbol{x}, \boldsymbol{y}, \boldsymbol{z} \in \mathbb{V}$ (*These four axioms mean that* $\mathbb{V}$ *is an abelian group with respect to addition*).

(5) *Interchange of multiplication:* $\alpha(\beta\boldsymbol{x}) = (\alpha\beta)\boldsymbol{x}$ for every $\alpha, \beta \in \mathbb{F}$, $\boldsymbol{x} \in \mathbb{V}$.

(6) *Action of unity:* $1\boldsymbol{x} = \boldsymbol{x}$ for every $\boldsymbol{x} \in \mathbb{V}$, where 1 is the unit element of $\mathbb{F}$.

(7) *Distributivity:* $\alpha(\boldsymbol{x}+\boldsymbol{y}) = \alpha\boldsymbol{x}+\alpha\boldsymbol{y}$ and $(\alpha+\beta)\boldsymbol{x} = \alpha\boldsymbol{x}+\beta\boldsymbol{x}$ for every $\alpha, \beta \in \mathbb{F}$, $\boldsymbol{x}, \boldsymbol{y} \in \mathbb{V}$.

If $\mathbb{V}_1, \mathbb{V}_2$ are both linear spaces over the same field $\mathbb{F}$ and $\mathbb{V}_1 \subseteq \mathbb{V}_2$, the space $\mathbb{V}_1$ is said to be a *subspace* of $\mathbb{V}_2$.

A.2.1.2 We say that a linear space is of *dimension* $d < \infty$ if it has a *basis* of d linearly independent elements (cf. A.1.1.2). If no such basis exists, the linear space is said to be of infinite dimension.

A.2.1.3 We assume for simplicity that $\mathbb{F} = \mathbb{R}$, although generalization to the complex field presents no difficulty.

A familiar example of a linear space is the d-dimensional vector space $\mathbb{R}^d$ (cf. A.1.1.1). Another example is the set $\mathbb{P}_\nu$ of all polynomials of degree $\leq \nu$ with real coefficients. It is not difficult to verify that $\dim \mathbb{P}_\nu = \nu + 1$.

More interesting examples of linear spaces are the set $C[0,1]$ of all continuous real functions in the interval $[0,1]$, the set of all real power series with a positive radius of convergence at the origin and the set of all *Fourier expansions* (that is, linear combinations of 1 and $\cos nx$, $\sin nx$ for $n = 1, 2, \ldots$). All these spaces are infinite-dimensional.

A.2.1.4 Inner products and norms over linear spaces are defined exactly as in A.1.3.1 and A.1.3.3 respectively.

A linear space equipped with a norm is called a *normed space.* If a normed space $\mathbb{V}$ is *closed* (that is, all Cauchy sequences converge in $\mathbb{V}$), it is said to be a *Banach space.*

An important example of a normed space is provided when a function is measured by a *p-norm.* The latter is defined by

$$\begin{aligned} \|f\|_p &= \left(\int_\Omega |f(\boldsymbol{x})|^p \, \mathrm{d}\boldsymbol{x} \right)^{1/p}, \qquad 1 \leq p < \infty, \\ \|f\|_\infty &= \sup_{\boldsymbol{x}\in\Omega} |f(\boldsymbol{x})|, \end{aligned} \qquad f \in \mathbb{V},$$

where Ω is the domain of definition of the functions (in general multivariate). A normed space equipped with the p-norm is denoted by $L_p(\Omega)$. If Ω is a closed set, $L_p(\Omega)$ is a Banach space.

A.2.1.5 A closed linear space equipped with an inner product is called a *Hilbert space.* An important example is provided by the inner product

$$\langle f, g \rangle = \int_\Omega f(\boldsymbol{x}) g(\boldsymbol{x}) \, \mathrm{d}\boldsymbol{x}, \qquad f, g \in \mathbb{V},$$

where Ω is a closed set (if the space is over complex numbers, rather than reals, $g(\boldsymbol{x})$ needs to be replaced by $\overline{g(\boldsymbol{x})}$). It induces the *Euclidean* norm (the 2-norm)

$$\|f\| = \left(\int_\Omega |f(\boldsymbol{x})|^2 \, \mathrm{d}\boldsymbol{x} \right)^{1/2}, \qquad f \in \mathbb{V}.$$

A.2.1.6 The Hilbert space $L_2(\Omega)$ is said to be *separable* if it has either a finite or a countable orthogonal basis, $\{\varphi_i\}$, say. (In infinite-dimensional spaces the set $\{\varphi_i\}$ is a basis if each element lies in the *closure* of linear combinations from $\{\varphi_i\}$. The closure is, of course, defined by the underlying norm.)

The space $L_2(\Omega)$ (denoted also by $L(\Omega)$ or $L[\Omega]$) is separable with a countable basis.

A.2.1.7 Let $\mathbb{V}$ be a Hilbert space. Two elements $f, g \in \mathbb{V}$ are *orthogonal* if $\langle f, g\rangle = 0$. If

$$\langle f, g\rangle = 0, \qquad g \in \mathbb{V}_1,$$

where $\mathbb{V}_1$ is a subspace of the Hilbert space $\mathbb{V}$ and $f \in \mathbb{V}$, then f is said to be *orthogonal* to $\mathbb{V}_1$.

A.2.1.8 A mapping from a linear space $\mathbb{V}_1$ to a linear space $\mathbb{V}_2$ is called an *operator.* An operator $\mathcal{T}$ is *linear* if

$$\mathcal{T}(\boldsymbol{x}+\boldsymbol{y}) = \mathcal{T}\boldsymbol{x} + \mathcal{T}\boldsymbol{y}, \qquad \mathcal{T}(\alpha\boldsymbol{x}) = \alpha\mathcal{T}\boldsymbol{x}, \qquad \alpha \in \mathbb{R}, \quad \boldsymbol{x}, \boldsymbol{y} \in \mathbb{V}.$$

Let $\mathcal{T}$ be a linear operator from a Banach space $\mathbb{V}$ to itself. The *norm* of $\mathcal{T}$ is defined as

$$\|\mathcal{T}\| = \sup_{\boldsymbol{x}\in\mathbb{V},\ \boldsymbol{x}\neq\boldsymbol{0}} \frac{\|\mathcal{T}\boldsymbol{x}\|}{\|\boldsymbol{x}\|} = \sup_{\boldsymbol{x}\in\mathbb{V},\ \|\boldsymbol{x}\|=1} \|\mathcal{T}\boldsymbol{x}\|.$$

It is always true that

$$\|\mathcal{T}\boldsymbol{x}\| \le \|\mathcal{T}\| \times \|\boldsymbol{x}\|, \qquad \boldsymbol{x} \in \mathbb{V},$$

and

$$\|\mathcal{T}\mathcal{S}\| \le \|\mathcal{T}\| \times \|\mathcal{S}\|,$$

where both $\mathcal{T}$ and $\mathcal{S}$ are linear operators from $\mathbb{V}$ to itself.

A.2.1.9 The *domain* of a linear operator $\mathcal{T} : \mathbb{V}_1 \to \mathbb{V}_2$ is the Banach space $\mathbb{V}_1$, while its *range* is

$$\mathcal{T}\,\mathbb{V}_1 = \{\mathcal{T}\boldsymbol{x} : \boldsymbol{x} \in \mathbb{V}_1\} \subseteq \mathbb{V}_2.$$

- If $\mathcal{T}\,\mathbb{V}_1 = \mathbb{V}_2$, the operator $\mathcal{T}$ is said to map $\mathbb{V}_1$ *onto* $\mathbb{V}_2$.
- If for every $\boldsymbol{y} \in \mathcal{T}\,\mathbb{V}_1$ there exists a unique $\boldsymbol{x} \in \mathbb{V}_1$ such that $\mathcal{T}\boldsymbol{x} = \boldsymbol{y}$ then $\mathcal{T}$ is said to be an *isomorphism* (or an *injection*) and the linear operator $\mathcal{T}^{-1}\boldsymbol{y} = \boldsymbol{x}$ is the *inverse* of $\mathcal{T}$.
- If $\mathcal{T}$ is an isomorphism and $\mathcal{T}\,\mathbb{V}_1 = \mathbb{V}_2$ then it is said to be an *isomorphism onto* $\mathbb{V}_2$ or a *bijection.*

A.2.1.10 A mapping from a linear space to the reals (or to the complex numbers) is called a *functional.* A functional L is *linear* if $L(\boldsymbol{x}+\boldsymbol{y}) = L\boldsymbol{x} + L\boldsymbol{y}$ and $L(\alpha\boldsymbol{x}) = \alpha L\boldsymbol{x}$ for all $\boldsymbol{x}, \boldsymbol{y} \in \mathbb{V}$ and scalar α.

A.2.2 Approximation theory

A.2.2.1 Denote by $\mathbb{P}_\nu$ the set of all polynomials with real coefficients of degree $\le \nu$. Given $\nu+1$ distinct points $\xi_0, \xi_1, \ldots, \xi_\nu$ and $f_0, f_1, \ldots, f_\nu$, there exists a unique polynomial $p \in \mathbb{P}_\nu$ such that

$$p(\xi_\ell) = f_\ell, \qquad \ell = 0, 1, \ldots, \nu.$$

It is called the *interpolation polynomial.*

A.2.2.2 Suppose that $f_\ell = f(\xi_\ell)$, $\ell = 0, 1, \ldots, \nu$, where f is a $\nu+1$ times differentiable function. Let $a = \min_{i=0,1,\ldots,\nu} \xi_i$ and $b = \max_{i=0,1,\ldots,\nu} \xi_i$. Then for every $x \in [a, b]$ there exists $\eta = \eta(x) \in [a, b]$ such that

$$p(x) - f(x) = \frac{1}{(\nu+1)!} f^{(\nu+1)}(\eta) \prod_{k=0}^{\nu} (x - \xi_k).$$

(*This is an extension of the familiar Taylor remainder formula and it reduces to the latter if* $\xi_0, \xi_1, \ldots, \xi_\nu \to \xi^* \in (a, b)$.)

A.2.2.3 An obvious way of evaluating an interpolation polynomial is by solving the *interpolation equations.* Let $p(x) = \sum_{k=0}^{\nu} p_k x^k$. The interpolation conditions can be written as

$$\sum_{k=0}^{\nu} p_k \xi_\ell^k = f_\ell, \qquad \ell = 0, 1, \ldots, \nu,$$

and this is a linear system with a nonsingular Vandermonde matrix (A.1.2.5).

An explicit means of writing down the interpolation polynomial p is provided by the *Lagrange interpolation formula*

$$p(x) = \sum_{k=0}^{\nu} p_k(x) f_k, \qquad x \in \mathbb{R},$$

where each *Lagrange polynomial* $p_k \in \mathbb{P}_\nu$ is defined by

$$p_k(x) = \prod_{j=0,\ j\neq k}^{\nu} \frac{x - x_j}{x_k - x_j}, \qquad k = 0, 1, \ldots, \nu, \quad x \in \mathbb{R}.$$

A.2.2.4 An alternative method of evaluating the interpolation polynomial is the *Newton formula*

$$p(x) = \sum_{k=0}^{\nu} f[\xi_0, \xi_1, \ldots, \xi_k] \prod_{j=0}^{k-1} (x - \xi_k), \qquad x \in \mathbb{R},$$

where the definition of the *divided differences* $f[\xi_{i_0}, \xi_{i_1}, \ldots, \xi_{i_k}]$ is given by recursion,

$$f[\xi_i] = f_i, \qquad i = 0, 1, \ldots, \nu,$$

$$f[\xi_{i_0}, \xi_{i_1}, \ldots, \xi_{i_k}] = \frac{f[\xi_{i_1}, \xi_{i_2}, \ldots, \xi_{i_k}] - f[\xi_{i_0}, \xi_{i_1}, \ldots, \xi_{i_{k-1}}]}{\xi_{i_k} - \xi_{i_0}};$$

here $i_0, i_1, \ldots, i_k \in \{0, 1, \ldots, \nu\}$ are pairwise distinct.

An equivalent definition of divided differences is that $f[\xi_0, \xi_1, \ldots, \xi_\nu]$ is the coefficient of x^ν in the interpolation polynomial p.

A.2.2.5 The practical evaluation of $f[\xi_0, \xi_1, \ldots, \xi_k]$, $k = 0, 1, \ldots, \nu$ is done in a recursive fashion and it employs a *table of divided differences*, shown below. Only the underlined divided differences are required for the Newton interpolation formula.

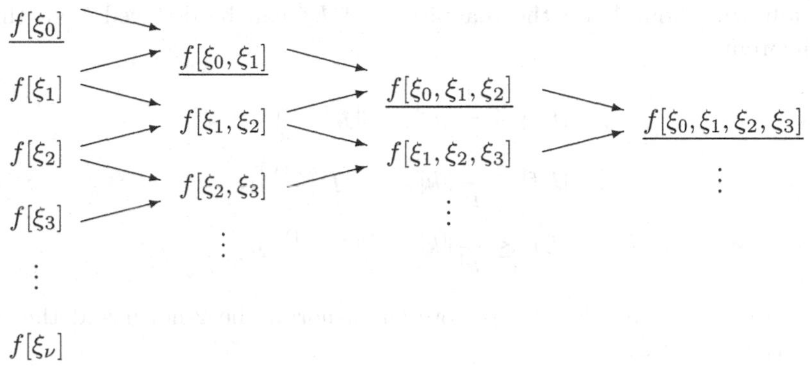

The cost is $\mathcal{O}(\nu^2)$ operations.

An added advantage of the above procedure is that only 2ν numbers need be stored, provided that overwriting is used.

A.2.2.6 A differentiable function f, defined for $x \in (a, b)$, is said to be of *variation*

$$V[f] = \int_a^b |f'(x)|\, \mathrm{d}x.$$

The set of all f's whose variation is bounded forms a linear space, denoted by $\mathcal{V}[a, b]$.

Let L be a linear functional from $\mathcal{V}[a, b]$. We assume that $f \in C^{\nu+1}[a, b]$, the linear space of functions that are defined in the interval $[a, b]$ and possess $\nu + 1$ continuous derivatives there. Let us further suppose that

$$L \int_a^b f(x, \xi)\, \mathrm{d}\xi = \int_a^b Lf(x, \xi)\, \mathrm{d}\xi$$

for any bivariate function f such that $f(\,\cdot\,, \xi), f(x, \,\cdot\,) \in C^{\nu+1}[a, b]$ and that L *annihilates* all polynomials of degree $\le \nu$,

$$Lp = 0, \qquad p \in \mathbb{P}_\nu.$$

The *Peano kernel* of L is the function

$$k(\xi) := L[(x - \xi)_+^\nu], \qquad \xi \in [a, b],$$

where

$$t_+^m := \begin{cases} t^m, & t \ge 0, \\ 0, & t < 0. \end{cases}$$

The *Peano kernel theorem* states that, as long as k is itself in $\mathcal{V}[a,b]$, it is true that

$$Lf = \frac{1}{\nu!}\int_a^b k(\xi) f^{(\nu+1)}(\xi)\,\mathrm{d}\xi, \qquad f \in C^{\nu+1}[a,b].$$

The following bounds on the magnitude of Lf can be deduced from the Peano kernel theorem:

$$|Lf| \le \frac{1}{\nu!}\|k\|_1 \times \|f^{(\nu+1)}\|_\infty,$$
$$|Lf| \le \frac{1}{\nu!}\|k\|_\infty \times \|f^{(\nu+1)}\|_1,$$
$$|Lf| \le \frac{1}{\nu!}\|k\|_2 \times \|f^{(\nu+1)}\|_2,$$

where $\|\cdot\|_1$, $\|\cdot\|_2$ and $\|\cdot\|_\infty$ denote the 1-norm, the 2-norm and the ∞-norm, respectively (A.2.1.4).

A.2.2.7 In practice, the main application of the Peano kernel theorem is in estimating approximation errors. For example, suppose that we wish to make the approximation

$$\int_0^1 f(\xi)\,\mathrm{d}\xi \approx \tfrac{1}{6}\left[f(0) + 4f(\tfrac{1}{2}) + f(1)\right].$$

Letting

$$Lf = \int_0^1 f(\xi)\,\mathrm{d}\xi - \tfrac{1}{6}\left[f(0) + 4f(\tfrac{1}{2}) + f(1)\right]$$

we verify that L annihilates $\mathbb{P}_3$, therefore $\nu = 3$. The Peano kernel is

$$\begin{aligned} k(\xi) = L[(x-\xi)_+^3] &= \int_\xi^1 (x-\xi)^3\,\mathrm{d}x - \tfrac{1}{6}\left[4(\tfrac{1}{2}-\xi)_+^3 + (1-\xi)^3\right] \\ &= \begin{cases} -\tfrac{1}{12}\xi^3(2-3\xi), & 0 \le \xi \le \tfrac{1}{2}, \\ -\tfrac{1}{12}(1-\xi)^3(3\xi-1), & \tfrac{1}{2} \le \xi \le 1. \end{cases} \end{aligned}$$

Therefore

$$\|k\|_1 = \frac{1}{480}, \qquad \|k\|_2 = \frac{\sqrt{14}}{1344}, \qquad \|k\|_\infty = \frac{1}{192}$$

and we derive the following upper bounds on the error,

$$|Lf| \le \frac{1}{1152}\|f^{(\mathrm{iv})}\|_1,\ \frac{\sqrt{14}}{8064}\|f^{(\mathrm{iv})}\|_2,\ \frac{1}{2880}\|f^{(\mathrm{iv})}\|_\infty, \qquad f \in C^4[0,1].$$

A.2.3 Ordinary differential equations

A.2.3.1 Let the function $\boldsymbol{f} : [t_0, t_0 + a] \times \mathbb{U} \to \mathbb{R}^d$, where $\mathbb{U} \subseteq \mathbb{R}^d$, be continuous in the cylinder

$$\mathbb{S} = \{(t, \boldsymbol{x}) : t \in [t_0, t_0 + a], \ \boldsymbol{x} \in \mathbb{R}^d, \ \|\boldsymbol{x} - \boldsymbol{y}_0\| \le b\}$$

where $a, b > 0$ and the vector norm $\|\cdot\|$ is given. Then, according to the *Peano theorem* (not to be confused with the Peano kernel theorem), the ordinary differential equation

$$\boldsymbol{y}' = \boldsymbol{f}(t, \boldsymbol{y}), \quad t \in [t_0, t_0 + \alpha], \qquad \boldsymbol{y}(t_0) = \boldsymbol{y}_0 \in \mathbb{R}^d,$$

where

$$\alpha = \min\left\{a, \frac{b}{\mu}\right\} \quad \text{and} \quad \mu = \sup_{(t,\boldsymbol{x}) \in \mathbb{S}} \|\boldsymbol{f}(t, \boldsymbol{x})\|,$$

possesses at least one solution.

A.2.3.2 We employ the same notation as in A.2.3.1.

A function $\boldsymbol{f} : [t_0, t_0 + a] \times \mathbb{U} \to \mathbb{R}^d$, where $\mathbb{U} \subseteq \mathbb{R}^d$, is said to be *Lipschitz continuous* (with respect to a vector norm $\|\cdot\|$ acting on $\mathbb{R}^d$) if there exists a number $\lambda \ge 0$, a *Lipschitz constant*, such that

$$\|\boldsymbol{f}(t, \boldsymbol{x}) - \boldsymbol{f}(t, \boldsymbol{y})\| \le \lambda \|\boldsymbol{x} - \boldsymbol{y}\|, \qquad \boldsymbol{x}, \boldsymbol{y} \in \mathbb{S}.$$

The *Picard–Lindelöf theorem* states that, subject to both continuity and Lipschitz continuity of $\boldsymbol{f}$ in the cylinder $\mathbb{S}$, the ordinary differential equation has a unique solution in $[t_0, t_0 + \alpha]$.

If $\boldsymbol{f}$ is smoothly differentiable in $\mathbb{S}$ then we may set

$$\lambda = \max_{(t,\boldsymbol{x}) \in \mathbb{S}} \left\| \frac{\partial \boldsymbol{f}(t, \boldsymbol{x})}{\partial \boldsymbol{x}} \right\|;$$

therefore smooth differentiability is sufficient for the existence and uniqueness of the solution.

A.2.3.3 The linear system

$$\boldsymbol{y}' = A\boldsymbol{y}, \quad t \ge t_0, \qquad \boldsymbol{y}(t_0) = \boldsymbol{y}_0,$$

always has a unique solution.

Suppose that the $d \times d$ matrix A possesses the spectral factorization $A = VDV^{-1}$ (A.1.5.4). Then there exist vectors $\boldsymbol{\alpha}_1, \boldsymbol{\alpha}_2, \ldots, \boldsymbol{\alpha}_d \in \mathbb{R}^d$ such that

$$\boldsymbol{y}(t) = \sum_{\ell=1}^{d} \mathrm{e}^{\lambda_\ell (t - t_0)} \boldsymbol{\alpha}_\ell, \qquad t \ge t_0,$$

where $\lambda_1, \lambda_2, \ldots, \lambda_d$ are the eigenvalues of A.

Bibliography

Birkhoff, G. and Rota, G.-C. (1989), *Ordinary Differential Equations* (4th edn), Wiley, New York.

Bollobás, B. (1990), *Linear Analysis: An Introductory Course,* Cambridge University Press, Cambridge.

Boyce, W.E. and DiPrima, R.C. (2001), *Elementary Differential Equations* (7th edn), Wiley, New York.

Davis, P.J. (1975), *Interpolation and Approximation,* Dover, New York.

Powell, M.J.D. (1981), *Approximation Theory and Methods,* Cambridge University Press, Cambridge.

Rudin, W. (1990), *Functional Analysis* (2nd edn), McGraw–Hill, New York.

Index

Printed in the United States
By Bookmasters

Printed in the United States
by Baker & Taylor Publisher Services